NF	noise figure
OC	open circuit
OCT	octal (base 8)
Ω	ohm, unit of resistance
op amp	operational amplifier
OS	one shot (monostable multivibrator)
OTA	operational transconductance amplifier
PIPO	parallel in, parallel out
PISO	parallel in, serial out
PLL	phase-locked loop
p-p	peak-to-peak
PROM	programable read-only memory
PSD	phase-sensitive detector = lock-in
Q	quality factor, of an inductor or band pass filter
R	a resistor
RAM	random access memory (volatile, in current usage)
RDY	ready
RE	real part of a complex number
Ref	reference: stable source of voltage or current
RF	radio frequency range (~0.5 MHz and above)
rms	root mean square
ROM	read-only memory
SC	short circuit
SCOPE	oscilloscope
SCR	silicon controlled rectifier
SEL	select
S/H or SH	sample and hold
SIPO	serial in, parallel out
SISO	serial in, serial out
S/N	signal to noise
SPDT	single pole, double throw 2-position switch
SR	shift register (sometimes, set-reset flip flop)
STROBE	timing signal in handshaking
SW	switch
SYNC	synchronization
TG	transmission gate
TRIG	trigger or device activation
TRI-STATE	tri-state logic: high, low or disconnected
TTL	transistor–transistor logic (bipolar)
UART	universal asynchronous receiver-transmitter
V	volts, unit of voltage
V_{cc}	collector power supply voltage
VCG	voltage-controlled gain
VCO	voltage-controlled oscillator
V_{dd}	drain power supply voltage
V/F	voltage-to-frequency converter
VOM	volt-ohm-meter
VTVM	vacuum tube voltmeter (obsolete)
WR	write
M	multiplier
X	multiplier
XOR	exclusive or gate

Abbreviations of size or scale

10^{-3}	m milli-
10^{-6}	μ micro-
10^{-9}	n nano-
10^{-12}	p pico-
10^{-15}	f femto-
10^{+3}	K kilo-
10^{+6}	M mega-
10^{+9}	G giga-
10^{+12}	T tera-

Electronics with Digital and Analog Integrated Circuits

RICHARD J. HIGGINS

University of Oregon, Eugene

PRENTICE-HALL, INC., *Englewood Cliffs, N.J.* 07632

Library of Congress Cataloging in Publication Data

Higgins, Richard J.
Electronics with digital and analog integrated circuits.

Bibliography: p.
Includes index.
1. Integrated circuits. 2. Digital electronics.
3. Linear integrated circuits. 4. Operational amplifiers. I. Title.
TK7874.H524 621.381'73 81-20987
ISBN 0-13-250704-8 AACR2

Editorial/production supervision: Karen J. Clemments
Interior design: Karen J. Clemments and Stan Spex
Editorial assistant: Susan Pintner
Page layout: Gail Collis
Cover design: Diane Saxe
Manufacturing buyer: Joyce Levatino

Printed in the United States of America

10 9 8 7 6 5 4 3 2

ISBN 0-13-250704-8

PRENTICE-HALL INTERNATIONAL, INC., *London*
PRENTICE-HALL OF AUSTRALIA PTY. LIMITED, *Sydney*
PRENTICE-HALL OF CANADA, LTD., *Toronto*
PRENTICE-HALL OF INDIA PRIVATE LIMITED, *New Delhi*
PRENTICE-HALL OF JAPAN, INC., *Tokyo*
PRENTICE-HALL OF SOUTHEAST ASIA PTE. LTD., *Singapore*
WHITEHALL BOOKS LIMITED, *Wellington, New Zealand*

Contents

Preface

Most people who feel at home with electronics pick it up by osmosis over a long period. When I was an undergraduate, I sometimes got a ride to class from a professor who lived near my home. He discovered that I'd never heard of negative feedback, and was astounded that an undergraduate education at MIT could leave out so elegant a concept. As a graduate student, I saw feedback put to work when a clumsy motor-generator power supply for a high-current electromagnet was replaced with an array of power transistors driven by operational amplifiers. Currents of microamps were controlling hundreds of amps! In my PhD project I found that I could enhance fine-structure in some data by running the signal through an op amp differentiator, and discovered the fun of signal processing. Later, in my own lab, we began some measurements that were extremely tedious. We interfaced the experiment to a minicomputer, which didn't get bored or tired like we did. Microprocessors make even more experiments amenable to automation.

There is a great satisfaction in putting one's own circuit together and having it work just like the design predicted. Debugging has the fun of detective work, once one gets over the feeling of being a complete klutz with electronics. There's just one problem: how to learn it. Now that everyone can afford a small computer, there's a wider need for a background course on how to connect the computer to the experiment. IC's have made all this much simpler. The *black box* approach gets results, so one rarely needs to know what's inside the box. One doesn't need the kind of course taught for electrical engineers, which takes too long for most scientists and doesn't adequately emphasize user applications. That's what this book is all about: *using electronics without fear!*

This book includes in one volume both digital and analog integrated circuit instrumentation. Many microcomputer interfacing examples are given. For example, the treatment of flip-flops in Chapter 4 ends with a discussion of tri-state logic and microcomputer bus interfacing. Microcomputers as such are not treated in this book. The emphasis here is on *applications* of IC electronics in measurement, control, signal generation, and signal processing.

The *approach* is to jump right into a given area, developing the necessary background as needed. The *focus* is on circuit principles and theorems, developing powerful "golden rules" which will work in many applications and which will not become obsolete with evolving technology. Details of circuit design are avoided whenever possible. On the other hand, the inside workings of specific industry-standard IC's are explained, and many applications with practical working circuits are included. Samples of IC specification sheets or tables are given, since it is important to learn how to read a "spec sheet" and select the right device.

The material in this book has served a primary *audience* of physics, computer science, chemistry, and biology students. It also has served students in other disciplines, attracted by the personal computer revolution, hi-fi electronics, or *Popular Electronics* tinkering projects. The material also can serve nonelectrical-engineering fields where electronic instrumentation is a present day necessity, or professionals whose electronics education came before IC's. The *level* of the material is matched to an audience of both graduate and upper-division undergraduate students. Half the audience has typically been physicists. The other half has been mostly computer scientists, providing valuable hands-on hardware background for this discipline. Students from other fields such as chemistry, biology, geology, psychology, and, occasionally, even music, business, architecture, urban planning, broadcasting, and physical education have used this material, with the math suitably scaled down.

The *background* of these students has varied widely. For many, this is the first time they have seen a differential equation, or have used complex numbers. The only necessary *electronics* background is Ohm's law, simple network theory, and a bit about discrete components like capacitors and transistors. A chapter on discrete component and semiconductor device electronics is provided for students without this background. The only math background needed for two-thirds of the book is enough introductory calculus to know how to recognize a derivative or an integral. Complex numbers are developed in an appendix. Students seem to pick up the math as it is introduced in context, and are motivated to learn more math by seeing its usefulness.

The text material has been developed into two independent one-term mini-courses. Chapters 1 to 9 form a manageable unit on digital IC electronics, with Chapter 9 (Digital to Analog and Analog to Digital Conversion) forming a bridge to the analog world. Chapters 10 to 20 form an analog IC electronics course. Enough additional material is provided to add flexibility for semester-long courses, varying background, or instructor's preference. For example, students without *any* prior electronics background should cover Chapter 1, The Basics: Discrete Component Circuits and Measurements.

The level varies greatly but systematically. I originally taught the analog material first because one then has the background to understand digital waveshaping and analog-to-digital and digital-to-analog conversion. However, the mathematics is much easier in the digital material, and the labs tend to be much more foolproof, so learning this part first is much more painless. The analog portion begins at a low math level, but includes several sections requiring some familiarity with differential equations, a willingness to become at home in the complex plane, and to learn the Laplace transform. These more difficult sections in Chapters 14, 16, 17, and 19 are marked with a † and are really only suitable for an upper division or graduate student audience of science or engineering majors.

This book came about after teaching an "electronics for scientists" course over a period of four years, using such standard texts as Brophy's *Basic Electronics for Scientists*, Diefenderfer's *Principles of Electronic Instrumentation*, or Malstadt *et al Electronic Measurements for Scientists.* Although written at the appropriate level for scientists, each had some disadvantages. Most did not go far enough in applications to bring the student to the point of working with state-of-the-art instrumentation. None of the standard texts used specific industry-standard IC devices, such as a 741 op amp or 7400 series TTL, to prepare the user to design and construct practical circuits. Recently, hobbyist-level publications have appeared which are organized as cookbooks. The user can jump right in and do surprisingly powerful tricks with very little previous electronics background. Examples include Larsen and Rony's *Bug Books*, Melen and Garland's *Understanding IC Op Amps* and Lancaster's *TTL Cookbook* (and *CMOS Cookbook*). Our trials of such material in the laboratory were successful in stimulating students' interest in what was really possible with IC's. Nonetheless, there were considerable gaps at the level of *why* things work (as to be expected from the cookbook approach) which were unsatisfying for an audience of college-level scientists. In addition, none of these books had enough applications for the sciences, particularly in measurement, control, and signal processing. This book is a combination of the best features of both the "electronics for scientists" and "hobbyist" or "cookbook" approach.

Laboratory experiments are the key to feeling at home with electronics. The textbook knowledge is useless unless one has wired together circuits and gotten them to work. A separate volume of lab experiments, *Experiments with Integrated Circuits*, is available to accompany this text. It is hard to convey to the newcomer how easy it is to get started in electronics these days. The inertia to learn about IC's disappears rapidly once one has had success with a few experiments that one can do on the kitchen table. IC's have made electronics both powerful and cheap. A kit of parts, illustrated in Chapter 2, which is sufficient to try out any of the ideas in this book, can be put together for about $200, and test instruments (except for an oscilliscope) cost little more. Buy it, try it, and have fun with electronics, rather than just being surrounded by it!

It was Henry Paynter of M.I.T. who first introduced me to what could be done with negative feedback, and J. H. Condon who, as a fellow grad student, encouraged me to try out an op amp in physics signal processing. The IC manufacturers have made valuable contributions in developing the state of the art with their applications literature, especially National Semiconductor, Texas Instruments, Signetics, RCA, Motorola, Analog Devices, Philbrick, and Burr Brown. Some of those contributions are extracted here. I have profited especially from D. H. Sheingold, Analog Devices, whose writing captures the sense of wonder one feels when an idea comes out of one's head and onto a circuit board and works well.

Nearly a decade of teaching assistants helped to develop the course on which this book is based. I am especially indebted to Hal Alles, Gary Karshner, and Tom Matheson. Margaret Graff gave editorial help with ruthless precision. Numerous typists (Liz Rachman, Sharon Robbins, Dolly Allen, Marc Baber, Linda Ficere, and Bev Jeness) put up with many revisions. The preparation of this manuscript, and especially the editing to incorporate helpful suggestions of several reviewers, was greatly facilitated by the Word Star (©Micropro International) word processing program.

Thanks to the many manufacturers who granted permission to use copyrighted material. These are specifically acknowledged in the figure or table captions. Material reprinted from *Electronics* magazine is copyrighted by McGraw-Hill, Inc., in the year of its publication. Material reprinted from *Scientific American* is copyrighted by Scientific American, Inc., 1977.

R. J. H.

Electronics with Digital and Analog Integrated Circuits

1

The Basics: Discrete Component Circuits and Measurements

This chapter is one person's view of what one still needs to know about discrete component circuits, plus a bit about basic measurement techniques. As electronics has evolved towards a "black box approach," circuit design has become simpler. One does not need to know a lot about equivalent circuit models for transistors if most transistor properties drop out of the final result, as they do for two-state binary devices or high-gain analog devices. But even though most electronics is now done with IC's, one still needs to know how to use a discrete transistor, for example to drive a light-emitting diode or other high-current device. In addition, there are some fundamental principles which allow you to make giant steps in understanding circuits, such as the notion that almost any two-terminal black box can be treated as if it were just a signal source and a resistor. The chapter closes with sections on what discrete devices look like, examples of how to convert a "real world" quantity into an electrical signal, and a discussion of what instruments one needs for basic electronic measurements.

1.1 MEASURING VOLTS, AMPS, AND OHMS

1.1.1 The Moving-Coil Meter

Inside the basic analog or moving-pointer meter [Fig. 1.1(a)] are a coil of wire forming an electromagnet and a permanent magnet at right angles. When a current I is passed through the coil, a torque T_I is created, which is balanced by a restoring torque T_S from the suspension spring. Since T_I = const × I and T_S =

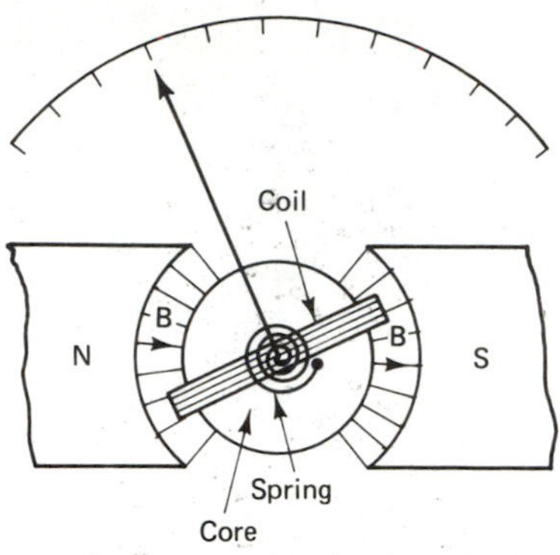

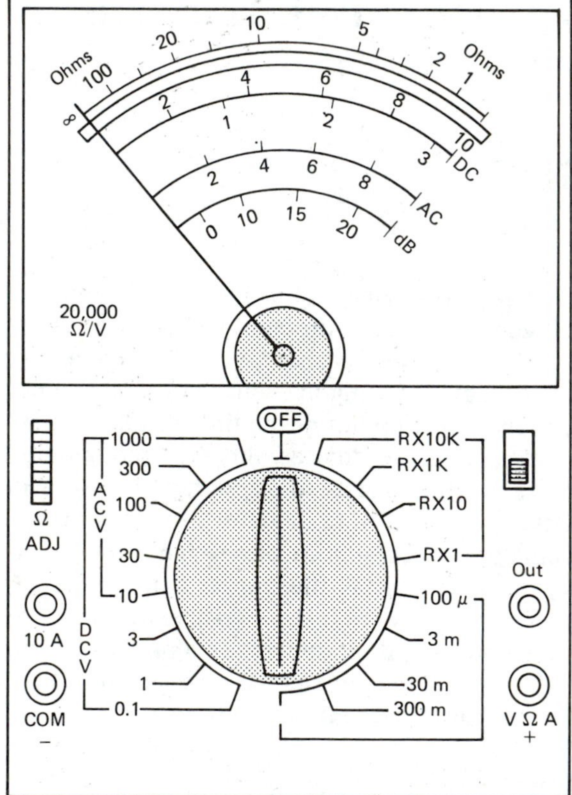

Figure 1.1 Instruments for measuring voltage, current, and resistance. (a) Inside the moving-coil analog meter. (b) Volt-ohm meter or VOM.

$k\theta$, the coil rotates and the two torques balance. The output angle is proportional to the input current being measured, and the basic meter circuit is a current sensor. Other applications require external circuitry to convert the quantity being measured to a current. The meter movement may be modeled as an ideal current meter, in series with a resistance equal in value to the resistance of the moving coil [Fig. 1.2(a)]. Typical parameter values are: full-scale meter deflection $I_m = 50\ \mu A$, and meter resistance $R_m = 100\ \Omega$.

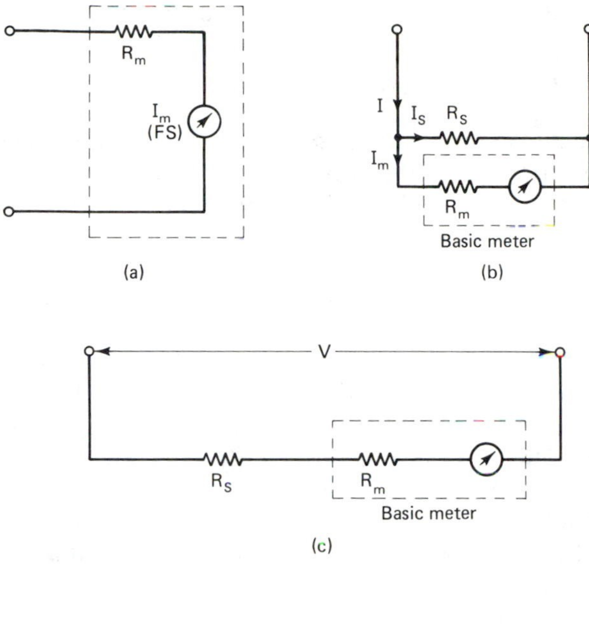

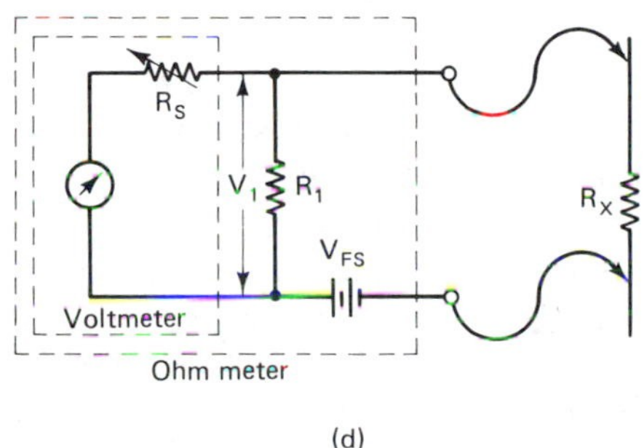

Figure 1.2 VOM circuits.
(a) Equivalent circuit of the moving-coil meter. (b) Ammeter circuit.
(c) Voltmeter circuit. (d) Ohmmeter circuit.

1.1.2 Volt-Ohm Meter Circuits

The moving-coil meter can measure current, voltage, or resistance. These functions are often combined in a basic test instrument called the *volt-ohm meter* (VOM). The measurement of currents larger than the basic meter's full-scale value requires diverting away most of the current (or else the meter will be destroyed) by placing in parallel a *shunt resistor* R_s [Fig. 1.2(b)]. R_s is selected so when the total input current equals the intended full-scale value, the meter is deflected to full-scale. Thus, for example, to measure 1-A full-scale requires a resistance $R_s = 0.5 \times 10^{-2}\ \Omega$ for the 50-μA, 100-Ω meter described above. (Why?)

The same basic meter may be used to measure a voltage, provided that an extra series resistance is inserted so the desired full-scale voltage results in a full-scale meter current [Fig. 1.2(c)]. To create a 1-V full-scale meter with the basic meter above requires a 20-KΩ resistor in series. (Why?)

The VOM measures resistance by a comparison method. The unknown resistor forms a voltage divider with an internal standard resistor R_1 [Fig. 1.2(d)].

When the unknown R_x is connected, the meter measures the fraction of the internal battery voltage that appears across R_1

$$V_1/V_{FS} = [1 + (R_x/R_1)]^{-1} \tag{1.1}$$

The nonlinear form explains the nonlinear ohms scale on a VOM. An initial calibration step with the leads shorted together ("zeroing" the meter) adjusts R_s to make the voltmeter combination read full-scale for a given value of the internal battery (note R_1 does not matter in this step).

1.1.3 Meter Errors

Since such meter circuits occur in test instruments such as the volt-ohm-meter, it is important to understand when they can be used without causing a measurement error. A current is measured by breaking the circuit at the desired point and inserting an ammeter in series. This will cause no error only if the effective resistance of the meter circuit [Fig. 1.2(b)] is small compared to the resistance of the circuit being measured. Otherwise, the meter will change the value of the current. The ammeter's resistance is dominated by the value of R_s. For circuit resistances $> 1\ \Omega$, there is usually no significant error. In the example above, $R_s = 0.5 \times 10^{-2}\ \Omega$. When operated as a voltmeter [Fig. 1.2(c)], the meter is connected in parallel with a portion of the circuit. No significant error is caused as long as the current flow into the voltmeter does not cause an appreciable voltage drop within the circuit under test. The meter series resistance R_s must therefore be large compared with circuit resistances. This can be a significant problem in high impedance circuits. For example, with circuit resistances of 100 KΩ, a voltmeter with $\sim$20 KΩ input resistance is not adequate. High-impedance voltmeters employ an isolation stage (typically a field-effect transistor or *FET*) to reduce the current flow into the meter by increasing its effective input impedance to $\sim 10^8\ \Omega$. Sometimes the term *VTVM* is used for such a meter, standing for the (obsolete) vacuum-tube voltmeter. Modern digital multimeters (*DMM*) all include FET-input stages.

1.2 PRINCIPLES FOR THE ANALYTICAL TOOLBOX

1.2.1 Ohm's Law

The potential drop across a circuit element is linearly proportional to the current flow in it.

$$V = IR \tag{1.2}$$

1.2.2 Resistors in Series Add [Fig. 1.3(a)]

$$R\,(\text{series}) = R_1 + R_2 \tag{1.3}$$

and the larger resistor dominates.

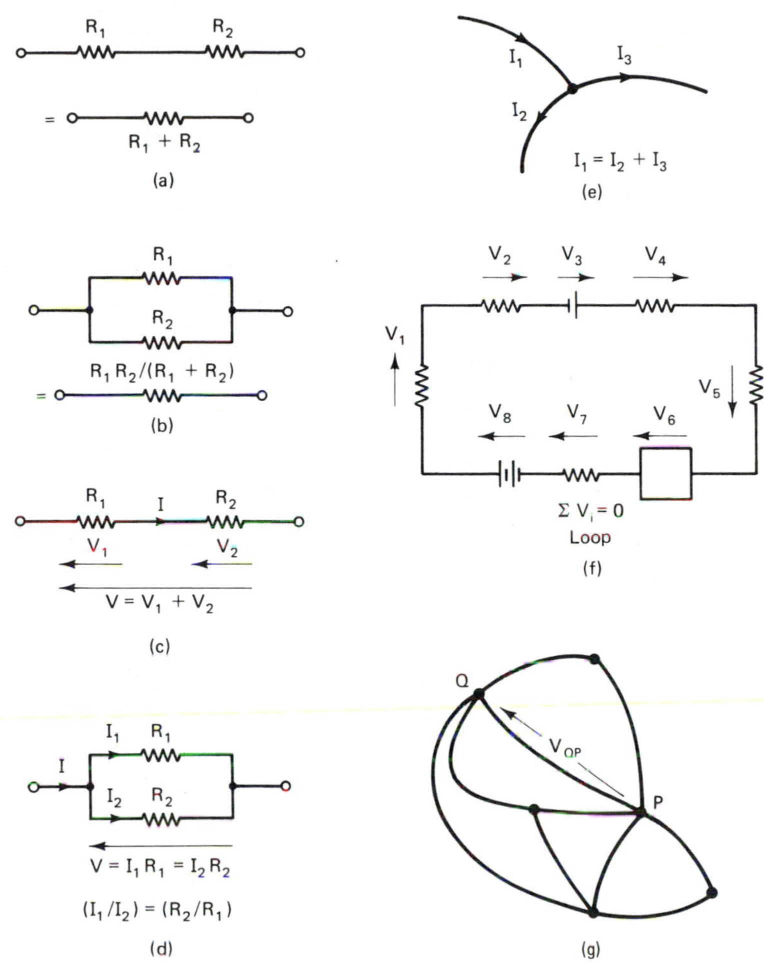

Figure 1.3 Some dc circuit principles. (a) Resistors in series. (b) Resistors in parallel. (c) Adding voltages. (d) Dividing currents. (e) Current nodes. (f) Voltage loops. (g) Path-independence of voltage loops.

1.2.3 Resistors in Parallel Add Inversely [Fig. 1.3(b)]

$$R\,(\text{parallel}) = (1/R_1 + 1/R_2)^{-1} = \frac{R_1 R_2}{R_1 + R_2} \tag{1.4}$$

and the smaller resistor dominates.

1.2.4 How Voltages Add in Series Branches

In a series branch of a circuit [Fig. 1.3(c)] the voltage drops add to produce the net voltage drop, and the same current flows in all elements. In this example,

$$V_1 = IR_1, V_2 = IR_2 \tag{1.5}$$

$$V = V_1 + V_2 = I(R_1 + R_2)$$

1.2.5 How Currents Divide in Parallel Branches

When two branches are in parallel, [Fig. 1.3(d)] the current divides inversely as the resistance, and the same voltage appears across all parallel elements.

$$V = I_1R_1 = I_2R_2 \tag{1.6}$$

$$I_1/I_2 = R_2/R_1$$

1.2.6 Current Is Conserved at a Node

When several wires are connected together into a *node* [Fig. 1.3(e)], the sum of the currents flowing into the node is zero.

$$\sum_{\text{node}} I_n = 0 \tag{1.7}$$

This follows from the fundamental principle of charge conservation: No electrons are lost as they flow around a circuit.

Question. Suppose it is not known which way an arrow of current I_1 should point. Will an error be made in the answer calculated?

Answer. No, the answer for I_1 will simply have a minus sign, to show that the initial guess was the wrong guess.

1.2.7 Voltage Loop Law

The sum of the voltages around a closed loop in a circuit is zero

$$\sum_{\text{loop}} V_n = 0 \tag{1.8}$$

This follows from the fundamental principle of energy conservation. Voltage is a form of potential energy just as height is potential energy. It takes as much energy to pump water up a series of hills as the potential energy available when the water falls down. Care must be taken with the algebraic signs of the voltages [Fig. 1.3(f)]. Voltage *sources* such as batteries should appear with opposite sign or be put on the opposite side of the equation from the voltage *drops* in resistors.

A useful corollary [Fig. 1.3(g)] is that the sum of the voltage drops in a closed loop starting from point P is always zero no matter which path is taken. It also follows that the voltage drop from point Q to point P equals the sum of the voltage drops around *any* of the possible paths connecting Q to P.

1.2.8 Moving Components in a Series Circuit

In a series circuit, the circuit elements can be redrawn in any order at will [Fig. 1.4(a)]. This follows from the Kirchoff voltage loop law, plus the fact that the order does not matter in adding numbers: $A + B = B + A$. *Caution:* This rule does not allow moving a circuit element across a three-branch node.

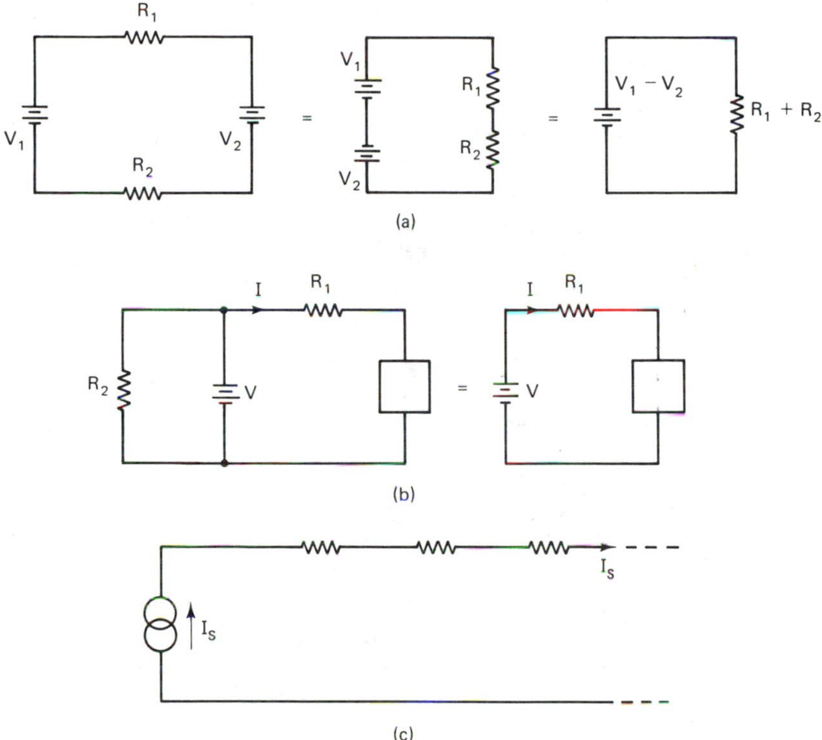

Figure 1.4 Some dc circuit theorems. (a) Movement of circuit elements and superposition of sources. (b) Voltage source. (c) Current source.

1.2.9 Superposition Principle

In a network with several independent sources, the resulting current is the sum of the currents caused by each source acting independently. This follows only for *linear* networks, where twice the voltage gives twice the current. The algebraic equations which govern linear networks are first order and linear. Examples of linear circuit elements are resistors, capacitors, and inductors. Transistors and diodes are not linear circuit elements. However, superposition and other linear principles can describe small changes even in nonlinear circuits.

1.2.10 Voltage Source

A voltage source produces the same voltage at its terminals no matter what current is being drawn. A battery is fairly close to an ideal voltage source. A useful result of this principle is shown in Fig. 1.4(b). The current I is unchanged if resistor R_2 is removed. As a result, resistors and other components in parallel with a voltage source are redundant as far as the rest of the circuit is concerned. Of course real sources are not perfectly ideal. A nonideal voltage source is best handled by the method of Thevenin's equivalent (next section).

1.2.11 Current Source

A current source produces the same output current independently of what load is connected to it. The voltage delivered by the current source will depend upon the total value of the resistances or impedances connected [Fig. 1.4(c)]. Of course this cannot be perfectly true, since a current source connected to an open circuit cannot deliver the same fixed current, since electrons do not spill out of wires. Although less familiar than ideal voltage sources, the current source is a useful idealization in solid state circuits, since transistors make excellent current sources.

1.3 THEOREMS FOR THE ANALYTICAL TOOLBOX

1.3.1 Thevenin's Equivalent Circuit

It is observed that the voltage measured across two terminals of a box (whose insides are unknown) depends upon the meter used for the measurement [Fig. 1.5(a)]. This is true even with meters calibrated against a reference battery. A model which can explain these results [Fig. 1.5(b)] specifies that inside the box is a battery V_T and a resistor R_T in series. If the meter resistance $R_m < \infty$, a current I flows when the meter is attached. A voltage drop IR_T will occur across the resistor in the box, so that $V < V_T$, and an error is made in the measurement. The error may be readily calculated. The circuit is redrawn in Fig. 1.5(c) as a *voltage divider.* This useful concept will come up many times in analyzing circuits. The voltage V_m seen at the meter can be calculated by subtracting the voltage drop in R_T caused by the current flow I.

$$V_m = V_T - IR_T, \tag{1.9}$$

where

$$I = V_T/(R_m + R_T) \tag{1.10}$$

Combining these two equations leads to

$$V_m = V_T[R_m/(R_m + R_T)] \tag{1.11}$$

Thus, a fraction $R_m/(R_m + R_T)$ of the source voltage actually appears across the meter. For an error of 1% or less in the measurement, the meter resistance must

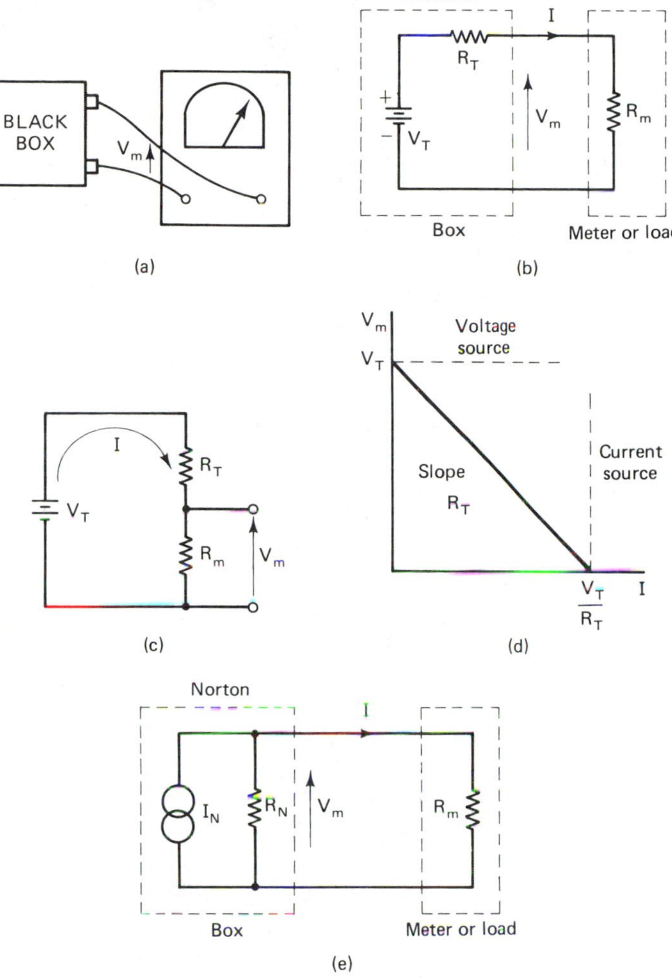

Figure 1.5 Thevenin and Norton equivalent circuits. (a) "Black box" whose voltage is being measured. (b) Loading by the meter alters the voltage of the box, now viewed as a Thevenin equivalent circuit. (c) Voltage divider circuit. (d) Voltage vs. current for the loaded circuit. (e) Norton equivalent circuit.

be at least $100R_T$. To display this idea of a loaded source [Fig. 1.5(d)], the output voltage is plotted as a function of current drawn from the black box, Eq. (1.9). The voltage falls as increasingly larger currents are drawn, with a slope equal to R_T.

Thevenin's theorem for dc circuits states that no matter what combination of batteries and resistors is actually inside the box, the output voltage as a function of current drawn from the box will have the form of Fig. 1.5(d) or Eq. (1.8). One can use black boxes without looking inside, since an observable characteristic curve is adequately described by the simple equivalent circuit [Fig. 1.5(b)], no matter what is actually inside the box. Although described here only for dc circuits, Thevenin's equivalent circuit is also a useful theorem in ac circuits. It is, however, restricted to circuits which are linear. Circuits with diodes and transistors are not linear, but if one's interest is in small changes about some operating point, Thevenin's equivalent circuit is a useful picture of *small signal* circuit operation.

Question. How can the parameters of Thevenin's equivalent circuit be measured when one cannot open the box?

Answer. The equivalent voltage V_T is measured as the "open-circuit" ($I = 0$) value of voltage using a meter which does not load the circuit, such as an FET input VOM or DMM. The equivalent resistance R_T may be measured in two ways. A variable resistance box may be connected across the output terminals and adjusted until the voltage measured across the box reaches half the open-circuit voltage. The adjusted value of the resistance box then equals R_T (Why?). This is called the *voltage divider method.* An alternative method, called the *short circuit method,* is to insert a current meter across the output terminals of the box, forcing V_m [Eq. (1.9)] to zero. The short-circuit current I_{sc} together with the open-circuit voltage V_{oc} yields the equivalent resistance

$$R_T = V_{oc}/I_{sc} \tag{1.12}$$

The most familiar resistive source is a battery. Why is a 12 V flashlight battery not adequate to start a car engine? The answer must involve the ability of the battery to deliver a large enough current, and it is a statement about the internal resistance of the battery. A typical 1.5 V flashlight battery can deliver about 1.0 A, and therefore has an internal resistance of about 1.5 Ω. A typical car battery can deliver approximately 100 A, and therefore must have an internal resistance 100 times smaller than this. An aging car battery which fails to start the engine properly is suffering from the buildup of oxides whose high resistance increases the equivalent resistance of the battery so it can no longer deliver the required current under load, even though the open-circuit voltage is adequate. Another low-impedance source is a mercury battery, whose internal resistance is at least 10 times smaller than that of a flashlight battery. It is inadvisable and even dangerous to measure the internal resistance of a mercury battery or other low-impedance source by the short circuit method, since the large current flow may cause explosive heating of the device.

1.3.2 Norton's Theorem

A graph identical with Fig. 1.5(d) would result if inside the box were a *current* source I_N in parallel with a resistor R_N [*Norton equivalent,* Fig. 1.5(e)]. As long as one does not open the box, there is no way to tell what is inside. The spirit of the black box approach is that it does not matter, as long as one has an adequate *equivalent* circuit to describe how the box will behave when connected to external components. The choice between the Thevenin and Norton equivalent circuits is arbitrary, based upon whether the equivalent resistance is less than (Thevenin) or greater than (Norton) the resistance of the circuit to which it is connected. It is readily shown that the correspondence between component values in the Thevenin and Norton equivalents is:

$$R_N = R_T \qquad (1.13)$$

$$I_N = V_T/R_T$$

1.3.3 Impedance Matching and Power Transfer

To avoid attenuating a signal, the input impedance of a normal *voltage* amplifier must be large compared to the impedance of the source. This is a consequence of Thevenin's theorem. However, there is another amplifier situation where the

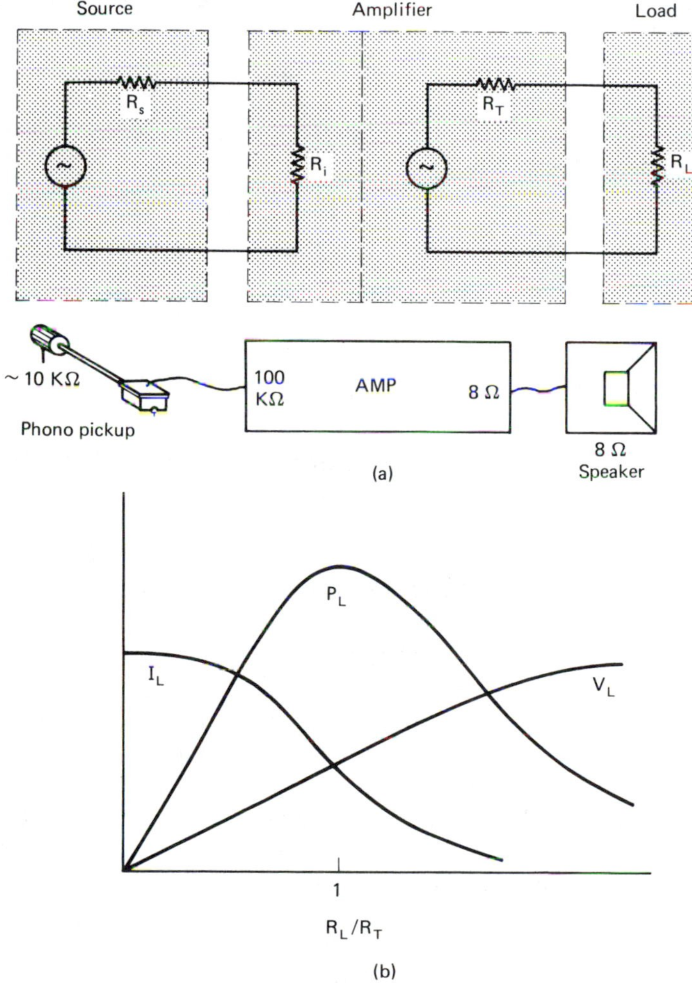

Figure 1.6 Impedance matching and optimum power transfer. (a) Typical impedance matching situation and its equivalent circuit. (b) Voltage, current, and power in the load as a function of load/source impedance ratio.

goal is to maximize the *power* delivered to the load. Maximum power transfer is achieved by making the load impedance *equal* to the source impedance. Both classes of amplifier are present in the familiar hi-fi amplifier [Fig. 1.6(a)]. A preamplifier stage is normally a voltage amplifier. Its input impedance ($\gtrsim$ 100 KΩ) is large compared to the typical source impedances ($\gtrsim$ 10 KΩ) of phono pickups or microphones. The final output stages are power amplifiers, producing both voltage gain and current gain, and are connected to a low-impedance load such as a loudspeaker (8 Ω typical). Maximum power transfer requires making $R_T \sim 8\ \Omega$ to match the load.

The conclusion that maximum power transfer to the load occurs when source and load impedances are matched follows from simple algebra. Referring to Fig. 1.6(b), there are two competing processes. The voltage delivered to the load rises as R_L/R_S increases, as a consequence of Thevenin's theorem.

$$V_L = V_S \frac{R_L/R_S}{1 + R_L/R_S} \tag{1.14}$$

However, the current delivered to the load decreases as the total impedance of the series circuit increases.

$$I_L = \frac{V_S}{R_S + R_L} = \frac{V_S}{R_S} \frac{1}{1 + R_L/R_S} \tag{1.15}$$

The power, which is the product $P_L = IV_L$, therefore has a peak as a function of R_L/R_S.

$$P_L = IV_L = \frac{V_S^2}{R_S} \frac{R_L/R_S}{(1 + R_L/R_S)^2} \tag{1.16}$$

It is straightforward to show by setting the derivative of the above function to zero that the peak occurs when $R_L = R_S$.

1.4 COMPLEX IMPEDANCE AND FILTER CIRCUITS

1.4.1 Ohm's Law for ac Circuits

A capacitor stores charge and builds up a voltage that depends upon how much current has been flowing for how long a time. It is a potential-energy storage reservoir in the same sense as a water reservoir on a mountain top. An inductor, on the other hand, generates a voltage opposing changes in the current flow in it, and may be said to store kinetic energy just as a moving car opposes change in its momentum. The *I-V* relationships are:

$$V_C = 1/C \int I\, dt \qquad V_L = L\, dI/dt \tag{1.17}$$

Suppose that each of these components is driven by a current source whose waveform is a sine wave of frequency f.

$$I = I_o \sin \omega t; \qquad \omega = 2\pi f \tag{1.18}$$

It will be useful for analysis to assume that the current has exponential form.

$$I = I_o \exp(j\omega t) = I_o(\cos \omega t + j \sin \omega t) \tag{1.19}$$

Of course the laboratory sine-wave generator does not generate a complex signal, but exponential notation simplifies problem solving:

1. The analysis is simplified, since it is easier to manipulate exponentials than sine functions, using algebra rather than trigonometry.
2. In many problems what matters is the ratio of output to input, or voltage to current. The precise form of the excitation will divide out anyway, and the unphysical exponential will disappear in the answer.
3. A useful picture of this complex excitation [Fig. 1.7(a)] is a vector in the complex plane whose magnitude is constant but which rotates once in a time

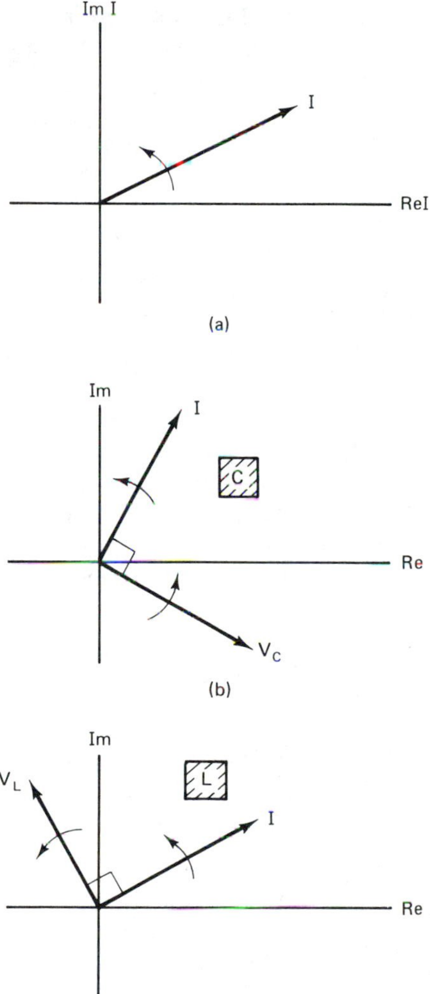

Figure 1.7 (a) Viewing ac signals as a rotating vector in the complex plane. (b) The voltage lags the current in a capacitor and (c) leads the current in an inductor.

$T = 2\pi/\omega$. A system's response may also be calculated in the complex plane and consists of a vector which also rotates at the same rate, but with a phase shift and a different magnitude. The response in the real world may be viewed as the projection on the real axis.

The voltage across each component as a result of this input current may readily be calculated, and leads to the version of Ohm's law for ac circuits.

Capacitor	Inductor
$V_C = (j\omega C)^{-1} I_o e^{j\omega t}$ $= Z_C I$	$V_L = j\omega L \; I_o e^{j\omega t}$ $= Z_L I$

(1.20)

The response may be put into the form of Ohm's law for ac circuits, except that a complex quantity called the *impedance* plays the role of the resistance.

$$Z_C = V/I = 1/j\omega C \qquad Z_L = V/I = j\omega L \tag{1.21}$$

Impedances are complex numbers, with both magnitude and phase.

$$V_C = (I_o/\omega C)(1/j) \exp(j\omega t) = (I_o/\omega C) \exp[j(\omega t - \pi/2)] \tag{1.22}$$

$$V_L = (I_o \omega L)(j) \exp(j\omega t) = I_o \omega L \exp[j(\omega t + \pi/2)]$$

The phase of Z_C is $-\pi/2$, while the phase of Z_L is $+\pi/2$. The significance may be seen by plotting equation 1.22 in the complex plane [Fig. 1.7(b) and (c)]. For a capacitor, the voltage *lags* the current by $\pi/2$, while for an inductor, the voltage *leads* the current by $\pi/2$.

1.4.2 Low-pass Filter or Integrator

The low-pass filter, used in signal processing to eliminate high-frequency noise, is a good way to practice manipulating RC combinations. The ratio V_o/V_i for the network shown in Fig. 1.8(a) may be calculated by viewing the circuit as a voltage divider.

$$\frac{V_o}{V_i} = \frac{Z_C}{R + Z_C} \tag{1.23}$$

With a little algebra, this may be rewritten as

$$\frac{V_o}{V_i} = (1 + j\omega RC)^{-1} \tag{1.24}$$

It is useful to bring such results into dimensionless form, to spot errors (if certain terms do not become dimensionless), and also to see graphically the essential behavior of the circuit. In this example, the ratio V_o/V_i (often called the *transfer function*) approaches 1 as ω approaches 0, and varies as $1/\omega$ in the high-frequency region ($\omega RC > 1$); hence the term *low-pass filter*. The magnitude of any transfer

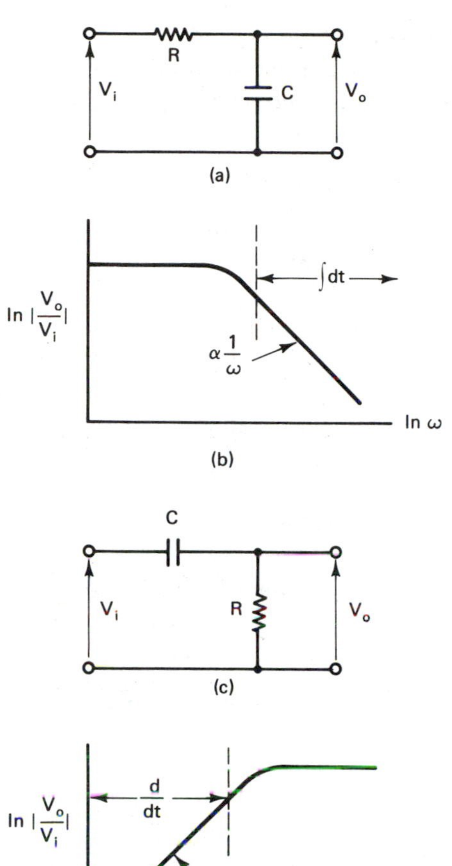

Figure 1.8 Two common RC filter circuits. (a) Low-pass filter (integrator) circuit and (b) transfer function. (c) High-pass filter (differentiator) circuit and (d) transfer function.

function is calculated from the magnitudes of the numerator N and denominator D.

$$\left| \frac{V_o}{V_i} \right| = \frac{|N|}{|D|} = [1 + (\omega RC)^2]^{-1/2} \tag{1.25}$$

As shown in Fig. 1.8(b), the transfer function is flat at low frequencies and falls off as $1/\omega$ at high frequencies, with the change occurring at a *cutoff frequency* $\omega RC = 1$.

The phase is calculated by bringing the transfer function (Eq. 1.24) into standard form: $X = a + jb$. This is done by rationalizing the denominator

$$\frac{V_o}{V_i} = \frac{1 - j\omega RC}{1 + (\omega RC)^2} \tag{1.26}$$

The phase shift θ is related to the ratio of the imaginary to real parts

$$\tan \theta = b/a = -\omega RC \tag{1.27}$$

The phase ranges from 0 at $\omega = 0$ ($Z_C = \infty$ and only the resistor matters) to $-\pi/2$ at $\omega = \infty$, and is always negative: the output is said to *lag* the input.

This circuit is often called a passive *integrator*. The integral of an input signal of the form $V = V_o \exp(j\omega t)$ is:

$$\int V\, dt = V_o\ (j\omega)^{-1}\ e^{j\omega t} = V\ (j\omega)^{-1} \tag{1.28}$$

A network whose output response multiplies the input by $1/j\omega$ and some real constant is therefore said to act as an integrator. In this example, the low-pass filter acts as an integrator for $\omega RC >> 1$. Although this has been made plausible only for a sine wave input, it is also true for inputs of any shape. An input square wave will come out looking like a triangular wave, if the square wave period is in the integrator regime of the filter. This follows because a square wave or any other periodic signal can be broken up into a sum of sine waves (*Fourier analysis*, Appendix 4), and the impedance formulas can be applied term by term.

1.4.3 High-pass Filter or Differentiator

The circuit shown in Fig. 1.8(c) is called a high pass filter. It is useful in removing dc or low-frequency components from high frequency signals of interest, and is used, for example, in decoupling stages in an ac amplifier. Since the analysis is similar to that of the low-pass filter, only the results will be quoted here. The circuit may be analyzed as a voltage divider

$$V_o/V_i = R/(R + Z_C) \tag{1.29}$$

giving the transfer function

$$V_o/V_i = j\omega RC/(1 + j\omega RC) \tag{1.30}$$

As ω approaches zero, the output approaches 0. As ω approaches ∞, the transfer function approaches 1, as expected from the name *high pass*. The magnitude is given by

$$|V_o/V_i| = |N|/|D| = \omega RC/[1 + (\omega RC)^2]^{1/2} \tag{1.31}$$

which is shown plotted in Fig. 1.8(d). To analyze for the phase, note that the phase of a ratio of two complex numbers is the difference between the phases of numerator and denominator. The phase of the numerator here is $+\pi/2$. The resulting phase of Eq. (1.30) is

$$\theta = \theta_N - \theta_D = \pi/2 - \tan^{-1}\omega RC \tag{1.32}$$

For this network, the response always *leads* the input. The phase shift approaches $+\pi/2$ as ω approaches 0, and approaches 0 as ω approaches ∞.

This network acts as a *differentiator* in the region of frequencies $\omega RC << 1$, where the response is linear in ω. The reasoning is essentially the same as for the integrator. A basic difference, however, is that the circuit will not do a good job of differentiating a signal whose *harmonics* extend into the region $\omega RC > 1$, where the network no longer follows the ideal differentiator transfer function.

Test your understanding. Suppose a square-wave input is applied to the high-pass filter. Show graphically the location of the first three terms (1ω, 3ω, 5ω) on Fig. 1.8(d) if the fundamental lies at $f_o/2$, where f_o is the cutoff frequency of the high-pass filter. Compare the output signal level for each of these first few harmonics with that obtained graphically for an ideal differentiator. [Refer to Appendix 4 to decide which harmonics are significant.]

1.4.4 Resonance

A basic understanding of resonance is necessary for tuned circuit filters, simulation, instability, and oscillators. Consider the series and parallel RLC resonant circuits, shown in Fig. 1.9(a) and (b). For the *series* resonance, suppose that the network is excited by an ideal voltage source. Since the impedances of the inductor and capacitor are 180° out of phase and vary in magnitude as ω and $1/\omega$, there will be some frequency (the resonant frequency) at which the two impedances cancel, as shown in a vector diagram [Fig. 1.9(c)]. The total network impedance passes through a minimum (equal to the value of the series resistor), and the current flow passes through a maximum.

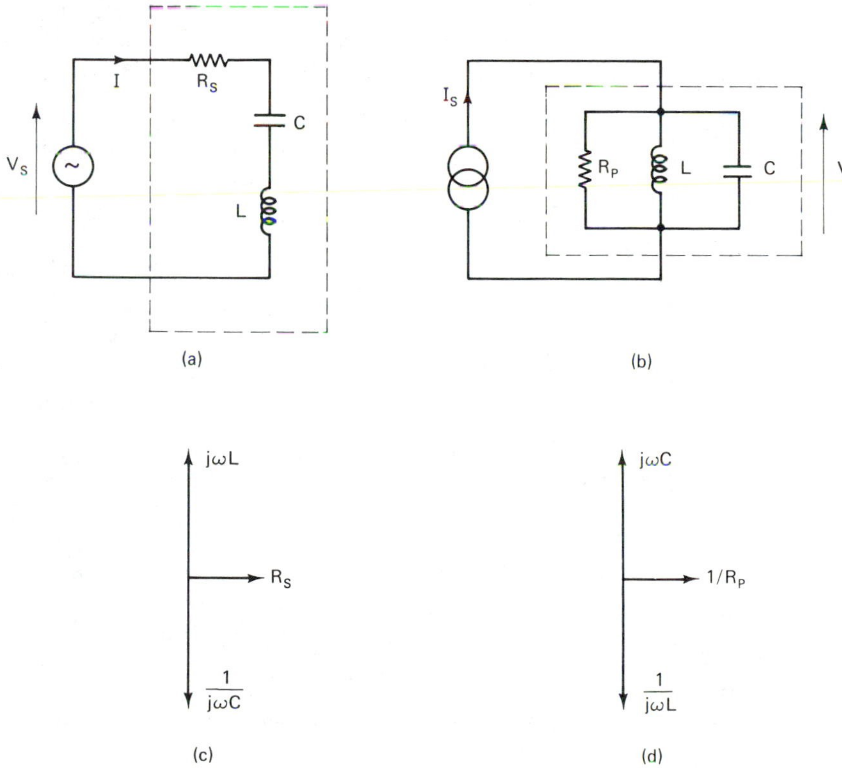

Figure 1.9 (a) Series and (b) parallel resonant circuits. (c) Vector diagrams of components of the series resonant impedance and (d) the parallel resonant admittance.

Parallel resonance is best understood by supposing that the network of Fig. 1.9 is excited by a current source. The current divides at the node joining the RLC combination, with the current in each branch proportional to the *admittance* (= impedance^{-1}) of the branch. There will be a frequency at which the inductive and capacitive admittances cancel, as shown in Fig. 1.9(d). The total parallel impedance and the voltage across the network pass through a maximum.

The height and sharpness of the resonant peak increase inversely as R_s in a series resonant circuit or directly as R_p in a parallel resonant circuit. These comparisons (which follow a relationship known as a *duality*) are summarized in Table 1.1. The transfer functions given there form the basis for the so-called universal resonance curve

$$F(\omega) = [(\omega^2 - \omega_o{}^2) + (j\omega_o)/(Q\omega)]^{-l} \tag{1.33}$$

whose amplitude and phase are plotted in Fig. 1.10.

TABLE 1.1 COMPARISON OF PARALLEL AND SERIES RESONANCE

Circuit	Parallel RLC	Series RLC
Excitation	Current source	Voltage source
Response	Voltage across network	Current in network
Impedance at resonance	Approaches infinity	Approaches zero
Transfer function	$\dfrac{V}{I} = \dfrac{R_p}{1 + jQ(\omega/\omega_o - \omega_o/\omega)}$	$\dfrac{I}{V} = \dfrac{1/R_s}{1 + jQ(\omega/\omega_o - \omega_o/\omega)}$
Resonant frequency	$\omega_o^2 = 1/LC$	$\omega_o^2 = 1/LC$
Q	$R_p/\omega_o L$	$\omega_o L/R_S$

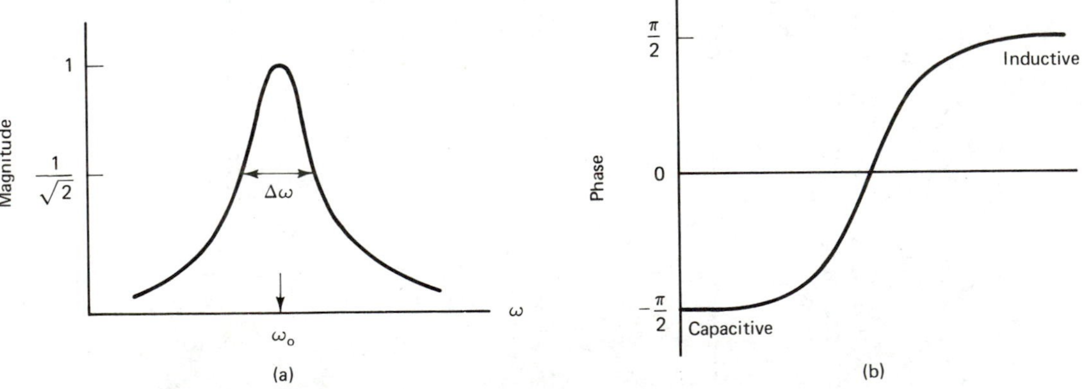

Figure 1.10 (a) Magnitude and (b) phase of the universal resonance curve as a function of frequency.

It may be difficult to see how the parallel resonant impedance could approach infinity while the impedance of the components is finite. During the initial excitation period, a large resonant circulating current is built up in the LC combination. The system acts as a large inertia resisting change. A large voltage is developed across a high-Q parallel resonant circuit. Thus, the divergence of the parallel impedance is not "getting something for nothing" but is a consequence of the fact that with high Q, a trickle of current in from the outside world gets stored up with little dissipation.

1.5 STEP RESPONSE OF PASSIVE NETWORKS

In the absence of more sophisticated methods, the transient response of networks unavoidably involves differential equations. The case of RC high-pass and low pass filters is sufficiently common to deserve analysis. Suppose the network is excited by a voltage step of size V. For either the high-pass or low-pass circuit [Fig. 1.8(a) or (c)], Kirchoff's voltage law requires that

$$V_i = V_R + V_C \tag{1.34}$$

Ohm's law is not sufficient to complete the analysis, since the $I-V$ relationship of the capacitor is not algebraic. Instead, differentiate both sides of Eq. (1.34)

$$0 = dV_R/dt + dV_C/dt = R\ dI/dt + C^{-1}I \tag{1.35}$$

The result is a first-order differential equation for $I(t)$. A suitable trial function is

$$I(t) = A + B \exp\,[-t/(RC)] \tag{1.36}$$

which, by direct substitution, is a solution. The constants A and B are determined by the initial condition resulting from the input step V_i

$$I(0_+) = V_i/R \tag{1.37}$$

where the subscript (+) designates the instant of time just after $t = 0$. If there was previously no voltage on the capacitor, it acts as a short circuit until some charge has built up (some current has flowed). The initial current is set by Ohm's law for the resistor alone. The resulting initial conditions are $A = 0$, $B = V_i/R$, and the complete solution is

$$I = \frac{V_i}{R} \exp\,[-t/(RC)] \tag{1.38}$$

The transfer function V_o/V_i for either the high-pass or the low-pass filter may be calculated from this result. Referring to the circuit diagrams [Fig. 1.8(c)], for the high-pass filter

$$\left(\frac{V_o}{V_i}\right)_{HP} = IR/V_i = \exp \frac{-t}{RC}, \tag{1.39}$$

which is plotted in Fig. 1.11(a). For the low-pass filter, the output voltage is proportional to the integral of the current

$$V_o/V_i = C^{-1}\ V_i \int_0^t I\ dt' = (RC)^{-1} \int_0^t \exp\,[-t'/(RC)]\ dt'$$

$$(V_o/V_i)_{LP} = \left\{1 - \exp\,[-t/(RC)]\right\} \tag{1.40}$$

which is plotted in Fig. 1.11(b).

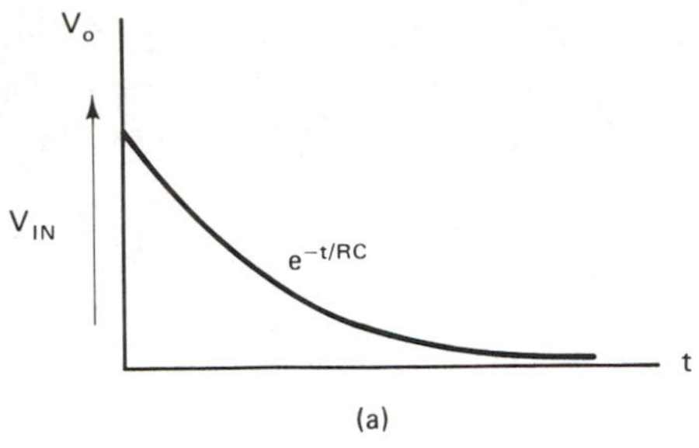

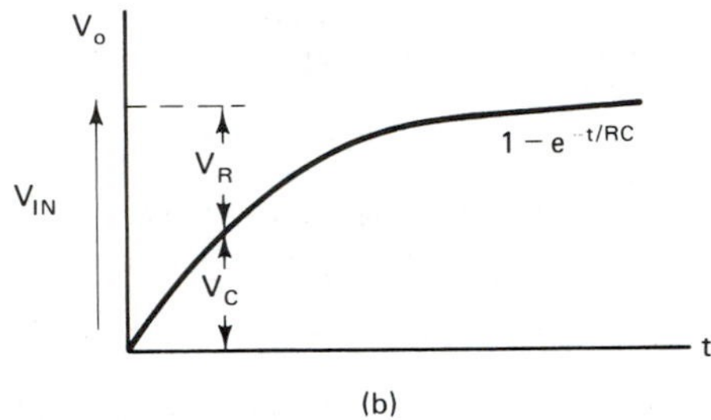

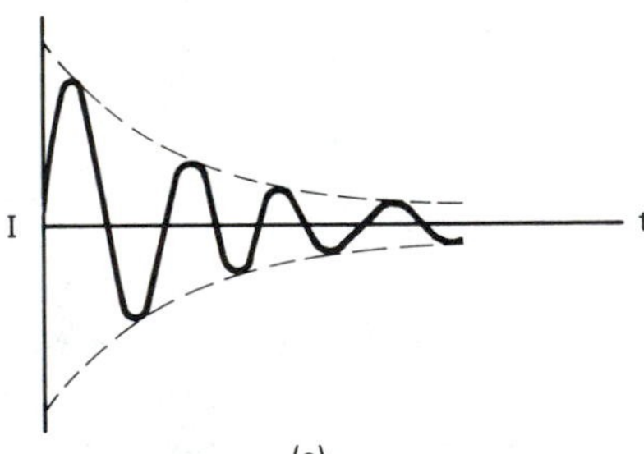

Figure 1.11 Response to a voltage step of the (a) high-pass filter; (b) low-pass filter; (c) RLC resonant circuit.

The step response of an RLC resonant circuit involves second-order differential equations. Since the method will not be used elsewhere in this book, only the result will be quoted. For a series RLC circuit, the current is

$$I = \frac{V_i}{R} e^{\left(\frac{-\omega_o t}{2Q}\right)} \sin \left[\omega_o t \left(1 - \frac{1}{4Q^2} \right)^{1/2} \right] \tag{1.41}$$

which is plotted in Fig. 1.11(c).

Note the close correspondence between circuit parameters which characterize the ac and the step response. For the RC high-pass or low-pass filter, the same time constant RC sets the cutoff frequency of the ac sine wave response and the *rise time* of the step response. For the RLC resonant circuit, the ringing frequency in Eq. (1.41) is nearly the same as the resonant frequency (identical, if Q is high), and the same circuit Q which sets the bandwidth of the sine wave response also determines the *decay time* of the ringing in the step response. The number of cycles it takes for the ringing to decay by a factor $1/e$ is just $n = Q/\pi$.

1.6 P-TYPE AND N-TYPE SEMICONDUCTORS

A silicon atom in a silicon crystal is surrounded by 4 other nearest neighbors, as shown schematically in Fig. 1.12(a). Each silicon atom has 4 outer electrons which bond with its 4 near neighbors. Since there are no leftover electrons, the bonds are complete, and the pure Si crystal is very nearly an insulator, called a *semiconductor.* It is possible to pull one of the electrons loose by thermal excitation, but since it takes an energy of about 1000°K to do so, pure silicon is a high-resistivity material. Suppose that one of the Si atoms is replaced by an atom one column to the right in the periodic table, such as arsenic [Fig. 1.12(b)]. Arsenic has one more electron than Si, so there is one unpaired electron after the 4 bonds to the Si are made. This extra electron is bound in an orbit similar to an electron in the hydrogen atom, except much more weakly. The extra electron is easily freed by thermal excitation and can then be moved by an electric field. This *doped* crystal can therefore conduct electricity readily. Similarly, suppose one of the atoms in the Si crystal is replaced by an atom such as gallium, which is one column to the left of Si in the periodic table. Since gallium has one less valence electron than Si, one electron will be absent in the bonding [Fig. 1.12(c)]. An electron which wanders by from somewhere else in the crystal could fill that vacancy. The vacancy, called a *hole*, acts in some ways like a bubble. A hole is a fictitious creation; there are no holes, only the absence of electrons. Nonetheless, when an electron moves to the left to fill a hole, there is real net motion of charge, which is described by saying the hole has moved to the right. A material doped with an electron *donor* such as arsenic is called *n-type*, n standing for negative charge carriers. A material doped with an electron *acceptor* such as gallium is called *p-type*, because the charge carriers act positively.

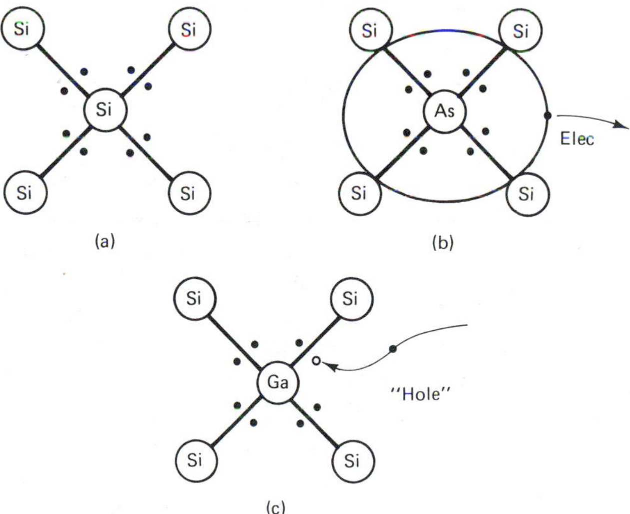

Figure 1.12 Semiconducting materials may be (a) intrinsic, with no unpaired electrons, (b) n-type, when electron donors are present, or (c) p-type, when electron acceptors create holes.

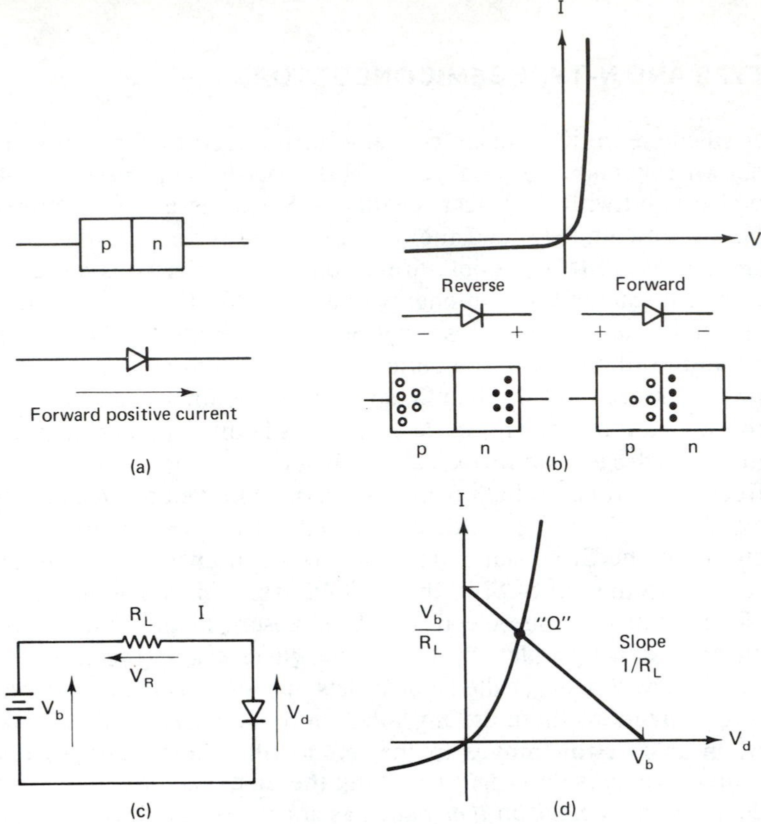

Figure 1.13 (a) Diode structure and circuit symbol. (b) Diode characteristic curve and its origin. (c) A diode circuit is a nonlinear problem best solved graphically, with (d) the load line concept.

1.7 DIODES AND POWER SUPPLIES

A diode is made by metallurgically joining together two pieces of p- and n-type material, as shown in Fig. 1.13(a). The symbol for a diode looks like an arrow, with the head of the arrow pointing from the p side to the n side of the junction. When a current is passed through the device, current flow as a function of the applied voltage [Fig. 1.13(b)] looks unlike that of a resistor. There is a low-resistance (large-current for small voltage) *forward* direction, and a high-resistance (low-current) direction when the device is *reverse*-biased. Qualitatively, the device has this asymmetric behavior because a forward bias applies a positive potential to the p side and a negative potential to the n side. The holes on the p side are repelled from the positive potential and move across the junction into the n side. The electrons in the n-side are repelled by the negative potential, and move to the left across the junction into the p side. These electrons in the p-side and holes in the n side, called *minority* carriers, recombine with holes and electrons normally present, but their net motion causes a large current flow in the device. When the device is reverse-biased, however, the holes on the p side move towards the nearby negative-biased electrode, and the electrons on the n side move towards the positive-biased electrode, leaving a region in between relatively free of charge carriers, and hence no net current flows across the junction. This is only a qualitative picture of the device action. The real explanation requires an understanding of *energy bands* in solids. But, from a circuit point of view, the transistor acts as a one-way switch or one-way valve, analogous to a check valve in plumbing.

In working with diodes or other nonlinear devices in electronic circuits [Fig. 1.13(c)], evaluating the current flow algebraically is awkward. The simplest way of evaluating the current flow in such a circuit is a graphical method using the *load line.* The current-voltage characteristic of the nonlinear device is plotted as in Fig. 1.13(d). On the same curve is plotted the I–V characteristic of the load resistor in series. The resistor's voltage drop IR_L is expressed in terms of the axes of the diode characteristic curve, using the Kirchoff voltage law to relate the sum of the voltages around the circuit.

$$V_d = V_b - IR_L \tag{1.42}$$

The current I is not yet known, so Eq. (1.42) is merely a statement of the possibilities for current flow in the resistor as a function of diode voltage drop V_d. This *load line*, also shown plotted in Fig. 1.13(d), is a straight line of negative slope $(-1/R_L)$ whose intercepts are the voltage on the battery [$I = 0$ in Eq. (1.42)] and V_b/R_L on the current axis ($V_d = 0$). But since the devices are in series, the current in the resistor and in the diode must be equal. The only point at which this happens is the intersection of the diode characteristic curve and the load line of the resistor. This graphical solution to the problem determines the operating point or Q point of the circuit.

1.7.1 Zener Diodes

In a normal diode, the low-current reverse-bias characteristic eventually breaks down at a high enough voltage, called the *reverse breakdown* voltage. The electric field becomes large enough so that carrier velocities are high enough to ionize atoms on impact. This creates many electron-hole pairs which also multiply, causing high-current *avalanche breakdown.* If this current is not limited, the pn junction will melt away, turning the diode into a resistor.

In the *Zener diode*, this liability is turned into an asset. The reverse breakdown region of the characteristic curve is made nearly vertical, so the same voltage appears across the device no matter what value the current takes on [Fig. 1.14(a)]. But that is the definition of a perfect voltage source. So, even though the Zener diode "supplies" nothing, it acts as a voltage regulator or voltage reference. The diode does not burn out because the current is limited by a source resistor R_S [Fig. 1.14(b); see also Prob. 1.14].

1.7.2 How to Make a Power Supply

To make a dc voltage source, feed the ac line voltage through a transformer to bring the voltage level close to the desired value, rectify the ac with diodes, and

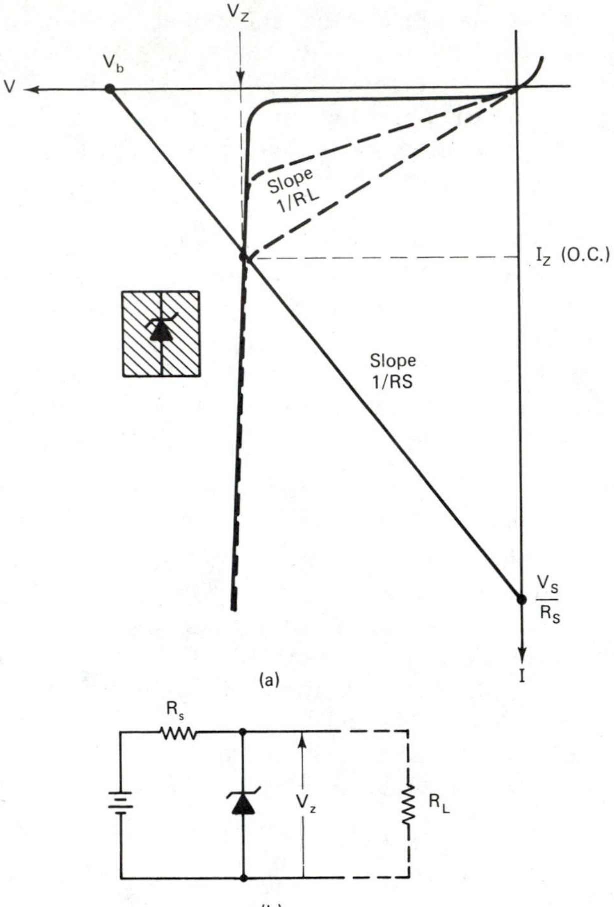

Figure 1.14 Zener diode.
(a) Characteristic curve and load line.
(b) Regulator circuit.

filter the rectified signal with a capacitor. The voltage value will wander if the load resistance or input voltage changes, but this can be eliminated with voltage regulation (Chapter 13). There are several standard rectifier circuits. The *half-wave* rectifier [Fig. 1.15(a)] blocks the half-cycle when the diode is reversed-biased and is a poor starting point to make a dc signal. The average dc output voltage is only 0.318 V_p, where V_p is the peak value of the ac input voltage. This number comes from Fourier analysis of the rectified waveform, which we will not do here. (A table of useful Fourier series is given in Appendix 4.)

Full-wave rectifiers steer both halves of the sine wave into the load [Fig. 1.15(b)]. The average value of output voltage is increased by a factor of 2 to 0.636 V_p. There are two full-wave rectifier circuits. One version uses only two diodes but requires a center-tapped transformer. The other version, called the *bridge rectifier*, uses four diodes but needs no center-tapped transformer. The bridge circuit is the more common. The transformer requirements are simpler, and the peak inverse voltage is only half the value which appears across a diode in the center-tap circuit.

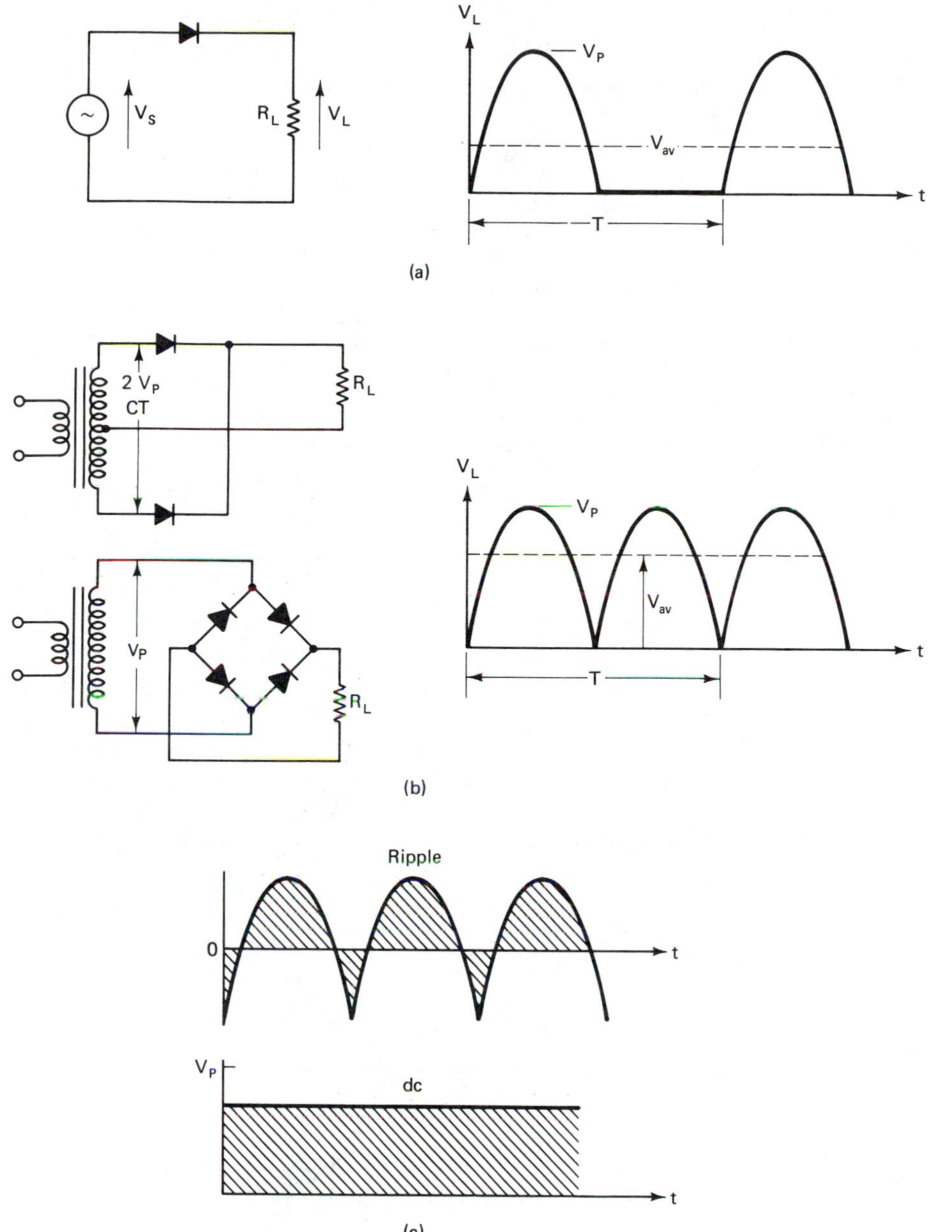

Figure 1.15 Rectifiers. (a) Half-wave rectifier circuit and output voltage waveform. (b) Full-wave rectified waveform and full wave rectifier circuits. (c) Separating the actual waveform into its dc and ac components.

Power supply filters. Even a full-wave rectified signal is far from constant [Fig. 1.15(c)]. In a conventional power supply, a large capacitor is added in parallel [Fig. 1.16(a)] to smooth out the ripple before the output goes to a regula-

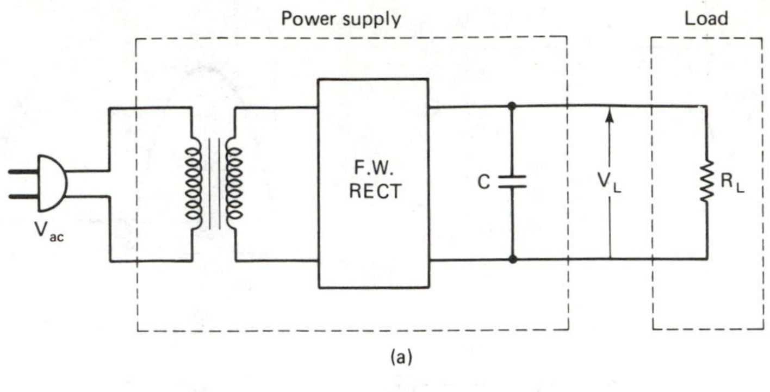

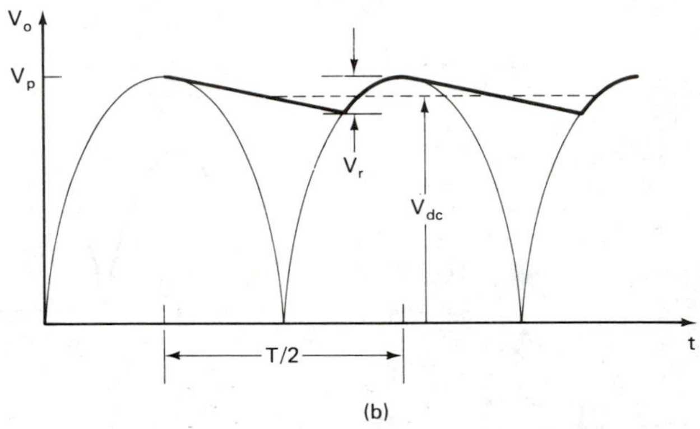

Figure 1.16 Power supply filtering. (a) Circuit. (b) Output voltage waveform.

tor section. The capacitor is charged up once per half-cycle of input and discharges only slightly in between [Fig. 1.16(b)]. The ripple V_r depends on how far the capacitor discharges through the load R_L before the next charging cycle $T/2$ s later ($T = f^{-1}$, where f is the ac line frequency). Although the discharge is exponential, the initial portion is nearly linear.

$$V = V_p \exp\,[-t/(R_L C)] \simeq V_p[1 - t/(R_L C)]$$

$$dV/dt = -V_p/(R_L C)$$

$$V_r = (T/2)(dV/dt) = V_p/(2R_L Cf) \quad \text{(peak-to-peak)}$$

The dc output voltage is V_p minus the average of the ripple, $V_r/2$ (including only the triangular portion and neglecting the short charging portion).

$$V_{dc} = V_p - V_p/(4R_L Cf)$$

Normally, one chooses C to match the load so that $V_{dc} \simeq V_p$, within about 20%. Within this approximation, the ripple fraction is

$$r = (V_r/2)/(V_{dc}) \simeq (4\, R_L Cf)^{-1}$$

This fraction can usually be brought to less than 10% (see Prob. 1.11) to serve as a filtered input for a regulator circuit.

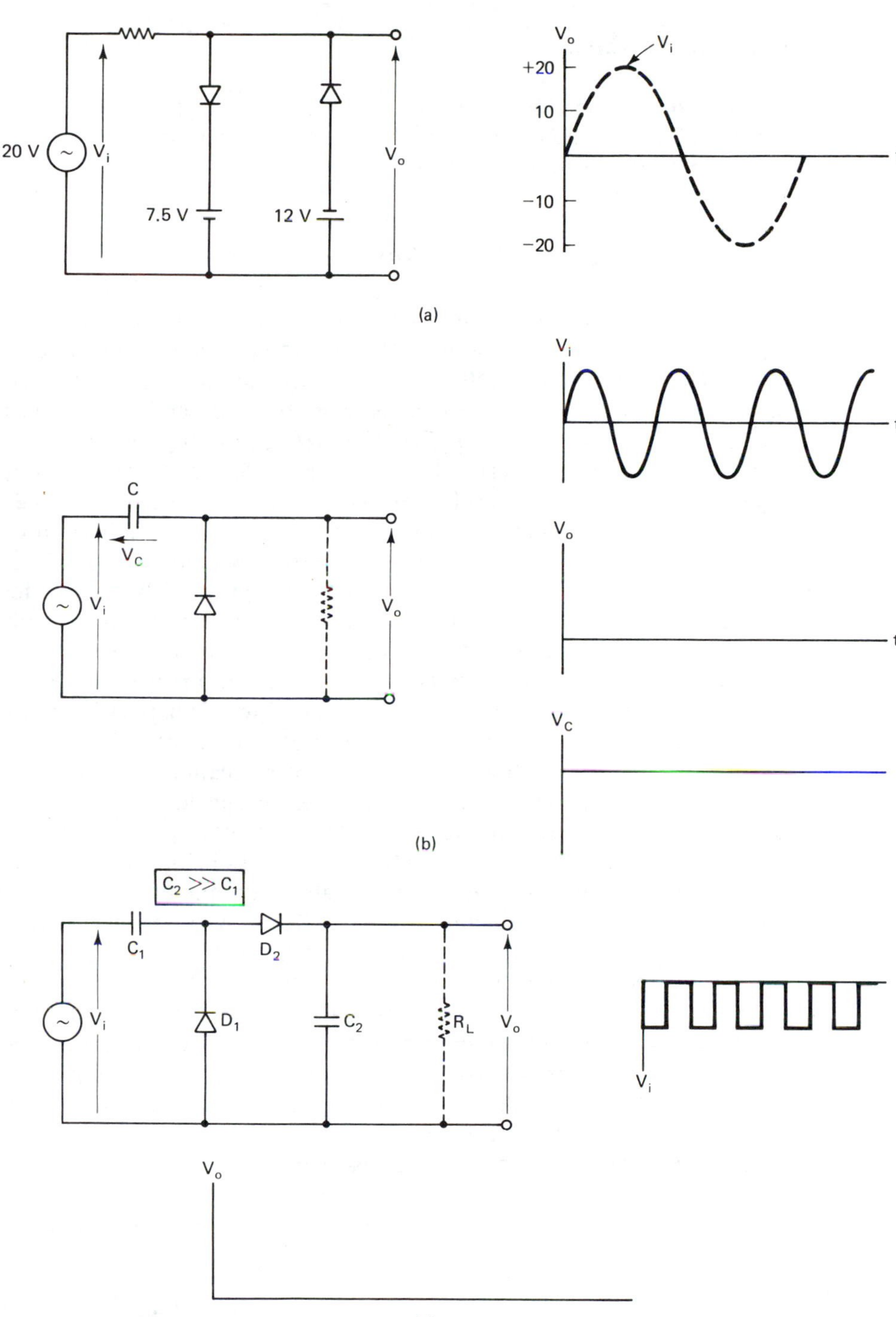

Figure 1.17 Useful diode circuits. (a) Clipper. (b) Clamp. (c) Pump. Complete the output waveform sketches in parts (b) and (c) as instructed in Problem 1.12.

1.7.3 Diode Circuits

In addition to rectifiers, diodes are also used to alter waveforms in other ways. We include in Fig. 1.17 a collection of the most common circuits: the clipper, the clamp, and the diode pump. See also Prob. 1.12.

1.8 JUNCTION (BIPOLAR) TRANSISTORS

The bipolar transistor consists of three slabs of semiconducting material, alternating pnp or npn. The term *junction* refers to the pn junctions in the device. The term *bipolar*, now preferred, refers to the participation of both holes and electrons in the action. The center segment (B) is called the *base*, and the others are called the *emitter* (E) and *collector* (C) [Fig. 1.18(a)]. The transistor structure looks at first like two pn diodes in series, back to back. A device actually made from two diodes would not be terribly useful because no current could flow from emitter to collector in either direction. The diode picture is, however, useful in understanding transistor biasing, and the difference between two diodes in series and a real junction transistor clarifies how transistors really do work. The symbol for an npn transistor is shown in Fig. 1.18(a). The arrow on the emitter is a reminder that there is a diode pointing in that same direction. The internal construction of the transistor has three slices of material, but one of them, the base, is very thin [Fig. 1.18(b)]. Current will flow between base and emitter if a battery is connnected so that the junction is biased in the forward direction. A second battery is placed between collector and emitter. By itself, the second battery will cause no current flow, because the collector battery back-biases the pn junction. When the base battery is connected, electrons are attracted from emitter into the base. If the base is thin enough, most of the electrons find themselves being swept across the base into the collector, and relatively few actually drift out of the base wire. This mechanism is shown by the arrow in Fig. 1.18(b). In what sense is this an amplifier? Most of the current flow actually comes from the collector battery, so no power is created. A fluid analogy is useful [Fig. 1.18(b)]. A relatively small amount of power applied to a control valve can control the flow of a large amount of fluid from a source. In the transistor, the collector battery acts as the source, and the base bias battery or base signal acts as the control turning on or off the flow of current. Although the current flow is initiated by the attraction generated by the base bias battery, it expends relatively little power in doing so, because very little current ends up flowing in the base lead.

Question. How can a large current flow across the base-collector junction when it is back-biased?

Answer. It is back-biased by the collector battery for the normal direction of majority carrier current flow in the collector-emitter pn diode, i.e., it is back-biased for holes to flow from p to n and for electrons to flow from n to p. However, the function of the emitter is to inject into the base region a concentration of

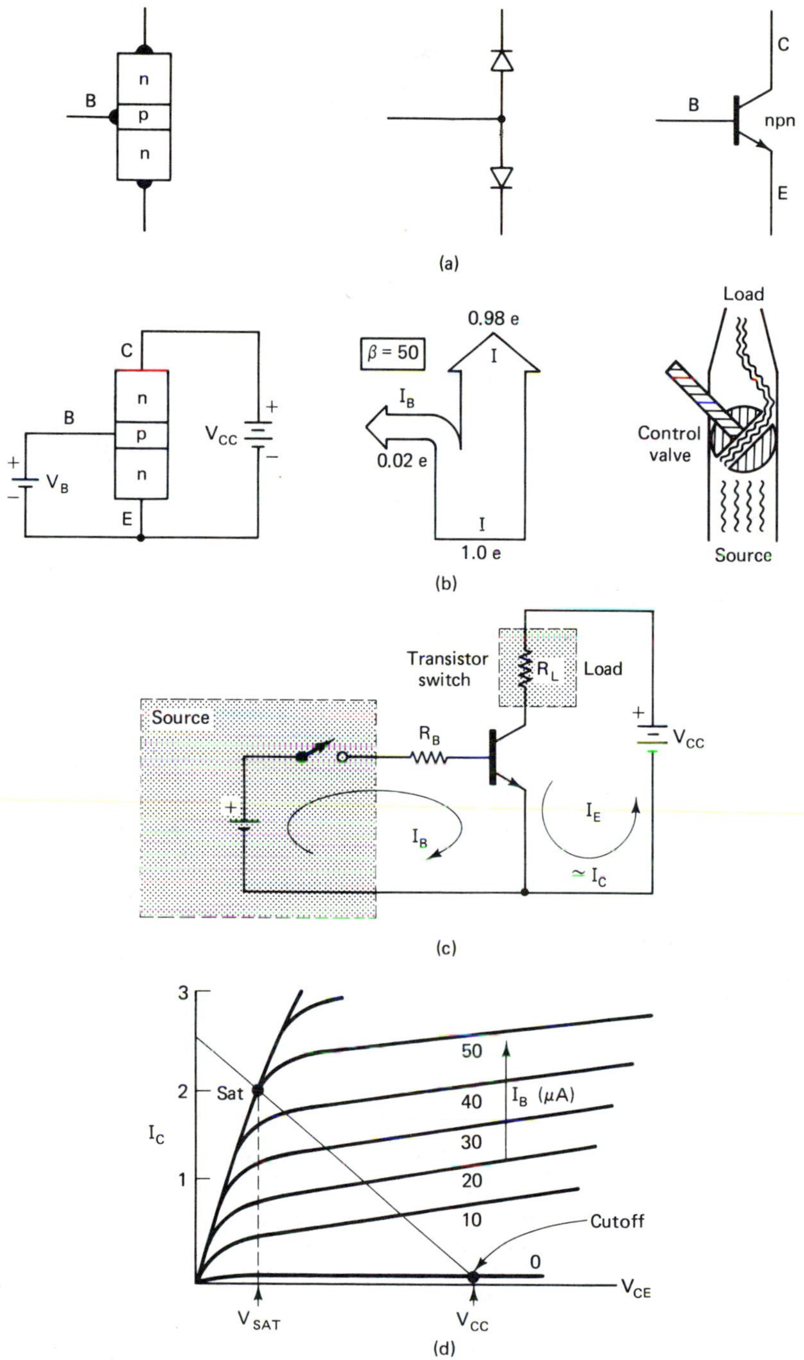

Figure 1.18 Junction or bipolar transistor, npn type. (a) Device structure, equivalent pn diode model, and circuit symbol. (b) In a circuit, most of the current flow is from emitter to collector, controlled by the base current, which acts like a control valve. (c) Transistor switch circuit. Typical values: R_L = 100 Ω to 10 KΩ; R_b = 3 KΩ to 10 KΩ. (d) Transistor characteristic curves, showing switching-mode saturation and cutoff states.

electrons (minority carriers) far in excess of the normal concentration in this p-type material. Once there, as long as they do not recombine with holes, these electrons find the base-collector junction a *forward* bias for their negative charge, and are readily accelerated across the junction into the collector.

> Definition of a junction (bipolar) transistor: a pnp or npn device whose base region is thin compared to the minority carrier *recombination length*, the distance over which an electron and hole pair will survive before recombining.

This is why two diodes back-to-back will not exhibit transistor action.

Such a device may be used as a switch [Fig. 1.18(c)]. A relatively small bias current I_B generates a relatively large current in the load resistor R_L. For a typical transistor, the ratio $I_E/I_B = \beta \simeq 100$. [Note that the direction of conventional current flow is opposite to the direction of electron flow shown above in Fig. 1.18(b).] Transistor operation may be described graphically using *characteristic curves* [Fig. 1.18(d)], with collector current plotted as a function of collector-emitter voltage for various values of the base current I_B. The curves are relatively flat, indicating that a bipolar transistor acts very nearly as a current source for a fixed base current. The curves allow one to estimate the current gain β. In this example, $\beta \simeq 50$.

To understand transistor operation as a switch, draw in a load line as for the diode circuit. Now, however, there is a family of curves, so the operating point moves along the load line as the base current is varied. A transistor *switch* is usually driven between two limits [Fig. 1.18(d)]: *cutoff*, the high-resistance off-state when $I_B = 0$, and *saturation*, the low-resistance on-state when I_B is large. The saturation voltage across a typical on-transistor is about 0.3 V. In analog circuits, care is taken to keep the transistor operating in the linear region in between.

1.9 FIELD EFFECT TRANSISTORS

1.9.1 MOSFETS and JFETS

For simplicity, the discussion of internal workings will be limited to one particular kind of field-effect transistor (FET), the metal oxide semiconductor FET, or *MOSFET*. This name refers to the layers of material found in the device. There are also FET's fabricated with bipolar junction technology (*JFET*), whose operation is similar but somewhat more complicated to understand. The MOSFET device structure [Fig. 1.19(a)] consists of a slab of semiconducting material, p-type in this example. Two small nearby regions called the *source* and the *drain* are doped to make them n-type. A wire is attached to each of these islands, and a layer of oxide applied over the device (conventionally prepared simply by oxidizing the silicon at an elevated temperature). A metal electrode is evaporated over the oxide, forming the *gate*. In typical circuit operation, the substrate and source are connected together, and that connection is tied to ground. The drain is connected to a positive voltage source. In the absence of a gate voltage, the structure

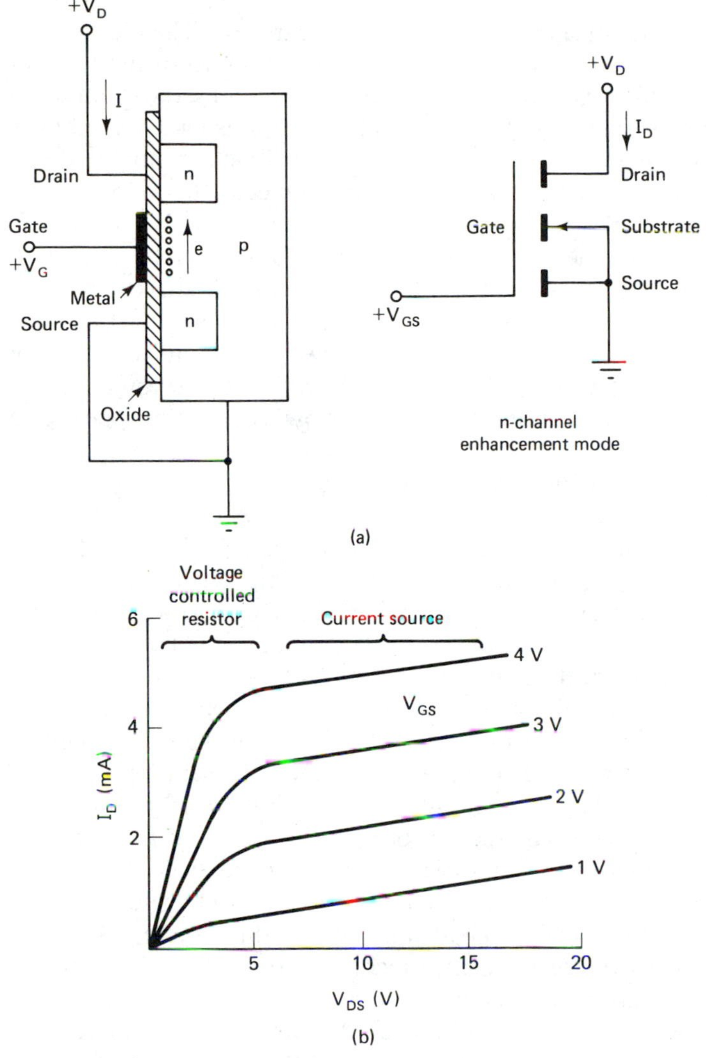

Figure 1.19 The metal-oxide-semiconductor field-effect transistor or MOSFET, n-channel enhancement mode example. (a) Device structure. (b) Circuit symbol. (c) Characteristic curves, showing two operating modes.

acts as a resistor, with a current flow between drain and source whose magnitude is determined by the amount of doping in the p-type substrate. Since the substrate is typically very lightly doped, the current flow is small. If, however, a positive bias is applied to the gate electrode, electrons are induced into the surface of the substrate under the gate.* These extra electrons can move under the influence

*Where do the electrons come from in a p-type material? The physics of the device operation actually involves the bending of energy bands by the gate bias to create a surface *inversion layer* which changes the material from p-type to n-type at the surface. The band bending can be large enough to create an inversion layer which is nearly metallic in its concentration of electrons, even though the starting material was lightly doped.

of the drain bias voltage and carry a current. (The electron flow has the opposite sign from the "conventional" current flow from drain to source.) There is no current flow across the pn junctions formed by the drain and substrate or source and substrate. The drain-substrate junction is back-biased by V_D, and the source-substrate junction does not have any positive potential to drive current flow, since source and substrate are electrically connected to one another.

The electronic symbol for this device is shown in Fig. 1.19(a). This example is an n-channel *enhancement mode* MOSFET, referring to the enhancement of conductivity produced when the gate voltage attracts charge carriers to the surface. The three bars in the electronic symbol stand for the drain, channel, and source. The arrow on the substrate wire points in the direction of the equivalent pn diode formed by the p-type substrate and the conducting n-channel. For a p-channel (n-substrate) device, the arrow faces in the opposite direction.

1.9.2 FET Characteristic Curves

The *characteristic curves* for such a device [Fig. 1.19(b)] look somewhat similar to those of a junction transistor, with two important differences. The control variable is not a current but a voltage, the gate-source voltage. An FET is therefore a voltage amplifier rather than a current amplifier. Also, because the oxide separating the gate from the rest of the device is an insulator, the gate current flow is virtually zero, and the input impedance R_i of a MOSFET is about 10^{12} Ω. These devices therefore function as very high-impedance voltage amplifiers with extremely high power gain. There are two regions of operation on the characteristic curve. For *high* values of V_{DS}, the curves are nearly flat, so the device operates as a current source. The reason for this is not obvious from the discussion given here, but originates in the variation with applied bias voltage of the *depletion region* between conducting channel and substrate. For *small* values of V_{DS}, the characteristic curves are straight lines whose slope increases as V_{GS} increases. This is physically reasonable, since increasing the value of V_{GS} creates increasingly large concentrations of charge carriers, and therefore larger current flow for a given voltage. In this region of operation, the MOSFET can be thought of as a voltage-controlled resistor, an extremely useful second mode of operation.

Another type of MOSFET uses what is called the *depletion mode.* Here, the channel between drain and source is already doped with change carriers, and the gate bias functions to push them away from the surface, reducing the conductivity. The symbol for these devices [Fig. 1.20(a)] has a solid bar between source and drain, designating conductivity even with no gate bias. The letter B on the symbol designates the substrate connection.

An FET can also be fabricated with bipolar technology. Such junction-FET's or *JFET's* have the advantage of toughness over MOSFET's, since no fragile oxide is present, but the disadvantage of lower input impedance (that of a back-biased diode) and biasing circuitry which must ensure that the gate-substrate junction never becomes forward-biased. An n-channel JFET is fabricated by incorporating a p-type controlling gate into an n-type bar. The gate bias alters the width of the insulating *depletion region* at the pn junction, altering the width of the

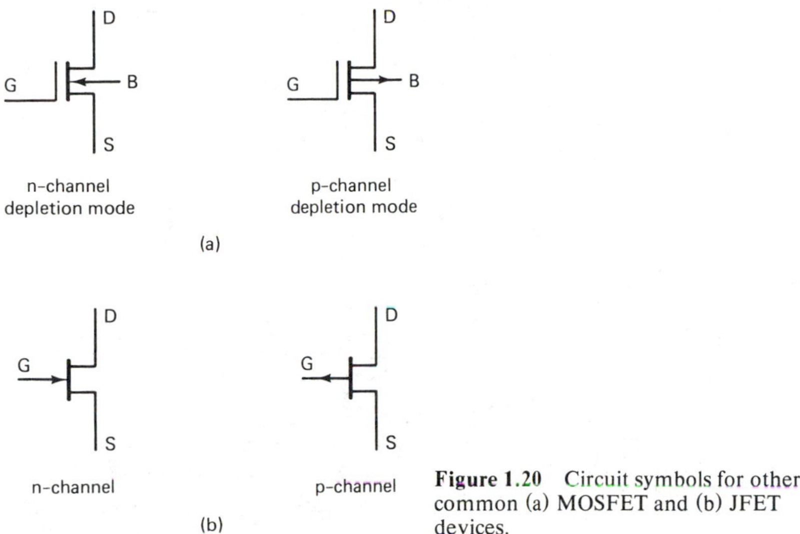

Figure 1.20 Circuit symbols for other common (a) MOSFET and (b) JFET devices.

conducting channel. Symbols for JFET's are shown in Fig. 1.20(b). JFET characteristic curves [Fig. 1.21] resemble those of a MOSFET, except that the voltage-controlled resistance region is not pronounced. The input characteristic curve is identical with that of a pn diode [Fig. 1.21(b)], so the input impedance of a JFET circuit, though high ($\sim 10^8$ Ω) is smaller than that of the insulated-gate MOSFET.

1.10 FET VOLTAGE AMPLIFIER

Because of the almost complete decoupling between input and output, FET circuits are easier to understand than the corresponding bipolar transistor equivalents. An FET voltage amplifier will be discussed as an example of the two principal design ideas used in analyzing discrete-component active circuits:

1. *Biasing*, or bringing the active device into a region of its characteristic curves where it can linearly amplify a signal
2. *Equivalent circuit* methods for analyzing circuit behavior such as gain

A JFET transistor circuit is chosen for generality. The treatment for a MOSFET is even simpler because input current flow is negligible.

1.10.1 FET Biasing

The goal is to turn on or *bias* the device with dc voltages so that it can then linearly amplify an input signal of either sign. Circuit component values are selected to bring the device to a chosen operating point [Q on Fig. 1.23]. Usually,

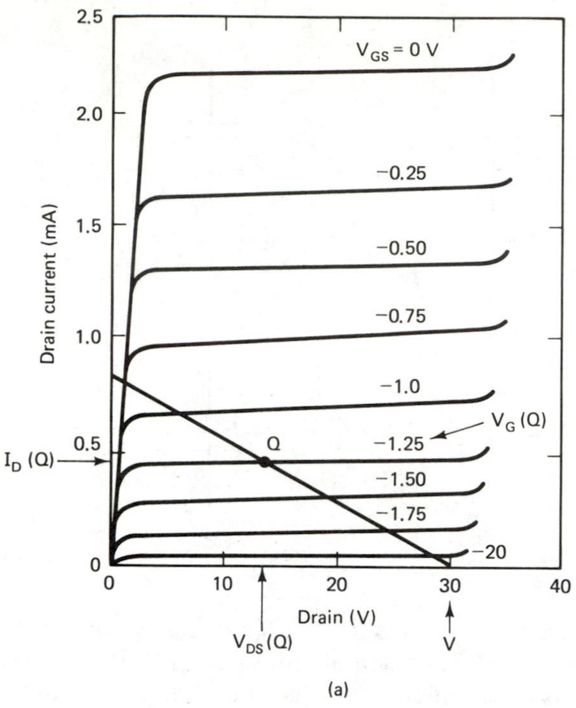

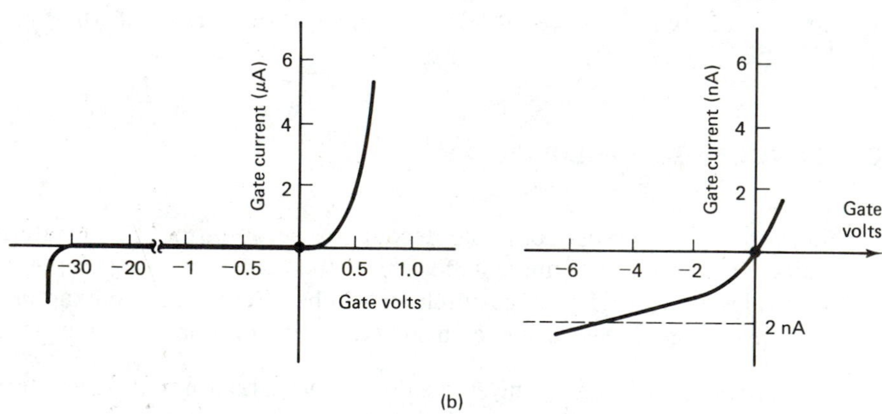

Figure 1.21 (a) Output and (b) input characteristic curves for an n-channel JFET. (*Source*: Alley and Atwood)

characteristic curves are not readily available in graphical form, and the key device specifications must be read from manufacturer's tables. Given these parameters and desired circuit properties (output current level; input power supply voltage), the design can be completed following the graphical model below.

Although the gate could be biased with a separate battery [Fig. 1.22(a)], it is inconvenient to require two separate power supplies. The conventional arrangement [Fig. 1.22(b)] employs a bias resistor R_S whose IR drop provides the needed gate bias when current flows in the drain-source circuit. First select a load

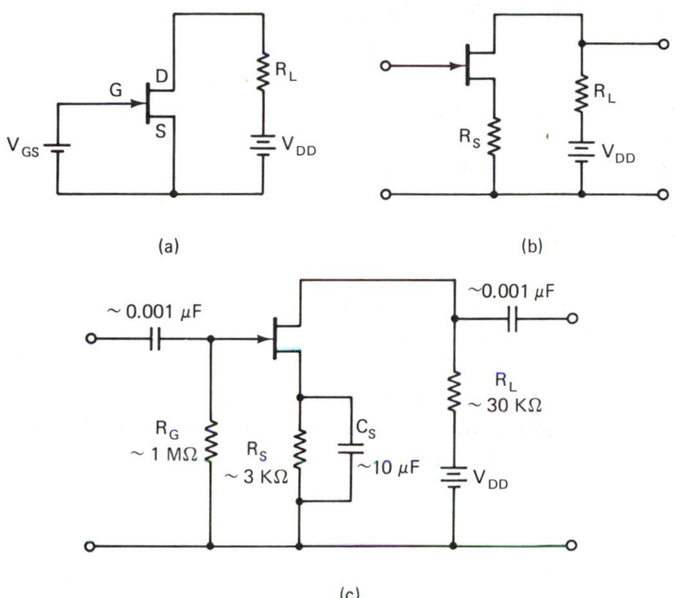

Figure 1.22 Circuits for a JFET voltage amplifier. (a) Idealized circuit. (b) Single bias-battery circuit. (c) Practical working circuit.

line containing the desired Q point. This fixes the total resistance $(R_S + R_L)$ as the slope of the triangle whose other two sides are $I_D(Q)$ and $[V_{DD} - V_{DS}(Q)]$.

$$R_L + R_S = \frac{V_{DD} - V_{DS}}{I_D} \quad \frac{30 - 14 \text{ V}}{0.5 \text{ mA}} = 32 \text{ K}\Omega \tag{1.43}$$

The value of R_S (and hence also R_L) is determined by the IR drop needed to create the desired bias voltage $V_{GS}(Q)$ when bias current $I_D(Q)$ flows.

$$R_S = \frac{|V_{GS}\,(Q)|}{I_D(Q)} = \frac{1.25 \text{ V}}{0.5 \text{ mA}} = 2.5 \text{ K}\Omega \simeq 3 \text{ K}\Omega \tag{1.44}$$

Hence, the load resistance is fixed at

$$R_L = (32 - 3) \text{ K}\Omega = 29 \text{ K}\Omega \tag{1.45}$$

In practice, several other components are added to make a working amplifier [Fig. 1.22(c)]. Resistor R_S is *bypassed* by capacitor C_S in order to prevent a deterioration of ac signal gain by the negative feedback which R_S provides. Resistor R_G is added in JFET amplifiers to prevent a shift in the operating point and nonlinear operation with large input signals. A JFET lacks the insulating oxide of a MOSFET, and a large enough ac input signal can forward bias the gate-substrate pn junction and cause a *diode clamping* effect as ac signals get rectified and charge up capacitors, shifting the operating point. R_S is chosen large enough not to load the input signal (1 MΩ to 10 MΩ is typical) but not so large as to shift the operating point as a consequence of the IR drop caused by the small gate current [~ 1 nA in a typical JFET; see Fig. 1.21(b)]. Finally, capacitors are added in input and output leads to decouple the dc bias from circuits on either side.

1.10.2 FET Equivalent Circuit

Amplifier action can be seen graphically by noting on Fig. 1.21 that an input voltage excursion of only about 0.5 V will cause an output voltage change of 12 V, or a gain of about 24. But characteristic curves are usually not available, and it is preferable to use an equivalent circuit model whose parameters can readily be read from manufacturer's specifications. The equivalent circuit model for an active device such as an FET is an extension of the Thevenin or Norton equivalent circuit. These are three terminal devices, so the equivalent circuit will have an input and an output side. One can extract from the (nonlinear) characteristic curves a set of (linear) *small signal equivalent circuit* parameters to characterize device operation.

As long as signal excursions are limited to a small portion of a characteristic curve [Fig. 1.23(a)], device response can be adequately characterized by a linear algebraic relationship, such as

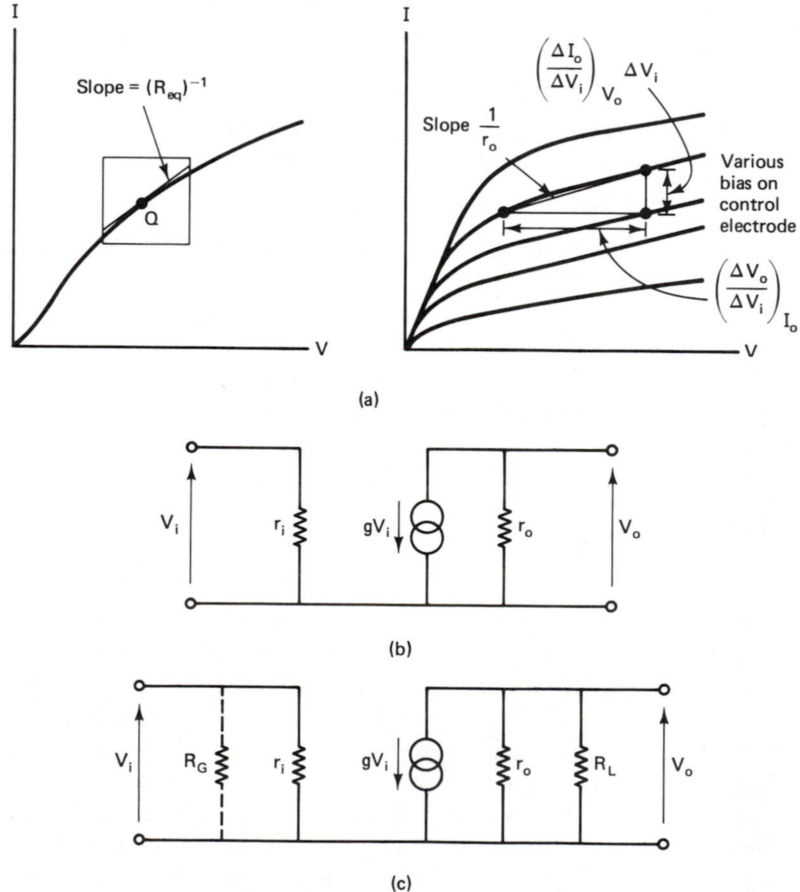

Figure 1.23 Equivalent circuit model for a JFET. (a) Extracting parameters from a small portion of the characteristic curve. (b) Small signal (linearized) JFET model. (c) Small-signal equivalent of the circuit of Fig. 1.22.

$$(R_{eq})^{-1} = (\Delta I/\Delta V)_Q \tag{1.46}$$

The subscript Q specifies that the value of the parameter will depend upon the operating point. From a black box point of view, an *active* device differs from a diode only in having a family of such curves [Fig. 1.23(b)], and requiring both input and output characteristics for its complete description. An adequate equivalent circuit model for most situations has a passive input resistor, r_i, and a nonideal current source output, as shown in Fig. 1.23(c). Here, R_o is the output resistance, and μ is the voltage gain $\Delta V_o/\Delta V_i$. The parameter g specifies the current gain of the output current source and has the units of conductance. Since g couples input to output, it is usually referred to as a *transconductance.* These parameters may be obtained graphically from the characteristic curves, as shown in Fig. 1.23(b). Although small-signal variables are often denoted by lowercase letters (v_i) to distinguish them from large-signal excursions, we will ignore this convention. So little small-signal analysis is done in this book that no confusion results. The three parameters are interrelated.

$$\mu = gr_o \tag{1.47}$$

These circuit parameters are obtainable from manufacturer's specification sheets for a given device. For FET's, it has become customary to use a somewhat different notation:

r_i	$\longrightarrow$	Y_{is}	input admittance
g	$\longrightarrow$	Y_{fs}	forward transfer admittance
$(1/r_o)$	$\longrightarrow$	Y_{os}	output admittance

Test your understanding. Estimate numerical values of the small-signal equivalent circuit parameters from the graphical characteristic curves [Fig. 1.21] and compare those numbers with tabulated manufacturer's small-signal parameters for a typical FET.

1.10.3 FET Amplifier Characteristics

The calculation of amplifier circuit properties is a straightforward extension of these equivalent circuit ideas. For an FET voltage amplifier, the main addition to the equivalent circuit [Fig. 1.23(d)] is the load resistor R_L. Other components in the actual circuit [Fig. 1.22(c)] do not appear. This follows because R_S is shorted out for ac signals by C_S, the input and output capacitors are also ac short circuits, and R_G is too large to matter. R_L appears in parallel with the output current source because (referring to the actual circuit) the bias battery V_{DD} is a shortcircuit from the ac point of view, in the sense that an ac signal applied to one end of it appears unchanged at the other end. The voltage gain is readily calculated from the output voltage

$$V_o = -gV_i[r_oR_L/(r_o + R_L)] \simeq gV_iR_L \tag{1.48}$$

Since $r_o \gtrsim 10^5\ \Omega$, the output is a good current source, and the parallel combination is dominated by $R_L \simeq 10^4\ \Omega$. The voltage gain is

$$a = V_o/V_i \simeq -gR_L \tag{1.49}$$

The voltage gain increases linearly as R_L increases. This can also be seen graphically (Fig. 1.21).

Test your understanding. Sketch graphically on the characteristic curves of Fig. 1.21 how the voltage gain increases as R_L increases. Estimate the magnitude of this gain, and compare it with values obtained from the small-signal equivalent circuit analysis.

1.11 COMMON-EMITTER AMPLIFIER

1.11.1 Junction Transistor Biasing

The two most common types of amplifiers made using bipolar transistors will now be discussed. The first, the *common emitter* circuit, is a general-purpose voltage amplifier [Fig. 1.24(a)]. R_B functions as a bias resistor, to generate a small amount of base current necessary to bring the transistor into a region of operation where it can respond to an input signal of either sign. This is shown as the point Q in Fig. 1.24(b). The value of R_L is determined by selecting a load line which will allow the device to produce the desired range of output current for a given power supply voltage V_{cc}. In this example, the load line shown requires that $R_L = V_C/I_C = 12\text{V}/4\text{mA} = 3\ K\Omega$. The last step in the design is to select the desired value of the base current in the absence of an input signal. The intersection with the load line of a particular characteristic curve at fixed I_B determines the quiescent operating point (Q point) of the amplifier [Fig. 1.24(b)]. The value of R_B may be selected by viewing the input circuit as a resistor in series with the diode formed by the base emitter junction. A forward-biased diode has a roughly constant voltage drop of $\sim$ 0.5 V across it. Once I_B has been decided upon, R_B is evaluated as

$$R_B = (V_{cc} - 0.5\ \text{V})/I_B \simeq 300\ \text{K}\Omega \tag{1.50}$$

An input signal causes the base current to vary up or down about this operating point, and as it does so the output circuit traverses the load line. Since an increase in I_B causes a decrease in the output voltage at the collector, the gain of this circuit is negative.

1.11.2 Amplifier Gain from Hybrid Parameters

Evaluating the voltage gain requires additional information not available on these characteristic curves, and is usually described in terms of the small-signal equivalent circuit. The simplest equivalent circuit for a common-emitter amplifier is shown in Fig. 1.25. There is a small ($\sim$ 2 KΩ) resistance in the forward-biased

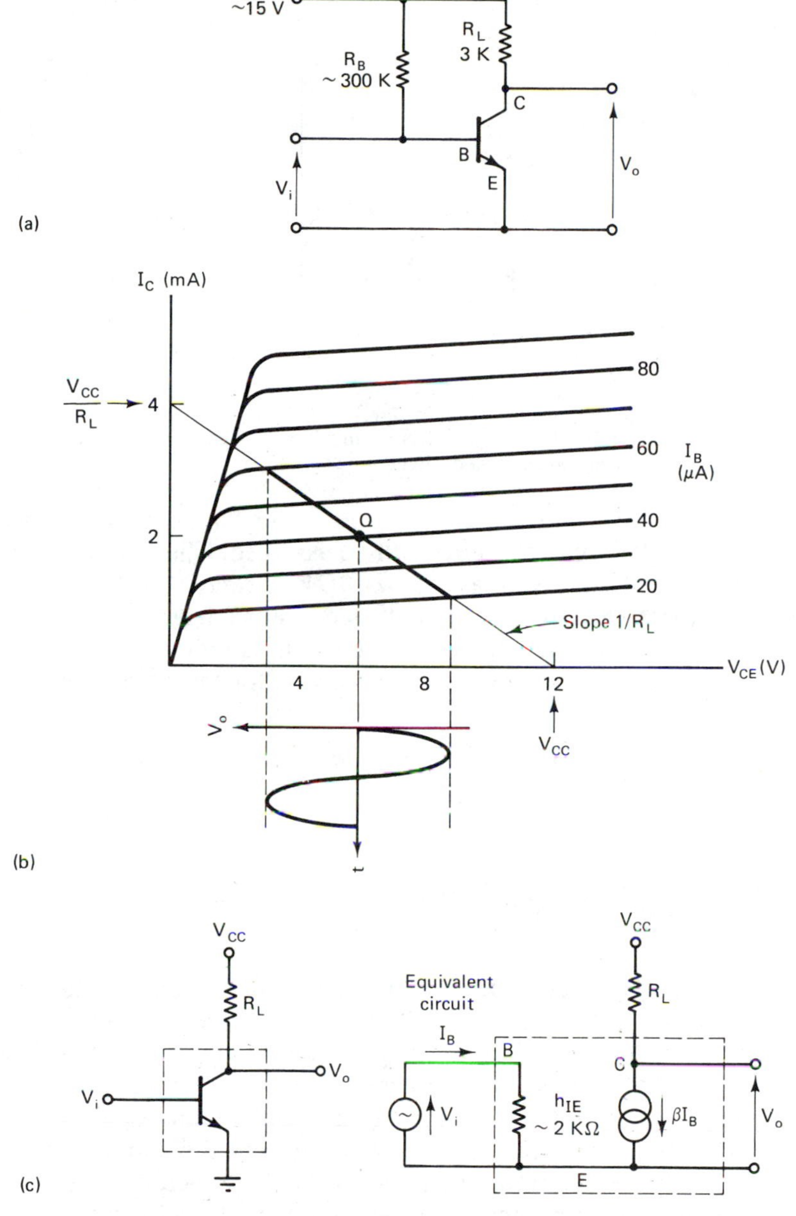

Figure 1.24 Common-emitter bipolar transistor amplifier. (a) Circuit. (b) Biasing and large-signal operation. (c) Small-signal equivalent circuit model.

base-emitter junction, labeled h_{ie}.* On the output side, there is a current source, whose magnitude equals βI_B. It is not quite a perfect current source, since the

*This labeling originates in the *hybrid parameter* analysis of transistor circuits. Here, h stands for hybrid, i stands for input, and e stands for the common emitter connection. This notation will be adopted here because it connects with the electrical engineering literature, even though it is unnecessary to use much hybrid parameter analysis in the understanding of IC circuit operation.

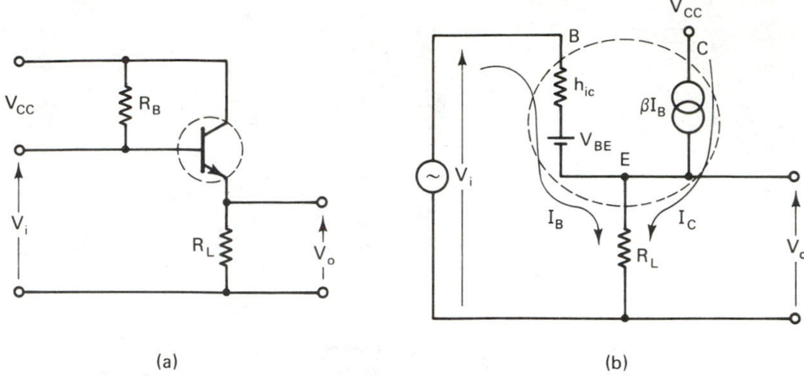

Figure 1.25 Common-collector amplifier or emitter-follower. (a) Circuit. (b) Small-signal equivalent circuit model.

slope of the characteristic curve is not quite flat, but for the purpose of this analysis an ideal current source will be assumed. The voltage gain of the circuit may be seen qualitatively as follows. If V_i increases, $I_C(=\beta I_B)$ also increases. This increased current in the output side causes a larger voltage drop in R_L, pulling V_o downward. The net gain of the circuit is therefore negative. The value of the voltage gain may be evaluated by solving for the ratio of output to input voltage change.

$$|A| = \frac{\Delta V_o}{\Delta V_i} = \frac{R_L \Delta I_C}{h_{ie} \Delta I_B} = \frac{\beta R_L}{h_{ie}} = \frac{(50)(3\ \mathrm{K\Omega})}{2\ \mathrm{K\Omega}} \simeq 75 \qquad (1.51)$$

1.12 COMMON-COLLECTOR AMPLIFIER (EMITTER FOLLOWER)

The second most useful connection of a junction transistor is the *common collector* amplifier, or *emitter follower.* It provides no voltage gain, but does provide a large current gain. It has a large input impedance and a low output impedance; it is therefore often used as a buffer stage to prevent a source from being loaded by a second circuit, and as a current booster. The common collector amplifier [Fig. 1.25(a)] has components identical with those in the common emitter amplifier, but arranged in a different order. The collector is tied directly to the power supply. The load resistor appears in the emitter lead, and the output voltage is also taken at this point. Qualitatively, circuit operation is as follows. If V_i increases, I_B increases, causing an increase in I_C. This increases the IR drop across resistor R_L, causing an increase in the output voltage. The gain of this circuit is therefore positive in sign. The subtlety is that the increased IR drop across $\mathrm{R_L}$ alters the base bias so as to oppose an increase in I_B. In fact, this effect, a form of *negative feedback,* is so strong as to keep I_B and the voltage drop across the base-emitter junction very nearly constant. The output voltage therefore follows the input voltage, except for the small constant drop across V_{BE}. This is why the circuit is called an emitter follower. Consider the simplified equivalent circuit shown in Fig.

1.25(b). A forward-biased base-emitter diode acts as a small constant battery V_{BE} in series with a small input resistance h_{ic}, where the *c* reminds us that this is the common collector connection. Because R_L is shared in both the input and output sides, it gives feedback to the input of what the output is doing, and acts like a much larger resistor, $\beta R_L >> h_{ic}$.

More quantitatively, the voltage gain may be evaluated by looking at the Kirchoff voltage loop on the input side.

$$\begin{aligned}\Delta V_i &= h_{ic}\Delta I_B + V_{BE} + (\Delta I_B + \Delta I_c)R_L \\ &\simeq h_{ic}\Delta I_B + \Delta I_c R_L\end{aligned} \tag{1.52}$$

The collector current is related to the base current

$$\Delta I_c = \beta \Delta I_B \tag{1.53}$$

which when substituted into Eq. (1.52) becomes

$$\Delta V_i \simeq (h_{ic} + \beta R_L)\Delta I_B \tag{1.54}$$

The output voltage is simply the IR drop across R_L

$$\Delta V_o = \beta R_L \Delta I_B \tag{1.55}$$

Combining the above results in the expression for the voltage gain.

$$\Delta V_o/\Delta V_i = \frac{1}{1 + (h_{ic}/\beta R_L)} = \frac{1}{1 + (2\ \text{K}\Omega)[(50)(5\ \text{K}\Omega)]^{-1}} \tag{1.56}$$

The expression in the denominator is nearly 1 because $\beta >> 1$, giving a voltage gain of unity for this amplifier. There is of course current gain, since the output current is β times bigger than the input current. This circuit therefore functions as a current booster.

The input impedance of this circuit is large, since the extra voltage drop across R_L caused by the current amplification has the correct sign to oppose input current flow in the base lead. Point E very nearly follows the voltage at point B because of the follower action [Eq. (1.56)] and the base current will be very much smaller than if the emitter were simply grounded. The input impedance is the input current flow for given input voltage change.

$$\begin{aligned}R_i &= \Delta I_i/\Delta V_i = \Delta I_B/\Delta V_i \simeq (h_{ic} + \beta R_L)(\Delta I_B/\Delta I_B) \\ R_i &\simeq h_{ic} + \beta R_L\end{aligned} \tag{1.57}$$

For typical parameter and component values ($h_{ic} \simeq 2\ \text{K}\Omega$, $R_L = 5\ \text{K}\Omega, \beta = 50$), $R_i \geqslant 250\ \text{K}\Omega$.

1.13 SCR's

The *silicon control rectifier* (SCR), also called the *thyristor*, is a high-current two-state switch [Fig. 1.26]. With no signal applied to the gate, the device is in the off-state, with cathode-anode resistance comparable to a back-biased diode. When a positive control signal I_G is applied to the gate, the device suddenly switches to a very low resistance on-state, and remains on even if the gate control signal is removed. The SCR will switch back to the off-state only if the power supply lowers the anode-cathode current I_{AC} below a threshold called the *holding current.* A closely related device is the *triac*, a bidirectional SCR which can be triggered into conduction in either direction, for ac power control.

Typical SCR parameter values are:

On-state current I_T (max)	10 A
Gate current I_G	50 mA
Gate voltage V_G	2 V
Off-state voltage V_{DRM} (max)	200 V
Holding current I_H	40 mA

The typical SCR or triac will switch a current 1000 times the value of the gate current in a time of only about 1 μs.

The physics of the SCR is subtle. The device has four layers, doped pnpn [Fig. 1.26(c)]. One may think of it as two transistors in series. The junction between layers 2 and 3 is reverse-biased and blocks current flow. When a gate current turns on the npn transistor (layers 2-3-4), emitter current in layer 2 acts as base current for the pnp transistor (layers 1-2-3) and turns it on as well. The current gains of the two transistors interact and multiply in a way which gives the latching action. See also the *solid state relay* in the next section.

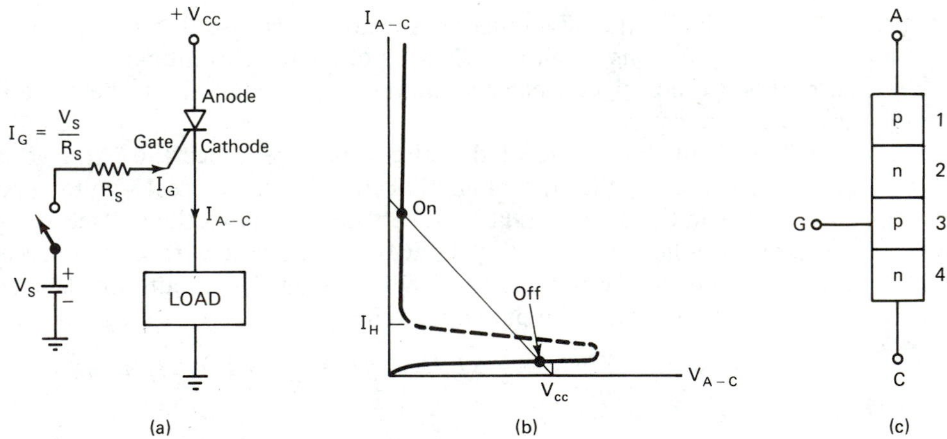

Figure 1.26 Silicon control rectifier. (a) Symbol. (b) Characteristic curve. (c) Internal structure.

1.14 TRANSDUCERS

A *transducer* [origin (Latin): to lead across] converts a "real world" quantity to an electrical quantity: a voltage, a current, or a resistance. We will discuss only a sample of the most common solutions to the most common problems. As more real world functions are controlled electronically, transducers have come to be a weak link. The microcomputer can only control what it can measure. For example, how would you cheaply instrument the measurement of CO/NO and other pollutants in automobile exhaust so a microcomputer could adjust the burning conditions to optimize catalytic converter efficiency? That problem has received an elegant solution, through a solid-state transducer (zirconium dioxide) whose conductivity changes by several orders of magnitude as the oxygen pressure varies and has a sharp change very near the optimum range for engine operation [Fig. 1.27]. Solid state IC electronics is revitalizing transducer design, with the development of IC's which can accurately measure temperature, light intensity, and force. These newer transducer types will be emphasized in the following section, with little discussion of more conventional transducers, which are amply discussed elsewhere.

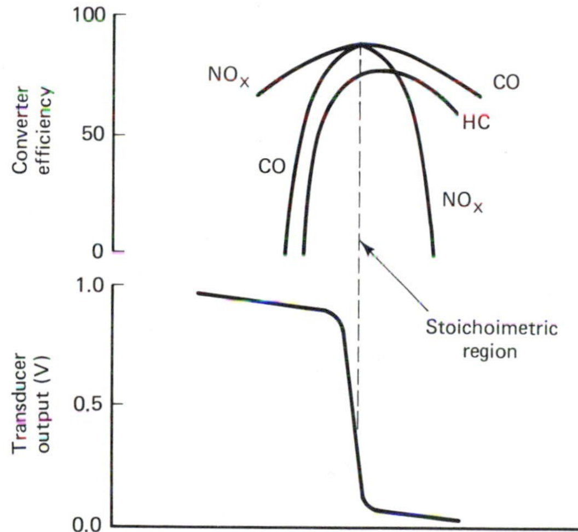

Figure 1.27 Transducer example: Oxygen partial pressure transducer for automotive exhaust. (*Courtesy* General Motors)

1.14.1 Temperature Transducers

The two older types are the thermocouple and the temperature-dependent resistor. The thermocouple is based on the thermal EMF developed when two dissimilar metals are bonded together. The thermocouple is moderately linear in T, but has a low output level (in the millivolt range) and a consequently small temperature coefficient of about 10 μV/°C. The most common resistive transducer is the thermistor, with a large but nonlinear temperature coefficient. Most industrial thermistors are made of SiC, which has a high melting point and is easy to fabricate as a ceramic. The thermistor's resistance varies approximately as $\exp(-a/T)$ and has a factor of 10 change in resistance from 0°C to 100°C.

A newer precise but inexpensive IC temperature transducer is the strongly preferred choice in the "ordinary" temperature range −50 to +150°C. Features of one example are quoted below.

Analog Devices AD590 IC Temperature Transducer

Linear output: precisely 1 μA/°K
Calibration error: ±1°C
Absolute error over entire range: ~ 2°C
Drift: < 0.1°C per month

These devices use a fundamental property of Si transistors: if two identical transistors are made to operate at a constant ratio r of collector currents, then the difference in their base-emitter voltages is $(kT/q)(\ln r)$. Since both k, the Boltzmann constant, and q, the charge of an electron, are known constants, the resulting voltage is fundamentally linear in the absolute temperature.

Circuit operation is simple [Fig. 1.28(a)]. The device is connected to a 5-V source, with a 1-KΩ series resistor to convert the output current [Fig. 1.28(b)] to a voltage. This circuit has an output of precisely 1 mV/°K.

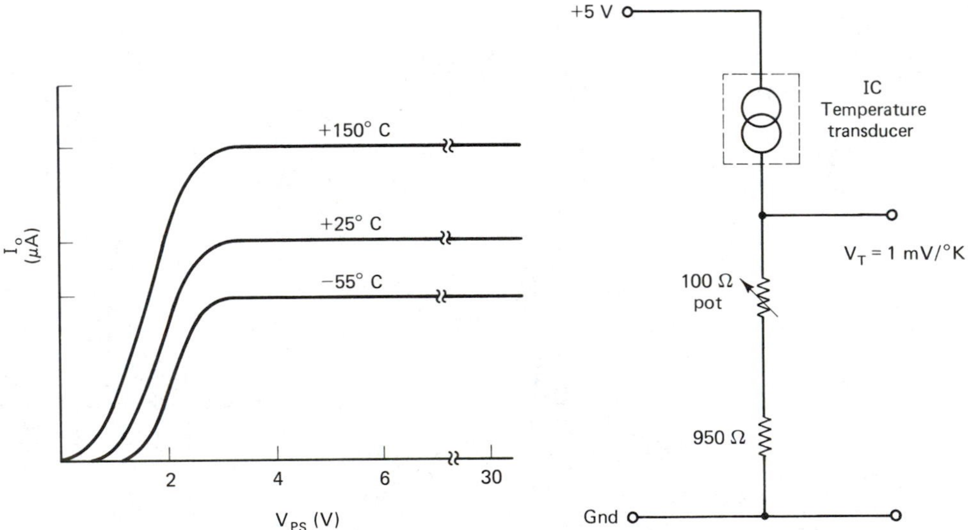

Figure 1.28 IC absolute temperature transducer. (a) Characteristic curve. (b) Operating circuit. (The example used is the AD590, *courtesy* Analog Devices, Inc.)

1.14.2 Light Transducers and Optoelectronic Devices

The older photodetector is the *photoresistor* [Fig. 1.29(a)], a bar of material such as cadmium selenide (CdSe), whose resistance changes by up to 4 orders of magnitude from darkness to room illumination. The resistance of one typical device falls from 10^8 Ω in the dark state, where the undoped material is a poor conductor, to about 10 KΩ in normal illumination of 2 foot-candles (I won't bother defining this bizarre unit) as the light creates electron-hole pairs. Photoconductive cells are slow devices, with response times in the millisecond range. Since one usually connects in a transistor to amplify the signal, it is simpler to use a *photo-*

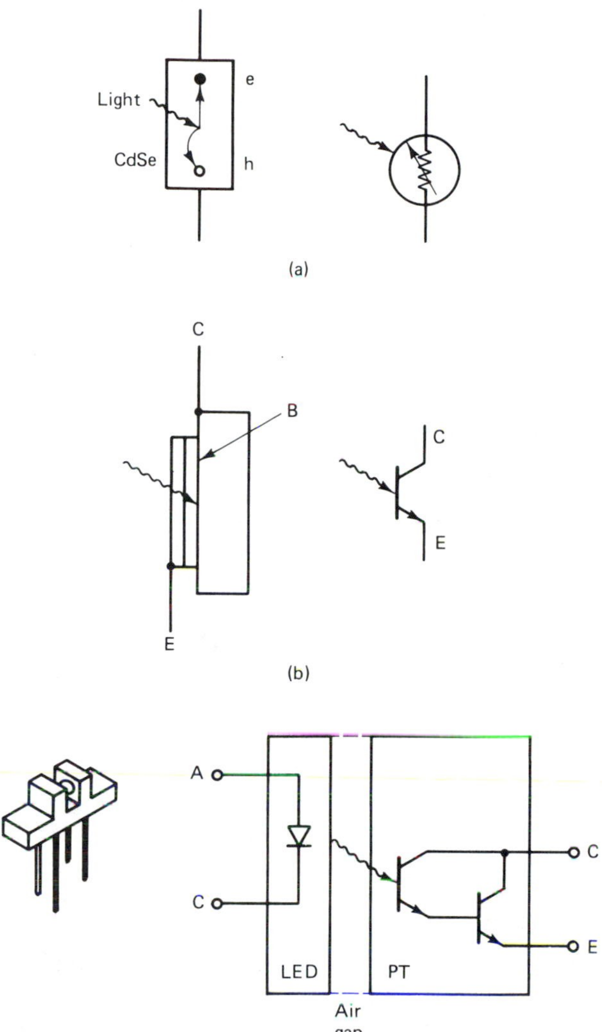

Figure 1.29 Light transducers.
(a) Photoresistor. (b) Phototransistor.
(c) Phototransistor–LED combination.

transistor [Fig. 1.29(b)]: a transistor with a window. Light creates minority carriers in the base; in fact, the base lead is usually left unconnected. The phototransistor is faster than the photoresistor because the light-induced carrier density change occurs in the narrow base region rather than in all of the bulk material. Response times of 1 μs to 10 μs are typical in phototransistors of moderate speed. Other specialized photodetectors such as the *PIN* diode (a pn diode with a narrow insulating layer at the junction) are optimized for higher speeds, in the nanosecond range. The phototransistor couples very efficiently with the *light-emitting diode* (LED), whose peak light intensity occurs just at the band gap energy where electron-hole pairs are most readily created in the detector. LED-phototransistor combinations in one package [Fig. 1.29(c)] facilitate object detection via light-beam interruption (thus permitting answers to questions such as "Is the cover

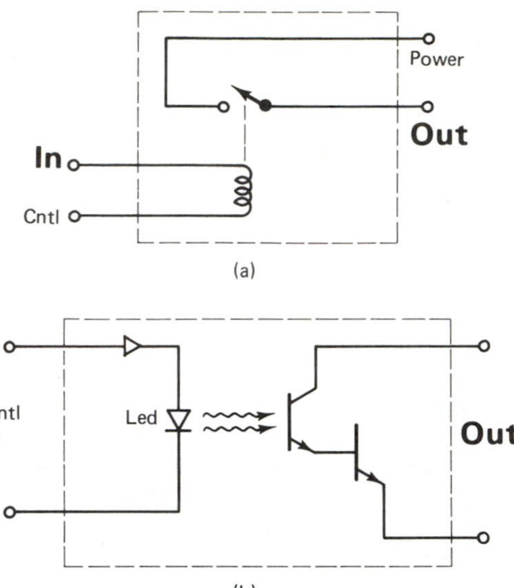

Figure 1.30 (a) Electromechanical relay. (b) Solid-state relay.

closed on the machine?'' or ''Is there a burglar in the house?'') and lend themselves to digital timing of events (Chapter 6). The device shown has a two-transistor configuration known as the *Darlington connection.* Since the phototransistor's emitter goes directly to the second transistor's base, the effective current gain of the Darlington combination is $(\beta)^2$, or nearly 10^4.

The common gallium arsenide (GaAs) LED emits mostly in the infrared range, so this combination lends itself to invisible communication on a beam of light. Optoelectronics is a rich and rapidly developing field, stimulated by the wide bandwidth of light for communications (e.g., solid state lasers and integrated optics), and by the unique electrical isolation it provides (e.g., *optoisolators* for decoupling two electronic systems).

The *solid-state relay* (Fig. 1.30) has the advantages of its electromechanical ancestor: the control signal is isolated from the system being switched. A tiny dc signal can switch high-power dc or even ac loads. A solid-state relay is TTL-compatible at the control or input side. Input and output are isolated (usually with an *optoisolator*, an LED-phototransistor pair), so the output and input can have independent grounds and no ac or noise from the output side will leak back to the input.

1.14.3 Force, Pressure, and Position

The classic stress transducer is the *resistance strain gauge* [Fig.1.31(a)]. A strain gauge is fabricated as a thin foil or evaporated film for easy bonding to the object under test. The observable quantity is the resistance change with elastic deformation of the material. Though the change is small, the strain gauge is used in a *resistance bridge* configuration so only differences are measured. Another class of deformation transducers is based on the *piezoelectric* effect. When an anisotropic insulator such as quartz is stressed, a proportional voltage appears

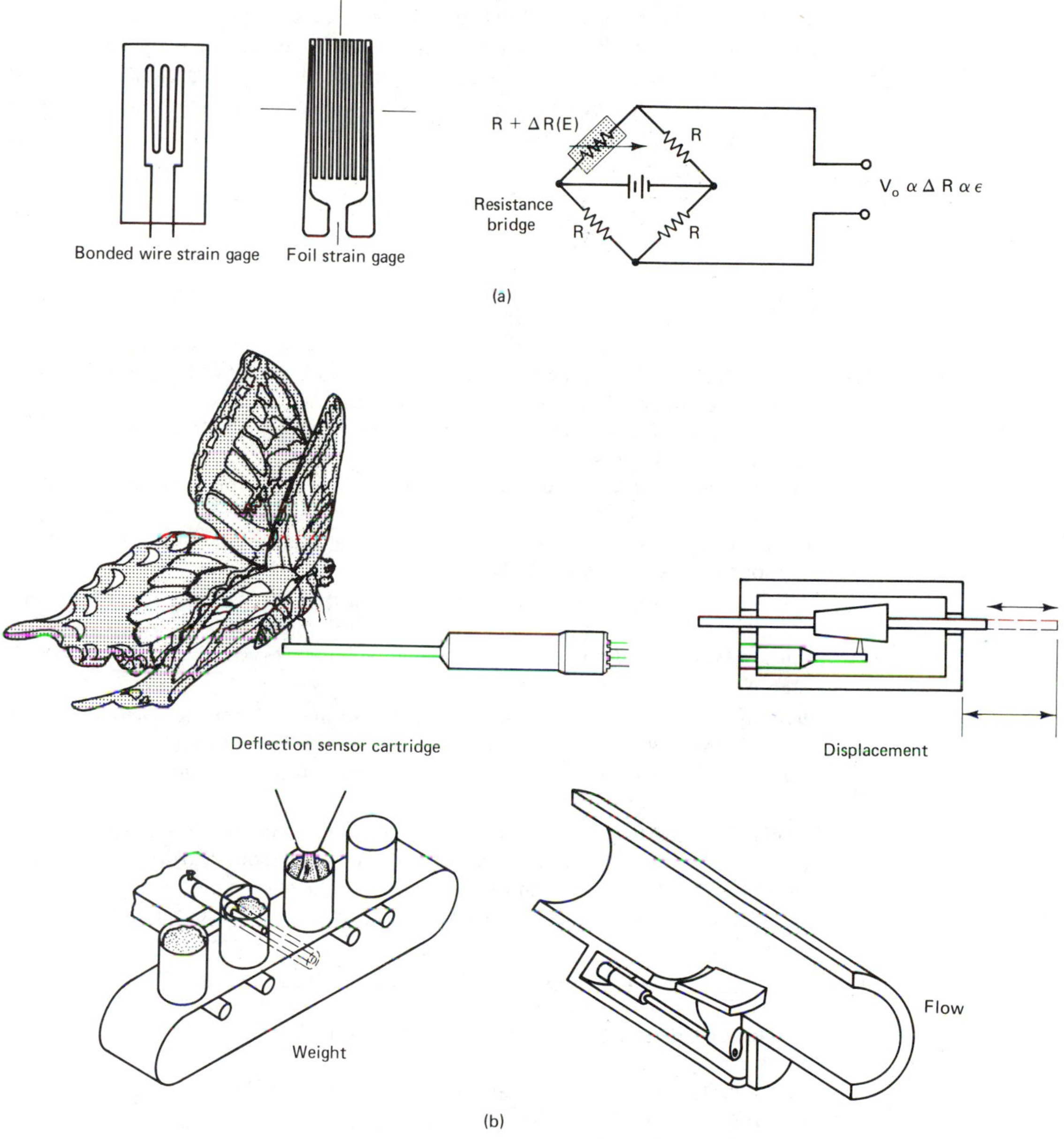

Figure 1.31 Mechanical transducers. (a) Strain gauge and resistance bridge readout. (b) Deflection sensor and its coupling methods.

across the material. This is the principle behind one kind of phonograph cartridge, as well as quartz crystal oscillators. Depending upon how it is coupled to the system, the same transducer can measure linear or angular displacement, thickness, weight, acceleration, flow, or even temperature [Fig. 1.31(b)].

IC devices are entering this arena also. There is no reason why one cannot fabricate a strain gauge deflection sensor right into an IC. National Semiconductor, for example, makes a series of inexpensive but accurate pressure transducers.

This is only a sampling of transducer types. Consult the references at the end of the chapter for transducers which fit the desired application, but bear in mind that the field is developing rapidly and manufacturers have manuals much more up to date than the textbooks.

PROBLEMS

1.1. A VOM is characterized by a quantity called *ohms-per-volt*, which is a measure of the series resistance, hence voltmeter loading effects.

- **(a)** Consider a 1-V full-scale voltmeter. Show that a 50 μa meter movement will have 20,000 ohms-per-volt.
- **(b)** The ohms-per-volt specification does not vary with the scale used. Explain, for a 10-V example.
- **(c)** How much error will be made in measuring a 5-V signal (10-V scale) whose source resistance is 1 MΩ?

1.2. A box whose open-circuit voltage (measured with a DVM whose input resistance is 10 MΩ) is 10.0 V is found to have an output voltage of 9.0 V when measured with a VOM (20,000 Ω/ V) on a 10 V scale. What is the Thevenin equivalent source resistance of the box?

1.3. Open up your flashlight, take out a battery, and measure the short-circuit current with a VOM. Set the meter on a current scale ($>$1 A full scale—careful!) and make the shorting connection only long enough to take a reading. What is the internal resistance of the battery?

1.4. Take a VOM out to your a car, if you have one. Connect the VOM, set on a volts scale ($>$12 V full scale), to the car battery. Have someone start the engine while you watch the meter and read the voltage while the starter motor is turning. Estimate the internal resistance of your car battery. A typical starter motor draws $\sim$ 50 A.

1.5.
- **(a)** Calculate the magnitude and phase of the impedance of a series RC combination (R = 10^3 Ω, C = 1 μF) when the frequency of excitation is 1600 Hz. Which component is the dominant one? Don't forget the 2π.
- **(b)** Repeat, for f = 16 Hz.
- **(c)** Repeat, but now make it a parallel RC at f = 1600 Hz.
- **(d)** Repeat (c) for f = 16 Hz.

1.6. A signal consisting of equal amplitudes of a 100 Hz and a 1000 Hz component is fed to the input of a filter. Calculate the *relative* amplitudes at these two frequencies at the output of the filter, under the following conditions:

- **(a)** Low-pass filter, cutoff frequency 300 Hz.
- **(b)** High-pass filter, cutoff frequency 300 Hz.
- **(c)** Low-pass filter, cutoff frequency 3000 Hz.
- **(d)** High-pass filter, cutoff frequency 3000 Hz.

1.7. Suppose the input to a low-pass filter is a square wave of frequency f. What will the output wave *shape* be for

(a) $2\pi fRC >> 1$?; **(b)** $2\pi fRC \sim 1$?; **(c)** $2\pi fRC << 1$?

In what regime is the filter acting as an integrator? Repeat for high-pass filter with the same frequency input square wave. In what regime is the filter acting as a differentiator?

1.8. Show that the current in the series RLC circuit is as given by Eq. (1.41). Set up an equation for the Kirchoff voltage loop. Take the time derivative of both sides to obtain a second-order differential equation. Use a trial solution of the form $I = A\exp(-\alpha t)\exp(j\beta t)$, and solve for the coefficients A, α, and β. The voltage input is a step, leading to initial conditions

$$V_C(t = 0_-) = 0 = V_C(t = 0_+)$$

$$I(t = 0_-) = 0 = I(t = 0_+)$$

Why?

1.9. Verify using complex impedances and Ohm's law the form of the transfer functions given in Table 1.1.

1.10. Select a transformer secondary voltage to drive a bridge rectifier circuit whose dc output voltage is to be 5 V. You can assume that there is enough RC filtering so the dc output is very nearly equal to the peak value of the secondary voltage; then add a volt or two for safety. When an ac voltage value is specified, the number given is the *rms* value. For example, 110 V ac has a peak value of 110/(0.707) = 156 V.

Transformer catalog Available values of secondary output voltage: 4.0; 7.5; 15.0; 30.0 V.

1.11. A 5-V RC-filtered power supply such as that shown in Fig. 1.15(a) delivers a current of 1 A to the load. How large a capacitor must be used to reduce the ripple to 1%?

1.12. Explain the operation of each of the diode circuits of Fig. 1.17. Show an output waveform for the input wave specified. It is necessary to consider what happens during the first few cycles after the signal is applied.

1.13. Why does the source impedance seen looking back at the secondary of a transformer scale as the turns ratio *squared*? **Hint :** Draw the Thevenin equivalent circuit of the actual source (V_T, R_T) plus transformer. Label it $V_{T'}$, $R_{T'}$. The value of $V_{T'}$ is $(N_2/N_1)V_T$. Since an ideal transformer neither dissipates nor creates power, the power dissipated in the actual source must be the same as the power dissipated in the equivalent source.

1.14. Explore how the Zener diode of Fig. 1.11 ceases to regulate if R_1 becomes too small. Consider the Zener and R_1 to form a parallel circuit component whose characteristic curve is the dashed line in Fig. 1.14(a). Calculate the critical R_1 below which the circuit goes out of regulation. Assume: $V_z = 10$ V; $V_b = 15$ V; $R_s = 50\ \Omega$. **Hint:** The current I_Z when $R_1 = \infty$ is the total current $I_L + I_Z$ when the load is connected. Regulation fails when I_Z goes to zero.

1.15. Design an LED driver circuit modeled after Fig. 1.18(c). The LED plus a 500-Ω series resistor is the load, and will need about 10 mA of current for adequate brightness. Pick the right R_b for a typical transistor ($\beta = 100$). Assume the source produces a 5-V signal level and $V_{cc} = 5$ V.

REFERENCES

More complete bibliographic information for the books listed below appears in the annotated bibliography at the end of the book.

ALLEY and ATWOOD, *Semiconductor Devices and Circuits*

BOYLESTEAD, *Introductory Circuit Analysis*

BROPHY, *Basic Electronics for Scientists*

DIEFENDERFER, *Principles of Electronic Instrumentation*

GIBBONS, *Semiconductor Electronics*

HIGGINS, *Experiments with Integrated Circuits*, Experiment 1

LION, *Instrumentation in Scientific Research*

MALMSTADT et al., *Electronic Measurements for Scientists*

Optoelectronic devices: see Hewlett-Packard,* *Optoelectronic Designer's Catalog*, and Monsanto,* *Solid State Optoelectronics Catalog*

Pressure transducers: see National Semiconductor,* *Pressure Transducers Data Book*

SHEINGOLD, Transducer Interfacing Handbook

SIMPSON, *Introductory Electronics for Scientists and Engineers*

SZE, *Physics of Semiconductor Devices*

*Manufacturer's data books are obtainable from local electronics distributors or manufacturer's representatives.

2

Binary Numbers and Digital Integrated Circuits

2.1 INTRODUCTION

This chapter is intended as an introduction to *using* digital integrated circuits (*ICs*). The concept of *integrating* many devices within one package has made electronics easier and also made it more powerful, because the device can be treated as a modular "black box" without knowing much about what goes on inside. Since the number of process steps is the same whether 100 or 1,000,000 devices are made at once, integrated circuits have resulted in device economy which has made it affordable to do clever things electronically. We begin with digital IC's, because digital electronics is easier to understand and easier to work with. A survey is given of some electronic devices which, because of their two-state nature, can represent binary numbers and binary operations. The next section discusses how the number of binary bits in a word is related to the resolution of a measurement or computation. Several common groupings or representations of binary numbers are then described. The chapter closes with some practical details of how to begin working nearly painlessly with IC's.

2.1.1 Binary Review

Most people learn about binary numbers in elementary mathematics. If you already understand the following two exercises, skim Chapter 2 and go on to Chapter 3.

TABLE 2.1 THE AGE GUESSING PUZZLE. IN WHICH COLUMNS A TO F DOES YOUR AGE APPEAR?

F	E	D	C	B	A
32	16	8	4	2	1
33	17	9	5	3	3
34	18	10	6	6	5
35	19	11	7	7	7
36	20	12	12	10	9
37	21	13	13	11	11
38	22	14	14	14	13
39	23	15	15	15	15
40	24	24	20	18	17
41	25	25	21	19	19
42	26	26	22	22	21
43	27	27	23	23	23
44	28	28	28	26	25
45	29	29	29	27	27
46	30	30	30	30	29
47	31	31	31	31	31
48	48	40	36	34	33
49	49	41	37	35	35
50	50	42	38	38	37
51	51	43	39	39	39
52	52	44	44	42	41
53	53	45	45	43	43
54	54	46	46	46	45
55	55	47	47	47	47
56	56	56	52	50	49
57	57	57	53	51	51
58	58	58	54	54	53
59	59	59	55	55	55
60	60	60	60	58	57
61	61	61	61	59	59
62	62	62	62	62	61
63	63	63	63	63	63

Exercise 1: Guessing Someone's Age

Show some people Table 2.1 and ask them to tell you in which of columns A through F their ages appear. The six pieces of yes/no information are sufficient to determine their ages uniquely.

The key to the puzzle is that when the number is represented in binary form, column A includes numbers whose binary representation includes a 1, column B has all numbers which include a 2, C has those numbers with a 4, etc. Someone whose age is 39 would reply: "My age is to be found in columns A, B, C, and F." This is enough to determine the person's age, as shown in Sec. 2.4.2.

Exercise 2: Optimum Random Number Guessing Strategy

What is the least number of guesses needed to identify a random number between 1 and 1000, if the only answers allowed are "it's higher than" or "it's lower than" your guess? The strategy uses binary numbers to successively narrow down the range by a factor of 2 with each guess. The optimum strategy therefore requires only N guesses, where 2^N is the first power of 2 larger than the allowed range. For a range of 1000, the first guess is 512, and it will take at most 10 guesses (since $2^{10} = 1024$).

Test your understanding. How many guesses are required if the number can be as large as 10^6?

2.2 BINARY ELECTRONIC DEVICES AND BINARY NUMBERS

2.2.1 Binary Electronic Devices

This introduction is intended to give an overview and introduce some terminology. These and other devices will be discussed in more detail in later chapters. Binary numbers may represent logic statements which are either true or false. The mathematical manipulation of such logic statements, called *Boolean algebra*, assigns the numbers 1 and 0 to quantities which are true and false, respectively. The light switches in a room are a familiar example, since the toggle switch has two states: on or off. The voltage on the light bulb also has two states: high or low. Up/down, on/off, high/low, true/false, one/zero are all examples of binary quantities. The use of binary numbers in electronic digital systems originates in the two-state nature of electronic switches used in *gates* (Chapter 3) and *flip flops* (Chapter 4).

There are other ways to manipulate signals or numbers electronically. For example, analog electronic systems and analog computers manipulate voltages which may take on any value. However, digital systems have advantages in speed, noise immunity, and cost. Digital switching transitions, which may be as fast as 10^{-9} s (10^{-11} s for some circuits under development), provide the speed advantages. Noise immunity results from a generous tolerance in the voltage values above which the circuit recognizes a 1 or below which it recognizes a 0. Noise will have no effect if it does not drive the circuit beyond these bounds. But it is economy which is most responsible for the increasing use of digital systems. The hand-held calculator, the digital watch, inexpensive digital measuring instruments, and the personal computer are familiar examples of this new digital technology. All of these have resulted from the development of integrated circuits which can squeeze thousands of circuit elements onto a slice or *chip* (see Fig. 2.1) of silicon less than ¼ in. square, without costing much more than a single device in the same package (Fig. 2.2). The number of components per circuit package has doubled about every two years since IC's were introduced! The cost per bit of memory storage elements (Chapter 7), for example, has fallen by more than a factor of 100 in the last decade, so that personal computers now may have more memory than a "huge" computer of the 60's.

Many electronic devices besides the toggle switch have only two states. Examples of two-state electronic devices are shown in Fig. 2.3. The basic electronic two-state device is the *flip flop* [Fig. 2.3(b)], which acts like a toggle switch. As Chapter 4 will explain, internal feedback inside the flip flop holds the circuit in

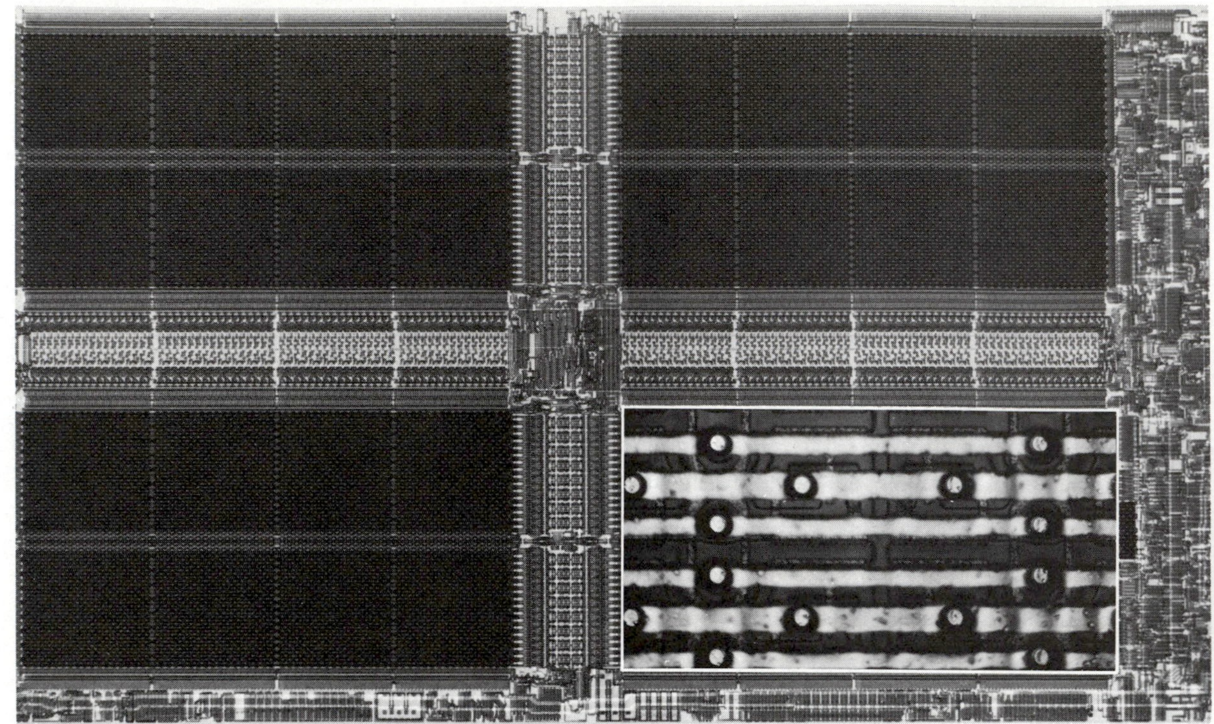

Figure 2.1 Scale of sizes in a 16-K memory integrated circuit (16-K RAM "chip"). About 1000 such chips are fabricated on one 4-in.-diameter silicon wafer, which therefore contains about 5 million bits of information. The chip is about 3×10^{-3} m wide. (Inset) The individual cells, about 2×10^{-5} m or 20 *micrometers* square, are smaller than the neurons in the brain. (Copyright Mostek 1980)

one of two stable states. A change of state is accomplished by forcing the circuit to a balance position, like a seesaw. While analog devices operate in this *linear* region where the response is proportional to the disturbance, the balance point is unstable for binary switching circuits. The flip flop was initially used for temporary data storage (called a *register*), but is now also widely used as the storage element in semiconductor memory. A flip flop constructed from discrete components cost several dollars in the 60's, but an integrated circuit flip flop now costs only a few cents in a *medium-scale integration* package, or only about 0.01 cent as part of a thousand-unit *large scale integration* package.

Two-state storage of information is also accomplished using magnetism [Fig. 2.3(c)]. An early example is *core memory:* a piece of magnetic material in the shape of a doughnut traps a magnetic field either in the clockwise or counterclockwise direction. In an IC magnetic *bubble memory*, the bits of information are part of a continuous sheet of material, which is easier to fabricate on a large scale. Another magnetic storage mechanism results from the fact that a closed loop of superconducting wire can store a current indefinitely. This, together with a fundamental quantization effect in superconductivity (the Josephson effect, Nobel

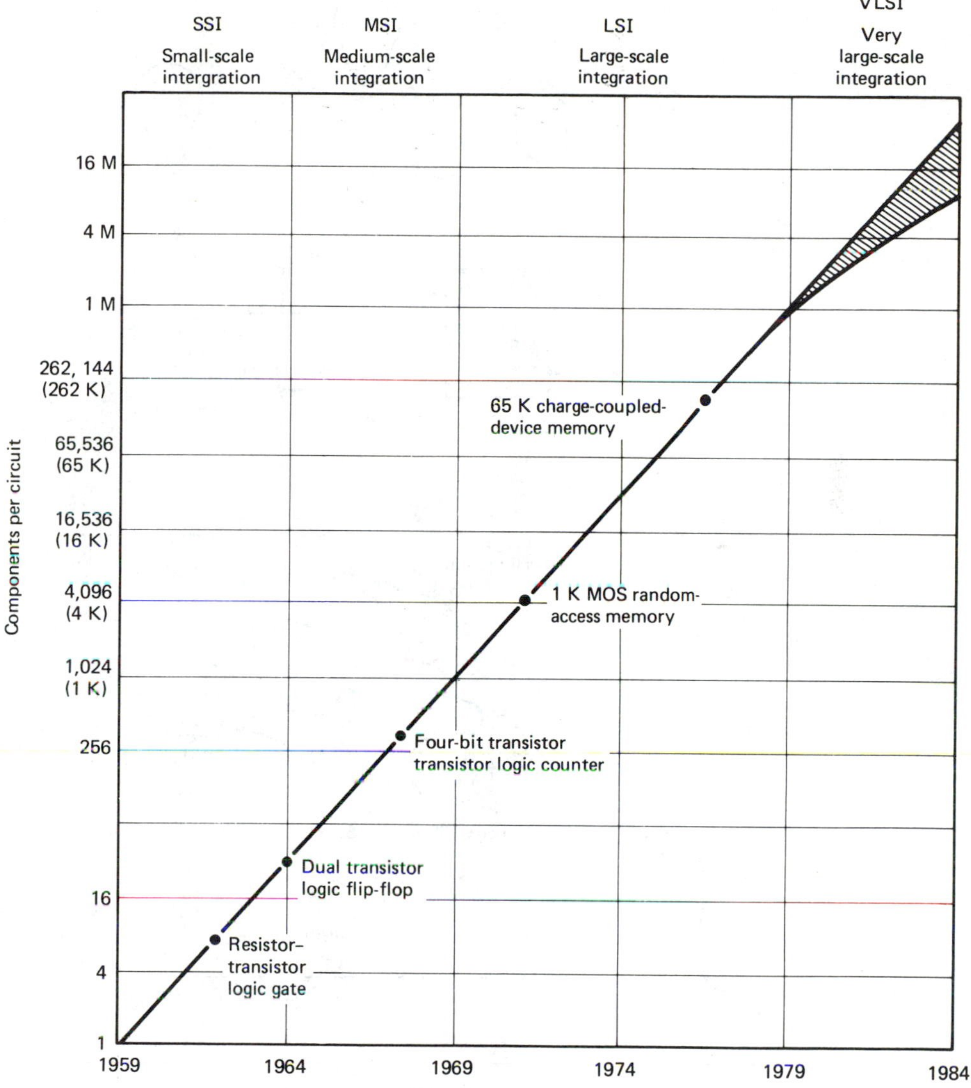

Figure 2.2 The number of devices which can be fabricated on a single IC chip is doubling about every two years. (*Extrapolated from* Robert N. Noyce, "Microelectronics," Copyright©1977 by *Scientific American*, 237:67; reproduced with permission.)

prize, 1973) is responsible for binary devices called *Josephson junction* logic elements [Fig. 2.3(d)].

Other two-state devices store electric charge. Memory devices based on *metal-oxide-semiconductor* technology store information by charging a capacitor (Chapter 7). The *charge-coupled device* memory [CCD, Fig. 2.3(e)] is an array of charge reservoirs, with information circulating in and out like sloshing water along a series of buckets.

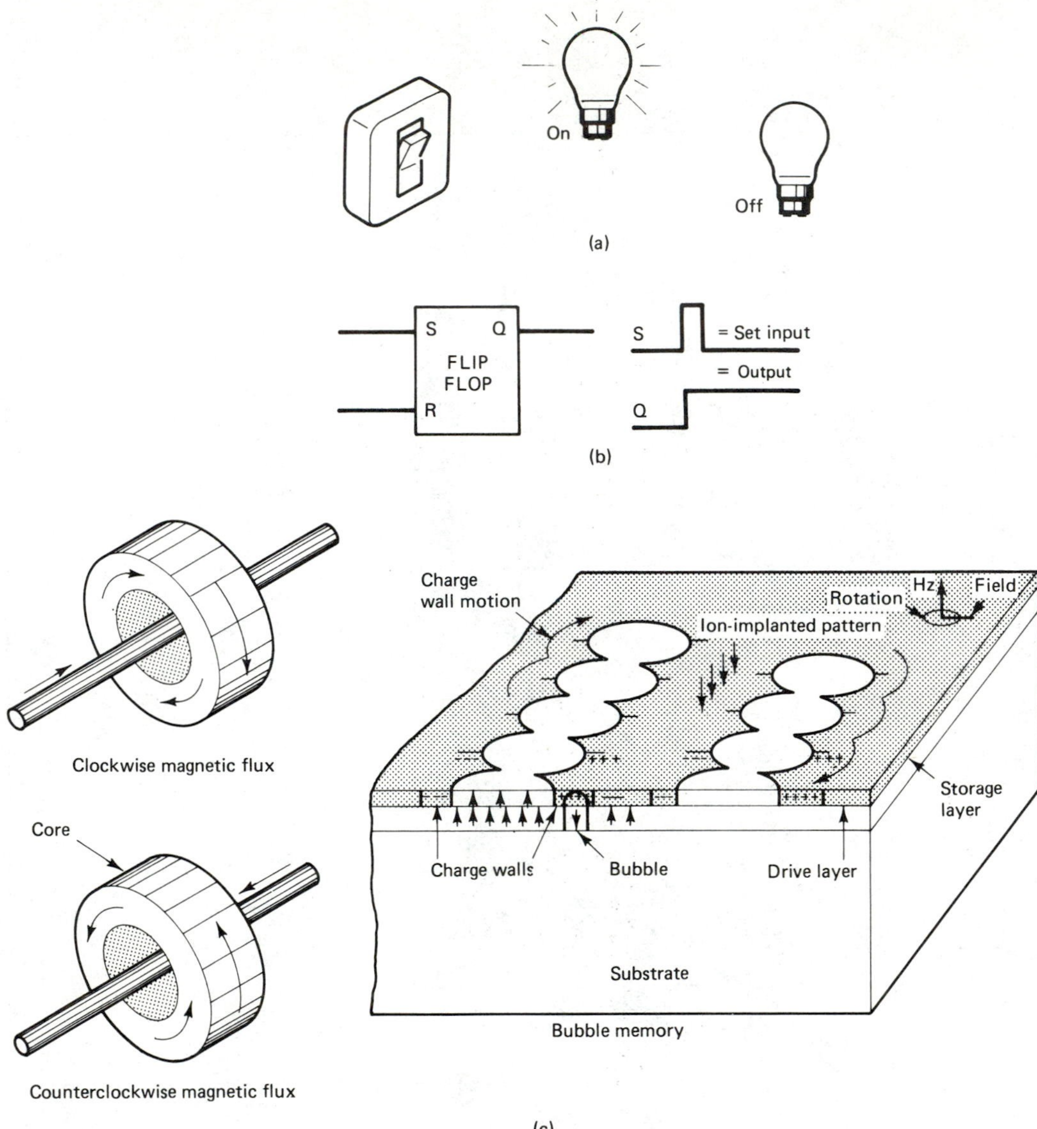

Figure 2.3 Examples of two state devices. (a) The familiar toggle switch is a two-state device. (b) A flip-flop is the electronic version of the toggle switch. (c) Binary storage elements based on magnetism. [*Source:* Electronics (August, 1979):100] (d) Josephson junction logic device. The dimpled areas in the upper center are "weak links" through which current can tunnel quantum-mechanically. (*Courtesy* IBM Corporation) (e) Binary storage elements based on charge storage; the charge-coupled device (CCD).

2.2.2 The Binary Representation

To understand the binary representation, consider what is meant when writing a number in the decimal representation. For example, consider 1984 (decimal):

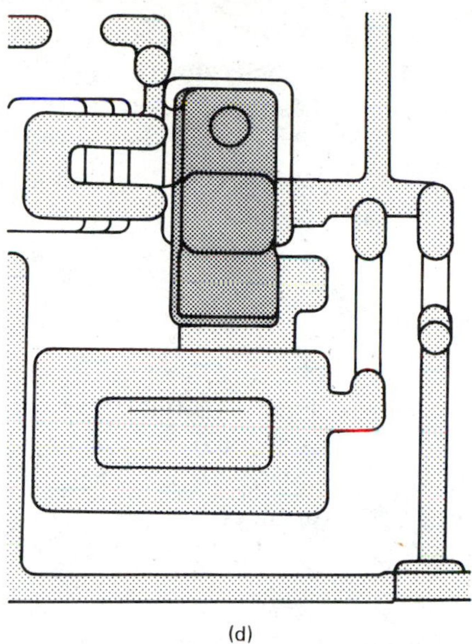

(d)

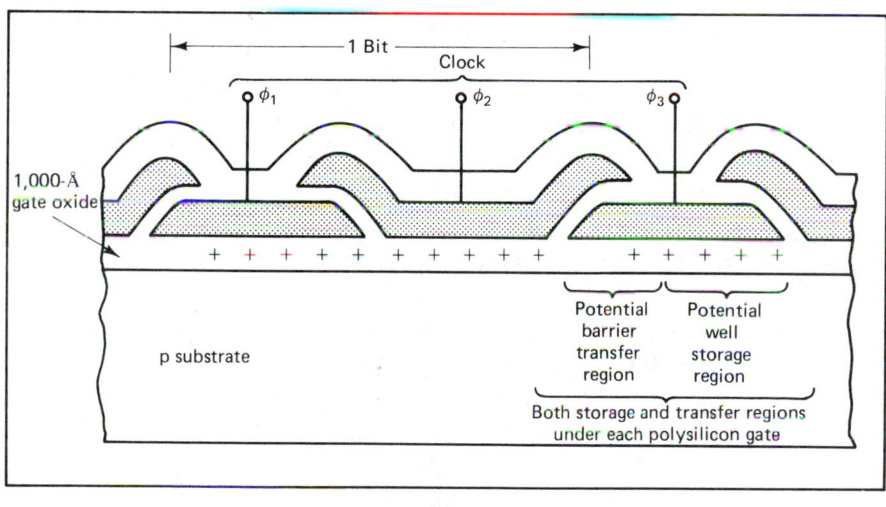

(e)

Figure 2.3 Continued

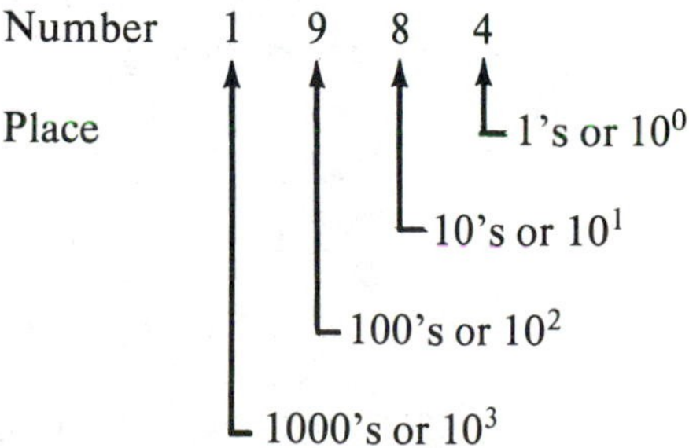

The numbers at each place specify how many times each power of 10 appears in the number. This example has four 1's, eight 10's, nine 100's, and one 1000. The same significance is attached to place values in the binary representation, with the contents of each place telling whether the number includes a given power of 2. For example, consider 10110100 (binary).

Number	1	0	1	1	0	1	0	0
Place	2^7	2^6	2^5	2^4	2^3	2^2	2^1	2^0

Each location in the binary number, called a bit, can take on only the values 1 or 0. To become adept at binary manipulations, it is essential to memorize some of the powers of 2. The range from $2^0 = 1$ through $2^{10} = 1024$ is an adequate subset, since others can be obtained from these and the rules for manipulating exponents (for a list, see Appendix 1).

Example

What power of 2 is closest to 1,000,000? The answer could of course be looked up in a table. But a close guess comes from knowing that 2^{10} is roughly 1000, since

$$10^6 = (10^3)^2 = (1024)^2 = (2^{10})^2 = 2^{20}$$

Exact answer: $2^{19} = 524\ 288$

$$2^{20} = 1\ 048\ 576$$

The approximate result is sufficiently close to the exact answer for purposes of estimation.

2.3 BINARY WORD LENGTH AND THE RESOLUTION OF MEASUREMENT OR COMPUTATION

There is a close connection between the number of bits in a binary word and the resolution possible in a measurement or computation using that word. A clear understanding of the connection is useful in selecting a computer or associated peripheral interfaces. For example, how many bits must an analog-to-digital converter have to encode a voltage to a desired resolution?

Measurement Example

To encode a voltage with 0.1% resolution, how many binary bits of information are required? To say 0.1% resolution is the same as saying that the number is known to one part in 1000. For example, $1023_{10} = 1022_{10}$ to within 0.1%. The subscript designates the *base* in which the number is being represented. The same numbers may be represented in binary form:

$$1023_{10} = 1111111111_2$$

$$1022_{10} = 1111111110_2$$

↑ MSB ↑ LSB

Here, MSB stands for *most significant bit*, and LSB stands for *least significant bit.* Since these two binary numbers are identical except for the least significant bit, a 10-bit word is necessary to represent a number to 0.1% resolution. More generally, the rule is:

> To represent a number to a resolution of one part in 2^n requires n bits of information in the corresponding binary word.

Computation Example

How large (i.e., how many digits) can two decimal numbers be without overflow or roundoff error when they are added? Consider two situations:

16 bits (typical minicomputer)

36 bits (typical large computer)

Applying the rule in the box above, $2^{16} = 64$ K. (The abbreviation K stands for kilo, or 1000, but in binary slang often stands for the nearest power of 2, i.e., "1K" = 1024.) As a result, with 16 bits, any two 4-digit numbers may be added without loss of precision, i.e., $9999 + 9999 = 19998 < 64\text{K}$. This is often sufficient for the results of measurement and control, if knowing physical variables to 0.01% is adequate. By contrast, the 8-bit word length of many microcomputers has a correspondingly smaller resolution in computation (see Prob. 1.1). Many computational or signal processing situations require taking differences of a series of numbers, and serious roundoff errors can accumulate. Such computations could be performed in *double precision*, using two 16-bit words to represent each number. But since such double precision operations take about twice as much time, machines designed for numerical processing generally use a longer word length.

What is the limit on the number of digits for accurate addition in a 36-bit computer?

Solution: $2^{36} = 2^6 \times (2^{10})^3 = 64 \times (1\text{K})^3 = 64 \times 10^9 = 6.4 \times 10^{10}$

Any two 10-digit numbers may therefore be added together in a 36-bit machine without loss of accuracy. (Note the use of exponential manipulation to avoid having to look up large powers of 2.)

Although 10 digits may seem like a large number, roundoff errors easily occur in multiplication. The product of two n-bit numbers is 2n bits long. With 16-bit word length, only words up to 8 bits long may be multiplied without loss of accuracy in this *fixed-point* method. Greater accuracy is achieved using scientific or *floating-point* notation. For example,

$$2.93 \times 10^8 \times 4.98 \times 10^9 = ??$$

Multiplication by this method requires special instructions to tell the machine how to carry out the operation. Such a *subroutine* is automatically used, for example, in hand calculators. Binary arithmetic is discussed more fully in Chapter 8. See also Prob. 2.3.

2.4 REPRESENTATION OF INTEGERS

2.4.1 Octal and Hexadecimal

When several different representations of integers are used, one must be careful to specify the *base*. Digital technology makes use of decimal, binary, octal, and hexadecimal representations, corresponding to bases 10, 2, 8, and 16, respectively. Octal and hexadecimal are used as shorthand for computer instructions and addresses. An octal digit can range from 0 to 7 and stands for 3 binary bits. The hexadecimal range is 0 to 15, written in single-digit form as 1,2,3,4,5,6,7,8, 9,A,B,C,D,E,F. A hexadecimal digit, representing 4 bits of information, is particularly convenient for microcomputers and digital communications, where the standard chunk of information is the 8-bit byte (two hexadecimal characters). Either of these is simply shorthand (Fig. 2.4), since it is more compact to write down $(177214)_8$ or $(FE8C)_{16}$ than the full binary number 1111111010001100.

Octal	1	7	7	2	1	4
Binary	1	1 1 1	1 1 1	0 1 0	0 0 1	1 0 0

Binary	1 1 1 1	1 1 1 0	1 0 0 0	1 1 0 0
Hexadecimal	F	E	8	C

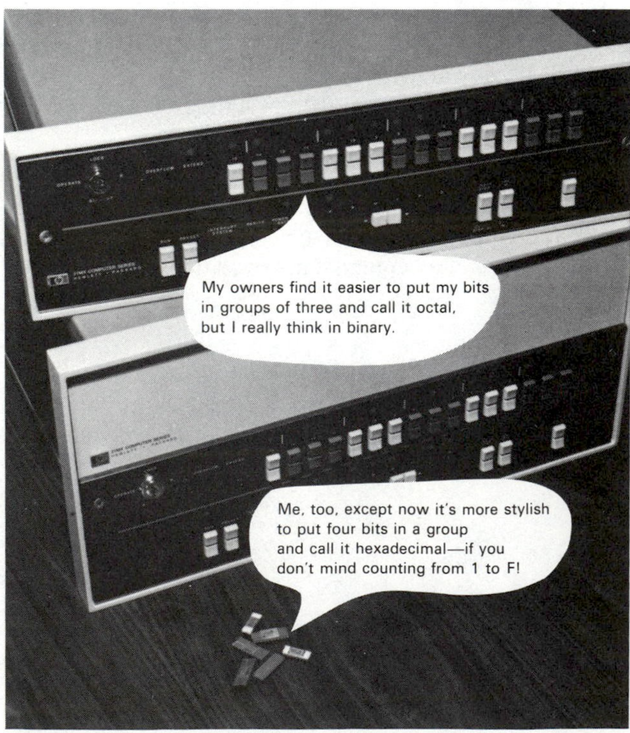

My owners find it easier to put my bits in groups of three and call it octal, but I really think in binary.
Me, too, except now it's more stylish to put four bits in a group and call it hexadecimal—if you don't mind counting from 1 to F!

Figure 2.4 Minicomputers and microcomputers use octal and hexadecimal codes as shorthand for binary numbers. (*Courtesy* Hewlett Packard Journal)

Confusion can result if the base is not carefully designated. For example, "30" can be an octal, decimal, or hexadecimal representation. Depending on the base assumed, it could be interpreted as 30_{10}, $30_8 = 24_{10}$, or $30_H = 48_{10}$. Similarly, a binary representation "101" could be interpreted as $101_2 = 5_{10}$, $101_8 = 65_{10}$, or simply 101_{10}.

Arithmetic in these different number systems involves learning to count all over again, with more or less fingers. Thus, counting in octal goes like

$$0,1,2,3,4,5,6,7,10,11,12,13,14,15,16,17,20,21. \; . \; . \; .$$

and an example of octal subtraction is $30_8 - 2_8 = 26_8$.

2.4.2 Conversion Between Bases

Converting from binary to decimal is straightforward.

Place	2_5	2_4	4_3	2_2	2_1	2_0	
Number	1	1	0	1	1	0	$= 32 + 16 + 0 + 4 + 2 + 0 = 54_{10}$

If the number contains all 1's, it is simpler to identify it in an alternative form

$$1111111 = 10000000 - 1$$

$$= 2^8 - 1 = (127)_{10}$$

Conversion from decimal to binary is done most simply by the *subtraction of powers* method. Find the largest power of 2 in the number, subtract it, then look for the largest power of 2 in the remainder, and continue until there is no remainder. For example,

$$39_{10} = 32 + 7$$

$$7 = 4 + 3$$

$$3 = 2 + 1$$

$$39_{10} = 1 \times 2^5 + 0 \times 2^4 + 0 \times 2^3 + 1 \times 2^2 + 1 \times 2^1 + 1 \times 2^0$$

$$39_{10} = (1\ 0\ 0\ 1\ 1\ 1)_2.$$

This is, of course, the trick behind the age guessing puzzle (Table 2.1) at the beginning of the chapter. The other method of decimal-to-binary conversion is called the *division method.* Divide the number by 2. If there's a remainder, the least significant bit (1's bit) is a 1; if not, the 1's bit is a 0. Divide by 2 successively, and each time the remainder (or lack of it) specifies whether successively more significant bits are 1 or 0.

Example

The number 9_{10}

9/2 = 4, remainder 1	1's bit = 1
4/2 = 2, remainder 0	2's bit = 0
2/2 = 1, remainder 0	4's bit = 0

$1/2 = 0$, remainder 1   8's bit = 1
$9_{10} = (1001)_2$

See Problem 2.8.

2.4.3 The BCD Representation

A representation used in digital measurements and in calculators is called BCD, which stands for *binary coded decimal.* This system is used for example to encode the lighted digits in calculators, digital watches, and digital measuring instruments. Each digit is driven by four wires containing binary information. Four binary bits are required, since $9_{10} = 1001_2$. Any device with decimal displays has this BCD information readily available, so interfacing such outputs to mini- or microcomputers is most simply done in BCD format. However, BCD is a less efficient code than binary, requiring more bits to send the same amount of information. Suppose the result of a measurement is the number 1023_{10}.

$$1023_{10} = (0001)(0000)(0010)(0011)_{BCD}$$
$$= (1111111111)_2$$

It takes 16 bits to represent this number in BCD, but only 10 bits in binary. The difference is more substantial with multidigit, high-precision measurements (see Prob. 2.4). However, as long as speed or the number of physical wires is not a problem, it is more convenient to transmit information already available in BCD directly to the computer and then do a code conversion in software.

Calculators and even some computers do arithmetic in BCD, because it resembles decimal. For example, the statement

$$9 + 1 = 10$$

becomes in BCD arithmetic

$$(1001)_{BCD} + (0001)_{BCD} = (10000)_{BCD}$$

rather than the binary sum $(1010)_2$.

2.5 WORKING WITH INTEGRATED CIRCUITS

The "user's" attitude of this book is to jump right in and try to use IC's, without necessarily understanding how they work, at least at first. Once the initial fear is lost and a certain confidence is gained, the user grows intrigued enough to go deeper into both understanding and applications. The background of this chapter is enough to begin using digital IC's. For example, the circuit shown schematically in Fig. 2.5 demonstrates how to represent binary numbers electronically, lighting a pattern of LED's which represents the binary number. A device called the *7-segment display*, familiar from calculators and digital watches, reads out the number in more familiar numeral format. The 4 binary bits are converted to a

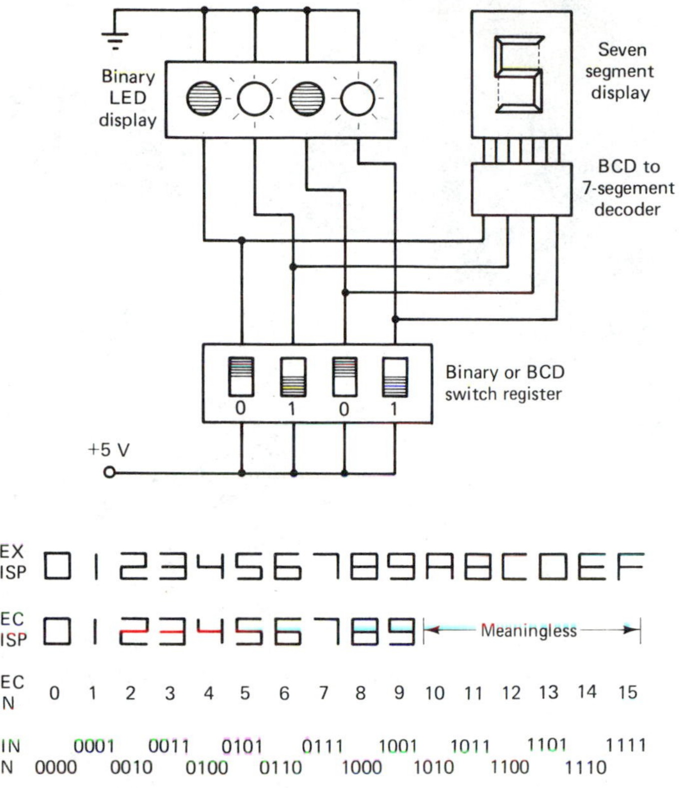

Figure 2.5 BCD or binary-coded decimal represents a decimal digit with four binary bits. A seven segment LED display represents the numbers with lighted bars, and may have decimal or hexadecimal range.

code which lights the proper segments by a device called a *decoder* (Chapter 3). A *decimal display* with range 0 to 9 has its segments driven by a *BCD to 7-segment* decoder. Binary numbers beyond 9 light up a meaningless segment pattern. However, if the decoder/display element has a *hexadecimal* range, all 16 of the 4-bit binary numbers light up a meaningful pattern. Hexadecimal displays are convenient for microcomputers and microcomputer-based instruments.

The example of Fig. 2.5 can be easily tried out using readily available IC breadboarding tools [Fig. 2.6]. IC circuits can be quickly wired together on a breadboard socket [Fig. 2.6(a)]. The IC's plug in along the center. Each IC pin is thereby connected to a string of other connection sockets located beneath the surface. Components or wires are pushed into these spring-connect sockets to complete the circuit. Although more permanent circuits should be made on more reliable printed circuit boards, the breadboard method facilitates getting circuit design tried out first.

Besides the breadboard, several other functions are basic enough to digital experimentation that they are incorporated into test boxes [Fig. 2.6(a)]. One needs binary readout lights, slide or toggle switches to input binary numbers, and pushbutton switches to input pulses. A variable-frequency square wave, called a

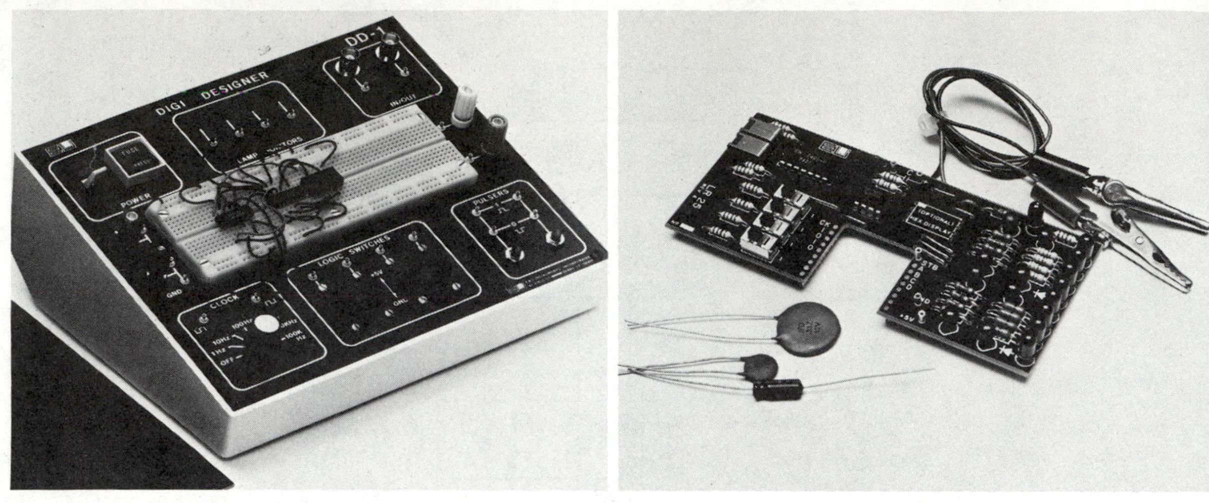

Figure 2.6 Integrated circuit breadboarding aids. (a) Breadboard socket and logic trainer or "Digidesigner." (b) "Outboard" plug-in with commonly used digital circuit test functions. (*Courtesy* E&L Instruments, Inc.; the original outboard modules were developed in coordination with Larsen, Rony, and Titus for the *Bugbook* series.)

clock, and one or more 7-segment decimal displays are also convenient. One particularly useful configuration has these functions prewired to plug into the breadboard socket [Fig. 2.6(b)].

PROBLEMS

2.1. **(a)** Estimate the resolution possible in additions or measurements using a microcomputer with 8-bit word length. Assume that the operation is performed on a single 8-bit word.

(b) What is the resolution easily visible in a graph plotted on standard 8 ½-in. by 11-in. paper? Is 8-bit resolution adequate?

2.2. Convert your telephone number (or social security number or bike lock combination) from decimal to binary, then from binary to octal and hexadecimal, then back to decimal again. Writing such a lock combination somewhere visible but in nondecimal format is useful if you tend to forget combinations easily, since fewer thieves are familiar with nondecimal formats!

2.3. On many scientific calculators, when two numbers are multiplied and the answer exceeds the width of the storage registers, the answer is returned in *scientific notation* or exponential format (e.g., 3×10^9), even though both numbers were entered and stored in *fixed-point* format (e.g., 2.02). Determine by experiment if this feature exists on your calculator. Find the number N such that N^2 comes back in exponential format. Based on this experiment, how many bits wide is the register used to store the product?

2.4. A voltage measurement yields 10.354 V. Neglecting the decimal point, how many wires are needed to send this result to a computer in **(a)** parallel BCD format; **(b)** parallel binary format?

2.5. Suppose a measurement of frequency is made using a counter with 8 decimal digits of precision at the output. How many bits are needed to send this information to a computer in BCD format? How many bits are needed to send the same amount of information in straight binary format?

2.6. Consider the numbers 11, 17, and 19. Which could be octal as well as decimal? Which could be binary, octal, or decimal? Write the decimal equivalent if the numbers are interpreted as binary or octal.

2.7. Practice converting from one base to another.

(a) To decimal from binary: 10110101; 111010011; 0100001101. Use the *subtraction of powers* method.

(b) To binary from decimal: 28; 57; 132; 1025.

2.8. Convert the number $(79)_{10}$ to binary by the *division* method.

2.9. Your computer has 64K bytes of memory. What is the last available word of memory **(a)** expressed in hexadecimal; **(b)** expressed in octal?

REFERENCES

More complete bibliographic information for the books listed below appears in the annotated bibliography at the end of the book.

HIGGINS, *Experiments with Integrated Circuits*, Experiment 2

FLOYD, *Digital Logic Fundamentals*

LARSEN and RONY, *Logic and Memory Experiments Using TTL Integrated Circuits*

MILLMAN, *Microelectronics*

Scientific American, *Microelectronics*

WILLIAMS, *Digital Technology*

Science, vol. 195, 18 March 1977, special issue on *The Electronics Revolution*

3

Gates and Digital Logic

3.1 ELECTRICAL SWITCHING AND MATHEMATICAL LOGIC

The connection between binary information and electronic circuits is the two-state switch. Just as a binary number can be either 0 or 1, a light switch can be either down (off) or up (on). One possible connection between binary, logical, and electrical representations of the same information is shown below.

Binary	1	0
Logical	True	False
Electrical voltage	High	Low
Switch	On	Off

The operations performed by electrical or electronic switches can be thought of as representations of logical operations. For example, for a relay (Fig. 3.1), a nonzero voltage will be present at the output C only if a nonzero voltage is simultaneously present at *both* inputs A and B. The relay therefore performs the function of an *AND* gate [Fig. 3.1(b)].

Mathematical logic is a convenient shorthand to represent verbal logic statements. The statement: "C is true if and only if both A and B are true" has an equivalent mathematical logic statement

$$C = A \cdot B \tag{3.1}$$

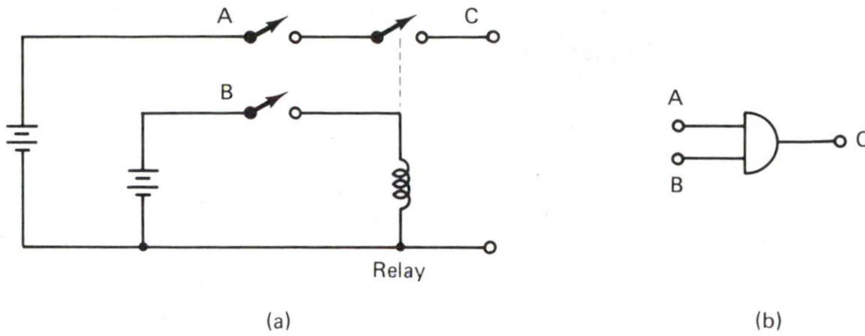

Figure 3.1 (a) The gate concept, illustrated by a relay logic circuit. (b) AND gate symbol for this logic function.

The dot is the symbol for AND function, and equals sign stands for logical *equivalence:* the right-hand side is true whenever the left-hand side is true. Similarly, the statement: "C is true whenever A or B (or both) are true" has a mathematical logic shorthand

$$\mathrm{C} = \mathrm{A} + \mathrm{B} \tag{3.2}$$

where the plus symbol stands for the logical OR function. This is called the *inclusive OR* function, to distinguish it from *exclusive* OR function: C is true if either A or B is true, but not if both are true. The symbol for the exclusive OR function is a plus sign with a circle around it:

$$\mathrm{C} = \mathrm{A} \oplus \mathrm{B} \tag{3.3}$$

For the circuit shown in Fig. 3.2(a), if the light bulb being *on* represents truth, what logical function is performed by the combination of switches A and B? Since the light will go on if either A or B is closed (true), this switching circuit performs the logical OR function. The switching network of Fig. 3.2(b) performs the logical AND function. Many other switching problems can be viewed as logical operations, such as the familiar circuit which turns a light on or off from either of two locations (Prob. 3.1).

Another logical operation is negation, symbolized by a bar above the quantity. The symbol $\overline{\mathrm{A}}$ represents the logical operation "not A". If A is false, then $\overline{\mathrm{A}}$ is true.

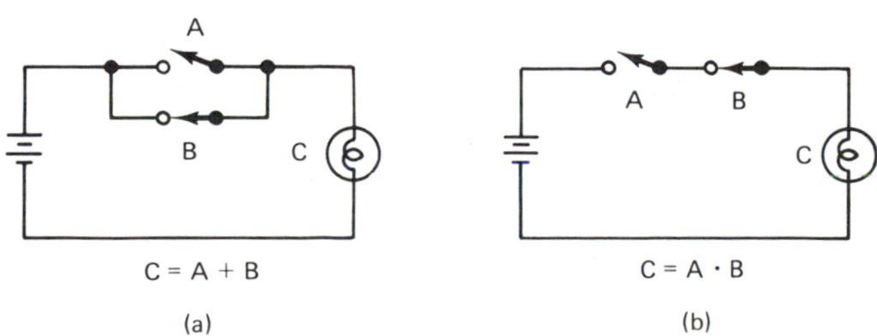

Figure 3.2 Switching circuits which perform (a) the OR and (b) the AND logic function.

An examination of logical combinations demonstrates the need for rules for manipulating mathematical logic. For example, are the expressions shown in Eq. 3.4 equivalent?

$$\overline{A} \cdot \overline{B} = \overline{A \cdot B} \tag{3.4}$$

They are not equivalent. Suppose A is true and B is false. Then the quantity $\overline{A} \cdot \overline{B}$ is false, but the quantity $\overline{A \cdot B}$ is true. The equivalence is therefore disproved. Testing such expressions verbally becomes very complicated. An improvement is to assign "values" 1 and 0 to represent true and false, then test the equivalence by example.

$$\overline{A} \cdot \overline{B} = 0 \cdot 1 = 0 \qquad \text{false} \tag{3.5}$$

$$\overline{A \cdot B} = \overline{1 \cdot 0} = \overline{0} = 1 \qquad \text{true}$$

Boolean algebra (Section 3.5) describes the mathematics of manipulating logic, including methods much more powerful than this proof by example.

3.2 HOW TO MAKE A GATE: DTL AND TTL GATE CIRCUITS

3.2.1 Diode Gates

Although the first computers were constructed using relays, speed limitations soon led to the use of electronic switching. Voltage levels perform gating functions in such circuits, using components whose resistance can be altered from a high to a low state. Early electronic gates used diodes. An example is shown in Fig. 3.3. Its operation is best understood as the "low man wins" gate. The output X is low if any of the inputs is connected to a low (ground) signal. This occurs because that diode becomes forward-biased, resulting in a current flow through resistor R which lowers the voltage at point X to nearly 0 V. This output voltage is somewhat higher than ground because of the diode forward voltage drop (typically 0.6 V). It is typical of solid-state logic circuits that the logic 0 and 1 states are not exactly at the ground and power supply potentials. Once any of the diodes is

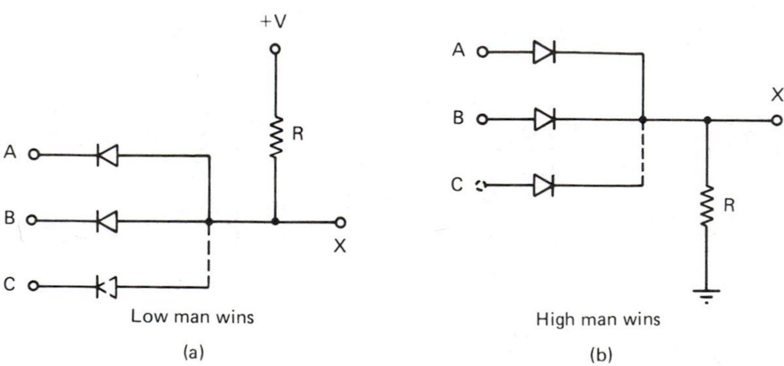

Figure 3.3 Diode gates. (a) "Low man wins." (b) "High man wins."

TABLE 3.1 VOLTAGE LEVELS AND TRUTH TABLES FOR THE "LOW MAN WINS" GATE

A	B	X
Voltage levels		
Low	Low	Low
Low	High	Low
High	Low	Low
High	High	High
Truth table if high = 1 and low = 0		
0	0	0
0	1	0
1	0	0
1	1	1
Truth table if high = 0 and low = 1		
1	1	1
1	0	1
0	1	1
0	0	0

forward-biased, the output state is independent of the status of the other inputs, since neither a reverse-biased diode nor a second forward-biased diode will alter the output voltage at X. This may be summarized by enumerating a table of possibilities [Table 3.1(a)].

If we identify the low-voltage state as the logic 0, and the high-voltage state as the logic 1, these states describe a *truth table* [Table 3.1(b)] which performs the logical AND function. If, on the other hand, we identify a low voltage as a logic 1 state and a high voltage as a logic 0, the truth table [Table 3.1(c)] matches that of the logical OR function. This change in the logic function due to a reversal of the logic polarity assignment occurs often, and will be dicussed in Section 3.3.

Another diode gate is called the "high man wins" gate [Fig. 3.3(b)]. Here, the output is high if any of the inputs are high. The explanation is left as Prob. 3.2. This gate performs the logical OR function if high = 1.

Diode gates formed the input side of a logic family called DTL (Diode-Transistor-Logic), now superseded by TTL (next section). Although occasionally used to construct switching matrices, diode gates are limited by slow switching speed and poor noise immunity. The diode, with junction capacitance C_J, stores charge and thereby limits switching speed to a time of the order of $C_J \cdot R$. With C_J typically 10 pF and $R > 10^3\ \Omega$, the switching time is as long as 10 μs. Diode gates have limited noise immunity because the transition from low to high states lacks "snap action." Improved gate circuits make this transition more sudden. Finally, diode gate inputs are poorly decoupled from one another, allowing noise to propagate through a circuit.

3.2.2 TTL Gates

Improved switching time (down to 10^{-8} s) and noise immunity result from replacing diodes with transistors. If the transistors are of the bipolar (junction) type, the logic family is called TTL, for *transistor-transistor-logic.* Transistors in digital circuits are operated as two-state devices, and are either driven into saturation or into cutoff (see characteristic curves, Fig. 1.18), with the linear active region traversed only during the switching transition. The switching characteristics used in TTL gates are summarized below.

Base-emitter	*Collector-emitter*
Forward bias	Saturation: *on*
Reverse bias	Cutoff: *off*

The circuit of a complete TTL NAND gate is shown in Fig. 3.4. The input transistor of TTL gates has multiple emitters. This is done by making a base region large enough for several emitter islands (Fig. 3.5). The circuitry of a TTL gate is designed to ensure a rapid transition between saturation and cutoff states, analogous to a toggle switch. The output stage (Fig. 3.6) contains two transistors

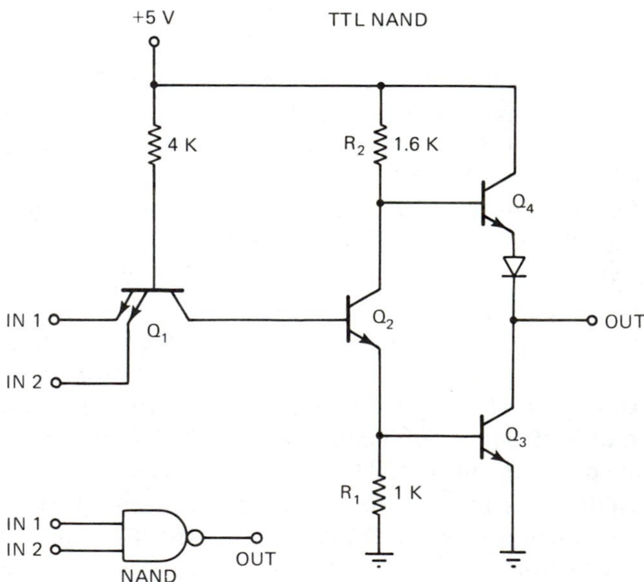

Figure 3.4 Complete circuit within a TTL IC NAND gate (7400).

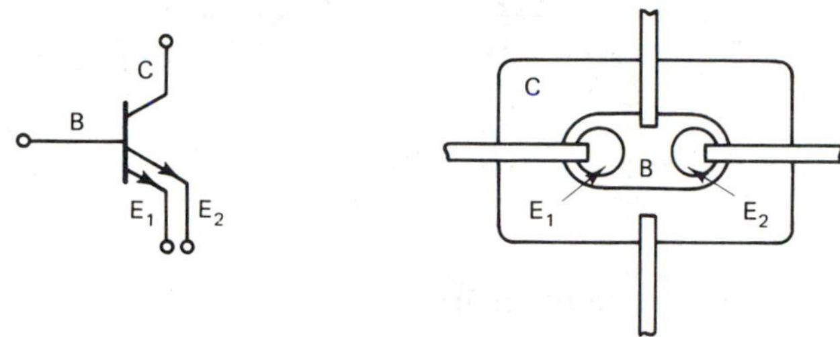

Figure 3.5 Fabrication of a multiple-emitter transistor.

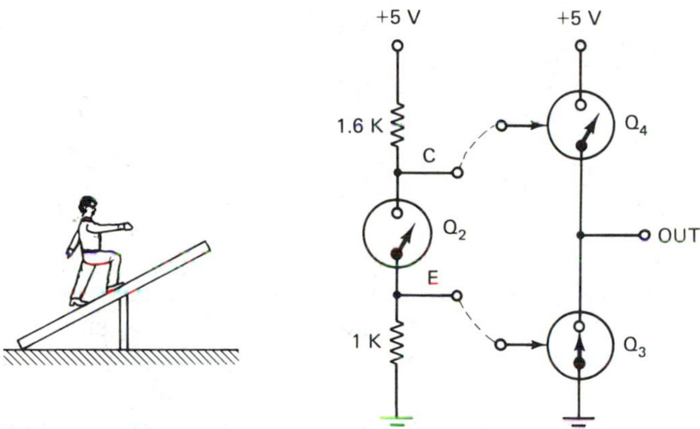

Figure 3.6 Output stage of Fig. 3.4, with snap action switching of output voltage level.

in series (Q_4 and Q_3), driven so that one of them will always be *on* and the other always *off*. The output is therefore always connected either to ground or to the power supply. When Q_2 is *off*, point C is at +5 V, and point E is at ground. This turns Q_4 *on* and Q_3 *off*. But when Q_2 is *on*, current flows through resistors R_1 and R_2, which form a voltage divider. Roughly one-third of the power supply voltage appears at point E, turning Q_3 *on*. The actual TTL NAND gate circuit (Fig. 3.6) includes a diode, whose forward drop is enough (0.6 V) to bias Q_4 *off*.

Transistor Q_2 is controlled by the multiple emitter transistor Q_1, which operates in an unusual manner. If both inputs are open, Q_1 functions as a diode rather than a transistor. Its collector-base junction becomes forward-biased, and causes a current flow into the base of Q_2, turning Q_2 *on*. This state is also maintained if the inputs are connected to voltages high enough so that neither base-emitter junction of Q_1 is forward-biased. The other state of the circuit occurs if either or both of the inputs are connected to a low voltage (ground). The collector current of Q_1 is drawn from the base of Q_2, and this reverse base current turns Q_2 *off*. (Alternatively, Q_1 becoming *on* results in its collector coming within 0.6 V of ground, which biases the base of Q_2 *off*.) The two stable states of this gate are summarized below.

Output	Q_4	Q_3	Q_2	Q_1	Input
Low	Off	On	On	Off	All High
High	On	Off	Off	On	Any Low

Since the output remains high as long as any of the inputs are low, the circuit functions as an AND gate whose output polarity has been inverted:

IN1	IN2	OUT
Low	Low	High
Low	High	High
High	Low	High
High	High	Low

IN1	IN2	OUT
0	0	1
0	1	1
1	0	1
1	1	0

This is called a NAND gate, where the N stands for negation, symbolized by a circle on the ouput side of the gate (Fig. 3.4). The circle represents an inverter whose output is the complement or inverse of its input.

TTL gate loading rules. This detailed examination of the TTL NAND gate demonstrates why the logic 0 state is not exactly 0.0 V, and the logic 1 state is not exactly equal to the power supply voltage. A TTL-compatible *input* stage is therefore designed to respond to any voltage less than 0.8 V as a low state or logic 0. In this state, the device supplying the logic signal must be able to accept (*sink*) enough current to turn Q_1 on fully, which requires $I_{IN} > 1.6$ mA. In practice, the input must be supplied from a source whose effective resistance (*source resistance*) is less than about 500 Ω, since anything greater will raise the input past the 0.8 V threshold. Since a TTL *output* comes from a saturated transistor like Q_3 with an effective source resistance under 100 Ω, a TTL output can easily drive other TTL inputs. However, since each TTL load requires about 1.6 mA in its low state, the saturation current of a TTL output can drive a maximum of about 10 devices. The maximum number of devices which can be driven, called the *fan-out*, ranges from 10 to 30 for devices in the TTL family.

In the output *high* state, the output voltage is smaller than 5 V by the forward drop across Q_4 plus the forward drop across the diode, or about 1.5 V less than the power supply voltage. TTL specifications guarantee that this state will produce an output voltage of at least 3.3 V. The circuitry on the *input* side is designed so that this voltage level is sufficient to keep Q_1 off. TTL specifications guarantee that an input greater than 2.4 V will be recognized as an input *high* state. A summary of these TTL specifications is given in Fig. 3.7.

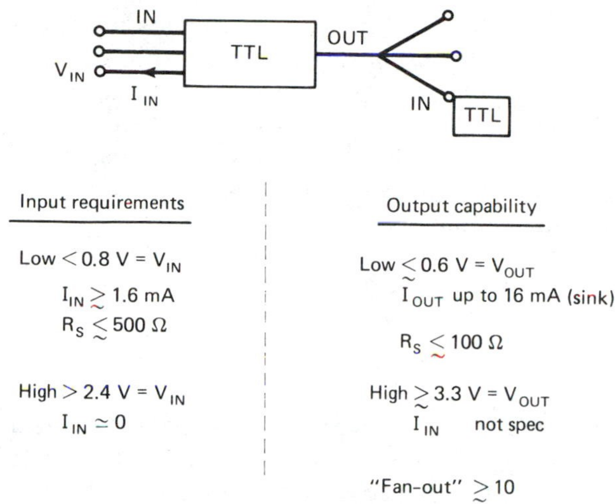

Figure 3.7 TTL voltage levels and terminology.

3.2.3 Open Collector Logic

Although TTL chips can be wired together at the input side, TTL outputs cannot be connected together directly. The output condition is indeterminate and damage could occur if one chip's output is trying to go high while another chip's output is trying to go low. *Open collector logic* gets around the problem, as shown in Fig. 3.8(a). Power connections to the output transistors are made *externally*, through a pull-up resistor. Used alone, an open collector circuit is indistinguishable from a normal TTL device, but when wired to a common bus, the open collector circuit

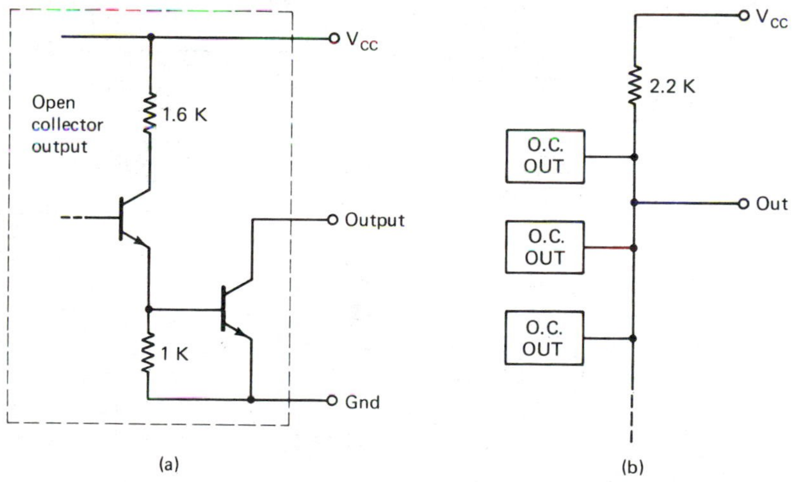

Figure 3.8 (a) An open collector output circuit. (b) Open collector devices tied to a common bus.

avoids the competition of outputs and provides an additional logic function called the *wired OR*. In a typical bus configuration [Fig. 3.8(b)], a given output is either connected to ground (the transistor is *on*) or is open (the transistor is *off*). But this is equivalent to the "low man wins" gate. This wire of the bus is pulled low if any transistor in *on*, and floats high only if all transistors are *off*. The "free" logic function is sometimes called a *wired* gate, performing the AND function in positive logic in this example.

3.3. MOS GATES

MOS gates employ the technology of the metal-oxide-semiconductor field-effect transistor (MOSFET) (see Chapter 1). The internal structure of a typical n-channel (p substrate) device is shown in Fig. 3.9(a). In a circuit [Fig. 3.9(b)], the gate is biased positively to draw electrons into the substrate, forming a conducting channel in otherwise high-resistance (weakly doped) material. Thus, the MOS device may be thought of as a voltage-variable resistor [Fig. 3.9(c)], whose resistance varies from about $10^3\ \Omega$ to $10^8\ \Omega$ from the *on* to the *off* states. In typical circuit operation, the substrate of an n-channel device is connected to the source [Fig. 3.9(d)] to short out the current which would otherwise flow through the

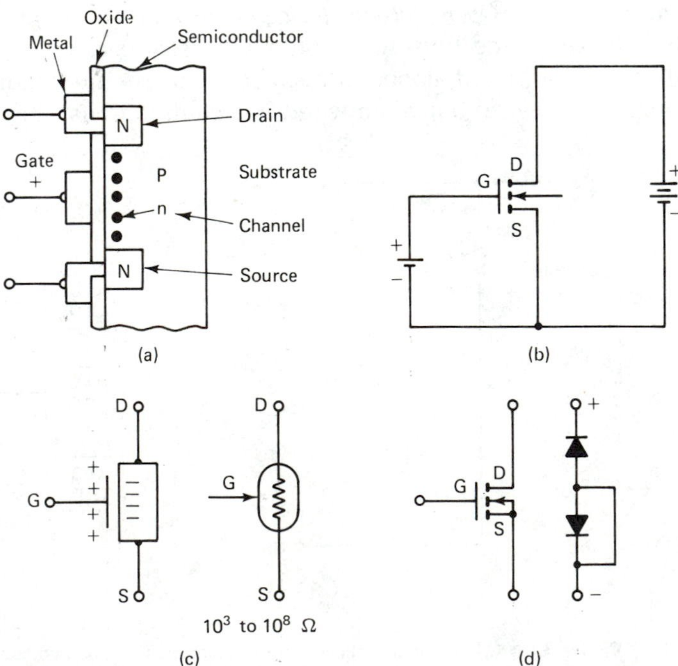

Figure 3.9 The metal-oxide-semiconductor (MOS) transistor. (a) Physical construction. (b) MOS circuit. (c) Voltage-controlled resistor concept. (d) External shorting of substrate to source.

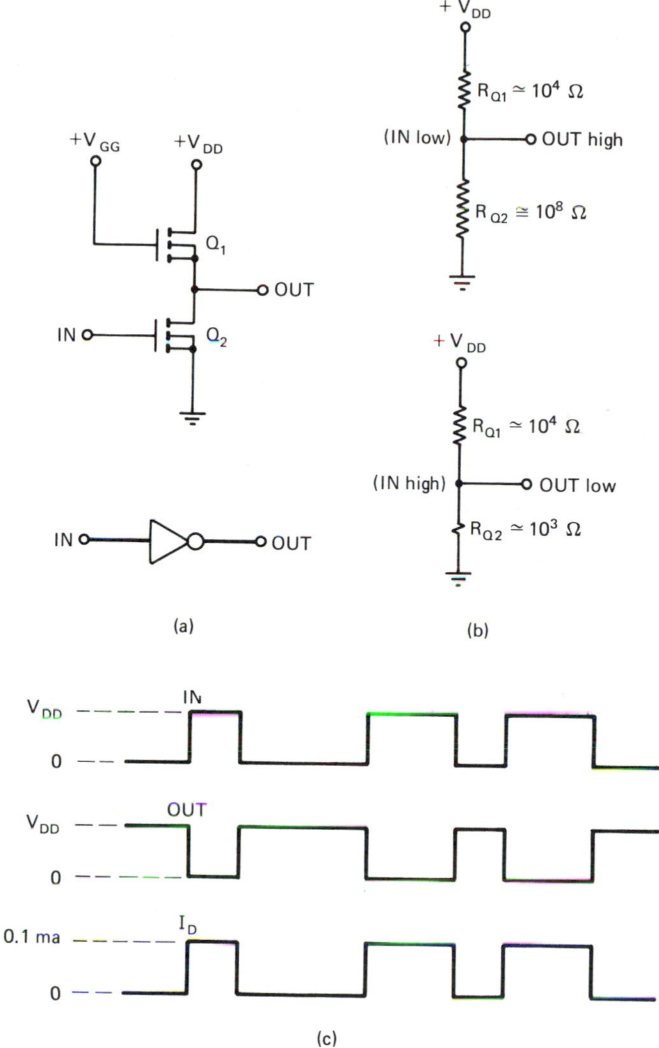

Figure 3.10 MOS inverter. (a) Circuit. (b) Equivalent circuit. (c) Waveforms, showing inversion of the signal.

substrate due to the forward-biased pn junction and interfere with the operation of other nearby devices. The complementary p-channel n-substrate device also exists, and is denoted schematically with the arrow reversed.

In a MOS inverter [Fig. 3.10(a)], one transistor Q_1 is connected to a constant bias voltage V_{GG} to form a constant load resistance of about 10^4 Ω [Fig. 3.10(b)]. This technique is often used in MOS IC's, since resistors are more expensive to fabricate and use more chip area than transistors. A second transistor Q_2 forms a voltage variable resistance which switches from a high value (10^8 ohms) when the input is low (no electrons being attracted into the channel) to a value of about 10^3 Ω when the input is high (electrons populating the conducting channel). The circuit therefore acts simply as a voltage divider, whose output is high (very near the value of V_{DD}) for a low input and low (very near ground) for a high input, since in either case R_{Q1} is very different from R_{Q2}. The circuit therefore functions as an inverter of data at the input terminal [Fig. 3.10(c)].

3.3.1 CMOS Inverter

The MOS inverter has a relatively large current flow when the input is high, a serious limitation in large-scale IC's with many gates on a single chip. CMOS devices overcome this difficulty by connecting two transistors with their source-drain current in series but their gate voltages in parallel. The two transistors are called *complementary* because their conducting channels have the opposite sense, and a given input signal turns one *on* and the other *off*. An example is the CMOS inverter of Fig. 3.11(a). Because of the large difference of on-state and off-state resistance, the circuit acts as two ideal switches [Fig. 3.11(b)] in series, arranged so that when one switch is closed, the other is open. For this reason, there is virtually no current flow in CMOS devices. Waveforms for the CMOS inverter [Fig. 3.11(d)] show a small current pulse only when the device changes its state. This results from unavoidable junction capacitance C_J, since a reverse-biased pn junction stores charge just as a capacitor does. The charging and discharging of C_J results in a current flow whenever the voltage across a junction changes.

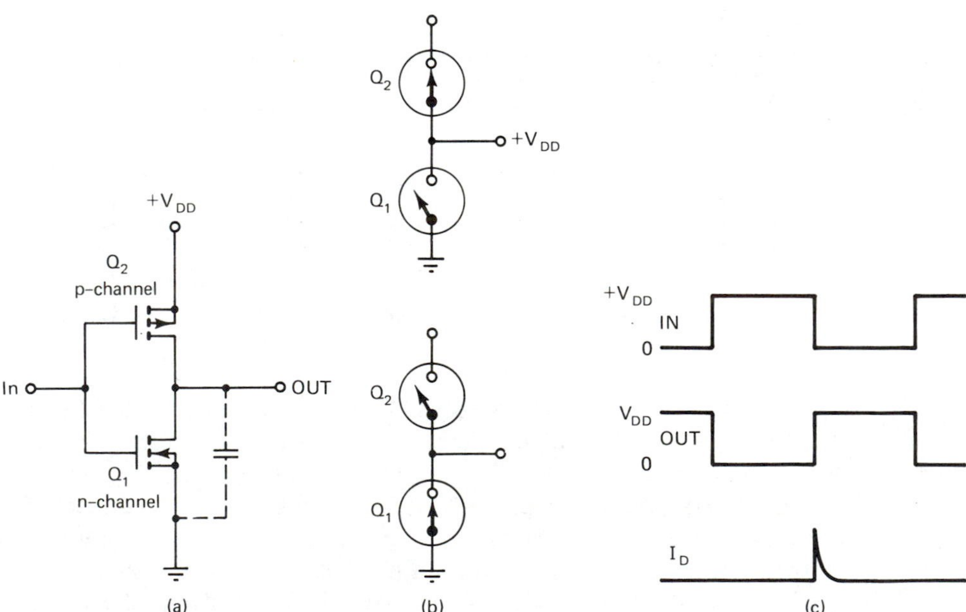

Figure 3.11 CMOS inverter. (a) Circuit. (b) Equivalent switching circuit. (c) Waveforms, showing inversion of the signal yet with current flow almost always zero (compare with previous figure).

3.3.2 CMOS Gates

The design of a CMOS gate [Fig. 3.12(a)] follows a similar strategy. Complementary devices are connected to each input with their source-drain currents in series and their gates in parallel. The circuit is analogous to the switching network of Fig. 3.12(b). The labels beside the switches indicate that a given switch is closed whenever the control signal applied to the gate is high, since a *high* closes an n-channel switch while a *low* closes a p-channel switch. The only way the output can be low is if both A and B are high, closing both switches A and B and opening switches $\overline{A}$ and $\overline{B}$. This circuit has the truth table of a NAND gate. Likewise, the series-parallel combination shown in Fig. 3.12(c) forms a NOR gate. The equivalent switching circuit analogous to Fig. 3.12(b) is left as an exercise (Problem 3.3).

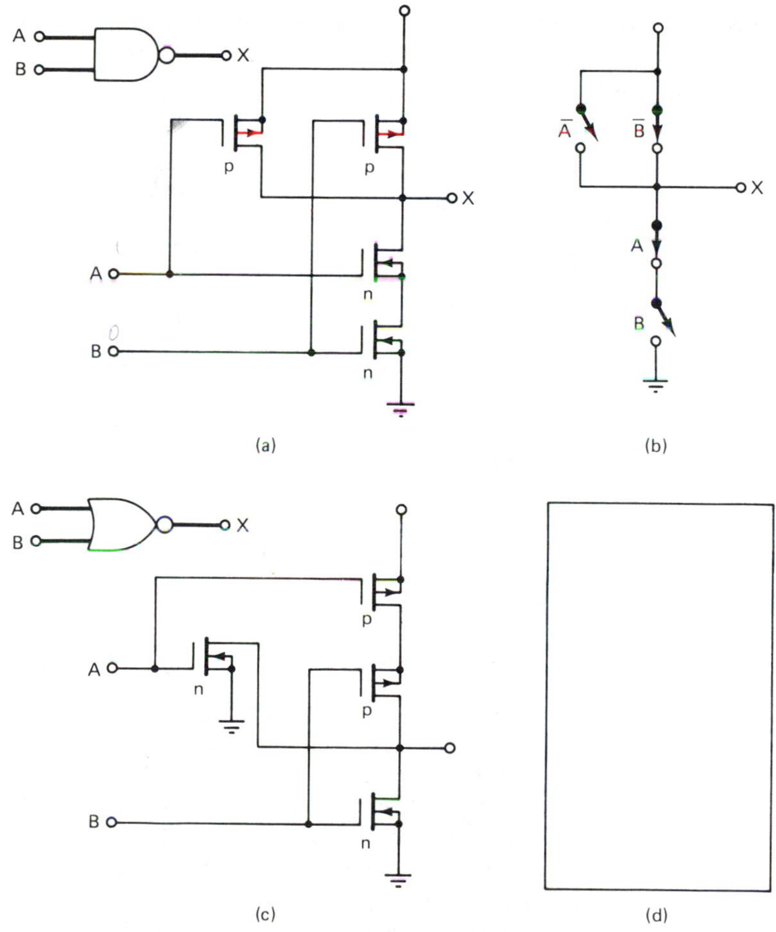

Figure 3.12 CMOS gate circuits and equivalent switching circuits. (a) and (b) NAND gate. (c) and (d) NOR gate.

In practice, protection diodes are connected between either input and both power leads, which protect the delicate oxide layer by turning on whenever input voltages exceed the diode forward turn on voltage. Without these diodes, the high impedance ($10^8\ \Omega$) of the gate electrodes may lead to the buildup of static voltages that could cause dielectric breakdown of the oxide. Even with these protection devices, CMOS devices are considerably more delicate than TTL devices. They are stored in conductive plastic foam and must be handled with extreme care (grounding soldering iron and grounded operator) when being installed into circuits.

3.4 LOGIC CIRCUITS USING GATES

Gates are used in measurement circuits, in interfaces to mini- and microcomputers, and inside computers, either to control the flow of information or to carry out logical and arithmetic operations. For example, the circuit shown in Fig. 3.13 allows the information in the data stream to appear at the output only when the *Enable* input is high.

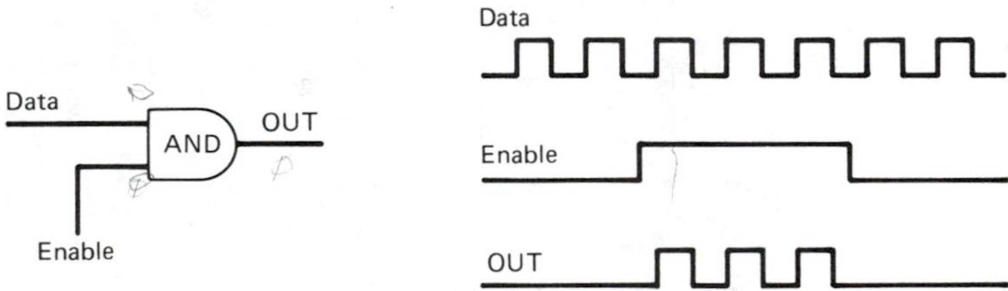

Figure 3.13 AND gate used to control the flow of information,

		Data	Output
Enable:	Low	0	0
		1	0
	High	0	0
		1	1

Viewed this way, the truth table for the AND gate falls into two parts, with only the "enable high" half of it allowing data transmission. This picture of gate operation resembles a faucet whose flow is controlled by the enable input.

This circuit is the basis for many measuring instruments. For example, a frequency counter connects an unknown signal to the data input, opens the enable input for a precisely timed interval, and connects the AND gate output to a counter to determine how many pulses fit within the timing interval. A precise measurement of a time interval may be made by reversing the role of data and en-

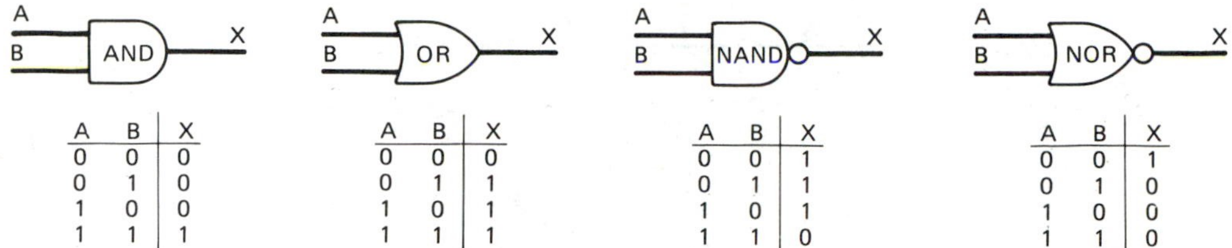

A	B	X (AND)
0	0	0
0	1	0
1	0	0
1	1	1

A	B	X (OR)
0	0	0
0	1	1
1	0	1
1	1	1

A	B	X (NAND)
0	0	1
0	1	1
1	0	1
1	1	0

A	B	X (NOR)
0	0	1
0	1	0
1	0	0
1	1	0

Figure 3.14 Comparison of gate symbols and truth tables.

able input, driving the data line with a fixed reference frequency (clock), and using a signal from the outside world event to open and close the gate at the enable input. Many other measuring instruments, for example the digital voltmeter (Chapter 9), also use this basic principle.

There are four basic logic gate functions (Fig. 3.14), whose effect on data differ greatly. AND and NAND gates allow data to flow when the enable input is high, but OR and NOR gates allow data to pass only when the enable input is low (see Prob. 3.4). The actual TTL circuit performs either a NAND (Fig. 3.7) or NOR operation, so it takes an extra inverter to perform AND or OR, requiring additional current and resulting in additional *propagation delay* between the application of input signals and the appearance of logically correct outputs. Consequently, only NAND and NOR gates are commonly chosen in circuit design.

	7400 NAND	7402 NOR	7408 AND	7432 OR	
Propagation delay	10	10	15	12	ns
Current per gate	3	3	4	4.5	mA

The added current requirements become significant in large circuits, and the small difference in propagation delay could cause logically incorrect decisions if the circuits attached to the output are fast enough to follow it or if two alternative pathways have slightly different propagation delays. Not all four kinds of gates are necessary, since the logic function performed by a given circuit can be flexibly reinterpreted using de Morgan's theorem.

3.4.1 de Morgan's Theorem

As described in Section 3.2, the logic function performed by a gate depends on how one chooses to interpret the high and low voltage levels. Thus, the "low man wins" gate (Fig. 3.3) functions as an AND gate if high equals 1, but is an OR gate if high equals 0 (see Table 3.1). As a reminder when a low signal is being interpreted as true, a bubble is placed on the input or output side of the gate [Fig. 3.15(a)]. The bubble stands for the logic function performed by an inverter, changing an input high to an output low, and vice versa. To describe the logic

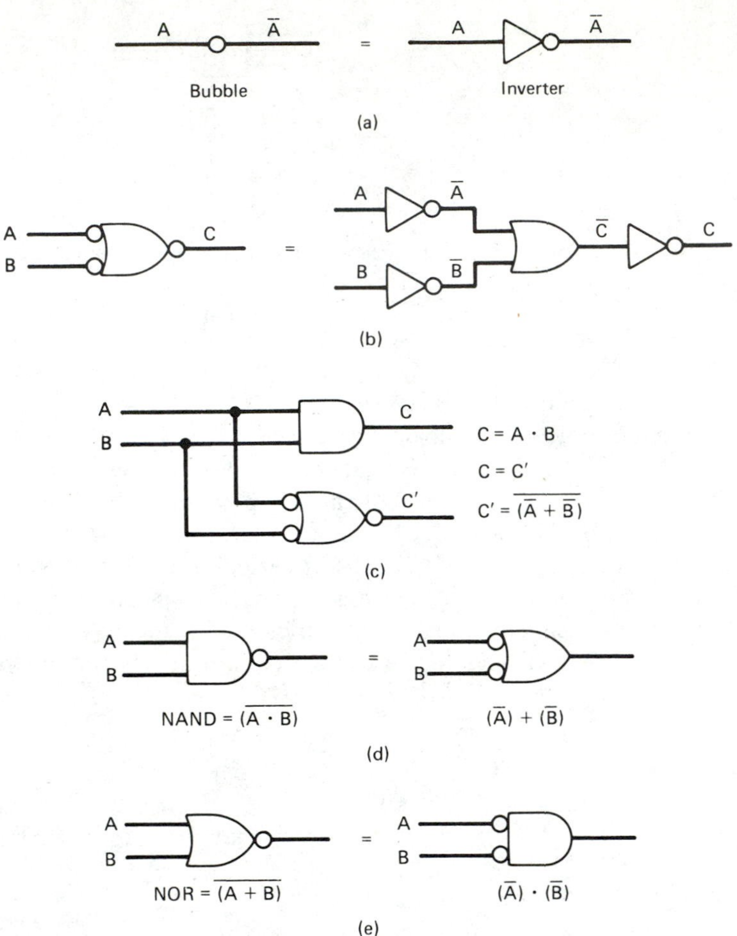

Figure 3.15 de Morgan's theorem. (a) Inverter symbol. (b) Designating inverted logic polarity using bubbles. (c) Changing of AND to OR when logic polarity is inverted. (d) Other examples of de Morgan's theorem.

function performed by the low man wins gate when both inputs and outputs are interpreted as "high equals zero," bubbles are attached to both the input and output sides of the gate [Fig. 3.15(b)]. The same gate therefore performs the AND function for *positive logic* (high equals true) or the OR function for *negative logic* (low equals true), as shown in Fig. 3.15(c). The two equivalent gate circuits in this figure represent a mathematical logical equivalence, which may be written in Boolean algebra as

$$C' = \overline{[(\bar{A}) + (\bar{B})]} = A \cdot B = C \tag{3.6}$$

This is a compact shorthand for saying in words: When the expression "A is not true or B is not true" is not true, then A and B are both true. This is one form of de Morgan's theorem. Other and more often used examples for NAND and NOR gates are shown in Fig. 3.15(d) and (e). de Morgan's theorem may be stated as

> **de Morgan's theorem.** Performing an operation on a complemented (inverted) input changes the logical operation performed from AND to OR, or from OR to AND, and complements the output.

Shorthand mnemonics useful in flexibly drawing circuit diagrams are:

> (1) Moving the bubble from output to input changes the logical operation performed from OR to AND, or from AND to OR.
> (2) Two inverters in series are equivalent to no inversion.

One consequence of de Morgan's theorem is the ability to construct any of the four basic gate functions using NAND gates only or using NOR gates only (Fig. 3.16). Although this involves different numbers of gates and hence different propagation delays and current demands, this feature is a convenience in minimizing the number of IC *packages* used, since (for example), a 7400 has four NAND's on a single chip. Inverter operations may be performed by wiring together the inputs of a NAND or NOR [Fig. 3.17(a)]. Thus, an OR gate may be constructed from NAND's as shown in Fig. 3.17(b). Two bubbles in a row cancel and may be replaced by a wire [Fig. 3.17(c)]. To avoid confusion and logic errors, one should avoid drawing circles which designate input and output connections the same size as the bubble for the inversion function.

3.4.2 The Active-Low Convention; Interfacing Examples

In interfacing measuring instruments or peripherals to a computer, it is necessary to provide control signals in addition to the data wires. For example, a printer [Fig. 3.18(a)] has a control line BUSY to indicate when it has received a piece of data, but is busy printing and not yet ready to receive more. Since electromechanical devices require milliseconds whereas computing devices can operate in microseconds, the *handshaking* provided by control signals is essential to avoid meaningless results due to the mismatch of natural speeds.

Sometimes, a device presents an output control line which when *active* or *asserted* is in the low voltage state. For example, the wire labeled $\overline{\text{BUSY}}$ in Fig. 3.18(a) means that this line is low when the printer is busy and not yet ready for more data. The bar on top designates an *active-low* line. It is becoming accepted to designate an active-low line with a bubble, as shown on the right side of Fig. 3.18(a). Alternatively, a bar over the pin identification label may indicate an active-low signal. For example, in a microcomputer system [Fig. 3.18(b)], there might be a control line ready (RDY) specifying when the memory is ready to accept information from the microprocessor unit (MPU, often written as CPU for central processing unit), and a control line $\overline{\text{WR}}$ which enables information to be

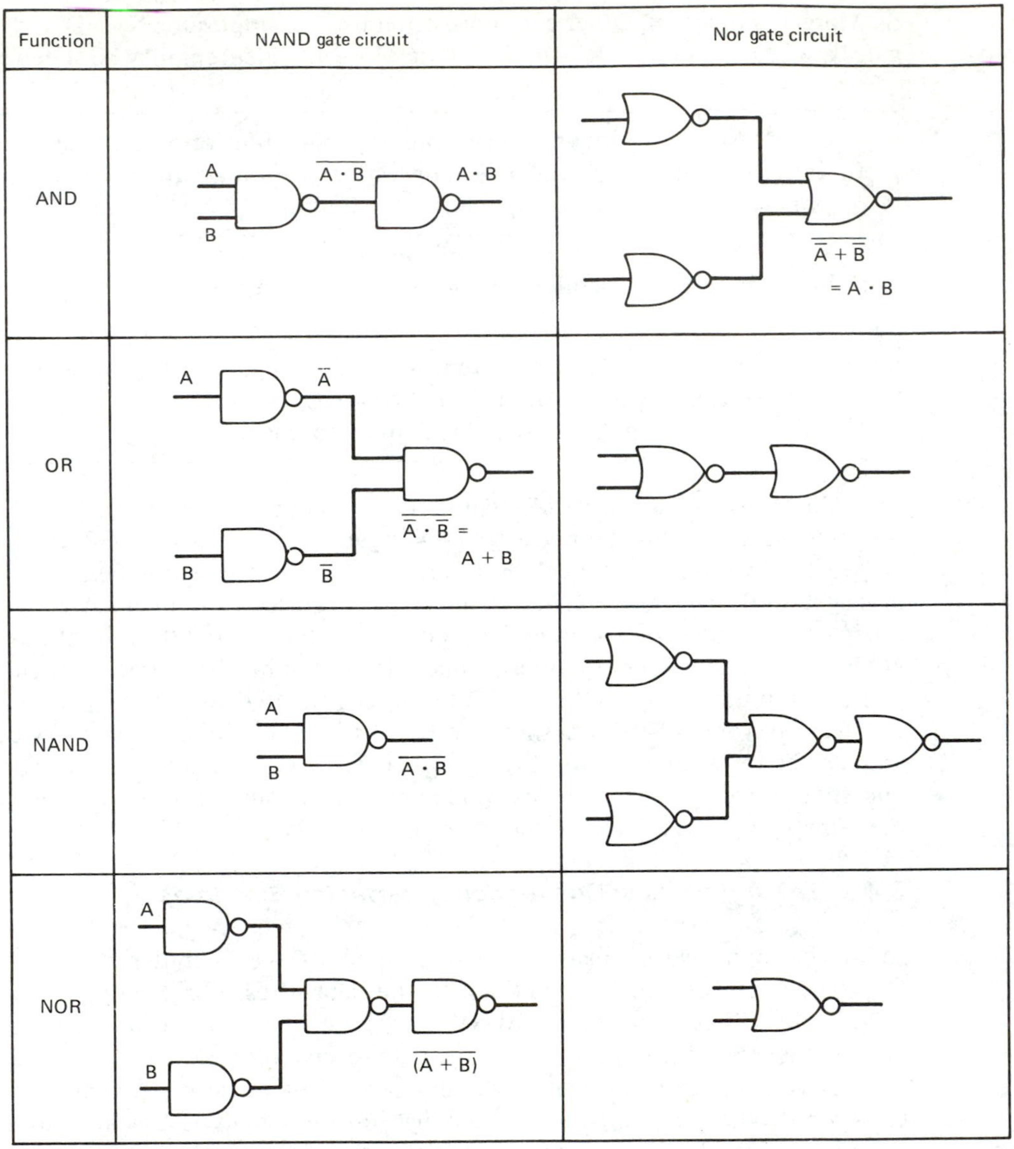

Function	NAND gate circuit	Nor gate circuit
AND	A, B, $\overline{A \cdot B}$, $A \cdot B$	$\overline{\overline{A} + \overline{B}}$ $= A \cdot B$
OR	A, $\overline{A}$, B, $\overline{B}$, $\overline{\overline{A} \cdot \overline{B}} =$ $A + B$	
NAND	A, B, $\overline{A \cdot B}$	
NOR	A, B, $\overline{(A + B)}$	

Figure 3.16 Synthesizing any kind of gate using NAND's only or NOR's only.

written into the random access memory (RAM). In this example, RDY is an active-high control line, and $\overline{WR}$ is an active-low control line. When control lines are interfaced to devices outside a given chip family, the logic polarity of control signals must be carefully matched, with additional inverters added as necessary.

Active-low logic requires a good understanding of De Morgan's theorem. Suppose device C is to be reset when device A signals that it has been selected, and device B signals that it is ready. Suppose both the select line $\overline{SEL}$ and the ready $\overline{RDY}$ are active low signals, and device C requires an active-low reset. An

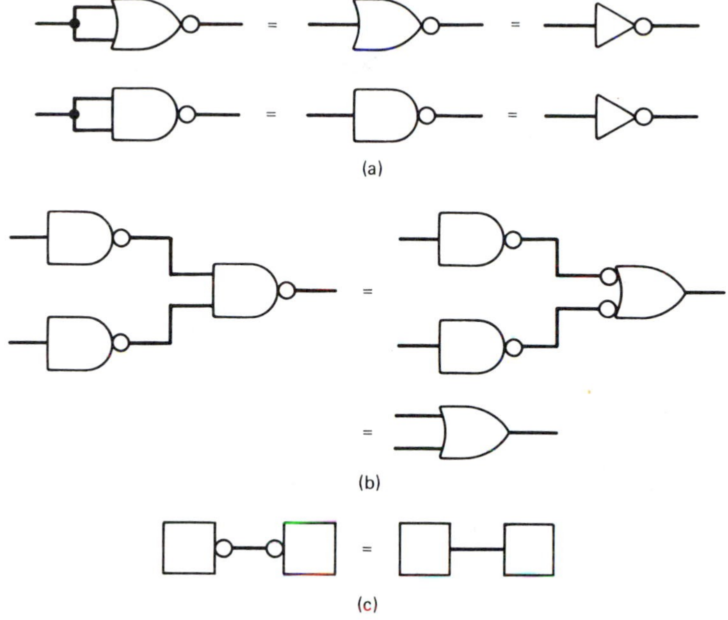

Figure 3.17 Rules for the manipulation of bubbles.

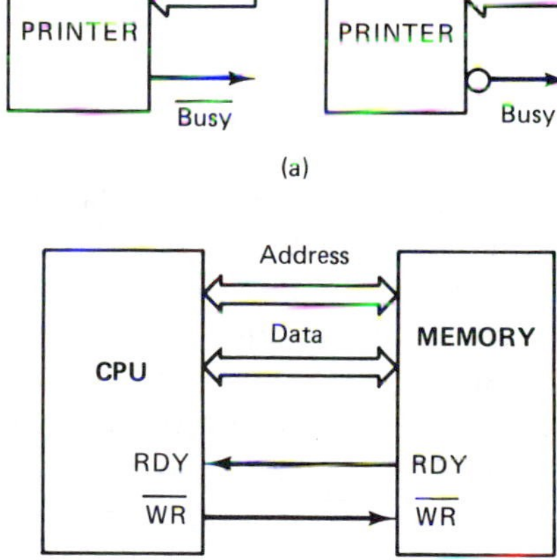

Figure 3.18 Interfacing examples. (a) Output control line; the symbol $\overline{\text{BUSY}}$ is equivalent to drawing a bubble on the device. (b) Microcomputer handshaking example.

interfacing circuit which performs the necessary AND function is shown in Fig. 3.19(a). The NOR gate output is high only when both the select and the ready lines are active because (moving the bubble through) it is equivalent to an AND gate for active-low signals. An inverter then provides the active-low signal needed for the device reset line. It is thus very useful to flexibly redraw gate circuits, using de Morgan's theorem, as AND's or OR's as convenient. Sometimes different gates within the same IC package may be used to perform different logical operations when interfacing control inputs which include both active-low and active-high lines [see Prob. 3.7 and Fig. 3.19(b)].

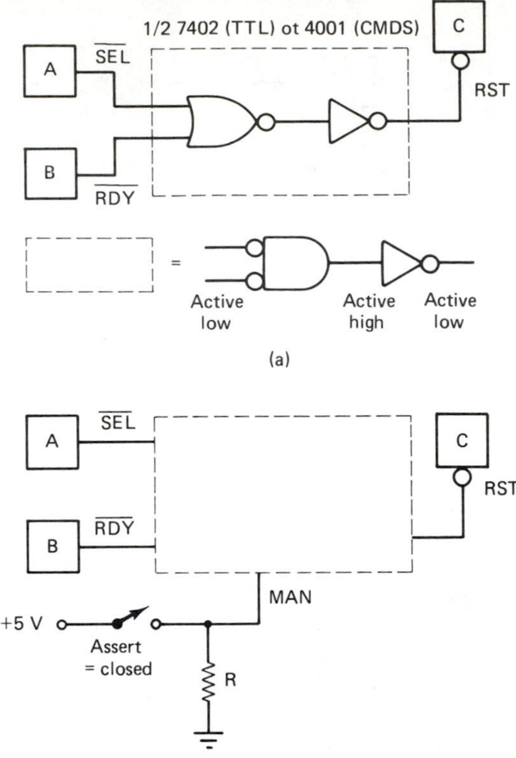

Figure 3.19 Interfacing control logic examples. (a) Device C is reset when both $\overline{SEL}$ and $\overline{RDY}$ lines are active. (b) Add a manual reset (see Prob. 3.7).

3.5 BOOLEAN ALGEBRA

In addition to de Morgan's theorem, there are a number of useful theorems in Boolean algebra, which are summarized in Table 3.2. Some are similar to algebra; all may be verified by writing truth tables. Boolean algebra is especially useful in determining the electronic circuit which requires the fewest gates to carry out a given logical operation. For example, the outputs in the circuits of Fig. 3.20 will be identical for both circuits for any combination of inputs A, B, and C, but the circuit on the left requires one less gate. This is a consequence of the distribution theorem

$$A + (B \cdot C) = D = (A + B) \cdot (A + C) \tag{3.7}$$

The equivalence may be proved by direct substitution (generate a truth table), or by using the simpler theorems of Boolean algebra.

$$(A + B) \cdot (A + C) = \underbrace{(A \cdot A)}_{A} + (A \cdot B) + (A \cdot C) + (B \cdot C) \tag{3.8}$$

$$A \cdot A = A \qquad \underbrace{A + (A \cdot B)}_{A}$$

$$A + A \cdot B = A \qquad \underbrace{A + (A \cdot C)}_{A}$$

$$A + A \cdot C = A \qquad \underbrace{A + (B \cdot C)}_{A + (B \cdot C)}$$

TABLE 3.2 THEOREMS OF BOOLEAN ALGEBRA

Commutation	$A + B = B + A$
Absorption	$A + (A \cdot B) = A$
	$A \cdot (A + B) = A$
de Morgan's theorem	$\overline{A + B} = \overline{A} \cdot \overline{B}$
	$\overline{A \cdot B} = \overline{A} + \overline{B}$
Association	$A + (B + C) = (A + B) + C$
	$A \cdot (B \cdot C) = (A \cdot B) \cdot C$
Distribution	$A + (B \cdot C) = (A + B) \cdot (A + C)$
	$A \cdot (B + C) = (A \cdot B) + (A \cdot C)$

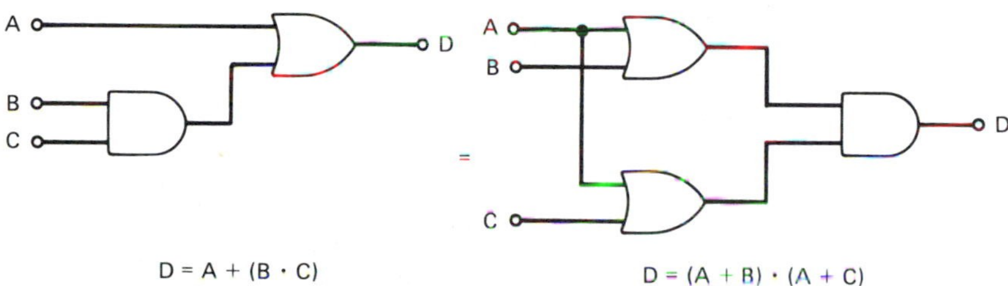

Figure 3.20 Boolean algebra example of two equivalent but dissimilar looking circuits.

Although gate minimization techniques were once used extensively, the low cost of IC's now makes it less important to design a circuit with the least number of gates. Other considerations sometimes take precedence. For example, the two equivalent circuits of Fig. 3.20 result in a subtle timing difference. In the minimized circuit, the B and C inputs pass through one more gate than the A input, whereas all inputs pass through an identical number of gates in the three-gate version. There is thus a period of time (1 *gate delay*, or about 10 ns in TTL) when the minimized circuit generates an incorrect output. This may cause difficulty if the circuit which follows it is fast enough to respond to the transient error. Nevertheless, the theorems of Boolean algebra remain very useful in circuit design, especially in redrawing gate circuits to make use of available gates.

3.6. THE EXCLUSIVE OR AND ITS APPLICATIONS

A final basic gate circuit which has many applications in data communications and in computer arithmetic is the *exclusive OR* (XOR) gate. For a two-input XOR gate, the output is true if either of the inputs is true, but not if both inputs are true. The Boolean symbol and truth table are shown on p. 86.

$$A \oplus B = C \tag{3.9}$$

A	B	C
0	0	0
0	1	1
1	0	1
1	1	0

The XOR function may be implemented in a number of ways from the four basic gates. For example,

$$A \oplus B = A \cdot \overline{B} + B \cdot \overline{A} \tag{3.10}$$

The equivalence may be proven simply by the truth tables

A	B	$(A \cdot \overline{B})$	$(B \cdot \overline{A})$	$(A \cdot \overline{B} + B \cdot \overline{A})$
0	0	0	0	0
0	1	0	1	1
1	0	1	0	1
1	1	0	0	0

This particular way of making the XOR function corresponds to the circuit used to control a light from either of two switches. The light will be *on* if either of the switches is up, but not if both are up or both are down (see Prob. 3.1). An electronic circuit for this logic combination requires two inverters, two AND gates, and one OR gate (Fig. 3.21). Using Boolean algebra, the number of gates may be reduced, or the XOR function may be performed using gates of only one kind. For example, a quantity ANDed to its complement is never true, and may therefore be ORed to any expression without changing it.

$$\begin{aligned} A \oplus B &= A \cdot \overline{B} + B \cdot \overline{A} \\ &= A \cdot \overline{B} + A \cdot \overline{A} + B \cdot \overline{A} + B \cdot \overline{B} \\ &= A \cdot (\overline{B} + \overline{A}) + B \cdot (\overline{A} + \overline{B}) \\ &= A \cdot \underbrace{(\overline{A \cdot B})}_{\text{NAND}} + B \cdot \underbrace{(\overline{A \cdot B})}_{\text{NAND}} \end{aligned} \tag{3.11}$$

A circuit which performs this logical combination is shown in Fig. 3.22(a). Application of de Morgan's theorem to the final OR gate changes it to an AND gate with bubbles on both input and output sides. The bubble may be slid along the wire to the left without changing the logical function, resulting in a four-gate XOR

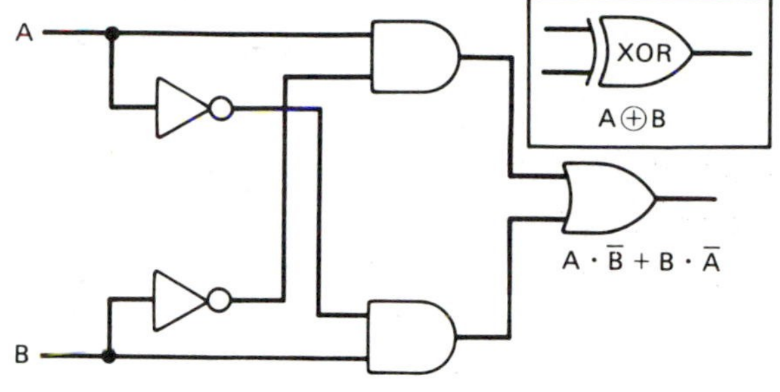

Figure 3.21 The exclusive OR gate, straightforward 5-gate version.

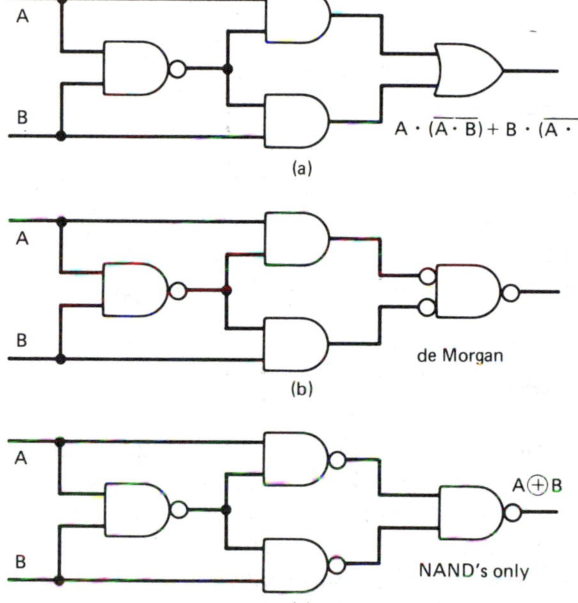

Figure 3.22 (a), (b), and (c) Evolution of a 4 NAND-gate XOR using Boolean algebra.

which uses NAND gates only. A NOR-gate version is also possible, but results in no savings in the number of gates (Prob. 3.12). A three-gate version of XOR is also possible, but requires gates of three different kinds (Prob. 3.13).

3.6.2 Exclusive OR Applications

Many applications of the XOR gate follow from its function as a controllable inverter. By the truth table (Eq. 3.9), if input A is viewed as a control input and input B is viewed as data, the output C follows the data when the control is low, and inverts the data when the control is high. This function is shown in Fig. 3.23. The controllable inverter may be used as a data scrambler for transmitting confidential messages (Fig. 3.24). If the control input is supplied by a pseudorandom noise signal, scrambled data results at the ouput. The scrambled data is transmitted by a phone line or radio link to a receiver which also contains an XOR gate. Its control input is fed by the same pseudorandom noise signal, either transmitted by an in-

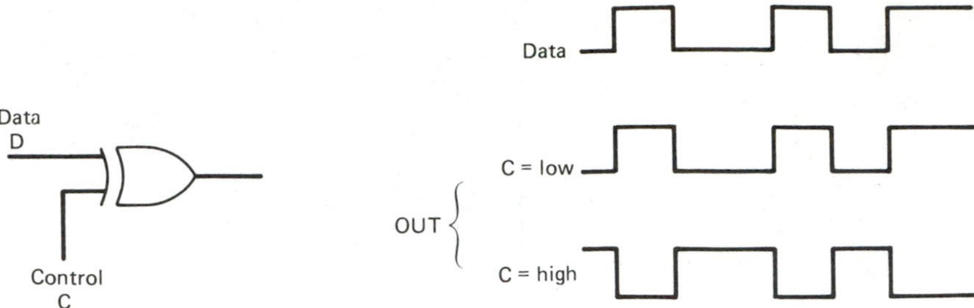

Figure 3.23 XOR application as controllable inverter.

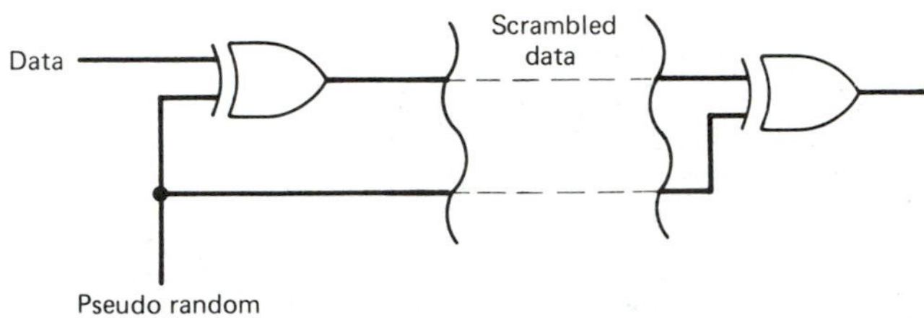

Figure 3.24 XOR application as data scrambler.

dependent link or generated locally using a clock synchronized by using edges present in the scrambled data, and a code generated by an identical pseudo-random circuit.

3.6.3 Parity Checking and Binary Comparisons

In the transmission of digital information, some means of detecting transmission errors is desirable. One method is a parity check. The *parity* of a binary word is defined as odd (1) if the word contains an odd number of 1's. In transmission, the parity bit is sent in addition to the data bits. At the receiver, the parity bit is once again generated, and compared with that which was sent [Fig. 3.25(a)]. If the two parity bits P and P′ are unequal, an error flag is generated. Parity checking is a useful way of eliminating the most common errors due to noise in transmission. If the probability of a noise error in a single word is $1/N$, the probability of two noise errors in the same word is $1/N^2$, which is much smaller if N is large. For the 4-bit example above, a parity generation circuit is constructed by combining XOR gates [Fig. 3.25(b)]. This may be shown most directly using truth tables (Prob. 3.14).

A *comparison* of two binary bits may be done using an XOR gate plus an inverter [Fig. 3.25(c)], since this circuit generates an output C which is high whenever the two bits are equal.

$$C = \overline{A \oplus B} \qquad (3.12)$$

A	B	C
0	0	1
0	1	0
1	0	0
1	1	1

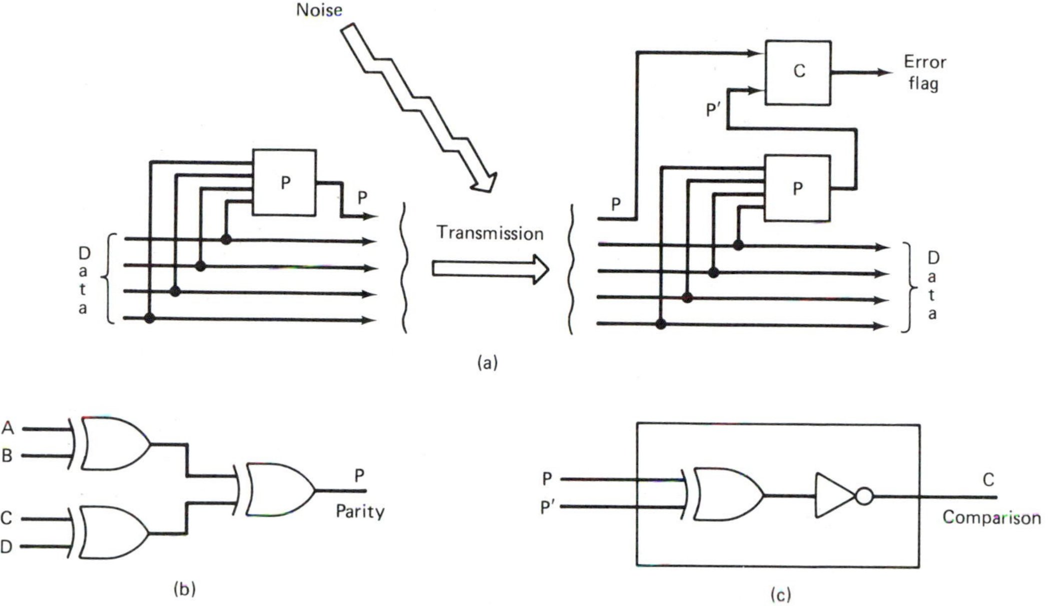

Figure 3.25 Use of parity to find errors in data transmission. (a) Concept of parity generation and comparison. (b) Parity generation circuit. (c) Parity comparison circuit.

Other more complicated comparisons form the basis for the inequality circuits in most computers and microcomputers. An example is shown in Fig. 3.26. The comparison of two 3-bit numbers is made using XOR gates. The algorithm for determining which of two binary words is larger involves a bit by bit comparison.

Comparison Algorithm. When two numbers are unequal, the one with the greater value is greater at the most significant unequal bit.

For example, on comparing (A = 111011) with (B = 110111), the fact that A > B shows up in comparison of the 2^3 bits. Once this is determined, the less significant bits should *not* be compared, since, in this example, the three least significant bits alone imply that B > A.

For a single bit, inequality testing is done by a single AND gate operating on A and the complement of B.

$$A \cdot \bar{B} = (A > B) \tag{3.13}$$

This may be proved by writing the truth table for the operation, and explains the uppermost AND operation in Fig. 3.26(b). Bit testing of the less significant bits is also done by AND gate inequalities, with extra inputs to disable the comparison if an inequality has already been found in testing a more significant bit (see Prob. 3.15).

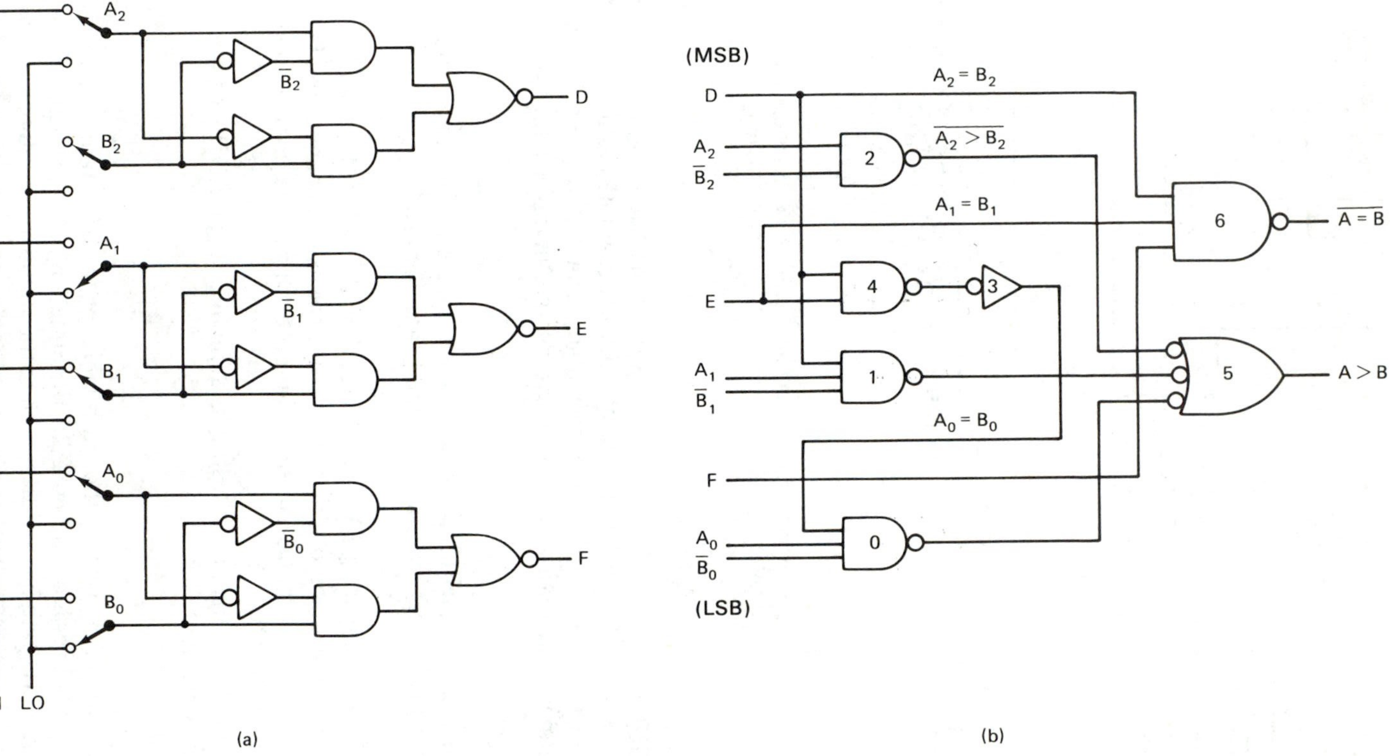

Figure 3.26 (a) Equality and (b) relative magnitude detector circuit.

3.7. GATES AS DECODERS, SELECTORS, AND MULTIPLEXERS

Gates are used extensively as decoders to translate a number from one representation to another; as selectors to enable one instrument at a time to listen to information being passed along a data bus; and as multiplexers and demultiplexers to allow a number of independent messages to be transmitted on the same line.

3.7.1 Code Conversion

Gate combinations may be used to convert a number between two different representations. For example, consider the binary-to-octal decoder (Fig. 3.27). The input on the left is a 3-bit parallel binary word. The output is a set of eight wires, only one of which is to be high. The wire which goes high is the octal representation of the input binary word. The logic design is made by noting which unique combination of binary bits must be true or false to specify a given octal number.

Octal	Binary	Logical
	C B A	
0	0 0 0	$\overline{C}\cdot\overline{B}\cdot\overline{A}$
1	0 0 1	$\overline{C}\cdot\overline{B}\cdot A$
2	0 1 0	$\overline{C}\cdot B\cdot\overline{A}$
3	0 1 1	etc.
4	1 0 0	
5	1 0 1	
6	1 1 0	
7	1 1 1	

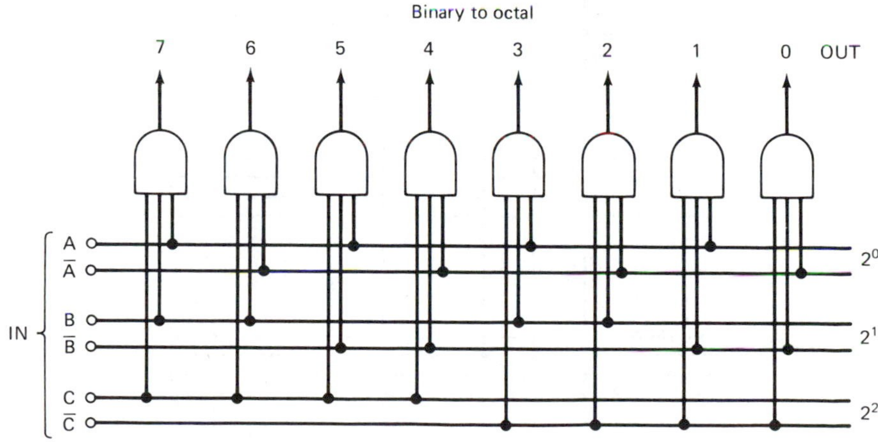

Figure 3.27 Binary to octal decoder circuit.

For example, octal 3 is true whenever binary bit C is not true and bits B and A are both true. To complete the logic, bit A (2^0) is true for all odd numbers, bit B (2^1) is true for numbers 2, 3, 6, and 7, and bit C (2^2) is true for numbers greater than 4. An extension for larger numbers is left for a problem (Prob. 3.16).

Question. Why is it necessary to include the inverters on the inputs? Why is it not sufficient just to AND the binary bits which are *true* in order to uniquely determine an octal output?

3.7.2 Octal to Binary Conversion

Conversion from octal to binary is also straightforward using gates (Fig. 3.28). On the input side, one and only one of eight wires is high, specifying an octal digit. The output is the corresponding parallel binary word.

Question. Why are OR gates used in this circuit, while AND gates were used in the binary to octal decoder?

Question. What is the algorithm determining which interconnections are made in the matrix formed by the octal input wires and the gate inputs in Fig. 3.28?

Question. Why use gates at all? Why not just interconnect the matrix of wires as shown, but in each box put a short circuit to the corresponding binary output?

It might at first seem adequate to connect together a group of wires to perform a logical function. But if the group of four wires inside each box in Fig. 3.28

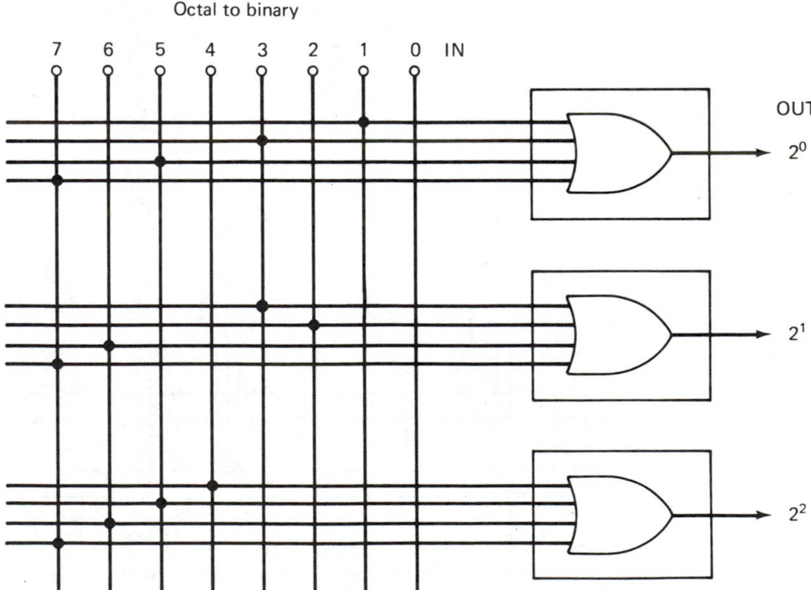

Figure 3.28 Octal to binary decoder circuit.

were simply shorted together, an input wire which was trying to go high could find itself connected to several other input wires which were trying to go low. A fight would result, the logical state would be indeterminate, and damage might occur to the devices shorted together. The OR gate lets the information through without connecting the input wires together.

General Rule. The outputs of gates cannot be wired together, or else contradictory messages, an indeterminate state, and possible circuit damage will result.

Exception. *Open collector* gates (Section 3.2) and *tri-state logic* (Section 4.7) are designed to be wired together at their outputs.

3.7.3 Data Selectors

Often a computer is connected to a number of devices along a single data bus (Fig. 3.29). These devices may be blocks of memory within the computer, or outside peripheral instruments interfaced to the computer. Normally data is to be sent between the computer and only one device at a time. All the other devices simultaneously connected to the data bus must be prevented from listening or talking along the data bus except when selected. This function is performed by a decoder whose input is a binary word specifying the address of the selected device. The decoder output contains a single active line which enables one device to interact at a time. An active high output decoder uses AND or NAND gates connected at the devices.

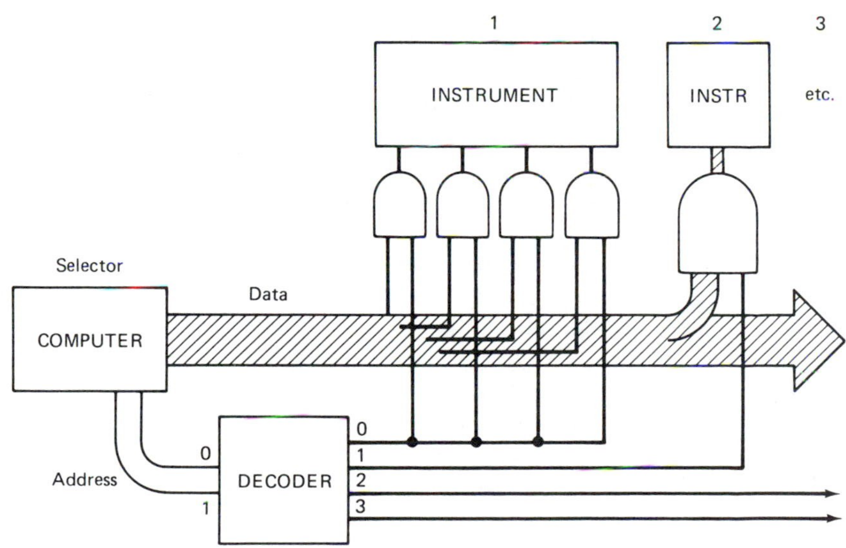

Figure 3.29 Decoder used as a selector.

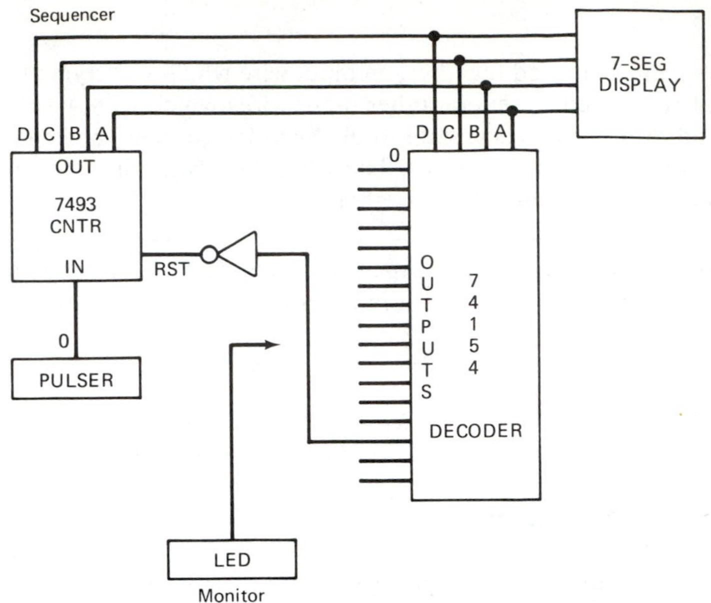

Figure 3.30 Selector circuit wired as a sequencer whose range N can be set arbitrarily.

Test your understanding. Which gate should be connected to the selected device if it is enabled by an active-low signal? Which gates when the decoder outputs are active-low?

When the address lines of a decoder are driven by a counter, the device is called a *sequencer*, since it advances the selected line each time the counter is clocked. If the number of devices connected to the sequencer is variable or less than the total number of lines, the device may be prevented from sequencing through unused outputs. The selector output itself resets the counter with the feedback line marked A in Fig. 3.30.

3.7.4 Multiplexing and Demultiplexing

It is often convenient to convert information present simultaneously in *parallel* form into *serial* information in a time sequence transmitted along a single wire. A multiplexer [Fig. 3.31(a)] performs this function. Its inputs are a parallel binary word and address lines to specify which input data bit is connected to the output wire. Multiplexing provides savings in the number of wires needed to transmit information. A number with 2^N bits may be transmitted in serial form using only N address wires, the data wire, and a timing signal called the *strobe* which specifies when the data appearing on the output is valid. The savings in interconnections may be substantial. For example,

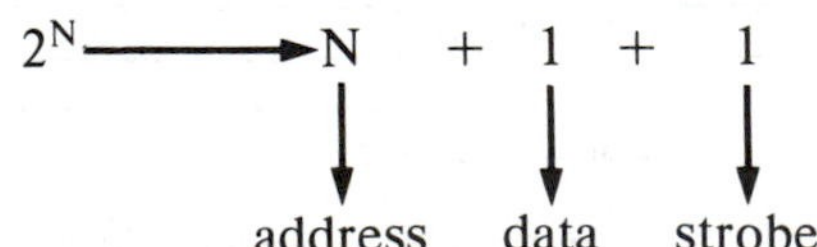

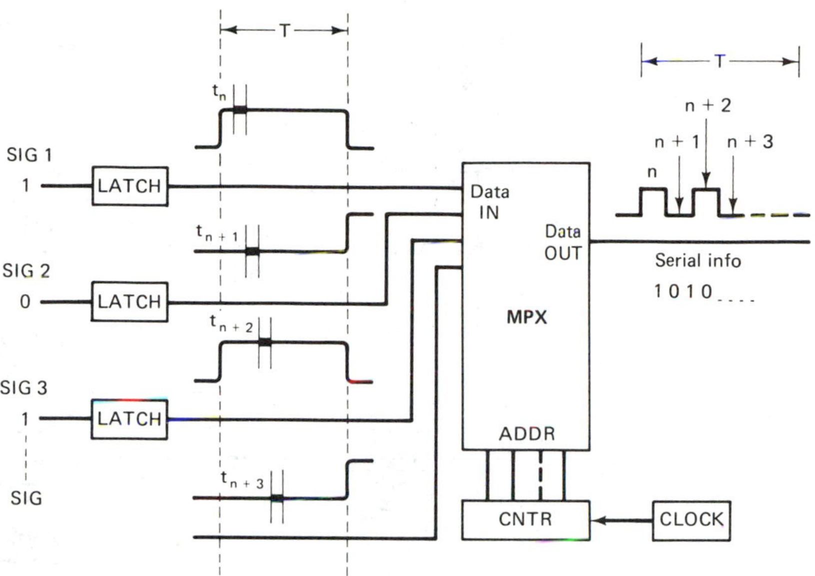

Figure 3.31 The parallel to serial multiplexer concept and its application to time sharing.

For example, 256 parallel bits reduce to 8 + 1 + 1 = 10. This technique is commonly used to strobe the output of multidigit instruments for encoding by a computer (Prob. 3.18). The address lines may be omitted, with timing information used to reassemble the word in software.

Multiplexing is also used in time-shared computers. Since CPU cycle times are less than 10^{-6} s, while response times range from 10^{-3} s (peripherals such as printers) to 10^{-1} s (humans), about 10^3 different functions could be serviced by a computer "at the same time." This is the principle behind time-sharing, although in practice the overhead time required to switch between independent computer programs decreases the number of individual functions by about an order of magnitude. The principle [Fig. 3.31(b)] assumes that information from the slow input devices will be present for a time T which is long compared to the time t_n necessary for the computer to read that information. A multiplexer driven by a counter sequentially scans a series of input devices, encoding the information present at times t_n, t_{n+1}, . . . into a time series of bits. The same counter information is also fed into the computer, so the data from each user can be placed into the proper location. Additional logic (not shown) is necessary to latch the data presented by a device until it has been received by the computer. Errors could otherwise result if the data were changing at the time the computer happened to read it. Translation back into a parallel word is accomplished by a demultiplexer (Fig. 3.32) whose function is similar to that of a decoder, with the addition of a data input line. The selected output line will go high and low, following the input data stream.

The logic inside multiplexers and demultiplexers is a straightforward extension of the logic of decoders. Examples of common TTL circuits are shown in Figs. 3.33 and 3.34 (see Prob. 3.20). Other common multiplexers allow multiple-

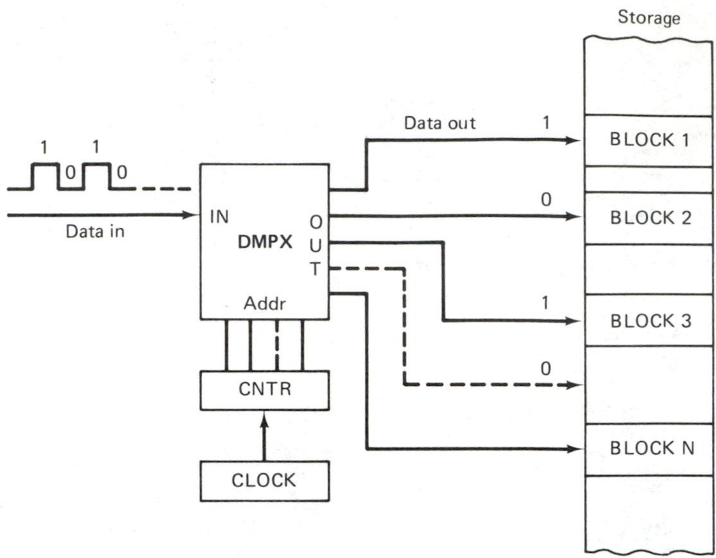

Figure 3.32 Demultiplexing serial data back into parallel form (simplified).

digit displays to be turned on one at a time, reducing power consumption and reducing wiring as well (see Prob. 3.22). The same multiplexing logic allows BCD outputs of digital measuring devices to be *strobed out* digit by digit, to interface simply to microcomputers with a minimum number of wires (see Problem 3.19).

3.7.5 Data Selector Logic and the Programmable Logic Array

A technique called data selector logic can be used to avoid extensive Boolean algebra. It allows an arbitrary truth table to be put into circuitry without needing to deduce from the truth table the necessary gate functions. For example, consider the four input–one output truth table shown in Table 3.3.

Data selector logic uses muliplexer IC's, with inputs permanently wired high or low to match the output column of the truth table (Fig. 3.35). A given input word ABCD applied to the address lines of the multiplexer selects a permanently wired high or low signal. Given the relatively low cost of IC'S compared to the cost of human time required for logic design, this technique is preferred whenever a complicated truth table is encountered.

The *programmable logic array* (PLA) is an extension of this idea. While data selector logic is limited to truth tables with a single bit output, a PLA can generate a truth table with parallel outputs. For example, encoding the correct segments on a seven segment LED display requires a truth table with 4 inputs and 7 outputs. A PLA contains a large number of gates which can be interconnected by the user. While some interesting applications exist for PLA's in microprocessor programming, the read-only-memory (ROM) covered in Chapter 9 is more often used in generating an arbitrary truth table.

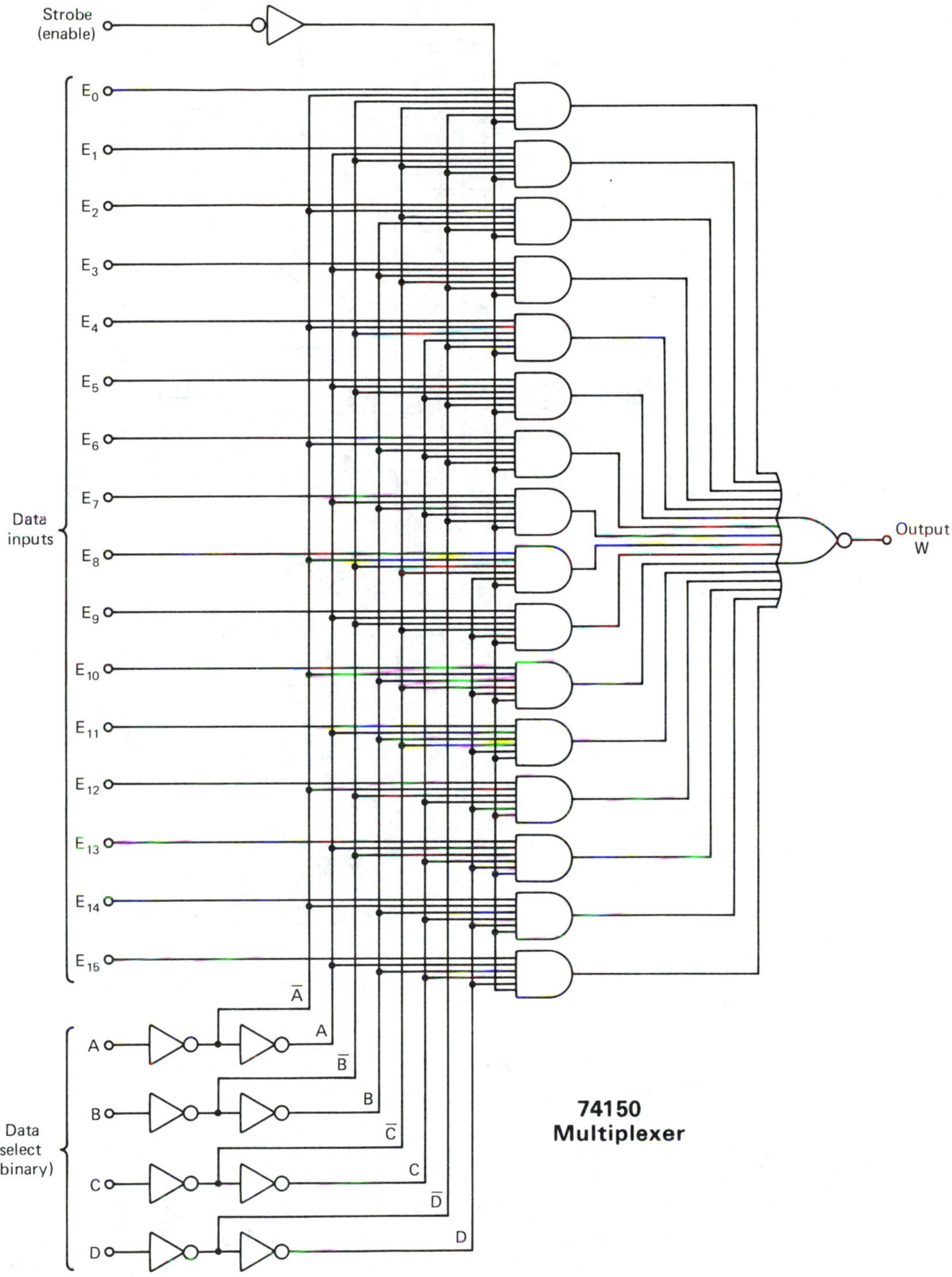

Figure 3.33 Circuit of the 74150 multiplexer. (*Courtesy* Texas Instruments Incorporated).

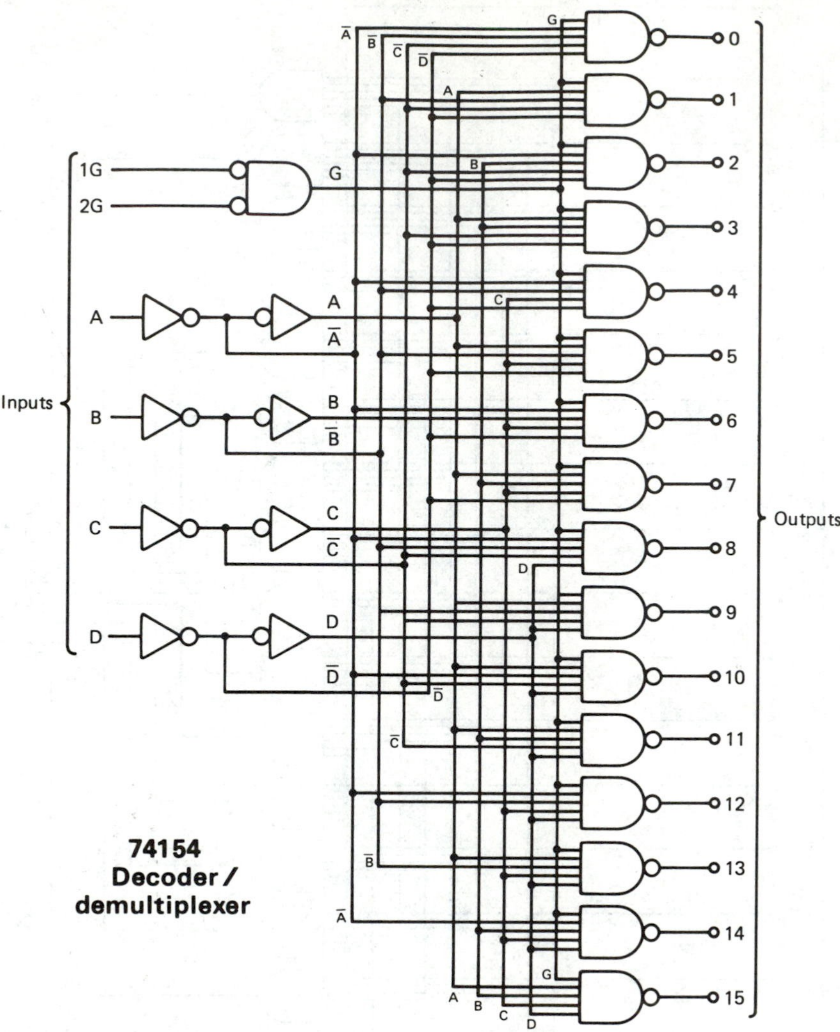

Figure 3.34 Circuit of the 74154 demultiplexer/decoder. (*Courtesy* Texas Instruments Incorporated).

3.8 TTL–CMOS GATE COMPARISON AND INTERFACING BETWEEN LOGIC FAMILIES

3.8.1 Comparison of Available Gate Circuits

CMOS and TTL differ very little in availability of the basic gate circuits discussed in this chapter. Standard part numbers for the most used gates in both families are listed in Table 3.4. Most TTL gates have a CMOS equivalent, often (though not always) pin-compatible. A summary of pin diagrams is given in the lab manual. Certain TTL circuits have no CMOS equivalent. An example is *open col-*

TABLE 3.3 TRUTH TABLE EXAMPLE FOR DATA SELECTOR LOGIC

D	C	B	A	X
0	0	0	0	1
0	0	0	1	1
0	0	1	0	1
0	0	1	1	0
0	1	0	0	1
0	1	0	1	0
0	1	1	0	0
0	1	1	1	0
1	0	0	0	1
1	0	0	1	0
1	0	1	0	0
1	0	1	1	1
1	1	0	0	1
1	1	0	1	1
1	1	1	0	0
1	1	1	1	0

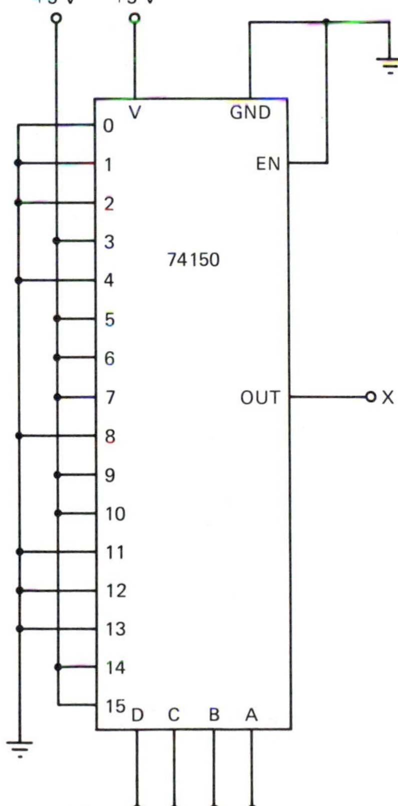

Figure 3.35 Data selector logic example.

TABLE 3.4 COMPARISON OF AVAILABLE GATE CIRCUITS IN TTL AND CMOS

			TTL	CMOS
Gates				
	NAND			
		2-input (4)[a]	7400	4011
		2-input (4) open collector	7401 7403	
		3-input (3)	7410	4023
		4-input (2)	7420	4012
		8-input (1)	7430	4068
	NOR			
		2-input (4)	7402	4001
		3-input (2)	7427	4000 4025
		4-input (2)	7425	4002
		8-input (1)		4078
	AND			
		2-input (4)	7408	4081
		3-input (3)	7411	4073
		4-input (2)	7421	4082
	OR			
		2-input (4)	7432	4071
		3-input (3)		4075
		4-input (2)		4072
	XOR			
		2-input (4)	7486	4070
	XNOR			
		2-input (4)		4077
Decoders, Selectors, Multiplexers, etc.				
	Decoders, demultiplexers			
		16-output DMPX[b]	74154	4514 4515 74C154
		10- or 8-output DECODE ONLY	7442	4028
		4-output (2) DMPX	74155	4555 4556
		BCD to 7-seg LED DECODE	7447	
	Selectors, multiplexers			
		16-input MPX	74150	74C150
		8-input MPX	74151	4512 74C151
		4-input (2) MPX	74153	4539
		2-input (4) MPX	74157	74C157
	Analog switches, MPX/DMPX			
		Analog switch		4016 etc.
		Analog MPX/DMPX		4051,2,3 etc.

TABLE 3.4 CONTINUED

		TTL	CMOS
Special Circuits			
	Parity generator/checker	74180 74280	40101
	Priority encoder, 8-input	74148	4532
	Magnitude comparator, 4-bit	7485	74C85

[a]Numbers in parentheses refer to the number of independent logic functions available on that chip.

[b]Any DMPX may also be used as a DECODE to enable a selected output line by wiring the data input appropriately high or low.

lector logic (e.g., 7401 NAND), available only in TTL. But the converse is also true, since the voltage-controlled-resistor nature of CMOS lends itself to functions not possible in TTL. An example is the family of *analog switches* and *analog multiplex-demultiplex* chips (e.g., 4051) available only in CMOS. Acting as a voltage-controlled valve, the inputs and outputs of these chips can actually be reversed. The term *analog* is used to specify such bidirectional linear chips, since most CMOS digital circuits do not have this capability.

3.8.2 Comparison of TTL and CMOS Specifications

The main features in which the TTL and CMOS logic families differ are summarized in Table 3.5. The dominant difference is power: CMOS gates can consume about 100,000 times less power than their TTL equivalents! The trade-off is speed: the present generation of CMOS circuits is about five times slower than TTL. In addition, TTL circuits are considerably more rugged than CMOS, and are therefore preferred in breadboarding. TTL circuits remain important in high-speed signal processing and in many interfacing applications to the outside world. CMOS circuits are natural in interfacing to microprocessors, which are also MOS

TABLE 3.5 COMPARISON OF TTL, LOW-POWER TTL, AND CMOS SPECIFICATIONS

	TTL 74	TTL 74L	CMOS 74C
Propagation delay (ns)	12	30	60 (5 V) 25 (10 V)
Minimum power dissipation per gate (mW)	15	1.5	10^{-5}
Minimum supply current (mA)	3	0.3	10^{-5}
Maximum output current (mA)	16	3.6	1.75
Input current demand, logic zero (mA)	1.6	0.18	10^{-5}

IC's. A hand calculator constructed from TTL would compute very fast, but would take about 5 A of current to power!

CMOS circuits have negligible *input current demand*, since the input drives an isolated gate, while a typical TTL gate requires 1.6 mA input current from a logic 0 source. On the other hand, a CMOS circuit cannot drive very much *output current*, compared to the 16 mA available from a TTL gate output. The 1.75 mA figure for CMOS output current capability is misleading, since drawing this much current from a logic 0 output would pull the output voltage past the logic 0 threshold of the next circuit. Of course, if a CMOS output is delivering 1 mA, the power dissipation is no longer nanowatts but becomes equivalent to a TTL gate; the power dissipation listed in Table 3.5 applies to the quiescent state with no output loads.

CMOS circuits have flexible power supply requirements, and will run on anywhere from 3 V to 15 V. This is advantageous in interfacing to analog circuits, which often run on 12 V to 15 V. In addition, doubling the supply voltage speeds up a CMOS gate to a 25 ns propagation delay (see Table 3.5), only a factor of 2 slower than TTL.

Useful intermediate logic families are *low-power* TTL, designated with an L (e.g., 74L02), *Schottky* TTL, designated with an S (e.g., 74S02), and the combination of the two (e.g., 74LS02). Low-power TTL is fabricated with resistor values 10 times larger than standard TTL, resulting in 10 times lower power dissipation, at the expense of a 3 times slower propagation delay. Schottky TTL employs Schottky barrier diodes (metal-semiconductor rather than pn) which turn on at a lower voltage, preventing saturation of the input transistors. This increases the switching speed by more than a factor of 2. The combination, the 74LS family, includes both Schottky barrier inputs and higher resistor values, and recovers the speed of standard TTL but with only one-fourth the power demands. The 74L and 74LS series are best for interfacing TTL to CMOS, but may be less available.

3.8.3 Interfacing TTL with CMOS

The principal interfacing problem between these families is the large input current demand of TTL coupled with the modest output current drive capability of CMOS. In a circuit with both CMOS and TTL powered by the same 5-V supply [Fig. 3.36(a)], a CMOS buffer (e.g., 4049) allows a CMOS output to drive up to two standard TTL loads. Although the CMOS output itself can supply more than 1 mA, the output voltage in a logic 0 state would then be pulled past the TTL input logic 0 threshold, giving an error. A CMOS output can, however, drive one *low-power* L or LS TTL gate directly.

On the CMOS input side, a pull-up resistor is added to the TTL output to ensure than the output 1 state will be pulled safely past the CMOS threshold for an input 1 state. This is related to the notion of logic thresholds and noise margins, which are compared in Chapter 6.

Often, the CMOS circuit will not be powered by a 5-V supply, so additional voltage level shifting circuits are added to the interface [Fig. 3.36(b)]. On the CMOS output side, the same CMOS buffer is used, but its power pin is wired over

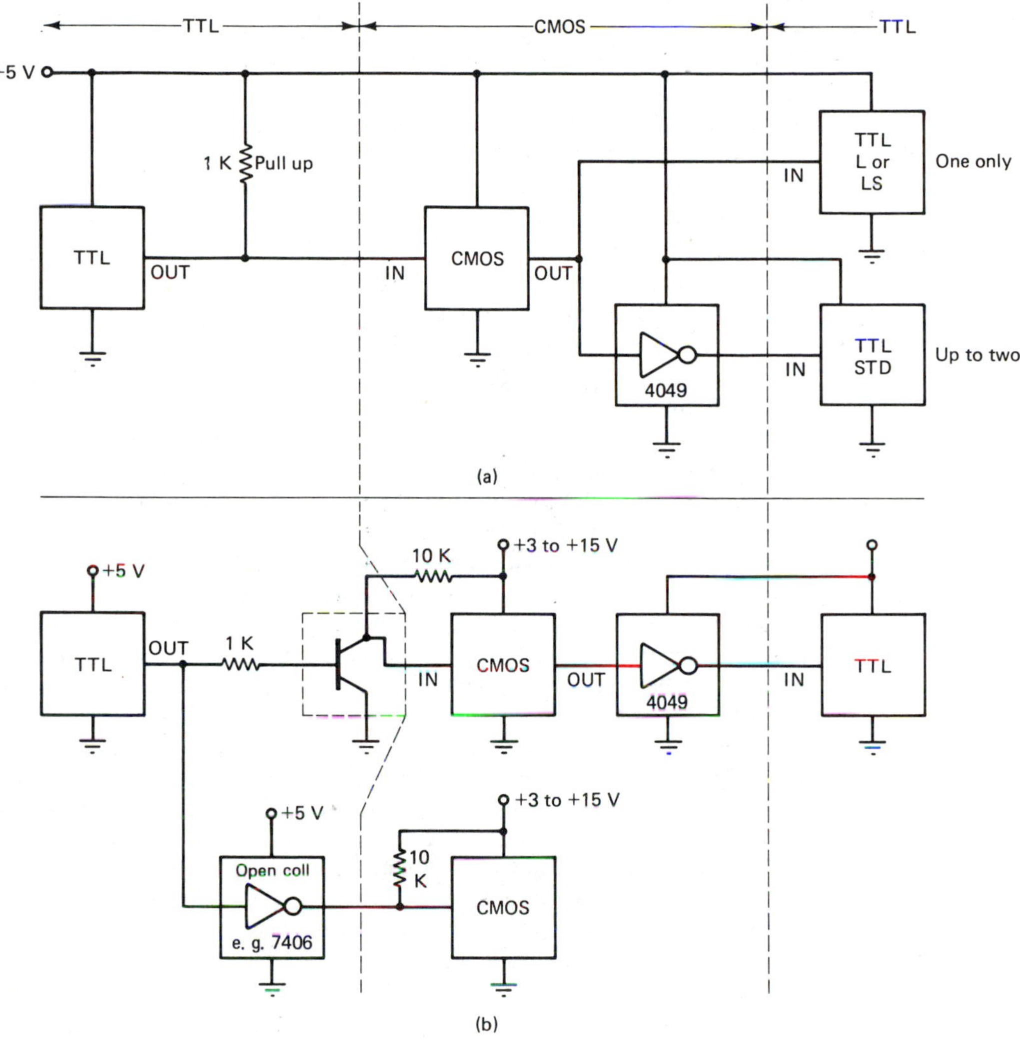

Figure 3.36 Interfacing between CMOS and TTL logic families. (a) Both CMOS and TTL running on 5 V power supplies; (b) CMOS running on independent power supply.

to the TTL 5-V supply to give TTL-compatible logic levels. On the CMOS input side, one can either let the TTL output drive a switching transistor whose collector is wired to the CMOS power supply, or use an open collector TTL buffer (e.g., 7406) capable of handling the CMOS voltage levels.

PROBLEMS

3.1. Suppose one wishes to be able to turn a light on or off from either of two locations. Given two switches, one light bulb, and a power source,

(a) What is the circuit?

(b) If the light being on is called "C is true" and one switch being closed is called "A or B is true," what is the logical operation performed by this switching circuit?

3.2. Explain how the "high man wins" gate [Fig. 3.3(b)] functions. With what logic polarity assignment does this circuit act as an OR gate?

3.3. Work out the switching circuit analogous to Fig. 3.12(b) for the CMOS NOR gate of Fig 3.12(d).

3.4. Sketch the output waveform you expect to see for each of the four gates shown in Fig. 3.14 when an arbitrary data stream enters the A input and the B input is used as an enable input. Follow the example of Fig. 3.13.

3.5. Write the truth tables for 2-input AND, NAND, OR, and NOR gates, assuming positive logic. Then change to negative logic. What happens to the logic function performed by each gate?

3.6. What is the difference between the logic function performed by an OR gate with bubbles on the inputs and an OR gate with a bubble on the output? Write the Boolean expressions for the functions performed. Show using both truth tables and de Morgan's theorem that the two are not equivalent, i.e., that it matters whether an inversion is performed before or after the logic gate.

3.7. Design an interfacing circuit [see Fig. 13.19(b)] using a single IC package (e.g., 7400 quad 2-input NOR) which will reset device C when both $\overline{\mathrm{SEL}}$ and $\overline{\mathrm{RDY}}$ are active, *or* when the manual reset switch MAN is depressed. Begin with the obvious circuit design and then use de Morgan's theorem to redraw it so only one kind of gate is needed.

3.8. An interface is needed to pass along pulses to a counter from a gravity wave detector but to block those pulses which occur during a time interval when a seismometer indicates that the pulse may be spurious due to building vibrations or earthquakes. Show that this *anticoincidence* circuit may be built using 2-input NAND's in a single quad package, or using a single quad 2-input NOR. Assume that both detectors generate active high outputs, and that the counter requires an active high input. Write Boolean expressions for both circuits, and show their equivalence.

3.9. Prove with truth tables the theorems of Boolean algebra (Table 3-2).

3.10. Verify the equivalence below using truth tables. Then show it using Boolean algebra. Draw the gate circuit for each version, and count the number of gates required.

$$A \cdot B + A \cdot C + \overline{B} \cdot C \overset{?}{=} A \cdot B + \overline{B} \cdot C$$

3.11. The symbols of Boolean algebra have an implied order of precedence, as in arithmetic. Thus,

$$A \cdot B + A \cdot C$$

implies

$$(A \cdot B) + (A \cdot C)$$

which is not the same as

$$A \cdot (B + A) \cdot C$$

Show this inequivalence, using truth tables.

3.12. Show how to implement the XOR function using only NOR gates. Show by Boolean algebra that $(A + B) \cdot (\overline{A} + \overline{B})$ is equivalent to $A \oplus B$, draw the circuit, and then using De Morgan's theorem convert the circuit to one using NOR's only.

3.13. Show that a three-gate implementation of the XOR function is possible. Write the Boolean expression for the operation performed by the circuit, and then show its equivalence to $A \cdot \overline{B} + B \cdot \overline{A}$. *Hint:* Gates of three different kinds are required.

3.14. Show using truth tables that the 3 XOR gate circuit of Fig. 3.25 generates the parity bit for a 4-bit binary word. Do this by working out successively the truth table for A B, C D, and (A B) + (C D), and then comparing the last result with the parity function.

3.15. Analyze the relative magnitude circuit (Fig. 3.26(b)]. Refer to the algorithm in the text for inequality testing. Show that the extra inputs on the lower two NAND's disable the comparison of those bits if the inequality has already shown up in more significant bits. Use Boolean algebra, truth tables, or examples.

3.16. Design a BCD-to-decimal decoder by analogy with the algorithm used for the 3-to-8 binary-to-octal decoder of Fig. 3.27.

3.17. The selector circuit of Fig. 3.29 enables a selected device to READ from the data bus. How would the logic be changed to allow the device to WRITE to the data bus?

3.18. The selector circuit of Fig. 3.29 assumes a decoder with active (selected) output *high*. How would the logic be changed if (as with devices such as the 74154) the selected line is *low*?

3.19. Design a multiplexing circuit which strobes the output of a 6-digit (BCD) digital voltmeter out onto four data lines plus address and control (strobe) lines. How many wires are needed? How many would be needed if the ouput were presented directly in parallel form?

3.20. Analyze the operation of the 74150 Data Selector-Multiplexer (Fig. 3.33) and the 74154 Decoder-Demultiplexer (Fig. 3.34). Relate them to the basic decoder idea (Fig. 3.27). How do the extra gates or extra gate inputs allow the IC's to pass a data stream?

3.21. Multiple-level decoding is sometimes used to bring data to or from many locations. Consider a 16-bit bus, with 8 bits reserved for data. Show how to decode the other 8 bits with 4-to-16 decoders to bring the data to or from as many as 256 different devices.

3.22. A 3-digit display is to be multiplexed to a 3-digit counter output so the digits are lit one at a time (though sequenced too fast to flicker). Draw a circuit to accomplish this. All digits of the counter are available in parallel simultaneously. The 3-digit display, however, presents only the 4 bits of a single digit (decoded to 7-segment format internally), plus three wires labeled *digit select.* Use a clock, a counter, and gating as needed.

REFERENCES

More complete bibliographic information for the books listed below appears in the annotated bibliography at the end of the book.

CRAWFORD, *MOSFET in Circuit Design*

DEMPSEY, *Basic Digital Electronics with MSI Applications*

FLOYD, *Digital Logic Fundamentals*

GROVE, *Physics and Technology of Semiconductor Devices*

HIGGINS, *Experiments with Integrated Circuits*, Experiments 3, 4, and 5

HUNTER, *CMOS Databook*

LANCASTER, *TTL Cookbook*

LANCASTER, *CMOS Cookbook*

LARSEN and RONY, *Logic and Memory Experiments Using TTL Integrated Circuits*

MUELLER and KAMINS, *Device Electronics for Integrated Circuits*

Scientific American, *Microelectronics*

SZE, *Physics of Semiconductor Devices*

WILLIAMS, *Digital Technology*

4

Flip Flops

A flip flop (often labeled FF for short) is a basic electronic storage element, analogous to the toggle switch which stores or *latches* information (the light is on or off) until told to do otherwise. In computers and microcomputers, flip flops are used as temporary data storage registers, and are also the basis for semiconductor random access memory (RAM). In measurement instruments, flip flops are used as latches to hold data for display or until it can be read by a computer. Flip flops are also used in measurement and signal processing. For example, flip flop arrays form the basis for frequency counters and for shift registers used in pipeline storage and parallel/serial conversion.

This chapter will trace the evolution of flip flop design, beginning with the discrete component version, to demonstrate what goes on inside an IC flip flop. We progress through several designs made from cross-coupled gates, and then discuss the most common designs: the master–slave flip flop and the edge-triggered flip flop. The last two circuits are complicated but important to understand in order to avoid serious errors in logic design.

4.1 DISCRETE COMPONENT FLIP FLOPS

A flip flop can be formed by cross-coupling inputs and outputs of two inverters and adding a shared emitter resistor R_E [Fig. 4.1(a)]. Because the circuit is so symmetric, one might think that the two transistors would perform identically. However, the cross coupling results in an instability. The slightest imbalance will cause the circuit to flop over into a state where one transistor is *on* hard and the

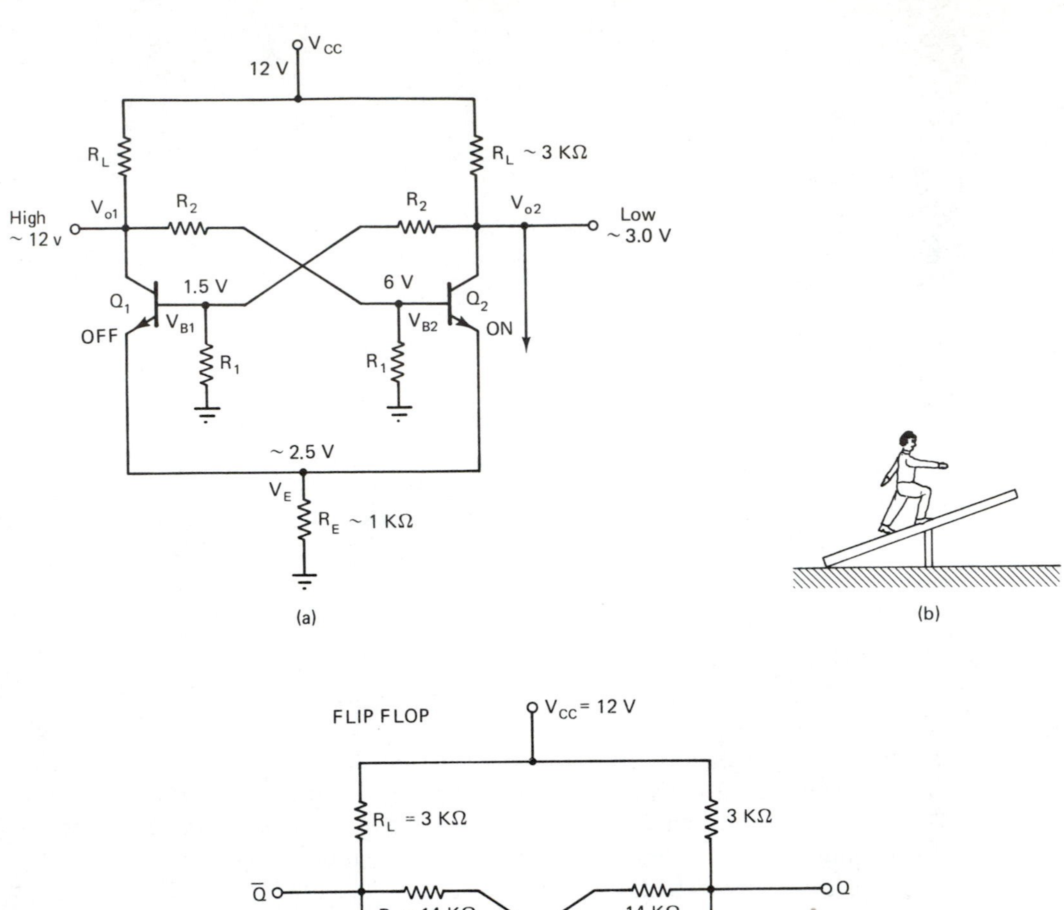

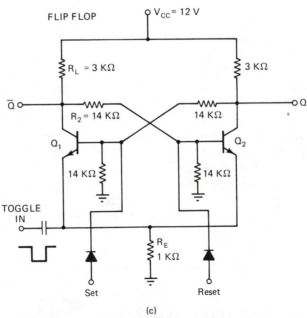

Figure 4.1 Discrete component flip flop. (a) Basic cross-coupled inverter circuit. (b) The bistability of a flip flop resembles a see-saw, with memory like a toggle switch. (c) Circuit for Toggle and Set-Reset change of state.

other transistor is *off* hard (saturated and cutoff, respectively), because each base bias network (R_2 and R_1) is connected to the collector of the opposite transistor. Suppose that when power is initially applied, Q_2 is on a little bit more than Q_1.

Since V_{O2} will be a bit lower, V_{B1} will also be a little bit lower. This will tend to turn Q_1 off. But this raises V_{O1}, which biases the base-resistor pair for Q_2. Q_2 will therefore turn on even more, lowering V_{O2}, which lowers V_{B1}, which turns it off even more, and so forth. The situation is analogous to someone walking past the pivot point on a seesaw [Fig. 4.1(b)]. The value of resistor R_E is selected to make sure that the *off* transistor is held off hard. For example, with the component values shown, when Q_2 is *on*, the voltage at the top of R_E is determined by the voltage divider formed by R_E and R_{L2}

$$V_E = [R_E/(R_E + R_{L2})]V_{CC} - 0.5 \text{ V} = 2.5 \text{ V} \tag{4.1}$$

Q_1 cannot be turned on unless its base voltage reaches this value, or unless V_E is somehow reduced. Both methods are used to induce a change of state.

A two-state circuit is not useful unless one can change the state upon command. Two ways to do this [Fig. 4.1(c)] illustrate two basic kinds of flip flop action. Either the circuit may flip into its opposite state (*toggle*) each time an input signal is applied, or else separate *Set* and *Reset* inputs may force the output into the Q = 1 or Q = 0 state, respectively, regardless of the previous state.

Toggle action results when a negative-going pulse is applied at the emitter connection. The pulse must be large to overcome the reverse bias of the *off* transistor. As soon as the *off* transistor begins to conduct, its collector voltage falls, lowering the bias voltage of the *on* transistor's base and tending to turn it off. If the input pulse is large enough and long enough to bring the circuit to the point where both transistors are conducting equally, the instability will result in a change of state.

The philosophy behind Set and Reset inputs is also to force the *off* transistor into conduction, but here inputs are applied separately to the base of the appropriate transistor. Suppose that the circuit is in the $Q = 0$ state, with Q_2 on and Q_1 off. A positive-going signal applied to the Set input large enough to overcome the negative bias of Q_1 will bring it into conduction. Once Q_1 begins to conduct, its collector voltage falls, and the instability propagates since the bias voltage holding Q_2 *on* also falls.

Practical discrete component flip flops have additional circuit components not illustrated here. It is nearly always preferable to use the IC flip flops to be described.

4.2 FLIP FLOPS FROM GATES

As with the discrete component FF, the simplest IC flip flop is formed by two inverters wired in a closed loop [Fig. 4.2(a)]. The cross coupling results in a two-state system. Suppose that the output state is a 0. The input to inverter 2 must have been a 1. The input to inverter 1 must therefore have been 0. But this is just the signal provided by the feedback wire from the ouput. By symmetry, an equally stable state would reverse the 1's and 0's. Changing the state of this flip flop requires forcing a change of state at some point in the circuit. For example, if the Go-to-1 switch is closed, the input of gate 2 is forced to 0, bringing its output

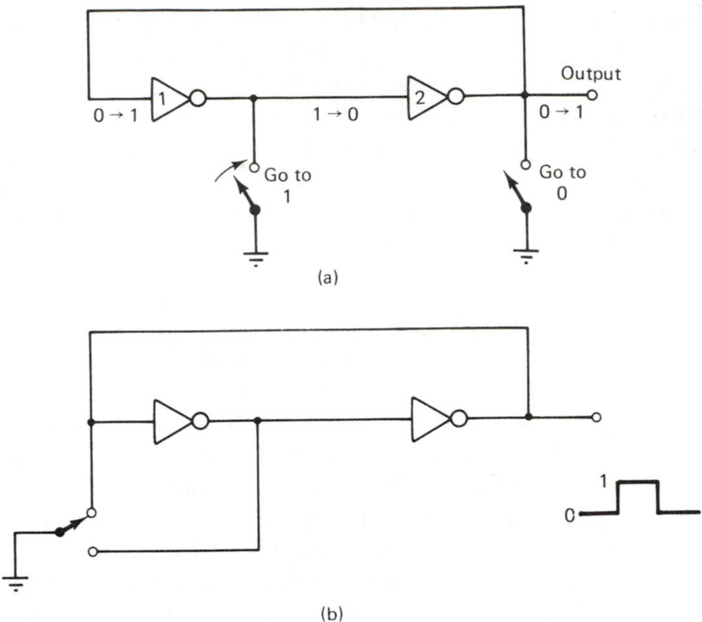

Figure 4.2 (a) Cross-coupled IC inverter flip flop circuit. (b) Application as switch contact debouncer.

high, giving gate 1 an input which will preserve the output high state even if the switch is then opened. If the two connections for changing the state are wired to a single pole–double throw switch [Fig. 4.2(b)], the circuit becomes a *switch debouncer.* Mechanical switches normally bounce for a period of milliseconds before establishing good contact. A debouncer is used to eliminate these multiple pulses which could cause serious errors in circuits connected downstream (e.g., a counter).

This circuit suffers from a limitation. If the Go To signal is supplied by an electronic switch whose impedance to ground is not 0, a change-of-state command may result in competition between the gate whose output is trying to be high and an input signal which is trying to drive it low. The winner is whichever circuit has the lower source impedance. This unpredictable behavior is unacceptable.

This indeterminacy can be overcome by using gates instead of inverters [Fig. 4.3(a)], since the change-of-state signal and the feedback connection do not interfere with one another when wired to different gate inputs. In the example shown (constructed from NAND's), the Q = 0 state shown is one of two stable possibilities. When inputs $\overline{S}$ and $\overline{R}$ are held high, the NAND gates function as inverters, and the stable states are identical with those of Fig. 4.2(a). However, one of the NAND's has both inputs high, while the other has only one input high. A change of state is effected by bringing *low* the input of the NAND gate which had both inputs high. In the example of Fig. 4.3(b), bringing the input labeled $\overline{S}$ low causes its output to go high. That will cause the lower NAND output to go low. This cross-couples back as an input to the upper NAND, keeping its output low even if

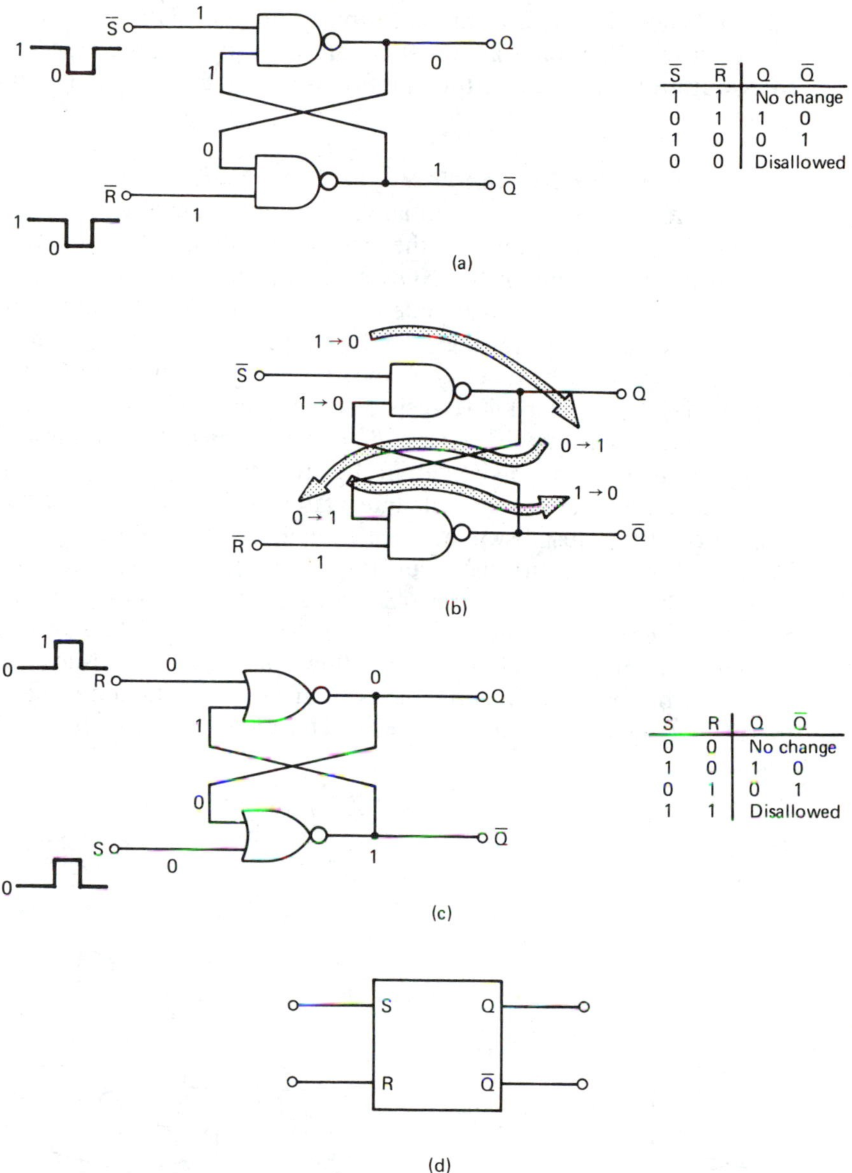

$\overline{S}$	$\overline{R}$	Q	$\overline{Q}$
1	1	No change	
0	1	1	0
1	0	0	1
0	0	Disallowed	

S	R	Q	$\overline{Q}$
0	0	No change	
1	0	1	0
0	1	0	1
1	1	Disallowed	

Figure 4.3 Flip flops from cross coupled IC gates. (a) NAND gate S-R FF circuit. (b) Change of state sequence in response to Set = 0. (c) NOR gate S-R FF circuit. (d) S-R flip flop symbol.

S should go high again. A change of state has been forced. Since the change of state is brought about by an input going low, the inputs are labeled with a bar on top to indicate their *active-low* nature. In the truth table [Fig. 4.3(a)] for this flip flop, the input combination $\overline{S} = 0$, $\overline{R} = 0$ is labeled *disallowed*, since the output state when the inputs went back high would be unpredictable.

A set-reset (SR) flip flop can also be constructed from NOR gates [Fig.

4.3(c)]. When the S and R inputs are held low, the NOR gates function as inverters, and two stable states are possible. Since 3 out of 4 input combinations bring a NOR output 0, the state of this flip flop is changed by an input signal which goes high.

Test your understanding. With the NOR gate FF of Fig. 4.3(d) in the Q = 0 state, and $S = R = 0$ initially, what input signal is required to bring it to the Q = 1 state? Trace through the changes in gate inputs and outputs.

In the truth table of the NOR gate FF, the state with both S and R high is disallowed, since it leads to an indeterminate state when S and R return low.

The logic symbol for these SR flip flops is shown in Fig. 4.3(d). These FF's are useful as switch contact debouncers (Fig. 4.4). In the NAND gate example [Fig. 4.4(a)] there is a 1-KΩ resistor connected between each of the inputs and the positive power supply. This *pull-up* resistor ensures that an ungrounded gate input will float reliably high even in the presence of noise. A pull-up resistor does not interfere with the change of state as long as its value is large compared to the resistance of the closed switch, since a reliable logic 0 input can still be generated. For a NOR gate debouncing circuit [Fig. 4.4(b)], the quiescent state requires both inputs to be low, so *pull-down* resistors are connected to ground. The value of pull-down resistors is critical in FF's constructed from TTL gates, since the input low current demand of about 1.6 mA flows through the pull-down resistor. There will therefore be a voltage drop across that resistor, which must be limited to less than the TTL *low* threshold of 0.8 V. The pull-down resistor must be therefore less than

$$R = V/I < 0.8 \text{ V}/1.6 \text{ mA} = 500 \ \Omega \tag{4.2}$$

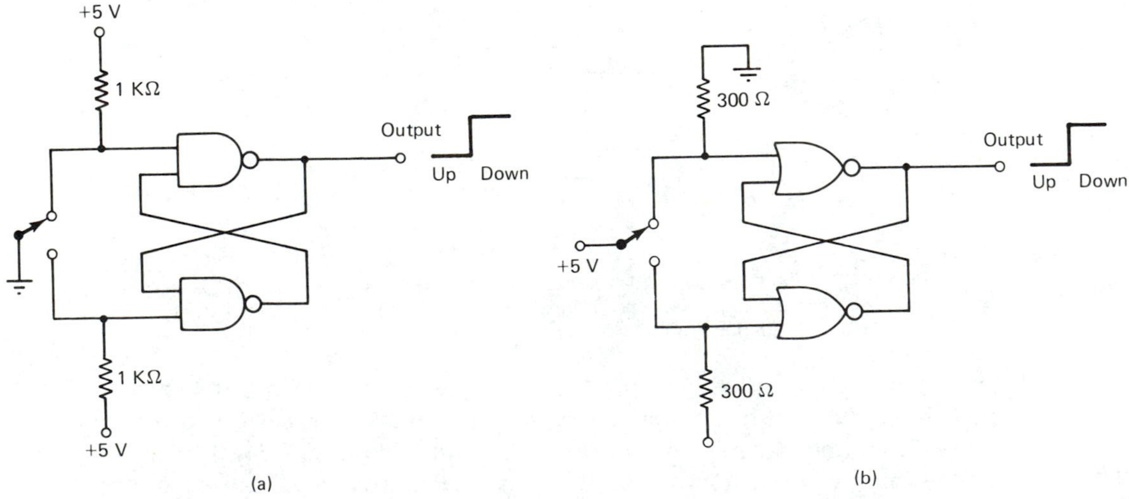

Figure 4.4 Switch contact debouncers using IC S-R flip flops. (a) NAND gate circuit. (b) NOR gate circuit. Note different input circuit.

On the other hand, if the pull-down resistor is too small, there will be an unnecessarily large current demand on the power supply. A value of 330 Ω is typical. Such constraints on resistor values are of course unnecessary in constructing FF's from CMOS gates.

4.3 CLOCKED FLIP FLOPS

Although the SR flip flop is useful as a switch debouncer, its utility as a data storage register is limited, because it does not discriminate between noise and meaningful data when responding to a change-of-state command. The addition of a clock input (Fig. 4.5) permits control over when the Set and Reset inputs are recognized. Only when the clock input is high will information present at S and R be able to change the state of the cross-coupled gate FF (dashed box). The truth table resembles the basic SR flip flop:

S	R	CLK	Q	$\bar{Q}$
0	0	0	No change	
0	1	0	No change	
1	0	0	No change	
1	1	0	No change	
0	0	1	No change	
0	1	1	0	1
1	0	1	1	0
1	1	1	Disallowed	

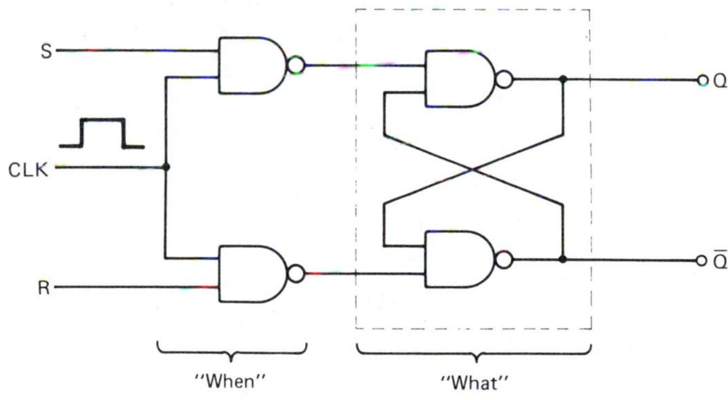

Figure 4.5 Clocked flip flop with additional gates enabling "when" a change of state may occur.

A useful modification of the clocked SR flip flop is the *D-type latch* (Fig. 4.6), in which the S input is inverted and fed to the R input. Only a subset of the SR truth table is accessed, and the disallowed state (S = R = 1) is never encountered. The truth table is:

D	S	R	CLK	Q	$\bar{Q}$
0	0	1	0 }	No change	
1	1	0	0 }		
0	0	1	1	0	1
1	1	0	1	1	0

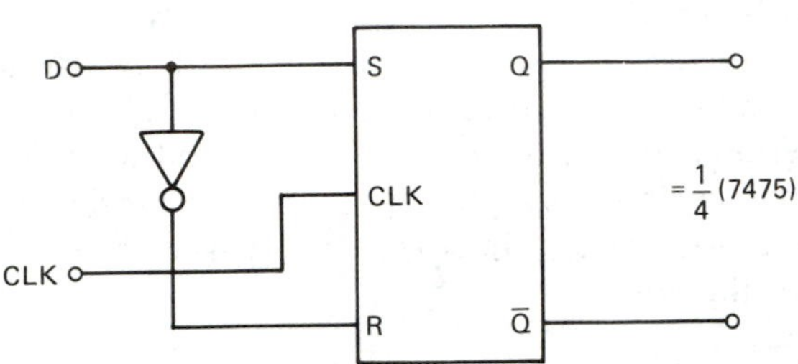

Figure 4.6 D-type flip flop or data latch made from an S-R FF.

The FF output follows whatever data is present at the D input as long as the clock is high, *latches* the information present when the clock falls, and holds it during the time the clock remains low. A latch is useful, for example, in storing the information fed to the display of a calculator or measuring instrument, keeping the most recent information until an update is made. Latches are also useful in encoding the output of a keyboard, holding the information ("this key was pressed") until it can be scanned by a decoding device.

The D-type latch is available in a multiple-device package, for example, the TTL 7475 or the CMOS 4042, with four latches per package.

Another useful circuit is the *toggle* or T flip flop, whose output changes state once each clock cycle. Since the output frequency is half the clock frequency, this is often referred to as the "divide by two" connection. Suppose we try to construct a T flip flop using the clocked SR flip flop (Fig. 4.7), wiring the $\bar{Q}$ output to the R input and the Q output to the S input. If the flip flop is initially in the Q = 0 output state, the inputs are prepared to shift into the Q = 1 state whenever the

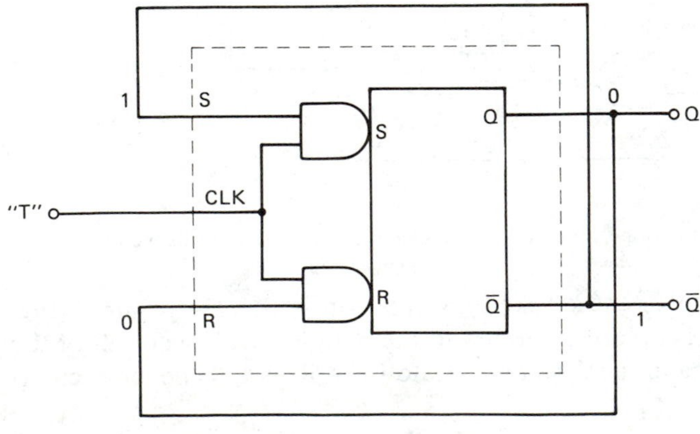

Figure 4.7 T or Toggle flip flop made from an S-R FF. The circuit fails to operate reliably when a simple S-R FF is used.

clock allows it. This feedback always gives instructions to change the output state and therefore provide toggling action. However, there is a problem. As soon as the clock goes high, the change of state to Q = 1 occurs. This feeds back a message to the inputs instructing them to change to the Q = 0 state ($S = 0,\ R = 1$). But then the output state gives the inputs a message to change back to Q = 1, etc. The result is a *race*, with the flip flop oscillating between states 1 and 0 when the clock is high, at a rate determined by the switching time of the gates. These race conditions, which often occur in simple clocked SR flip flops, are a reason for the development of the more sophisticated master–slave and edge-triggered flip flops.

4.4 MASTER–SLAVE TTL FLIP FLOPS

Simple clocked R-S or D-type flip flops suffer from a number of problems. Since the output will follow the input as long as the clock enables it (Fig. 4.8), one must be sure that the data was present just before the clock fell. Often, FF's are chained together to form a counter or shift register (Fig. 4.9), in which the information present advances one stage to the right each time the clock goes through one cycle, like a theater ticket line. Unless the flip flop is prevented from changing state more than once during each clock cycle, each data transfer will propagate through the circuit in an uncontrolled way. As another example, the inputs to a logic

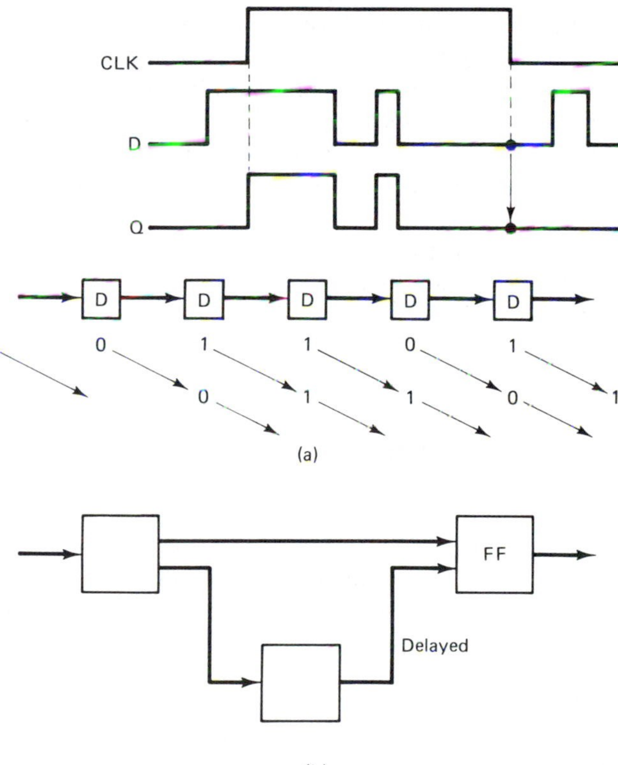

Figure 4.8 The output of a D-latch is transparent to the input while enabled by the clock.

Figure 4.9 Two inadequacies of simple flip flops. (a) Data racing uncontrollably through a string of FF's. (b) Unpredictable response as a result of unmatched gate delays.

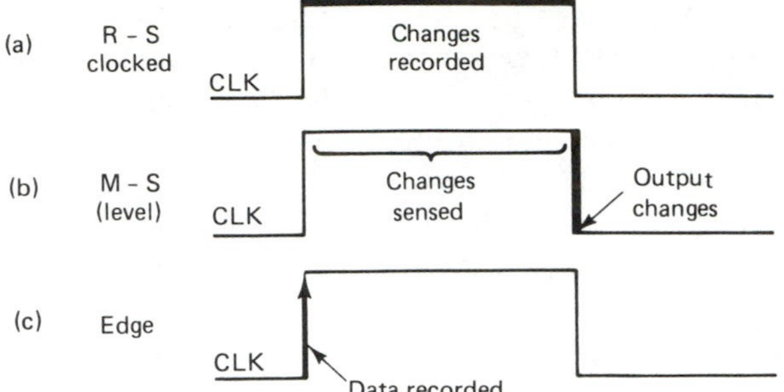

Figure 4.10 Timing diagrams for the (a) clocked Set–Reset, (b) level triggered Master–Slave, and (c) edge-triggered flip flops, showing when data is recorded.

block may be delayed different amounts [Fig. 4.9(b)], due to differing numbers of gate delays. A logic error may result if the information is stored at the wrong time.

For these reasons, several more error-free FF circuits have evolved. The most important types are the *master–slave* (level-triggered) and the *edge-triggered* flip flops. Their responses are compared to those of the RS clocked flip flop in Fig. 4.10. In the case of a *master–slave* FF, the inputs during half of the clock cycle cause action to occur *within* the flip flop, but no *output* change is seen until the end of the clock cycle. With *edge triggering*, data is recorded only during the brief time interval while the clock is changing its state. In the example shown, the active time period is when the clock goes from low to high. This is called *positive* edge triggering, denoted by a positive-going arrow. The active time period is quite short, for example 20 ns for a TTL 7474. Although output changes occur only when the clock level changes in both the master–slave (M–S) and edge-triggered designs, the M–S type is not an edge-triggered device, since input action is sensed during one half of the clock cycle. Another possible confusion in terminology exists because such a M–S flip flop is called *negative level triggered*, since changes in the output are not seen until the clock falls from high to low. However, only input changes which occur while the clock is high or positive are sensed and recorded.

A master–slave flip flop constructed from NAND gates (Fig. 4.11) consists of two pairs of cross-coupled gate FF's, each gated by an enabling circuit. There are two basic ideas in the design.

(1) Because of the extra inverter, when the master is sensing input changes, the slave is not, and vice versa.

(2) The slave is isolated from the master during the time in which the master's state may change.

The second point is carried out in Fig. 4.11 by the resistor at the inverter input. With the current flow of the input low state, the voltage drop across this resistor shifts the inverter slightly closer to the point of transition to the input high state. As the clock input rises, the inverted clock signal closes off the slave before the master inputs become active (Fig. 4.12). As a result, the slave is isolated from the master during the period of time (clock high) when the master's state may change. As the clock signal falls, the same difference in voltage level sensitivity ensures that the master inputs are disabled before the slave senses the

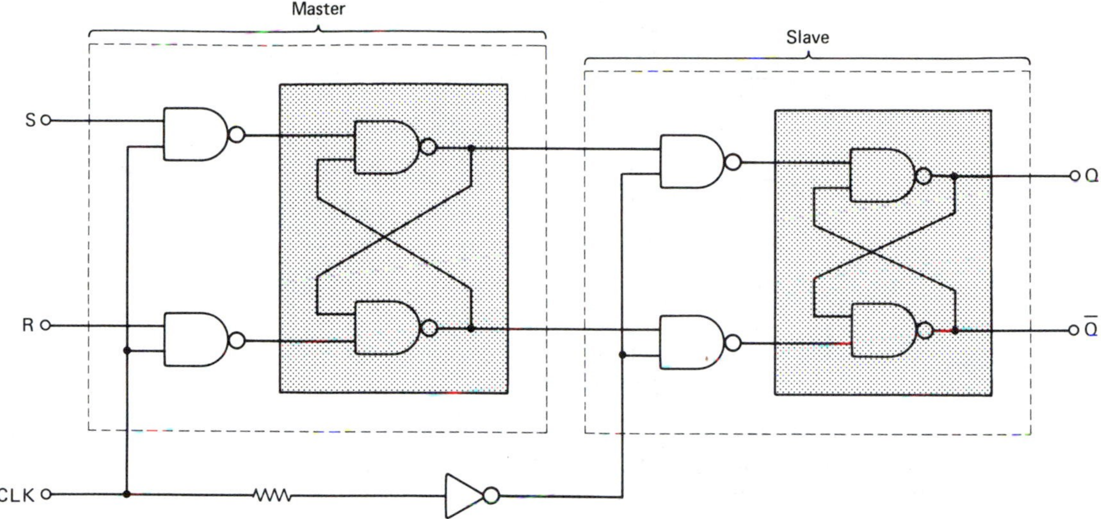

Figure 4.11 S-R Master–Slave flip flop circuit.

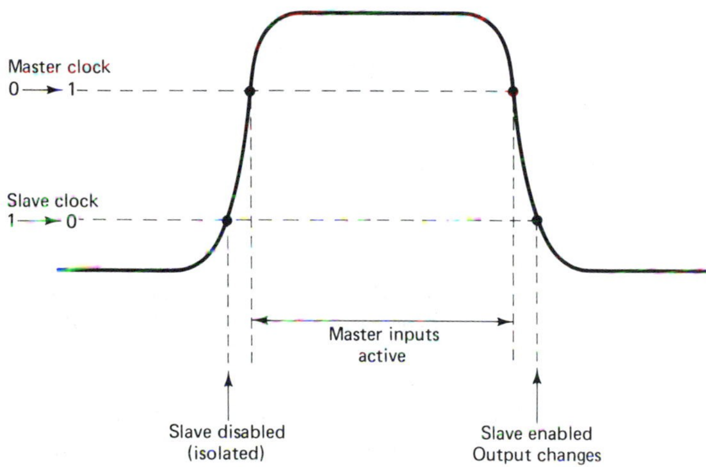

Figure 4.12 Timing diagram for a Master–Slave FF circuit. This is called "negative level triggering" since output changes do not occur until the clock falls.

state of the master and transfers it to the output. Although the circuit of Fig. 4.11 illustrates the key features of master–slave flip flop operation, actual circuits do not contain the inverter shown, since the extra gate delay would complicate circuit operation at high frequencies.

4.4.1 JK Master–Slave Flip Flops

The circuit of Fig. 4.11 still permits the disallowed state of the original SR flip flop. The JK flip flop (Fig. 4.13) removes this disallowed state and replaces it with a useful truth table entry. Internal connections feed the outputs back to the in-

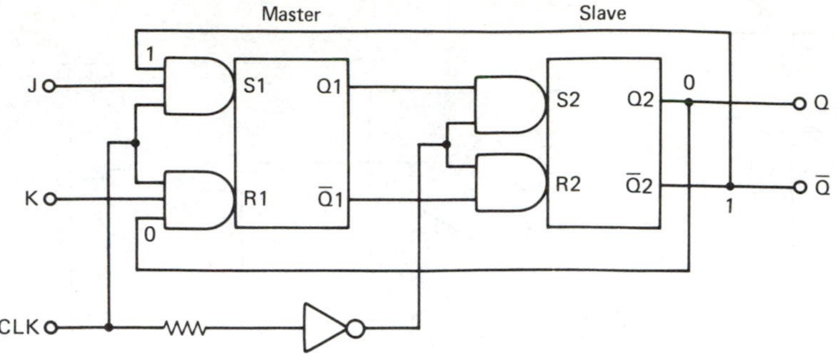

Figure 4.13 A JK Master–Slave FF circuit is created by gating feedback from the outputs to the inputs.

puts, connecting Q to a Reset input, and $\overline{Q}$ to a Set input. This toggle connection does not allow racing as in Fig. 4.7, since the master–slave connection allows an output change at most once per clock cycle. The truth table for the JK master–slave flip flop therefore has no disallowed states.

Inputs (t_n) J	K	CLK	Outputs (t_{n+1}) Q	$\overline{Q}$
0	0	⌐L	Q_n	$\overline{Q}_n$
0	1	⌐L	0	1
1	0	⌐L	1	0
1	1	⌐L	$\overline{Q}_n$	Q_n

The subscript n reflects the fact that inputs present during one clock cycle do not appear at the output until the next clock cycle. If J and K are held high, the state will change once each clock cycle, dividing the clock frequency by 2.

There is a subtlety in JK circuit operation which, unless carefully understood, will cause logic errors in circuit design. The feedback wiring of outputs back to inputs disables one of the inputs at all times. If the FF is in $Q = 0$ state (Fig. 4.13), the K input can have no effect on changing the state of the master. Only the J input will be sensed, and if J is a 1 at any time while the clock is high, the state will change from $Q = 0$ to $Q = 1$ at the end of a clock cycle. This may be summarized by writing the truth table slightly differently.

Q_n	J	K	CLK	Q_{n+1}
0	0	X	⌐L	0
0	1	X	⌐L	1
1	X	0	⌐L	1
1	X	1	⌐L	0

Here, X stands for "Don't care": it does not matter whether the input is high or low. A change of state of the master while the clock is high does not change this feedback enabling, since the state of the slave remains unchanged until the clock falls. Also, once an input signal condition has produced a change in the master, no further changes are possible during this clock cycle.

> The master–slave flip flop may change its state only once during a clock cycle, and the change recorded is the first change to occur.

In the example of Fig. 4.14(a), the K input is disabled, and the J input can only act to set the master flip flop. Once set, the master cannot be reset during this cycle. The value of K is irrelevant because the FF began the clock-high cycle in the $Q = 0$ state. In the example shown in Fig. 4.14(b), the change to the $Q = 1$ state occurs because J was high when the clock went high, and all later changes at the J input are irrelevant.

Reliable operation of JK negative level-triggered M–S flip flops may be summarized as follows.

4.4.2 Rules for JK Master–Slave Flip Flops

(1) A change of the output state (slave) may occur only at the end of the clock cycle, as the clock falls.

(2) A change of the internal state (master) may occur any time while clock is high.

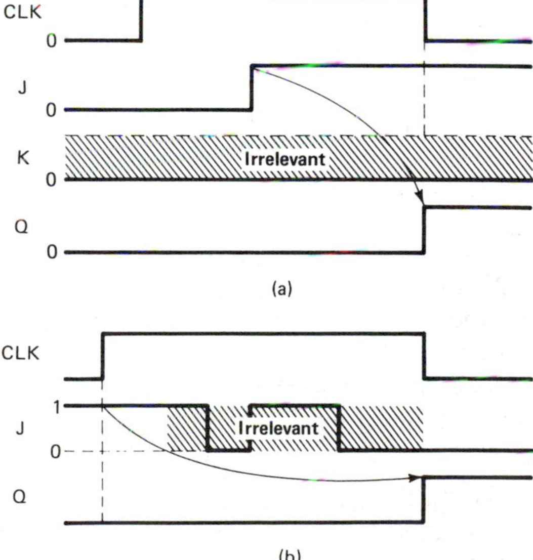

Figure 4.14 Timing diagrams for a negative level triggered JK M-S flip flop. (a) The output change caused by J going to 1 does not appear at the output until the clock falls. K input data is irrelevant because Q was low. (b) Further changes in J are irrelevant after the first change of state has been recorded.

(3) The only changes in the master which are enabled are those which will cause a change in the slave output. The first such input condition will be stored, and later changes in inputs are irrelevant.

(4) The master will change its state at most once during a clock cycle.

4.5 EDGE-TRIGGERED TTL D FLIP FLOPS

Like the master–slave FF, the edge-triggered D flip flop also has two FF's in series (Fig. 4.15). The circuit is enabled to listen to a signal at the D input only during a brief instant of time while the clock changes its state from low to high in a *positive* edge-triggered FF or from high to low in a *negative* edge-triggered FF. This circuit is particularly useful as a data storage register where spurious noise

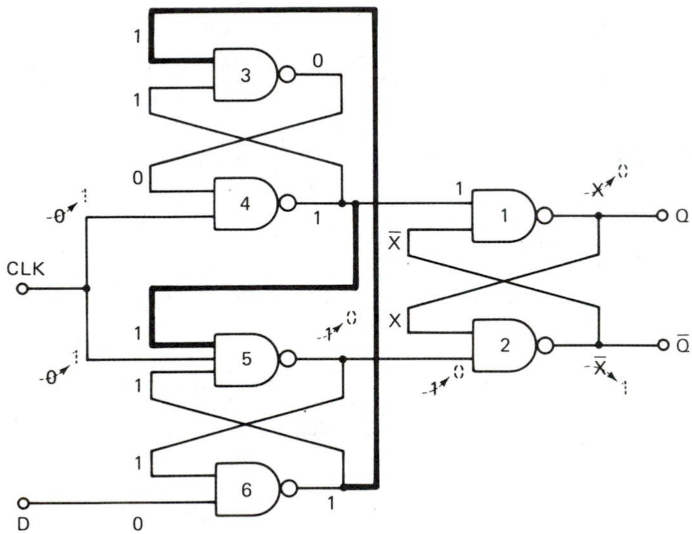

Figure 4.15 D-type edge-triggered flip flop constructed from NAND gates. The heavy lines designate feedback connections which cause the edge-triggered action.

may occur during circuit transitions, since clock timing can be arranged so that the data at D settles to a reliable value before being latched. The circuit may be viewed as three SR flip flops, the two on the left driving the one on the right. In addition, each of the outputs on the left is fed back as an input to the other FF (shown as heavy lines in Fig. 4.15). The states labeled in Fig. 4.15 result if initially the D input is low and the clock input is low.

Exercise

Verify this initial state by directly tracing through the gate logic. Keep in mind that as long as one input to a NAND gate is 0, the output will be 1. The label X designates a "don't care" situation.

4.5.1 Change of State

Initially, both outputs of the lower left FF (gates 5 and 6) are simultaneously high. This state was disallowed for the simple SR flip flop (Fig. 4.3) because it leads to a race in which the final state depends upon which input returns high first. However, this unusual situation is responsible for edge-triggered action, and leads to no race since the basic design rule of edge triggering demands that the data at the D input be settled and waiting at the time the clock input goes high. Thus, as soon as CLK reaches 1, gate 5 flips to a 0 state, which is seen as a Reset To 0 command by the right-hand FF (gates 1 and 2). Further changes on the D input while the clock is high will have no effect, since neither flip flop is "primed" for action and gate 6 is disabled by the 0 output from gate 5. A similar chain of events occurs if D is high at the time the clock goes high, except that in this case the upper left flip flop (gates 3 and 4) is primed with both outputs high, and will trigger a change to the $Q = 1$ state when the clock goes high.

TRUTH TABLE FOR THE D-TYPE EDGE-TRIGGERED FLIP FLOP

D	CLK	Q
0	↑	0
1	↑	1

Edge-triggered flip flops are therefore able to listen only to data present when the clock goes high. Data changes occurring before or after this transition are not seen, hence the term *edge-triggered.* Design rules for applying edge-triggered flip flops are therefore somewhat simpler than those for JK flip flops.

4.5.2 Design Rules for Edge-Triggered Flip Flops

(1) The data which will be latched and appear at the output is that which was present at the time the clock went high (for positive edge triggering) or low (for negative edge triggering).

(2) The data D must be held steady during the time period of the clock low-high transition, plus a period of about one gate delay.

A comparison between JK-MS and D-edge flip-flop characteristics is given below.

	JK-MS-level	D-edge
Trigger	Level	Edge
"No change" state	Yes	Impossible

	JK-MS-level	D-edge
Toggle	Intrinsic; wire $J = K = 1$	Only if hard-wired (extra inverter)
Noise immunity	Requires data be stable	Excellent
TTL example	7476	7474

4.6 CMOS FLIP FLOPS

Like their TTL counterparts, CMOS flip flops also come in three basic types:

(1) The clocked latch, useful in routine temporary data storage;
(2) The D-type edge-triggered latch, for precise data latching and synchronizing;
(3) The JK master–slave, with its flexible truth table useful in counters and logical manipulations.

Important differences exist:

(1) More simple latches can be squeezed into a single package in CMOS, making this family very useful for bus-oriented design, multiplexing, etc.;
(2) The D-edge latch is a master–slave design in CMOS FF's;
(3) The CMOS JK M-S flip flop is edge-triggered rather than level-triggered as in TTL.

Difference (3) makes the CMOS JK M-S flip flop more reliable. Only the data present when the clock changes gets stored, removing the possibility of recording changes which occur during the clock-high phase, as in TTL level clocked M-S designs [Fig. 4.14(a)]. In most FF applications, data is present when the clock enables storage, so this design change is not a restriction and enhances noise immunity.

CMOS IC flip flops are based around the *transmission gate* rather than cross-coupled gates alone. A CMOS transmission gate [Fig. 4.16(a)] is a four-terminal device which functions as a single-pole–single-throw switch. When C is low, input and output are connected by nearly a short circuit. When C is high, input and output are isolated from one another. A transmission gate (TG) is thus a MOS "relay" switch, and forms the basis not only for CMOS FF circuits but also for tri-State logic and for CMOS analog switches. The diamond-shaped symbol [Fig. 4.16(a)] is equivalent to the switch shown, and is often drawn more simply by a box labeled TG. The control connections may be left off. A box labeled TG is a normally closed switch, with In and Out connected together as long as control in-

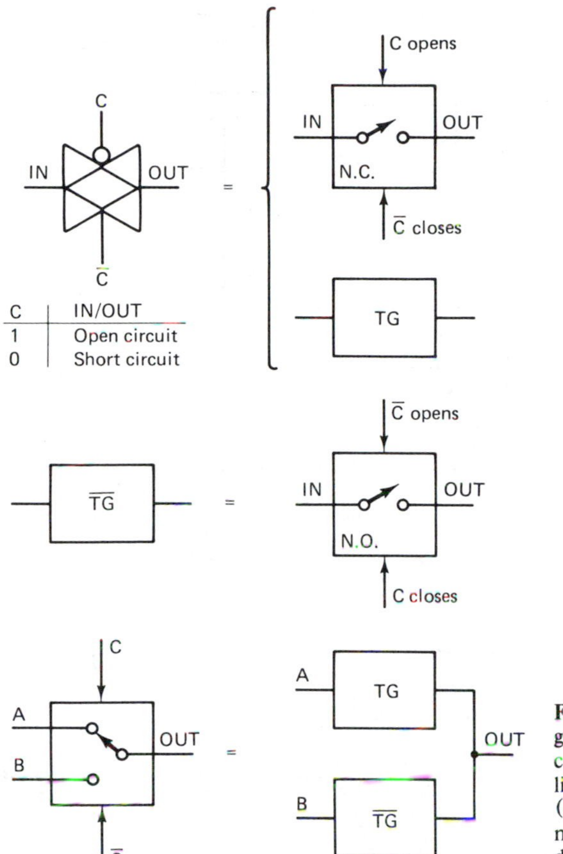

Figure 4.16 The MOS transmission gate. (a) Circuit symbols for a normally closed TG. (b) Reversing the control lines makes a normally open switch $\overline{\text{TG}}$. (c) Combining a normally open and normally closed TG makes a single pole, double throw switch.

put C stays low. A bar on top of the TG symbol [Fig. 4.16(b)] indicates that inputs C and $\bar{C}$ have been reversed to make a normally open switch. A single-pole–double-throw switch which selects one of two inputs is made using two complementary transmission gates TG and $\overline{\text{TG}}$ [Fig. 4.16(c)]. The connection of these two outputs together results in no indeterminacy since one of the boxes is always an open circuit.

4.6.1 The CMOS D-Type Edge-Triggered Flip Flop

Transmission gates simplify flip flop design. An example is the CMOS D-type flip flop (Fig. 4.17). This is a master–slave design, with two cross-coupled gate FF's in series. In addition to transmission gates connecting the Data input to the Master FF and the Master to the Slave, TG's are also within the feedback loop of each FF. Opening one of these TG's breaks the FF feedback loop, so the NOR's act as two inverters in series, with output following input. When the clock is low, TG 1 and TG 4 are short circuits, and TG 2 and TG 4 are open circuits. The Slave acts as an FF, retaining whatever was previously stored, but is closed off from the Master. Input data is passed along to the Master, which continuously follows but

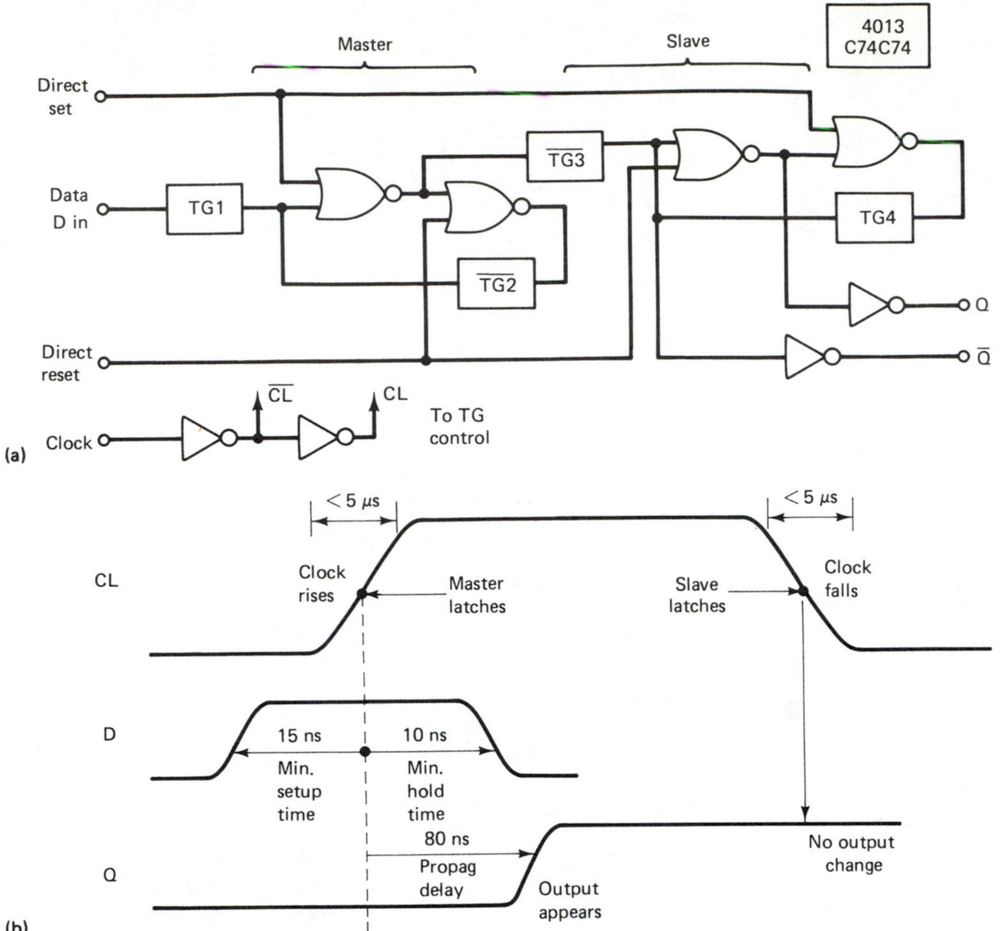

Figure 4.17 CMOS D-type edge-triggered flip flop. (a) Circuit, typical of type 4013 or 74C74. (b) Timing diagram, showing immediate edge-triggered response although the internal circuit is a Master-Slave type.

does not store it. As the clock goes high, the roles of all transmission gates are reversed. The Master becomes an FF, storing whatever had been present at D when the clock went high. TG 1 opens, preventing any further input data changes from being recorded. TG 3 closes, passing the Master's state along to the Slave and also directly to the output. The Slave is not an FF now, since TG 4 is open. When the clock falls again, the Slave latches the Master's state before TG 3 opens to disconnect the two FF's. Outputs Q and $\overline{Q}$ are buffered to prevent output loading from changing the FF state. The CMOS D-type FF has several unusual features.

(1) Only one FF is storing data at any time. While the clock is high, it is the Master's state which is seen at the output. When the clock falls low, no output change is seen, since the Slave latches the same information.

(2) Data present at D is immediately latched and appears at the output when the clock goes high, after a brief propagation delay. Although a Master–Slave design, the CMOS D-type FF is edge-triggered rather than level-triggered.

The internal changes occurring when the clock falls again are not apparent to the user.

(3) As with any edge-triggered design, the clock transition must occur *faster than* a certain period to ensure reliable operation. The clock must rise in less than 5 μs in most CMOS designs. The clock fall time must be just as fast, even though no output changes are apparent during this phase.

4.6.2 The CMOS JK M-S Flip Flop

JK M-S operation is obtained by a modification of the D-type input (Fig. 4.18). A signal equivalent to the Q output is fed back to point D but is inverted or not inverted depending upon the state of the J and K inputs. If J and K are both low, the AND gate is disabled, and the NOR gates pass along to D the previous state Q. This state is reentered during the clocking cycle, but looks like a "no change" event to an observer. If K is high and J is low, both the AND gate and input NOR gate 1 are enabled. NOR 2's inputs are Q and $\overline{Q}$, so its output passes along to D a zero no matter what Q had been, forcing a change to the output 0 state. If J is high and K is low, the AND is disabled and the output of NOR 1 is low regardless of Q, passing along to D a Set to 1 command. Finally, if both J and K are high, NOR 2's inputs are 0 (from NOR 1) and Q. NOR 2's output is $\overline{Q}$, passing along to D the

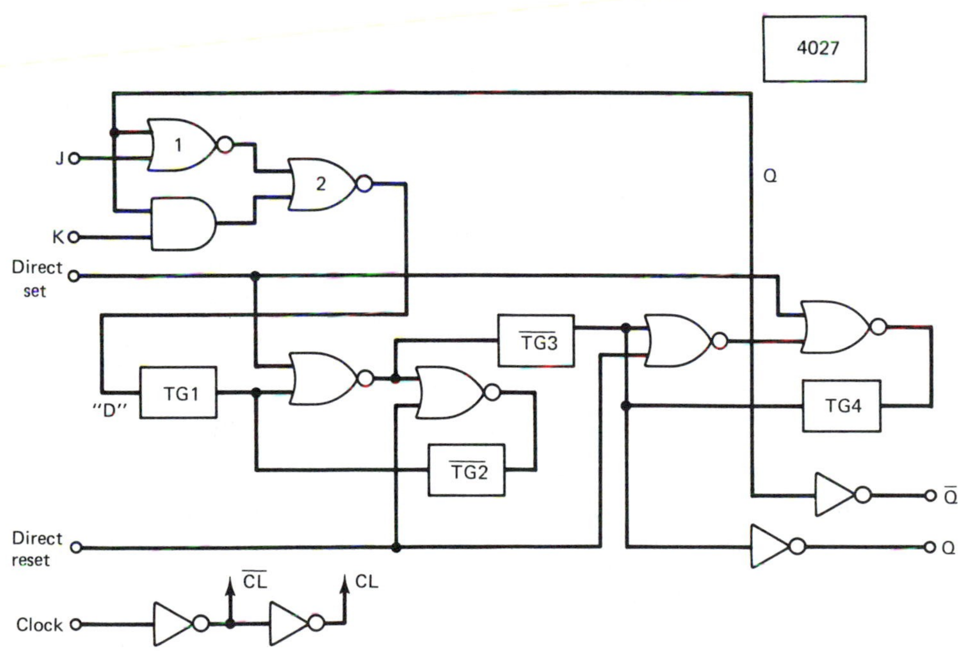

Figure 4.18 A CMOS JK flip flop is made by adding gating and feedback to a D-edge FF.

command Change State or Toggle. Thus, the addition of three gates to the D-FF gives the full JK truth table.

TABLE 4.1 SUMMARY OF MOST COMMONLY USED FLIP FLOPS AND LATCHES IN TTL AND CMOS VERSIONS

	TTL	CMOS
D-edge		
Dual D, with direct Set and direct Reset	7474	74C74 4013
JK M-S		
Dual, JK, with direct Set and direct Reset	7473 7476 (neg level)	74C73, 74C76 (neg edge) 4027 (pos edge)
4-Bit Latch		
Quad latch	7475	4042
Dual, Quad latch, tri-state output		4508
Dual Quad addressable latch (MPX inputs)		4723
Quad D-edge FF	74175	74C175, 40175/4175
Quad D-edge FF, tri-state output	74173	74C173
Quad D-edge FF, tri-state output *and input*		4076
8-Bit Latch		
Octal latch, tri-state output	74373	74C373
Octal D-edge FF, tri-state output	74374	74C374
Octal addressable latch (MPX inputs)		4099 4724
Octal bidirectional bus register, tri-state inputs and outputs		4034

As with the D-type, the CMOS JK M-S is an edge-triggered FF. The only data which counts is that present while the clock is going high, and changes of state appear at the output immediately after the propagation delay. This is very different from the most common TTL JK M-S, which is level-triggered: the changes recorded occur while the clock is high, with no change seen at the output until the clock falls. But in either the TTL or the CMOS JK, only one change can occur per clock cycle, and the two types may be used nearly interchangeably in logic design. Rules 1, 2, and 3 for the level-triggered TTL JK are removed, while the truth table is unchanged. The CMOS JK FF is therefore simpler to use. In addition to the positive edge-triggered FF of Fig. 4.18, negative edge-triggered

CMOS JK's also exist, which are pin-compatible with their TTL equivalents. A summary of commonly used TTL and CMOS flip flops is given in Table 4.1.

4.7 TRI-STATE LOGIC AND MICROCOMPUTER BUS INTERFACING

It is common in microcomputer system organization for many devices to be connected along a common set of wires called a *bus*. Information transmitted along a bus may be data or the address of a device or memory cell. A bus is often shown symbolically as in Fig. 4.19 to avoid drawing all the wires. Some devices such as line printers or graphic display units will only read data from the bus, and are called *listeners*. Other devices such as A/D converters and keyboard inputs will only write data to the bus and are called *talkers*. There will also be devices which both read and write at different times, such as terminals and memory arrays. The normal output of either a TTL or a CMOS device has a strong will to be either 1 or a 0, with current sourcing or sinking capability to force circuits connected to it to

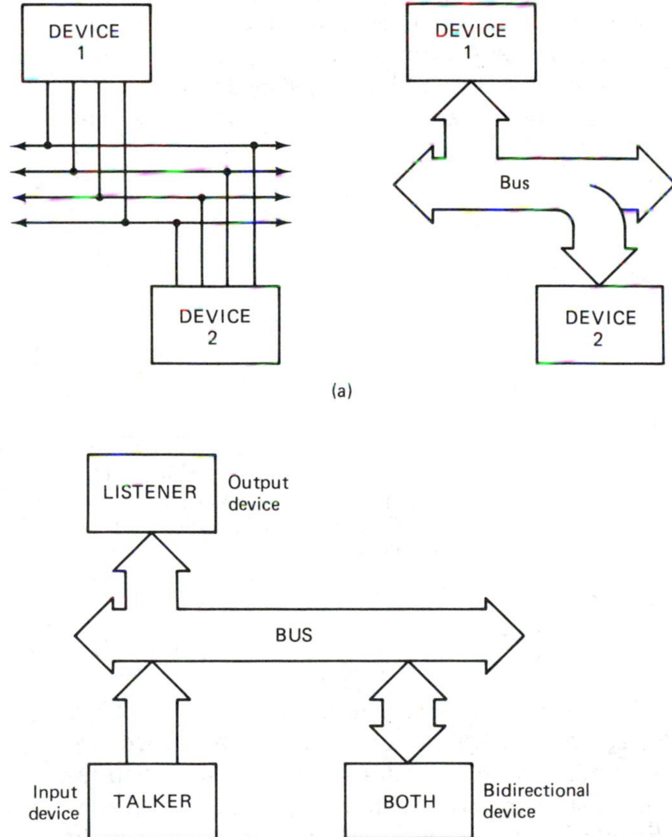

Figure 4.19 (a) The bus concept simplifies conceptual circuit drawings. (b) Output devices "listen" only, input devices are "talkers," and some devices are bidirectional.

follow. Two devices cannot ordinarily have their outputs connected to a common bus. Open-collector TTL devices allow output interconnection, but lose flexibility; and there is no equivalent of open-collector logic in CMOS. A superior solution called *tri-state* logic has evolved in both logic families. A tri-state output has the usual 1 and 0 states, plus a third high-impedance state which functionally disconnects the device from the bus.

4.7.1 TTL Tri-State

The normal output circuit (e.g., Fig. 3.4) of a TTL device always has one output transistor with a low impedance path to ground or to V_{CC}. Two devices with outputs connected together can therefore short V_{CC} to ground *through the devices*, causing catastrophic failure. The TTL tri-state *Enable* input (Fig. 4.20) forces *both* output transistors into the OFF state, and has priority over all other inputs. When the Enable input is high, the multiple emitter transistor Q_1 is turned on. This is the "low man wins" state, so the signal on the *Data* input is irrelevant. Q_2 is forced Off, turning Q_4 Off as well, because, with no current flow in R_1, Q_4's base is grounded. So far, this is no different from a normal TTL gate with an extra input. Tri-state action comes from the extra wire connecting Q_3's base to the collector of Q_0. When the Enable input is high, Q_3 is *also* forced off, because reverse-bias base current is drawn from it by *On* transistor Q_0. Since both output transistors are in the off-state, the output acts as a high impedance to either V_{CC} or to ground, and the circuit is isolated from other circuits connected to its output.

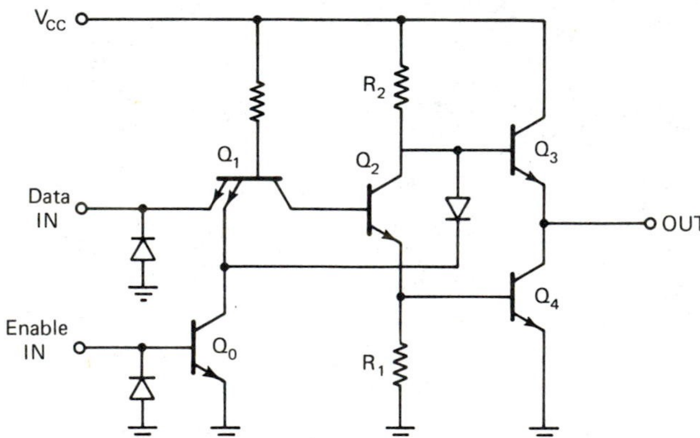

Figure 4.20 Tri-state inverter in TTL logic. The state of the enable input determines whether data will appear at the output.

4.7.2 CMOS Tri-State

Although the inputs to CMOS devices always see a high impedance, one of the output transistors is always ON. The output is therefore always a low impedance connection either to V_{CC} or to ground, as with TTL. A CMOS tri-state output requires only the addition of two transistors in series with the V_{CC} and ground connections (Fig. 4.21). With the Enable (EN) input held low, both transistors are in a low resistance state, and the circuit operates normally. When EN is brought

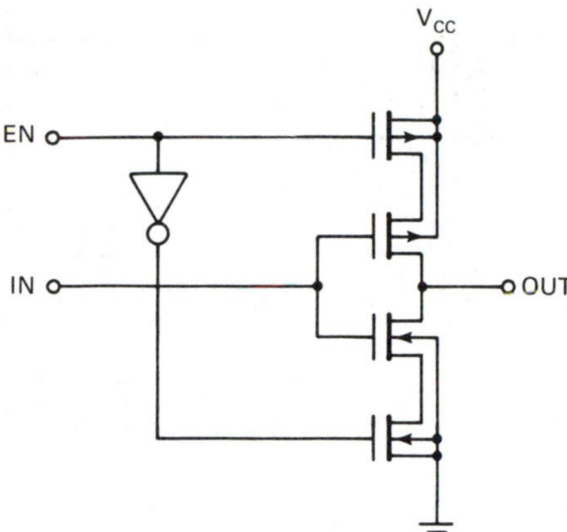

Figure 4.21 Tri-state inverter in the CMOS logic family. When EN is high, Out is functionally disconnected.

high, both transistors become high resistances, disconnecting the output from both V_{CC} and ground.

4.7.3 Tri-State Devices and Bus Design

A gate symbol with an extra wire on its side is a tri-state device [Fig. 4.22(a)]. In drawing the circuit, an extra gate with a bubble on the input is often added to indicate that *a low on the EN input enables normal circuit operation.* This matches naturally with the low output of a 1-of-*n* data selector. It is essential that a data selector be included in a bus circuit (Fig. 4.23) so that one and only one data *source* is enabled at a time, or else competition for the bus and circuit damage will result. There is no such limitation for listeners, so any number of devices may have their *inputs* connected to the bus at one time.

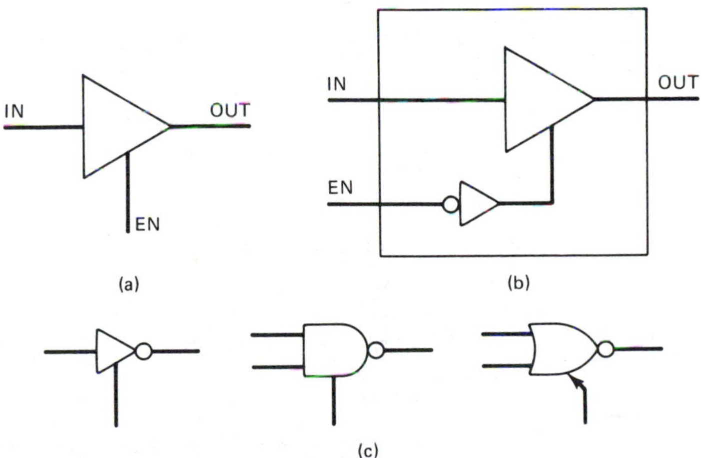

Figure 4.22 Symbols for tri-state components. (a) Buffer. (b) Input bubble to specify that a low signal on EN enables normal transmission of data. (c) Other tri-state gate symbols.

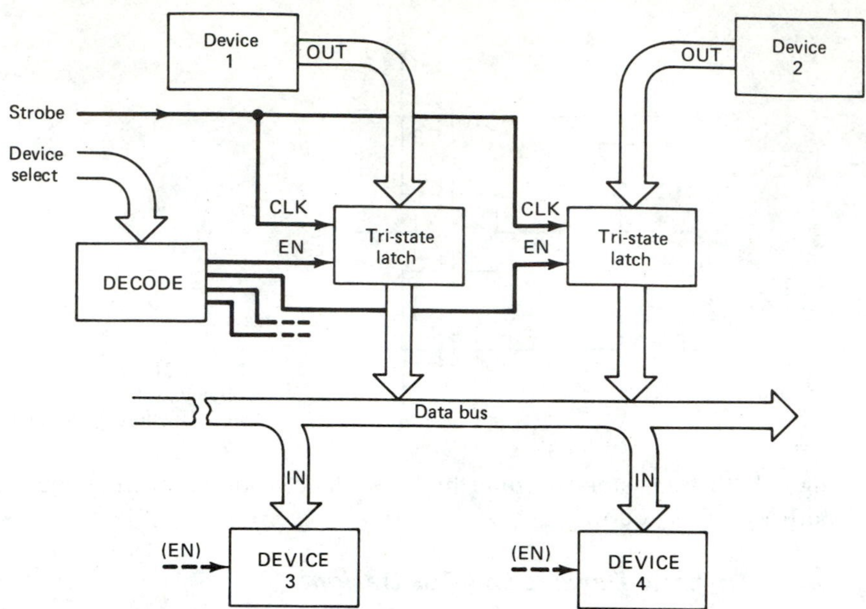

Figure 4.23 Typical microcomputer data bus circuit with uniquely decoded tri-state buffers on devices which can output data to the bus.

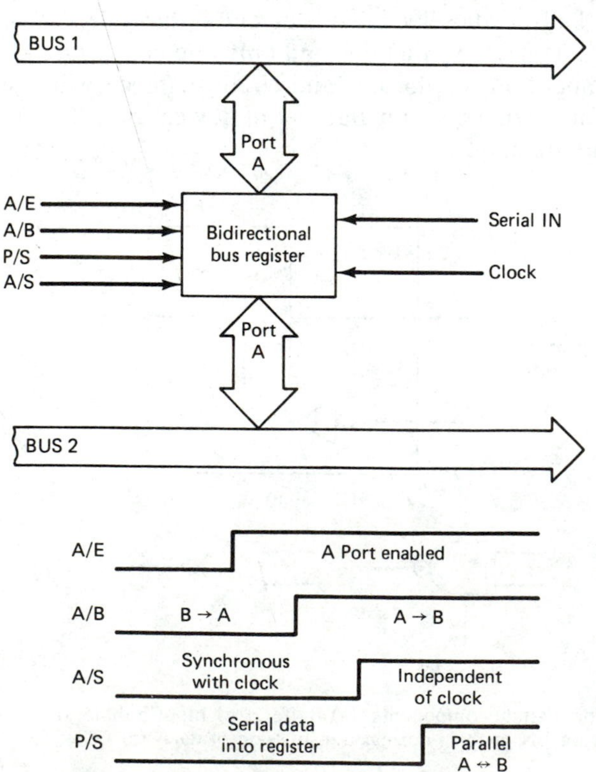

Figure 4.24 Bidirectional bus register IC allows tri-state buffered data transfer between two busses. Control signals determine the mode and direction of transmission.

A number of flip flop arrays intended for bus-latching applications are available with tri-state outputs, as summarized in Table 4.1. Both D-edge and simpler D-latches are available. The nomenclature is confusing. If the term "D" is used, it usually means D-edge or D-edge M-S. Examples are the 74173/74C173 (4-bit) and the 74374/74C374 (8-bit). Simpler latches are also sometimes called D-type, and resemble the TTL 7475, with output following input as long as the clock enables it to. This type is sometimes called the "transparent" or "gated" latch to distinguish it from the D-edge FF. Examples are the 4508 (4-bit) and the 74373/74C373 (8-bit). In most microcomputer applications, the gated latch is adequate, since both data and control signals are correctly timed by the system clocking. In data transmission or interfacing in noisy environments or between asynchronous systems, the D-edge-type latch may be preferred.

Sometimes bus interconnection must be *bidirectional*, with data going to or from two buses (Fig. 4.24). Although this can be accomplished using two tri-state latches enabled one at a time, self-contained bidirectional bus register IC's exist, such as the CMOS 4034. Control lines enable the ports (line A/E), define the direction of data transfer (line A/B), latch data synchronous or asynchronous with the clock (line A/S), and even allow a separate serial bit stream to be loaded into the register (line P/S).

PROBLEMS

4.1. Work through the change of state from Q = 0 to Q = 1 for the NOR gate FF of Fig. 4.3 when a *Set* pulse is received. Follow the procedure used in the text for the NAND gate FF, putting 1's and 0's on all gate inputs and outputs. What happens if a second Set pulse is received? What happens when a Reset pulse is received?

4.2. A toggle FF can be made from a D FF by wiring the $\overline{Q}$ output back to the D input. How does this work? Will a race occur if the D FF is a TTL D-edge or CMOS D-edge M-S type? Use timing specifications for actual IC's if needed.

4.3. Show the output waveforms at Q for each of the flip flop circuits of Fig. 4.25 when (a) data is set up and waiting when the clock enables storage; (b) data is changing within a clock cycle. Discuss similarities and differences in the responses. Indicate with arrows *when* the output change was stored for each M-S type.

4.4. Show the output waveforms at Q for each different type of JK flip flop in Fig. 4.26. Indicate with arrows *when* a given output change was stored. What requirement must be put on the data for a JK negative level-triggered device (typical TTL) to act the same as a JK negative edge-triggered FF (typical CMOS)? Assume each FF begins in the Q = 0 state for case (a) and in the $Q = 1$ state for case (b).

4.5. Sketch the outputs Q1 and Q2 in Fig. 4.27, showing phase and frequency relation to the clock.

4.6. The circuit shown in Fig. 4.28 functions as a phase meter whose output duty cycle T_H/T is proportional to the phase difference between the two input signals. How does it work? Most phase meters only cover a 180° range, but this one covers a full 360° range. The resistor is for timing purposes and can be ignored in the analysis.

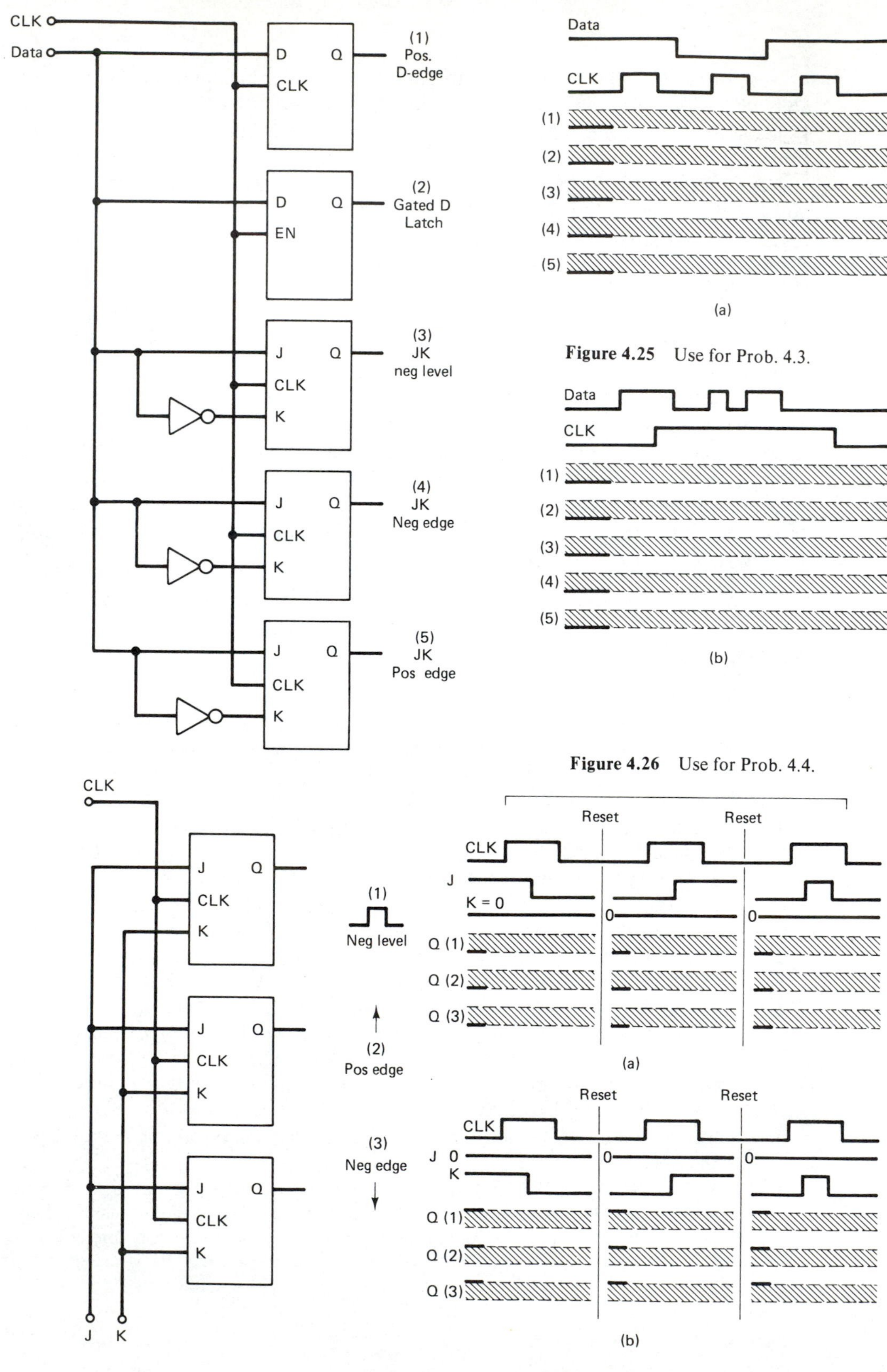

Figure 4.25 Use for Prob. 4.3.

Figure 4.26 Use for Prob. 4.4.

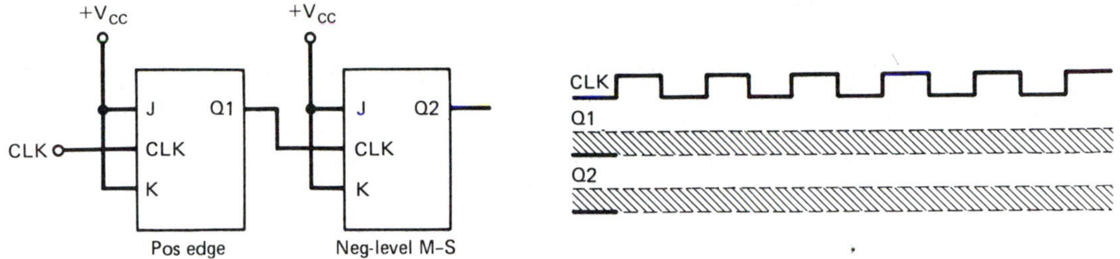

Figure 4.27 Use for Prob. 4.5.

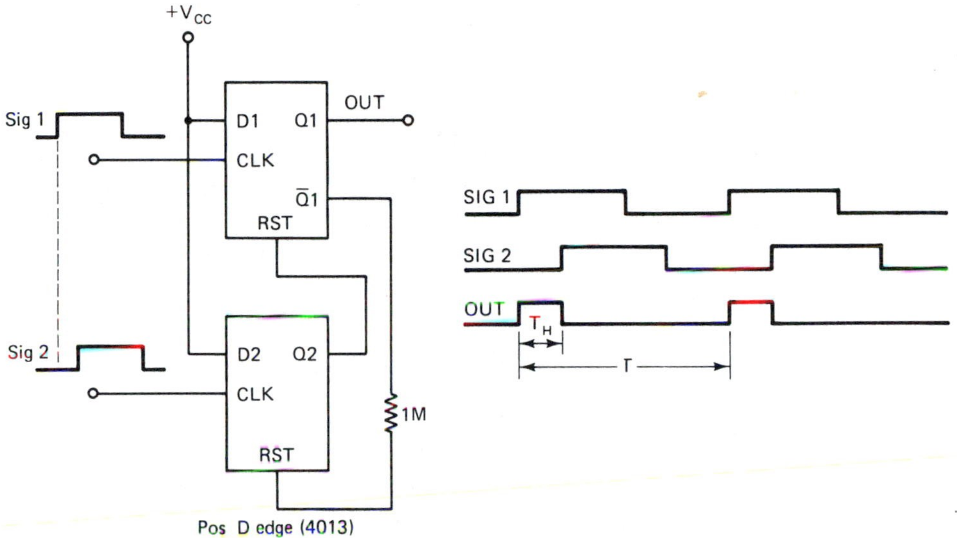

Figure 4.28 Phase meter for Prob. 4.6.

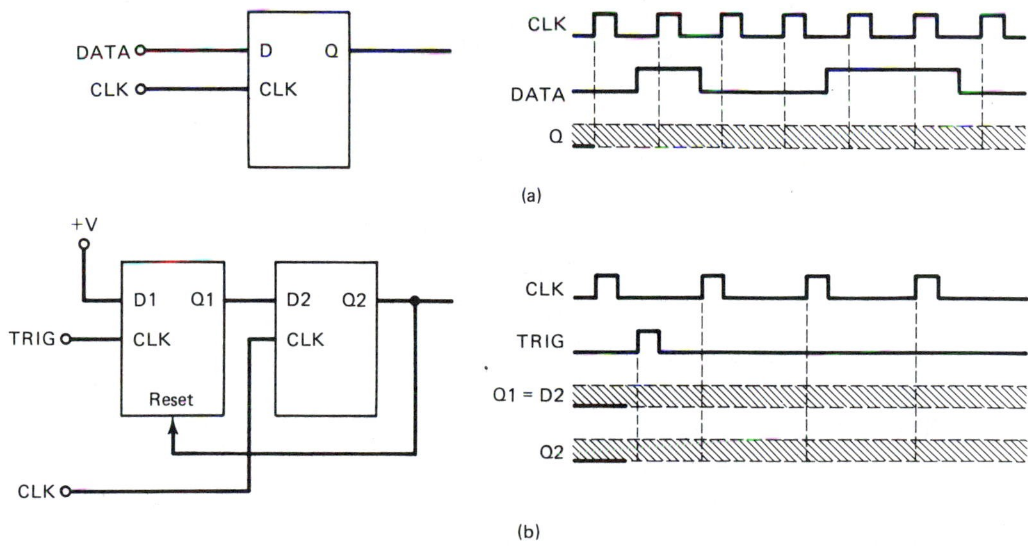

Figure 4.29 (a) Data synchronizer and (b) framing pulse generator for Prob. 4.7.

4.7. The flip flop circuits shown in Fig. 4.29 act as synchronizers. Circuit (a) replicates the input data stream but with rising and falling edges coherent with the clock. Circuit (b) issues a *single* "framing" pulse equal in width to the clock frequency within one clock cycle after a trigger pulse is received. Sketch output waveforms and explain circuit operation. The FF's are positive edge-type, and reset on a positive Reset input.

4.8. Match the following FF types to the list of features below. More than one match is possible for a given type.

(a) JK Master-Slave (level)	(d) R-S (NAND)
(b) D-type gated latch	(e) Clocked R-S (NAND)
(c) D-edge FF	(f) JK Master-Slave (edge)

Features: Available (1) only in TTL; (2) only in CMOS; (3) either in TTL or CMOS. Triggers on (4) positive transition; (5) negative transition; (6) edge. Can be used with appropriate wiring for (7) R-S; (8) D; (9) T. Best choice for storing (10) unpredictably spaced input signals; (11) controllably timed input signals. Used (12) frequently; (13) infrequently.

4.9. Specify which types of flip flop are good choices for each of the following applications.

Available Flip Flops

(1) D-type, either negative or positive edge
(2) D, gated (transparent latch)
(3) JK Master-Slave, negative level-triggered
(4) RS Master-Slave, negative level-triggered

(a) The data to be latched consists of pulses much shorter than the clock cycle.
(b) The flip flop is to toggle immediately when the input rises from 0 to 1. There is no way to predict when this input signal will be received.
(c) The FF is to act as a *unit delay*, passing input data to its output one clock cycle later.
(d) Data bits are to be latched after a settling time.
(e) The output of the FF is to be a square wave whose frequency is exactly one-half the input signal frequency.

REFERENCES

More complete bibliographic information for the books listed below appears in the annotated bibliography at the end of the book.

DEMPSEY, *Basic Digital Electronics with MSI Applications*

DIEFENDERFER, *Principles of Electronic Instrumentation*

FLOYD, *Digital Logic Fundamentals*

GROVE, *Physics and Technology of Semiconductor Devices*

FEMILING, D., *Tri-State Logic in Modular Syatems*, National Semiconductor Application Note AN-43

HIGGINS, *Experiments with Integrated Circuits*, Experiment 6

HUNTER, *CMOS Databook*

LANCASTER, *TTL Cookbook*

LANCASTER, *CMOS Cookbook*

LARSEN & RONY, *Logic and Memory Experiments using TTL Integrated Circuits*

Manufacturers' Data Books: see especially: Texas Instruments, *TTL Data Book*, and the National and Motorola *MOS Data Book*

MALMSTADT, et al. *Electronic Measurements for Scientists*

MUELLER and KAMINS, *Device Electronics for Integrated Circuits*

Scientific American, *Microelectronics*

SZE, *Physics of Semiconductor Devices*

WILLIAMS, *Digital Technology*

5

Counters and Shift Registers

Counters and shift registers are similar in that both are strings of flip flops where the output of one feeds the input of the next. In a counter, a sequence of input pulses causes the pattern of flip flop outputs to increment or count over the binary, decimal, or other sequence. In a shift register, a clock sequence moves data within the device, effecting conversion from parallel to serial format (or the reverse), or generating unusual sequences, or even multiplying numbers.

Counters are among the most useful medium-scale integration (MSI) and

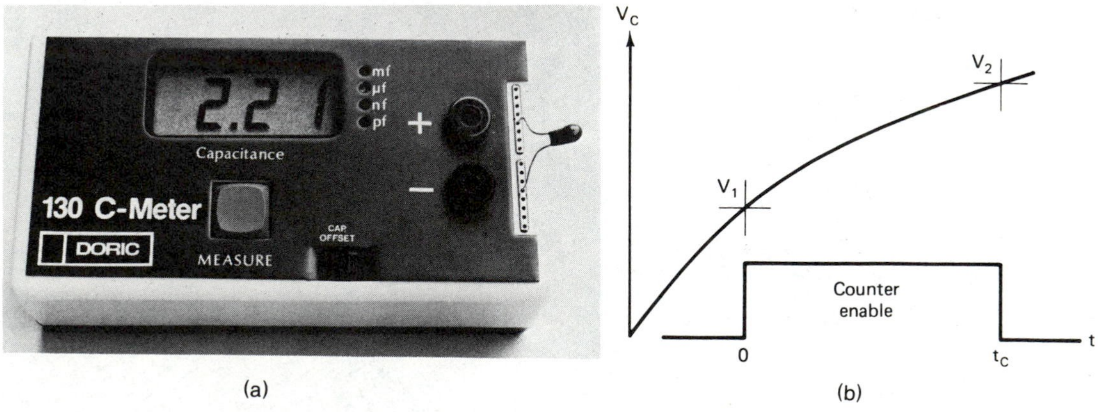

Figure 5.1 The digital capacitance meter is basically a counter, which measures the time interval t_c to charge the unknown capacitor. (*Courtesy* Doric Scientific Division, Emerson Electric Company)

large-scale integration (LSI) circuits. They carry out frequency and period measurement in a digital frequency counter and are also used in many other digital measuring instruments where the unknown quantity can be converted to a frequency or time. For example, in a digital capacitance meter (Fig. 5.1), the unknown capacitor becomes part of an RC time interval circuit, and the number of clock cycles which fit within the timing interval is displayed directly in units of capacitance. A dual slope digital voltmeter (Chapter 9) measures the time required for an unknown voltage to ramp an integrator through a given voltage change. Chapter 5 introduces counter principles, including some subtle sources of error, and describes the properties of a number of standard TTL and CMOS counter devices.

5.1. BASIC COUNTER CIRCUITS AND SPECIFICATIONS

5.1.1 Ripple Counters

Asynchronous or *ripple* counters are composed of a chain of JK flip flops, arranged so that the output of one changes the state of the next [Fig. 5.2(a)]. The J and K inputs are wired high, and the input to the clock on each state is provided by the output Q of the previous stage. A toggle occurs once per cycle of the waveform at the CLK input, so each successive stage divides the input frequency by 2 [Fig. 5.2(b)]. This is called a *counter*, because as the clock goes through a number of cycles, the pattern of parallel output states goes through the binary counting sequence and may be therefore assigned an appropriate *weight:* $A = 2^0$, $B = 2^1 \ldots$

Clock cycle N	---------Output lines--------			
	$D = 2^3$	$C = 2^2$	$B = 2^1$	$A = 2^0$
0	0	0	0	0
1	0	0	0	1
2	0	0	1	0
3	0	0	1	1
4	0	1	0	0
5	0	1	0	1
6	0	1	1	0
7	0	1	1	1
8	1	0	0	0
.	.	.	.	.
.	.	.	.	.
15	1	1	1	1
0	0	0	0	0

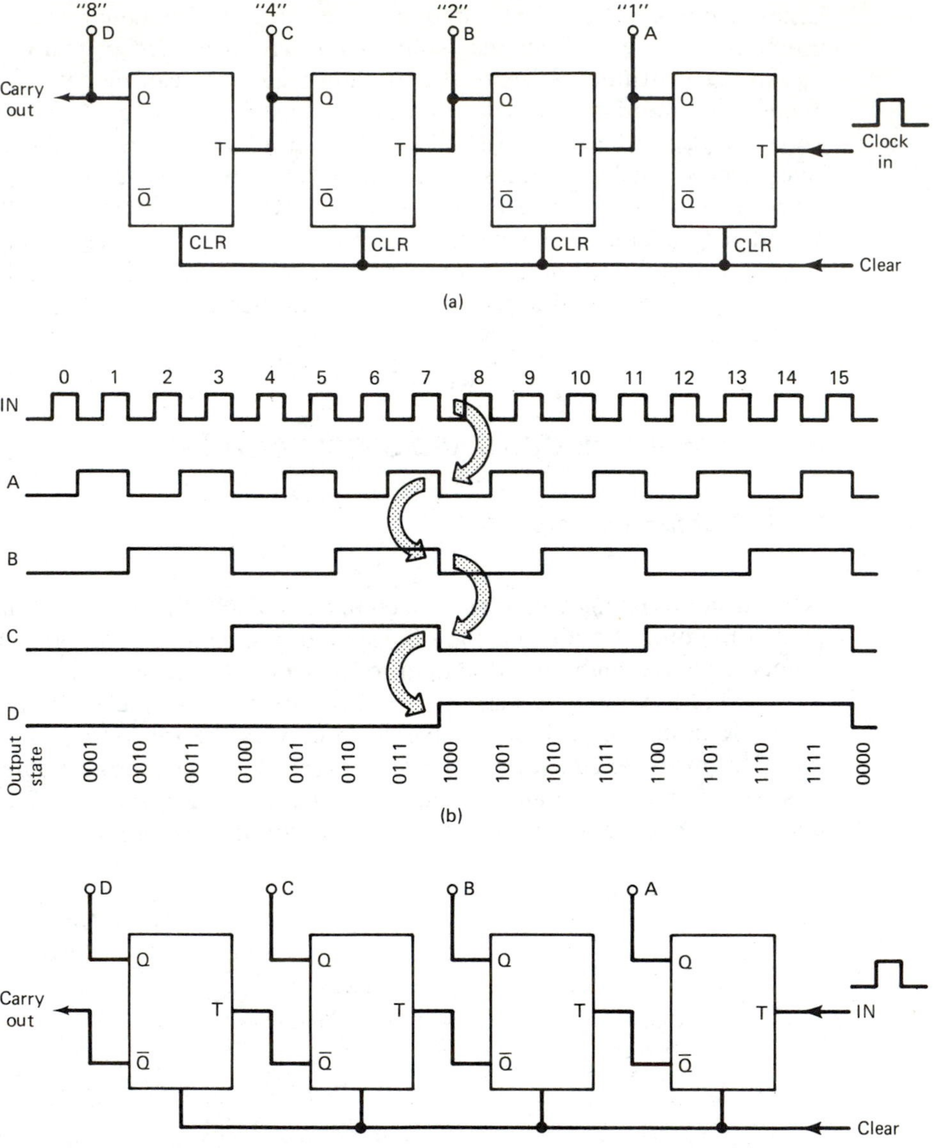

Figure 5.2 Four bit binary ripple counter. (a) Circuit of up-counter. (b) Waveforms of up-counter, with an example of a toggling transition rippling through. (c) Circuit of down-counter.

If the toggle input of each stage is connected instead to the complement $\overline{\text{Q}}$ output of the previous stage, the output lines will count down through the same sequence [Fig. 5.2(c); see also Prob. 5.2]. A programmable up/down counter may be designed by using gates to select the routing of outputs to inputs (see Prob. 5.3).

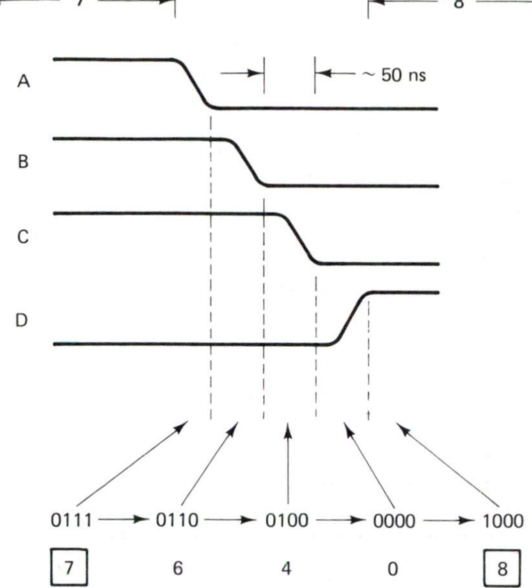

Figure 5.3 Intermediate state example for the ripple counter of Fig. 4.1(a). The transition from state 7 to state 8 goes briefly through states 6, 4, and 0.

In a long chain of ripple counter stages, the last FF changes its state considerably later than the first FF due to propagation delays in each stage. For example, the maximum toggle frequency of the TTL 7490 or 7493 is 18 MHz or a propagation delay of about 50 ns per bit, so the ripple delay in a 16-bit counter will be nearly a microsecond. This is much longer than the response time of other logic circuits. An even more serious source of error results from the fact that the ouputs of a ripple counter during changes of state often correspond to intermediate states outside of the intended sequence. The transition between state 7_{10} and state 8_{10} for the ripple counter of Fig. 5.2(a) is diagrammed in Fig. 5.3. When the output of FF_A falls, there is a brief interval before the outputs of FF_B will toggle in response. During this time, the output state is $0110 = 6_{10}$. Similarly, the changes in C and D will not occur immediately. As a result, the change from state 7 to state 8 actually includes brief time intervals with outputs 6, 4, and then 0. These unexpected intermediate states change too fast to be seen on readout devices such as LED's. However, if the output lines are connected to logic which can respond to these in-between states, serious errors may occur. Because the *glitch* is so fast (about 50 ns), the error may be difficult to track down.

The occurrence of these glitches may be readily detected using a combination of a JK flip flop and an LED, as shown in Fig. 5.4. This is the basis for inexpensive "latched" pulse detectors in commercial *logic probes.* The flip flop is set to toggle whenever its clock input falls. This happens whenever all the inputs of the NAND gate go high. The NAND inputs are wired to the unique combination of counter outputs which correspond to the glitch. For example, to detect the intermediate state 0100 which occurs in between states 7 and 8 of Fig. 5.3, the gate inputs are connected to counter outputs $\overline{Q}_D$, $\overline{Q}_C$, $\overline{Q}_B$, and $\overline{Q}_A$. An LED connected directly to the gate output is included to demonstrate that the glitch is too fast to be seen without being latched.

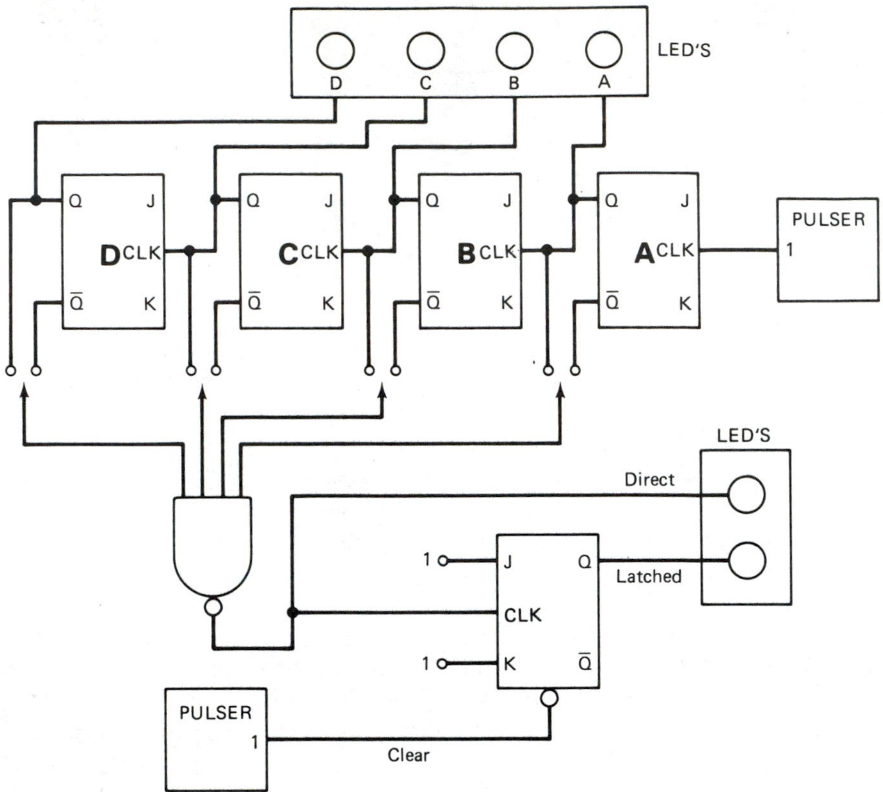

Figure 5.4 Latching an intermediate state or "glitch" of a ripple counter. The J and K inputs will float high if left unconnected as shown, but it is preferable to wire them high.

5.1.2 Synchronous Counters

The problems of propagation delay and erroneous intermediate states are overcome in synchronous counters (Fig. 5.5). Since all FF clock inputs are wired together, the transitions of all stages occur simultaneously. JK flip flops are used, with J and K wired together so the FF will toggle when the inputs are held high. Gating is used to program the correct sequence of states. For example, the 4's bit (FF_C) should go high in the next clock cycle after the 2's and the 1's bits are high. This is done by ANDing the J and K inputs to the Q output of both preceding bits. The master flip flop of FF_C is set while the clock is high, and toggling occurs when the clock falls. Similarly, FF_D is wired to receive a toggle instruction only after state 0111.

5.1.3 Counter Specifications

There are some important counter functions which must be specified for each application. The list below applies both to TTL and CMOS counters.

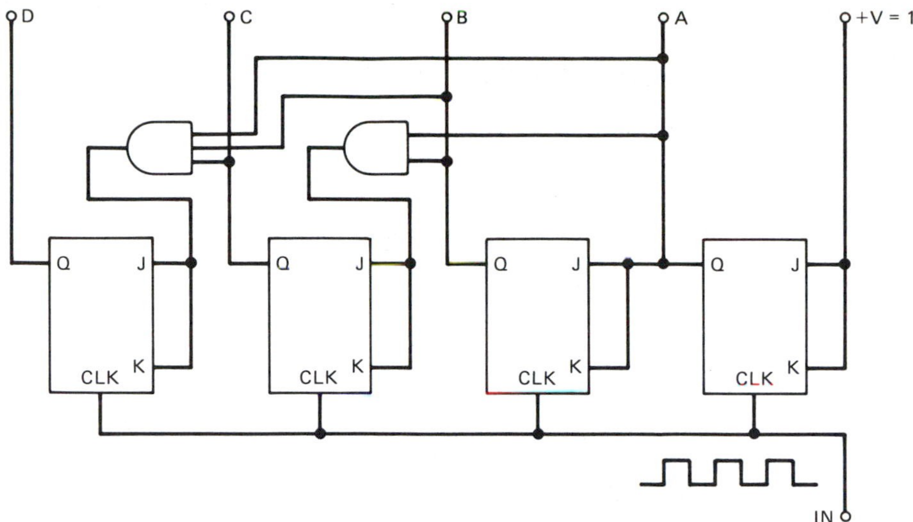

Figure 5.5 Synchronous counter circuit.

Modulus The *modulus* is the number of distinct states the counter goes through before repeating. The modulus of a 4-bit binary counter is 16.

Symmetry A counter is *symmetric* if the outputs all are square waves, as in Fig. 5.3(b). A binary counter is symmetric, but a BCD counter is not, since the D output is high for only 2 of the 10 states.

Weighting The outputs are *weighted* if a number may be assigned to each output bit so the output word goes through a sequence of values as the counter advances. 1-2-4-8 binary or BCD weighting is the most common. By contrast, a 1-output-low selector counter is unweighted.

Count direction The sequence of (weighted) output states may go up, down, or selectably.

Synchronism A transition between states may propagate along the FF chain (*asynchronous* or *ripple*), or take place at the same time for each FF (*synchronous* or clocked).

Reset and Parallel Load All counters need a *reset*, usually to the 0 state. Some counters (e.g., 74193) also have *parallel load* capability to preset the counter to an arbitrary binary word. This is used, for example, in generating a counter of arbitrary modulus.

Cascadability and Unit Cascadability Most counters can be *cascaded*, with the most significant bit output of one connected to the input of the next. This allows more precision, with a modulus equal to the product of moduli in the chain. For example, a digital clock cascades 60 s/min × 60 min/h × 24 h/day = 86,400 counts per modulus. Sometimes a multidigit counter's modulus is to be set to the *sum* of moduli. For example, a timer circuit may be set to go off after 1000 + 400 + 60 + 3 = 1463 s. Counters suitable for summed moduli are called *unit cascadable.*

TABLE 5.1 TTL COUNTER SELECTION GUIDE

Main feature	Device	Comments		
Binary (4-bit)				
Ripple	7493			
Ripple	74197	Parallel load		
Dual ripple	74493			
Synchronous	74161,74163	Parallel load		
Synchronous	74191	Parallel load	up/down	no clear
	74193	Parallel load	up/down	
Decade (4-bit)				
Ripple	7490			
Ripple	74196	parallel load		
Dual ripple	74490			
Synchronous	74160,74162	parallel load		
Synchronous	74190	parallel load	up/down	no clear
	74192	Parallel load	up/down	

5.2 TTL COUNTERS

Specifications for important TTL counters are summarized in Table 5.1. Circuits, specifications, and pin diagrams for several of these are given in Fig. 5.6.

5.2.1 *The 7493*

This counter is nearly identical to the 4-bit binary ripple counter of Fig. 5.2(a). However, two FF's in the chain are left unconnected, so for normal 4-bit operation an external jumper is added connecting Q_A to input B. The *Reset to zero* circuitry includes a NAND for making arbitrary modulus counters. In normal operation, one of these NAND inputs is wired high, while the other is held low for counting and brought high to reset the counter (note the bubble at the FF Reset inputs). In normal operation, the *Set to 9* inputs are not used and must be wired to ground.

5.2.2 *The 7490*

Operation of the 7490 BCD counter is similar, except that a modulus 5 stage is provided by three FF's (B, C, and D). When the output of FF_A is externally wired to the input of FF_B, a modulus 10 counter results with 1-2-4-8 BCD weighting. This counter may be cascaded to make a digital counter with multidigit resolution.

Exercise

Explain the divide-by-10 operation of the 7490 counter, following the simplified circuit of Fig. 5.7.

TTL
MSI

TYPES SN5490A, SN5492A, SN5493A, SN54L90, SN54L93, SN7490A, SN7492A, SN7493A, SN74L90, SN74L93 DECADE, DIVIDE-BY-TWELVE, AND BINARY COUNTERS

BULLETIN NO. DL-S 7211807, DECEMBER 1972

'90A, 'L90 . . . DECADE COUNTERS

'92A . . . DIVIDE-BY-TWELVE COUNTER

'93A, 'L93 . . . 4-BIT BINARY COUNTERS

description

Each of these monolithic counters contains four master-slave flip-flops and additional gating to provide a divide-by-two counter and a three-stage binary counter for which the count cycle length is divide-by-five for the '90A and 'L90, divide-by-six for the '92A, and divide-by-eight for the '93A and 'L93.

All of these counters have a gated zero reset and the '90A and 'L90 also have gated set-to-nine inputs for use in BCD nine's complement applications.

To use their maximum count length (decade, divide-by-twelve, or four-bit binary) of these counters, the B input is connected to the Q_A output. The input count pulses are applied to input A and the outputs are as described in the appropriate function table. A symmetrical divide-by-ten count can be obtained from the '90A or 'L90 counters by connecting the Q_D output to the A input and applying the input count to the B input which gives a divide-by-ten square wave at output Q_A.

'90A . . . J, N, OR W PACKAGE
'L90 . . . J, N, OR T PACKAGE
(TOP VIEW)

'92A . . . J, N, OR W PACKAGE
(TOP VIEW)

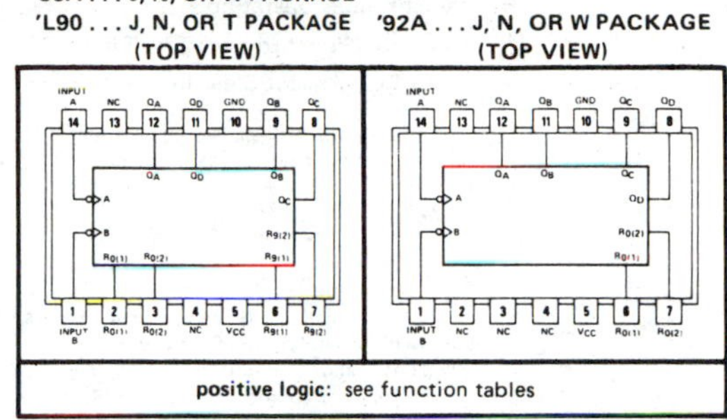

positive logic: see function tables

'93A . . . J, N, OR W PACKAGE
(TOP VIEW)

'L93 . . . J, N, OR T PACKAGE
(TOP VIEW)

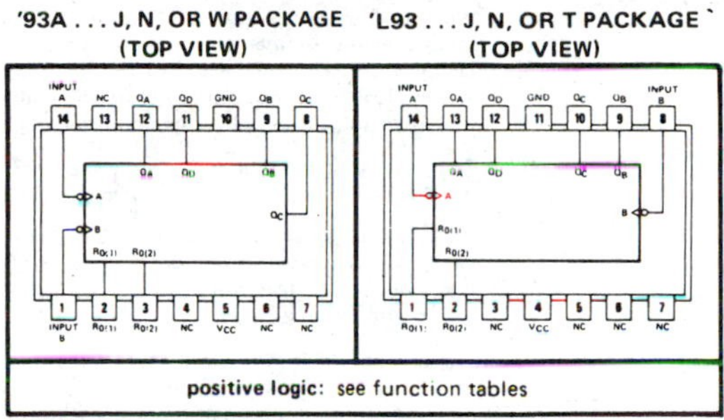

positive logic: see function tables

NC—No internal connection

functional block diagrams

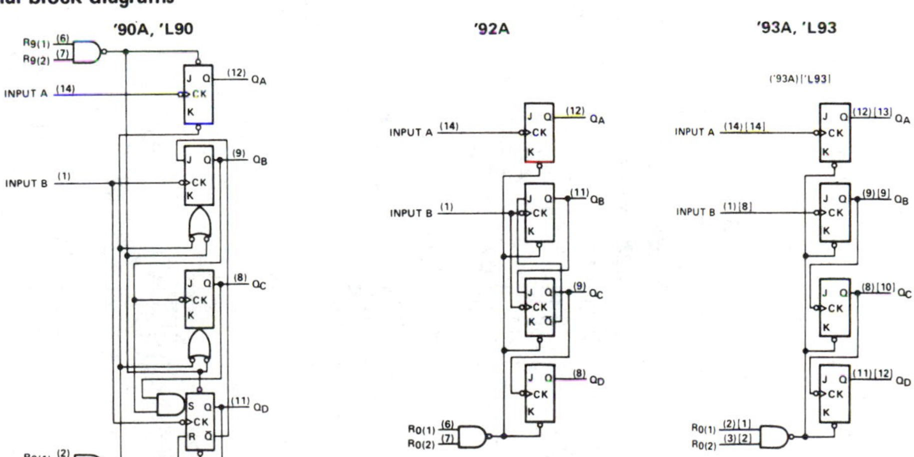

. . . dynamic input activated by transition from a high level to a low level.

The J and K inputs shown without connection are for reference only and are functionally at a high level.

Figure 5.6 Manufacturer's specifications for common TTL counters (7490, 7493, 74193). (*Courtesy* Texas Instruments, Incorporated)

TYPES SN54192, SN54193, SN54L192, SN54L193, SN54LS192, SN54LS193 SN74192, SN74193, SN74L192, SN74L193, SN74LS192, SN74LS193 SYNCHRONOUS 4-BIT UP/DOWN COUNTERS (DUAL CLOCK WITH CLEAR)

BULLETIN NO. DL-S 7211828, DECEMBER 1972

- Cascading Circuitry Provided Internally
- Synchronous Operation
- Individual Preset to Each Flip-Flop
- Fully Independent Clear Input

'192, '193 . . . J, N, OR W PACKAGE
'L192, 'L193 . . . J OR N PACKAGE
'LS192, 'LS193 . . . J, N, OR W PACKAGE
(TOP VIEW)

logic: Low input to load sets $Q_A = A$, $Q_B = B$, $Q_C = C$, and $Q_D = D$

TYPES	TYPICAL MAXIMUM COUNT FREQUENCY	TYPICAL POWER DISSIPATION
'192, '193	32 MHz	325 mW
'L192, 'L193	7 MHz	43 mW
'LS192, 'LS193	32 MHz	85 mW

description

These monolithic circuits are synchronous reversible (up/down) counters having a complexity of 55 equivalent gates. The '192, 'L192, and 'LS192 circuits are BCD counters and the '193, 'L193 and 'LS193 are 4-bit binary counters. Synchronous operation is provided by having all flip-flops clocked simultaneously so that the outputs change coincidently with each other when so instructed by the steering logic. This mode of operation eliminates the output counting spikes which are normally associated with asynchronous (ripple-clock) counters.

The outputs of the four master-slave flip-flops are triggered by a low-to-high-level transition of either count (clock) input. The direction of counting is determined by which count inputs is pulsed while the other count input is high.

All four counters are fully programmable; that is, each output may be preset to either level by entering the desired data at the data inputs while the load input is low. The output will change to agree with the data inputs independently of the count pulses. This feature allows the counters to be used as modulo-N dividers by simply modifying the count length with the preset inputs.

A clear input has been provided which forces all outputs to the low level when a high level is applied. The clear function is independent of the count and load inputs. The clear, count, and load inputs are buffered to lower the drive requirements. This reduces the number of clock drivers, etc., required for long words.

These counters were designed to be cascaded without the need for external circuitry. Both borrow and carry outputs are available to cascade both the up- and down-counting functions. The borrow output produces a pulse equal in width to the count-down input when the counter underflows. Similarly, the carry output produces a pulse equal in width to the count-down input when an overflow condition exists. The counters can then be easily cascaded by feeding the borrow and carry outputs to the count-down and count-up inputs respectively of the succeeding counter.

functional block diagrams

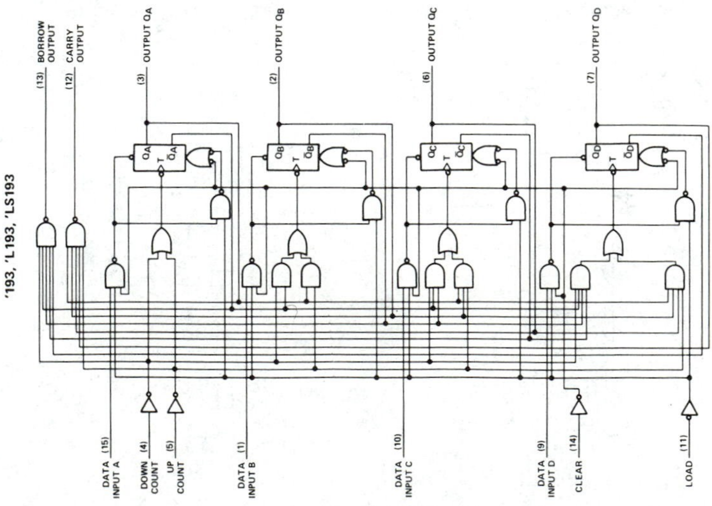

Figure 5.6 Continued

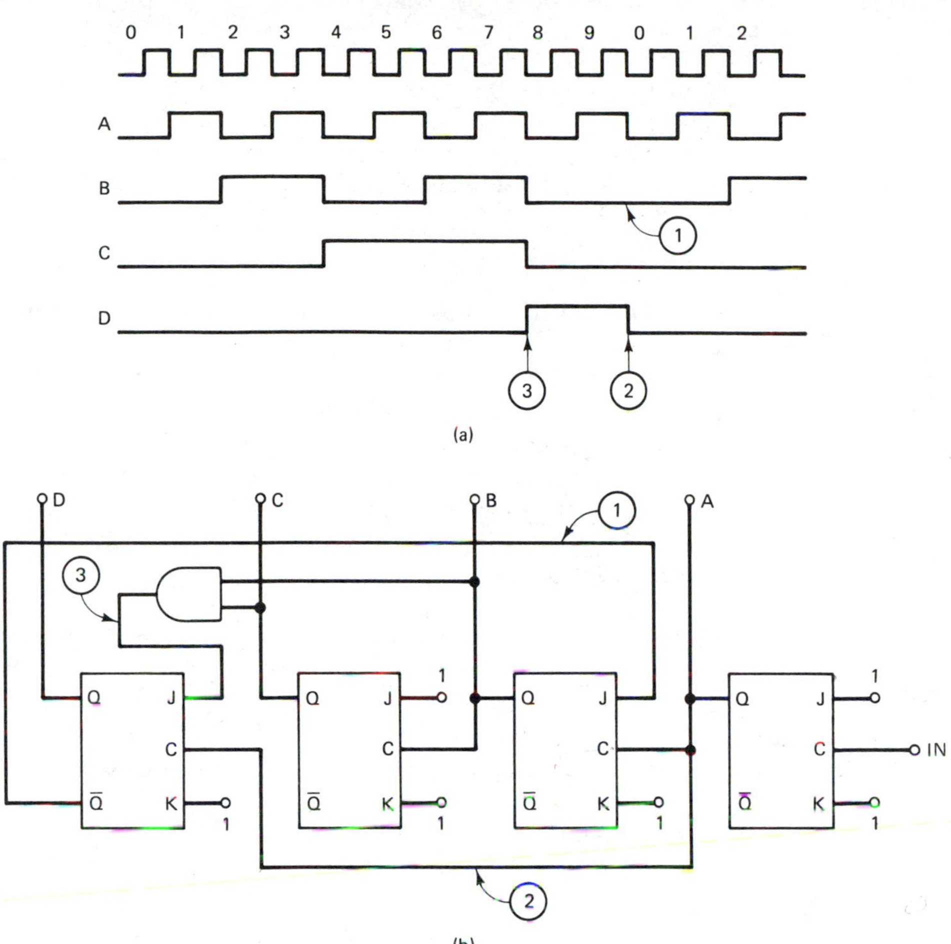

Figure 5.7 BCD counter (simplified).

Solution

Operation of FF_A and FF_C is normal. With their J and K inputs wired high, each toggles once per cycle of their input clock signal, which is the external clock input for FF_A and the output of stage B for FF_C. The special circuitry of the BCD counter controls the times when events which are different than in the binary counter occur [Fig. 5.7(a)]. Each of these special events corresponds to a specific circuit connection in Fig. 5.7(b).

(1) *Desired result:* FF_B must not toggle when FF_D has just been high.
Method: Wire the J input of FF_B to the $\overline{Q}$ output of FF_D. Since $\overline{Q}_D = 0$ when D is high, FF_B is not allowed to toggle high when the next clock pulse comes from FF_A.

(2) *Desired result: Reset* FF_D to 0 after the 9-state.
Method: FF_D is clocked by the output of FF_A, rather than the adjacent stage FF_C, since FF_D should be high for only two out of ten states or one full clock cycle of FF_A. The reset to 0 occurs after n = 9 because $J_D = 0$ and $K_D = 1$.

(3) *Desired result:* Set FF_D high after state n = 7.
Method: Although this occurs automatically for the binary counter, the unusual clocking for FF_D requires that J_D be allowed to go high only when FF_C and FF_D have just been high, requiring AND-gate decoding of this state.

5.2.3 The 74193

Operation of the 74193 synchronous, presettable up/down counter is not as complex as the circuit diagram [Fig. 5.6(b)] suggests. The count up/down function is provided by gates which route the coupling between successive stages (see Prob. 4.13). The *Preset* function loads 4 bits of data into the *direct set* FF inputs. This data is ANDed so that loading occurs when the *Load* input is brought to 0. A logic 1 on the *Clear* input resets the counter to the 0 state. For normal counting operation, the *Load* input is held high, and the *Clear* is held low.

5.2.4 Setting the Modulus

Many waveform-generating applications require a counter whose modulus can be varied. For factors of 2 this is accomplished by varying the number of FF stages. A modulus 5 or 10 example was shown in the 7490 counter. A modulus 3 counter can be made with two JK flip flops, as shown in Fig. 5.8 (see Prob. 5.4). *Arbitrary* modulus counters can be made using MSI counters and several basic tricks.

(1) Use of the *Reset* inputs, plus logic to decode the outputs.
(2) Use of the *Preset* inputs available on some counters.

A modulus 6 counter can be made using a 7490 or 7493 [Fig. 5.9(a)]. The Q_2 and Q_4 outputs are wired to the Reset inputs, activating the internally NANDed reset function when state 6 is reached. Other moduli may require external gating to uniquely decode the desired reset state. A counter like this actually has an extra state outside the modulus for a brief time. For modulus 6 the *Reset* is wired to look for the state X11X (X = "don't care"). As a result, the actual cycle of states is:

0	0000 ←	
1	0001	
2	0010	
3	0011	
4	0100	
5	0101	
(6)	0110 →	*Reset*

The seventh state 0110 exists briefly (about 30 ns) before the reset has time to function. This momentary glitch can cause difficulties in some applications.

Test your understanding. How would the glitch detector of Fig. 5.4 be wired to detect the momentary extra state of the modulo 6 counter?

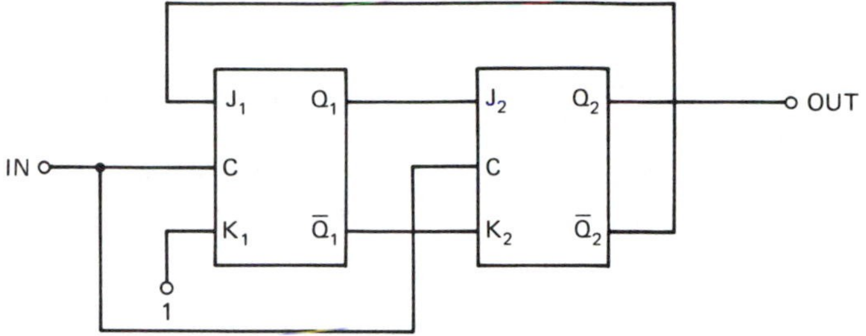

Figure 5.8 Divide by 3 counter.

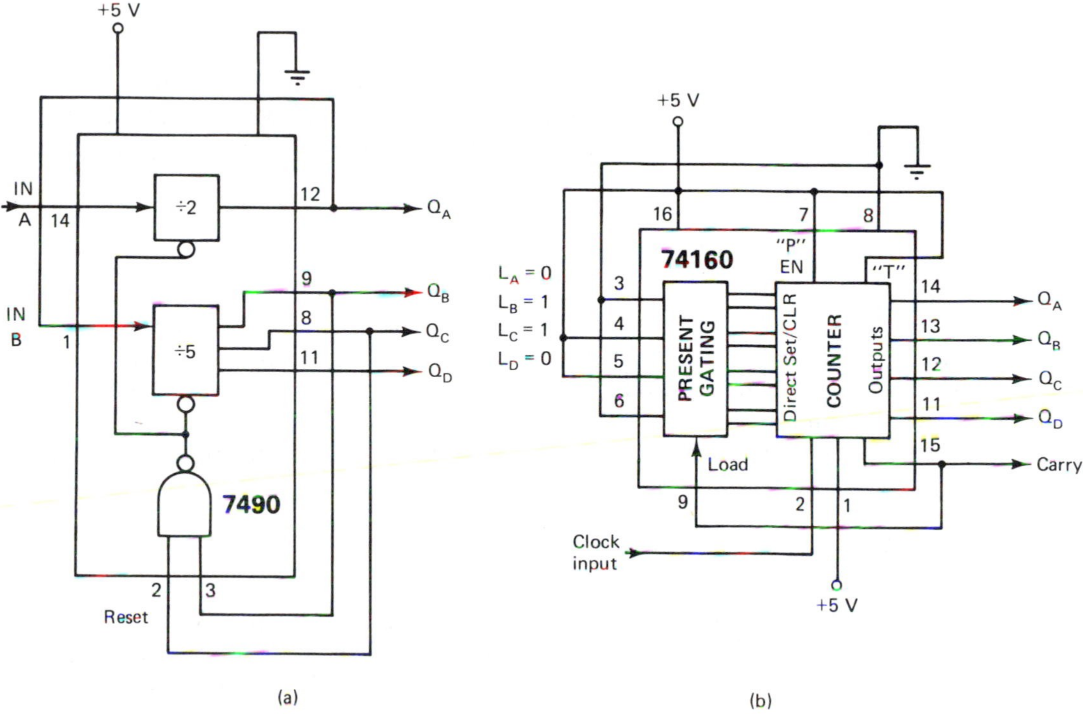

Figure 5.9 (a) Divide by 6 counter, using reset. (b) Divide by 9 counter, using preset.

The *Preset* inputs available on some counters may be used to make a counter of arbitrary modulus. The counter is allowed to count to the top of its range, where the *Carry* output generates the *Load* command. This causes the counter to jump to an initial state determined by the number wired into the *Preset* inputs. The Preset data is the *complement* of the desired modulus. In the example of Fig. 5.9(b), a modulus of 9 is generated by a Preset of 6, since 0110 is the complement of 1001. See Prob. 4.5

Test your understanding. The above trick is correct for an *up* counter. Show that for a *down* counter, the preset is wired to the desired modulus rather than the complement.

5.3 MOS COUNTERS

MOS counter circuits show significant advantages over their TTL counterparts. Some CMOS counters are designed to be TTL-equivalent, but offer lower power dissipation. Other CMOS counters offer more functions, such as easier preset or variable modulus. The real advantage of MOS counters, however, is their capability for more devices on a chip, with many useful applications in signal generation and measurement. For example, one CMOS counter has a range of $2^{16} = 16{,}384$. LSI MOS chips are available for high precision counter/timer applications. Finally, a class of LSI chips is available for the generation of precisely related frequencies from a single fixed clock frequency. One application is frequency synthesis: generating any frequency precisely from a single stable fixed-frequency crystal. A familiar consumer example is CB radio; 40 channels are synthesized from one crystal. Laboratory frequency synthesizers use the same principle. Another example is the top octave synthesizers for music, which will be discussed in detail as an example of complex waveform generation using the principles of this chapter.

5.3.1 CMOS Counters

A comparison of commonly used CMOS counters is given in Table 5.2. Most small-modulus (i.e., 4-bit) CMOS counters are synchronous and therefore glitch-free. One binary and one decimal counter in the table have functions corresponding exactly to a TTL equivalent, including pin-compatibility. Other counters have additional features not available in TTL. There are several characteristic features of CMOS counters:

(1) Several counters such as the 4520 offers two independent counters on a single chip [Fig. 5.10(a)].

(2) Many CMOS counters have a separate *enable* input for external gating of counter action [Fig. 5.10(b)]. Input IN is the signal to be counted, while input EN controls when to start and stop counting. This is useful in stopwatch or timer applications (see Chapter 6).

(3) Most up/down counters in CMOS have a single *count* input with another input which controls the direction of counting [4516, Fig. 5.10(c)]. This is useful, for example, in positioning applications with stepping motors. TTL and TTL-compatible up/down counters (e.g., 45193) have less convenient separate inputs for count up and count down pulses.

(4) There is a significant difference in the convention used for the *Preset* function in TTL and CMOS counters [Fig. 5.10(d)]. Most CMOS counters load the number to be preset when a high reaches the *Load* input, opposite to the convention used in TTL. The *Reset* command is a *high* in both logic families.

(5) In the case of *divide by N* counters such as the 4526, the number wired into the *preset* inputs is the modulus itself [Fig. 5.10(e)], not the complement as in TTL. The device is a down counter, and the signal appearing at the *zero*

TABLE 5.2 CMOS COUNTER SELECTION GUIDE

Main feature	Device	Decription
Binary (4-bit)		
TTL—equivalent (Presettable, up/down)	4193	Pin compatible with 74193. Separate count up/down inputs.
Dual	4520	Two independent counters. Count up only.
Presettable, up/down	4516	Single count input plus up/down input.
Programmable ÷n	4526	Modulus entered on *load* inputs.
Decimal (BCD)		
TTL—equivalent (Presettable, up/down)	4192	Pin compatible with 74192. Separate count up/down inputs.
Dual	4518	Two independent counters. Count up only.
Presettable, up/down	4510	Single count input plus up/down input.
Programmable ÷n	4522	Modulus entered on *load* inputs.
Ubiquitous		
BCD or binary	4029	Also includes up/down and preset.
Ring or 1-of-*n* selector		
Decimal	4017	Fully decoded 1-of-*n* selector. Includes internal *enable* gating.
Binary	4022	
2-through-10 counter	4018	Divides by n = 2 to 10, selected by wired feedback.
Large modulus binary		*Note:* These are ripple counters.
7-stage	4024	Divide by 12: All 6 output bits available.
12-stage	4040	Divide by 4096. All 11 output bits available.
14-stage	4020	Outputs for bits 2 and 3 missing.
14-stage	4060	Includes internal oscillator. Outputs for bits 1, 2, 3, and 11 missing.

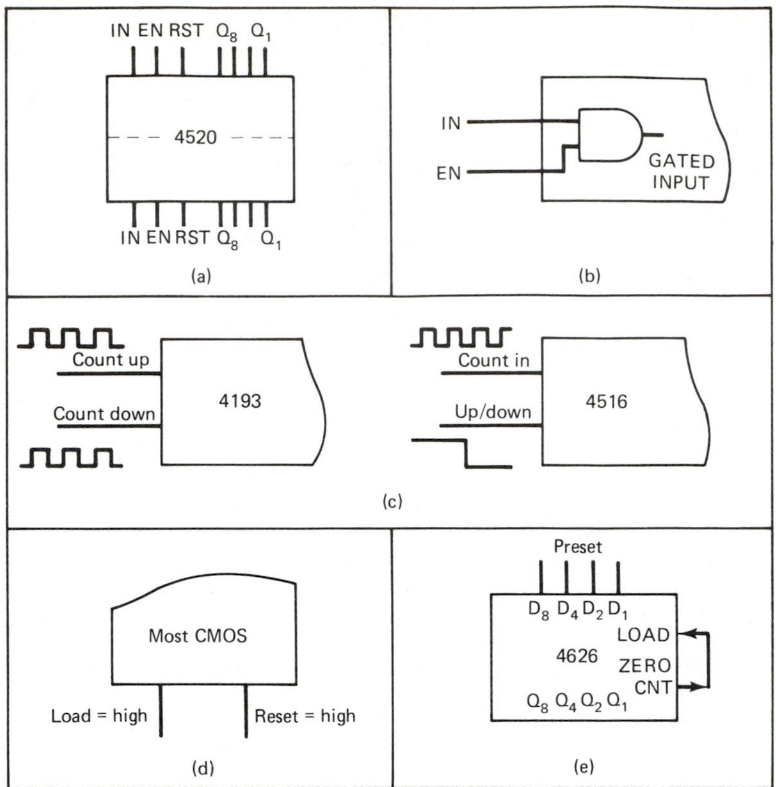

Figure 5.10 CMOS counter features. (a) Dual counters. (b) Enable input. (c) Up/down control. (d) Preset convention. (e) Divide by *N*.

count output is wired externally to the *load* command input. Compare the TTL example, Fig. 5.9(b).

(6) There are several differences in the output signals available in variable modulus divide-by-N counters. This is irrelevant if only a single output line with a divide by N signal on it is needed, but is important if parallel outputs are to be used. In the case of binary counters, such as the 4522 [Fig. 5.11(a)], the output has the usual binary weighting. However, *ring* counters, such as the 4018 [Fig. 5.11(b)], are also available, which employ a series of flip flops wired as a shift register (Section 5.5). All outputs display the same (divided) signal, but each is phase-shifted from the next. This can be useful in generating multiphase signals or complex waveforms (see Section 5.6).

(7) The modulus of the 4018 counter is set by a feedback connection from one or more outputs to the *data* input. A separate *Reset* input is also available. Even-modulus counters may be constructed with a single feedback wire, while odd-modulus counters require two feedback connections and a gate, as shown in the table on p. 151.

(8) A class of large-modulus ripple counters is available only in CMOS. Because of the large number of stages (up to 14 for the examples in Table 5.2), a long time is required for the output to settle to a reliable state. The glitch-

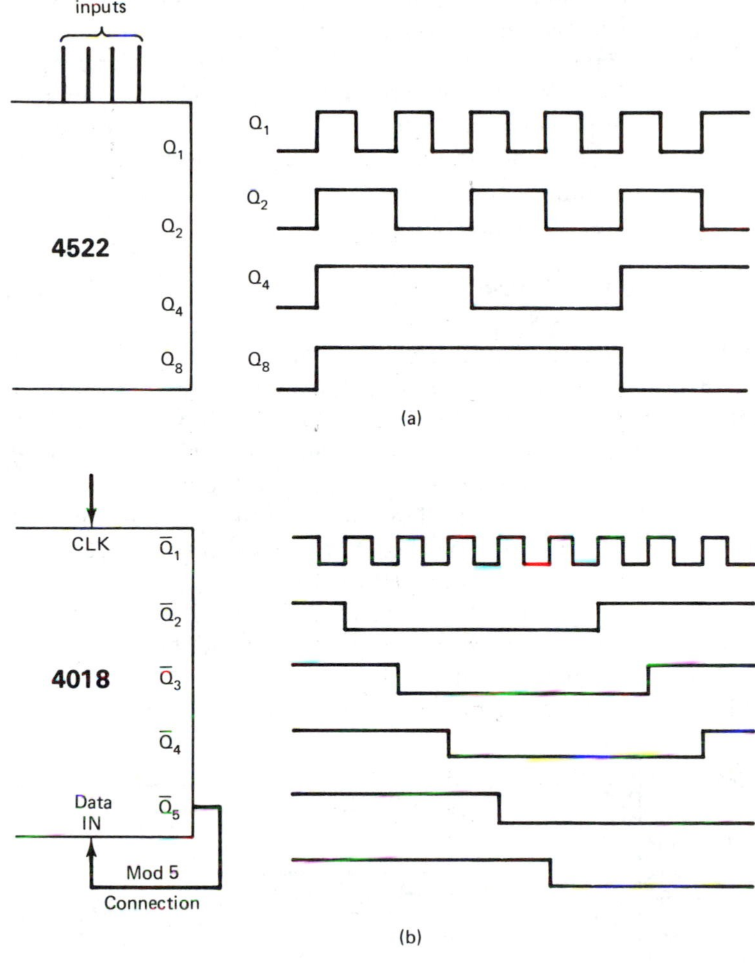

Figure 5.11 Divide by *N*: (a) binary/BCD versus (b) ring counter.

Modulus	Output to Input Connection
10	Q_5
8	Q_4
6	Q_3
4	Q_2
2	Q_1
9	Q_4 ANDed to Q_5
7	Q_3 ANDed to Q_4
5	Q_2 ANDed to Q_3
3	Q_1 ANDed to Q_2

free operation of *synchronous* counters is therefore preferred in fast timing and precision frequency-synthesis applications, despite the larger cost of cascading many chips. However, glitch problems in large-modulus counters can be easily eliminated in audiofrequency applications by passive filtering. Because of pin limitations, some low-order bits are missing on these chips (see Table 5.2). Although not serious in divide-by-N applications or when only a limited number of outputs are needed, these counters are not useful in driving high-precision parallel binary outputs.

5.3.2 Counter-Timers

A lovely example of the complexity of functions which can be performed with MOS-LSI hardware is the *universal timer*. One example (Fig. 5.12) includes two independent 4-digit counters. Either counter output may be time-multiplexed out to a 4-digit display via a 4-bit BCD data output plus a 4-bit *Digit Select* output. In addition to control inputs to start and stop the clock, the user can select one of seven different functions. Two of the functions are stopwatches, where one counter is actively counting the time interval since the *Start* pulse, and the other counter keeps track of either the total elapsed time (previous event) or the total accumulated time (all events since the *Reset* pulse). There is also a separate rally timer function. Two other functions enable the device to be used as a programmable up or down counter, counting to a preset number and then issuing a "done" pulse for precision timing of other devices. This universal timer therefore

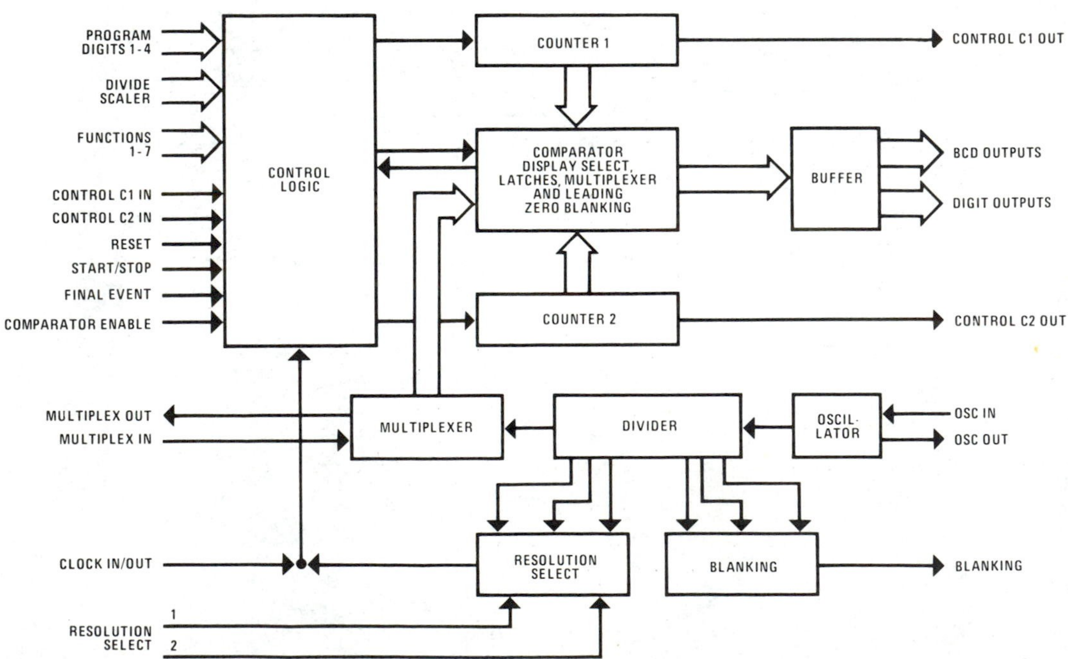

Figure 5.12 Universal timer IC, MM5865. (Copyright 1980 National Semiconductor Corporation)

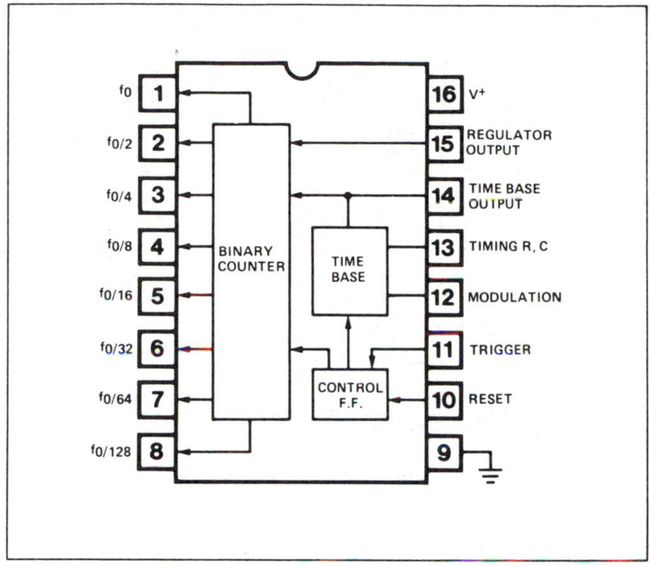

Figure 5.13 Programmable counter/timer, XR 2440. (*Courtesy* EXAR Integrated Systems)

has applications both in measuring time intervals of events and as a timer to signal when a preset interval is over.

Another example of complex timer chips is the programmable timer/counter (Fig. 5.13). This chip combines an 8-stage binary counter with an adjustable RC timer (Chapter 6). In addition to a divide-by-N function, an external clock waveform can synchronize the internal timer to oscillate on a frequency M times higher. As a result, the XR2440 is a frequency synthesizer whose output f_o can be any rational multiple of an input frequency f_R:

$$f_o = \frac{M}{(1 + N)} f_R$$

This capability allows generating a signal which is coherently locked to but not harmonically related to another signal. For example, a 100-Hz signal locked to the 60-Hz line frequency is generated by setting $M = 5$ and $N = 2$. This could be useful in coherent detection instrumentation (Chapter 18) to avoid beats with the line frequency.

5.3.3 The Top-Octave Synthesizer

This chip, called TOS for short, includes within one LSI-MOS chip 13 independent divide-by-N counters whose moduli are preset to an approximation of the *even-tempered* musical scale, which divides a single octave into 12 notes separated by the frequency ratio $2^{1/12}$. An example of a top-octave circuit is shown in Fig. 5.14. The accuracy of the TOS binary approximation depends on the number of bits used. Some examples are shown in Table 5.3. The 9 and 10-bit versions are

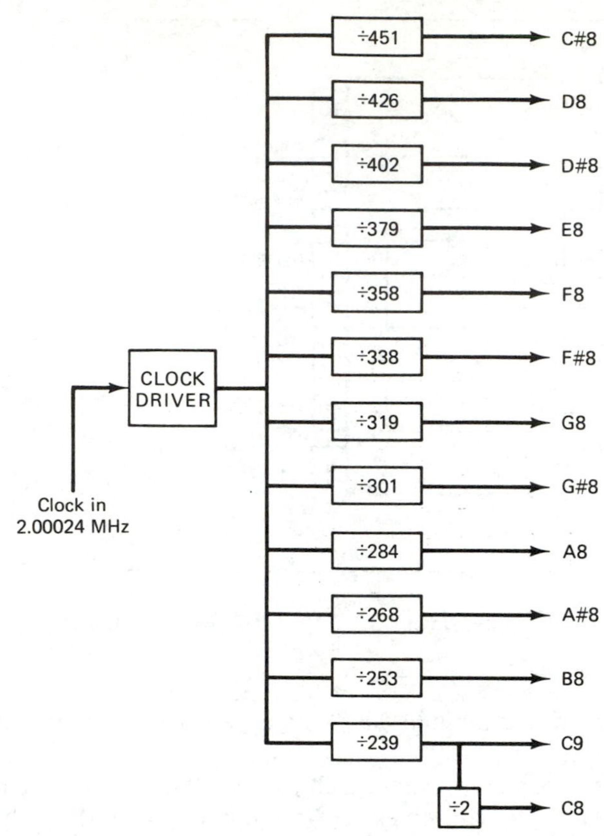

Figure 5.14 Top octave synthesizer (T.O.S.) IC.

TABLE 5.3 APPROXIMATIONS TO THE EVEN-TEMPERED SCALE

Note	Even-tempered frequency	8-bit		9-bit		10-bit	
		Divisor	Error%	Divisor	Error%	Divisor	Error%
C9	8372.02	116	+.036	239	−.034	254	−.023
B8	7902.13	123	−.048	253	+.050	269	+.0151
A#8	7458.62	130	+.194	268	+.067	285	+.014
A8	7040.00	138	−.002	284	+.004	302	−.004
G#8	6644.88	146	+.139	301	+.007	320	−.017
G8	6271.93	155	−.067	319	−.025	339	−.009
F#8	5919.91	164	−.065	338	−.034	359	+.035
F8	5587.65	174	−.007	358	−.007	380	+.127
E8	5274.04	184	+.111	379	+.069	403	+.026
D#8	4978.03	195	+.081	402	−.046	427	+.018
D8	4698.64	207	−.115	426	−.069	452.5	−.006
D#8	4434.92	219	+.026	415	−.000	479.5	−.026
C8	4186.01	232	+.036	478	−.034	508	−.023
Input frequency (MHZ)		0.9715		2.00024		2.126	
Examples		TTL or CMOS preset counters		National 5891		National 5555/5556	

found in commercial top-octave chips. The 8-bit version demonstrates the TOS idea using 4-bit TTL circuits, but is musically unsatisfactory, since the notes are out of pitch even to a relatively unmusical listener. The absolute errors in Table 5.3 are often expressed by musicians as the percentage of the interval between adjacent notes. The worst-case notes in these approximations are in error by:

	8-bit	9-bit	10-bit
	$G^{\#}8$	E8	$F^{\#}8$
Percent frequency error	0.194	0.069	0.035
Percent of one-note interval	3	1.1	0.6

Notes in other octaves may be generated by binary division. For a monophonic (one note at a time) synthesizer (Fig. 5.15), 3 bits of information select the octave, and an additional 4 bits driving a 16-input multiplexer select the note within the octave. Note select inputs may be provided either by matrix encoding a keyboard or by signals from a microcomputer output port.

The circuit of Fig. 5.15 does not utilize the capability of the TOS to generate many notes simultaneously. A polyphonic circuit is shown in Fig. 5.16. Each TOS output is routed to a separate octave divider to cover a full 6 × 12 = 72 note

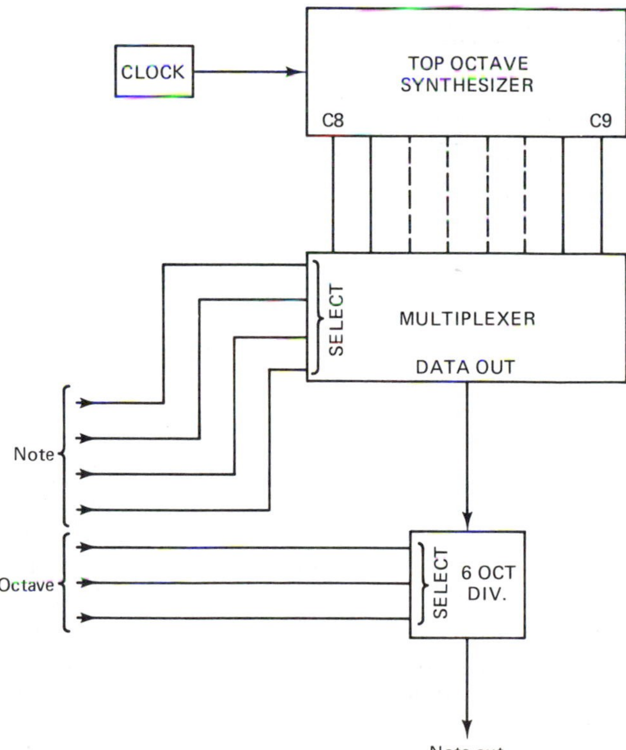

Figure 5.15 Monophonic TOS note generating circuit.

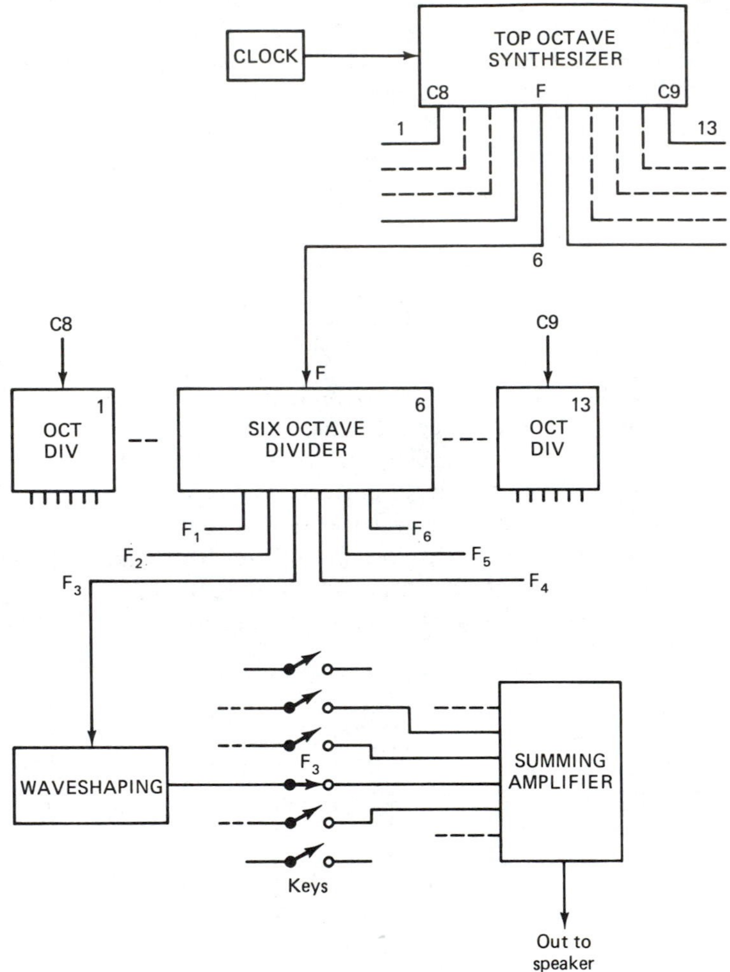

Figure 5.16 Polyphonic TOS note generating circuit (simplified).

range. All notes are available simultaneously. The keys pressed connect the corresponding notes to a summing amplifier. The square waveform generates a reed-like sound. Further waveshaping is necessary to generate the variety of sounds in an electronic organ, and further shaping of the sound intensity envelope is added in electronic music synthesizers. A square wave is not an adequate basis for producing other tone qualities, since it contains only odd harmonics. A better starting point is to add harmonically related square waves [Fig. 5.17(a)] to approximate a ramp function, which contains all harmonics. The approximation in Fig. 5.17(b) deviates from an ideal triangular wave only above the sixteenth harmonic, which is adequate for most musical purposes and can be smoothed by passive filtering.

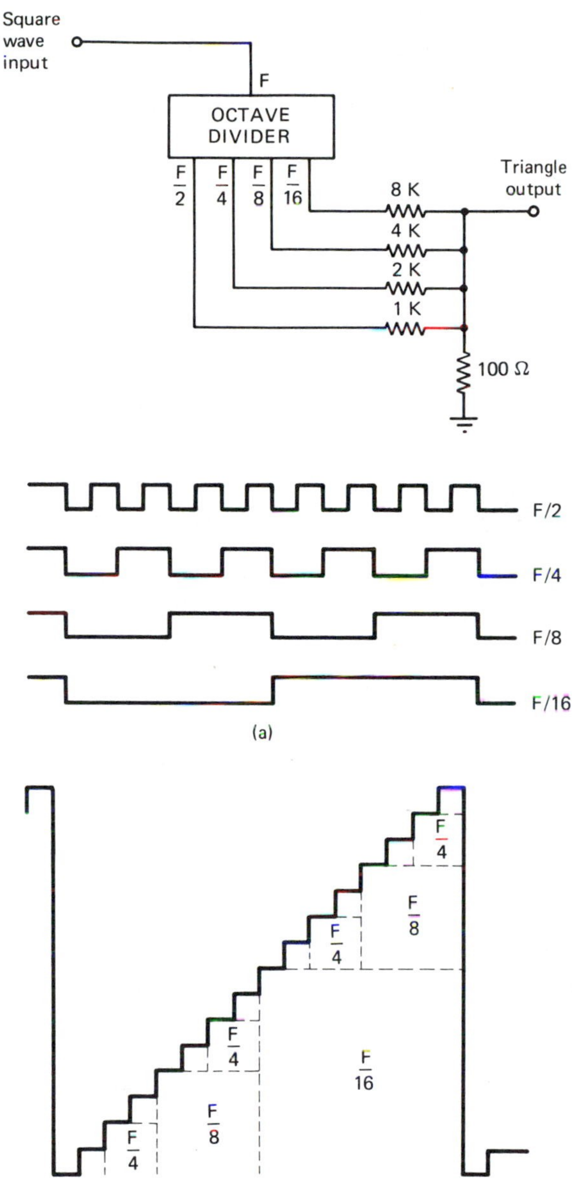

Figure 5.17 Ramp waveform generation using an approximation of four square waves.

5.4 SHIFT REGISTER CIRCUITS

A *shift register* (SR) is a chain of flip flops connected so that binary data moves down the chain on command, at a rate set by an external clock. A shift register, often labeled SR for short, resembles a linear chain of buckets [Fig. 5.18(a)], with the information in each bucket shifting to its neighbor upon command. In *binary* shift registers, the information is 1's or 0's. In *analog* shift registers (Chapter 19), the unit of information is a voltage which may have any value.

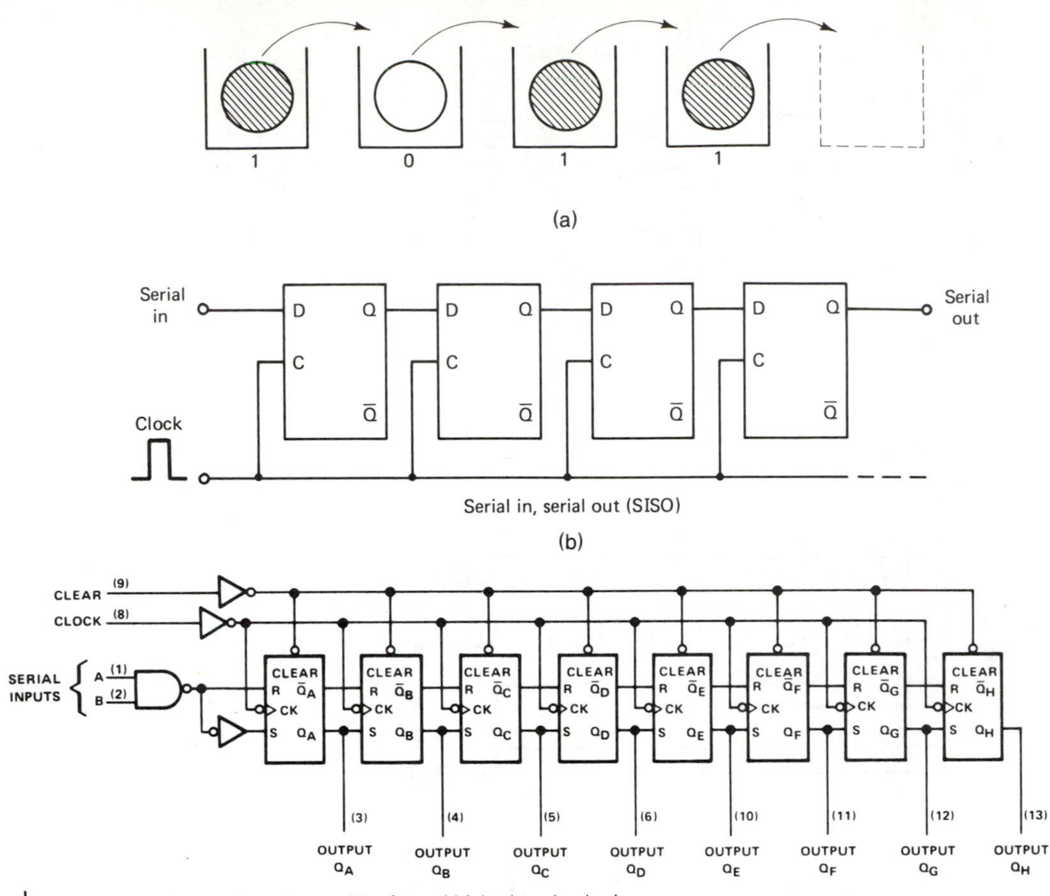

Figure 5.18 Shift register circuits. (a) Bucket brigade analogy. (b) Serial input, serial output (SISO). (c) Serial input, parallel output (SIPO) = 74164. (d) Parallel input, serial output (PISO) = 74165. (e) Universal shift register (PIPO) = 74194. [Parts (c), (d), and (e) *courtesy* Texas Instruments Incorporated]

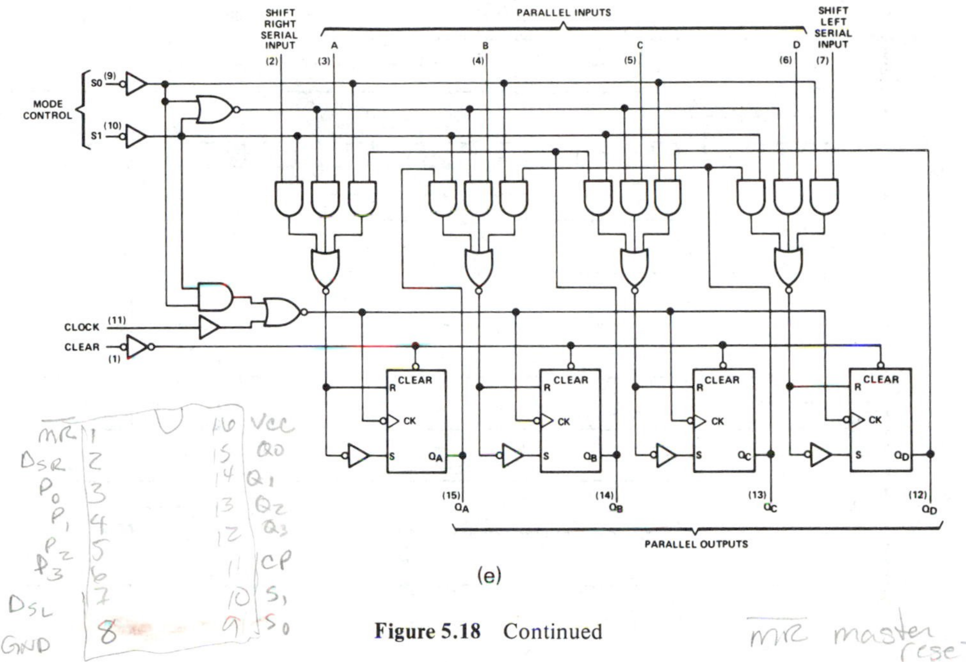

(e)

Figure 5.18 Continued

Data may be input to shift registers (SR) serially, one bit at a time, or in parallel form, all bits simultaneously. Shift registers have applications as temporary data storage registers, and are often used to convert data between serial and parallel format. When a key is pressed on a terminal, a parallel data word is generated. An SR is used in data communications to convert this parallel data into a serial time series transmitted down a single wire. SR's are also used in computation, since binary multiplication is a series of add and shift operations (Chapter 8). SR's have applications in signal processing. A shift register wired to recirculate data around it can generate nearly random numbers or white noise. The delay capability of SR's may be used to advantage in digital filters (Chapter 19) or in audio "reverb" units.

The simplest shift register [Fig. 5.18(b)] accepts a bit of data at its input each time it is clocked. The same data appears at the serial output after N clock cycles, where N is the number of FF's in the chain. This operation is called serial input–serial output (SISO). The basic TTL unit, shown as a D-type FF, is actually constructed from a R-S master–slave FF with an extra inverter connected to the S input [Fig. 5.18(c)]. The master–slave FF acts as a race-free unit delay element: the data present at D appears at the output Q one clock cycle later. A shift operation usually occurs when the clock makes a transition from low to high.

If connections are made to each of the FF outputs [Fig. 5.18(c)], the shift register is said to have a *parallel* output. A time series of input bits will appear in parallel form at these outputs after the number of clock cycles equal to the number of FF's in the chain [Fig. 5.19(a), SIPO]. The last of these outputs may also be used as a serial output.

Conversion of parallel data to serial form (PISO) may be accomplished [Fig.

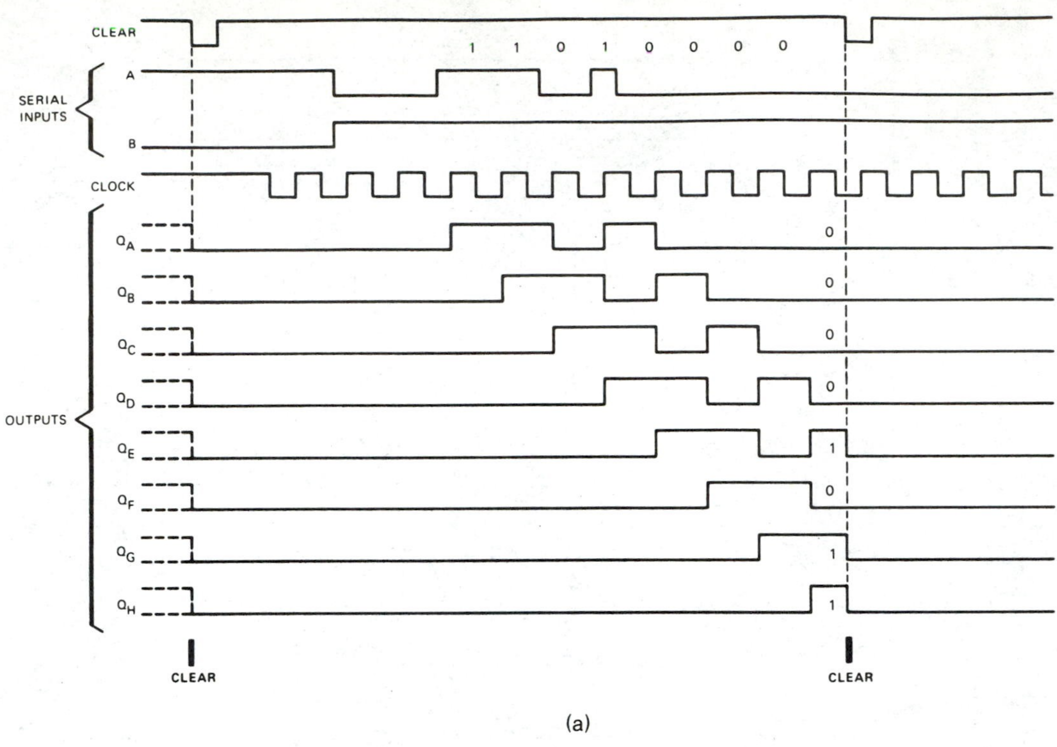

(a)

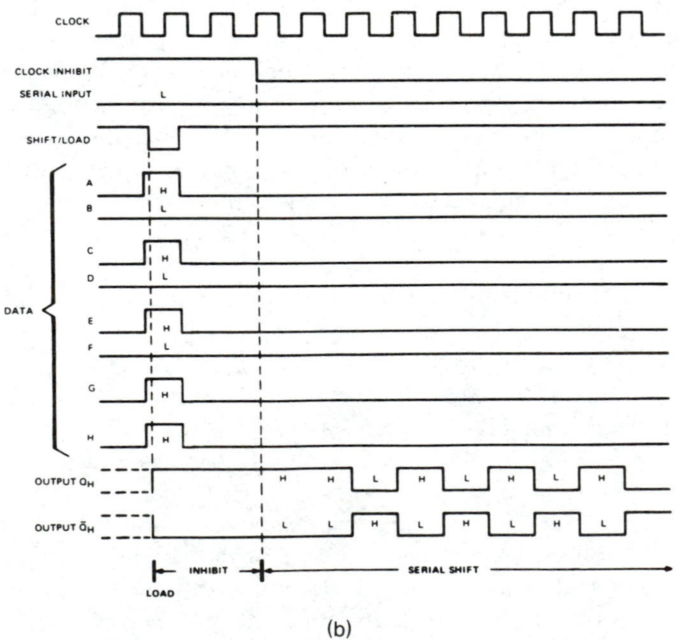

(b)

Figure 5.19 (a) Data sequences in the serial in, parallel out shift register of Fig. 4.17(c). (b) Data sequences in the parallel in, serial out shift register of Fig. 4.17(d). (*Courtesy* Texas Instruments Incorporated)

5.18(d)] by connecting parallel inputs to the direct Set or direct Clear FF inputs, with the addition of gating to route each bit either to the set or to the clear input, depending upon whether the bit is high or low. An additional *load* control line is provided to enable information to be loaded only after it is certain that the input data has settled reliably. In the example shown in Fig. 5.19(b), parallel data is latched when the load command is given (shift/load line falls). Data is clocked out serially only after the clock/inhibit line falls. In Fig. 5.18(d), a gate minimization trick eliminates an extra inverter needed in the obvious method of routing input data to the Preset or Clear inputs (see Prob 5.8).

The two functions of Fig. 5.18(c) and (d) can be combined to create a parallel input–parallel output SR. The maximum flexibility exists in the *universal* shift register [Fig. 5.18(e)] in which additional *mode control* inputs and internal gating allow the user to configure the SR to parallel load data or to serially shift data left or right.

5.5 TTL AND CMOS SHIFT REGISTER EXAMPLES

5.5.1 CMOS vs. TTL

As in the case of counters, CMOS SR's can hold more bits than TTL SR's (Table 5.4). In addition, CMOS SR's dissipate far less power than their TTL equivalents, yet they have enough current-driving capability to drive one low-power TTL circuit. On the other hand, they are slower by about a factor of 5.

5.5.2 Functions vs. Bits

Although the extra functions of parallel inputs and parallel outputs make a quite modest increase in SR design complexity, there are disadvantages in providing these functions on any SR chip. Each function requires one or more pins for connection to other IC's, leading to less SR stages within a given package. This trade-off is demonstrated in Table 5.4. Shift registers available in a 16-pin package, for example, range from a 4-bit universal device to a 64-bit device with only serial inputs and serial outputs. The 8-bit devices are a useful compromise where parallel access is necessary and the 8-bit length is natural, as in ASCII data communications.

5.5.3 The 74194

This TTL chip [circuit in Fig. 5.18(e)] does any possible SR manipulation on 4 bits of data. There are both parallel inputs and parallel outputs, and two serial in-

TABLE 5.4 SHIFT REGISTER EXAMPLES

Device	74194	4035	74164	74165	4031
Logic family	TTL	CMOS	TTL and CMOS	TTL and CMOS	CMOS
Application	Universal	Universal	SIPO	PISO	SISO
Bits	4	4	8	8	64
Parallel inputs	4	4	0	8	0
Parallel inputs	4	4	8	0	0
Serial inputs	2 (Left/right)	2 (J,K)	1 plus Enable	1	2: Data and Recirc.
Serial outputs	Use Q_D or Q_A	Use Q_D	Use Q_H	Q_H	2: OUT and complement
Other[a]: mode	2: S_0, S_1	1: P/S	0	0	1: Ext/Rec
Pins	16	16	14	16	16
Power dissipation (quiescent)	195 mW	0.015 mW	168 mW (TTL) 50 nW (CMOS)	100 mW (TTL) 50 nW (CMOS)	
Maximum clock frequency	36 MHz	8 MHz	36 MHz (TTL)	26 MHz	8 MHZ
Current sink capability (number of TTL loads)	20	1	20 (TTL) 1 (CMOS)	20 (TTL) 2 (CMOS)	1

[a]All devices have clock, clear (= reset), and power pins.

puts, since data may be shifted in either from the left or from the right of the chain. Serial output may be taken at either Q_D or Q_A. Functions are selected by logic levels at pins S_1 and S_0, shown in the chart on p. 162.

S_1	S_0	Mode selected
0	0	No change
1	0	Shift right
0	1	Shift left
1	1	Parallel load

A given operation is executed when the clock signal rises high.

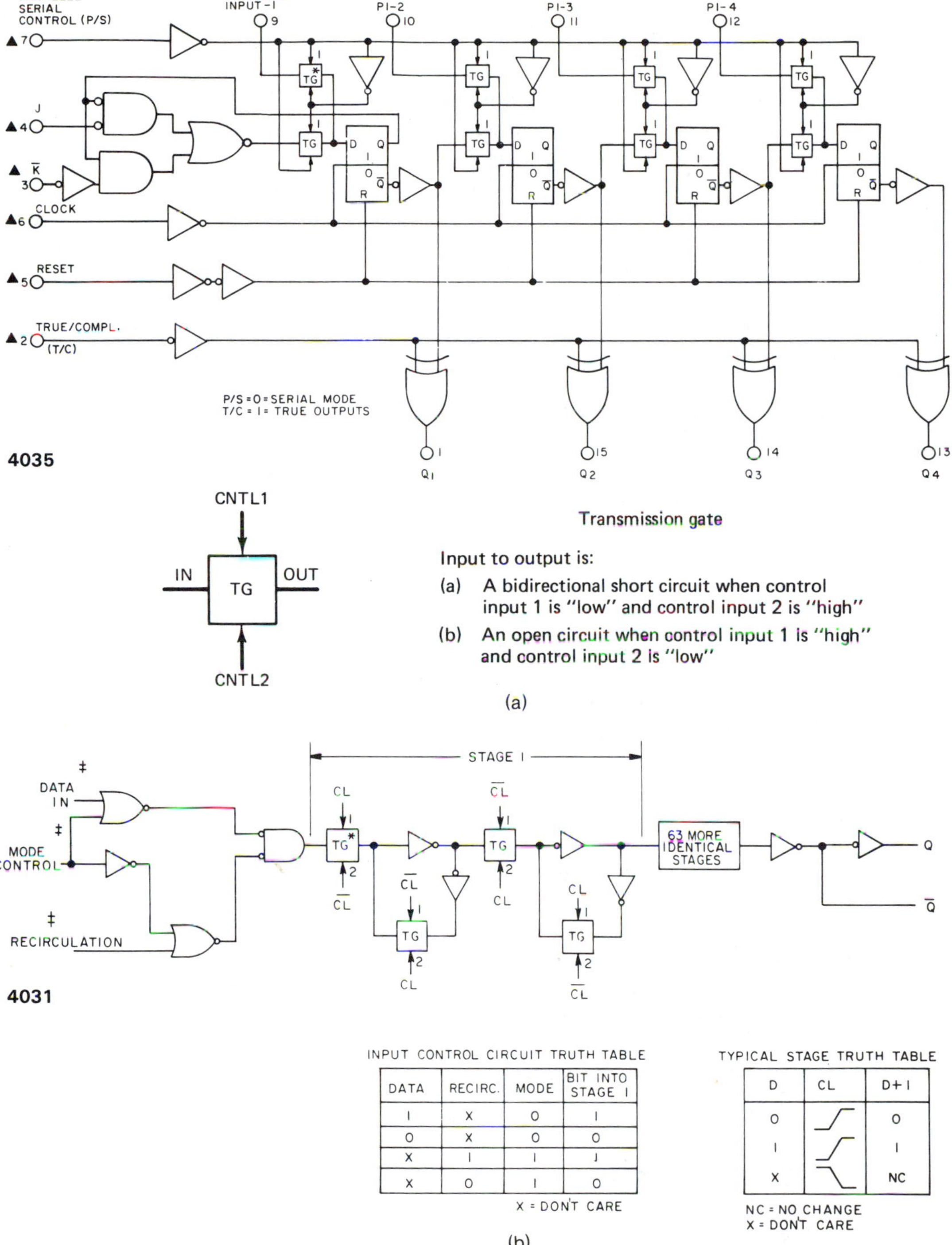

INPUT CONTROL CIRCUIT TRUTH TABLE

DATA	RECIRC.	MODE	BIT INTO STAGE 1
1	X	0	1
0	X	0	0
X	1	1	1
X	0	1	0

X = DON'T CARE

TYPICAL STAGE TRUTH TABLE

D	CL	D+1
0	⎽⎽╱⎺⎺	0
1	⎽⎽╱⎺⎺	1
X	⎺⎺╲⎽⎽	NC

NC = NO CHANGE
X = DON'T CARE

(b)

Figure 5.20 Circuit diagrams of two CMOS shift registers. (a) 4035 4-bit PIPO. Note the use of FET transmission gates (TG). (b) 4031 64-bit SISO. (*Courtesy* RCA Solid State Division)

5.5.4 The CD4035

This CMOS chip has almost the number of functions of the TTL 74194. The circuit diagram [Fig. 5.20(a)] illustrates the major differences between CMOS shift registers and their TTL equivalents. D-Type FF's are used, with only a single input. The two transmission gates on each FF function as a single pole–double throw switch, and determine whether the signal appearing at D comes from the preceding FF stage or from the external parallel inputs. The mode control input P/S serially shifts data to the right when $P/S = 0$, and parallel-loads data when $P/S = 1$. Separate J and K inputs are provided on the first stage to facilitate certain counting and sequence generation operations (Section 5.6). With J and K inputs connected together, the first stage becomes a normal D-type FF.

5.5.5 The 74164 and 74165

This complementary pair of SR's is used in data transmission, with the 74165 acting as a parallel-to-serial data transmitter, and the 74164 acting as a serial-to-parallel receiver. Both TTL and CMOS versions are available. The circuit diagrams are logically identical to Fig. 5.18(c) (74164) and 5.18(d) (74165). Typical waveforms are shown in Fig. 5.19(a) and (b). The 74164 serial input has an additional gated *enable* line to prevent data from loading until the desired bit string arrives at the serial input [Fig. 5.19(a)]. Likewise, the 74165 has a Clock Inhibit line which prevents data in the SR from shifting out until the correct time. These Enable/Inhibit controls are essential in noise-free data transmission, and intelligent use requires understanding methods of noise elimination and of handshaking.

5.5.6 The 4031

This circuit [Fig. 5.20(b) and Table 5.4] has 64 bits on one 16-pin chip, with only serial inputs and outputs available. A second serial input, called *recirculate*, is provided to facilitate ring counter applications (Section 5.6). When in use, the recirculate input is wired externally to the output, and recirculation of data already in the SR is effected by bringing the *mode control* input high. The circuit uses a CMOS master–slave D-FF. Control signals from the clock turn off the slave before the master is allowed to change its state, and vice versa. This provides the one clock-cycle delay necessary for race-free operation as a shift register (see Prob. 5.9).

5.6 SHIFT REGISTER APPLICATIONS

5.6.1 Data Communications and the UART

A parallel-to-serial shift register, such as the 74165, may be used to convert parallel binary information into a time series sent along a single wire. The corresponding receiver, the 74164, converts that same information back to parallel form, so

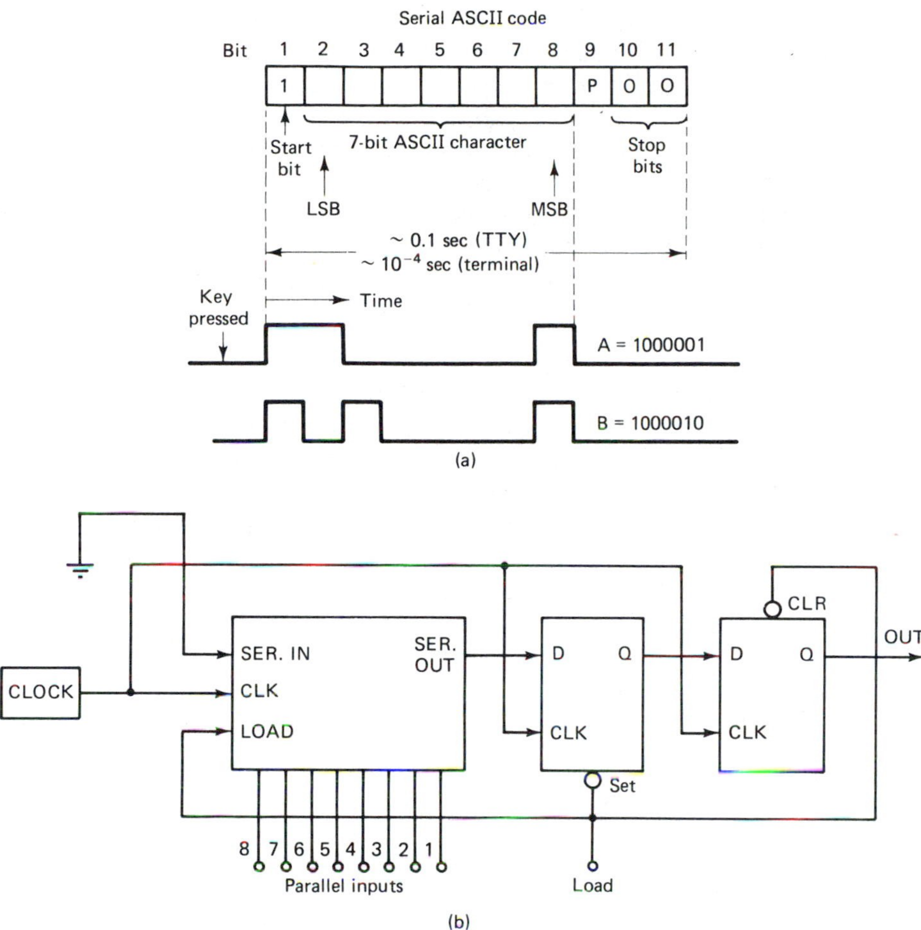

Figure 5.21 Parallel to serial code conversion. (a) The serial ASCII code convention. (b) Parallel to serial shift register circuit.

it may be read, for example, by an 8-bit microcomputer port. Parallel-to-serial conversion is accomplished as shown in Fig. 5.21.

The serial ASCII* code is used to represent characters generated by a teletype or terminal [Fig. 5.21(a)]. Six bits of information are enough to represent both upper and lower case for the 26 letters of the alphabet. A seventh bit enables the transmission of control characters which specify nonprinting special functions either locally at the terminal (ring the bell; clear the screen) or at the receiving device or computer (*control*-C to regain the attention of a computer). In addition, a *start bit* (always 1) signals the receiving device that a character is coming, and one or more *stop bits* (always 0) ensure that adjacent characters do not overlap and

* ASCII = American Standard Code for Information Interchange. In serial *asynchronous* transmission, sender and receiver clocks are not synchronized, although they must be set to the same frequency. The receiver uses the start bit to generate timing information for grabbing data bits at the right time.

become garbled. There is also an optional parity bit which may be used to determine if the message has been received without errors. The ASCII code is arbitrary and bears no particular correspondence to the order of letters in the alphabet or on the typewriter. Two examples are shown in Fig. 5.21(a).

A circuit for an ASCII transmitter, shown in Fig 5.21(b), includes an 8-bit PISO shift register and two additional FF's. The first FF, which is preset high, generates the start bit. The second FF, which is preset low, generates a leading 0 to ensure that the start bit can last a full clock cycle, since there is no way to know when a key will be pressed during a clock cycle [see arrow in Fig. 5.21(a)]. The FF's are triggered and the SR is loaded when a key is pressed.

The UART. It has become more common to use the MOS-LSI *Universal Asynchronous Receiver/Transmitter* (UART), shown in Fig. 5.22, for both transmitter and receiver functions. This flexible circuit is a good example of the versatility of MOS-LSI hardware. The UART contains two 8-bit shift registers (the receiver and the transmitter), each with an additional buffer register for temporary data storage. The mode of operation of each SR and the timing are set by control circuits. The UART is most often used as a bidirectional communications device, with parallel data received from and transmitted to 8-bit ports of a microcomputer

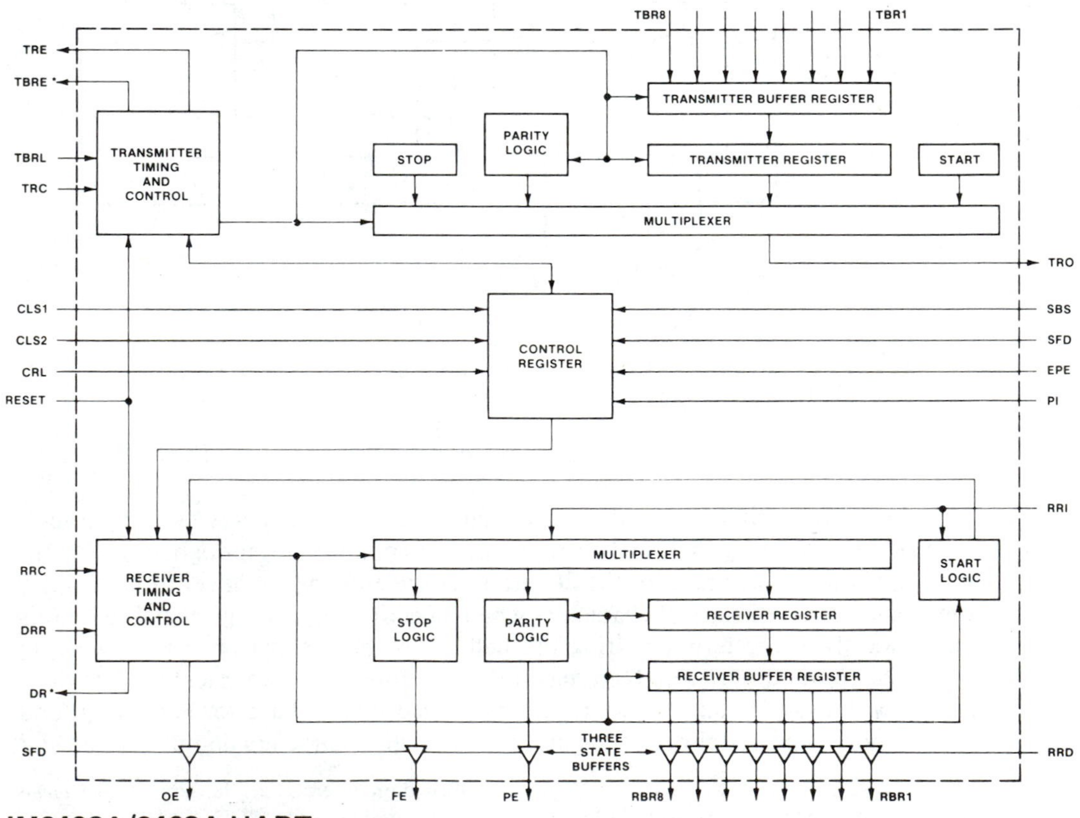

Figure 5.22 Universal Asynchronous Receiver/Transmitter (UART). (*Courtesy* Intersil, Inc.)

via RBR (receiver buffer register) and TBR (transmitter buffer register), and serial data received and sent via lines RRI (receiver register input) and TRO (transmitter register output).

The logic of the control functions is beyond the scope of this book, but will be touched on briefly. Control lines CLS1 and CLS2 (character length) determine whether the characters will have 5, 6, 7, or 8 bits. PI (parity inhibit) and EPE (even parity enable) determine if parity will be checked, and, if so, whether odd or even. Control line TBRL loads a character to be transmitted into TBR, and TBRE and TRE provide handshaking to signal that the transmit buffer is empty or that serial tranmission is complete. Line DR (data ready) indicates that a character has been received and is waiting in the receiver buffer register. Tri-state control line RRD makes the data in the receiver buffer available. The transmitter and receiver clocks TRC and RRC run at 16 times the serial bit rate. The actual latching of data in receiver or transmitter occurs near the center of a given time slot to allow noise or ringing to settle.

5.6.2 Ring Counters

A *ring counter* is a shift register whose serial output is connected to its serial input (Fig. 5.23). These circuits are useful for storing a predetermined number to be clocked out repetitively, as in sequencing and in waveform generation. When the circulation control gate input is low, external data is clocked into the SR as usual. When the circulation input is high, the output is fed back to the input, forming a circulating ring counter or ring register. If a single binary bit is loaded in, the sequence of states is the 1-2-4-8 counting sequence below.

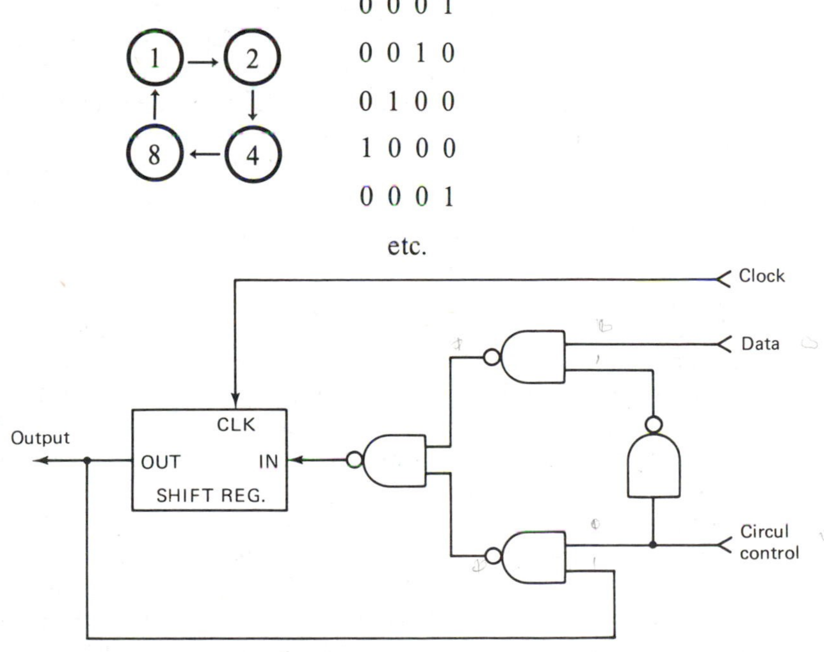

Figure 5.23 Basic ring counter.

Some CMOS SR's such as the 4031 (Table 5.4) include ring counter gating internally. A large number of ring counter stages connected in parallel can be useful as memory storage elements. Both bubble memory and charge-coupled device memories (Chapter 7) are functionally wired as shift register rings.

Ring counters have *disallowed states* which if entered (as in the power turn-on) cause an unintended sequence. For example, the disallowed states shown below cycle nowhere or through two instead of four states.

Disallowed states must be avoided by including a power-on reset to the starting state or by self-correcting circuitry.

The basic ring counter is used as a sequencer to enable a series of devices or to turn on bits in a successive-approximation analog-to-digital converter (Chapter 9). However, it is a relatively inefficient way of generating N states, since N FF's are required. By comparison, a binary counter-plus-decoder sequences through 2^N states with N FF's. The ring counter's advantage comes in high-speed applications. Since only one FF is changing state at any time, the sequencer output is glitch-free by comparison with the counter-decoder alternative. The Johnson counter improves upon efficiency by generating $2N$ states with N FF's, and retains glitch-free operation.

5.6.3 Johnson Counter

If the input to the first stage of a ring counter is taken from the complement $\overline{Q}$ of the final-stage output, the *Johnson counter* or *twisted ring* results [Fig. 5.24(a)]. The output waveforms in this example [Fig. 5.24(b)] are symmetric square waves of frequency eight times lower than the clock.

What gets the Johnson counter going? Assume the FF's are preset to 0. Then $\overline{Q}_D = 1$ provides a high input to D_A, which sets this FF high at the next clock pulse. This 1 state propagates down the chain until all FF's are set high, and $\overline{Q}_D = 0$ resets FF_A at the next clock pulse. The 0 propagates down the chain until the cycle repeats.

Digital waveform generation is a Johnson counter application. If the output voltages are summed up as in Fig. 5.25, the square-wave contributions add up to give a reasonable approximation to a sine wave. With an N flip-flop shift register, harmonic distortion occurs only at frequencies $2N - 1$, $2N + 1$, and above. This example generates a sine and cosine wave in precise 90° phase relationship. With the resistor values shown in this 6-stage approximation, the waves have 0 harmonic distortion up to the eleventh and thirteenth harmonic. Digital sine wave generation has a number of advantages over analog methods, in addition to precise sine and cosine output phase relationship. The signal amplitude is constant, independent of frequency. The output frequency can be very stable if a crystal clock is used, yet can be precisely adjusted using digital divide-by-N circuits on the input.

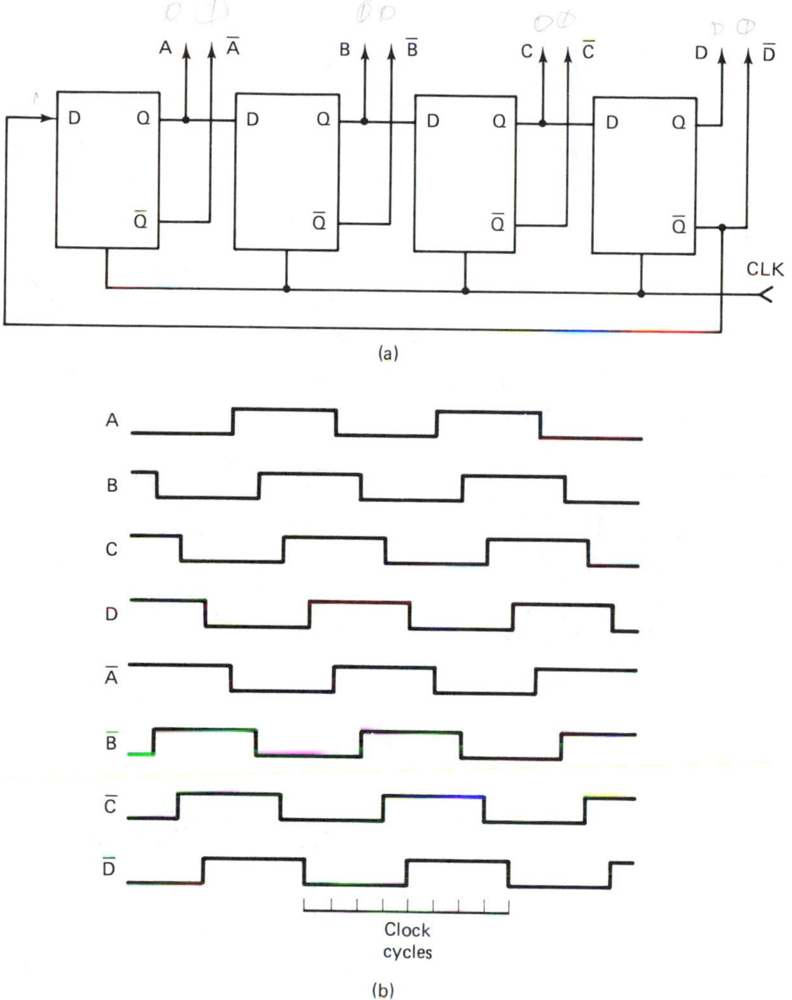

Figure 5.24 Johnson counter or walking ring counter. (a) Circuit of 4-bit version. (b) Waveforms; note the symmetric divide-by-8 result.

Johnson counters can begin in a disallowed state, with a cycle of states very different than intended. To solve the problem, one could load the correct starting state. This is inconvenient if it requires parallel inputs, although an adequate solution if a reset to 0 starts the desired sequence. A second solution is a *self-correcting* Johnson counter which uses gating between several stages to selectively eliminate disallowed states. CMOS Johnson counters usually have such gating built in (NOR gates in Fig. 5.26; see also Prob 5.10).

Johnson counters are also used as glitch-free "one-out-of-N" data selectors. The phase-shifted square-wave outputs are tied to AND gates to provide the decoding. With the $2N$ outputs of an N-stage Johnson counter, a single two-input NAND is sufficient to decode each of the $2N$ states. Fig. 5.26 shows 8 and 10 state examples. The decoding is synchronous and therefore glitch-free, since identical gate delays occur on all inputs.

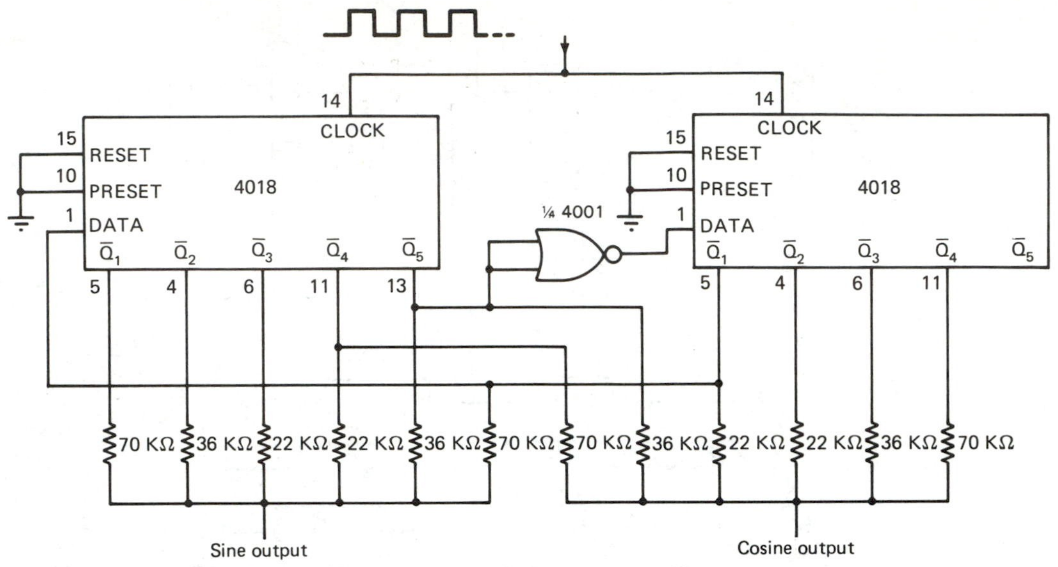

Figure 5.25 Digital sine wave generator using a Johnson counter. (*Adapted from* Timothy E. Jordan, *Electronics.* (Aug. 18, 1977): 115.)

5.6.4 Random Noise Generators

A final example of shift register applications is the generation of *pseudorandom* noise. Although each number in the sequence appears unrelated to the next, the identical sequence occurs each time the circuit is reset to the starting point. Eventually the cycle will repeat itself, although the modulus may be made very long. Pseudorandom noise is useful in code generation for protected data transmission, white noise generators for testing the response of circuits or systems, and the production of a sequence of random numbers.

The circuit is constructed by connecting the output of two or more shift register stages back to the input through an exclusive OR gate. Table 5.5 shows the sequence of states resulting from a 4-bit example whose circuit is shown in Fig. 5.27. The same 15-number sequence would result if any number in the sequence were used as the preset. The number 0000 is disallowed, because it results in no sequence. Any pseudorandom number generator has one or more disallowed states.

A circuit constructed from N FF's will generate a sequence $(2^N - 1)$ numbers long. The gating feedback depends upon the modulus, and several gates may be required. If the goal is to construct a white noise or long-sequence random number generator and the modulus does not matter, the subset shown in Table 5.6 can be selected which requires only one 2-input XOR gate. For example, a $(2^{33} - 1)$ sequence generator is shown in Fig. 5.28. When used as an audio white noise generator, passive filtering or a clock frequency above the audio range gives a signal indistinguishable from analog white noise. This circuit does not repeat itself for nearly 10 h when the input clock frequency is 250 kHz!

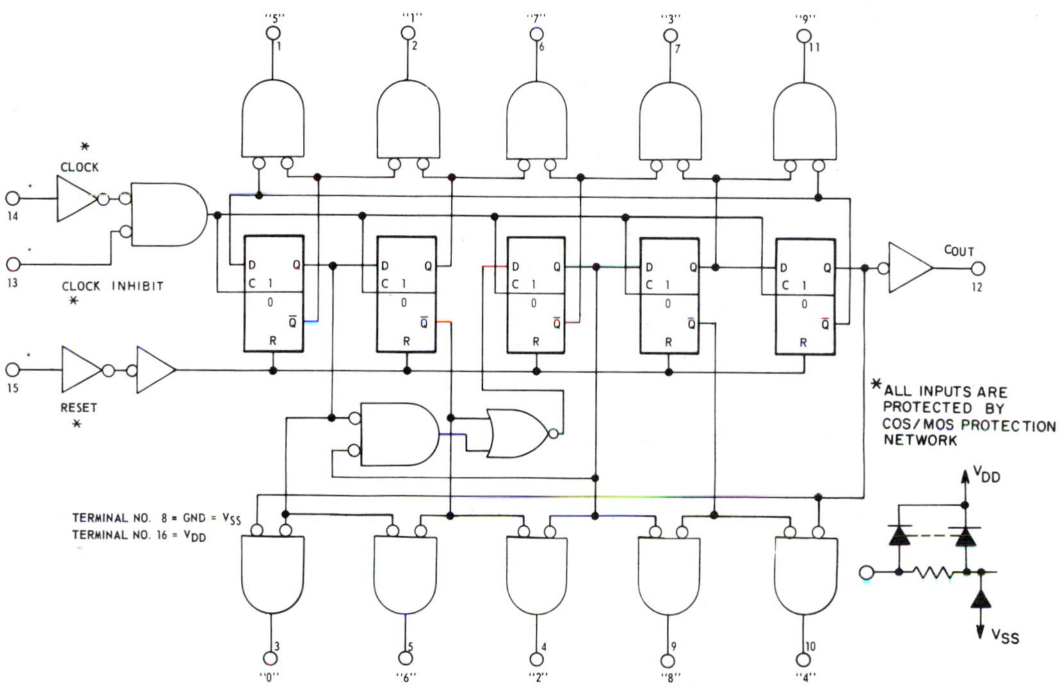

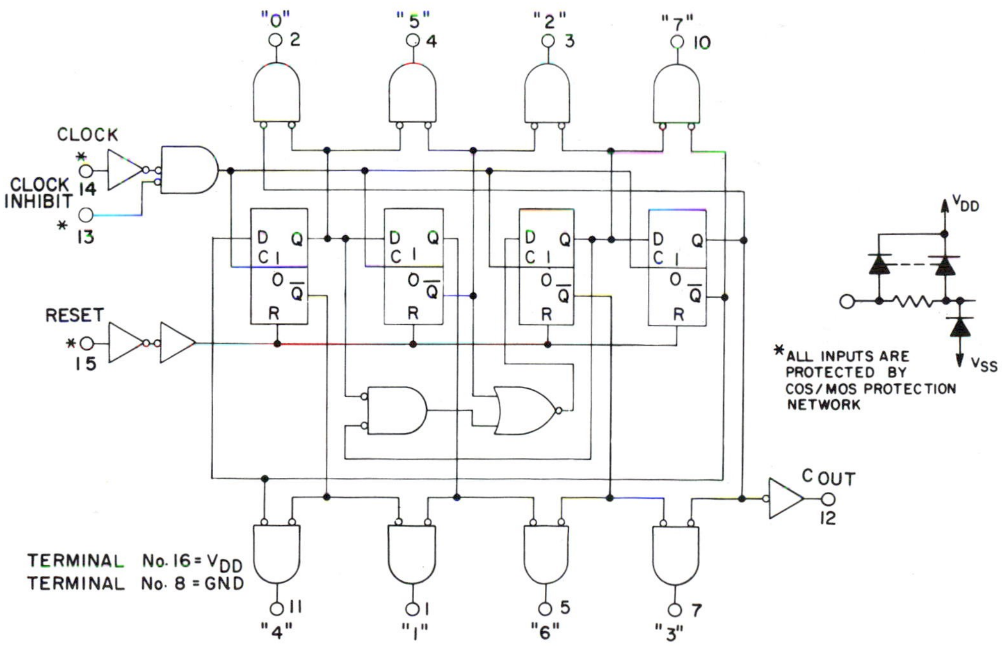

Figure 5.26 Type 4017 and 4022 Johnson counter with NOR gating to eliminate disallowed states, and NAND decoding of outputs. (*Courtesy* RCA Solid State Division)

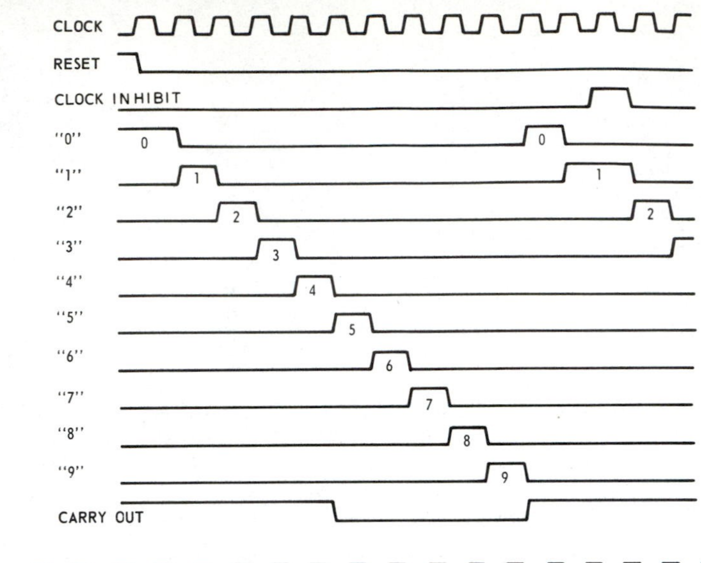

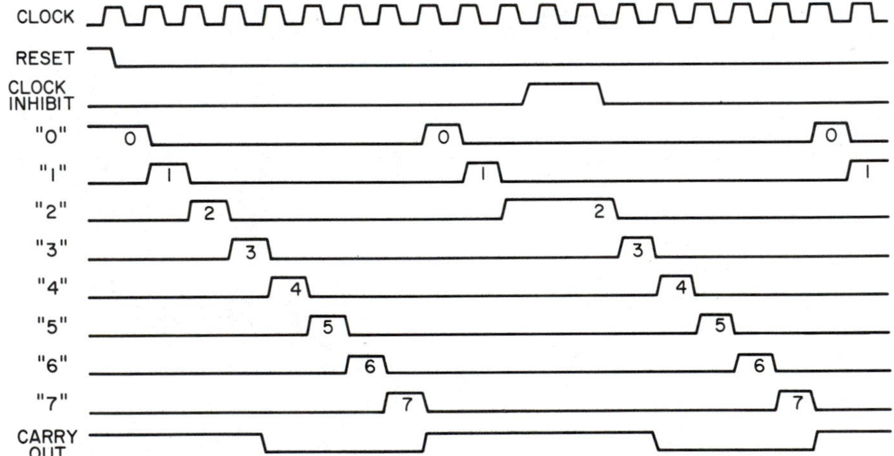

Figure 5.26 (cont.)

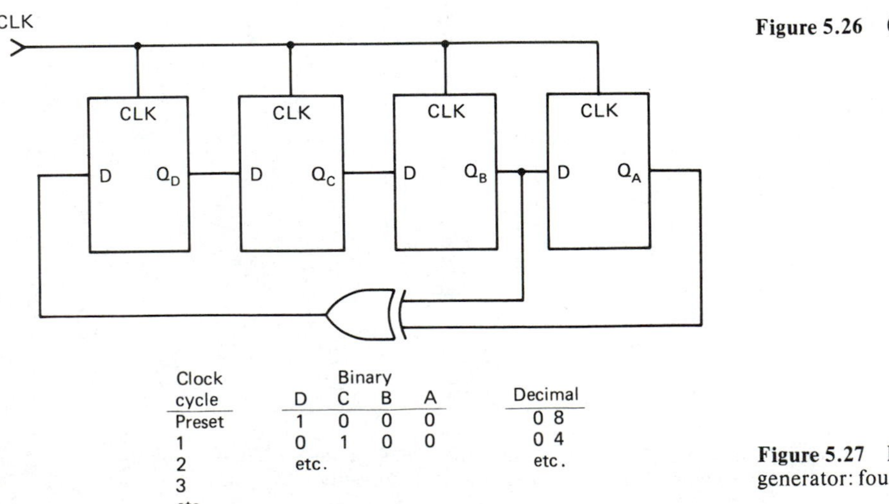

Clock cycle	Binary D	C	B	A	Decimal
Preset	1	0	0	0	0 8
1	0	1	0	0	0 4
2	etc.				etc.
3					
etc.					

Figure 5.27 Pseudorandom noise generator: four bit example.

TABLE 5.5 PSEUDO RANDOM NUMBER GENERATOR

	Binary	Decimal
Preset:	1000	8
Clock 1	0100	4
2	0010	2
3	1001	9
4	1100	12
5	0110	6
6	1011	11
7	0101	5
8	1010	10
9	1101	13
10	1110	14
11	1111	15
12	0111	7
13	0011	3
14	0001	1
—	—	—
15	1000	
16	etc.	

TABLE 5.6 MAXIMUM LENGTH WHITE NOISE SEQUENCES THAT REQUIRE ONLY TWO FEEDBACK TAPS ON THE SHIFT REGISTER[a]

No. of stages	Stages with taps	Sequence length	Duration of sequence using 250-kHz clock
7	1,7 or 3,7	127	0.51 ms
15	15 and 1, 4, or 7	32,767	131 ms
20	3,20	1,048,575	4.2 s
28	28 and 3, 9 or 13	268,435,455	18 m
33	13,33	8,589,934,591	9.5 h
39	39 and 4, 8, or 14	5.5×10^{11}	25 d

[a](Selected from the list by Danashek M., *Electronics*, May 27. 1976, p. 107.)

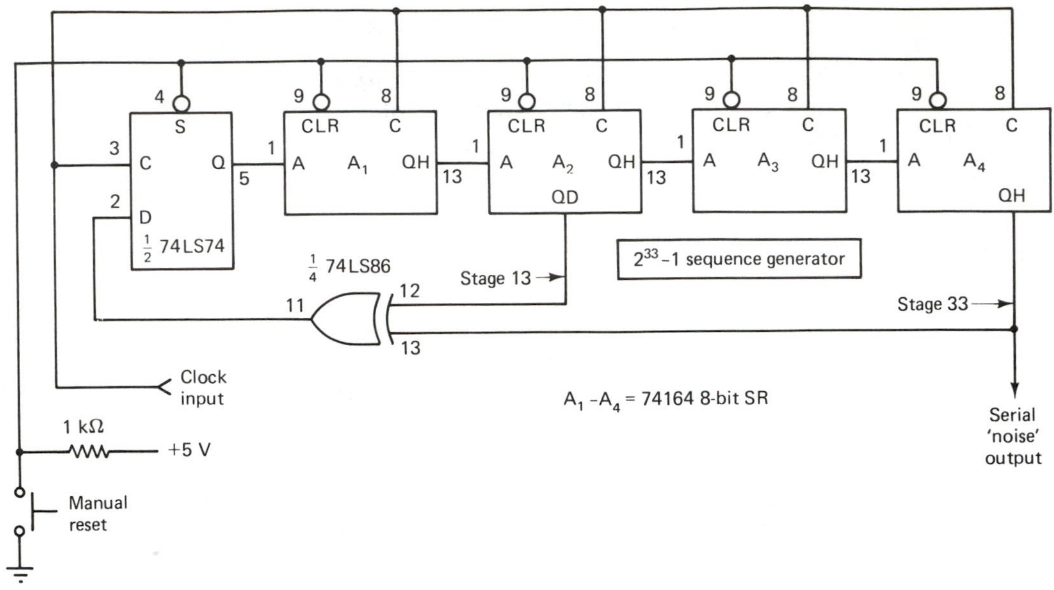

Figure 5.28 ($2^{33} - 1$) random sequence generator circuit. (*Adapted from* Leonard H. Anderson, *Electronics.* (Nov. 9, 1978): 134.)

PROBLEMS

5.1. (Flip flop review) Show how a toggle flip flop for a ripple counter (Fig. 5.2) may be constructed from either a JK-M–S or a D-edge flip flop.

5.2. Sketch the pattern of waveforms at the outputs of the counter shown in Fig. 5.2(c) during a complete count cycle. Show that if the outputs are assigned the usual binary weights, the cycle corresponds to counting down.

5.3. Design a 4-bit up/down ripple counter. There are two inputs: a data input which receives clock pulses, and a count direction input which when high causes counting up and when low causes counting down. Use a minimum of extra gating.

5.4. Analyze the operation of the divide-by-3 counter shown in Fig. 5.8 using the truth table for the JK flip flop. Sketch the waveforms seen at Q_1, Q_2, and Q_2. Begin with the clock high and assume that both Q_1 and Q_2 start out low.

5.5. Explain the two unusual modulus counter examples given in Fig. 5.9. (a) Drawing waveforms may be helpful in the case of the divide-by-6. What glitch (short-lived state) will occur? (b) In the case of the divide-by-9, writing the pattern of binary numbers seen at the outputs should explain why the counter is wired to load the complement of the desired modulus. Note the use of the Carry, which goes low as the count of 15 is reached.

5.6. The presence of internal gating and two independent counters (divide-by-2 and divide-by-5) in the 7490 allows construction of *any* modulus between 2 and 10 using external jumpers only, as shown in the table on p. 175. Show how this works, using the 7490 circuit diagram and waveform timing diagrams as needed.

Divisor N	Input on pin	Output on pin	External connections
2	14	12	Pin 2 or 3 low
3	1	8	Pin 8 to pin 2 Pin 9 to pin 3
4	1	8	Pin 11 to pins 2 and 3
5	1	11	Pin 2 or 3 low
6	14	8	Pin 12 to pin 1 pin 9 to pin 2 pin 8 to pin 3
7	1	12	Pin 11 to pin 14 Pin 12 to pin 2 Pin 9 to pin 3
8	14	8	Pin 12 to pin 1 Pin 11 to pins 2 and 3
9	14	11	Pin 12 to pins 1 and 2 Pin 11 to pin 3
10	14	11	Pin 12 to pin 1 Pin 2 or 3 low

(*Adapted from:* T. Durgavich and D. Abrams, *Electronics*, July 8, 1976, p. 90)

5.7. The circuit shown in Fig. 5.29 generates a graduated-scale "ruler line" when displayed on an input channel of an oscilloscope, and is useful, for example, in locating channels of a histogram display. How does it work?

Figure 5.29 Circuit for putting a graduated scale "ruler" line on an oscilloscope (see Problem 5.7). (*Source:* Ken E. Anderson, *Electronics.* (Jan. 8, 1976): 108.)

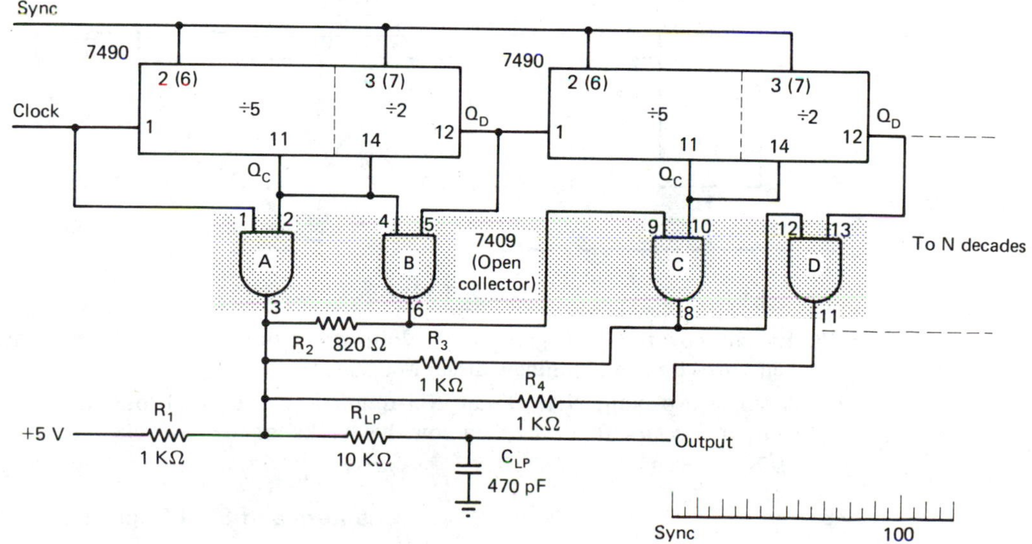

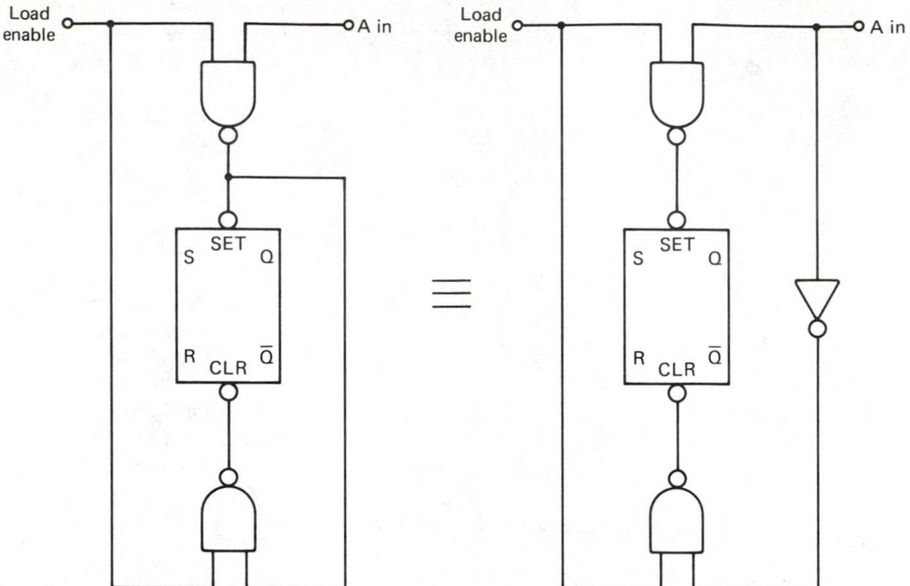

Figure 5.30 Gate minimization for a parallel load shift register.

5.8. Explain how the gating of the input data in the parallel input SR [Fig. 5.18(d)] is accomplished. Gate minimization (Fig. 5.30) has been used to eliminate an inverter needed in the obvious method. Work out the truth tables for both the minimized and the obvious circuit to show the equivalence. Truth table inputs are A and LOAD, and outputs are SET and CLR.

5.9. The design of race-free shift registers depends upon the D flip flops acting as a unit delay. How is this done when D-edge FF's are used in CMOS SR's such as the 4031 [Fig. 5.20(b)]? Consider two CMOS D FF's in series (Fig. 5.31). Fill in the waveform diagrams. What requirements must the basic CMOS D FF satisfy so that the *previous* state of FF2 gets latched by FF1? (The triangular symbol on the CLK input is a reminder that the FF is edge-triggered.)

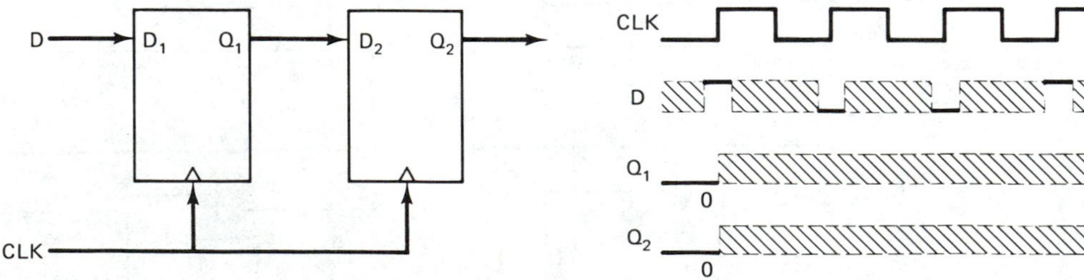

Figure 5.31 Use for Prob. 5.9.

5.10. Explain how the NOR gates on the Johnson counter of Fig. 5.26 make the circuit self-correcting and eliminate disallowed states.

5.11. A single two-input NAND gate will uniquely decode each output of a Johnson counter *selector* circuit. Explain how this works for the example illustrated in Fig. 3.25.

5.12. Explain the operation of the 4-bit random number generator of Fig. 5.27. Assume one state of Table 5.5 to start, and show what transitions occur as the cycle goes on. What will happen if the initial state is 0000?

5.13. Explain how the decoding of the 74193 counter [Fig. 5.6(b)] provides synchronous up/down counting action.

REFERENCES

More complete bibliographic information for the books listed below appears in the annotated bibliography at the end of the book.

DANASHEK, M., *Electronics*, May 27, 1976, p. 107; maximal length white noise generators

DEMPSEY, *Basic Digital Electronics with MSI Applications*

DIEFENDERFER, *Principles of Electronic Instrumentation*

DURGAVICH, T., and D. ABRAMS, *Electronics*, July 8, 1976, p. 90; divide-by-N with any modulus

HIGGINS, *Experiments with Integrated Circuits*, Experiment 7

LANCASTER, *TTL Cookbook*

LANCASTER, *CMOS Cookbook*

LARSEN & RONY, *Logic and Memory Experiments using TTL Integrated Circuits*

MALMSTADT et al., *Electronic Measurements for Scientists*

Manufacturer's literature: see, especially, National, EXAR, Intersil, Motorola

WILLIAMS, *Digital Technology*

6

Digital Waveshaping and Instrumentation

6.1 INTRODUCTION

Signals encountered in real systems rarely are idealized two-state binary numbers. This is particularly true in measurement, where there is no reason to expect that the transducer or signal detector will produce TTL logic levels. Signal processing problems are also encountered in data transmission, where noise can prevent an initially clean digital signal from being reliably received. Within a given circuit, large current pulses in one part of the circuit may cause erroneous operation elsewhere through fluctuations in the power supply voltage.

Several examples of waveshaping have been introduced earlier. Switch debouncing, using gates or SR FF's, is universally used to eliminate mechanical contact bounce (Fig. 4.4). Storage latches for data transmission and microcomputer interfacing require waveshaping in order to distinguish the data from noise introduced in transmission. Anticoincidence gate circuits (Chapter 3 problems) help select which events to count as real data, for example: Was that pulse really a gravity wave, or just a truck going by outside?

Signal processing circuits for measurements often combine digital and analog techniques. For example, an analog amplitude-sensing circuitry must determine not only the occurrence of a pulse, but also its amplitude. Other analog circuits might determine exactly when the peak of a pulse occurred. These techniques must be deferred until after analog background has been developed. Chapter 6 will be limited to techniques of *digital* signal conditioning to make a waveform cleaner for reliable triggering of digital logic devices. The chapter closes with some examples of digital instrumentation.

6.1.1 What is Digital Signal Conditioning?

A number of waveshaping problems (Fig. 6.1) are solved by standard circuits.

(1) *Level Restoration.* The output signal from a detector or transducer may have voltage levels very different from those needed to drive digital logic circuits [Fig. 6.1(a)]. Level restoration means adjusting the range of the signal by amplification or attenuation, adjusting a bias voltage V_b to keep a signal from crossing through 0, and clipping the result to produce a two-state waveform.

(2) *Bounce Elimination.* If the input signal is noisy, it may cross the threshold from logic 0 to logic 1 several times on the way to a "real" transition [Fig.

Figure 6.1 Examples of digital signal conditioning problems. (a) Level restoration. (b) Bounce elimination. (c) Threshold setting. (d) Event timing. (e) Pulse stretching. (f) Pulse generation.

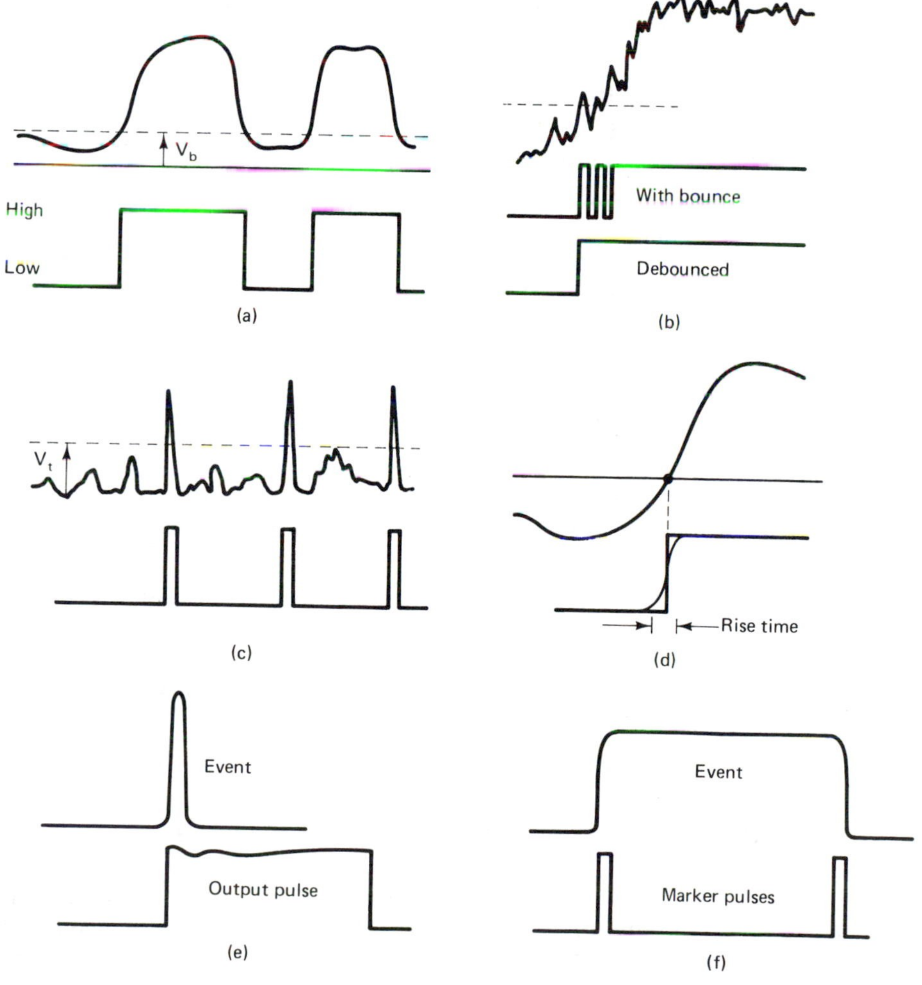

6.1(b)]. Contact bounce from a switch is a familiar example. If the signal is connected directly to a digital buffer amplifier, multiple triggering will result, giving errors if, for example, the signal is then fed to a counter. Debouncing is the cure, and examples of flip-flop debouncers have been encountered earlier (Chapter 4).

(3) *Threshold Setting.* In a measurement of pulses from photons or particles, real events may be accompanied by noise pulses, introduced, for example, by vibration (nearby moving vehicles) or by stray pulses from other sources. Often there exists a threshold V_t above which one can be reasonably certain the pulse is from a "real" event [Fig. 6.1(c)]. Setting a threshold to define meaningful data significantly increases the signal to noise ratio of the measurement.

(4) *Event Timing or Edge Restoration.* In timing applications it is necessary to know *when* a given voltage level was crossed [Fig. 6.1(d)]. The signal may be varying slowly across the threshold, as in the precise digital measurement of the phase of a sine wave. In other cases the events may have started with a sharp edge but become rounded in transmission through a long signal path. Edge restoration may also be required if digital devices (e.g., edge-triggered FF's) require the input signal to rise faster than a given time interval. In such cases, use waveshaping circuits to give an output signal that rises quickly once an input threshold has been crossed.

(5) *Pulse Stretching.* An input signal can occur faster than digital logic can follow it. For example, a photon burst from a laser pulse may last only picoseconds. Pulse stretching holds the event long enough for conventional logic circuits to detect it [Fig. 6.1(e)].

(6) *Pulse Generation.* Sometimes an event occurs over a long time interval, and one needs to generate a marker pulse to indicate when the event started or stopped [Fig. 6.1(f)]. The marker pulses are made short compared to the event, allowing high-resolution measurement of the time interval.

All these problems may be solved with waveshaping circuits such as familiar FF debouncers plus two new circuits: the Schmitt trigger and the *one-shot* or monostable multivibrator. Basic circuitry and IC versions of these devices will be discussed in this chapter, together with a closely related circuit, the astable multivibrator used as a basic clock. Common to these circuits are two gates cross-coupled with feedback to hold a given state. The circuits differ only in the type of feedback: direct (flip flop), resistive (Schmitt trigger), capacitive (astable multivibrator), or resistive and capacitive (one-shot).

6.1.2 Noise Immunity: Comparison of TTL and CMOS Logic Families

CMOS circuits have a higher noise immunity than their TTL counterparts. This is demonstrated in Fig. 6.2 for an inverter. The input of a CMOS circuit identifies a logic 0 as any signal up to about 0.3 V_{cc} (or 1.5 V when V_{cc} = 5 V), treating any-

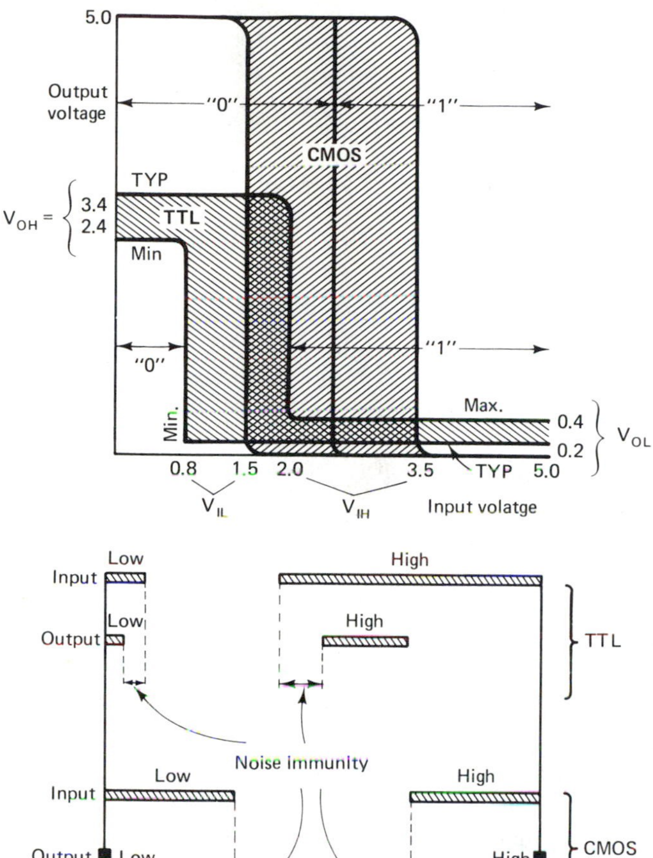

Figure 6.2 TTL and CMOS logic thresholds.

thing larger as a logic 1. By contrast, a TTL device only accepts signals up to 0.8 V as an input 0. A noisy 0 signal can trigger a TTL device erroneously if the noise exceeds 0.8 V. The noise can be almost twice as large without causing an error for a CMOS device. A more striking difference exists at the output. CMOS output voltage levels are pulled to the power supply voltage (1) or to ground (0). By contrast, the TTL logic 1 output voltage may range from 2.4 V to 3.4 V, and a logic 0 output may be anywhere from 0.2 V to 0.4 V. This difference is most significant when the output of one device drives the input of another (lower half of Fig. 6.2). A noise spike of as little as 0.4 V (worst case) riding on a TTL signal may cause an erroneous change of state at the input of the next device. By contrast, since the CMOS outputs are at the maximum or minimum voltage of the circuit, the noise must be at least four times larger (1.5 V) to cause an erroneous transition. In practice, the difference is not quite this extreme, since TTL output highs and lows are usually at the top and the bottom of the shaded range in Fig. 6.2. Noise margins in system design are estimated using guaranteed logic threshold values shown on Fig. 6.2 and listed below. The CMOS values assume a 5-V power supply for comparison purposes.

Parameter	Definition	CMOS Value	TTL Value
V_{IL}	Upper limit voltage value which input circuit will recognize as low state	1.5 (max)	0.8 (max)
V_{IH}	Lower limit voltage value which input circuit will recognize as high state	3.5 (min)	2.0 (min)
V_{OL}	Voltage value of output in low state	0.0	0.2 (typ) 0.4 (max)
V_{OH}	Voltage value of output in high state	5.0	2.4 (min) 3.4 (typ)

6.2 DIGITAL WAVESHAPING CIRCUITS

6.2.1 Schmitt Trigger

A Schmitt trigger circuit converts a signal that is noisy or slowly varying (or both) into clean digital form [Fig. 6.3(a)]. As the input signal rises, the output changes from the low to the high state when the input first crosses the *On threshold* [Fig. 6.3(a)]. The output then remains high as the signal falls past the on threshold, until the input crosses the *Off threshold.* The On and Off thresholds are separated by a region called the *deadband,* within which the circuit does not respond to input signal changes. The input-output transfer function [Fig. 6.3(b)] displays a kind of memory called *hysteresis;* that is, the input voltage at which the output changes depends upon whether the output is presently high or low [arrows on Fig. 6.3(b)].

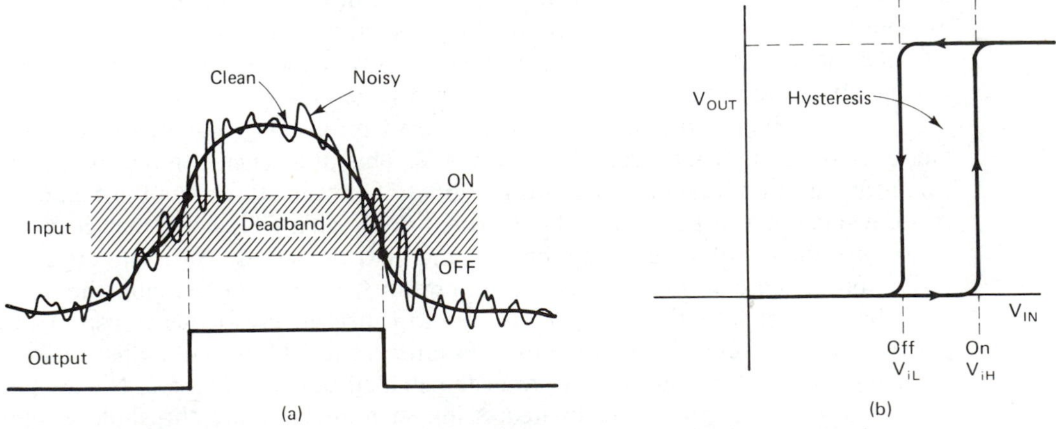

Figure 6.3 (a) Schmitt trigger functioning to clean up a waveform. (b) Transfer function showing hysteresis.

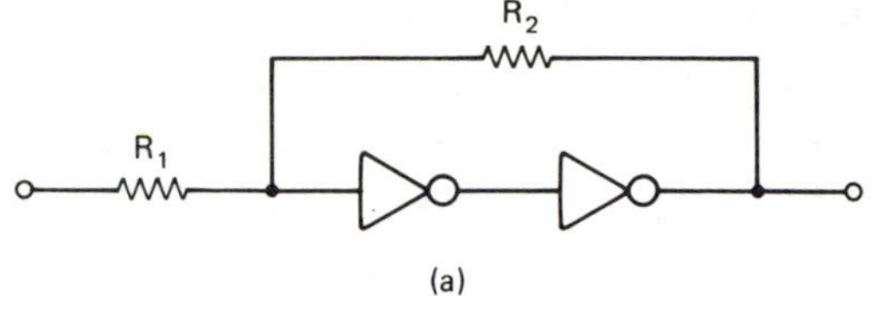

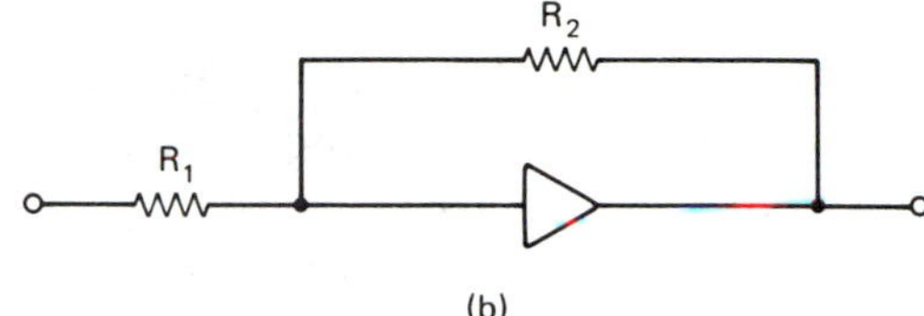

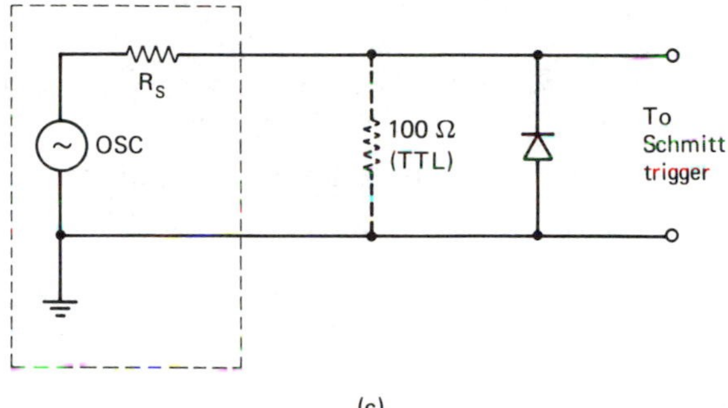

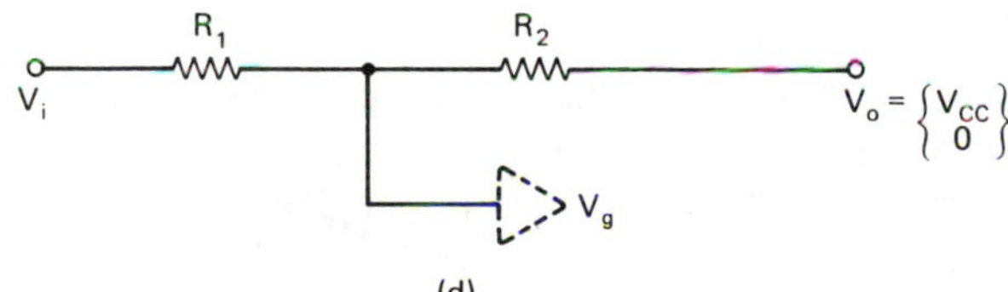

Figure 6.4 Schmitt trigger circuit from inverters, constructed with (a) TTL logic or (b) CMOS logic. (c) Sine wave input test signal circuit. (d) Principle behind Schmitt operation as a weighted voltage divider.

Hysteresis or deadband is useful in waveshaping a very noisy input, since only the first crossing of a threshold is effective. Later excursions due to noise have no effect as long as they do not exceed the deadband. In Fig. 6.3(a), both the clean and the noisy signal examples produce a single pulse at the output.

A Schmitt trigger may be constructed using two inverters or a single noninverting buffer amplifier [Fig. 6.4(a) and (b)]. The hysteresis may be demonstrated on an oscilloscope using the circuit of Fig. 6.4(c). The turn-on and turn-off points will occur at different amplitudes of the sine wave. The 100-Ω resistor ensures that the source has a resistance to ground low enough to adequately define the TTL low state. The *clamp diode*, which allows only positive portions of the ac signal to be applied to the logic circuitry, is necessary unless the logic family provides clamping internally.

The Schmitt trigger generates a deadband by making the input voltage for the 0-to-1 transition depend upon the value of the output voltage [Fig. 6.4(d)].

The following analysis assumes the CMOS version, since the output voltage is then well-defined (0 V or power supply voltage V_{cc}). The voltage V_g at the gate is the weighted sum of the input voltage V_i and the output voltage V_o. Resistors R_1 and R_2 form a voltage divider, and the resulting voltage at the gate is, by the superposition principle

$$V_g = \frac{R_2}{R_1 + R_2} V_i + \frac{R_1}{R_1 + R_2} V_o \tag{6.1}$$

A change in the output state occurs when the gate input crosses a threshold voltage V_T. This occurs for two input voltages, V_{iL} and V_{iH}, depending upon whether V_o was high or low. If the output was high, a transition occurs when

$$V_g = V_T = \frac{R_2}{R_1 + R_2} V_{iL} + \frac{R_1}{R_1 + R_2} V_{cc} \tag{6.2a}$$

If the output was low,

$$V_g = \frac{R_2}{R_1 + R_2} V_{iH} + 0 = V_T \tag{6.2b}$$

These equations may be solved to determine the two values of the input threshold voltage:

$$V_{iH} = \frac{R_1 + R_2}{R_2} V_T$$

$$V_{iL} = \frac{R_1 + R_2}{R_2} V_T - \frac{R_1}{R_2} V_{cc} \tag{6.3}$$

$$\text{Hysteresis } \Delta V = V_{iH} - V_{iL} = (R_1/R_2) V_{cc}$$

The two transition points have been shifted above and below the normal value V_T, with a hysteresis fraction set by the ratio R_1/R_2. With $R_2 = 10\ R_1$, the circuit exhibits a deadband $\Delta V = 0.1\ V_{cc}$. Although shown shaded in Fig. 6.2, the parameters V_{IL} and V_{IH} listed there do *not* indicate hysteresis for ordinary gates, but rather the variation from one unit to another.

The calculation is identical for TTL circuits, except that V_o becomes the TTL output high voltage (typically 3.4 V) and the voltage V_T at which an input transition will take place is less well defined (Fig. 6.2). Although the hysteresis is readily calculated, the upper and lower threshold points are best determined experimentally. Circuit details also differ substantially. Since a CMOS input is an open circuit, practically any resistor values may be used. With a TTL device, however, the input resistor must be less than about 500 Ω. The TTL Schmitt trigger, while faster than CMOS, may therefore be used only when the signal has a low source impedance or is suitably buffered. A summary of likely IC choices and resistor values is given in Table 6.1.

Analog versions of Schmitt triggers use linear amplifiers rather than two-state devices, and allow both the hysteresis and the threshold to be adjusted continuously. This is particularly useful in control circuits or in measurements when the input signal from a transducer or detector is offset from ground or has a vol-

TABLE 6.1 SCHMITT TRIGGER CIRCUIT DETAILS

CMOS	
Two-inverter circuit:	4041 (will drive TTL)
	4069 (will not drive TTL)
Single buffer:	4050 (best choice in CMOS)
Resistors:	R_1 = 10 K; R_2 = 100 K (for 10% hysteresis)
TTL	
Two-inverter circuit:	7404 (best choice in TTL)
Single buffer:	7407, 7417 (requires additional pull-up)
Resistors:	R_1 = 330 Ω, R_2 = 3.3 K Ω

tage range different from typical logic circuits. Analog Schmitt triggers are discussed in Chapter 19.

6.2.2 The One-Shot Multivibrator

The *one-shot* produces upon command a single clean output pulse of adjustable width. The term *monostable multivibrator* is a leftover from older terminology that classes as multivibrators the one-shot, flip flops (bistable multivibrator), and square-wave oscillators or clocks (astable multivibrator).

One-shots from gates. Monostable circuits constructed from gates [Fig. 6.5] resemble switch contact debouncers, except that the closed loop includes an RC combination whose time constant sets the output pulse width. The NAND gates with only one input shown function as inverters, to facilitate making the circuit from a single IC package. In viewing these circuits, we will assume that the output of a gate is "active" and the input side is "passive." That is, the state of the gate depends upon the voltage level at its input, but the gate input will not affect circuit components connected to it. This is not exactly true for TTL, since the input low state is a current sink. As a result, the following analysis is only approximately true for TTL (though accurate for CMOS).

The circuit shown in Fig. 6.5(a) converts long pulses into short ones. In the steady state, with the signal at INPUT held low, the output of gate 1 is high and C is charged. Gate 2 has one high and one low input, so its output is high. When a positive-going input signal brings the output of gate 1 low, the charge on C finds a path to ground through R. During the time that C is first discharging, both inputs of NAND gate 2 are high, and a low signal is generated at the output. When the capacitor voltage drops to the turnoff threshold of gate 2, the output returns to the high state to end the pulse. If the circuit is constructed from TTL gates, R must be limited to values less than about 500 Ω, or a reliable input low state will not be achieved. This limitation is not present in CMOS versions, but their speed limitation must also be considered. The circuit of Fig. 6.5(a) is most useful in generating a trigger pulse whose output width is shorter than the input pulse width. If the

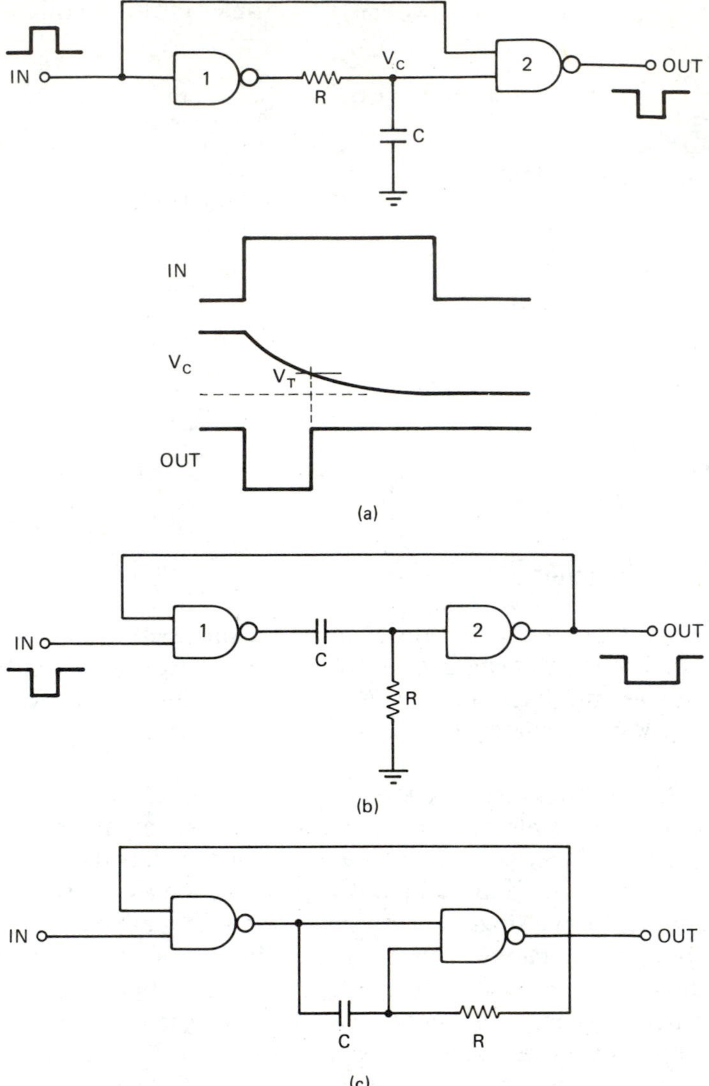

Figure 6.5 One-shot circuits constructed from gates. (a) Input pulse longer than output pulse. (b) Input pulse shorter than output pulse. (c) Another version.

input pulse falls before a time $\sim RC$ has passed, the upper input of gate 2 will be 0, and the output will return to the normally high state prematurely.

The circuit of Fig. 6.5(b) does not have this limitation, and is used in making short pulses into longer ones. This circuit requires a normally high input and has a normally high output, with capacitor C initially discharged. When the input signal goes low, the output of gate 1 goes high, and the capacitor will begin to charge. Since the capacitor is initially uncharged, most of the output voltage from gate 1 appears initially across R, driving the output of gate 2 low. The coupling of

Output back to Input ensures that the output of gate 1 will remain high no matter what the input state is, since NAND 1 has one input low. After a period approximately RC, C will have charged to the point that the input voltage for gate 2 falls below threshold, returning the system to its initial state.

A final example is shown in Fig. 6.5(c). The circuit is most easily understood if viewed as an SR FF. The input pulse changes the state using the *set* input, and the output *resets* the FF after a time $\sim RC$. The analysis is left as Prob. 6.1.

Integrated circuit monostables. Although one-shot circuits constructed from gates perform adequately in generating short pulses, they have serious limitations. The range of pulse widths available in TTL versions is limited by input loading, since R must be kept lower than 500 Ω. In addition, TTL gate inputs will load the RC combination, so the relationship between component values RC and pulse width will not be quite linear. Finally, since the output pulse width depends upon the time to charge or discharge a capacitor to a gate input threshold, the pulse width will change if the supply voltage varies. Special-purpose monostable IC's overcome all these limitations.

A basic IC monostable circuit is shown in Fig. 6.6. The triangular component is a *comparator*. It compares the voltages at its two inputs, producing an output voltage that is high (power supply voltage) or low (ground) when input V_+ is greater or less than V_-, respectively. The comparator is a *hybrid* device; its output is a binary signal, but its inputs are continuous analog signals. The comparator symbol is identical with the symbol for the operational amplifier (second half of this book), and any operational amplifier in analog electronics may be used as a comparator. However, special-purpose comparators have been developed (Chapter 19) for fastest switching speeds.

Since comparators have a high input impedance, R and C can take on a wide range of values without loading the circuit. Other circuit components include an SR FF and a transistor switch to discharge the capacitor. In the quiescent state, Input is held high and Output is low. The transistor is held on, keeping C discharged, so the comparator output is in the *high* state ($V_+ = V_{\text{Ref}} > V_- = 0$). (Note the procedure used in analysis: assume an input/output combination and see if the consequences are consistent around the circuit.) When the input signal falls from high to low, the FF changes state, driving Output high and also switching the transistor *off*. Capacitor C begins charging towards V_{cc} at a rate determined by RC. When V_C reaches the reference voltage V_{ref}, the comparator switches to the low output state, resetting the FF and terminating the output pulse. Since V_{ref} is a fixed fraction of V_{cc}, the time at which the comparator switches (and therefore the output pulse width) is independent of the supply voltage.

A common monostable IC is the 74121, whose timing graphs are given in Fig 6.7. The pulse width extends over 6 decades using easily available R and C values. The pulse width is accurately linear in R, and nearly linear in C. The curvature at the low-C end is due to internal parallel capacitance. The relationship

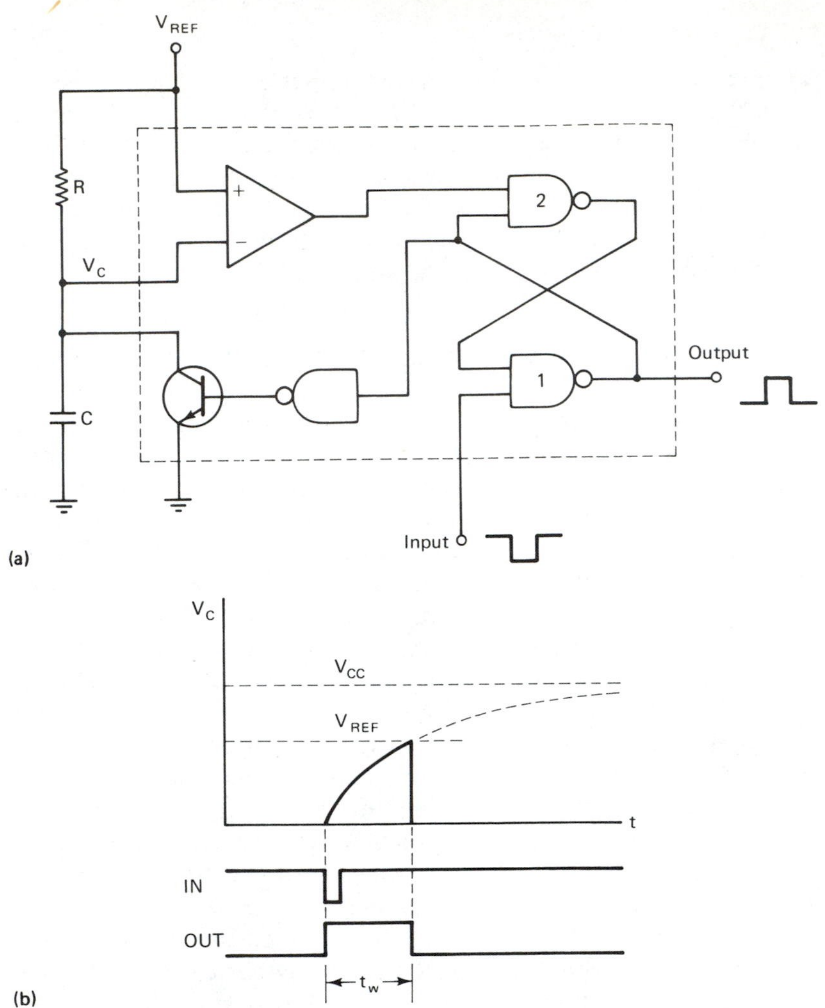

Figure 6.6 (a) One-shot integrated circuit (similar to 74121). (b) Waveform of charging cycle sets t_w.

between output pulse width and circuit component values is approximately

$$t_w \simeq 0.3RC \tag{6.4}$$

Although TTL one-shots like the 74121 are the best general-purpose solution for short pulses, C values become excessive for pulses longer than about 1 s. Although an electrolytic capacitor could be used, ordinary electrolytics have too much leakage, so high-quality tantalum electrolytics must be used. Other IC monostables are listed in Table 6.2. A CMOS one-shot is a good solution for long time intervals because larger values of R are tolerable. The 555 timer is popular for very long time intervals, though inadequate for pulse widths under a microsecond. This general-purpose timing device will be discussed with astable multivibrator circuits.

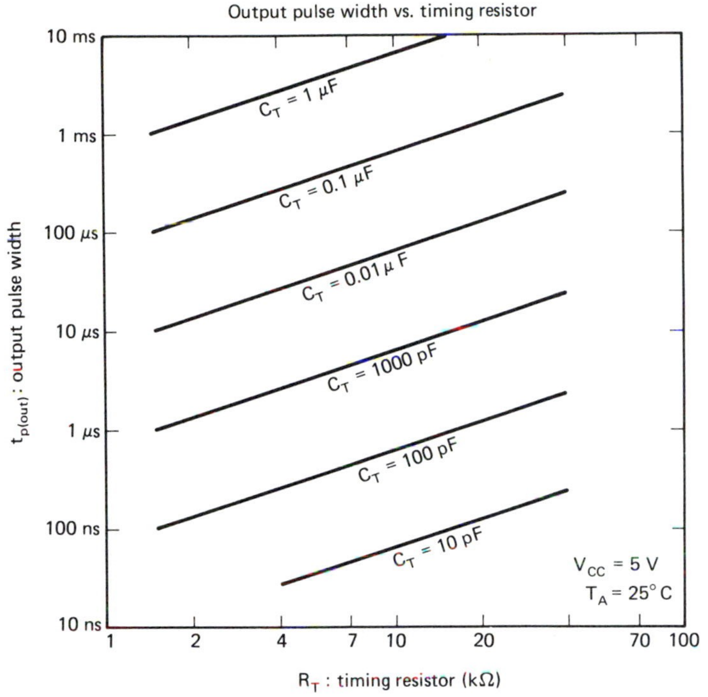

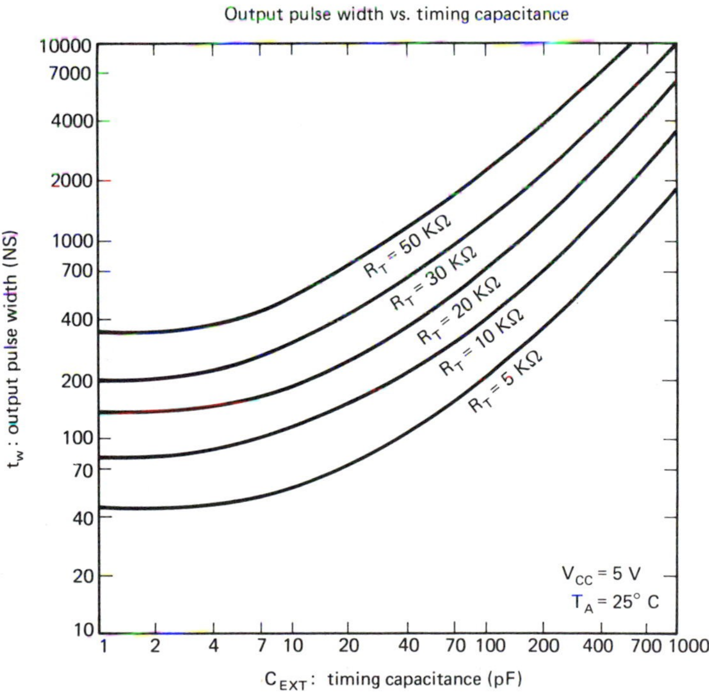

Figure 6.7 74121 one-shot timing graphs.

TABLE 6.2 IC MONOSTABLE COMPARISON

IC	Logic family	R-range min K Ω	R-range max K Ω	C-range min μF	C-range max μF	Inputs	Retrig?	Reset?	Special features
74121	TTL	1.4	40	None	1000	2A,1B	No	No	Schmitt input
74122,9600	TTL	5	50	None	None	2A,3B	Yes	Yes	
74123,9602	TTL	5	50	None	None	1A,1B	Yes	Yes	Dual
9601	TTL	5	50	None	None	2A,2B	Yes	No	
74L5221	Low power Schottky TTL	1.4	1100	None	1000	1A,1B			Dual Schmitt input
74C221	CMOS	5	350	None	None	1A,1B	No	Yes	Dual
4098B 4528B MC14528CP	CMOS	5	1000	None	None	1A,1B	Yes	Yes	Dual
555	Linear	1K	10,000	5×10^{-4}	None		No	No	Warning: $t_{in} < t_w$

A comparison of devices in Table 6.2 illustrates several other monostable features.

Inputs. Most one-shots include more than one independent input. In TTL devices, a pulse is generated when the voltage at an A input falls from high to low, while a B input indicates the reverse input logic. This convention is often reversed in CMOS. A variety of input circuits is available. For example, the two A inputs in the 74121 (Table 6.2) are separately gated to allow additional control over enabling or disabling pulse generation [Fig. 6.8(a)]. Input gate logic differs considerably among the IC's available. The 74123 and the 9602 [Fig. 6.8(b) and (c)] have both A and B inputs and follow the same input pulse convention but require different treatment of unused inputs. This is summarized in one-shot *input tables* that specify the combination of input signals that will produce a pulse:

Input table Examples	74123 A	74123 B	9602 A	9602 B
	↓	1	↓	0
	0	↑	1	↑

When using the A input, B must be wired high for the 74123 but low for the 9602.

The convention on inputs is reversed in most CMOS one-shots, except for TTL replacements carrying a 74C part number. The CMOS A input triggers on a

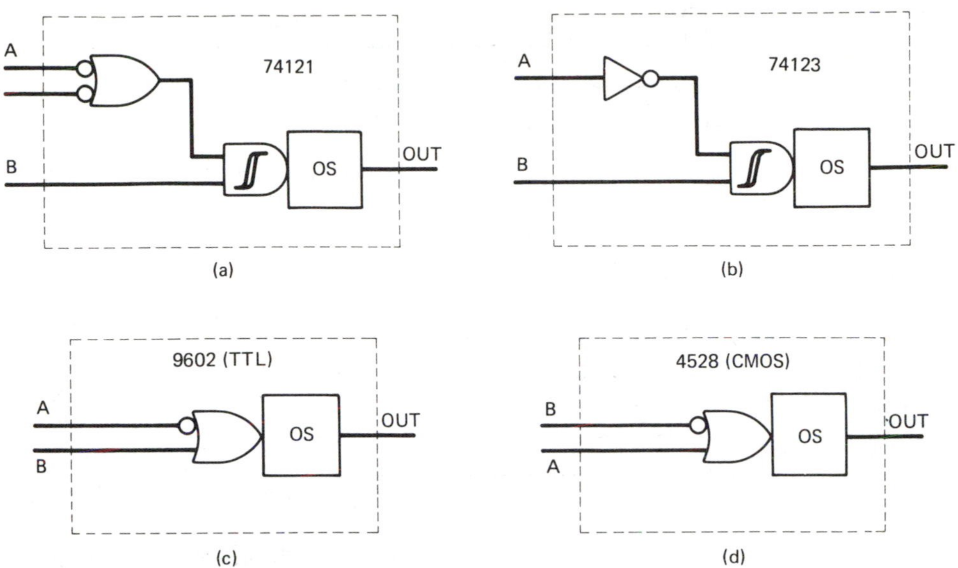

Figure 6.8 One-shot IC input circuitry. (a) 74121. (b) 74123. (c) 9602. (d) 4528.

positive-going transition, while B triggers on a negative-going transition [Fig. 6.8(d)]. The gating circuitry of Fig. 6.8 demonstrates the required state of unused inputs (see Problem 6.2). The unusual OR gate symbol in Fig. 6.8(c) and (d) designates what input signals will enable the circuit. The one-shot in Fig. 6.8(c) is enabled either by B going high or by A going low (note the bubble).

Triggering. Although trigger signals are shown as vertical arrows in input-state tables, one-shot inputs are not *edge*-triggered. A pulse is generated when an input signal crosses a *voltage* threshold. The emphasis is on edges because in one-shot applications for timing, it is the time at which the pulse appears that matters. Noise at the input will alter the time when the input signal crosses the logic threshold. A slow threshold crossing may cause a large error in timing. This shift of pulse appearance time is called *jitter* when the input is repetitive. If the input is a fixed waveform such as a sine wave but of variable frequency, jitter becomes worse as the input frequency is lowered. The uncertainty in crossing of the voltage threshold translates to an uncertainty in when the pulse appears that depends on the rate of change (hence the frequency) of the input signal. To reduce this problem, some one-shots (the 74121 and 74221) include a Schmitt trigger input, so only the first crossing of the logic threshold can fire the device. For example, the 74121 is guaranteed to be jitter-free for input signals as slow as 1 V/s, with a guaranteed noise immunity of 1.2 V. The Schmitt input is the B input in these circuits and is designated on the gate circuit of Fig. 6.8(a) and (b) by the hysteresis loop symbol.

Retriggering. Monostables listed in Table 6.2 with *retriggering capability* fire for another period t_w if a second input pulse is received during the period of

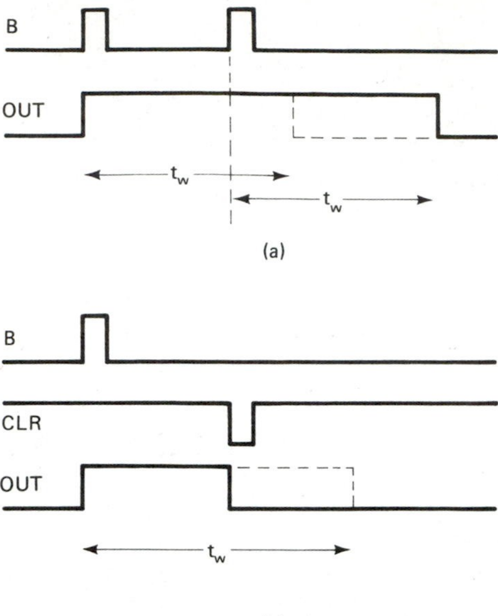

Figure 6.9 (a) Retriggering capability. (b) Clear input.

an output pulse [Fig. 6.9(a)], extending the output pulse beyond the normal period $t_w \simeq RC$. Applications will be discussed in Section 5.4. Retriggering may be eliminated when undesired by connecting the Q or $\overline{Q}$ output to an appropriate input gate. Further inputs are disabled during the entire period when the one-shot is firing. The precise wiring depends upon which circuit input receives the start pulse.

Clear. An output pulse is terminated before its normal time t_w if a pulse is received on the *Clear* input [Fig. 6.9(b)]. The Clear input is used to reset (manually or by computer) a timing circuit for another cycle, when the previous cycle has not finished. Retriggering is available on all common one-shots (Table 6.2) except the 74121 and the 555.

Monostables for very long time periods. Time periods from seconds to hours are achieved using circuits like the 555 timer. A wiring diagram and time period graph are shown in Fig. 6.10. The pulse width is approximately 1.1 *RC*s. Because the RC combination is connected to a high-impedance input, *R* values as large as 10 M Ω may be used, resulting in a timing range of over 7 decades. The output circuitry has a large current-driving capability (100 mA), adequate to drive a small speaker. The 555 is limited to a 10 μs lower limit. In addition, there is a very important application rule that does not appear clearly in most commercial application notes:

> **555 Warning.** When operated as a monostable, the 555 output will remain high as long as the input is low, so that erroneous timing will occur if the input pulse is longer than the output pulse.

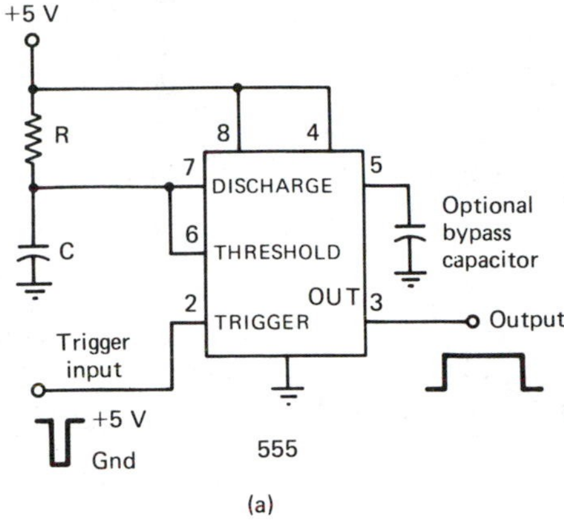

(a)

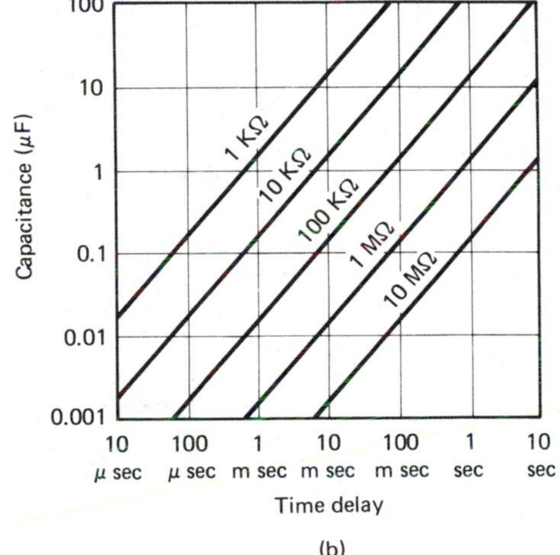

(b)

Figure 6.10 555 timer as monostable multivibrator. (a) Wiring diagram. (b) Timing graph.

This can be circumvented, when necessary, by shortening the input pulse using a passive RC differentiator (see Prob. 6.5).

6.2.3 Astable Multivibrator Circuits

An astable multivibrator generates a repetitive waveform whose frequency can be adjusted by an RC combination. This circuit is an example of digital wave generation. Circuit operation is very similar to the one-shot, and some IC's can serve both functions. Timing applications in digital instrumentation require an astable circuit as a frequency reference. These devices are also used as clocks in digital computers and data communications. Oscillator circuits based upon linear (rather than two-state) switching concepts will be presented after some analog concepts have been developed (see Chapter 17).

Integrated circuit astable multivibrator. Although an astable circuit can be constructed from gates, the price and performance of IC astable circuits are

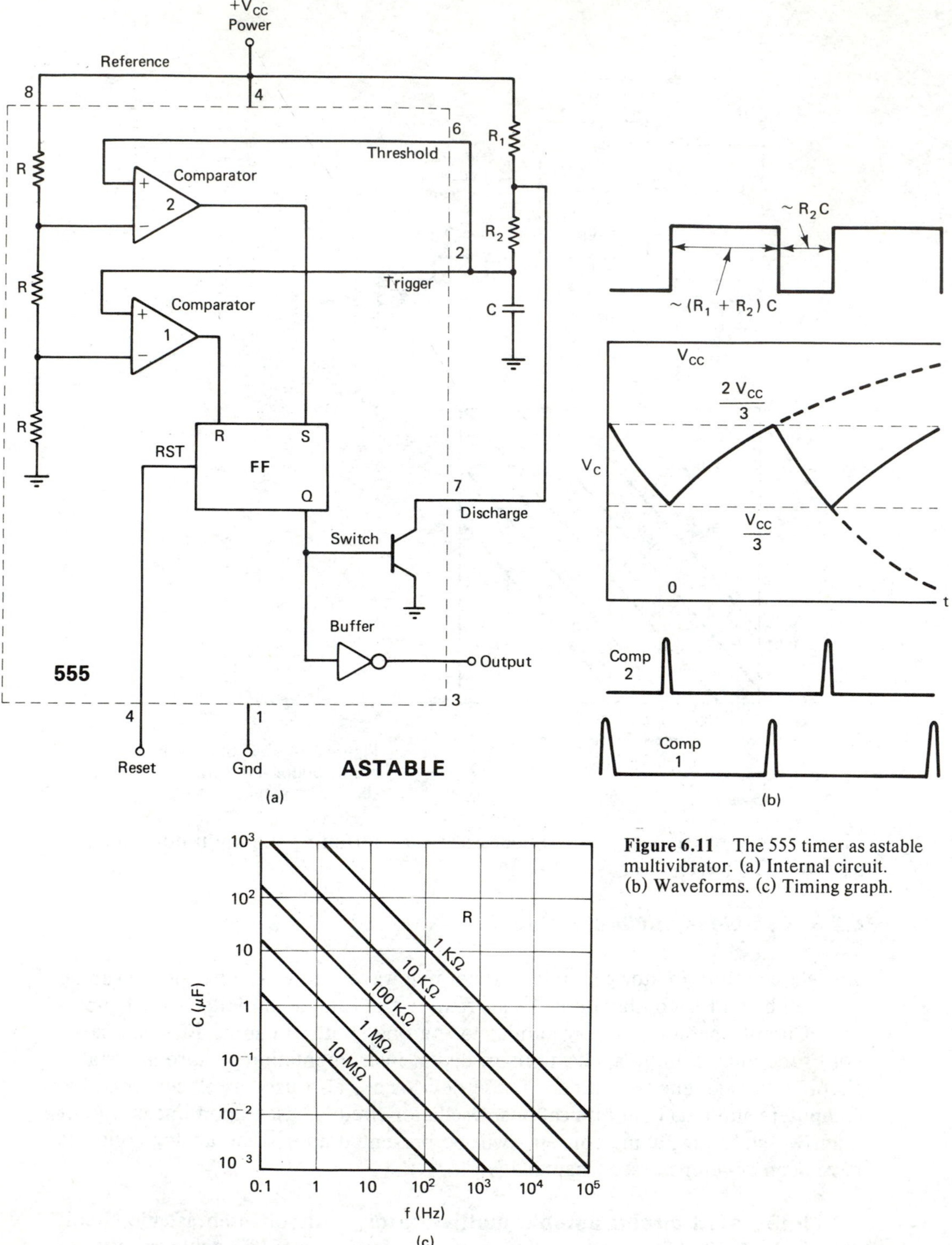

Figure 6.11 The 555 timer as astable multivibrator. (a) Internal circuit. (b) Waveforms. (c) Timing graph.

more attractive. An industry standard for all but the most stable applications is the 555 timer. The internal circuit is shown in Fig. 6.11(a). Like the IC one-shot, the 555 is a hybrid circuit containing both analog and digital components. There are two comparators, with one input of each connected to an internal resistor ladder. Comparator 1 resets the FF when the voltage on its plus terminal reaches $V_{Ref}/3$, while comparator 2 sets the FF when the voltage on its plus terminal reaches $2V_{Ref}/3$. In operation as an astable circuit, the other inputs of both comparators are wired together to sense voltage V_C across an external capacitor C. When the transistor switch is off, C is charging exponentially towards V_{cc} through $(R_1 + R_2)$, as shown in Fig. 6.11(b). When V_C reaches $2V_{cc}/3$, the output of comparator 2 switches high, setting the FF and turning on the transistor. This grounds the junction between R_1 and R_2, discharging C through R_2. When V_C falls below $V_{cc}/3$, the output of comparator 1 switches to the high state, resetting the FF. This turns the transistor off, and the charging cycle begins again. Each comparator is high only momentarily, long enough to change the FF state.

The relationship between component values and operating characteristics is

$$\text{Charging time} = 0.685\,(R_1 + R_2)C \qquad \text{(output high)}$$

$$\text{Discharge time} = 0.685\,R_2C \qquad \text{(output low)}$$

$$\text{Frequency} = 1.46/[(R_1 + 2R_2)C]$$

$$\text{Duty cycle} = \frac{\text{time high}}{\text{total period}} = \frac{1 + R_2/R_1}{1 + 2\,R_2/R_1}$$

Limits $\quad 1\text{ K}\,\Omega < (R_1,R_2) < 3\text{ M}\,\Omega$
$\quad 500\text{ pF} < C <$ any value (limited by leakage)

The 555 has a number of useful features.

(1) Since high-input-impedance comparators are used, external timing resistors can vary over a wide range. External capacitance can vary over 6 decades. This circuit is capable of periods from microseconds to hours.

(2) The waveshape can range from nearly a square wave ($R_2 >> R_1$) to a sharp pulse ($R_2 << R_1$).

(3) Since the voltage that charges the capacitor is also the internal comparison voltage on the resistor ladder, the output frequency change due to power supply voltage fluctuation is small: frequency drift is 0.1% per volt of supply voltage change. Drift with temperature is more serious, approaching 0.01% per °C. While the 555 is adequate for most data communications and instrumentation, computer clocks and precise instrumentation require more stable crystal clocks.

(4) The internal buffer enables the 555 circuit to provide a large output current (up to 100 mA), which can drive many TTL loads and is powerful enough to drive a small speaker directly.

Fig. 6.12 shows the 555 circuit with external connections as a one-shot (previous section). The analysis is left for Prob. 6.4.

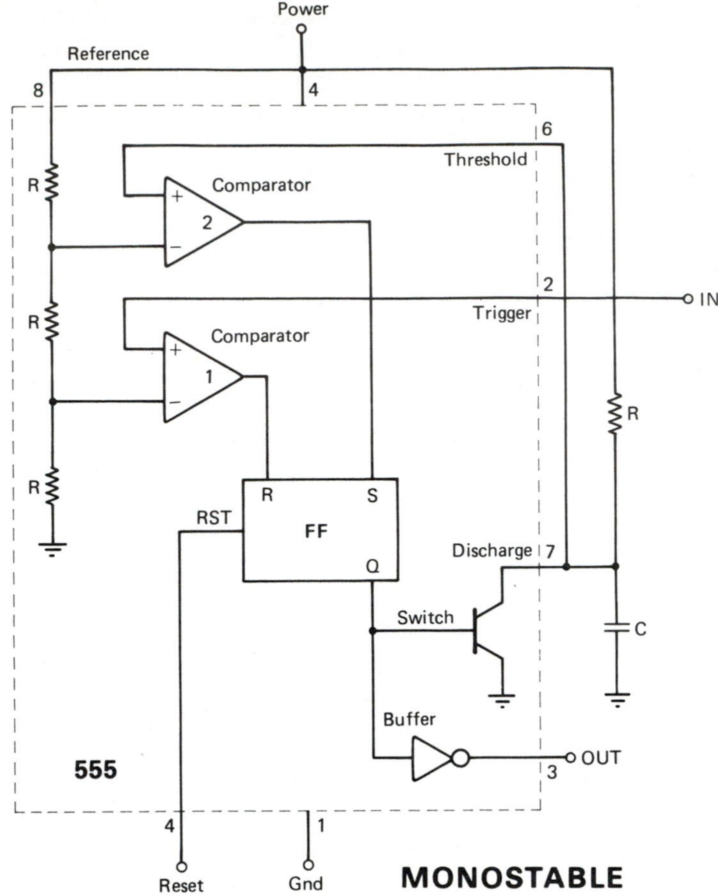

Figure 6.12 Internal circuit of 555 timer wired as a one-shot (see Prob. 6.4).

6.3 WAVESHAPING APPLICATIONS IN MICROCOMPUTER INTERFACING

6.3.1 Strobing vs. Waiting

In interfacing an experiment or in transmitting data to a mini- or microcomputer, binary data may be far from ideal when they reach the receiver. Noise with glitches large enough to cross a logic threshold may be present [Fig. 6.13(b)]. A binary edge that was clean when transmitted may be rounded or have overshoot or ringing at the receiver if the transmission path is very long. There are two approaches to ensure that data is latched reliably. The first approach [Fig. 6.13(c)] is to clean up the data, restoring it to nearly binary form using a Schmitt trigger or debounce circuit. The second approach [Fig. 6.13(d)] is to wait until the worst of the ringing on the line has settled out before reading the bit.

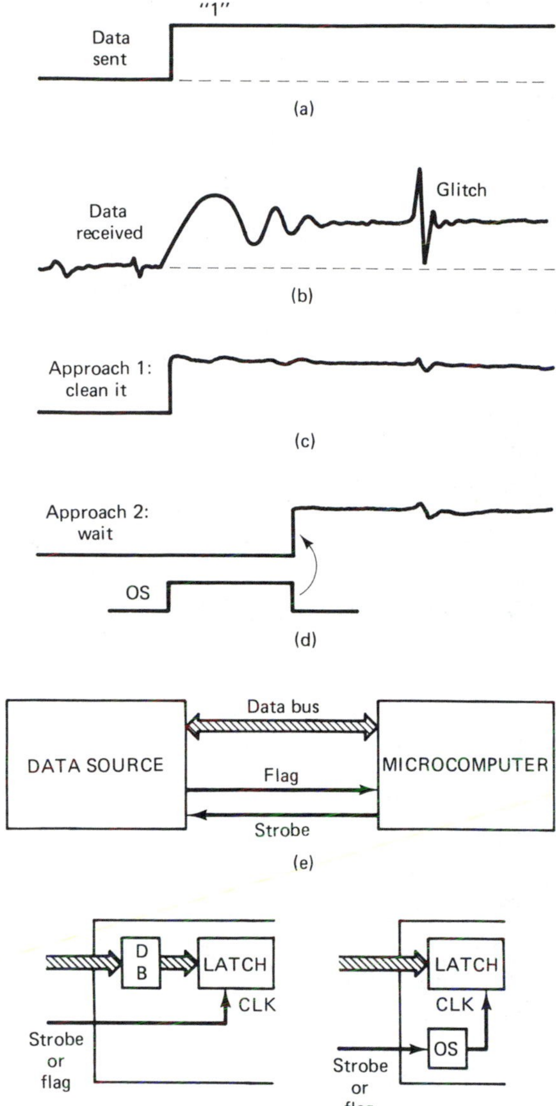

Figure 6.13 Reliable latching of data in a microcomputer interface. Data (a) as sent; (b) as received; (c) after waveshaping; (d) after settling. Microcomputer interface (e) circuit and (f) use of control lines for data latching with waveshaping (DB) or settling (OS).

Consider a parallel interface or *bus* between data source and computer [Fig. 6.13(e)]. In addition to the data bus, there are *control lines*, such as a *strobe* to tell a data source when it should take a data point and a *flag* to let the computer know the data source is ready to act or has data available. Although the term *data source* is used here, the handshaking logic is similar if the peripheral is a listener such as a printer or graphics device. One has the choice [Fig. 6.13(f)] of including debounce circuitry in the interface or using a control line to drive a one-shot to clock the data into a latch after an appropriate settling time.

The waiting approach is simpler if time is available, but when the settling time is comparable to the data rate, information might be lost. This method is best for latching a voltage *level*, but is poor if the data is a pulse whose width or amplitude needs to be characterized. It can also not be used if the settling time is unknown in advance or if the signal is nearly always noisy. Nonetheless, waiting is a simple and adequate solution in most microcomputer interfacing where delays of a few microseconds are tolerable.

The approach of cleaning the data has the advantage of preserving information that tells *when* the event happened. It is the only method that works in pulse experiments where the data are events or counts. If one were to wait to latch the data, the signal would be gone. This approach is also best in real-time signal processing, since it does not delay the measuring system. On the other hand, this method is poor in extremely noisy situations where the noise may reach the logic threshold or deadband width of the debouncing circuitry. Noise amplitude must be carefully estimated when setting the deadband.

With binary data that can be synchronized, the D-type edge flip flop is the usual choice as a data latch for either approach, since it can change its state only once per clock cycle and it allows control over *when* the latching occurs. If a level-triggered JK master–slave FF is used, a noise event might be latched erroneously during the clock high period when the master is listening. On the other hand, the level-triggered JK is the preferred way to latch an asynchronous pulse whose arrival time is unknown.

6.3.2 Comparison of Debounce Circuitry

Here are three common waveshaping problems. It is left to the reader to determine which waveshaping solution is best in each case (see Prob. 6.7), and to compare and contrast the output waveforms.

(1) Switch Contact Bounce. The signal in Fig. 6.14(a) represents a noisy signal coming from contact bounce of a mechanical switch [Fig. 6.14(a)]. For particularly noisy data, there is a one-shot debouncer circuit that waits until the transient has settled, using a one-shot plus flip flop [Fig. 6.15].

(2) Sine Wave Input. One often wants to convert a sine wave input into binary form, preserving timing information such as the zero crossings to measure phase or frequency [Fig. 6.14(b)]. Any noise or variations in signal amplitude may cause undesirable jitter in the output binary waveform.

(3) Square Wave or Pulse Input. A signal that started out as a clean binary waveform may be noisy by the time it reaches the detection circuitry [Fig. 6.14(c)]. The solution will depend upon whether the information of interest is only the number of pulses, or also includes the amplitude, width, or arrival time of the pulse.

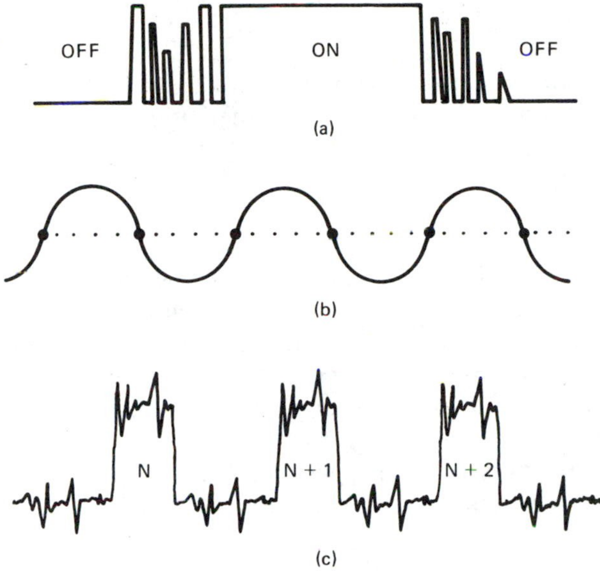

Figure 6.14 Data examples for cleaning up by the waveshaping techniques discussed in this chapter. (a) Switch contact bounce. (b) Sine wave. (c) Noisy pulses. (See Prob. 6.7.)

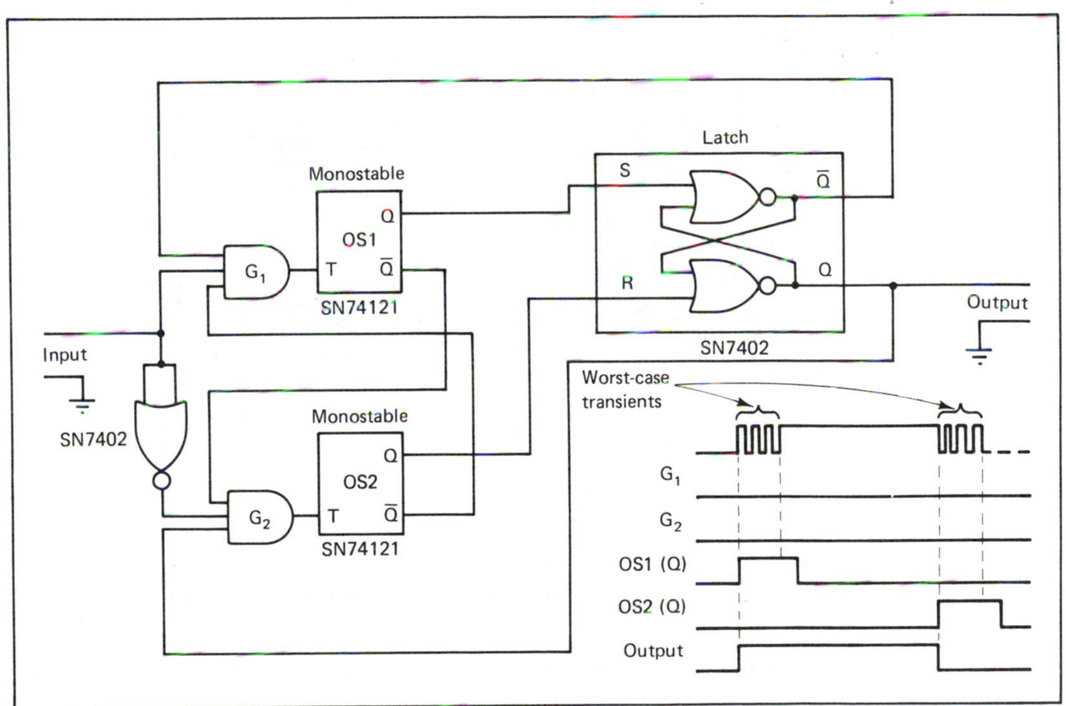

Figure 6.15 Debouncer for data. (*From* C. Strangio, *Electronics.* (Aug. 2 1978): 98)

6.4 DIGITAL INSTRUMENTATION

This section includes some examples illustrating how the digital waveshaping circuits and concepts of this chapter may be applied to instrumentation and measurement.

6.4.1 Measurement of Elapsed Time

It is surprisingly easy to make measurements with digital techniques. The measurement may be made very accurate, involving a comparison against a fixed reference or stable clock frequency rather than relying on the linearity of an analog voltmeter. The measurements may also be made very precise, since digital instrumentation circuits lend themselves to multiple digits with resolution that far exceeds that of an analog meter.

An elapsed time measurement is a typical application [Fig. 6.16]. The object being measured could be as slow as someone running across a room, or as fast as a

Figure 6.16 Digital measurement of time period. (a) Physical situation. (b) Block diagram of measuring electronics. (c) Waveforms.

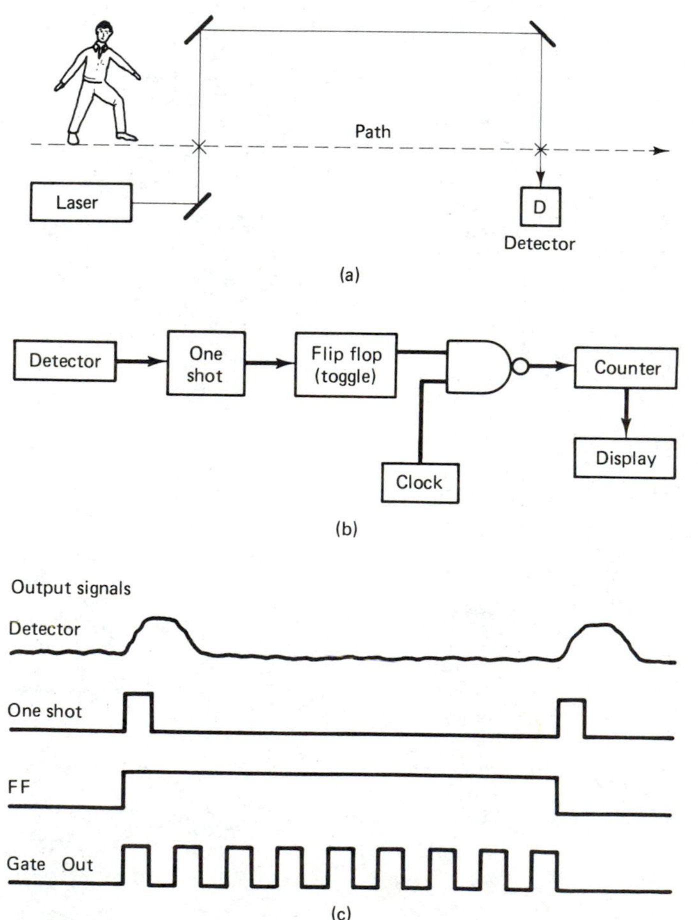

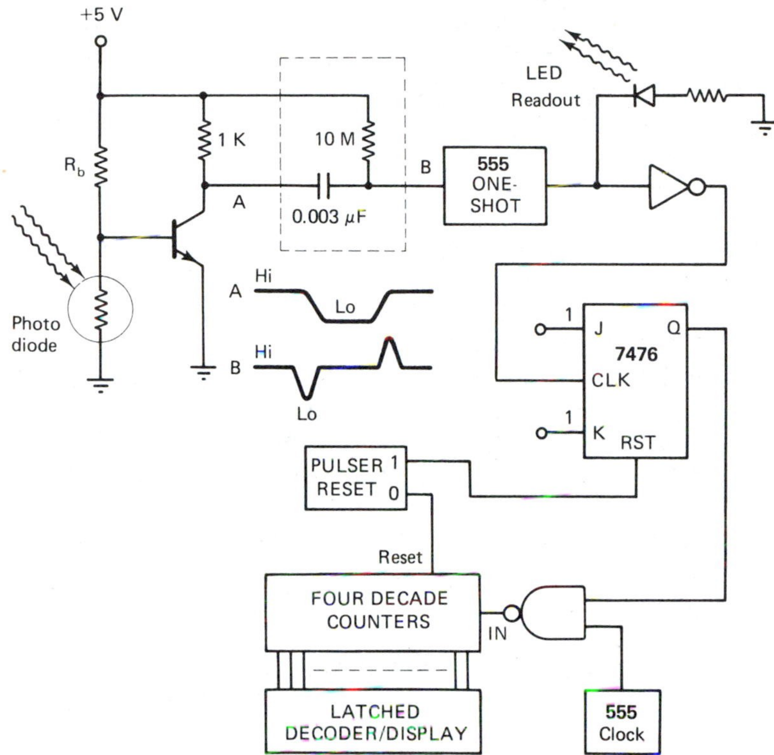

Figure 6.17 Detailed circuit for the digital period measurement of Fig. 6.16.

particle approaching the speed of light. The measurement requires a signal when the event starts and stops, such as breaking a laser beam [Fig. 6.16(a)]. Elapsed time is measured with a gated counter. The reference clock pulses are counted for the time period between laser detector pulses. The first break in the laser signal causes the detector to fire a one-shot, which toggles a flip flop high and thus enables the counter. When the beam is broken a second time, the one-shot toggles the FF again, disabling the counter. The high resolution of digital measurements originates here. If a million pulses pass during the counting period, the time has been measured with 6-digit resolution.

A detailed circuit diagram for an elapsed time measurement is shown in Fig. 6.17. The light signal is converted into an electrical signal by using a photodiode to bias a transistor. For the experiment in Fig. 6.16, the laser beam is normally falling on the photodiode, making it a low resistance. The low voltage on the base turns the transistor off, so point A is high. When the light beam is interrupted, the photodiode becomes a high resistance. The transistor is turned on by bias current through resistor R_b, and the voltage at output A falls. Additional waveshaping is provided by the 10 M Ω, 0.003 μF RC combination in Fig. 6.17, which acts as a high-pass filter or differentiator (see Prob. 6.5), providing a sharp digital pulse to turn on the one-shot. The one-shot generates a positive-going pulse which is inverted to provide a negative-going clock signal to toggle the JK flip flop. Each pulse successively starts and stops the counter through the NAND gate. The multiple-digit counter may be constructed by cascading MSI IC's such as 7490's or using an LSI multidigit counter IC.

6.4.2 Digital Instrumentation of Other Physical Variables

Many other quantities besides time may be measured digitally by conversion to a time interval.

Digital capacitance meter. The capacitance meter shown in Fig. 5.1 illustrates how digital measurement techniques can provide powerful features in a relatively inexpensive commercial instrument. The unknown capacitor is made part of an RC time interval which is then measured. By using a high-frequency reference clock, 3 digits of precision are possible even for small capacitance values. The circuit is autoranging, changing scales as needed. Actual circuit details are beyond the scope of this book, but the concept resembles the 555 timer, sensing when the capacitor voltage crosses two voltage reference thresholds and counting the time interval between them ($T_2 - T_3$ in Fig. 5.1).

Frequency meter or tachometer. This circuit is used in determining the frequency of a motor or other rotating device, displaying the output on a conventional moving-coil meter. The circuit [Fig. 6.18] samples the device under test (for example, spark plug firing or shaft rotation) and first cleans up the signal into binary form. This feeds a one-shot whose output drives a current meter in parallel with a large capacitor. The current meter forms a resistive load for the one-shot, and together with C it averages the asymmetric square wave so the meter reading is the average value of the waveform.

$$I = \frac{V_{cc}}{R_m} \frac{t_W}{t_M} = \text{const} \times f_M \tag{6.5}$$

Here, t_W is the pulse width of the one-shot, while t_M is the period of the motor rotation. The *duty cycle*, or fractional On time of the output waveform, is a linear measure of the unknown frequency. The full-scale range is adjusted using the RC value of the one-shot. The meter reading will be in error if the power supply voltage fluctuates, since the average current follows the one-shot output amplitude even if the duty cycle is unchanged. The error can be eliminated by driving the meter-capacitor combination with a current source or FET current-pump circuit that is turned on whenever its input (in this case the one-shot output) is high.

Simple digitizer or position encoder. A one-shot provides a particularly simple method of digitizing the position of a variable resistor [Fig. 6.19]. There are many ways to convert this analog information into digital form (Chapter 9); however, the one-shot is a particularly useful solution when many resistor values are to be measured. The R value to be digitized sets the output pulse width. When a microcomputer needs to measure a potentiometer setting, it

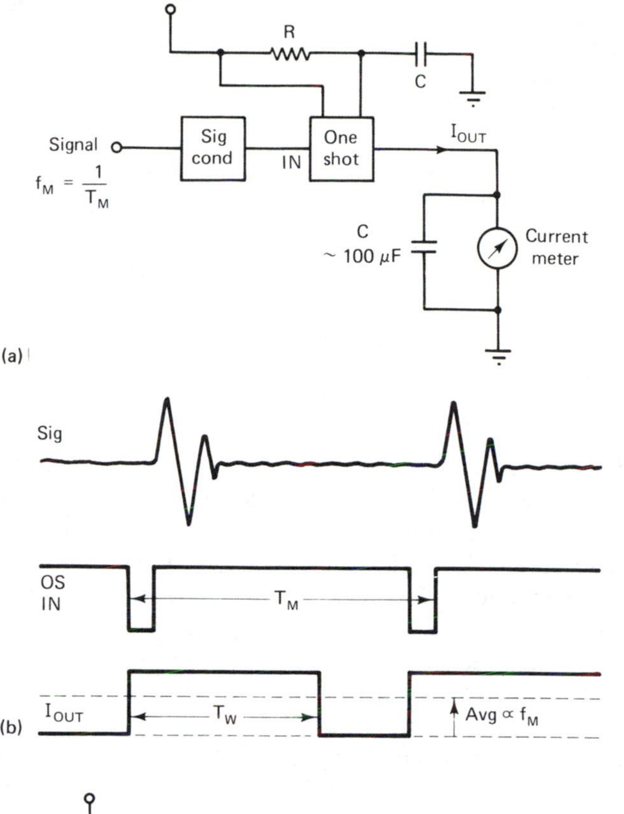

Figure 6.18 Frequency meter or tachometer. (a) Circuit. (b) Waveforms.

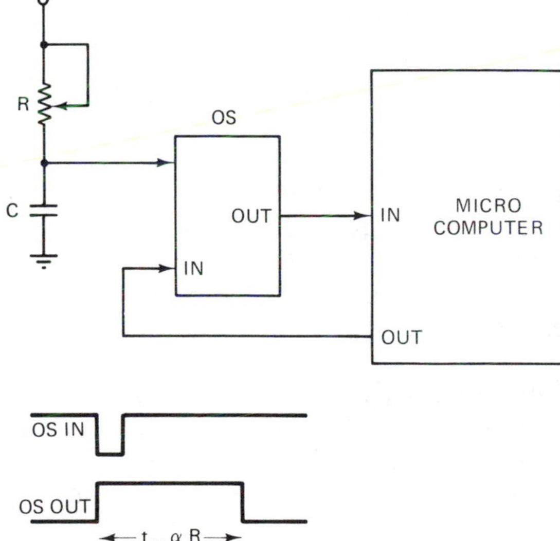

Figure 6.19 Digitizer or position encoder.

issues a pulse to fire the one-shot and starts a software timing loop. A computer input line connected to the one-shot output is enabled to interrupt the program when the pulse ends, giving t_w as a program output. This method is convenient for measuring many independent potentiometers, since one-shots are inexpensive and each interface uses only two wires of a microcomputer I/O port. The accuracy is adequate for many situations, although accuracies in excess of 8 bits demand excessive amounts of the microcomputer's time.

Digital thermometer. Differential or comparison techniques increase measurement resolution when the quantity measured varies over only a small range. Measuring *changes* in a signal also increases the number of usable transducers, since many nonlinear transducers are adequately linear over a small range.

An example is a digital thermometer [Fig. 6.20]. The circuit uses two 555 timers. The first 555 is connected as an astable multivibrator. One of its timing resistors is a temperature-sensing thermistor, selected to have a linear output over the 0° to 100 °C temperature range. The thermistor R_T sets the discharge time R_TC at about 0.01 s. The second resistor is about 100 times bigger, and sets the charging time to 1 s. Output OS 1 becomes an input signal for both the counter and the second 555, which is connected as a one-shot. Its pulse width is adjusted to equal the time interval $t_1 = R_TC$ of OS 1 when the temperature is 0° (selectable either in Celsius or in Fahrenheit). For temperatures above 0°, the thermistor resistance falls, so that $t_1 < t_2$. The counter is reset to zero on the positive edge of OS 1's output after time t_1. The counter then counts until halted by the negative edge of OUT_2 [Fig. 6.20(b)]. The RC combinations at the 555 outputs convert these edges into sharp pulses of the sense appropriate to control the counter cycle. During the long time period $t_H = 1$ s, the counter holds the most recent temperature measurement for easy viewing. An update occurs about once per second, with each measurement requiring only about 0.01 s. The number of counts is a measure of the time interval $t_2 - t_1$, and therefore of the temperature difference from 0°. Circuit parameters have been selected so the counter output reads directly in degrees, with a calibration 1 count = 0.1°.

The next application utilizes one-shots with retriggering capability, particularly useful in decision-making circuits.

Missing pulse and rate-sensitive detectors. If a one-shot is continuously retriggered by pulses with a period t_R slightly shorter than the one-shot pulse width t_w, the output falls only when the input is interrupted for a time longer than t_w. This may be used to sound an alarm. One application of this circuit is burglar alarms: input pulses from an infrared or ultrasonic signal are transmitted across a room, and breaking the beam for a time longer than t_w will cause the alarm to sound. Other applications include monitoring patients in hospital intensive care units with the input derived from a heartbeat EKG signal. The failure of this signal could be used to begin emergency life-saving measures.

Sometimes one needs a circuit to detect the opposite condition, with pulses occurring faster than a given rate. Applications include intensive care monitoring (some heart attacks signal their onset by ultrafast "fibrillation"), but also radiation monitoring, to sound an alarm whenever the pulse rate exceeds a "safe" background level. This circuit may be constructed using one-shots with retriggering capability. Two identical one-shots are used, with identical pulse width t_w that sets the threshold between safe and dangerous count rates [Fig. 6.21]. OS 1 normally falls or *times out* if t_w has passed since the last radiation pulse counted, but is

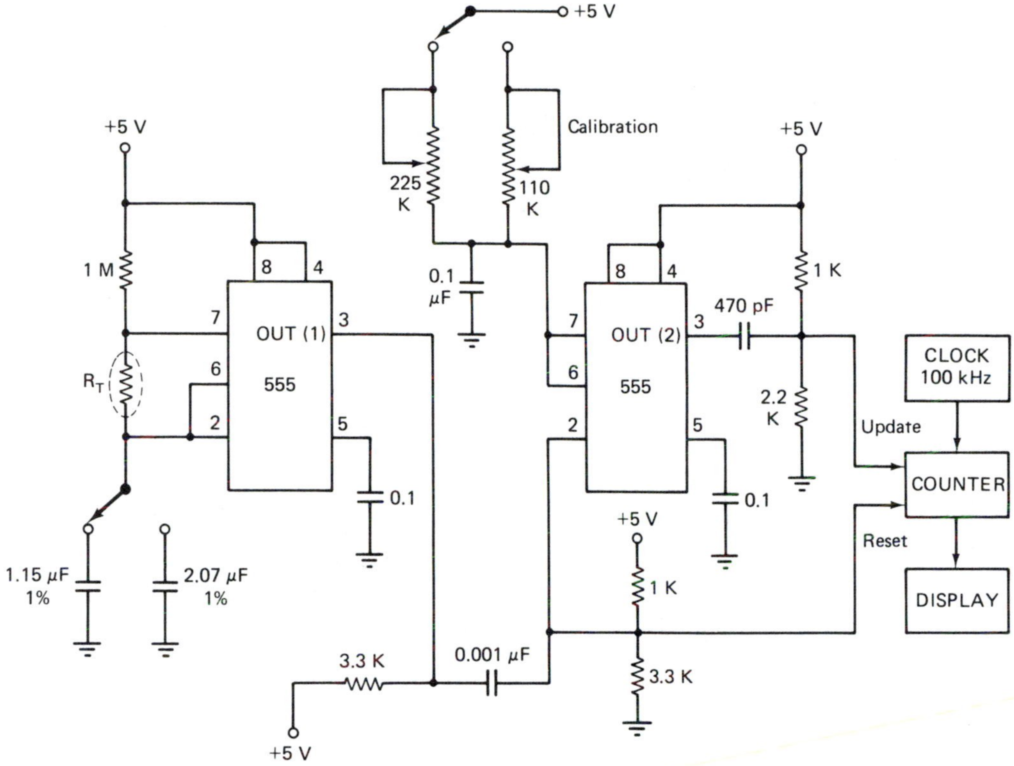

(a)

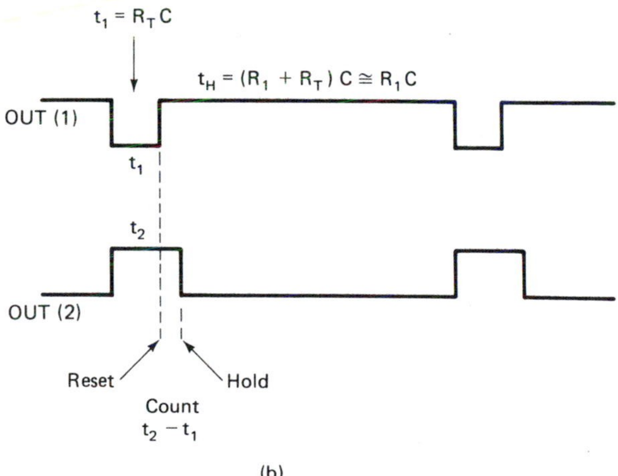

(b)

Figure 6.20 Digital thermometer using 555 timers. (a) Circuit. (b) Waveforms. The linear thermistor is Yellow Springs #44203 or equivalent, $R = 12.2\ \text{K}\ \Omega$ (0°C) and $dR/dT = -127\ \Omega/°\text{C}$. (*Circuit adapted from* D. Lancaster, *TTL Cookbook. Courtesy* Howard W. Sams & Co., Indianapolis, Indiana)

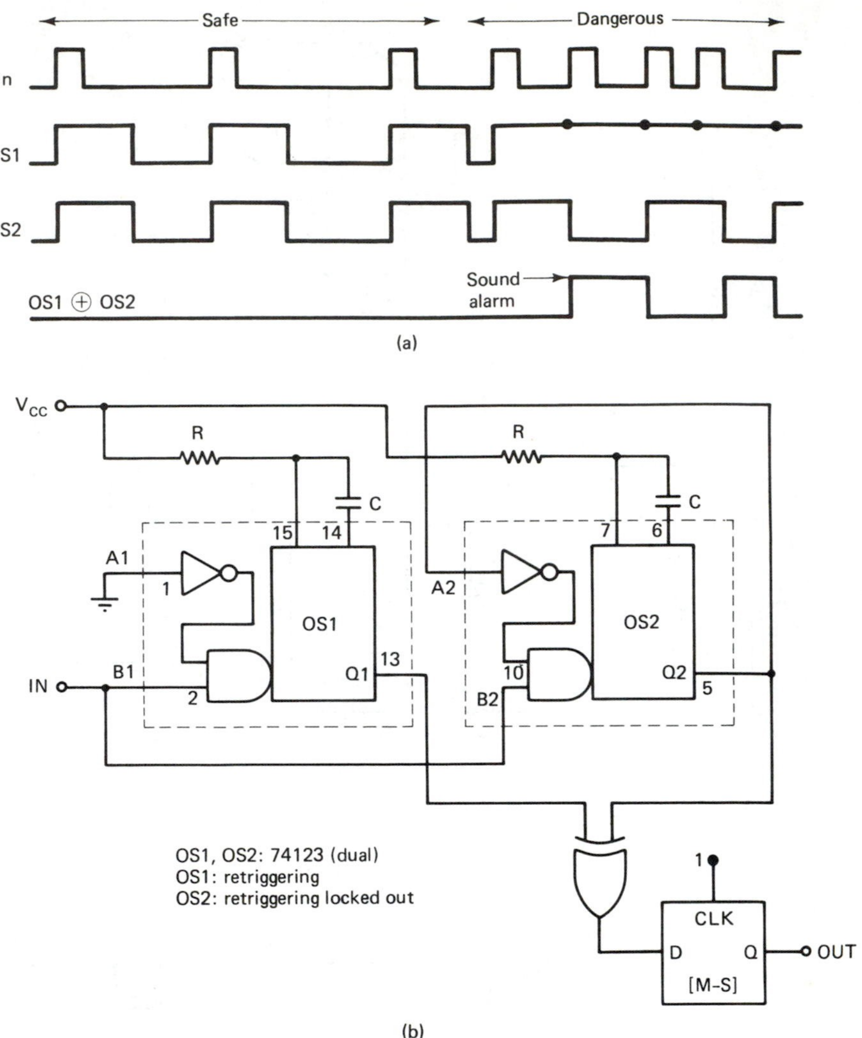

Figure 6.21 Radiation rate monitoring. (a) Pulse stream. (b) Circuit.

retriggered and stays high whenever the count rate becomes dangerous. The retriggering capability of OS 2 is locked out by feedback: connecting output Q2 to input A2 disables any input pulses during the time that OS 2 is firing, and its output therefore ignores many of the radiation pulses [Fig. 6.21(a)]. Both OS outputs become inputs to an exclusive OR gate, so an alarm signal may be generated [Fig. 6.21(b)] when the one-shots disagree. This circuit has a possible bug in it. If the periods of the two RC combinations differ slightly, the XOR output will contain alarm glitches of short duration even during periods of safe count rate. It is straightforward to add circuitry to ignore these short pulses.

PROBLEMS

6.1. Explain the operation of the one-shot of Fig. 6.5(c). Draw waveform diagrams and define gate states as was done in the text for the one-shots of Fig. 6.5(a) and (b).

6.2. For the one-shot circuit of Fig. 6.8(d), explain why an unused A input must be wired low and an unused B input must be wired high. Construct an input state table.

6.3. Which of the one-shots of Fig. 6.8 can be retriggered? Show in each case how retriggering may be locked out using feedback from an output to an unused gate input. First assume the input start pulse comes in on the A input, and then repeat with the B input. No additional gates may be used.

6.4. Describe the operation of the 555 timer as a one-shot using the circuit diagram of Fig. 6.12. Follow the transitions of all circuit components through one cycle of operation and plot waveforms [analogous to Fig. 6.11(b)] at the input, the capacitor, and the output. Explain why the output pulse will not return to the quiescent state if the input is held low past a time RC.

6.5. Explain the operation and show input and output waveforms V_A and V_B of the passive differentiator of Fig. 6.22. Let the input be the falling edge of a normally high signal. This circuit is used to generate short pulses for triggering purposes (e.g., Figs. 6.17 and 6.20). **Hint:** The capacitor is normally uncharged. The lower resistor 2R establishes an output dc level conveniently close to the logic threshold of the device on the right, but may be ignored in the analysis.

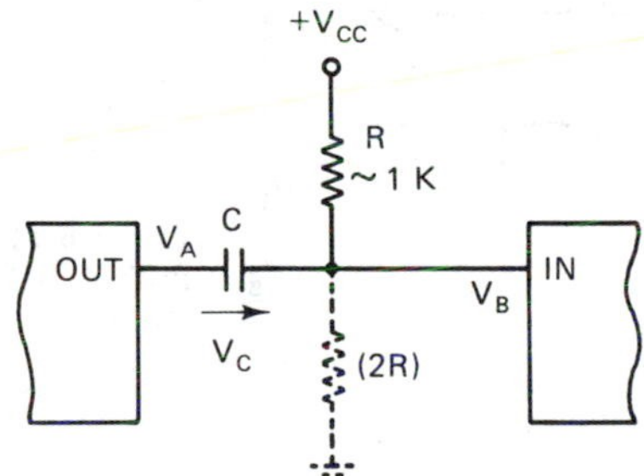

Figure 6.22 Passive waveshaping to turn an edge into a short pulse (see Problem 6.5).

6.6. The three circuits in Fig. 6.23 are known as edge-detectors. The box marked *d* is an inverter with delay-time *d* nanoseconds. Sketch waveform B and the output waveforms C1, C2, and C3 in the spaces provided. Ignore the delay in all gates besides inverter *d*. Which circuit functions as a frequency doubler?

6.7. Compare and contrast waveshaping solutions for converting the signals shown in Fig. 5.14 into clean binary form. Discuss as possible circuits: (a) debouncer; (b) Schmitt trigger; (c) one-shot debouncer [Fig. 5.15]. Display output waveforms to facilitate comparison of solutions.

6.8. The circuit of Fig. 6.24 is a clock constructed from two CMOS gates. Explain its operation, and sketch waveforms at points A, B, and C.

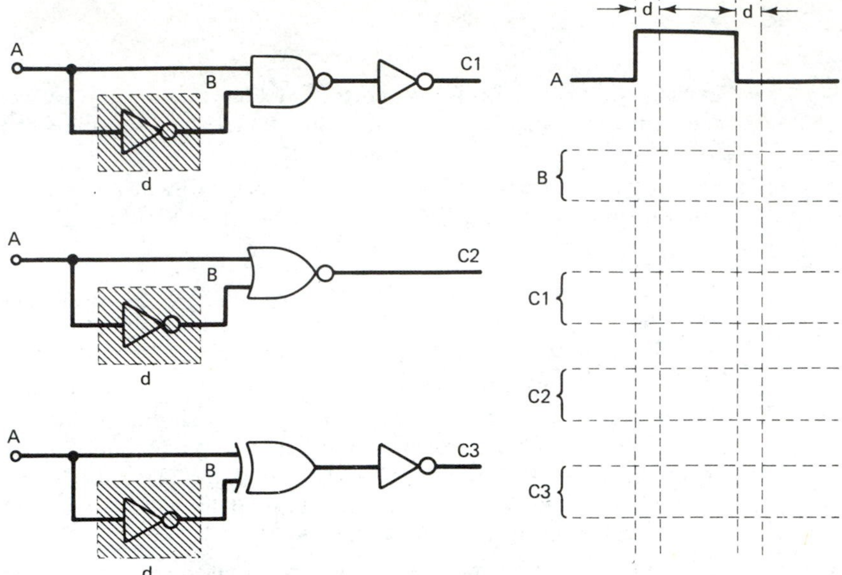

Figure 6.23 Edge-detector circuits (see Prob. 6.6).

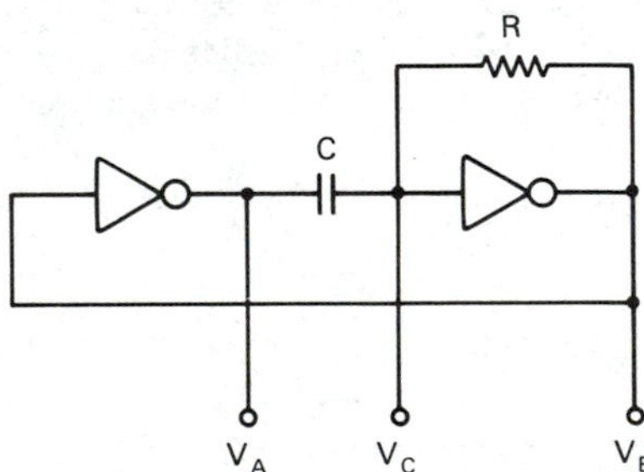

Figure 6.24 Two-gate clock circuit (see Prob. 6.8).

REFERENCES

More complete bibliographic information for the books listed below appears in the annotated bibliography at the end of the book.

BROPHY, *Basic Electronics for Scientists*

DEMPSEY, *Basic Digital Electronics with MSI Applications*

DIEFENDERFER, *Principles of Electronic Instrumentation*

HIGGINS, *Experiments with Integrated Circuits*, Experiments 8 and 9

LANCASTER, *TTL Cookbook*

LANCASTER, *CMOS Cookbook*

LARSEN & RONY, *Logic and Memory Experiments Using TTL Integrated Circuits*

MALMSTADT et al., *Electronic Measurements for Scientists*

7

Memory

7.1 INTRODUCTION: THE OLDER I GET THE MORE I CAN AFFORD TO REMEMBER

Memory is finding its way into applications far beyond computers because of rapidly falling costs and improvements in technology. A typical "smart" terminal may have several thousand (K) words of memory containing an image of the characters on the screen; a laboratory data-gathering instrument may have room for 16K words of information to tell it what to do; the permanently resident programs which control a personal computer, speech synthesizer, digital voltmeter, or electronic music synthesizer may occupy 50K *bytes* (8 bits = 1 byte) of information. The all-in-one microcomputer (CPU, memory, I/O) will fit within a paper clip (Fig. 7.1). This chapter introduces the reader to the newer families of memory devices, providing the background for choosing intelligently between RAM and ROM, and for understanding, from a user's point of view, the distinctions between core, semiconductor, bubble, and charge-coupled device memory types.

The cost of memory has fallen rapidly since the introduction of large-scale integration (LSI) semiconductor random-access memory (RAM) chips, as shown in Fig 7.2. The cost of a given chip falls rapidly after its introduction, because the economy of scale spreads development costs out over a larger number of units, and because the fraction of chips which work, called the *yield*, increases with experience. Newer chip families with more bits per chip are significantly less expensive, because the same amount of silicon crystal is used, and the number of manufacturing process steps is very nearly unchanged. The cost of semiconductor

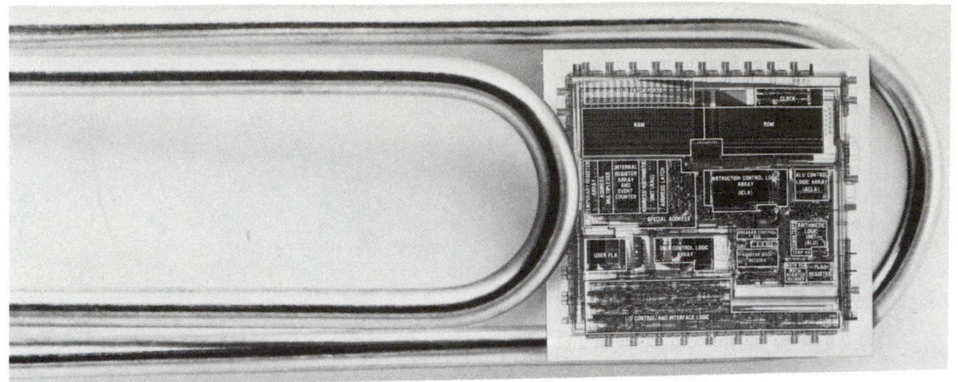

Figure 7.1 A modern microcomputer fits well within the end of a paper clip. (The MAC-4, complete with CPU, memory, and input/output. *Courtesy* Bell Laboratories.)

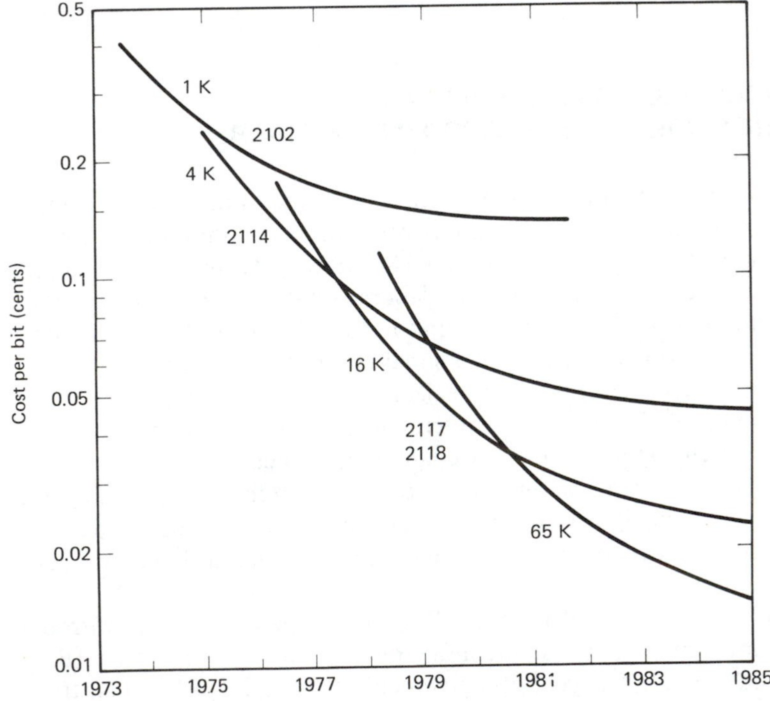

Figure 7.2 The cost per bit of memory is falling with production experience and the evolution of larger chips. (*Source:* Robert N. Noyce, "Microelectronics." Copyright © 1977 by *Scientific American*, 237:69)

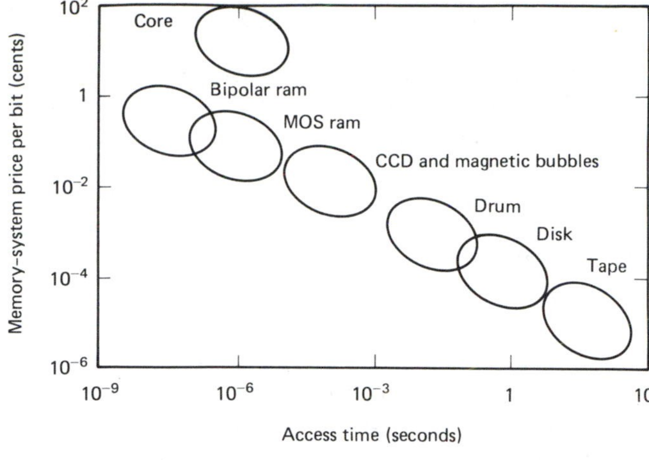

Figure 7.3 Trade-off between access time and price per bit for various memory technologies. The spectrum ranges from bipolar (expensive but good for signal processing) to tape (cheap solution for mass storage). (*Source:* David A. Hodges, "Microelectronic Memories." Copyright © 1977 by *Scientific American*, 237: 144)

RAM is now less than 0.1 cent per bit, and will soon approach 0.01 cent per bit. By contrast, the cost of core-type memory in older computers was much higher: at a typical price of 5 cents per bit, a large timesharing system of the early 70's with 128K words of 36 bits might have had an investment of $150,000 in memory. That same amount of memory purchased now in semiconductor RAM chips would cost $2000 or less. Personal microcomputers now often have more memory than older central computers or large minicomputer installations.

Types of memory devices form a spectrum, with a trade-off between access time and price (Fig. 7.3). Older mass-storage devices such as magnetic tape and disk cover the range of access times from a few milliseconds to minutes. Magnetic tape is slow, because it is inherently a *sequential-access* device, as opposed to a *random-access* device. A sequential access device is like getting from point A to point B by train, but a random-access device is like doing it by car: one can choose the closest route. If a piece of information labeled File N is buried somewhere on a reel of tape [Fig. 7.4(a)], it will take time for the tape to wind to the right posi-

Figure 7.4 Comparison of (a) sequential access device with (b) random access device.

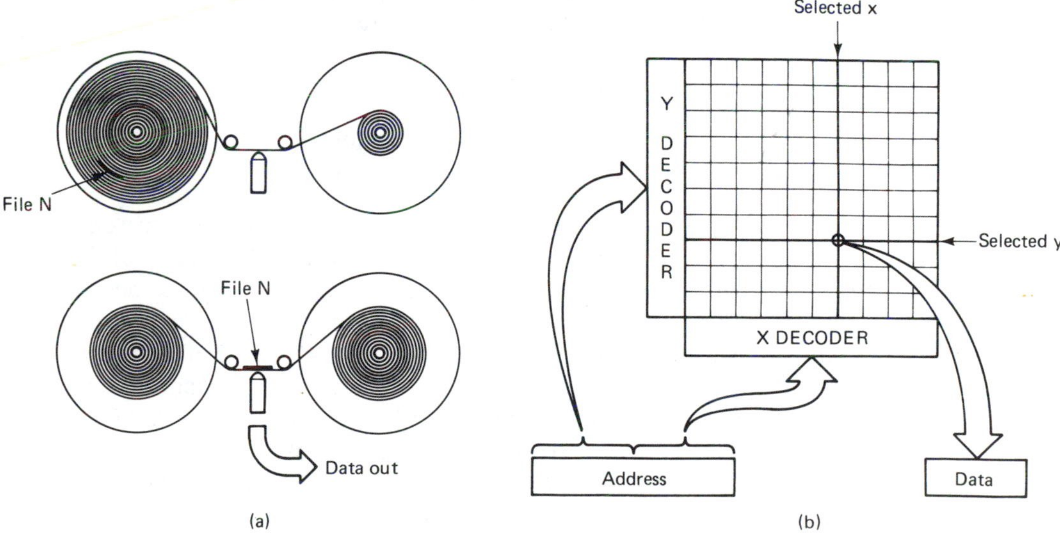

tion for accessing the file. Random-access devices, by contrast, assign to each piece of information a unique address which may be reached without passing sequentially through all of the files in between [Fig. 7.4(b)]. The time required to find and read a specific piece of data does not depend on the location of the file.

7.2 MEMORY TYPES AND TERMINOLOGY

7.2.1 Magnetic Core Memory

Although largely superseded by semiconductor memory, core illustrates many principles of memory organization, and the magnetic storage principle remains in use in the IC version, *magnetic bubble* memory. Core memory is still found in older large computer systems where its *nonvolatile* nature is particularly valuable: the information remains stored even when the power is turned off. For example, in Defense Department and NASA installations, key information must not be lost in a power failure. Core memory uses the storage capability of permanent magnets [Fig. 7.5(a)]. Fine magnetic powder is baked like clay in the shape of a doughnut to keep the magnetic flux lines within the material. This allows the magnetization to be altered by a single wire threading the core. The key feature which makes magnetic memory useful for storage is *remanence:* the material remembers the direction of the magnetizing current I_m even after the current has been removed [Fig. 7.5(b)]. A magnetic core is a two-state storage element, since the material may be in one of two possible states depending upon the previous sign of I_m. Magnetic cores are typically wired into a planar array [Fig. 7.5(c)]. Each core is threaded by wires which uniquely specify its X and Y address and two additional wires, the *sense line* and the *inhibit line*, which thread through all cores in a plane. The delicate assembly process of a core plane is the principal reason that the bit price of core memory has saturated. A given bit is uniquely specified by breaking up the binary address to drive row and column decoder circuits as in Fig. 7.4(b). In an 8-bit example, address 00100011 might specify X-line 3 and Y-line 2.

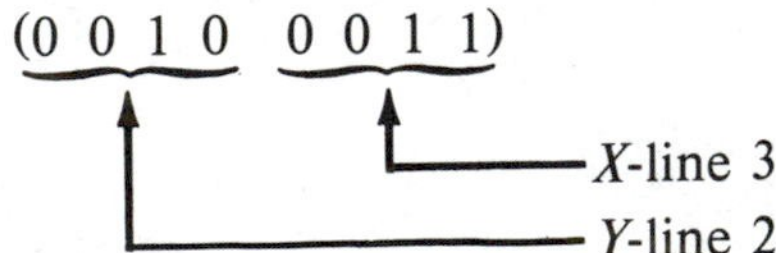

To read information, the selected X and Y lines are each driven by a current $I_m/2$. This alone is not enough to bring cores on the X or Y line past the knee of the hysteresis curve [see Fig. 7.5(b)]. Only the selected bit at the X-Y intersection sees the total magnetizing current I_m which is sufficient to switch that core into the 1 state. Nothing happens if the core was in the 1 state originally, but if the core was in the 0 state, a change in magnetization state occurs. The corresponding change in flux causes an induced voltage which is detected in the sense wire. Thus, the observation of voltage pulse on the sense wire indicates that the selected core used to be in a 0 state. This *read cycle* is destructive, and a complete memory cycle must include a writing phase to restore those bits that were originally in a 0 state.

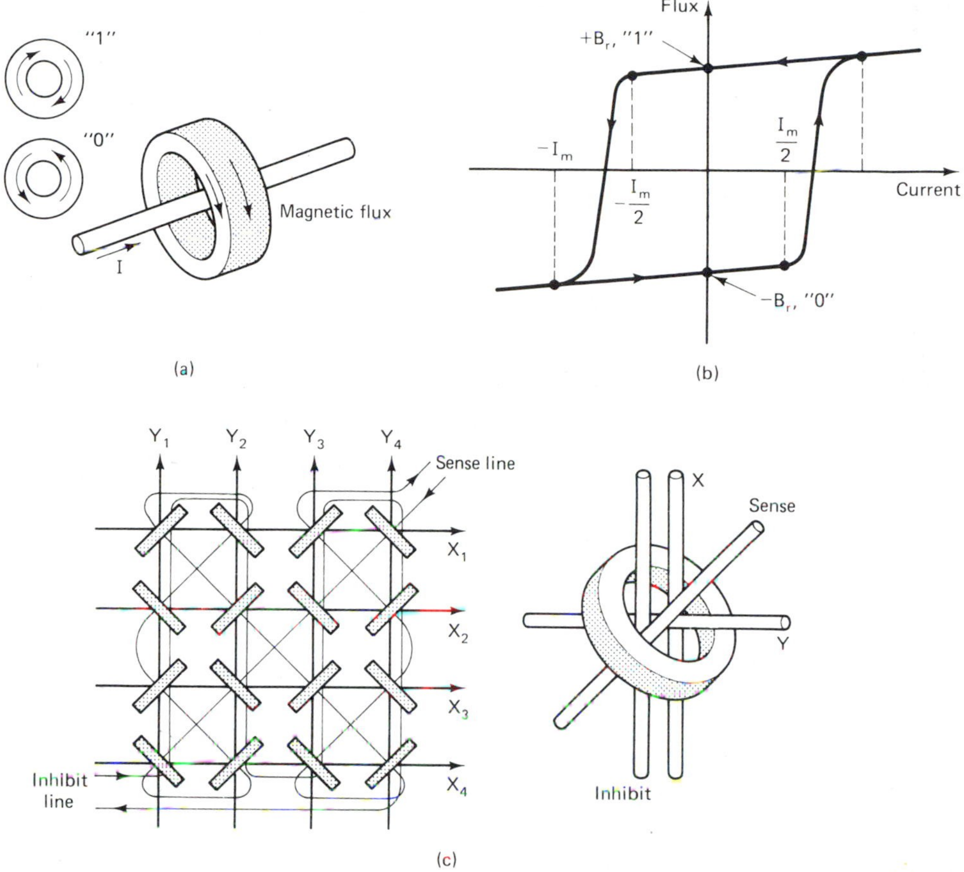

Figure 7.5 Magnetic core storage. (a) The toroid is threaded by current-carrying wires which switch its magnetization state. (b) Hysteresis loop showing critical magnetizing current I_m. (c) Core plane array showing X and Y drive lines, Inhibit line, and voltage Sense line. The same magnetic storage idea is used in bubble memories, though the organization is different and the fabrication is simpler.

The *write cycle* is similar. A current $I_m/2$ is applied to each of the selected X and Y address lines to write a 1 into the selected location. If the selected bit is supposed to have a 0 written in it, a current $- I_m/2$ is fed along the inhibit line to prevent the total current from getting past the knee of the hysteresis curve. To write a 0, a core must have previously been a 0, so the Write cycle must be preceded by a clear.

7.2.2 Bipolar Semiconductor Memory

Since a flip flop can store information, semiconductor memory superseded core memory as soon as it became possible to make many flip flops inexpensively on a single chip. Advantages include economy of scale (Fig. 7.2), increased speed, and simpler application, since the read-process is nondestructive. The major tradeoff

TABLE 7.1 BIPOLAR RAM EXAMPLE SPECIFICATIONS[a]

TTL
LSI

TYPE SN7489
64-BIT READ/WRITE MEMORY

- For Application as a "Scratch Pad" Memory with Nondestructive Read-Out
- Fully Decoded Memory Organized as 16 Words of Four Bits Each
- Fast Access Time . . . 33 ns Typical
- Diode-Clamped, Buffered Inputs
- Open-Collector Outputs Provide Wire-AND Capability
- Typical Power Dissipation . . . 375 mW
- Compatible with Most TTL and DTL Circuits

J OR N DUAL-IN-LINE
OR W FLAT PACKAGE (TOP VIEW)†

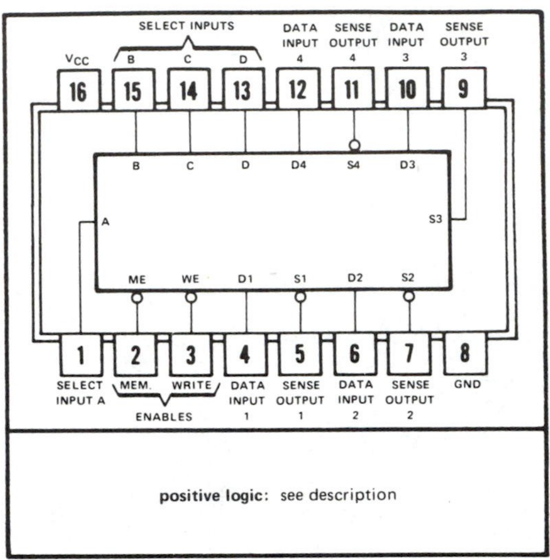

positive logic: see description

†Pin assignments for these circuits are the same for all packages.

description

This 64-bit active-element memory is a monolithic, high-speed, transistor-transistor logic (TTL) array of 64 flip-flop memory cells organized in a matrix to provide 16 words of four bits each. Each of the 16 words is addressed in straight binary with full on-chip decoding.

FUNCTION TABLE

ME	WE	OPERATION	CONDITION OF OUTPUTS
L	L	Write	Complement of Data Inputs
L	H	Read	Complement of Selected Word
H	L	Inhibit Storage	Complement of Data Inputs
H	H	Do Nothing	High

The buffered memory inputs consist of four address lines, four data inputs, a write enable, and a memory enable for controlling the entry and access of data. The memory has open-collector outputs which may be wire-AND connected to permit expansion up to 4704 words of N-bit length without additional output buffering. The open-collector outputs may be utilized to drive external loads directly; however, dynamic reponse of an output can, in most cases, be improved by using an external pull-up resistor in conjunction with a partially loaded output. Access time is typically 33 nanoseconds; power dissipation is typically 375 milliwatts.

write operation

Information present at the data inputs is written into the memory by addressing the desired word and holding both the memory enable and write enable low. Since the internal output of the data input gate is common to the input of the sense amplifier, the sense output will assume the opposite state of the information at the data inputs when the write enable is low.

read operation

The complement of the information which has been written into the memory is nondestructively read out at the four sense outputs. This is accomplished by holding the memory enable low, the write enable high, and selecting the desired address.

TEXAS INSTRUMENTS
INCORPORATED
POST OFFICE BOX 5012 • DALLAS, TEXAS 75222

7489, *courtesy* Texas Instruments Incorporated

is that semiconductor random-access memory (RAM) is *volatile:* the data is lost when the power is turned off. This is not usually a serious problem, since nonvolatile read-only memory (ROM) can be used as a "monitor" to tell a computer what to do when it wakes up in the morning, and data can then be read into RAM from slower nonvolatile storage units such as tape or disk. In *bipolar* RAM, bipolar or junction transistors are used, as opposed to the FET's used in metal-oxide-semiconductor (MOS) memories. Although bipolar RAM's have been superseded by less expensive MOS RAM's in most program and data storage applications, fast bipolar RAM retains its usefulness in real-time signal processing applications and high-speed computing.

An example of a TTL bipolar RAM chip is shown in Table 7.1. This chip offers fast access time (33 ns), but has fairly high power dissipation (375 mW). The cost is high: 2.5 cents per bit, nearly 100 times higher than the cost of LSI MOS RAM. Bipolar memory chips are therefore used only in relatively small scratch-pad memories where rapid speed is essential. For example, a Fourier transform (FT) processor requires many multiplications of a signal $Y(t_n)$ by sine waves:

$$\mathrm{FT}(f) = \sum Y(t_n) \cdot \sin(2\pi f \cdot t_n) \tag{7.1}$$

With values of the sine function in a look-up table (Fig 7.6), the calculation proceeds much faster than if the sine function had to be calculated each time. Bipolar RAM chips retain advantages if the signal processing is to be done in *real time*, i.e., as it is occurring, as in laboratory data acquisition. In real-time waveform synthesis, a laboratory function generator may be replaced by a far more general digital device (Fig. 7.7). A waveform stored in RAM is clocked out by a counter. The real time frequency of the signal is adjusted to arbitrary precision by selecting the value of N in the divider circuit, loading the desired value of the *frequency control word.* The signal is brought to the analog world through a digital-to-analog converter (Chapter 9). The waveform may be changed at will by reloading the look-up table stored in RAM. Bipolar RAM allows fast real time signals to be generated. For example, if a 7489 is used to store a sine wave, the full

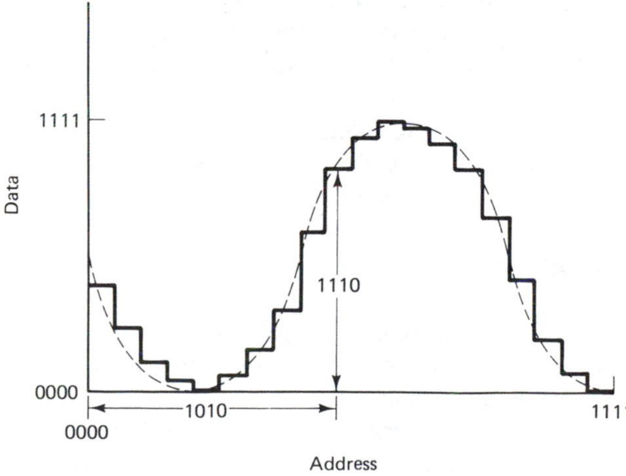

Figure 7.6 Real time application of random access memory. A sine wave is stored in RAM. The data-address array becomes $\sin(2\pi f \cdot t)$ if the address field is scanned by a counter.

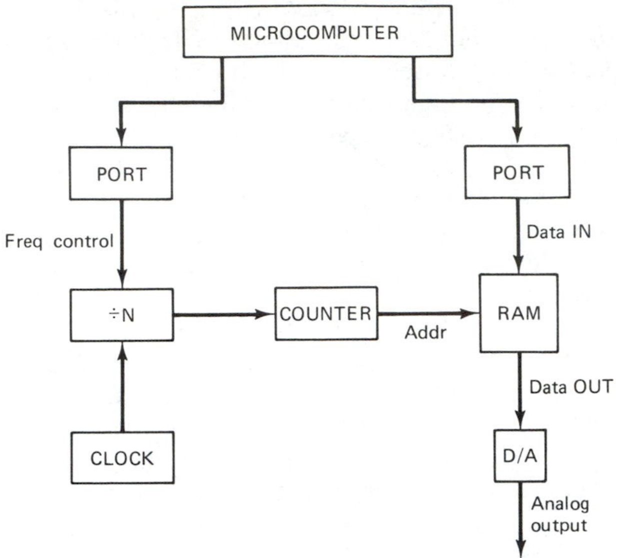

Figure 7.7 Circuit for clocking out a real-time sine wave or other function from RAM storage.

contents of the memory may be strobed out in about 1 μs, allowing real-time waveform generation up to about 50 KHz (see Prob. 7.2).

The internal circuitry of this bipolar RAM chip is shown in Fig. 7.8. Each bit of information is stored in a simple flip-flop cell. The inputs and outputs of all cells are connected in common. Gate arrays act as address decoders to uniquely select a single row containing a 4-bit word. The row of gates at the bottom enable or disable data being written to a selected row. The organization of the RAM chip is shown in block form in Fig. 7.9. A single row selected by address bus decoding is enabled to read or write data. Two other control lines common to nearly all memory chips are the *chip select* ($\overline{\text{CS}}$) and the *write enable* ($\overline{\text{WR}}$). The write enable line determines whether or not a given word in memory will have its contents changed by data present on the data bus. This is sometimes called the *read/write enable* line, to indicate which of the two memory functions is currently being performed. In this case, a low signal allows data to pass into the memory (note the bubble on the front of the write enable buffer in Fig. 7.8). The chip select control line determines whether information stored in the chip may be put on the data output lines, which allows many chips in large memory arrays to share the same data bus. The bar on top ($\overline{\text{CS}}$) designates that a low signal allows a chip to output to the data bus.

7.2.3 Organization of Larger Memories

Chip organization is usually not interchangeable. A chip with sixteen 4-bit words cannot be used to make four 16-bit words, since address and data lines are in general not interchangeable. However, a given chip can be easily stacked with others to make longer address words, longer data words, or both. The following examples, illustrated for a small bipolar RAM chip, are also common to any of the

TYPE SN7489
64-BIT READ/WRITE MEMORY

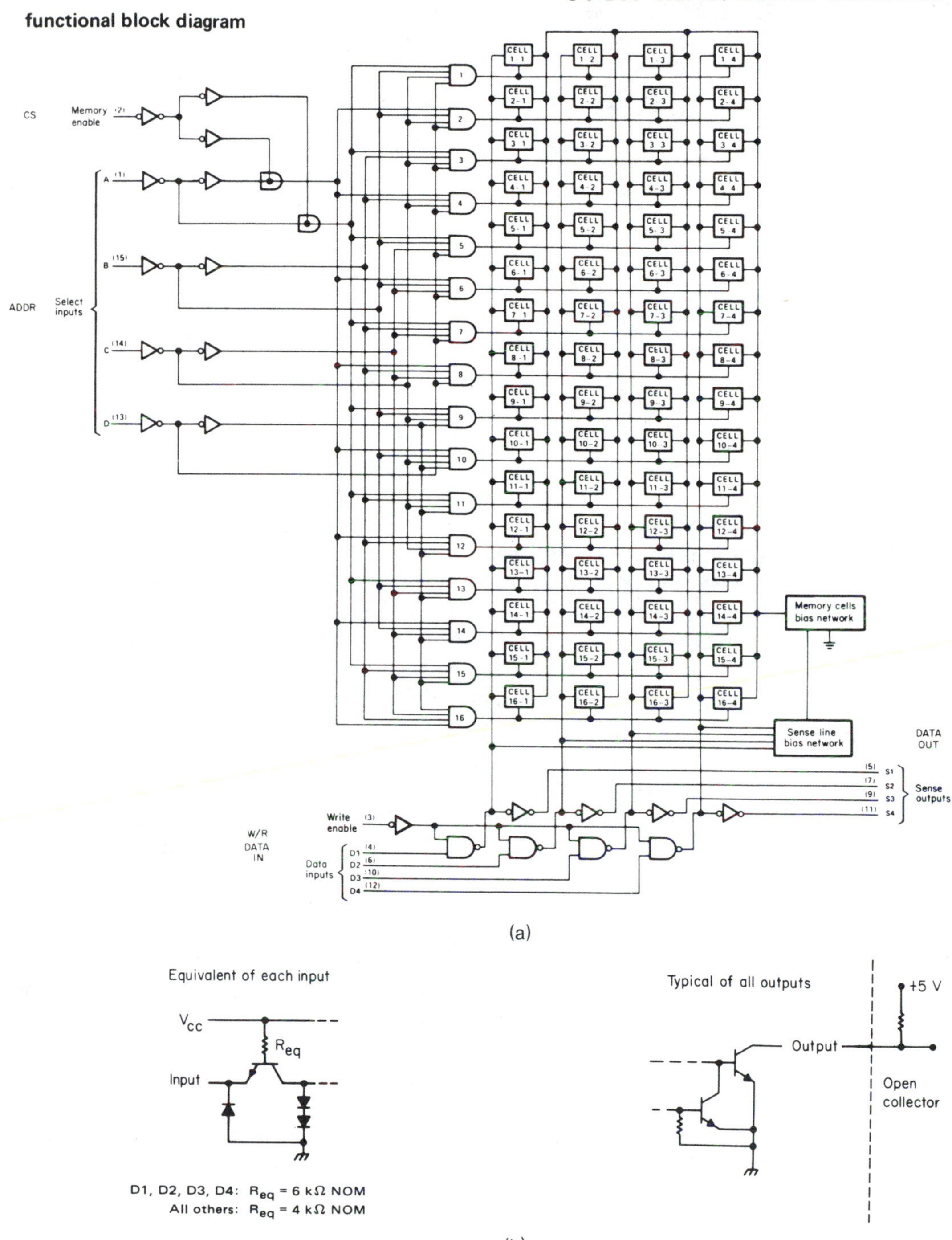

Figure 7.8 Example of a bipolar RAM circuit. (a) Functional block diagram. (b) Schematics of inputs and outputs. (7489, *courtesy* Texas Instruments Incorporated).

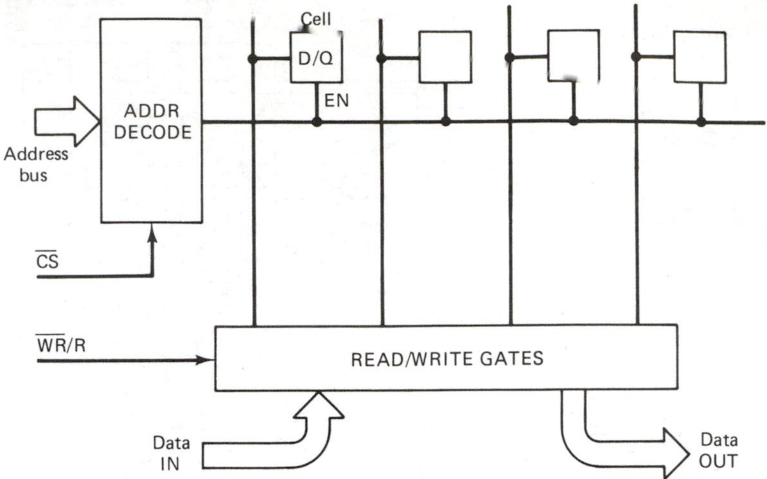

Figure 7.9 Organization of the bipolar RAM of Fig. 7.7.

larger MOS RAM family. Four-bit data has inadequate measurement or computational resolution, but wider data words can be made by stacking chips together as in Fig. 7.10(a). The data word is broken into chunks which fit the word length of the chip itself. Since the devices that put data on the bus have no way of knowing in which chip a given chunk resides, the effect is a memory chip with wider data words. This figure also illustrates the address bus concept: a number of devices are connected in parallel to the same address lines. No indeterminacy results, since

(1) Device inputs can be connected together without difficulty;
(2) A given word on the address bus is intended to select several chips, whose data outputs are brought out separately to make the wider data bus.

An address field wider than 4 bits can be created using extra address bits as inputs to external decoding to uniquely select one chip at a time [Fig. 7.10(b)]. Devices talking to the address bus have no way to know whether the decoding which uniquely selects a single data word is done on the chip or externally. Some address bits select a cell within a chip, while others select a particular chip by *chip select* (CS) lines. This figure illustrates the data bus concept: data lines from a number of devices are connected in parallel to a common data bus. But the outputs of digital devices cannot be connected together without a nonuniqueness problem. If one flip flop is in a 1 output state while another is in a 0 output state, it is not clear which will win when they are connected together on a data bus. Memory device data outputs therefore include a *disconnected mode.* In older TTL RAM chips, this is done using *open-collector* outputs. Address decoding allows only one cell at a time to pull the data bus low. The other transistors are all off, leaving their collectors open. MOS RAM devices ensure one-at-a-time data bus access with *tri-state* outputs enabled by the chip select line.

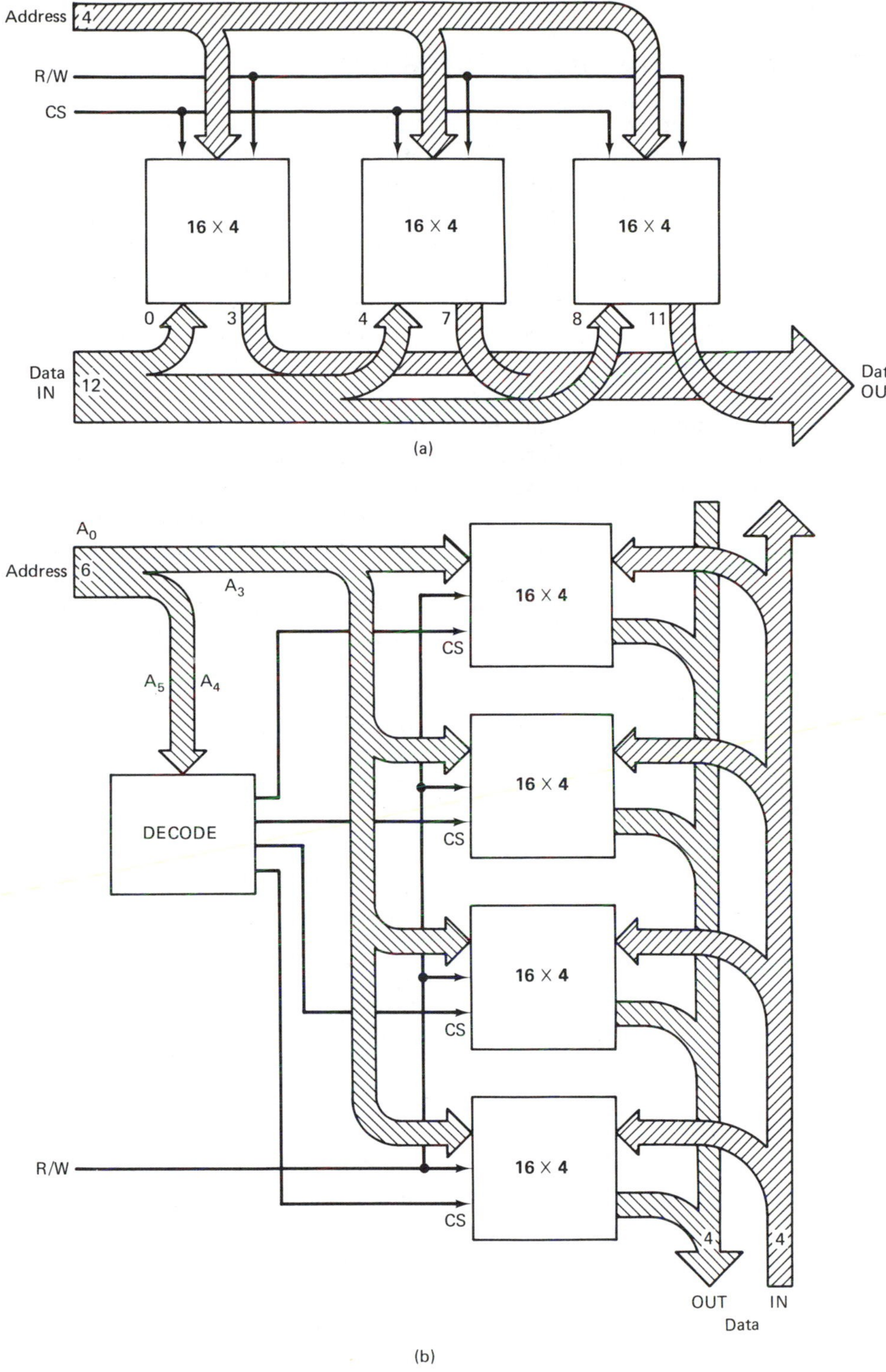

Figure 7.10 Making larger memory blocks. (a) Expanding the data field. (b) Expanding the address field.

7.3 METAL-OXIDE-SEMICONDUCTOR RAM

Dramatic improvements in the availability of small, inexpensive, and reliable semiconductor memory have come with the development of metal-oxide-semiconductor (MOS) RAM. The inside of one type of MOS RAM cell (Fig. 7.11) is a simple flip flop equivalent to two cross-coupled inverters. The gates of transistors Q_3 and Q_4 are permanently biased, forming the equivalent of load resistors. Each of transistors Q_2 and Q_1 forms an inverter, cross-coupled to ensure that when one transistor is in the low resistance state, the other transistor is held in the high resistance state. Transistors Q_5 and Q_6 act as switches to enable measurement or change in the state of the cell. When their channels are in the high resistance state, the cell is effectively isolated from the outside world. Switches Q_5 and Q_6 are closed by the bit select control signal which comes from external decoding of address lines. The cell may then *read* by sensing voltage levels at the flip-flop outputs. The state may be *changed* by forcing a given *Write* line to ground, as with cross-coupled inverter flip flops. This configuration is typical of a *static* RAM cell where information remains stored only as long as the power supply is on.

Another form of MOS RAM cell [Fig. 7.12(a)] stores information as a voltage on a capacitor. Transistor Q_1 acts as a switch connected to capacitor C. Q1 has a low resistance when enabled by the *Write Select* line. The voltage level present on the *Write Data* line is then stored on C. When the Write Select line is not enabled, Q_1 has such a large resistance (10^{10} Ω) that the voltage level on C decays

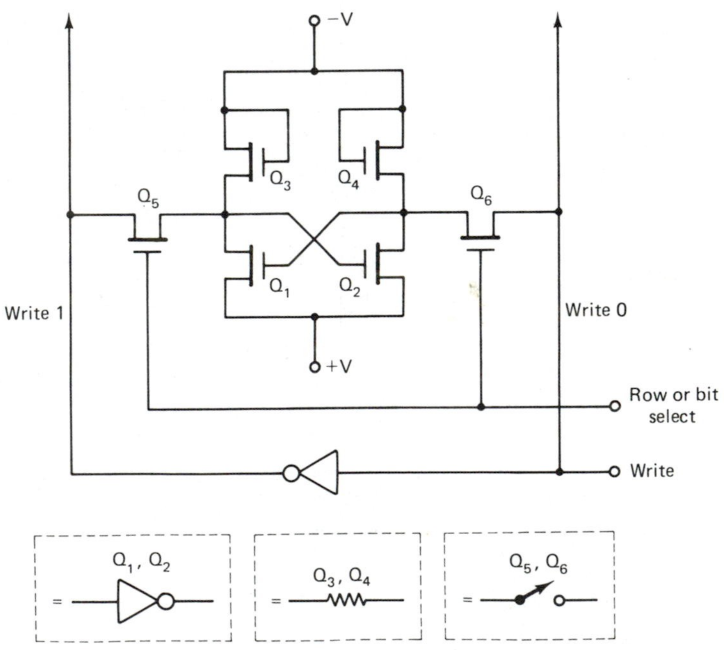

Figure 7.11 Static MOS RAM cell circuit. The six transistors play three different roles.

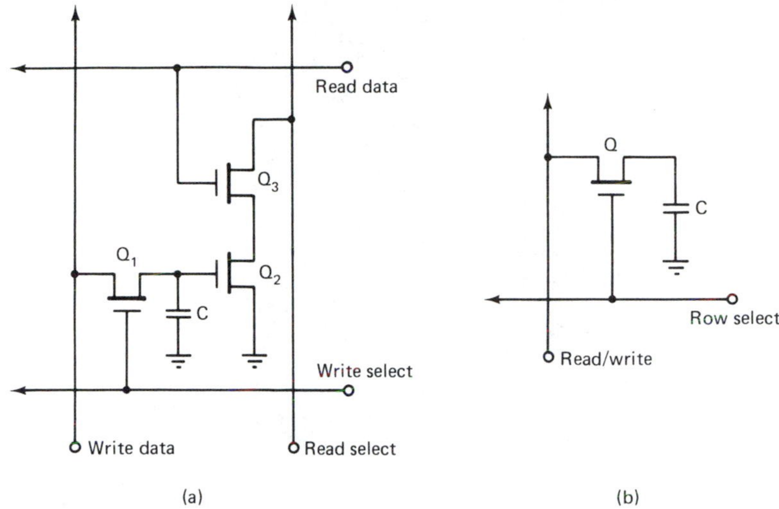

Figure 7.12 Dynamic MOS RAM cells, with information stored on a capacitor. (a) Three transistor cell. (b) One transistor cell.

rather slowly. Since C is also connected to the gate of Q_2, the information stored determines whether Q_2 will be a large or small resistance. Information is read out by opening switch Q_3 using the *Read Select* line. The *Read Data* line is then pulled to ground if Q_2 is conducting, and floats high otherwise. This design uses half the transistors per cell (3 rather than 6) of the standard static RAM cell (Fig. 7.11), so more bits of information can fit on a chip. A disadvantage is that information stored on C decays and must be refreshed periodically. This design is therefore called the *dynamic* RAM. Refresh times of milliseconds are typical, and the process is usually accomplished automatically by additional circuitry built onto a memory board. The inconvenience of refreshing dynamic RAM has not proved to be significant, especially when balanced against further circuit improvements resulting in a one-transistor dynamic RAM cell [Fig. 7.12(b)] which packs more than 10^4 bits onto a single 16-pin chip. In addition to increased memory capacity and decreased cost, increased speed results from higher packing density: the switching speed of this circuit is about 200 ns.

Figure 7.13 shows how the one-transistor dynamic RAM cell of Fig. 7.12(b) is fabricated in an integrated circuit. This is the same 16K bit array shown in Fig. 2.1. The exploded view shows how all of the circuitry is packed into a 15 μm by 30 μm area. The capacitor is also fabricated on the chip: the bottom plate is the substrate itself. Such cells may be packed very densely onto the silicon chip. In Fig. 2.1, 16,384 bits of information are packed onto a chip only 3 mm by 6 mm.

This improvement in technology is resulting in a major advance in readily available computing power: single board computers with up to 64K bytes of storage, minicomputer or microcomputer memory boards with 128K bytes or more, and moderately priced microcomputer systems with more memory than a large central computer of 10 years ago.

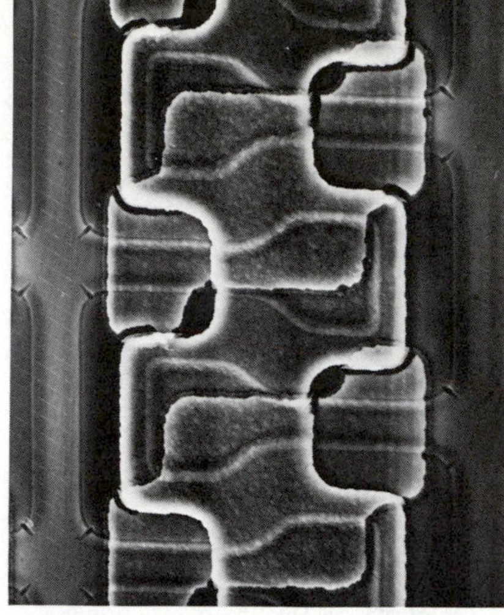

Figure 7.13 Fabrication on an integrated circuit of the one transistor cell of Fig. 7.12. (*Courtesy* W. G. Oldham, University of California at Berkeley)

A comparison of typical examples from various memory families is given in Table 7.2. Newer technologies have made core memory obsolete. Bubble memory and charge-coupled devices retain some advantages in slower speed mass storage devices where semiconductor RAM still is economically unfeasible. Bipolar memory retains the advantage of speed, but as MOS RAM cells continue to shrink, the speed difference is diminishing. MOS devices offer significantly lower power dissipation than bipolar technology and a factor of 100 lower cost. CMOS memory is the best choice where extremely low power dissipation is essential, since it allows nonvolatile storage using a battery backup power supply.

TABLE 7.2 MEMORY COMPARISON TABLE

Logic family	Core	TTL MSI	ECL LSI	MOS LSI	MOS LSI	CMOS LSI
Type	Nonvolatile	RAM	RAM	RAM Static	RAM Dynamic	RAM Static
Size (bits per chip)	1	64	1024	4K	16K	1K
Cost per chip (dollars)		1.75	15	8	10	8
Cost per bit (cents) [a]	5	3	15	0.2	0.06	0.8
Speed (access time)	1 μs	33 ns	24 ns	450 ns	350 ns	650 ns
Power dissipation (per bit)	mW	6 nW	0.5 mW	0.12 mW	0.04 mW	1 nW
Examples	Obsolete	7489	10146	2114	2117	5101

[a]Price is approximate, intended for comparisons only, and will vary greatly with number of chips purchased, the source, and date of purchase.

7.3.1 Organization of Typical MOS RAM Chips

An early example of MOS RAM, the 2102, is shown in Table 7.3. The internal organization is (10 address bits) × (1 data bit). Multiple chips are stacked together following the ideas of Fig. 7.8(a) to make a data word. The 10-bit address word is

TABLE 7.3 EARLY MOS RAM SPECIFICATIONS[a]

Silicon Gate MOS 2102

1024 BIT FULLY DECODED STATIC MOS RANDOM ACCESS MEMORY

- **Single +5 Volts Supply Voltage**
- **Directly TTL Compatible — All Inputs and Output**
- **Static MOS — No Clocks or Refreshing Required**
- **Low Power — Typically 150 mW**
- **Access Time — Typically 500 nsec**
- **Three-State Output — OR-Tie Capability**
- **Simple Memory Expansion — Chip Enable Input**
- **Fully Decoded — On Chip Address Decode**
- **Inputs Protected — All Inputs Have Protection Against Static Charge**
- **Low Cost Packaging — 16 Pin Plastic Dual-In-Line Configuration**

The Intel® 2102 is a 1024 word by one bit static random access memory element using normally off N-channel MOS devices integrated on a monolithic array. It uses fully DC stable (static) circuitry and therefore requires no clocks or refreshing to operate. The data is read out nondestructively and has the same polarity as the input data.

The 2102 is designed for memory applications where high performance, low cost, large bit storage, and simple interfacing are important design objectives.

It is directly TTL compatible in all respects: inputs, output, and a single +5 volt supply. A separate chip enable ($\overline{CE}$) lead allows easy selection of an individual package when outputs are OR-tied.

The Intel 2102 is fabricated with N-channel silicon gate technology. This technology allows the design and production of high performance easy to use MOS circuits and provides a higher functional density on a monolithic chip than either conventional MOS technology or P-channel silicon gate technology.

Intel's silicon gate technology also provides excellent protection against contamination. This permits the use of low cost silicone packaging.

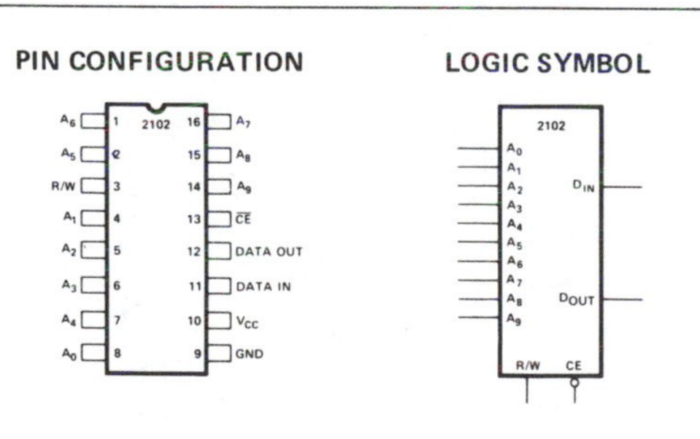

PIN NAMES

D_{IN}	DATA INPUT	$\overline{CE}$	CHIP ENABLE
$A_0 - A_9$	ADDRESS INPUTS	D_{OUT}	DATA OUTPUT
R/W	READ/WRITE INPUT	V_{CC}	POWER (+5V)

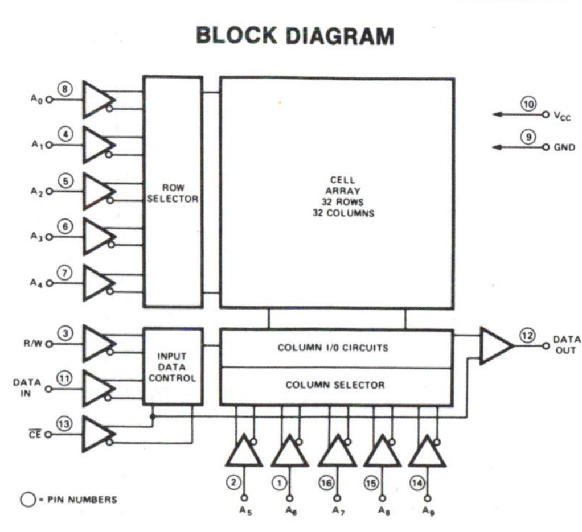

[a]2102, *courtesy* Intel Corporation

TABLE 7.4 MOS STATIC RAM EXAMPLE SPECIFICATIONS[a]

2114
1024 X 4 BIT STATIC RAM

	2114-2	2114-3	2114	2114L2	2114L3	2114L
Max. Access Time (ns)	200	300	450	200	300	450
Max. Power Dissipation (mw)	525	525	525	370	370	370

- **High Density 18 Pin Package**
- **Identical Cycle and Access Times**
- **Single +5V Supply**
- **No Clock or Timing Strobe Required**
- **Completely Static Memory**
- **Directly TTL Compatible: All Inputs and Outputs**
- **Common Data Input and Output Using Three-State Outputs**
- **Pin-Out Compatible with 3605 and 3625 Bipolar PROMs**

The Intel® 2114 is a 4096-bit static Random Access Memory organized as 1024 words by 4-bits using N-channel Silicon-Gate MOS technology. It uses fully DC stable (static) circuitry throughout — in both the array and the decoding — and therefore requires no clocks or refreshing to operate. Data access is particularly simple since address setup times are not required. The data is read out nondestructively and has the same polarity as the input data. Common input/output pins are provided.

The 2114 is designed for memory applications where high performance, low cost, large bit storage, and simple interfacing are important design objectives. The 2114 is placed in an 18-pin package for the highest possible density.

It is directly TTL compatible in all respects: inputs, outputs, and a single +5V supply. A separate Chip Select ($\overline{CS}$) lead allows easy selection of an individual package when outputs are or-tied.

The 2114 is fabricated with Intel's N-channel Silicon-Gate technology — a technology providing excellent protection against contamination permitting the use of low cost plastic packaging.

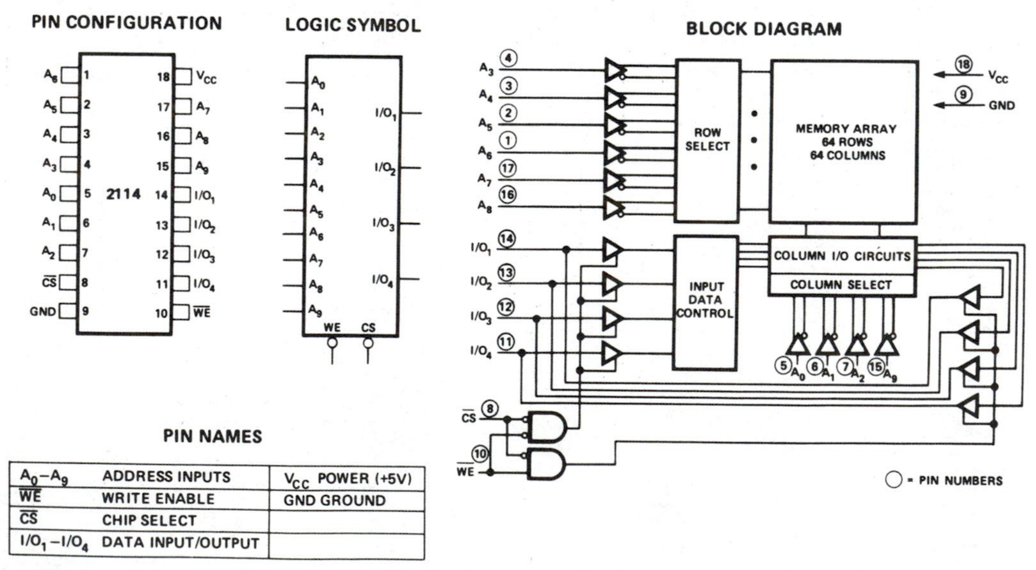

A_0–A_9	ADDRESS INPUTS	V_{CC}	POWER (+5V)
$\overline{WE}$	WRITE ENABLE	GND	GROUND
$\overline{CS}$	CHIP SELECT		
I/O_1–I/O_4	DATA INPUT/OUTPUT		

[a]2114, *courtesy* Intel Corporation

broken up internally so that 5 bits select rows and 5 bits select columns. The internal organization of the cell keeps data input and output lines separate as in older bipolar designs. The data output line includes a tri-state buffer to facilitate wiring many chips to a data bus.

More modern MOS memory chips increase the number of bits per package by replacing separate input and output lines with a single *bidirectional* data I/0 line, using tri-state logic plus I/O control signals. This is possible in memory chips because no instruction tries to read and write to the chip at the same time. Control signals and decoding allow a device to write data via the bus onto the RAM chip. A change in the control signals allows the RAM data to be read by another device via the same I/O lines on the same data bus. An example of a MOS static RAM chip is shown in Table 7.4. Tri-state logic allows data input and output lines to share common wires, as shown in simplified form in Fig. 7.14.

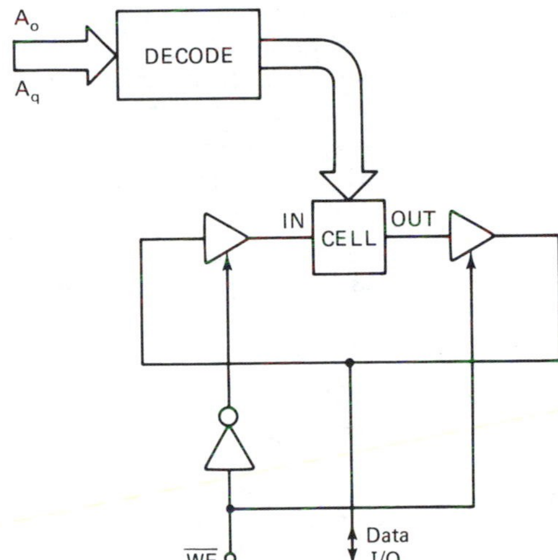

Figure 7.14 Bidirectional data I/O lines on a memory chip using tri-state logic.

Still larger RAM's multiplex 2 address bits on a single pin, which results in very effective use of "real estate" on a memory board, since 16K bits of information are packed into a single 16-pin package. An example is shown in Table 7.5. During a memory cycle, the address lines first contain row and then column information, with a latch to retain the row information. Two external strobe signals, derived from two phases of the system clock, are therefore required. The multiplexing of addresses from an unmultiplexed bus requires external tri-state logic.

7.3.2 CMOS Memory

Although still less common in microcomputer systems than MOS memory, CMOS memory has unique low power demands and will become more popular as costs fall and speed increases. An example of a CMOS memory chip is described in Table 7.6. Since a typical MOS microcomputer system needs several amps of current, only CMOS is feasible for a remote or portable system powered by bat-

TABLE 7.5 MOS DYNAMIC RAM EXAMPLE SPECIFICATIONS[a]

2117 FAMILY
16,384 x 1 BIT DYNAMIC RAM

	2117-2	2117-3	2117-4
Maximum Access Time (ns)	150	200	250
Read, Write Cycle (ns)	320	375	410
Read-Modify-Write Cycle (ns)	330	375	475

- **Industry Standard 16-Pin Configuration**
- **±10% Tolerance on All Power Supplies: +12V, +5V, -5V**
- **Low Power: 462mW Max. Operating, 20mW Max. Standby**
- **Low I_{DD} Current Transients**
- **All Inputs, Including Clocks, TTL Compatible**
- **Non-Latched Output is Three-State, TTL Compatible**
- **$\overline{RAS}$ Only Refresh**
- **128 Refresh Cycles Required Every 2ms**
- **Page Mode Capability**
- **$\overline{CAS}$ Controlled Output Allows Hidden Refresh**

The Intel® 2117 is a 16,384 word by 1-bit Dynamic MOS RAM fabricated with Intel's standard two layer polysilicon NMOS technology — a production proven process for high performance, high reliability, and high storage density.

The 2117 uses a single transistor dynamic storage cell and advanced dynamic circuitry to achieve high speed with low power dissipation. The circuit design minimizes the current transients typical of dynamic RAM operation. These low current transients and ±10% tolerance on all power supplies contribute to the high noise immunity of the 2117 in a system environment.

Multiplexing the 14 address bits into the 7 address input pins allows the 2117 to be packaged in the industry standard 16-pin DIP. The two 7-bit address words are latched into the 2117 by the two TTL clocks, Row Address Strobe ($\overline{RAS}$) and Column Address Strobe ($\overline{CAS}$). Non-critical timing requirements for $\overline{RAS}$ and $\overline{CAS}$ allow use of the address multiplexing technique while maintaining high performance.

The 2117 three-state output is controlled by $\overline{CAS}$, independent of $\overline{RAS}$. After a valid read or read-modify-write cycle, data is latched on the output by holding $\overline{CAS}$ low. The data out pin is returned to the high impedance state by returning $\overline{CAS}$ to a high state. The 2117 hidden refresh feature allows $\overline{CAS}$ to be held low to maintain latched data while $\overline{RAS}$ is used to execute $\overline{RAS}$-only refresh cycles.

The single transistor storage cell requires refreshing for data retention. Refreshing is accomplished by performing $\overline{RAS}$-only refresh cycles, hidden refresh cycles, or normal read or write cycles on the 128 address combinations of A_0 through A_6 during a 2ms period. A write cycle will refresh stored data on all bits of the selected row except the bit which is addressed.

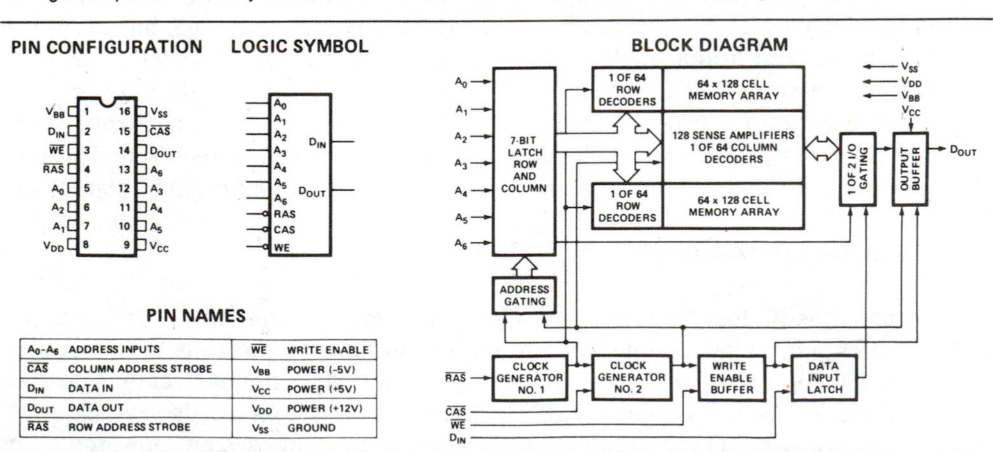

A_0-A_6	ADDRESS INPUTS	$\overline{WE}$	WRITE ENABLE
$\overline{CAS}$	COLUMN ADDRESS STROBE	V_{BB}	POWER (-5V)
D_{IN}	DATA IN	V_{CC}	POWER (+5V)
D_{OUT}	DATA OUT	V_{DD}	POWER (+12V)
$\overline{RAS}$	ROW ADDRESS STROBE	V_{SS}	GROUND

[a]2117, *courtesy* Intel Corporation. See also the 2118, which requires only a single 5-V power supply.

5101 FAMILY
256 x 4 BIT STATIC CMOS RAM

P/N	Typ. Current @ 2V (μA)	Typ. Current @ 5V (μA)	Max Access (ns)
5101L	0.14	0.2	650
5101L-1	0.14	0.2	450
5101L-3	0.70	1.0	650

- **Single +5V Power Supply**
- **Ideal for Battery Operation (5101L)**
- **Directly TTL Compatible: All Inputs and Outputs**
- **Three-State Output**

The Intel® 5101 is an ultra-low power 1024-bit (256 words X 4 bits) static RAM fabricated with an advanced ion-implanted silicon gate CMOS technology. The device has two chip enable inputs. Minimum standby current is drawn by this device when CE2 is at a low level. When deselected the 5101 draws from the single 5-volt supply only 10 microamps. This device is ideally suited for low power applications where battery operation or battery backup for non-volatility are required.

The 5101 uses fully DC stable (static) circuitry; it is not necessary to pulse chip select for each address transition. The data is read out non-destructively and has the same polarity as the input data. All inputs and outputs are directly TTL compatible. The 5101 has separate data input and data output terminals. An output disable function is provided so that the data inputs and outputs may be wire OR-ed for use in common data I/O systems.

The 5101L has the additional feature of guaranteed data retention at a power supply voltage as low as 2.0 volts.

A pin compatible N-channel static RAM, the Intel® 2101A, is also available for low cost applications where a 256 X 4 organization is needed.

The Intel ion-implanted, silicon gate, Complementary MOS (CMOS) process allows the design and production of ultra-low power, high performance memories.

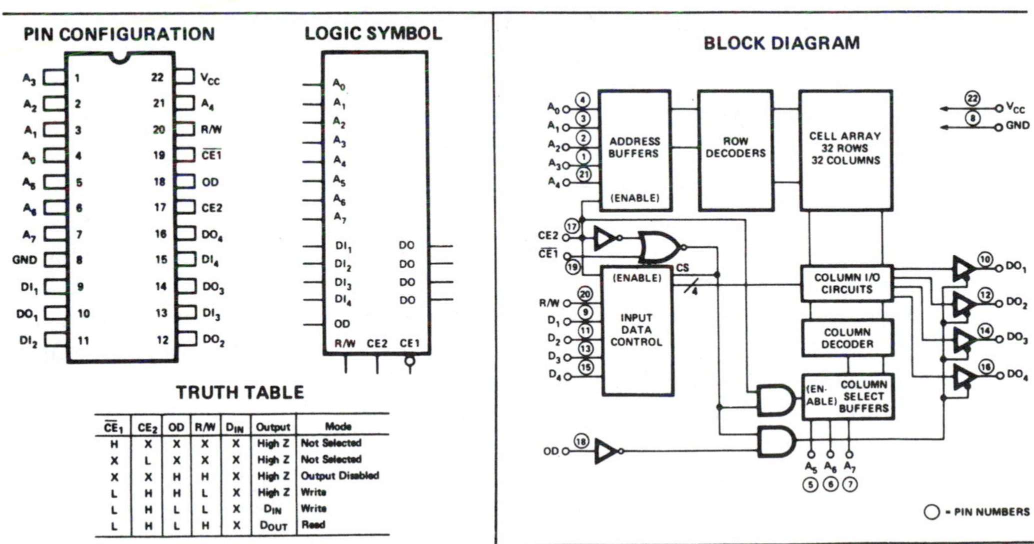

$\overline{CE}_1$	CE_2	OD	R/W	D_{IN}	Output	Mode
H	X	X	X	X	High Z	Not Selected
X	L	X	X	X	High Z	Not Selected
X	X	H	H	X	High Z	Output Disabled
L	H	H	L	X	High Z	Write
L	H	L	L	X	D_{IN}	Write
L	H	L	H	X	D_{OUT}	Read

[a](5101, *courtesy* Intel Corporation)

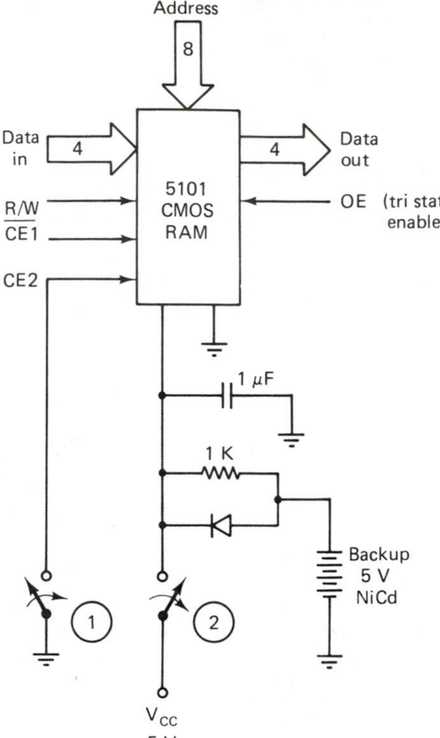

Figure 7.15 CMOS memory with battery-powered backup system.

teries. Several CMOS microprocessors exist, for example, the RCA 1800 series. Even for microcomputers containing an MOS processor, CMOS memory plays a role in fail-safe battery backup memory systems which retain programs and data in the event of a power failure. An example is shown in Fig. 7.15.

7.4 READ-ONLY MEMORIES

7.4.1 ROM, Uses and Types

A read-only memory (ROM) contains permanently stored information which can be read but not altered. A familiar example of ROM-like action is present in any 7-segment numerical display. A decoder translates the 4-bit binary input into a code which lights up the appropriate segments for the corresponding decimal or hexadecimal character. Thus, the 7447 decoder acts as a kind of ROM. Although these MSI decoders are actually specific gate arrays, permanent ROM memory makes designing wired gate arrays unnecessary. If the desired function can be written as a truth table, one can use a ROM instead of designing a gate array.

ROM is also used to replace RAM in signal processing applications. Because ROM is nonvolatile, a stored function will not be lost when the power is turned off. The amount of storage required depends upon the amplitude and frequency resolution needed. For example, consider a digital sine wave generator using ROM in a look-up table. The ROM size requirement is reduced somewhat using the symmetry of a sine wave: only one of four quadrants needs to be stored (see Fig. 7.6), since the full wave can be reconstructed using simple external logic (Prob. 7.3). A 16K ROM is adequate to provide 1% amplitude resolution and 1% frequency resolution:

Amplitude resolution	1 in 64	6 bits wide
Frequency resolution	1 in 1024	+ 10 bits wide
Sine wave symmetry	1 of 4 quadrants	− 2 bits wide
Address + data bus		14 bits wide
ROM storage	$2^{14} = 16\text{K}$ bits	

Here, it is sufficient by symmetry to store only ¼ cycle of the sine wave. Without this, 64K bits would be required rather than 16K bits.

ROM's are also used in calculators, although for trigonometric calculation it is not the function itself that is stored, but rather the method or *algorithm* for calculating a sine wave based on a power series approximation. A ROM is also used in *microprogrammed* computers, which retain the instruction set of an older but popular computer and execute it in a new and faster way. The actual computer, which is constructed using the best available technology, may not resemble the original either in architecture or in instruction set. A read-only memory translates the instructions of the older computer into the actual instructions needed, and another ROM translates the results back into the older format. Such *emulation* of an older machine by a newer one can save many person-years of software investment.

A ROM is present in the *monitor* of any microcomputer and most newer minicomputers to remember permanently the machine's basic operations. A monitor includes not only instructions for reading characters from a teletype or terminal, but software debugging features, and very likely a high-level-language translator such as BASIC. With the development of *programmable* but *erasable* ROM's called *EPROM's*, information may be "permanently" stored by a user, so any computer program may be stored in a nonvolatile way. This is particularly useful in fixed-purpose applications where microcomputers are used as controllers or laboratory data acquisition instruments.

How is a ROM made if information can only be read and not written? The ROM is an outgrowth of the *diode matrix* method of generating an arbitrary truth table. For example, one could construct the binary 4-bit-to-7-segment display decoder shown in Fig. 7.16. The outputs of the 4-to-16 decoder form the rows of the matrix, and the columns are terminated at the top by pull-up resistors to the positive power supply. The rows and columns are unconnected except by diodes at selected junctions. The diodes on a given column form a "low person wins" diode gate. Since the selected row corresponding to the input address word is pulled low by the decoder, output lines A through H at the bottom of the columns

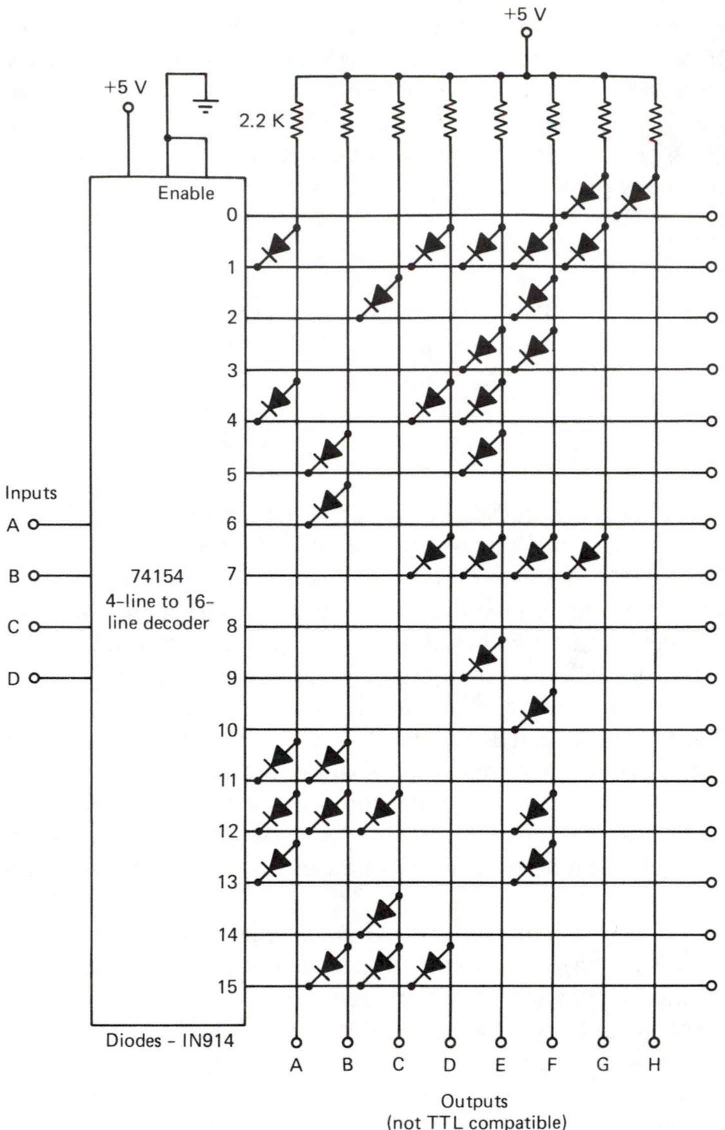

Figure 7.16 A "home-made" 7-segment display decoder circuit using a diode matrix to illustrate the ROM idea.

will be low wherever a column has a diode connection to the selected row, and float high otherwise because of the pull-up resistor. The truth table for a binary 4-bit-to-7-segment display decoder is shown in Table 7.7. The diode connections on a given row correspond to the complement of the desired binary word, since the diodes generate zeros on the selected row. For example, on row 4, diodes are connected at columns A, D, and E in order to generate the binary output 01100111.

TABLE 7.7 7-SEGMENT DISPLAY TRUTH TABLE

Input				Output								Pattern
D	C	B	A	A	B	C	D	E	F	G	H	
0	0	0	0	1	1	1	1	1	1	0	0	0
0	0	0	1	0	1	1	0	0	0	0	1	1
0	0	1	0	1	1	0	1	1	0	1	1	2
0	0	1	1	1	1	1	1	0	0	1	1	3
0	1	0	0	0	1	1	0	0	1	1	1	4
0	1	0	1	1	0	1	1	0	1	1	1	5
0	1	1	0	1	0	1	1	1	1	1	1	6
0	1	1	1	1	1	1	0	0	0	0	1	7
1	0	0	0	1	1	1	1	1	1	1	1	8
1	0	0	1	1	1	1	1	0	1	1	1	9
1	0	1	0	1	1	1	1	1	0	1	1	a
1	0	1	1	0	0	1	1	1	1	1	1	b
1	1	0	0	0	0	0	1	1	0	1	1	c
1	1	0	1	0	1	1	1	1	0	1	1	d
1	1	1	0	1	1	0	1	1	1	1	1	e
1	1	1	1	1	0	0	0	1	1	1	1	F

A ROM resembles a gate array rather than the assembly of flip flops in a RAM. A matrix of selected connections replaces storage cells. There are three kinds of read-only memories which differ in the way matrix connections are made: the ROM, the PROM, and the EPROM.

ROM. This label usually refers to a *mask-programmable* ROM. The matrix array is stored at the time the integrated circuit is manufactured by selectively drawing connections on the photolithographic mask. Because of the high cost of this step, mask-programmable ROM's are practical only when thousands of a given kind are used, as in pocket calculators.

PROM. This label refers to a *programmable* ROM. Active elements (actually transistors rather than diodes) are fabricated at *all* row-column intersections at the time of manufacture. A weak or fusible link is in series with each device. Links may be selectively fused or evaporated by the user to program the device. Fusing is done by applying a large current in excess of what is normally used in reading the PROM. A PROM may be programmed only once.

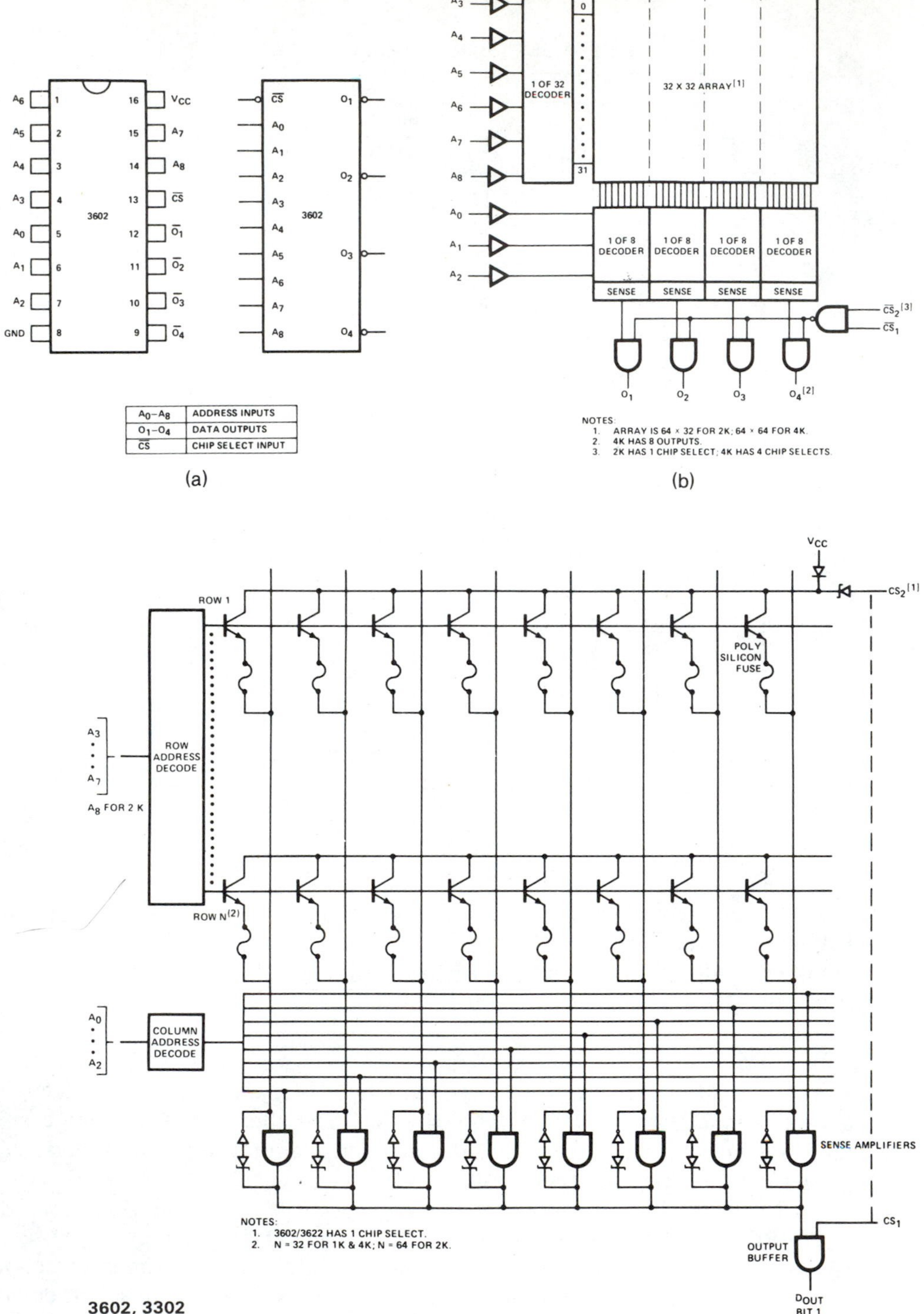

(c)

Figure 7.17 Bipolar PROM example. (a) Simplified bipolar PROM block diagram. (b) 3602/3622 and 3302/3322 pin configuration and logic symbol. (c) Typical bipolar PROM schematic, e.g., 3602, 3302. (3603, 3302; *courtesy* Intel Corporation)

EPROM. This is an erasable PROM, whose contents may be changed as many times as desired. Programming is done by storing charge on an extra *floating gate.* The gate is electrically isolated in the oxide layer so information remains stored even when the power is off. Erasure is effected optically by making the oxide photoconductive. In some EPROM devices, erasure can be done electrically. See Table 7.8 for examples of EPROM's.

An example of a PROM is shown in Fig. 7.17. This is a bipolar PROM, since the active elements are junction transistors. The horizontal line passing through all transistors on a given line indicates that all transistors in a row share a common base connection and are turned on in parallel when a given row is selected. The emitters are connected to the columns by a fusible link, typically a thin strip of polycrystalline Si (*polysilicon*). A given matrix intersection is programmed by passing a current large enough to destroy this link.

In addition to the usual row address decoding, the columns are further decoded. This results from a decision to store the information in a 16-pin chip. The decision is not a trivial one, since it ultimately affects the amount of memory which can be packed onto a given computer board. Suppose a chip is organized internally with 2^n rows and 2^m columns, i.e., contains 2^{n+m} bits of information (Fig. 7.18). The number of wires required to bring 2^m bits out directly is excessively large for any reasonable value of m. For example, for a 2K-bit chip organized with 32 columns ($m = 5$) and 64 rows ($n = 6$), the data may be brought out on 4 lines rather than the full 32. A total of n address bits can uniquely select one of 2^n distinct rows via the row address decoder (RAD in Fig. 7.18). To reduce the number of data wires, the columns are also decoded (CAD in Fig. 7.18). This adds k additional bits to the address word, whose width W_A becomes

$$W_A = n + k = 6 + 3 \tag{7.2}$$

The number of bits W_D in the data word is reduced from 2^m to

$$W_D = 2^{m-k} = 2^{5-3} = 2^2 \tag{7.3}$$

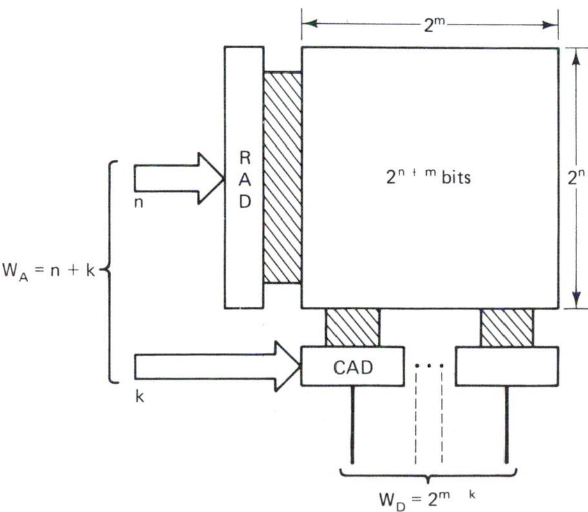

Figure 7.18 Decoding of the PROM of Fig. 7.17 includes columns as well as rows to reduce the number of data wires brought out of the chip.

TABLE 7.8 EPROM EXAMPLE SPECIFICATIONS[a]

2708/8708*
8K AND 4K UV ERASABLE PROM

	Max. Power	Max. Access	Organization
2708	800 mW	450 ns	1K × 8
2708L	425 mW	450 ns	1K × 8
2708-1	800 mW	350 ns	1K × 8
2704	800 mW	450 ns	512 × 8

- **Low Power Dissipation — 425 mW Max. (2708L)**
- **Fast Access Time — 350 ns Max. (2708-1)**
- **Static — No Clocks Required**
- **Data Inputs and Outputs TTL Compatible during both Read and Program Modes**
- **Three-State Outputs — OR-Tie Capability**

The Intel® 2708 is a 8192-bit ultraviolet light erasable and electrically reprogrammable EPROM, ideally suited where fast turnaround and pattern experimentation are important requirements. All data inputs and outputs are TTL compatible during both the read and program modes. The outputs are three-state, allowing direct interface with common system bus structures.

The 2708L at 425 mW is available for systems requiring lower power dissipation than from the 2708. A power dissipation savings of over 50%, without any sacrifice in speed, is obtained with the 2708L. The 2708L has high input noise immunity and is specified at 10% power supply tolerance. A high-speed 2708-1 is also available at 350 ns for microprocessors requiring fast access times. For smaller size systems there is the 4096-bit 2704 which is organized as 512 words by 8 bits. All these devices have the same programming and erasing specifications of the 2708. The 2704 electrical specifications are the same as the 2708.

The 2708 family is fabricated with the N-channel silicon gate FAMOS technology and is available in a 24-pin dual in-line package.

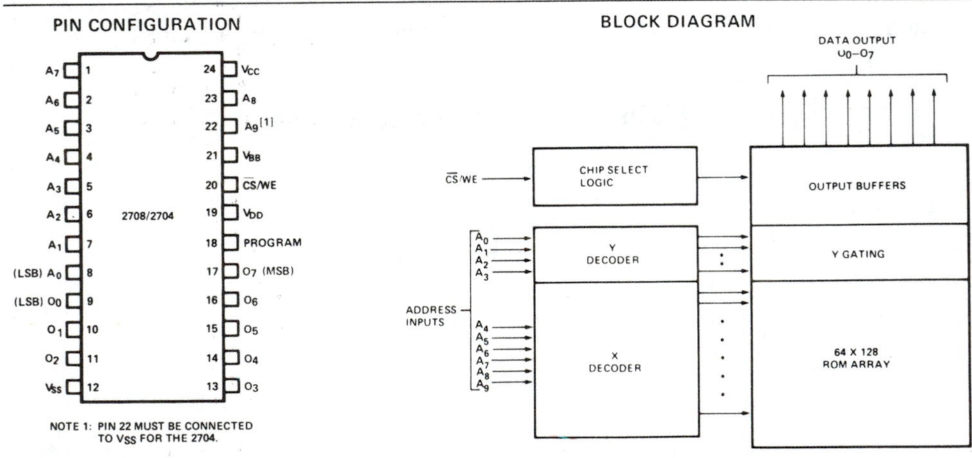

PIN NAMES

A_0-A_9	ADDRESS INPUTS
O_1-O_8	DATA OUTPUTS/INPUTS
$\overline{CS}$/WE	CHIP SELECT/WRITE ENABLE INPUT

PIN CONNECTION DURING READ OR PROGRAM

	PIN NUMBER							
MODE	DATA I/O 9-11, 13-17	ADDRESS INPUTS 1-8, 22, 23	V_{SS} 12	PROGRAM 18	V_{DD} 19	$\overline{CS}$/WE 20	V_{BB} 21	V_{CC} 24
READ	D_{OUT}	A_{IN}	GND	GND	+12	V_{IL}	-5	+5
DESELECT	HIGH IMPEDANCE	DON'T CARE	GND	GND	+12	V_{IH}	-5	+5
PROGRAM	D_{IN}	A_{IN}	GND	PULSED 26V	+12	V_{IHW}	-5	+5

***All 8708 specifications are identical to the 2708 specifications.**

[a]2708, *courtesy* Intel Corporation. See also the 2718, with twice the number of bits, a single power supply, and simpler programming.

The column address decoding requires 3 extra address bits, but reduces the number of data lines from 32 to 4. With $W_D = 4$ and $W_A = 9$, the total number of address and data lines equals 13, leaving room for power pins within a 16-pin package. Far more pins and a bigger package would have been required to bring the data out simultaneously without multiplexing, and even 8-bit bytes would exceed the 16-pin package in this case (see problems).

7.4.2 Erasable Read-Only Memories

The *erasable programmable read-only memory* (EPROM) allows nonvolatile storage of information, yet the contents can be changed when desired. This is particularly useful during the development phase of a microcomputer program, when changes and improvements occasionally need to be made. The most common version of this device is the optically erasable PROM, shown in Fig. 7.19. The configuration resembles a normal MOSFET with an additional *floating gate* which is not electrically connected to the rest of the device. Information is stored in the form of charge trapped on the floating gates, and is read out nondestructively since the stored electrons modify the electric field in the channel, altering the characteristic curves. The device is programmed by applying a high electric field (25 V across a 1000-Å oxide layer) between control gate and drain. This allows a few charge carriers to migrate through the insulating oxide layer and become trapped on the floating gate. Since normal operating voltages are significantly lower than this, no changes occur in normal operation. The oxide conductivity is sufficiently low that the decay time is years long in typical devices.

An EPROM is erased optically by shining ultraviolet light through a window on the device. This creates electron–hole pairs in the oxide, temporarily providing a path for electrons to leak away from the floating gate. The circuit symbol [Fig. 7.20(a)] resembles the MOSFET except for the extra floating gate. The device's characteristic curves [Fig. 7.20(b)] resemble an enhancement-mode MOSFET. Little current flows until a gate bias V_T is reached which is sufficient to populate the channel with electrons. Electrons on the floating gate shift this conduction threshold, providing a readout mechanism. With no electrons on the floating gate

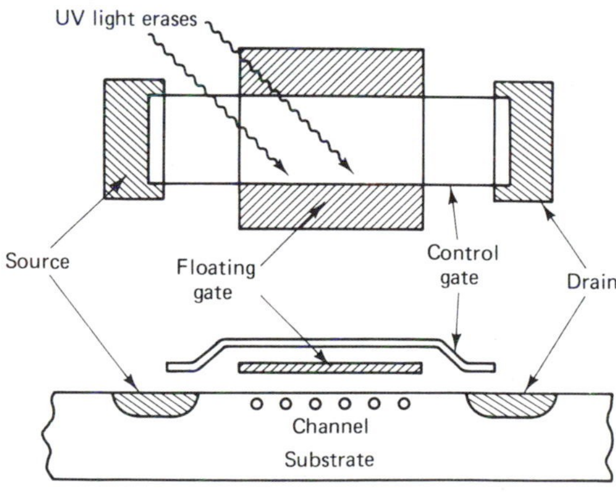

Figure 7.19 Optically erasable, programmable, read-only memory (EPROM) fabrication.

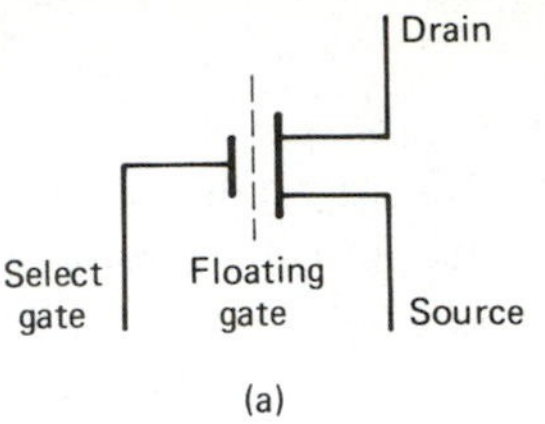

Drain
Select gate
Floating gate
Source
(a)

Figure 7.20 EPROM circuit operation. (a) Circuit symbol; note the floating gate. (b) Characteristic curves shift toward zero current when cell is programmed.

Current through transistor cell
Not programmed
1
I(1)
Programmed
0
V_{T_1} (not programmed)
V_{T_0} (programmed)
Voltage on gate of cell
(b)

Figure 7.21 Internal organization of the 2708 EPROM. (*Courtesy* Intel Corporation)

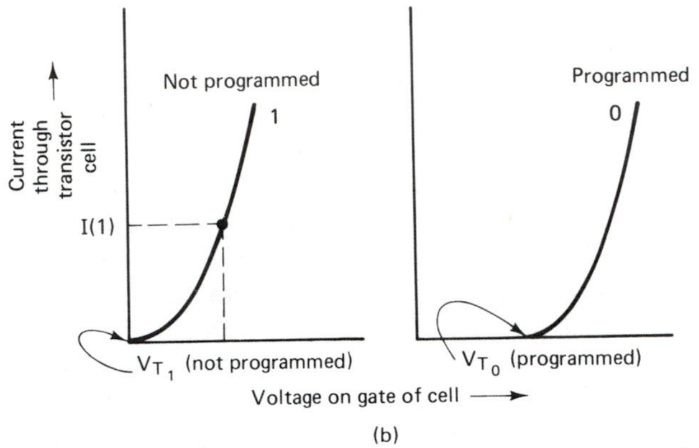

D_0 D_1 D_2 D_3 D_4 D_5 D_6 D_7
$\overline{CS}$/WE
OUTPUT BUFFER
PROG
READ SENSE AMPLIFIER & PROGRAM DATA BUFFER
A_0 A_1 A_2 A_3
COLUMN ADDRESS BUFFERS
COLUMN DECODER
A_4 A_5 A_6 A_7 A_8 A_9
ROW ADDRESS BUFFERS
ROW DECODERS
64 X 128 CELL ARRAY

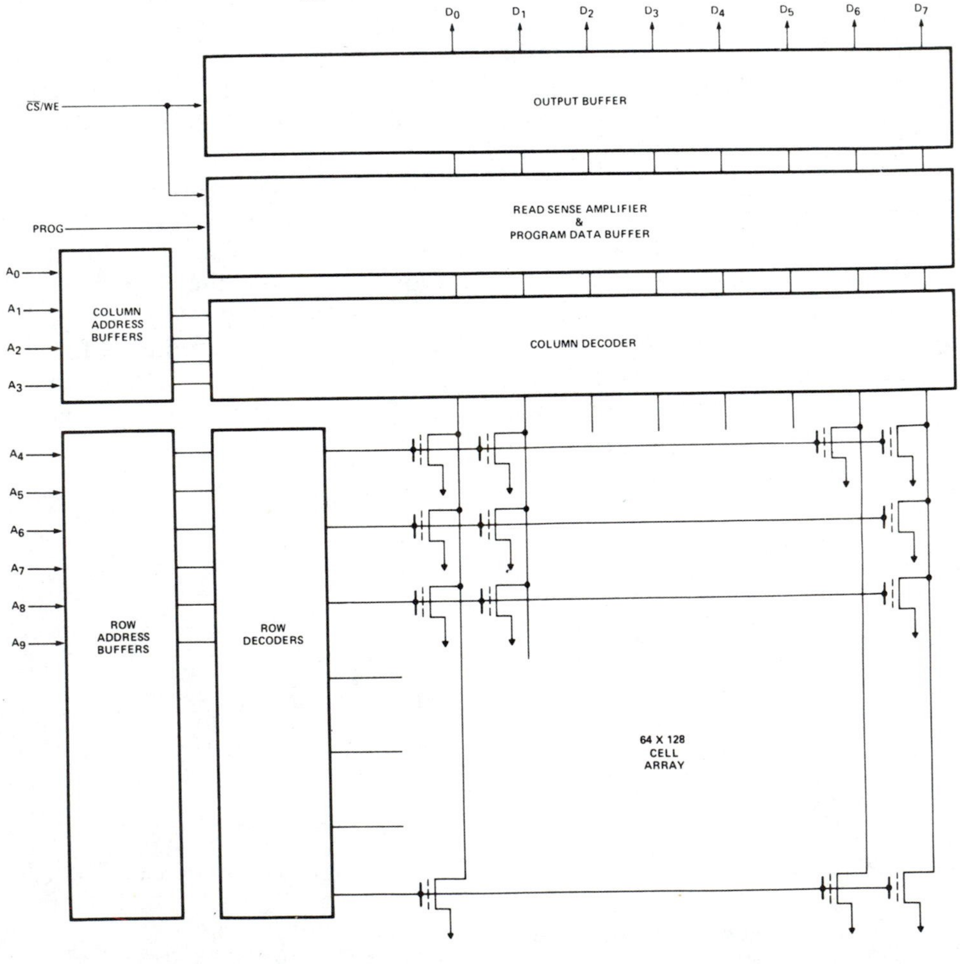

[left side Fig. 7.20(b)], a given *read voltage* V_R on the control gate results in a current I_R between source and drain. However, with electrons on the floating gate, the characteristic curve is shifted to the right and a larger gate bias is required to overcome the repulsion felt by electrons in the channel. Thus, bias voltage V_R results in nearly zero output current for a *programmed* cell.

The organization of a typical EPROM chip is shown in Fig. 7.21. This configuration provides 8K bits of storage organized as 1024 8-bit words (1K bytes). The internal organization is 64 rows by 128 columns. The columns are partially decoded, using 4 address bits to bring out the 8 data lines. The output buffer allows this MOS device to drive about 1 mA to the data bus. A pair of transistors connected to the chip select (CS) line form a tri-state output, disconnecting the internal data when CS is held high.

Further evolution of EPROM devices is in the direction of

(1) Single power supply connections. Early devices required two power supplies.
(2) Simpler programming cycles. The device shown above must be "hit" a number of times before it remembers.
(3) More bits ($\geqslant$ 64K) per chip.
(4) Electrical erasing capability. Simple changes can be made without reprogramming the whole device when erasure is done bit by bit instead of the whole chip at one time.

PROBLEMS

7.1. Estimate how much memory you could afford to buy for your microcomputer in 1990 if you had \$1000 to spend. Use the data of Fig. 7.2, and make plausible extrapolations about larger chip sizes. Neglect inflation, as does the figure.

7.2. Estimate the maximum-frequency sine wave signal which can be generated using the circuit of Fig. 7.7. Assume that the full sine wave is stored in 16 adjacent channels of address space. First assume typical MOS RAM (500 ns access time), and then repeat using TTL RAM (30 ns access time).

7.3. Design external logic so that a higher-resolution sine wave can be stored in the circuit of Fig. 7.7. Store 1 of 4 quadrants (i.e., 90°), so the full sine wave will occupy 64 time slices when reconstructed. You may use gates, flip flops, counters, and a clock. The reconstruction should occur automatically as the clock advances.

7.4. Design a memory board for your microcomputer. The data bus is 8 bits wide, and the address bus is 16 bits wide. A single memory board should hold 16K *bytes*. Use MOS static RAM chips which hold 1024, 4-bit words (e.g., 2114, Table 7.4). Use decoders and data selectors as needed, and show how a single 8-bit word with 16 bit address finds its way to a unique physical location. Draw a complete circuit diagram and also specify how the 16-bit address is broken up to select a given board and then a given chip.

7.5. Does fusing a link (the wire becomes *open*) on the bipolar PROM of Fig. 7.17 make that bit a 1 or a 0?

7.6. What would be the consequences of omitting column decoding in larger PROM's? For the example of Fig. 7.18, how many address and data pins would be needed if all data were brought out directly? If package size is set by the 0.10-in. pin spacing, what amount of memory could then fit on a 5-in. by 10-in. board?

7.7. Repeat Prob. 7.6 and redesign the chip with a compromise: 8 data bits brought out directly. Use the internal cell array of Fig. 7.17, and check your solution against Fig. 7.21, which also uses 8-bit words in a different size chip.

REFERENCES

More complete bibliographic information for the books listed below appears in the annotated bibliography at the end of the book.

HIGGINS, *Experiments with Integrated Circuits*, Experiment 11

Intel, *Memory Design Handbook*

LARSEN & RONY, *Logic and Memory Experiments Using TTL Integrated Circuits*

MALMSTADT et al., *Electronic Measurements for Scientists*

Scientific American, *Microelectronics*

WILLIAMS, *Digital Technology*

8

Binary Arithmetic

Binary arithmetic is used not only for numerical calculations in calculators and computers, but also for signal-processing devices such as digital filters and spectrum analyzers. This chapter provides the necessary background for both classes of applications. These days no one constructs a binary adder from gates as described in the following section; one uses a microprocessor instead. However, this material provides the necessary background for understanding how microprocessors do it, in addition to the limitations and strengths of several alternative methods.

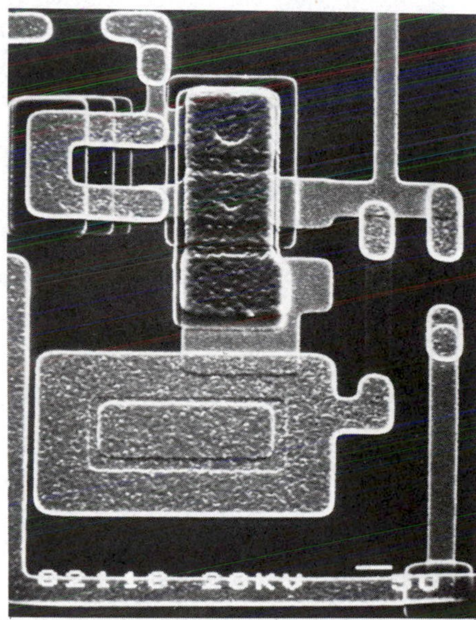

Figure 8.1 Logic element of a superconducting computer, based on the Josephson effect. Circuits in this family can switch in less than 20 picoseconds. (*Courtesy* IBM Corporation)

Relatively few microprocessors include multiplication hardware. Although this is changing, the special-purpose hardware-multiply IC'S described in Section 8.3 are the method of choice for fast signal processing and for extending microcomputer systems where the speed of number crunching is important.

Even though the principles of binary number manipulation remain the same, the methods are changing rapidly. An example is the rise of quantized information in the superconducting Josephson-effect computer (Fig. 8.1).

8.1 BINARY ADDITION SOFTWARE AND HARDWARE

8.1.1 Adder Logic

Consider the following example of binary addition:

	MSB							LSB		
	0	0	1	1	0	1	0	1	=	53_{10}
+	0	1	1	1	0	1	1	0	=	$+118_{10}$
								1		
							1			
					1	0				Partial
					0					sums
			1	0						
		1	0							
	1	1								
	0									
	1	0	1	0	1	0	1	1		$=171_{10}$

The partial sums indicate the possibilities for the carry bit generated by overflow when two 1's are added.

To determine the electronic circuit required for addition, we first examine the truth table for adding two single-bit numbers (Table 8.1).

TABLE 8.1 TRUTH TABLE FOR ADDING SINGLE BITS

A	B	C	S
0	0	0	0
0	1	0	1
1	0	0	1
1	1	1	0

Two bits are needed to specify the result: the *sum bit* S and the *carry bit* C. The truth tables for S and for C are identical to the simple logic combinations:

$$S = A \oplus B \tag{8.1}$$
$$C = A \cdot B$$

An adder circuit for a single bit number therefore consists of an exclusive OR gate and an AND gate (Fig. 8.2). This circuit is called a *half adder* because it ignores the possibility of a carry bit left over from a less significant bit (e.g., the second bit from the left in the above example). The truth table therefore must be expanded to include this possibility.

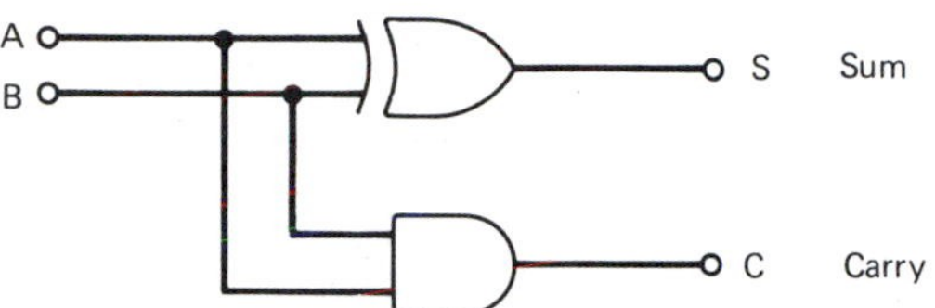

Figure 8.2 Half-adder circuit.

One may test one's understanding of Table 8.2 by trying Prob. 8.1. How can this truth table be implemented using gate circuits? The sum bit S_i is 1 whenever (C_{i-1}, A_i, B_i) contain an odd number of 1's. This is just the definition of parity, so S_i is logically equivalent to

$$S_i = A_i \oplus B_i \oplus C_{i-1} \tag{8.2}$$

TABLE 8.2 TRUTH TABLE FOR THE FULL ADDER

C_{i-1}	A_i	B_i	C_i	S_i
0	0	0	0	0
0	0	1	0	1
0	1	0	0	1
0	1	1	1	0
1	0	0	0	1
1	0	1	1	0
1	1	0	1	0
1	1	1	1	1

A circuit for the carry bit C_i is less obvious. A portion of the truth table for C_i is carried out by an AND gate with A_i and B_i as inputs. The balance of the truth table for C_i is just the exclusive OR function whenever the previous carry bit was a 1. The complete carry function ORs these two partial results together.

$$C_i = C_{i-1} \cdot (A_i + B_i) + A_i \cdot B_i \tag{8.3}$$

Figure 8.3 is the circuit for the full adder which implements the logical functions of Eqs. (8.2) and (8.3). The AND functions have been replaced by NAND gates to decrease the number of different kinds of gates required (see Prob. 8.2).

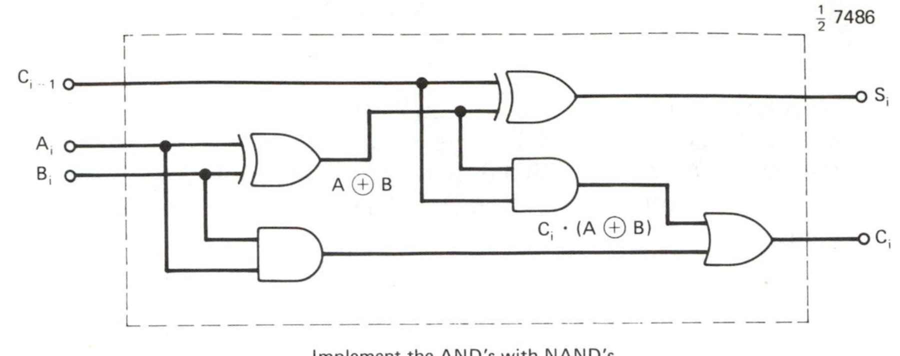

Implement the AND's with NAND's

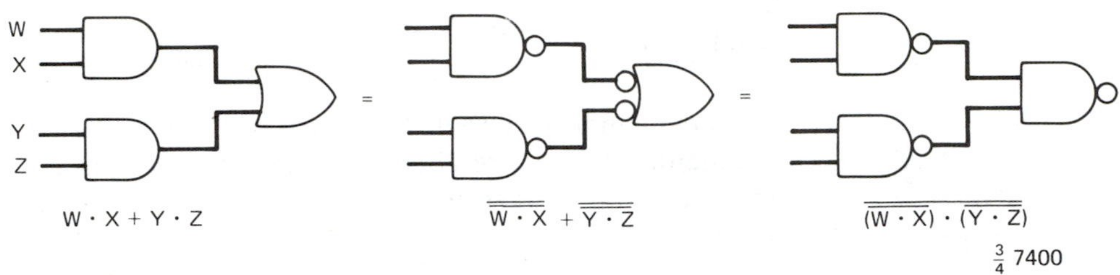

Figure 8.3 Full-adder circuit.

8.1.2 Adder Terminology: Asynchronous vs. Synchronous and Serial vs. Parallel

Large numbers may be added by combining several full-adder functions. Since adders require gates only, the sum is present as soon as gate delays are over. This type of addition is *asynchronous,* since there is no clock. However, the basic adder also has no memory, since the sum remains only as long as the inputs are present. A practical adder circuit must also include latches for temporary storage of inputs and outputs.

Full-adder units may be combined in parallel or serial configuration. A *parallel adder* is constructed by wiring together one full-adder gate array per bit (Fig. 8.4). By contrast, the *serial adder* (Fig. 8.5) uses only one full-adder gate array, and performs additions bit by bit on numbers stored in shift registers. A D-type flip flop stores the carry output of one step, thereby delaying it by one clock cycle to become the carry input for the next step. The serial adder requires the minimum amount of gate hardware but it requires $(N + 1)$ clock pulses for an N-bit addition. Serial addition is used where high precision is necessary but slow speed may be tolerated, as in calculators. The serial circuit also minimizes the number of interconnections necessary between packages. The parallel adder (Fig. 8.4) requires more gate hardware but it is much faster. Parallel addition is the most common method in micro- and minicomputers. The time required for a valid answer equals the propagation delay of a single full adder times the number

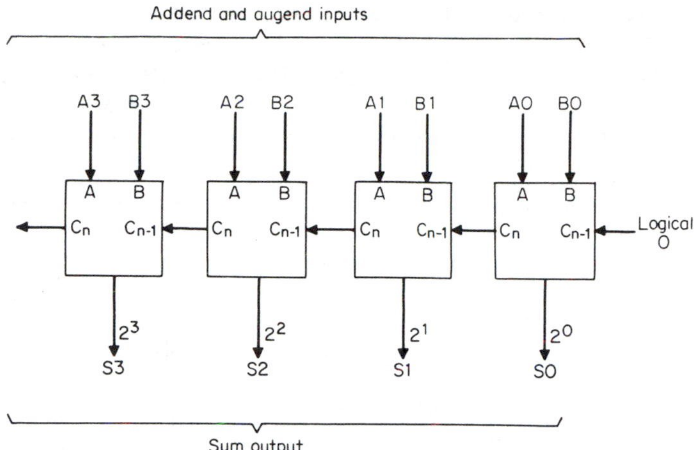

Figure 8.4 Four bit parallel-adder circuit. (*Courtesy* Texas Instruments Incorporated)

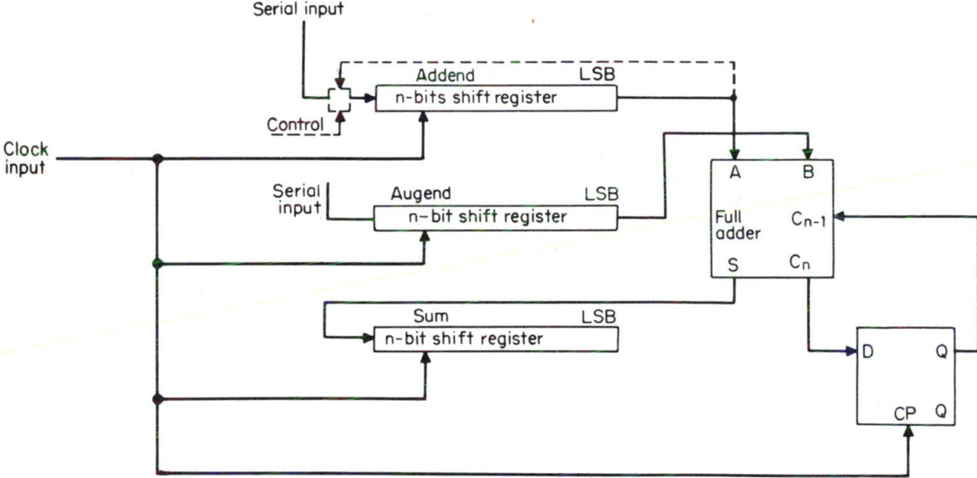

Figure 8.5 Serial-adder circuit. (*Courtesy* Texas Instruments Incorporated)

of bits being added. If the number of bits is large enough for this delay to become unacceptable, one uses an improved algorithm called the *look-ahead carry*, which presents the appropriate carry bit to all stages simultaneously.

8.1.3 IC's for Arithmetic and Logic

Constructing adder circuits with basic gates is unnecessary, since there are available MSI IC's which perform addition with fewer packages and fewer interconnections. For example, here is the number of IC's needed to implement a 4-bit adder using gates:

TABLE 8.3 FOUR BIT-FULL ADDER SPECIFICATIONS[a]

TTL MSI — TYPES SN5483A, SN54LS83, SN7483A, SN74LS83 4-BIT BINARY FULL ADDERS

- For applications in:
  - Digital Computer Systems
  - Data-Handling Systems
  - Control Systems
- SN54283/SN74283 Are Recommended For New Designs as They Feature Supply Voltage and Ground on Corner Pins to Simplify Board Layout

TYPE	TYPICAL ADD TIMES TWO 8-BIT WORDS	TYPICAL ADD TIMES TWO 16-BIT WORDS	TYPICAL POWER DISSIPATION PER 4-BIT ADDER
'83A	23 ns	43 ns	310 mW
'LS83	89 ns	165 ns	75 mW

J OR N DUAL-IN-LINE OR W FLAT PACKAGE (TOP VIEW)

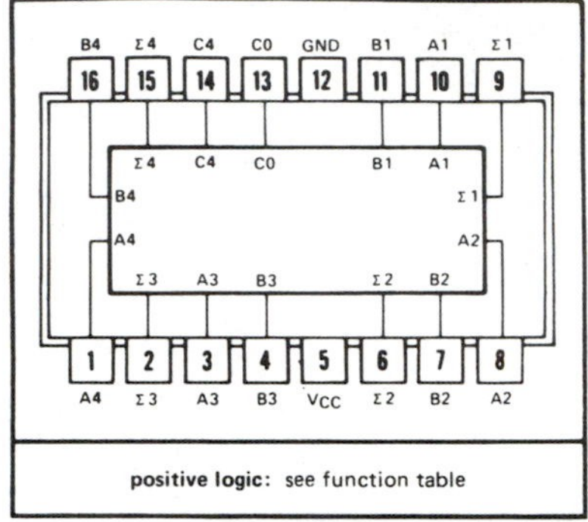

positive logic: see function table

description

These full adders perform the addition of two 4-bit binary numbers. The sum (Σ) outputs are provided for each bit and the resultant carry (C4) is obtained from the fourth bit. The adders are designed so that logic levels of the input and output, including the carry, are in their true form. Thus the end-around carry is accomplished without the need for level inversion. Designed for medium-to-high-speed, the circuits utilize high-speed, high-fan-out transistor-transistor logic (TTL) but are compatible with both DTL and TTL families.

The '83A circuits feature full look ahead across four bits to generate the carry term in typically 10 nanoseconds to achieve partial look-ahead performance with the economy of ripple carry.

The 'LS83 can reduce power requirements to less than 20 mW/bit for power-sensitive applications. These circuits are implemented with single-inversion, high-speed, Darlington-connected serial-carry circuits within each bit.

Series 54 and 54LS circuits are characterized for operation over the full military temperature range of −55°C to 125°C; Series 74 and 74LS are characterized for 0°C to 70°C operation.

FUNCTION TABLE

INPUT				OUTPUT WHEN C0 = L / WHEN C2 = L			OUTPUT WHEN C0 = H / WHEN C2 = H		
A1 / A3	B1 / B3	A2 / A4	B2 / B4	Σ1 / Σ3	Σ2 / Σ4	C2 / C4	Σ1 / Σ3	Σ2 / Σ4	C2 / C4
L	L	L	L	L	L	L	H	L	L
H	L	L	L	H	L	L	L	H	L
L	H	L	L	H	L	L	L	H	L
H	H	L	L	L	H	L	H	H	L
L	L	H	L	L	H	L	H	H	L
H	L	H	L	H	H	L	L	L	H
L	H	H	L	H	H	L	L	L	H
H	H	H	L	L	L	H	H	L	H
L	L	L	H	L	H	L	H	H	L
H	L	L	H	H	H	L	L	L	H
L	H	L	H	H	H	L	L	L	H
H	H	L	H	L	L	H	H	L	H
L	L	H	H	L	L	H	H	L	H
H	L	H	H	H	L	H	L	H	H
L	H	H	H	H	L	H	L	H	H
H	H	H	H	L	H	H	H	H	H

H = high level, L = low level

NOTE: Input conditions at A3, A2, B2, and C0 are used to determine outputs Σ1 and Σ2 and the value of the internal carry C2. The values at C2, A3, B3, A4, and B4 are then used to determine outputs Σ3, Σ4, and C4.

absolute maximum ratings over operating free-air temperature range (unless otherwise noted)

Supply voltage, V_{CC} (see Note 1)	7 V
Input voltage	5.5 V
Interemitter voltage (see Note 2)	5.5 V
Operating free-air temperature range: SN54', SN54LS' Circuits	−55°C to 125°C
SN74', SN74LS' Circuits	0°C to 70°C
Storage temperature range	−65°C to 150°C

NOTES: 1. Voltage values, except interemitter voltage, are with respect to network ground terminal.
2. This is the voltage between two emitters of a multiple-emitter transistor. For the '83A, this rating applies between the following pairs: A1 and B1, A2 and B2, A3 and B3, A4 and B4. For the 'LS83, this rating applies between the following pairs: A1 and B1, A1 and C0, B1 and C0, A3 and B3.

TEXAS INSTRUMENTS
INCORPORATED
POST OFFICE BOX 5012 • DALLAS, TEXAS 75222

TABLE 8.3 CONTINUED

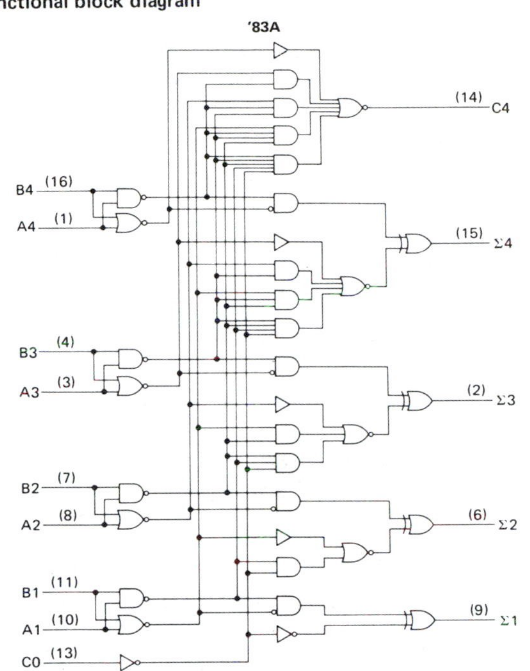

[a] *Source:* Texas Instruments Incorporated

	TTL	CMOS
XOR	2-7486	2-4507
NAND	3-7400	3-4011

Only one IC is required using a complete adder package

	TTL	CMOS
ADDER	1-7483	1-4008

An example of an MSI adder is shown in Table 8.3

Adder gate logic, combined with control circuits, forms the heart of a microprocessor. In practice one uses a microcomputer, rather than adder IC's, to perform arithmetic functions. However, TTL adder hardware is used in constructing mockups of new microprocessor designs, and because of speed advantages may still be useful in some high speed signal processing applications.

It is straightforward to extend adder gating logic to provide other logical functions. For example, an IC called an arithmetic and logical unit (ALU) per-

TABLE 8.4 ARITHMETIC AND LOGICAL UNIT SPECIFICATIONS[a]

TTL
MSI

TYPES SN54181, SN54LS181, SN54S181, SN74181, SN74LS181, SN74S181 ARITHMETIC LOGIC UNITS/FUNCTION GENERATORS

PIN DESIGNATIONS

DESIGNATION	PIN NOS.	FUNCTION
A3, A2, A1, A0	19, 21, 23, 2	WORD A INPUTS
B3, B2, B1, B0	18, 20, 22, 1	WORD B INPUTS
S3, S2, S1, S0	3, 4, 5, 6	FUNCTION-SELECT INPUTS
C_n	7	INV. CARRY INPUT
M	8	MODE CONTROL INPUT
F3, F2, F1, F0	13, 11, 10, 9	FUNCTION OUTPUTS
A = B	14	COMPARATOR OUTPUT
P	15	CARRY PROPAGATE OUTPUT
C_{n+4}	16	INV. CARRY OUTPUT
G	17	CARRY GENERATE OUTPUT
V_{CC}	24	SUPPLY VOLTAGE
GND	12	GROUND

'181, 'LS181 . . . J, N, OR W PACKAGE
SN54S181 . . . J OR W PACKAGE
SN74S181 . . . J, N, OR W PACKAGE
(TOP VIEW)

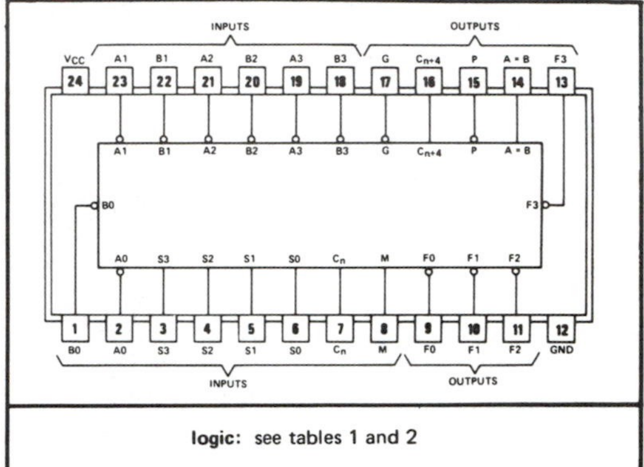

logic: see tables 1 and 2

- **Full Look-Ahead for High-Speed Operations on Long Words**
- **Input Clamping Diodes Minimize Transmission-Line Effects**
- **Darlington Outputs Reduce Turn-Off Time**
- **Arithmetic Operating Modes:**
  - **Addition**
  - **Subtraction**
  - **Shift Operand A One Position**
  - **Magnitude Comparison**
  - **Plus Twelve Other Arithmetic Operations**
- **Logic Function Modes:**
  - **Exclusive-OR**
  - **Comparator**
  - **AND, NAND, OR, NOR**
  - **Plus Ten Other Logic Operations**

TYPICAL ADDITION TIMES

NUMBER OF BITS	ADDITION TIMES			PACKAGE COUNT		CARRY METHOD BETWEEN ALU's
	USING '181 AND '182	USING 'LS181 AND '182	USING 'S181 AND 'S182	ARITHMETIC/ LOGIC UNITS	LOOK-AHEAD CARRY GENERATORS	
1 to 4	24 ns	24 ns	11 ns	1		NONE
5 to 8	36 ns	40 ns	18 ns	2		RIPPLE
9 to 16	36 ns	44 ns	19 ns	3 or 4	1	FULL LOOK-AHEAD
17 to 64	60 ns	68 ns	28 ns	5 to 16	2 to 5	FULL LOOK-AHEAD

description

The '181, 'LS181, and 'S181 are arithmetic logic units (ALU)/function generators which have a complexity of 75 equivalent gates on a monolithic chip. These circuits perform 16 binary arithmetic operations on two 4-bit words as shown in Tables 1 and 2. These operations are selected by the four function-select lines (S0, S1, S2, S3) and include addition, subtraction, decrement, and straight transfer. When performing arithmetic manipulations, the internal carries must be enabled by applying a low-level voltage to the mode control input (M). A full carry look-ahead scheme is made available in these devices for fast, simultaneous carry generation by means of two cascade-outputs (pins 15 and 17) for the four bits in the package. When used in conjunction with the SN54182, SN54S182, SN74182, or SN74S182, full carry look-ahead circuits, high-speed arithmetic operations can be performed. The typical addition times shown above

TABLE 8.4 CONTINUED

The signal designations shown below result in the logic functions and arithmetic operations shown in the truth table.

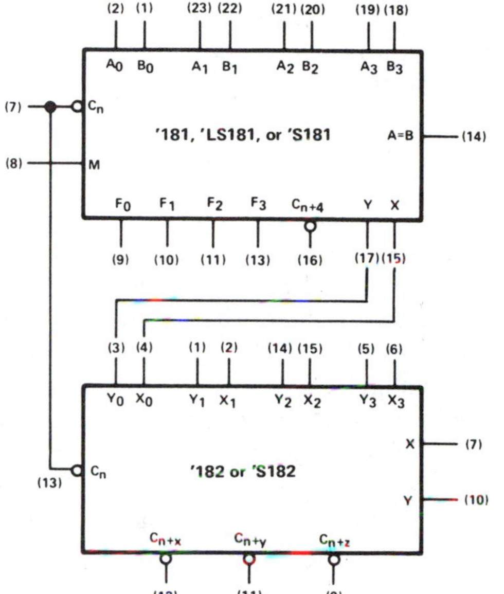

SELECTION	ACTIVE-HIGH DATA		
	M = H LOGIC FUNCTIONS	M = L; ARITHMETIC OPERATIONS	
S3 S2 S1 S0		C_n = H (no carry)	C_n = L (with carry)
L L L L	$F = \overline{A}$	F = A	F = A PLUS 1
L L L H	$F = \overline{A + B}$	F = A + B	F = (A + B) PLUS 1
L L H L	$F = \overline{A}B$	$F = A + \overline{B}$	$F = (A + \overline{B})$ PLUS 1
L L H H	F = 0	F = MINUS 1 (2's COMPL)	F = ZERO
L H L L	$F = \overline{AB}$	$F = A$ PLUS $A\overline{B}$	$F = A$ PLUS $A\overline{B}$ PLUS 1
L H L H	$F = \overline{B}$	$F = (A + B)$ PLUS $A\overline{B}$	$F = (A + B)$ PLUS $A\overline{B}$ PLUS 1
L H H L	$F = A \oplus B$	F = A MINUS B MINUS 1	F = A MINUS B
L H H H	$F = A\overline{B}$	$F = A\overline{B}$ MINUS 1	$F = A\overline{B}$
H L L L	$F = \overline{A} + B$	F = A PLUS AB	F = A PLUS AB PLUS 1
H L L H	$F = \overline{A \oplus B}$	F = A PLUS B	F = A PLUS B PLUS 1
H L H L	F = B	$F = (A + \overline{B})$ PLUS AB	$F = (A + \overline{B})$ PLUS AB PLUS 1
H L H H	F = AB	F = AB MINUS 1	F = AB
H H L L	F = 1	F = A PLUS A*	F = A PLUS A PLUS 1
H H L H	$F = A + \overline{B}$	F = (A + B) PLUS A	F = (A + B) PLUS A PLUS 1
H H H L	F = A + B	$F = (A + \overline{B})$ PLUS A	$F = (A + \overline{B})$ PLUS A PLUS 1
H H H H	F = A	F = A MINUS 1	F = A

*Each bit is shifted to the next more significant position.

illustrate the little additional time required for addition of longer words when full carry look-ahead is employed. The method of cascading '182 or 'S182 circuits with these ALU's to provide multi-level full carry look ahead is illustrated under typical applications data for the '182 and 'S182.

Subtraction is accomplished by 1's complement addition where the 1's complement of the subtrahend is generated internally. The resultant output is A–B–1 which requires an end-around or forced carry to provide A–B.

The '181, 'LS181 or 'S181 can also be utilized as a comparator. The A = B output is internally decoded from the function outputs (F0, F1, F2, F3) so that when two words of equal magnitude are applied at the A and B inputs, it will assume a high level to indicate equality (A = B). The ALU should be in the subtract mode with C_n = H when performing this comparison. The A = B output is open-collector so that is can be wire-AND connected to give a comparison for more than four bits. The carry output (C_{n+4}) can also be used to supply relative magnitude information. Again, the ALU should be placed in the subtract mode by placing the function select inputs S3, S2, S1, S0 at L, H, H, L, respectively.

INPUT C_n	OUTPUT C_{n+4}	ACTIVE-HIGH DATA (FIGURE 1)	ACTIVE-LOW DATA (FIGURE 2)
H	H	$A \leq B$	$A \geq B$
H	L	$A > B$	$A < B$
L	H	$A < B$	$A > B$
L	L	$A \geq B$	$A \leq B$

These circuits have been designed to not only incorporate all of the designer's requirements for arithmetic operations, but also to provide 16 possible functions of two Boolean variables without the use of external circuitry. These logic functions are selected by use of the four function-select inputs (S0, S1, S2, S3) with the mode-control input (M) at a high level to disable the internal carry. The 16 logic functions are detailed in Tables 1 and 2 and include exclusive-OR, NAND, AND, NOR, and OR functions.

TEXAS INSTRUMENTS
INCORPORATED
POST OFFICE BOX 5012 • DALLAS, TEXAS 75222

[a]*Source:* Texas Instruments Incorporated

forms not only addition and subtraction, but also incrementation, decrementation, and combined operations. The 74181 ALU (the 40181 in CMOS) is described in Table 8.4. The bit pattern applied to control inputs S_0 through S_3 determines which of 16 operations will be performed. The set covers most standard microprocessor operations, so this chip is suited for prototyping new computer architecture. Logical inequalities are performed by magnitude comparator chips such as the 7485 and its CMOS equivalent, the 14585.

8.2 2's COMPLEMENT SUBTRACTION

Binary subtraction requires a way to represent a minus sign. Consider an analogy with measurement. If 10 bits of measurement resolution are available, one has the choice of measuring numbers ranging from 0 to 1023, or, if the numbers may be either positive or negative, measuring from −512 to +511. Ten bits of resolution cover the same range as nine bits plus a sign bit.

To see the idea behind the commonly used 2's complement method of representing negative numbers (Fig. 8.6), the binary number field is written on a strip of paper, which is wrapped into a circle and glued and then cut at a new place halfway from the original ends. When the strip is opened and laid flat, the

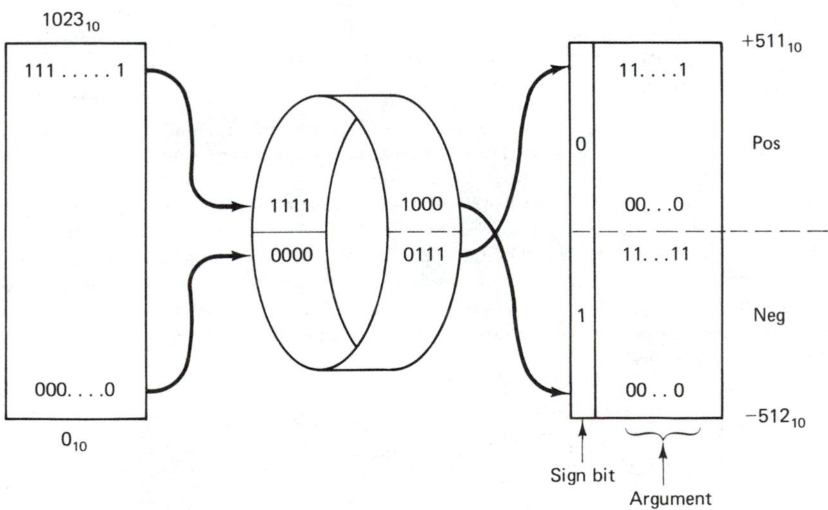

Figure 8.6 Graphic view of two's complement negative numbers.

numbers written on it will be in *2's complement* notation. The leftmost bit is designated as the sign bit: 0 for positive numbers and 1 for negative numbers. In this representation a small negative number is mostly 1's, while a large negative number is mostly 0's. The rule for writing 2's complement negative numbers may be seen by an example. Apply the circular number strip of Fig. 8.6 to 3-bit numbers:

2's COMPLEMENT NUMBERS: 3-BIT EXAMPLE

Positive only		2's Complement	
7	111	011	3
6	110	010	2
5	101	001	1
4	100	000	0
3	011	111	−1
2	010	110	−2
1	001	101	−3
0	000	100	−4

These negative numbers give the right answers in simple arithmetic. For example, $(-N)$ is the number which when added to N makes 0.

$$\begin{array}{rr} -1 & 111 \\ +1 & 001 \\ \hline 0 & 000 \end{array} \quad \text{(ignore carry)}$$

The rule for finding a 2's complement negative number is:

2's Complement. Reverse all 1's and 0's in the original number (*1's complement*) and then add a 1.

Test Your Understanding. Apply the rule to the examples below and verify that the results make sense.

Example

Number			
Decimal		Digital	Negative of the number
−1	=	111	000 + 001 = 001 = +1
+3	=	001	etc.
−3	=	101	etc.
0	=	000	
−4	=	100	

These 3-bit examples demonstrate that there is one more negative number than positive number in 2's complement notation. This results from the fact that 0 becomes a positive number in this notation, since its leftmost bit is a 0.

Exercise

What happens to the number 0 if negative numbers are formed by taking the complement alone?

The rules for 2's complement addition and subtraction are particularly simple and require no changes in the hardware developed in the previous section, because the sign bit may be treated as just another binary bit.

Example 2's Complement Addition

Decimal	8-bit 2's complement		
(−13)	11110011	=	− (00001101)
+(+ 6)	00000110		
− 7	11111001	=	− (00000111)

↑ Sign

Subtraction is done simply by taking the 2's complement of the number to be subtracted and adding it to the first number.

This discussion has ignored the *carry bit:* this is used when the sum of two numbers will not fit on the circular strip. Also only one of several algorithms for binary addition and subtraction has been presented. Further discussion of these topics may be found in many computer science texts.

8.3 THE SOFTWARE AND HARDWARE OF BINARY MULTIPLICATION

8.3.1 Simple Multiply Algorithms

Recall the simplest way of multiplying:

Multiplication is Repetitive Addition

$$\begin{array}{r} 3 \times 4 = 4 \\ +4 \\ +4 \\ \hline 12 \end{array}$$

This method may be converted to a binary version (Prob. 8.5). Repetitive addition requires no additional hardware besides an adder, but it obviously becomes very time-consuming if the numbers are very large.

For Longer Numbers, Use Long Multiplication

372	*Algorithm*
× 102	
744	Multiply
000	Shift + multiply
372	Shift + multiply
37944	Add partial products

The binary version of this procedure is constructed by applying the same algorithm.

$$
\begin{array}{lrcr}
\text{Multiplier} & Y = 1011 & = & (11)_{10} \\
\text{Multiplicand} & X = 1001 & = & (9)_{10} \\
\hline
 & 1011 & & \\
 & 0000 & & \\
 & 0000 & & \\
 & 1011 & & \\
\hline
X \cdot Y = & 1100011 & = & (99)_{10}
\end{array}
$$

Binary multiplication is easier than decimal multiplication because the partial products are formed simply by including or not including the (shifted) multiplicand in the partial product, depending upon whether the corresponding bit of the multiplier is a 1 or a 0. Since binary multiplication is just repetitive shifting and adding, it can be done in software using any computing device. However, a large number of steps is required to carry out software-only multiplication: the multiply time is on the order of milliseconds for a memory cycle time on the order of microseconds. This is inefficient use of machine time for numerical computations, and in digital signal processing it seriously limits the bandwidth of signals which may be processed (under 1 KHz if the multiply time is 1 ms). The problem is more serious when the operation to be performed requires processing an array. For example, consider the Fourier transform of a signal $y(nt)$ sampled at time intervals nt:

$$Y(f) = \sum_{n=1}^{N} y(nt) \cdot \exp(2\pi \text{ jf} \cdot nt) \tag{8.4}$$

If the signal $y(nt)$ is stored as N sampled data points, the Fourier transform at a single frequency f requires N multiplications. The Fourier spectrum at N different frequencies requires approximately N^2 multiplications. Although improved algorithms, such as the *FFT* (fast Fourier transform), reduce this number to about $N \log N$, the slow speed of software-only multiplication would make *real-time* Fourier transform signal processing virtually impossible. Special-purpose hardware multipliers have been devised, and newer generations of microprocessors and minicomputers often include multiply hardware. We will now consider briefly the hardware implementation of binary multiplication, and then discuss some newer LSI multipliers which achieve even higher speeds.

8.3.2 TTL Hardware Multiplier

The repetitive *shift* and *add* algorithm may be carried out at high speed using the TTL medium-scale integration (MSI) circuit shown in Fig. 8.7. Since even faster LSI multipliers are now available, the circuit is included here only for illustration. Adders A_1 and A_2, which contain gates only, have no memory and are used for

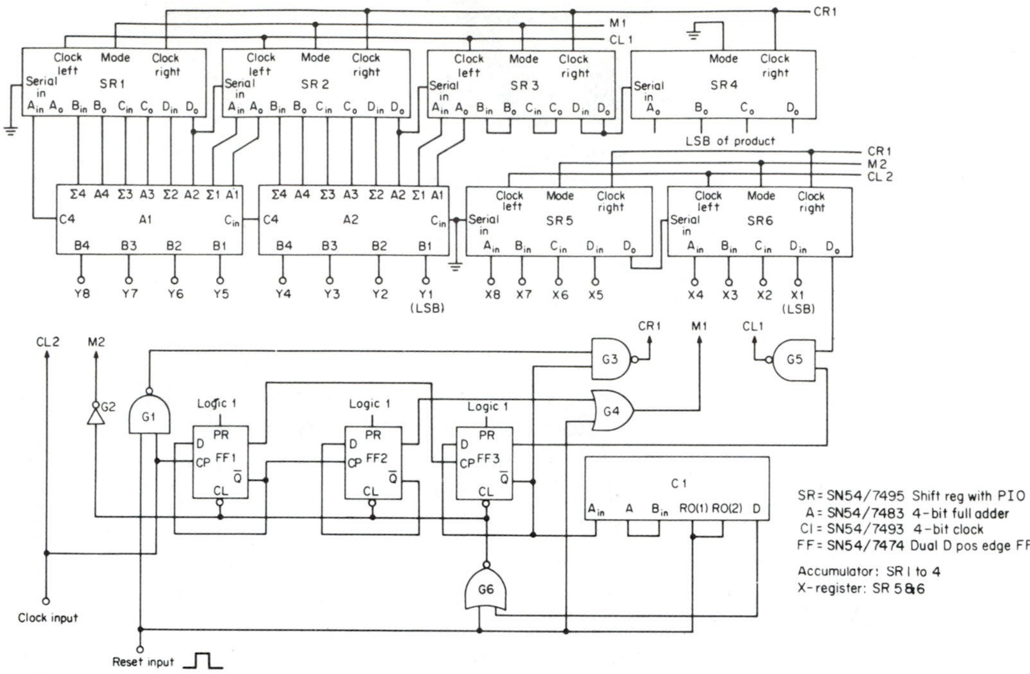

Figure 8.7 Hardware multiplier constructed from TTL MSI integrated circuits. (*Courtesy* Texas Instruments Incorporated)

the partial sums. The components labeled SR are universal shift registers, allowing both parallel and serial input and output, which are controlled by signals applied to the *mode*, *clock left*, and *clock right* inputs. These mode control signals are generated by the flip flop–gate combination at the bottom half of the figure. FF1 divides the input clock pulse by 2 (ensuring that the first cycle of operation actually gets a full clock cycle), and drives FF2 and FF3, thereby generating mode control signals CR1, M1, and CL1. Counter C1 keeps track of the multiply cycle and halts the process after 16 clock cycles (8 adds and 8 shifts).

During the first clock cycle, the mode controls are set to parallel-load the multiplicand X into SR5 and SR6. During the clock cycles which follow, X is shifted out serially. Its 1 and 0 bits generate the command Add or Do Not Add by forming the mode control signal CL1 applied to the accumulator (SR1, SR2, and SR3). The Y input is always applied to A1 and A2, and the current bit of X determines if the current partial product in SR1 through SR3 will be parallel loaded to the adder and the new partial product parallel loaded back in to SR1 through SR3. If the current bit of X is 0, the partial product will only shift. If the current bit of X is a 1, the partial product will have Y added to it. The sequence of operations and partial products is shown below for the example Y = 1011, X = 1001.

X-bit	Operation	SR Mode	Partial product
1	Add Y	Parallel load	00001011
	Shift	Shift right	00010110
0	Null	Do not load	
	Shift	Shift right	00101100
0	Null	Do not load	
	Shift	Shift right	01011000
1	Add Y	Parallel load	01100011

The answer matches the one obtained in the above long multiplication example.

8.3.3 Modern LSI Multiplier IC's

The effort of constructing and debugging the hardware multiplier of Fig. 8.7 is unnecessary, because self-contained high-speed LSI multipliers have been developed. Hardware multiplier chips offer higher speed due to improved algorithms, easy interfacing to microcomputers, simplified programming, and programmable functions other than multiplication. The specifications of several such devices are summarized in Table 8.5.

A *combinatorial* multiplier such as the MPY-8 (TRW) is a good choice for real-time signal processing because of its high speed: an 8 × 8 parallel multiply requires less than 0.2 μs. A calculator-like multiplier such as the AM9511 (Advanced Micro Devices) has somewhat slower speed but offers much higher precision, executing 16-bit or 32-bit operations in either fixed- or floating-point format. The AM9511 can also divide and can evaluate trigonometric functions.

TABLE 8.5 MULTIPLIER IC COMPARISON

Method	Example	Speed	Cost
Serial pipeline	AM 25LS14 (Advanced Microdevices)	$8 \times n$; 40 MHz clock (max)	~ \$15
Full parallel ("combinatorial")	MPY-8 (TRW)	8 × 8 170 ns	~ \$100
Calculator-like (stack; floating-point; special functions)	AM 9511	4 MHz clock Add = 20 cycles FP MPY = 200 cycles Trig = 10^3 cycles	~ \$200
	MM 57109 (National)	0.5–10^3 ms	~ \$20

Multiplication speeds range from 10 μs to 100 μs, depending upon the precision, and the time required to evaluate trigonometric functions is about a millisecond. This is still considerably faster than the typical pocket calculator. A device such as the AM9511 is a good general-purpose arithmetic processor for microcomputer applications.

If the high cost of either of these devices is a problem, there are several alternatives (Table 8.5) which are much cheaper. However, these involve tradeoffs in speed or in complexity of interfacing and programming. For example, the MM 57109 (National) does essentially all of the calculator-like arithmetic functions of the AM 9511 at one-tenth of the cost, but it is about 100 times slower. This is an excellent choice for slow data-logger applications where one can afford to wait a second for an answer. By contrast, the serial pipeline multiplier AM25LS14 (Advanced Micro Devices) is as fast as the MPY-8 at one-tenth the price. Its high speed is enhanced by a clever algorithm which shortcuts the multiplication whenever one of the inputs contains a string of 1's or a string of 0's. However, much of the apparent cost saving is eliminated by interfacing complexity, since one of the inputs must be fed in serially. This requires the user to supply additional hardware (a shift register) and/or considerable software control over the multiplication process.

The combinatorial approach of multipliers such as the MPY-8 exploits the logical operations performed in forming a product of binary numbers:

$$
\begin{array}{cccccccc l}
 & & & & X_3 & X_2 & X_1 & X_0 & \text{Multiplicand} \\
 & & & & Y_3 & Y_2 & Y_1 & Y_0 & \text{Multiplier} \\
\cline{5-8}
 & & & & A_3 & A_2 & A_1 & A_0 & \text{A partial product} \\
 & & & B_3 & B_2 & B_1 & B_0 & & \\
 & & C_3 & C_2 & C_1 & C_0 & & & \\
 & D_3 & D_2 & D_1 & D_0 & & & & \\
\hline
S_7 & S_6 & S_5 & S_4 & S_3 & S_2 & S_1 & S_0 & \text{Final product}
\end{array}
$$

Each entry in a partial product is just a single bit of the multiplier and a single bit of the multiplicand ANDed together [Fig. 8.8(a)]. The final product is obtained by adding together all terms using single-bit full adders [Fig. 8.8(b)]. Each of the square boxes is shorthand for a full adder [Fig. 8.8(c)]. Extremely high speed results, since only asynchronous logic (AND gates and adders) are used; no shift registers, temporary latches, or internal clocking are required. The cost is high because approximately N^2 AND gates and adders are required to perform an N-bit multiplication. This approach is reasonable for 8-bit signal processing, but becomes prohibitive for high-precision applications. For example, a 32-bit multiplication would require about 1000 full adders.

An example of a fast calculator-like arithmetic processor such as the AM9511 is shown in Fig. 8.9. Data is entered in 8-bit bytes through the data buffer into the data stack. Also entered through the same buffer is a command word which selects the function to be performed. The processor then operates independently, looking up the appropriate subroutines in its ROM, placing the answer back in an internal register stack, and signaling the controlling microprocessor when done.

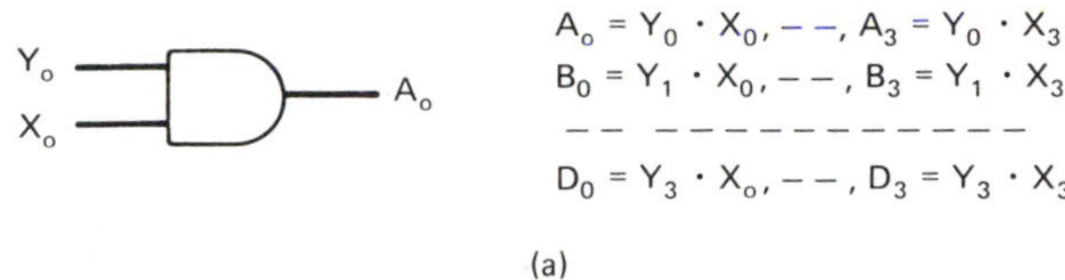

Y₀
X₀
A₀
$A_o = Y_0 \cdot X_0, --, A_3 = Y_0 \cdot X_3$
$B_0 = Y_1 \cdot X_0, --, B_3 = Y_1 \cdot X_3$
$D_0 = Y_3 \cdot X_o, --, D_3 = Y_3 \cdot X_3$
(a)

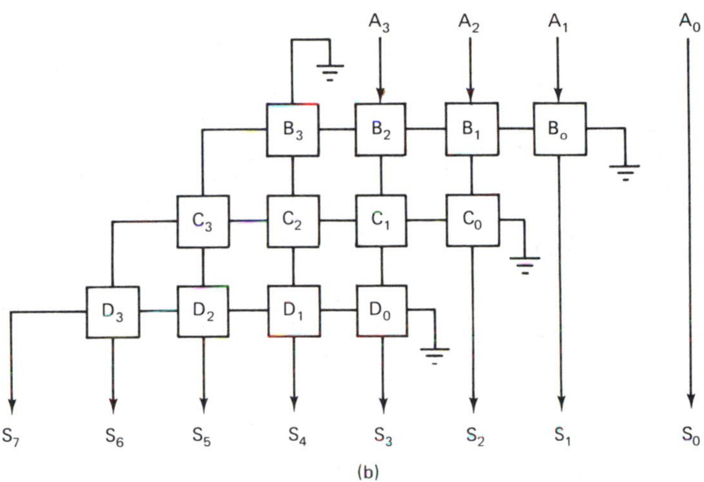

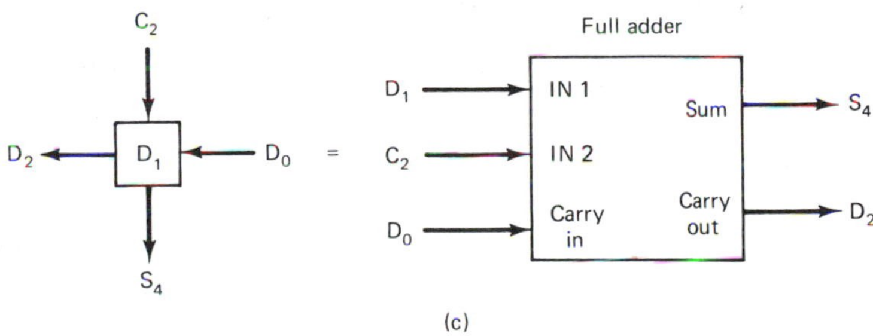

Figure 8.8 Combinatorial multiplier design. (a) Partial products are sums of AND's. (b) Network of full adders forms a multiplier. (c) Adder notation shorthand. (*Source:* TRW and Waser article)

There are both speed and cost advantages in transferring the responsibility for arithmetic computations to a satellite processor. The number of clock cycles needed to execute arithmetic and special functions in these devices is summarized in Table 8.6. Although the number of cycles for a given operation may seem large, the speed is still about a factor of 100 faster than the same operation done in software using only a microprocessor. Although the cost ($\simeq$\$150) is large, these devices may actually save costs where considerable numerical computation is required, because there is no need to develop and test relatively complex arithmetic subroutines.

Am9511A

Arithmetic Processor
Advanced Micro Devices
Advanced MOS/LSI

DISTINCTIVE CHARACTERISTICS

- Replaces Am9511
- Fixed point 16 and 32 bit operations
- Floating point 32 bit operations
- Binary data formats
- Add, Subtract, Multiply and Divide
- Trigonometric and inverse trigonometric functions
- Square roots, logarithms, exponentiation
- Float to fixed and fixed to float conversions
- Stack-oriented operand storage
- DMA or programmed I/O data transfers
- End signal simplifies concurrent processing
- Synchronous/Asynchronous operations
- General purpose 8-bit data bus interface
- Standard 24 pin package
- +12 volt and +5 volt power supplies
- Advanced N-channel silicon gate MOS technology
- 100% MIL-STD-883 reliability assurance testing

GENERAL DESCRIPTION

The Am9511A Arithmetic Processing Unit (APU) is a monolithic MOS/LSI device that provides high performance fixed and floating point arithmetic and a variety of floating point trigonometric and mathematical operations. It may be used to enhance the computational capability of a wide variety of processor-oriented systems.

All transfers, including operand, result, status and command information, take place over an 8-bit bidirectional data bus. Operands are pushed onto an internal stack and a command is issued to perform operations on the data in the stack. Results are then available to be retrieved from the stack, or additional commands may be entered.

Transfers to and from the APU may be handled by the associated processor using conventional programmed I/O, or may be handled by a direct memory access controller for improved performance. Upon completion of each command, the APU issues an end of execution signal that may be used as an interrupt by the CPU to help coordinate program execution.

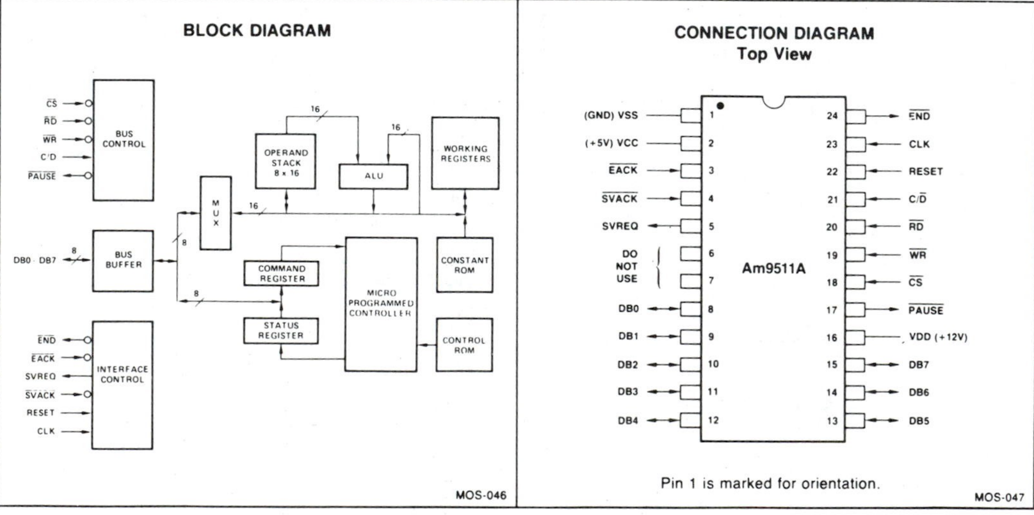

Figure 8.9 Calculator-like multiplier circuit. (Copyright © 1980 Advanced Micro Devices, Inc. Reproduced with permission of copyright owner.)

TABLE 8.6 NUMBER OF CLOCK CYCLES TO EXECUTE ARITHMETIC AND SPECIAL FUNCTIONS IN THE AM 9511[a]

	Add	Subtract	Multiply	Divide
16 × 16 Fixed-point	15	30	90	90
32 × 32 Fixed-point	20	40	200	200
32 × 32 Floating-point	50–30	70–400	150–170	150–180
32-bit special functions	Sq. rt.	Sin/cos, etc.		Log/exp
	800	4000		5000–7000

[a]Maximum clock frequency 2 MHz (regular version) or 4 MHz (high-speed version).

PROBLEMS

8.1. Carry out the binary addition below. Use the truth table for the full adder (Table 8.2). Check your answer by converting both the original numbers and your answer to base 10.

$$0\,1\,0\,1\,0\,0\,1\,1 = A = (\quad)_{10}$$

$$1\,0\,1\,1\,0\,1\,1\,0 = B = (\quad)_{10}$$

$$= A + B = (\quad)_{10}$$

8.2. Show how the function of the two AND and one OR gate of the full adder (Fig. 8.3) may be replaced by three NOR gates, so that the full adder could be implemented if necessary using ½ 7486 and ¾ 7400. Use De Morgan's theorem.

8.3. Sketch the circuit for a 4-bit full adder constructed using XOR and NAND gates. Assume you have XOR's (4 per package) and NAND's (4 per package). Show your result as a wiring diagram for the actual minimum number of packages required (e.g., 2-7486's and 3-7400's). Now aren't you glad that complete adder chips are available to make this unnecessary?

8.4. Examine the gate circuit diagram for the 74181 ALU in Table 8.4. Select 4 of the 16 functions in the table, and show how the select lines S_0 to S_3 cause a given function to be performed. You may ignore the carry bit section.

8.5. Try binary multiplication by the two basic methods: repetitive addition and long multiplication. Write out the various partial sums along the way, and compare the number of steps required. Try $(11)_{10} \times (5)_{10}$.

REFERENCES

More complete bibliographic information for the books listed below appears in the annotated bibliography at the end of the book.

DEMPSEY, *Basic Digital Electronics with MSI Applications*

HIGGINS, *Experiments with Integrated Circuits*, Experiments 12 and 13

Scientific American, *Microelectronics*

WILLIAMS, *Digital Technology*

ADDITIONAL REFERENCES

BRYANT, JACK, and MANOT SWASDEE, "How to Multiply in a Wet Climate," *Byte* (April 1978), pp. 28–35, 100–110 (Part 1); (May 1978), pp. 104–114 (Part 2). Application example for S-100 bus computer; uses TRW MPY 8.

MICK, JOHN R., and JOHN SPRINGER, "Single Chip Multiplier Expands Digital Role in Signal Processing," *Electronics*, May 13, 1976, pp. 103–108. Serial pipeline Am 25LS14; Booth's algorithm.

WASER, SCHLOMO, "High Speed Monolithic Multipliers for Real Time Digital Signal Processing," *IEEE Computer* (October 1978), pp. 19–29. Compares algorithms and specific hardware; emphasis on full parallel methods.

WEISSBERGER, ALAN J., and TED TOAL, "Tough Mathematical Tasks Are Child's Play for Number Cruncher," *Electronics*, Feb. 17, 1977, pp. 102–107. Calculator-like micro chip; Independent processor: National 57109, includes trig functions.

9

Digital To Analog and Analog to Digital Conversion

9.1 INTRODUCTION: BRIDGING THE GAP TO THE REAL WORLD

As the adaptation of Leonardo's classic drawing illustrates (Fig. 9.1), analog and digital functions in any data gathering or control system form a continuous closed loop with information flow in both directions. This chapter bridges the gap between the digital and analog worlds, and in doing so sets the stage for the analog half of the book. Signals from the analog system (Fig. 9.2)—physical variables such as voltage, resistance, or capacitance—are converted from linear to binary form using an *analog-to-digital converter* (A/D). A digital computer can adjust parameters in the experiment, generating control voltages from binary numbers using a *digital-to-analog converter* (D/A). Although direct digital pathways of stimulus or response are also possible, A/D and D/A methods are essential whenever the parameters to be measured or contolled are linear.

The analog world has problems that are absent in the binary world. Since signals are continuous rather than just true or false, noise and grounding become a greater problem. Digital experiments often work right the first time; this is less often true for analog experiments. A balanced knowledge of instrumentation, however, must include powerful analog signal processing techniques to condition analog signals to or from the binary world.

This bridge chapter is placed at this part of the book since digital electronics is incomplete until one knows how to bring a voltage in (A/D) or out (D/A). On the other hand, we have not yet discussed a basic analog component, the *operational amplifier* (op amp), used in both D/A and A/D circuits. We include enough

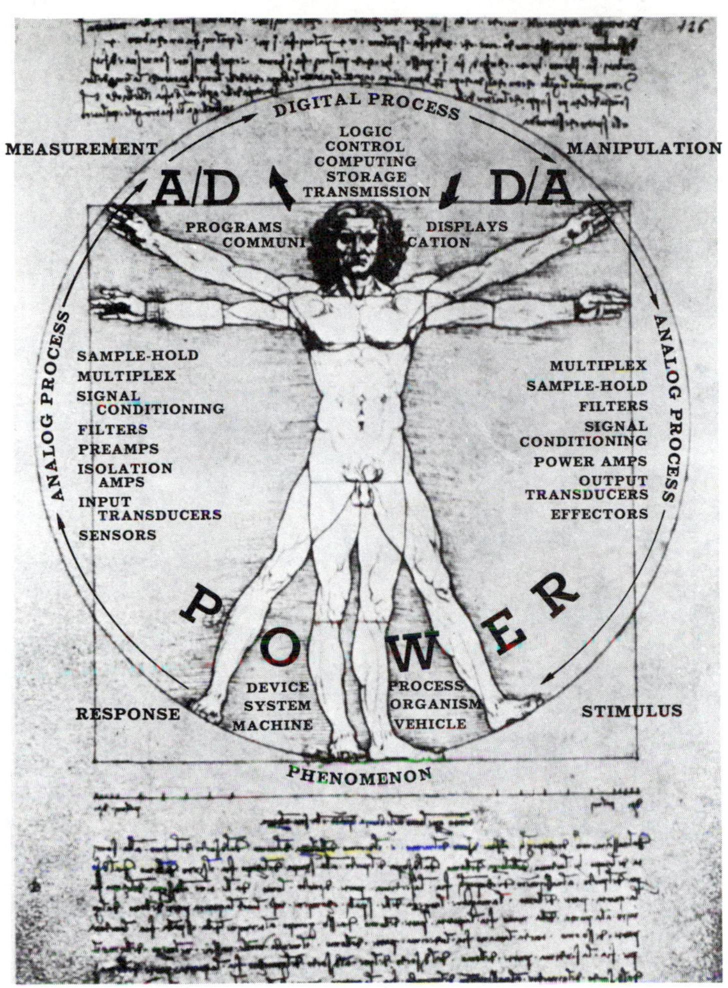

Figure 9.1 The interaction between a "real world" phenomenon and the digital image of it involves a cycle going both *to* (D/A) and *from* (A/D) the real world. (*Courtesy* D. Sheingold, Analog Devices Inc., with apologies to Leonardo da Vinci)

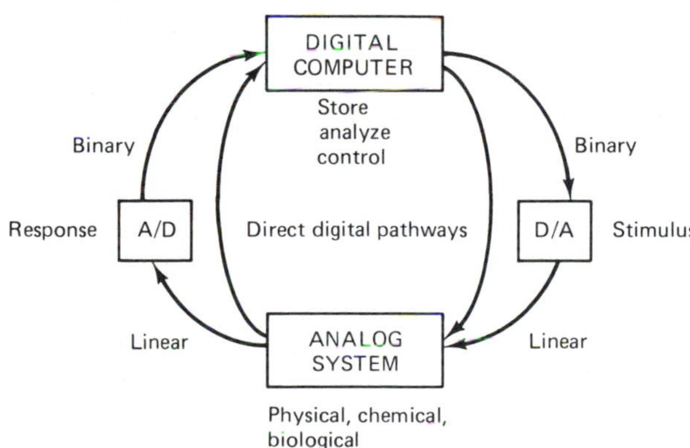

Figure 9.2 A/D and D/A converters form a bridge between analog and digital worlds.

about op amps in Chapter 8 to see how A/D and D/A circuits work. Another component, the CMOS *analog* switch, is mentioned and used, but not explained until Chapter 20.

9.1.1 Digital-to-Analog Converter Applications

Graphics. Analog information may be drawn graphically on an oscilloscope or with a pen on a hard copy plotter. Often a conventional analog plotter or oscilloscope is interfaced to the computer with D/A converters (Fig. 9.3). If the discrete points are drawn close enough together with line segments between them, the curves look continuous. Ten or twelve bits of resolution is adequate for most graphs, and even 8 bits is enough for the small screen of a typical oscilloscope.

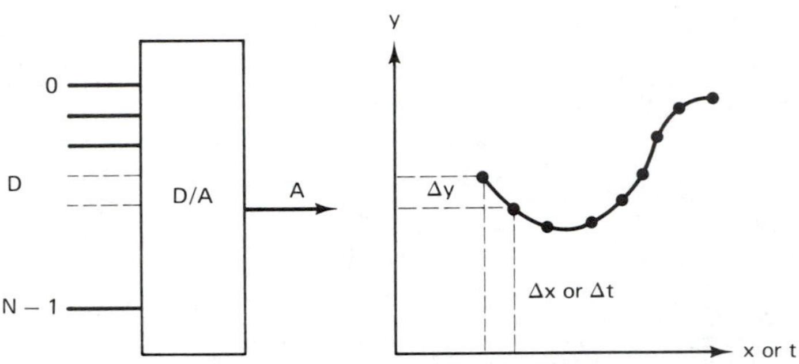

Figure 9.3 D/A converter as a graphic plotting device.

Control. Most analog power supplies have an external control voltage input, making computer control straightforward using a D/A. The analog control signal may adjust the temperature of a furnace, locate a position using a servomotor, or control any physical variable.

Automated testing. D/A converters provide the basic stimulus for many automated measurement systems (Fig. 9.2): physical, chemical, biological, industrial, etc.

Function generation. The conventional oscillator or function generator can be replaced by a real-time signal generated by a D/A. This method has the advantage that an arbitrary programmable waveform can be generated from a computer memory or permanently stored in read-only memory (ROM). A sequence of numbers clocked out of memory (Fig. 9.4) can, for example, generate a ramp approximation which, after simple analog filtering, is indistinguishable from that of a function generator.

Multiplying D/A. Scale factors or gains can be set digitally using a *multiplying* D/A converter. One input is the signal to be controlled, and the other input is a binary number which sets the amplitude. This brings the advantages of digital control to analog signals. For example, a binary "volume control" can adjust the amplitude of an audio test signal. The multiplying D/A has superseded the cumbersome, expensive, and slow servomotors or mechanical switches formerly used in this application.

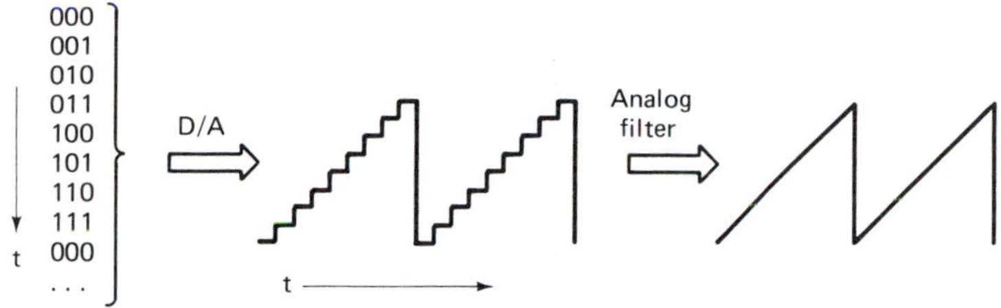

Figure 9.4 D/A converter as a function generator.

A/D converters. Several kinds of A/D converters use a D/A together with a comparator as a feedback element, comparing the signal with the computer's most recent approximation of it.

9.2 DIGITAL TO ANALOG CONVERTER CIRCUITS

9.2.1 The D/A Concept

A D/A converter is an amplifier whose gain can be programmed digitally. The gain of the operational amplifier circuit of Fig. 9.5 can be adjusted with passive circuit component values. The negative feedback from op amp output and input acts to keep the voltage across the op amp's input terminals zero (Chapter 10). The op amp has essentially infinite input impedance, so negligible current flows into the input terminals. Application of Ohm's law then gives:

$$V_o = -(R_f/R_i)\,V_R = -R_f I_i \tag{9.1}$$

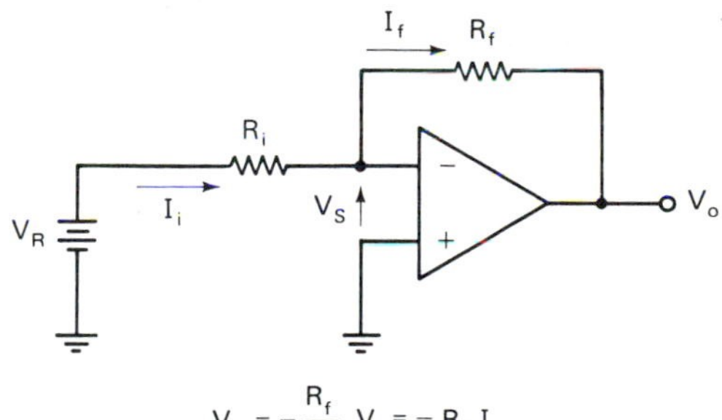

Figure 9.5 Variable gain amplifier circuit.

If V_R is a fixed reference voltage, the output voltage may be varied by adjusting the gain R_f/R_i. A D/A adjusts input current I_i by switching the value of R_i digitally (Fig. 9.6). Resistors may be connected in parallel, depending upon which switches are closed by binary control signals. The total current I_i flowing into the circuit is the sum of the currents flowing through individual resistors:

$$I_i = \sum I_n = V_R \sum 1/R_n \tag{9.2}$$

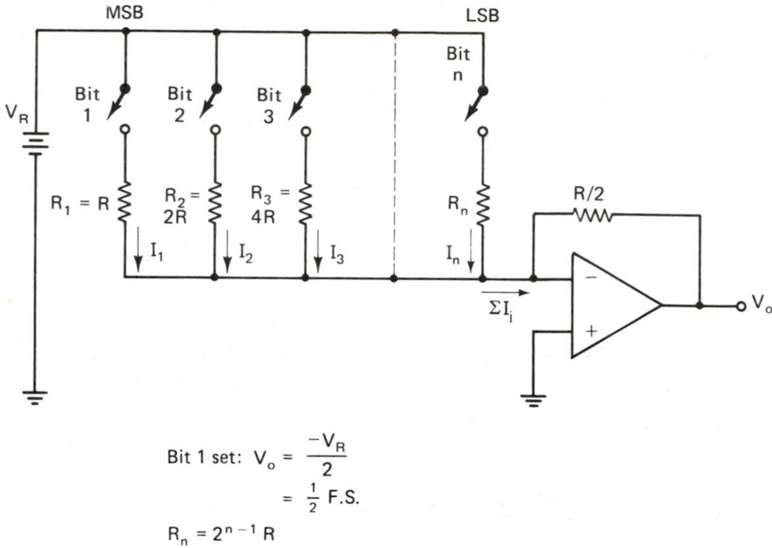

Figure 9.6 D/A converter concept as binary gain adjustment.

and by Eq. (9.1), this determines the output voltage. Each resistor is twice as large as its neighbor:

$$R_n = 2^{n-1}R \tag{9.3}$$

If only switch 1 is closed, the output voltage is half the reference voltage or half of the full scale output:

$$\text{Bit } 1 = 1 \qquad V_o = -\ V_R/2$$

If switch 2 alone is closed, the output is one-quarter full scale:

$$\text{Bit } 2 = 1 \qquad V_o = -\ V_R/4$$

Successively smaller contributions are made by each successive bit:

$$\text{Bit } N = 1 \qquad V_o = -\ V_R/(2^N)$$

Thus, a D/A generates a voltage in a way which resembles putting a series of binary weights onto a chemical balance. A 5-bit example is shown below.

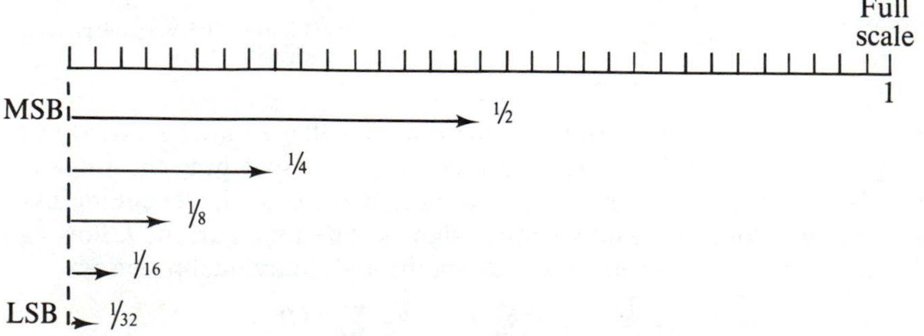

If all of the bits are *set* (all switches closed), the output is very nearly the full-scale reference voltage:

$$\frac{V_o}{V_R} = \frac{1}{2} + \frac{1}{4} + \frac{1}{8} + \frac{1}{16} + \frac{1}{32} = \frac{31}{32} = 1 - \frac{1}{32}$$

or, more generally,

$$\frac{V_o}{V_R} = \sum \begin{bmatrix} \text{all} \\ \text{bits} \\ 1 \end{bmatrix} = 1 - 2^{-N}$$

Generalizing from this example, the output voltage of a D/A is:

$$V_o = -V_R \sum_{m=1}^{N} x_m (\tfrac{1}{2})^m \tag{9.4}$$

where x_m takes on the values 0 or 1. The maximum output voltage with all bits set is:

$$(V_o)_{\text{max}} = -V_R(1 - 2^{-N}) \tag{9.5}$$

for binary input (1111....1)

This becomes closer to the full-scale value V_R as N becomes larger. For example, V_o is within 0.1% of V_R for $N = 10$.

The circuit of Fig. 9.6 is not used in practice because of limitations in resistance values, resistor adjustment, and settling time. For example, R might be chosen as 10 KΩ, to keep the current drain low. For a 10-bit D/A, $R_{10} = 2^9 R =$ 5 MΩ. Such large resistance values are difficult to achieve on IC's. In addition, for R_N to be meaningful, R_1 must be precise to 1 part in 2^N. This requires precision trimming of many different resistor values. Most serious is the switching time limitation, set by the LSB resistor and by stray capacitance, which can easily approach 100 pF. In the 10-bit example, the settling time T_s would be:

$$T_s = (5 \times 10^6 \Omega) \times (10^{-10}\text{ F})$$
$$= (\tfrac{1}{2}) \times 10^{-3}\text{ s} \tag{9.6}$$

A millisecond settling time is very slow for a device whose digital components operate 1000 times faster. This also limits the practical number of bits, since each additional bit reduces the speed by a factor of 2.

9.2.2 The R/2R Ladder Network D/A

The *R/2R ladder* (Fig. 9.7), which is the most common D/A circuit, overcomes the problems of the previous circuit. Only two resistor values are required, regardless of the number of bits, which simplifies resistor trimming. The circuit has no voltage *changes* at the switch terminals. With no voltage transients, RC settling problems are reduced. The principle of the R/2R ladder [Fig. 9.8(a)] is that at each node of the ladder, current flow from V_{ref} is divided in two. The current divides equally because the resistor values are $2R$ for each leg [Fig. 9.8(b)]. No

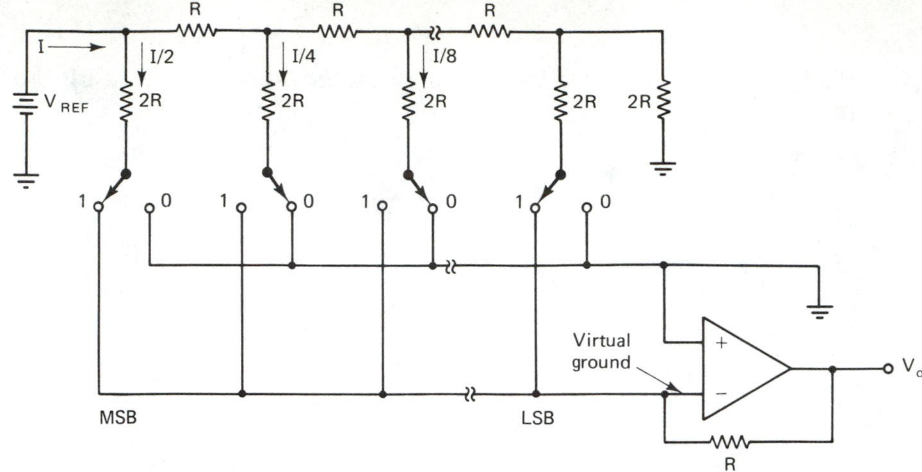

Figure 9.7 The R/2R ladder D/A circuit. The bits set in this example correspond to the binary word 101...1.

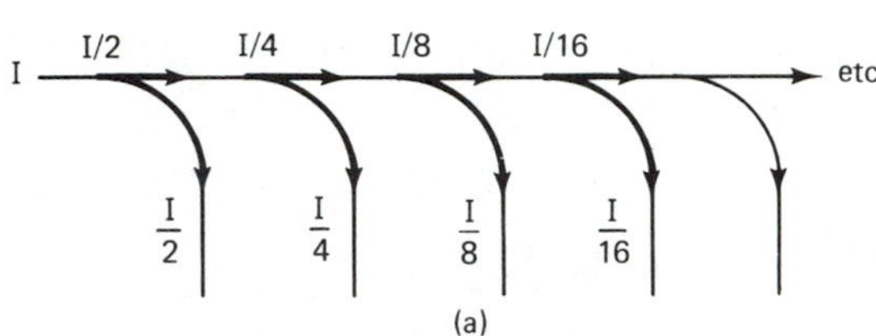

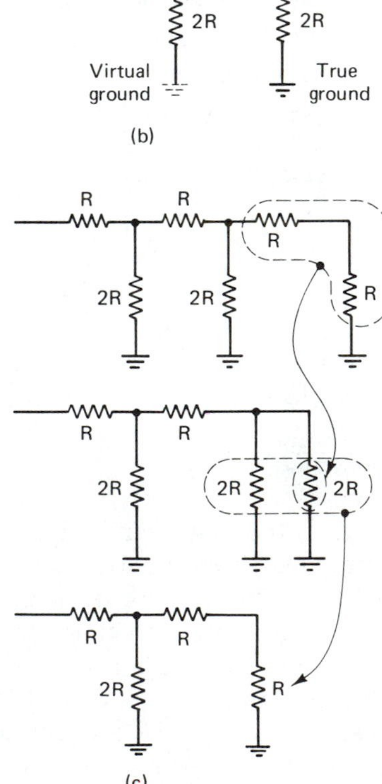

Figure 9.8 Principles of the R/2R ladder. (a) Binary current divider. (b) Each node sees two 2R legs to ground. (c) Successive nodes are equivalent.

matter what switch connections are made in Fig. 9.7, current finds a path either to a true ground or to a *virtual ground* (essentially 0 volts) at the op amp's negative input terminal (see Chapter 10). Each node gives two paths of resistance $2R$ because of the geometry of the ladder circuit [Fig. 9.8(c)]. Beginning at the right side of Fig. 9.8(c), two resistors $2R$ in parallel are equivalent to one resistor of value R. Successive series–parallel resistor combinations reduce to the R/2R circuit of Fig. 9.8(b) at each node. The current flow from V_{ref} is thus constant, but is reduced by a factor of 2 at each node. Binary bits set to 1 direct currents into the op amp and thus become contributions to the output voltage, while the currents from bits set to 0 are shunted to ground.

An example of a commercial D/A is given in Table 9.1. Formerly an expensive component, 8-bit D/A's now cost only a few dollars, greatly expanding D/A utilization.

9.2.3 The Multiplying D/A Converter

This circuit [Fig. 9.9(a)] is an R/2R ladder D/A with the reference voltage supplied externally. Although the digital inputs are typical TTL logic levels (0 V to 5 V), V_{ref} can take on any value up to the D/A *analog* power supply, and can even be negative. This is because the digital inputs are isolated at MOSFET gates, while V_{ref} creates current flow between source and drain of an FET analog switch. This portion is just a bar of semiconducting material, so current can flow in either direction. As a result, the multiplying D/A is useful as a digital volume control even for ac signals.

In normal operation [Fig. 9.9(b)], an op amp is wired externally to complete the circuit. The circuit acts as a multiplier:

$$V_o = -(A \times D) \tag{9.7}$$

where A is the analog input signal and D, the magnitude of the binary input, sets the output amplitude. This is called a *two-quadrant multiplier*, because A can be either positive or negative, but D is only positive. Input coding for this circuit is shown in Table 9.2.

A circuit modification [Fig. 9.9(c)] allows D also to have either sign, giving a *four-quadrant multiplier*. As with binary negative numbers (Chapter 8), the MSB is designated as the sign bit. Input coding for this circuit is *offset binary* (modified 2's complement), also shown in Table 9.2. Current from a switch whose binary input is a 1 is steered to output I_{OUT1}. Current resulting from an input bit 0 is fed to a second amplifier which causes a current contribution of opposite sign at I_{OUT1}. The total current I_{OUT1} is then the difference of the 1 and 0 bit contributions. Resistors R_1 and R_2 need not match the internal resistors in the D/A, but they should match each other to ensure that negative bits will cause the same magnitude output signal as positive bits. With the MSB a logic 1 and all other bits a logic 0, a ½ LSB difference current exists between I_{OUT1} and I_{OUT2}, creating an offset of ½ LSB. To shift the output voltage to 0 as in Table 9.2, resistor R_3 adds an additional ½ LSB of current into terminal I_{OUT2}.

Specifications of a representative multiplying D/A are given in Table 9.3.

TABLE 9.1 D/A EXAMPLE SPECIFICATIONS[a]

DAC-100

10 BIT DIGITAL-TO-ANALOG CONVERTER

FEATURES

- **Complete.......................... Internal Reference**
- **Flexible .... 0 to 2mA Output**
- **Fast Settling......... 225nsec (8 Bits), 375nsec (10 Bits)**
- **Stable .... Tempcos to ±15ppm/°C Maximum**
- **0°C/+70°C, −25°C/+85°C, −55°C/+125°C Models Available**
- **TTL and DTL Compatible Logic Inputs**
- **Wide Supply Range .... ±6V to ±18V**
- **8 and 10 Bit Versions Available**
- **MIL-STD-883 Class B Processing Models Available**
- **Low Cost Q3, Q4 Series**

GENERAL DESCRIPTION

The DAC-100 is a complete 10-bit resolution digital-to-analog converter constructed on two monolithic chips in a single 16-pin DIP. Featuring excellent linearity vs. temperature performance, the DAC-100 includes a low tempco voltage reference, ten current source/switches and a high stability thin-film R-2R ladder network. Maximum application flexibility is provided by the fast current output and by matched bipolar offset and feedback resistors which are included for use with an external op amp for voltage output applications. Although all units have 10-bit resolution, a wide choice of linearity and tempco options is provided to allow price/performance optimization.

The small size, wide operating temperature range, low power consumption and high reliability construction make the DAC-100 ideal for aerospace applications. Other applications include use in servo-positioning systems, X-Y plotters, CRT displays, programmable power supplies, analog meter movement drivers, waveform generators and high speed analog-to-digital converters.

PIN CONNECTIONS

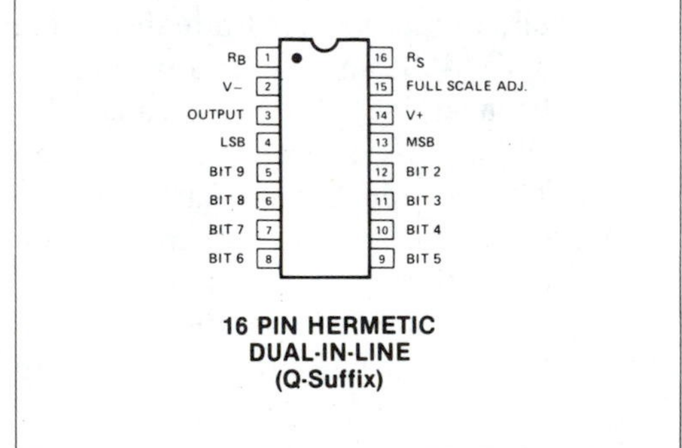

SIMPLIFIED SCHEMATIC

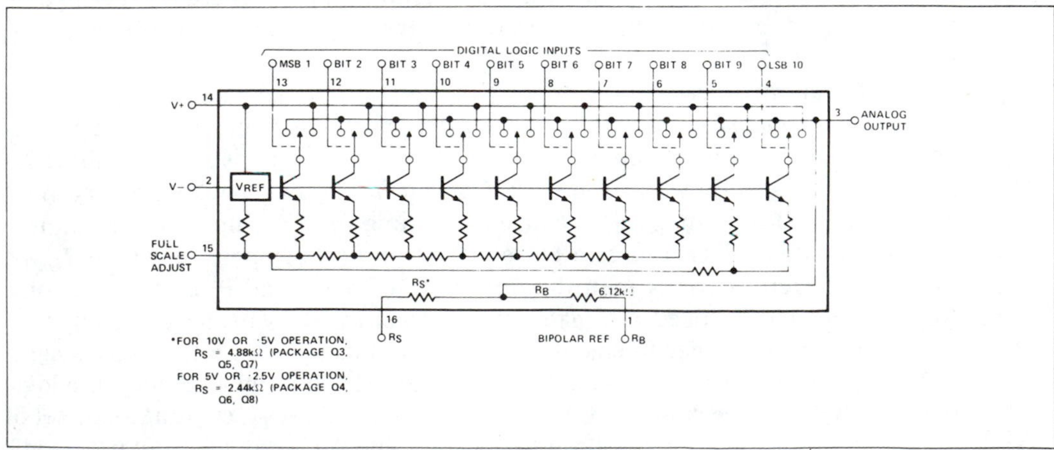

[a] DAC 100. *Source:* Precision Monolithics Inc.

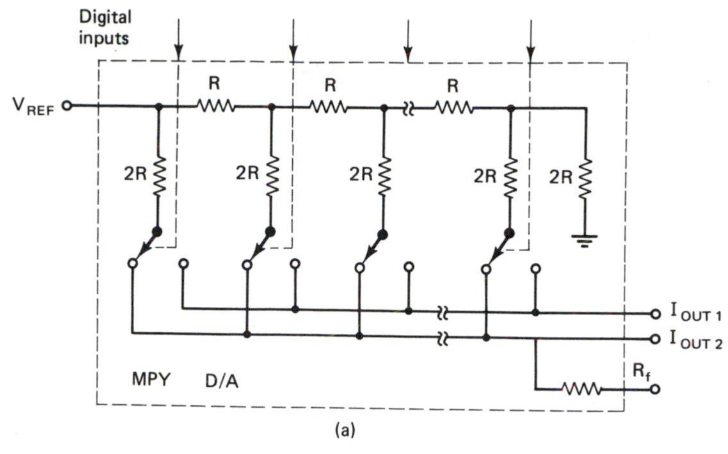

(a)

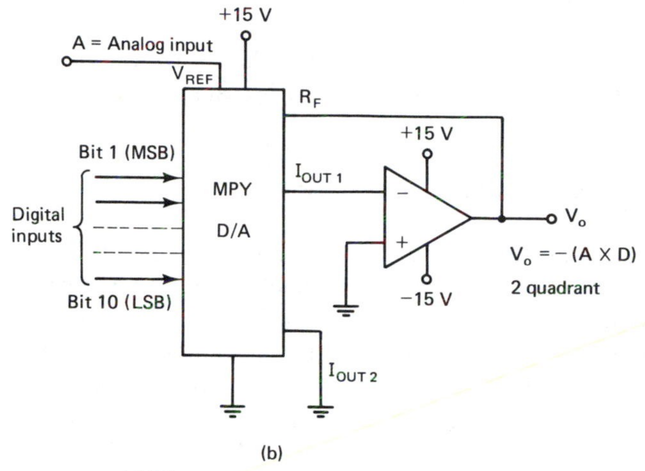

(b)

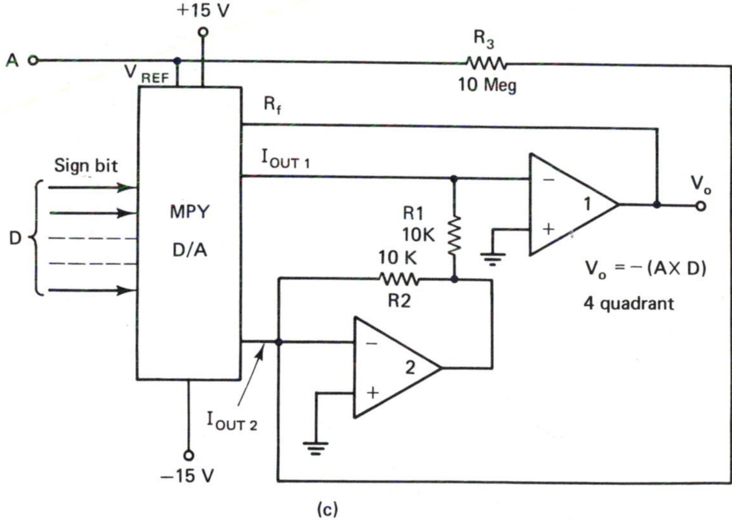

(c)

Figure 9.9 Multiplying D/A converter. (a) Internal IC configuration. (b) Two quadrant multiplier. (c) Four quadrant multiplier.

TABLE 9.2 MULTIPLYING D/A CONVERTER CODING TABLE FOR THE CIRCUIT OF FIG. 9.9(b) AND (c).

Input binary word										Output voltage, fraction of full scale: Unipolar binary, two-quadrant	Offset binary, four-quadrant
1	1	1	1	1	1	1	1	1	1	$-(1-2^{-10})$	$-(1-2^{-9})$
1	0	0	0	0	0	0	0	0	1	$-(\frac{1}{2}+2^{-10})$	$-(2^{-9})$
1	0	0	0	0	0	0	0	0	0	$-\frac{1}{2}$	0
0	1	1	1	1	1	1	1	1	1	$-(\frac{1}{2}-2^{-10})$	$+(2^{+9})$
0	0	0	0	0	0	0	0	0	1	-2^{-10}	$+(1-2^{-9})$
0	0	0	0	0	0	0	0	0	0	0	$+1$

9.3 D/A APPLICATIONS

In addition to generating control voltages D/A converters have numerous real-time signal generation and signal processing applications. Two examples are given here.

9.3.1 D/A Function Generator

The familiar analog oscillator or function generator may be replaced by a D/A driven by a binary counter or a waveform stored in memory. The resulting analog waveform has little harmonic distortion if a sufficient number of bits is used. The method offers programmability in frequency, waveform, and amplitude. A ramp wave is generated by driving the binary inputs of the D/A with a binary counter [Fig. 9.10(a)]. The waveform repeats itself when counter overflow brings the binary output back to zero (Fig. 9.4). The frequency can also be adjusted digitally using a programmable counter.

A symmetric triangular wave is generated by driving the D/A with an up/down counter [Fig. 9.10(b)]. The *borrow* and *carry* outputs generate control signals to steer clock pulses into the *count-up* or *count-down* inputs when the counter reaches either end of it range. The 7470 FF (a positive edge-triggered JK) toggles each time either a borrow or a carry pulse occurs, thereby enabling either gate A or gate B (see Prob. 9.3). Triangular wave generation using CMOS counters is even simpler: the Count Input line and Up/Down Select line eliminate gates A and B in Fig. 9.10(b) (see Prob. 9.2). Although the counter preset inputs in Fig. 9.10(b) are wired to cover the full-scale range, other modulus waveforms are possible (see Prob. 9.3).

An *arbitrary* waveform may be generated digitally by driving a D/A with data stored in memory [Fig. 9.11(a)]. A counter sequences through the binary address lines of the RAM, strobing out stored data as a binary input to the D/A. A waveform is initially loaded into the RAM by cycling through the counter states, bringing the Write Enable line low as each new data word is ready at the data input lines. Although shown as a RAM, a waveform can be stored permanently using a ROM.

TABLE 9.3 MULTIPLYING D/A EXAMPLE SPECIFICATIONS[a]

CMOS Low Cost 10-Bit Multiplying DAC

AD7533

PRELIMINARY TECHNICAL DATA

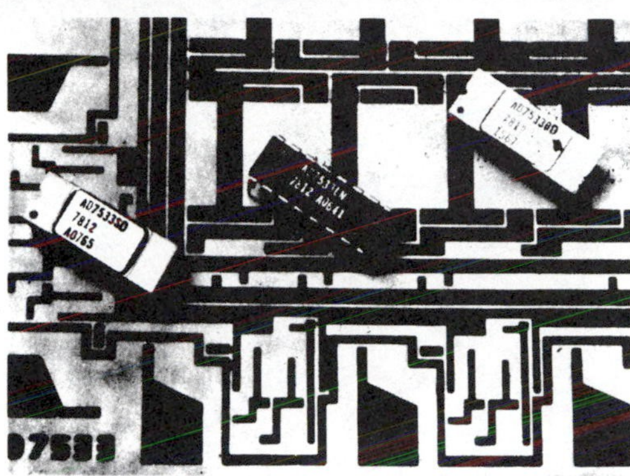

FEATURES

- Lowest Cost 10 - Bit DAC
- Direct AD7520 Equivalent
- Linearity: ½, 1 or 2 LSB
- Low Power Dissipation
- Full Four - Quadrant Multiplying DAC
- CMOS/TTL Direct Interface

APPLICATIONS

- Digitally Controlled Attenuators
- Programmable Gain Amplifiers
- Function Generation
- Linear Automatic Gain Control

GENERAL DESCRIPTION

The AD7533 is a low cost 10 - bit, 4 - Quadrant multiplying DAC manufactured using an advanced thin - film - on - monolithic - CMOS wafer fabrication process.

Pin and function equivalent to the industry standard AD7520, the AD7533 is recommended as a lower cost alternative for old AD7520 sockets or new 10 - bit DAC designs.

AD7533 application flexibility is demonstrated by its ability to interface to TTL or CMOS, operate on +5 to +15V power, and provide proper binary scaling for reference inputs of either positive or negative polarity.

FUNCTIONAL DIAGRAM

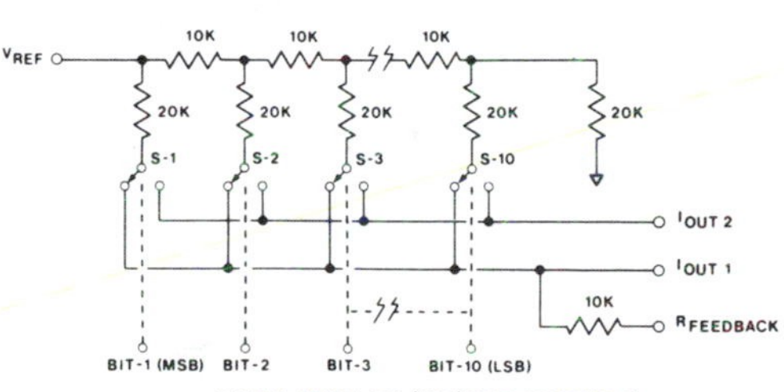

ORDERING INFORMATION

Nonlinearity	Temperature Range and Package		
	Plastic 0 to +70°C	Ceramic -25 to +85°C	Ceramic -55 to +125°C
±2 LSB	AD7533JN	AD7533AD[1]	AD7533SD[1]
±1 LSB	AD7533KN	AD7533BD[1]	AD7533TD[1]
±½ LSB	AD7533LN	AD7533CD[1]	AD7533UD[1]

Note 1: 883B Version is available. To order add "/883B" to part number shown. See note 7, page 2.

PIN CONFIGURATION

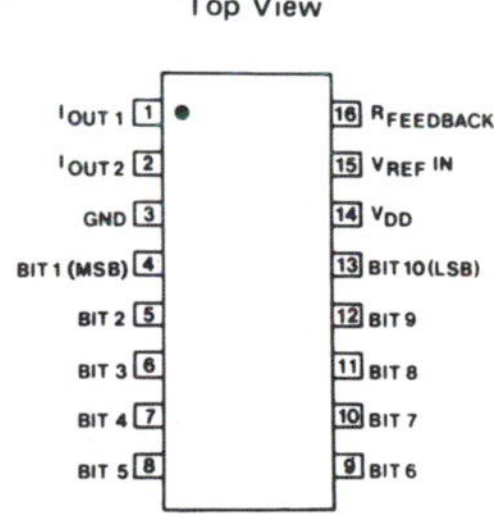

[a]AD 7533. *Source:* Analog Devices, Inc.

The vertical resolution of this waveform generator is limited by the number of bits per data word in the memory. The horizontal resolution is set by the number of words of memory or number of address lines. Horizontal resolution can be increased by a factor of 2 or 4 if the waveform is symmetric. This is particularly convenient in real-time function generation if the 16-channel resolution of

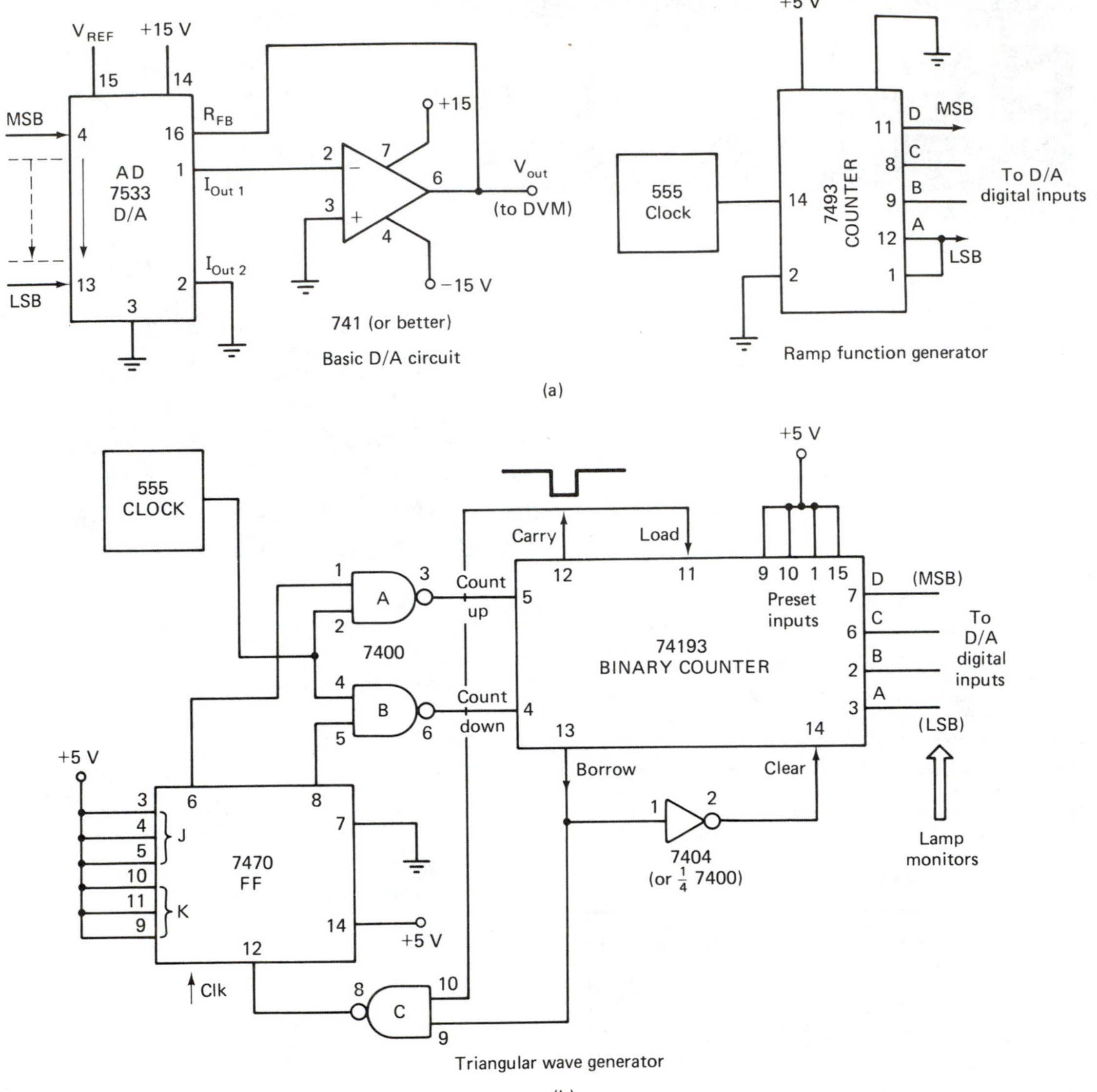

Figure 9.10 D/A function generators. (a) Ramp waveform. (b) Triangular waveform.

conventional 4-bit counters does not generate a sufficiently smooth waveform. For example, to generate a sine wave [Fig. 9.11(b)] it is sufficient to store in ROM only the first of four quadrants, generating the other quadrants by symmetry. The sine wave is strobed out from the origin towards the peak and then back down again when the count direction of the address lines is reversed. This generates a *full wave rectified* sine wave. Each time the counter reaches zero, a flip flop is toggled to set the sign bit of a *sign/magnitude* D/A [similar to Fig. 9.9(c)], giving a full ac sine wave.

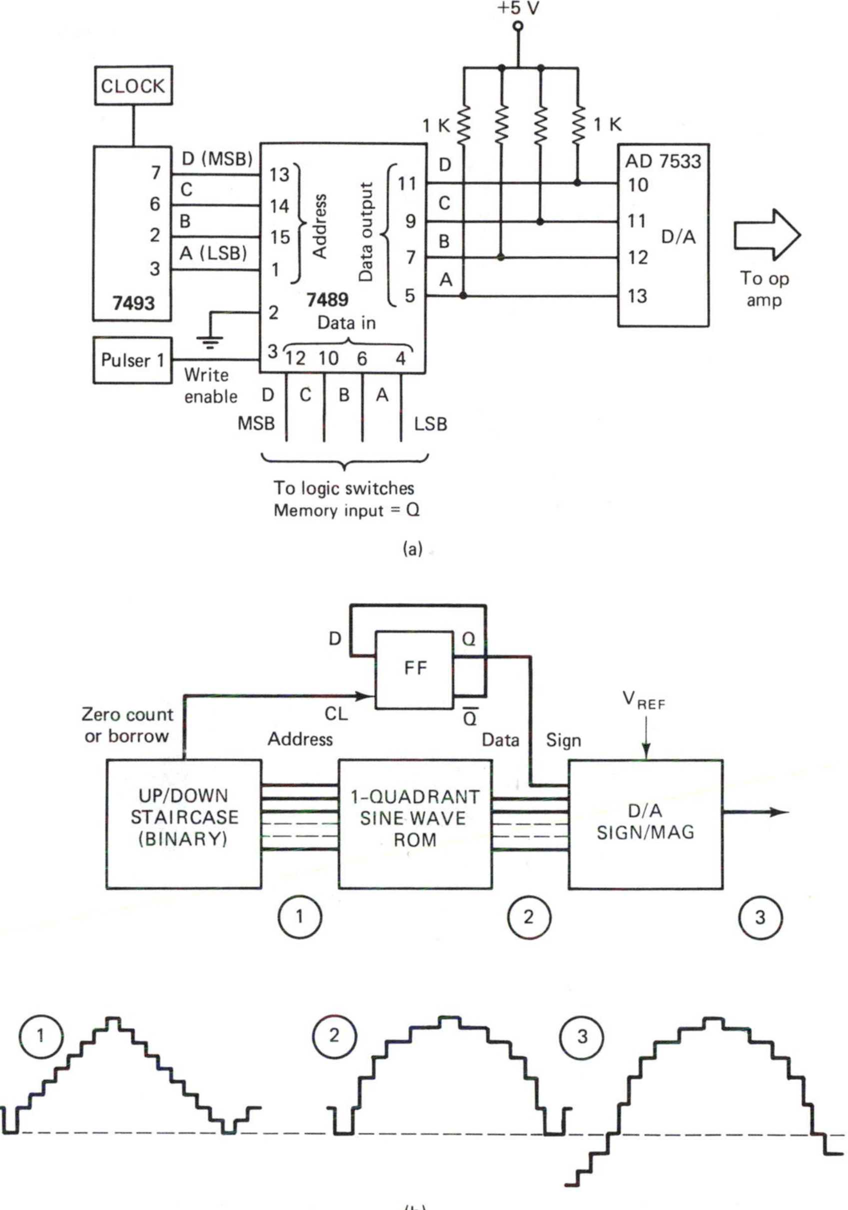

Figure 9.11 D/A function generators. (a) Arbitrary waveform using RAM or ROM. (b) Increased horizontal resolution using waveform symmetry. The D/A is a special sign/magnitude circuit whose MSB sets the sign of the voltage generated by the data bits.

9.3.2 Dot Matrix Graphic Display Using a D/A

A high-resolution graphic display terminal uses D/A's to steer the beam of an oscilloscope (Fig. 9.12). D/A's 1 and 2, driven by binary signals from a computer, position the beam with high precision. The other circuitry allows an arbitrary al-

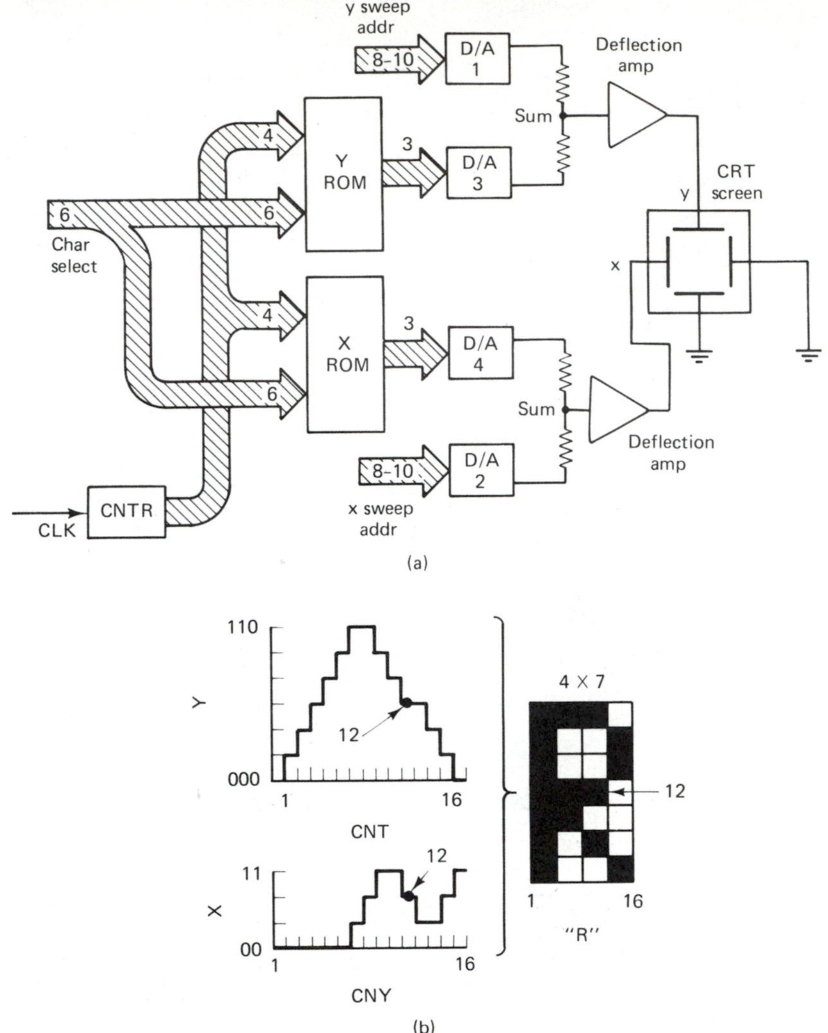

Figure 9.12 Dot matrix display using D/A's. (a) Circuit. (b) How ROM contents draw the letter *R*. (*Courtesy* Analog Devices Inc.)

phanumeric or graphic character to be generated. Thus, D/A's 1 and 2 locate the beam and D/A's 3 and 4 draw the characters. The "where" and "what" signals are added together by the resistors feeding the horizontal and vertical amplifiers. The *dot matrix* scheme approximates a character by tracing dots within a matrix. In this example, the letter R is traced within a matrix 4 dots high by 7 dots wide. However, any alphanumeric character can be adequately approximated by tracing 16 points. A 4-bit counter is therefore sufficient to sequence through the address lines of ROM's which steer data to the low-resolution X and Y D/A's drawing the character. The sequence for drawing the letter R is shown [Fig. 9.12(b)] in graphs of ΔX and ΔY, which map ROM data contents. A different ROM pattern is stored for each character, specified by the additional 6-bit character address lines. This

vector-addressed method is used for graphic-type terminals; lower resolution terminals which draw only the characters on a typewriter use fewer D/A's and a *raster scan* (a time sequence of video information) as in a TV set.

9.4 COMPARING THE FEATURES OF A/D CONVERSION METHODS

In selecting a method to convert a voltage into a binary number, a choice must be made between resolution and speed. Sometimes one needs to measure a quantity accurately but high speed is not required, as in slow temperature measurements. At the other extreme are the high-speed demands of real-time digital signal processing. The two methods which meet most of these needs are compared in Table 9.4. The *dual slope* method is used in most digital voltmeters and multimeters, because they demand high resolution and high accuracy but only moderate speed. A resolution of 3 to 6 digits is accomplished at the expense of speed: 1 to 10 measurements per second is typical. The *successive approximation* method achieves high speed (10^4 to 10^5 measurements per second) at the expense of only moderate resolution: 8 to 12 bits. The methods differ also in their accuracy; each gives an accuracy comparable to its resolution. Either method is available in inexpensive LSI IC's easily interfaced to microcomputer systems.

9.4.1 Signal Processing Subtleties and the Nyquist Theorem

The A/D method chosen depends on the characteristics of the events to be measured. The successive approximation method will encode the data present at the instant a sample is taken [Fig. 9.13(a)]. If a noise spike occurs at one of the sampling points, it will be stored as data. The dual slope method, on the other hand,

TABLE 9.4 COMPARISON OF A/D CONVERTER SPECIFICATIONS

Method	Dual slope	Successive approximation
Speed	Low 1–10/s	High 10–10^5/s
Resolution	High 3–6 digits 10–20 bits	Medium 8–12 bits
Accuracy	Excellent; 0.01% or better	Good; 0.1% is typical
Noise immunity	Excellent (integration averaging)	None; must prefilter
Typical application	Digital multimeter	Digital signal processing

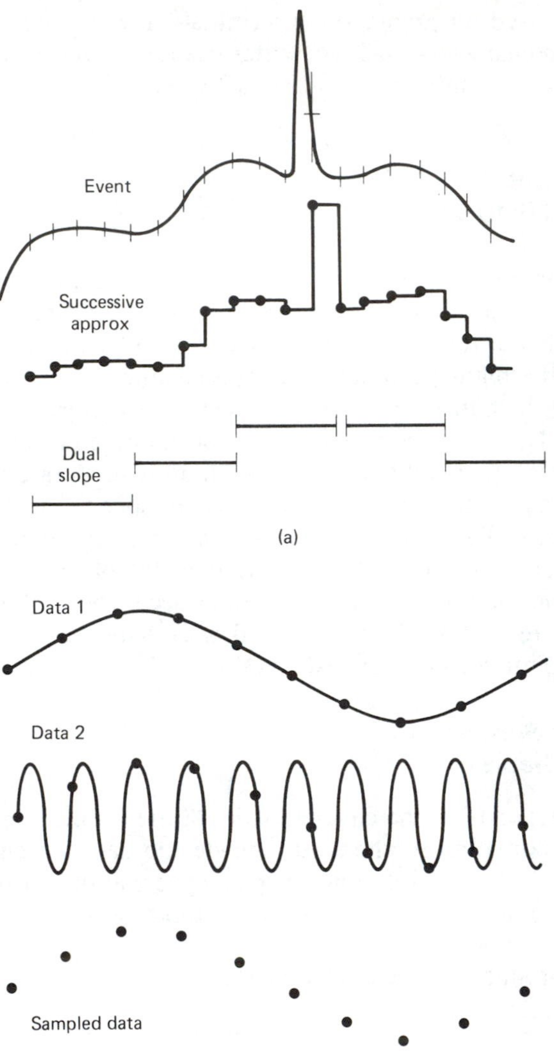

Figure 9.13 (a) Comparison of noise immunity with successive approximation and dual slope methods. (b) Nyquist images in sampled data.

averages data over the sampling interval, smoothing out momentary fluctuations. If the spike in Fig. 9.13(a) is the event of interest, the dual slope method fails and only the successive approximation method can encode it.

An important subtlety in high-speed data acquisition comes from the theory of sampled data. If many data samples are taken within one cycle of a sine wave [data 1 in Fig. 9.13(b)], the sampled data builds an accurate image of the event. But if the sampling frequency is not fast enough to follow the data [data 2 in Fig. 9.13(b)], the sampled data will be a spurious image of the real event. Both data 1 and data 2 examples lead to *identical* sampled data at the sampling rates shown. This is a consequence of the *Nyquist theorem*, which states that data must be sampled at a rate high enough to include at least two points within one cycle of the

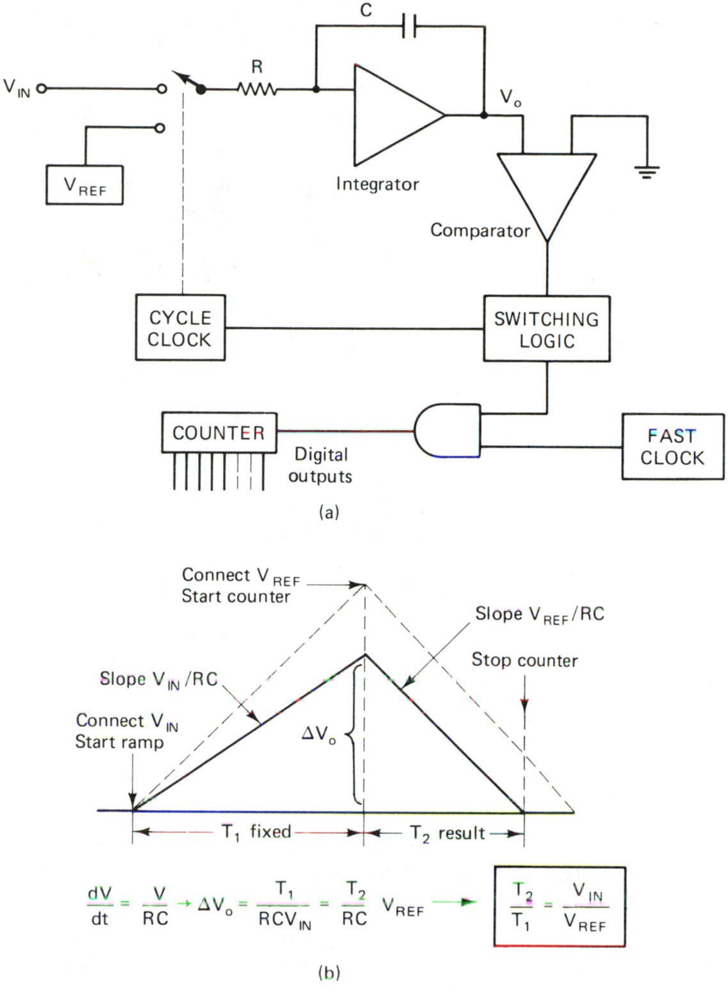

Figure 9.14 Dual slope A/D converter. (a) Circuit. (b) Integrator output V_o during one measuring cycle.

highest signal frequency. Since spurious *Nyquist images* actually generate peaks in a frequency spectrum calculated from sampled data, a high-speed data acquisition system must include a low-pass filter with a cutoff frequency set to eliminate frequencies greater than half the sampling frequency.

9.5 DUAL SLOPE A/D CONVERTER

The concept of the dual slope A/D method is shown in Fig. 9.14. The circuit includes an analog integrator whose output voltage is the time integral of the input voltage (Chapter 11). With a constant input voltage, the output of an integrator increases at a rate proportional to the size of the input voltage:

$$dV_o/dt = V_{in}/RC \tag{9.8}$$

An analog switch (Chapter 19) alternately connects the integrator to the unknown input voltage V_{in} and a fixed reference voltage V_{ref}. At the start of a cycle, the switch connects V_{in} to the integrator, whose output starts ramping upwards. At the end of a *fixed* time interval T_1, the cycle clock switches V_{ref} to the integrator. V_{ref} has the opposite polarity, so the integrator starts ramping downwards. Meanwhile, the switching logic enables the gate to start the counter. The switching logic disables the counter when the comparator senses that V_o has crossed zero. A larger input voltage [dashed line in Fig. 9.14(b)], would cause a larger integrated change ΔV_o in the fixed interval T_1. The integrator would take longer to ramp back down to zero in the second phase of the cycle, and more counts would accumulate during time T_2. Thus, the dual slope A/D measures a voltage by converting it to a time interval T_2. High resolution is easily achieved using a fast clock and a multidigit counter. The relationship between the unknown V_{in} and the measured T_2 follows from Fig. 9.14(b). The triangle with base T_1 shares a common vertical side with the triangle of base T_2.

$$\Delta V_o = (T_1/RC)\ V_{in} = N\ V_{in}/(f_c RC) \tag{9.9}$$

$$\Delta V_o = (T_2/RC)\ V_{ref} = n\ V_{ref}/(f_c RC)$$

where $N = T_1 f_c$, a fixed number of counts used to set T_1, and $n = T_2 f_c$, the number of counts occurring in the second half of the cycle. Equating the two expressions,

$$n = (N/V_{ref})\ V_{in} \tag{9.10}$$

An important feature of this method is that the result is independent of clock frequency f_c. The *accuracy* of the dual slope method depends only upon V_{ref}, though the *precision* depends upon f_c and T_1 (see Prob. 9.4).

The dual slope circuit has inherent noise rejection which increases linearly with frequency (Fig. 9.15). Noise which goes through several cycles during the sampling period T_1 makes positive and negative contributions to the integral which tend to average out [Fig. 9.15, waveform examples (a) and (c)]. The noise rejection increases with noise frequency f_N because the integral of a sine wave falls off inversely with frequency:

$$\left|\int_0^{T_1} \sin\,(2\pi ft)\ dt\right| = 1/(2\pi\ f_N\ T_1) \tag{9.11}$$

$$\frac{\left|\int \text{Signal}\right|}{\left|\int \text{Noise}\right|} = \frac{V_{sig}\cdot T_1}{V_N\cdot 2\pi\ f_N T_1} = \frac{V_{sig}}{2\pi\ f_N V_N} \tag{9.12}$$

In addition, noise components whose periods are integral multiples of the sampling time T_1 are totally rejected, since the integral averages precisely to zero [Fig. 9.15, waveform example (b)]. For this reason, T_1 is often chosen as s to eliminate ac power line noise components. Dual slope noise rejection capability is demonstrated in Fig. 9.16. In this dual-trace photograph, noise comparable in

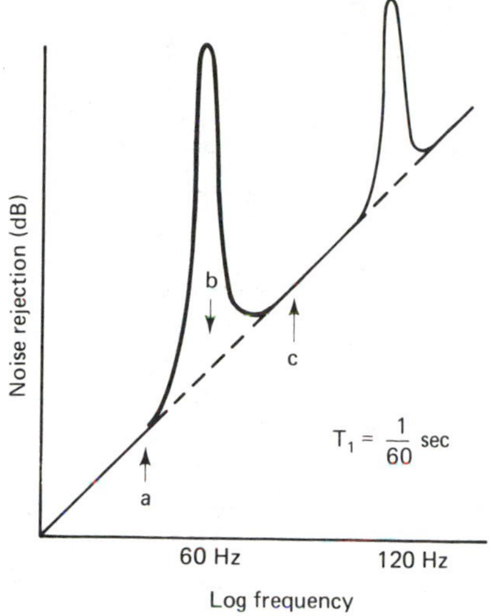

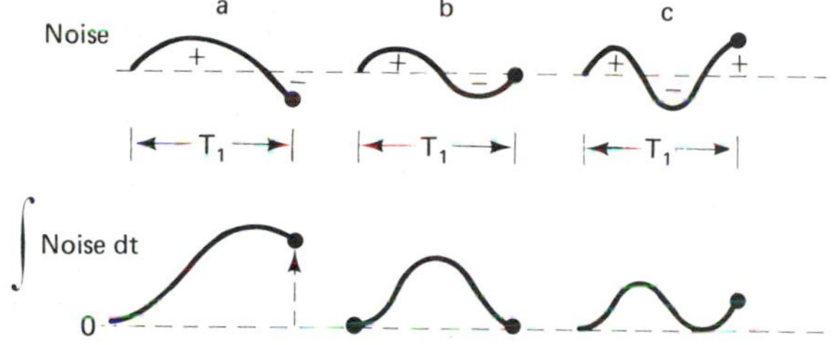

Figure 9.15 Dual slope noise rejection.

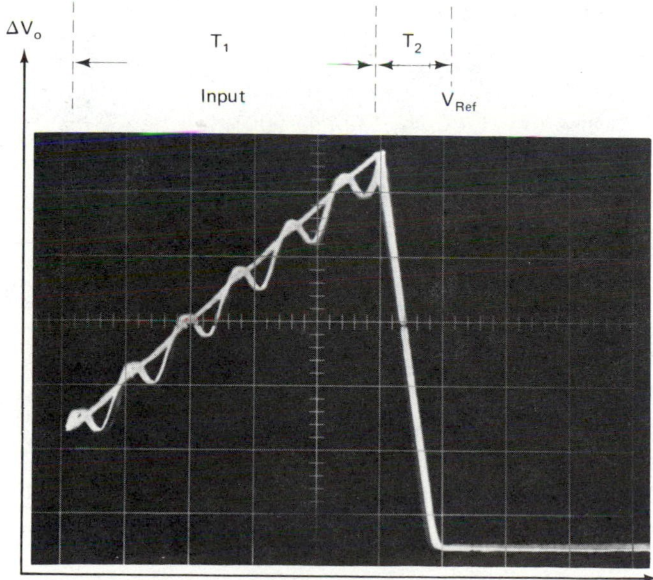

Figure 9.16 Example of noise averaging in the dual slope A/D. The input is a constant dc voltage plus (in one of the traces) an ac noise component. The signal shown is the integrator output V_o.

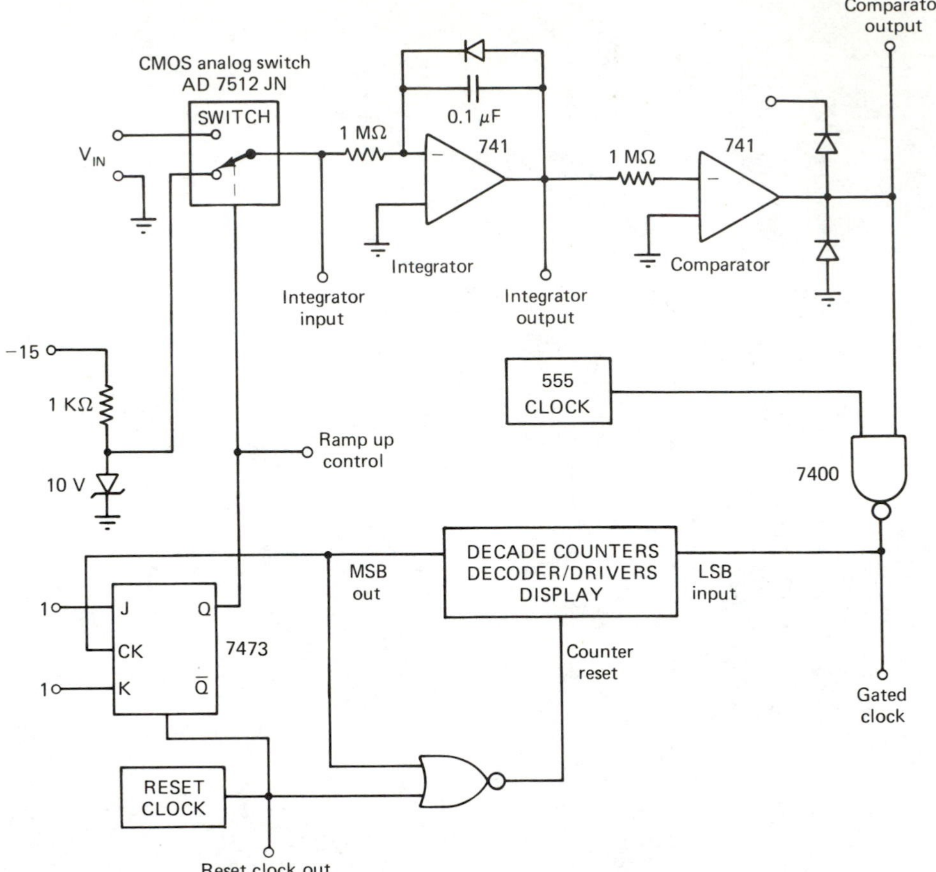

Figure 9.17 Practical working dual slope A/D test circuit. Note the labeled test points for examining circuit operation. (Circuit due to T. G. Matheson)

size to the dc voltage of interest has been added in one case, yet the integral ΔV_o (height of the triangle) is little changed.

A practical dual slope converter may be constructed from readily available circuit components (Fig. 9.17). Key features of circuit operation are:

(1) The cycle is started by the reset clock pulse, generated by a one-shot.

(2) The clock frequency in hertz is chosen to be 10 times the number of states of the counter. For example, for a 3-decade counter, choose

$$f = 10 \times 1000 = 10^4 \text{Hz}$$

This choice allows the MSB of the decade counter to control the switching logic.

(3) The counter is enabled during *both* time T_1 and T_2. Switching between V_{in} and V_{ref} occurs when the count reaches the number N where the most significant bit of the decade counter changes from 0 to 1. (The display rapidly se-

quences to full scale during T_1.) The next count brings the display to 000 (full scale plus 1) and triggers the second phase. When the full cycle is complete, the number n is held in the display.

Inexpensive IC dual slope A/D converters make excellent analog input ports for a microcomputer. An example is shown in Table 9.5. The circuit includes a multiplexer which strobes out the measured result digit by digit to minimize the number of connections required for interfacing. A microcomputer interface requires only one 8-bit port: 4 bits for a data digit, 4 bits to identify which digit is being transmitted, plus 2 extra bits, *display update* and *end of conversion*, for handshaking. This IC also makes an inexpensive digital voltmeter: the BCD data lines may be brought in parallel to four 7-segment displays, with the *digit strobe* lines used to enable the displays one digit at a time.

9.6 SUCCESSIVE APPROXIMATION A/D CONVERTER

The successive approximation method is the standard way to do *fast* A/D conversion. Only n clock cycles are required to encode a number to n bits of resolution, so data throughput rates of 1 MHz are readily possible. Although very high resolution (16-bit) conversions are also possible by this method, the inherent lack of noise averaging makes other methods (e.g., dual slope) more attractive when high resolution is required.

Successive approximation is a comparison method (Fig. 9.18) in which the input voltage V_{in} is compared with the output V_{appr} of a D/A converter. The *sign* of the error signal ($V_{in} - V_{appr}$) sets successive bits of the digital input to the D/A. The bits are set and tested in turn, starting with the most significant bit, one-half full scale. The process resembles weighing an object on a chemical balance with a set of binary weights. The digital logic must perform the following functions. When the *Encode* command is received, the most significant bit (MSB) of the D/A is set. If the input voltage is larger than one-half full scale, the sign of the comparator voltage results in the MSB being "kept." In the next clock cycle, the next most significant bit is set and this new approximation is compared with V_{in}. In the example of Fig. 9.18(b), the second approximation is too large, resulting in an opposite sign to the comparator output, so that bit is turned off. The third bit does not tip the scales too far, however, so it is kept. The process continues until all bits of the D/A have been set and tested in turn. At the end of the last test, a status line is set to indicate that the encoding is done.

One method to implement these logic functions is shown as a 4-bit example in Fig. 9.19. A shift register such as a TTL 74194 sets each bit in turn. The SR is parallel-loaded with a 0111 which is shifted with each clock cycle. An output register such as the TTL 7474 (positive edge D-type) provides the D/A input which is also the binary output of the circuit. The register is preset with the contents of the SR at each clock cycle. This SR word matches the negative logic requirements of the output register's preset inputs, and acts to turn on successively less significant bits of the D/A. The comparator output is wired in parallel to all the data (D) in-

TABLE 9.5 DUAL SLOPE IC A/D EXAMPLE SPECIFICATIONS[a]

MOTOROLA

SEMICONDUCTORS

3501 ED BLUESTEIN BLVD., AUSTIN, TEXAS 78721

MC14433

3½ DIGIT A/D CONVERTER

The MC14433 is a high performance, low power, 3½ digit A/D converter combining both linear CMOS and digital CMOS circuits on a single monolithic IC. The MC14433 is designed to minimize use of external components. With two external resistors and two external capacitors, the system forms a dual slope A/D converter with automatic zero correction and automatic polarity.

The MC14433 is ratiometric and may be used over a full-scale range from 1.999 volts to 199.9 millivolts. Systems using the MC14433 may operate over a wide range of power supply voltages for ease of use with batteries, or with standard 5 volt supplies. The output drive conforms with standard B-Series CMOS specifications and can drive a low-power Schottky TTL load.

The high impedance MOS inputs allow applications in current and resistance meters as well as voltmeters. In addition to DVM/DPM applications, the MC14433 finds use in digital thermometers, digital scales, remote A/D, A/D control systems, and in MPU systems.

- Accuracy: ±0.05% of Reading ±1 Count
- Two Voltage Ranges: 1.999 V and 199.9 mV
- Up to 25 Conversions/s
- $Z_{in} > 1000$ M ohm
- Auto-Polarity and Auto-Zero
- Single Positive Voltage Reference
- Standard B-Series CMOS Outputs—Drives One Low Power Schottky Load
- Uses On-Chip System Clock, or External Clock
- Low Power Consumption: 8.0 mW typical @ ±5.0 V
- Wide Supply Range: e.g., ±4.5 V to ±8.0 V
- Overrange and Underrange Signals Available
- Operates in Auto Ranging Circuits
- Operates with LED and LCD Displays
- Low External Component Count

CMOS LSI

(LOW-POWER COMPLEMENTARY MOS)

3½ DIGIT A/D CONVERTER

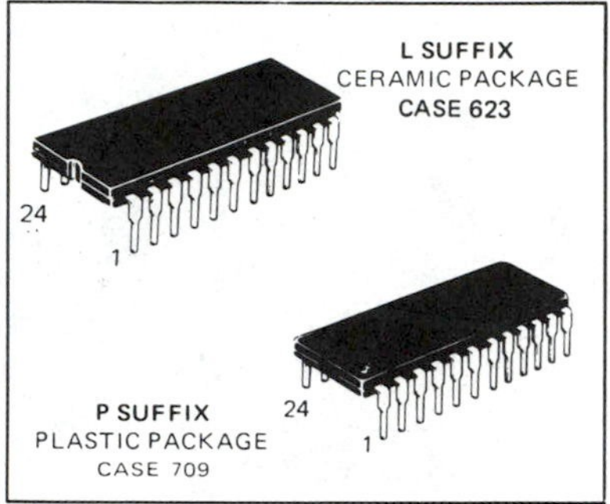

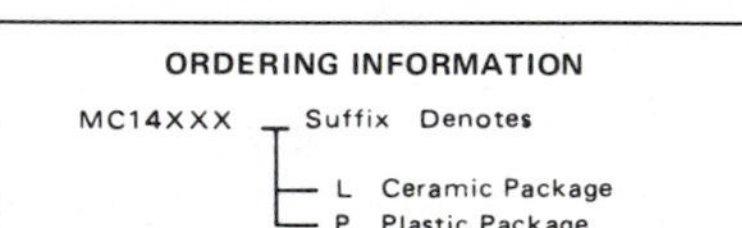

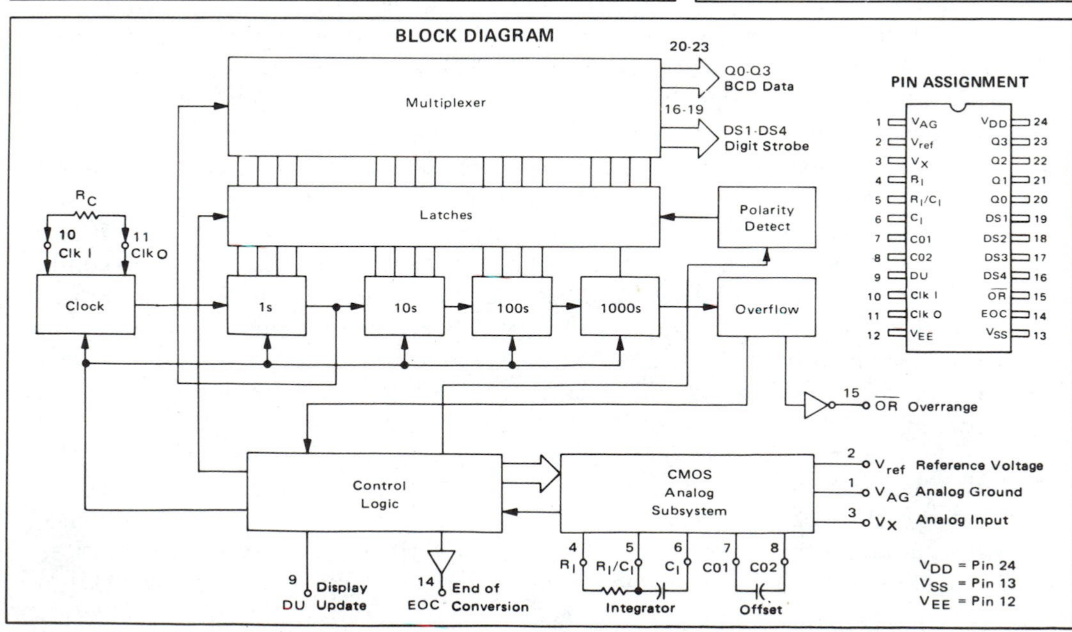

[a]MC 14433. *Source:* Motorola Semiconductor, Inc.

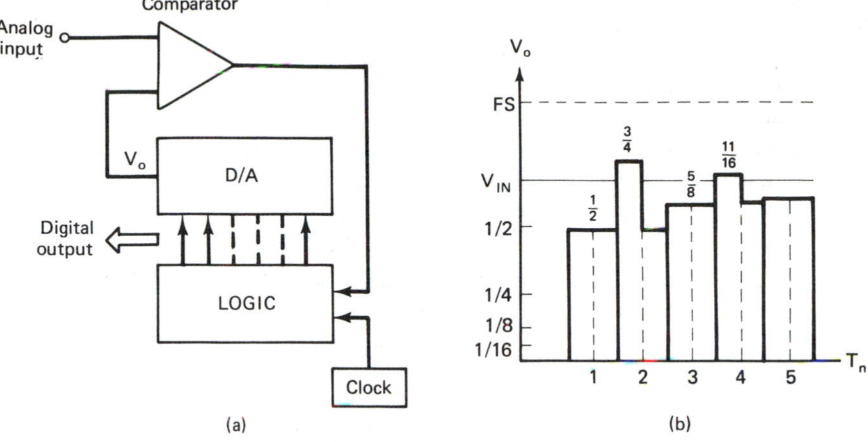

Figure 9.18 Successive approximation A/D converter concept. (a) Circuit. (b) Measurement example.

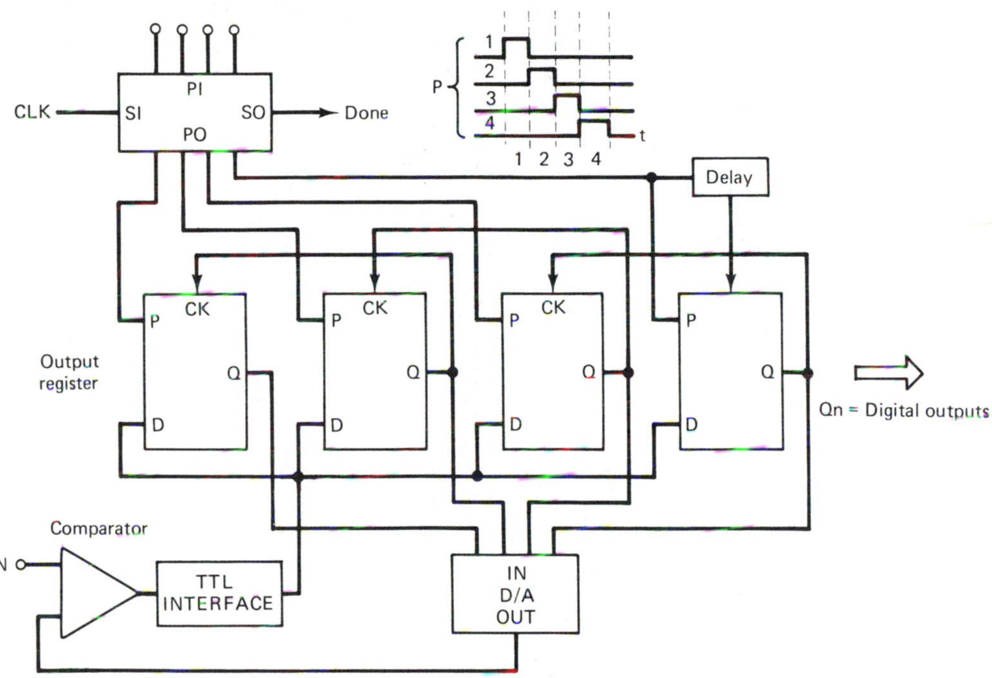

Figure 9.19 Successive approximation A/D, detailed 8-bit circuit. The inset shows waveforms applied to the direct Preset (P) inputs of the FF output register.

puts of the flip flops, and sets a bit high if the D/A output is less than the input voltage. Since the FF's are edge-triggered, they will not change state until the leading edge of a clock pulse is received. The clock for a given bit comes from the output of the next less significant bit. As a result, the command to latch a given bit to 1 or 0 (depending on the sign of the comparator output) is received at the instant the next less significant bit is preset by the shift register. This preset causes the D/A output to change, and the comparator may change state if the bit is too large. No race or indeterminacy occurs in the FF output, since the previous FF has already latched (30 ns) by the time the D/A responds to the new digital input (one μs). Each FF receives only one clock pulse during an encode cycle, so each successive bit remains stored while the remaining bits are set and tested. To complete the logic, the arrival of the test bit out the end of the SR alters a status line, indicating that the measurement is complete and binary data is waiting.

Other circuit details complete the picture. A pulser commands initial parallel loading of the shift register. The stage labeled *TTL interface* shifts the voltage range of the analog comparator to the 0 V to 5 V digital range. The clock input for the last FF is obtained from the preset bit from the SR. A delay is inserted to ensure that the comparator signal has time to respond to that last preset bit before the last data bit is latched.

The logic of the successive approximation method has become standard, and inexpensive commercial LSI IC's are available which include all the digital logic of Fig. 9.19 in a single package. An example of a complete 8-bit A/D is shown in Fig. 9.20. The circuit uses an AM2502 (Advanced Micro Devices) successive approximation logic chip described in Table 9.6. The total chip cost is less than $30 for this 8-bit A/D, which will run at data rates up to 100 KHz.

Figure 9.20 Successive approximation logic IC (AM 2502) used in a three chip A/D system.

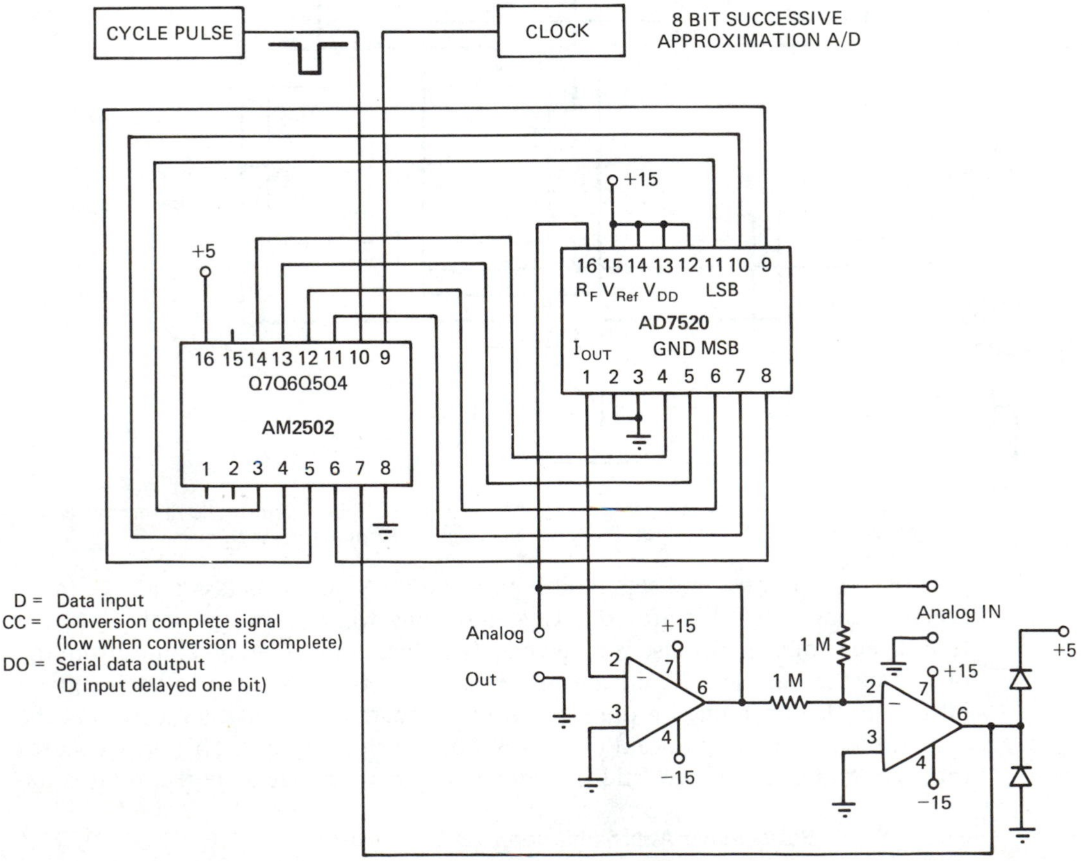

D = Data input
CC = Conversion complete signal (low when conversion is complete)
DO = Serial data output (D input delayed one bit)

Am2502/2503/2504

Eight-Bit/Twelve-Bit Successive Approximation Registers
Advanced Micro Devices
TTL-MSI Integrated Circuits

Distinctive Characteristics

- Contains all the storage and control for successive approximation A-to-D converters.
- Provision for register extension or truncation.
- Can be operated in START-STOP or continuous conversion mode.
- 100% reliability assurance testing in compliance with MIL-STD-883.
- Can be used as serial-to-parallel converter or ring counters.
- Electrically tested and optically inspected dice for the assemblers of hybrid products.

FUNCTIONAL DESCRIPTION

The Am2502, Am2503 and Am2504 are 8-bit and 12-bit TTL Successive Approximation Registers. The registers contain all the digital control and storage necessary for successive approximation analog-to-digital conversion. They can also be used in digital systems as the control and storage element in recursive digital routines.

The registers consist of a set of master latches that act as the control elements in the device and change state when the input clock is LOW, and a set of slave latches that hold the register data and change on the input clock LOW-to-HIGH transition. Externally the device acts as a special purpose serial-to-parallel converter that accepts data at the D input of the register and sends the data to the appropriate slave latch to appear at the register output and the DO output on the Am2502 and Am2504 when the clock goes from LOW-to-HIGH. There are no restrictions on the data input; it can change state at any time except during the set-up time just prior to the clock transition. At the same time that data enters the register bit the next less significant bit is set to a LOW ready for the next iteration.

The register is reset by holding the $\overline{S}$ (Start) signal LOW during the clock LOW-to-HIGH transition. The register synchronously resets to the state $Q_7(11)$ LOW, (Note 2) and all the remaining register outputs HIGH. The $\overline{CC}$ (Conversion Complete) signal is also set HIGH at this time. The $\overline{S}$ signal should not be brought back HIGH until after the clock LOW-to-HIGH transition in order to guarantee correct resetting. After the clock has gone HIGH resetting the register, the $\overline{S}$ signal is removed. On the next clock LOW-to-HIGH transition the data on the D input is set into the $Q_7(11)$ register bit and the $Q_6(10)$ register bit is set to a LOW ready for the next clock cycle. On the next clock LOW-to-HIGH transition data enters the $Q_6(10)$ register bit and $Q_5(9)$ is set to a LOW. This operation is repeated for each register bit in turn until the register has been filled. When the data goes into Q_0, the $\overline{CC}$ signal goes LOW, and the register is inhibited from further change until reset by a Start signal.

In order to allow complementary conversion the complementary output of the most significant register bit is made available. An active LOW enable input, $\overline{E}$, on the Am2503 and Am2504 allows devices to be connected together to form a longer register by connecting the clock, D, and $\overline{S}$ inputs together and connecting the $\overline{CC}$ output of one device to the $\overline{E}$ input of the next less significant device. When the Start signal resets the register, the $\overline{E}$ signal goes HIGH, forcing the $Q_7(11)$ bit HIGH and inhibiting the device from accepting data until the previous device is full and its $\overline{CC}$ goes LOW. If only one device is used the $\overline{E}$ input should be held at a LOW logic level (Ground). If all the bits are not required, the register may be truncated and conversion time saved by using a register output going LOW rather than the $\overline{CC}$ signal to indicate the end of conversion.

LOGIC DIAGRAM/SYMBOLS

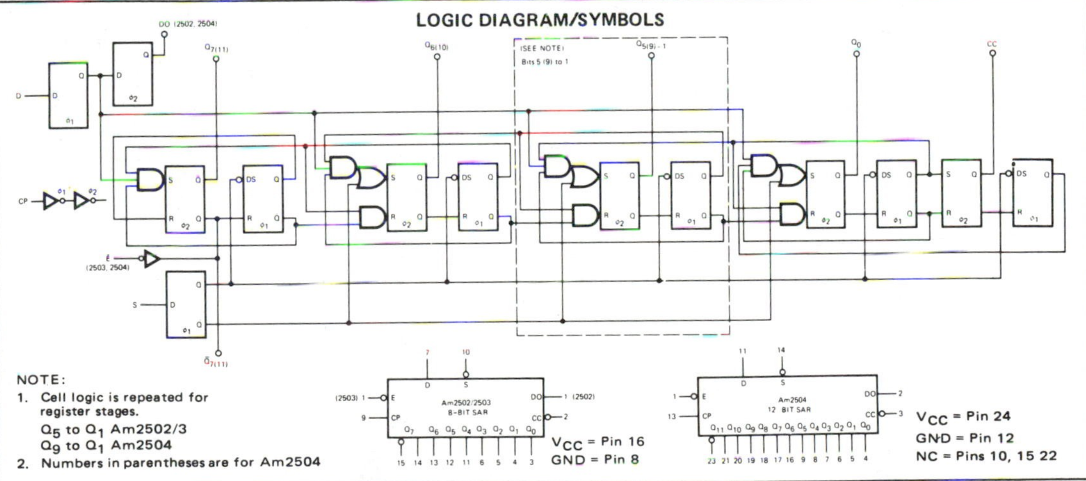

ORDERING INFORMATION

Package Type	Temperature Range	Am2502 Order Number	Am2503 Order Number	Am2504 Order Number
Hermetic DIP	0°C to +75°C	AM2502PC	AM2503PC	AM2504PC
Molded DIP	0°C to +75°C	AM2502DC	AM2503DC	AM2504DC
Dice	0°C to +75°C	AM2502XC	AM2503XC	AM2504XC
Hermetic DIP	–55°C to +125°C	AM2502DM	AM2503DM	AM2504DM
Hermetic Flat Pak	–55°C to +125°C	AM2502FM	AM2503FM	AM2504FM
Dice	–55°C to +125°C	AM2502XM	AM2503XM	AM2504XM

CONNECTION DIAGRAMS

Top View

Am2502 Am2503 (16-Pin) — Am2504 (24-Pin)

NOTE: PIN 1 is marked for orientation.

9.6.1 Successive Approximation Subtleties

The successive approximation method has limitations which result both from the concept and from the response time of the D/A.

(1) If the input signal is changing relatively fast during the conversion time interval [Fig. 9.21(a)], a given bit may be too much at the beginning of the encode cycle yet belongs there by the end. Successive bits are all turned on in an attempt to balance the input (data point $n-1$ in the figure). The digitized output is the same for both data point $n-1$ and data point n, despite the very different input voltage levels. The algorithm cannot track changes in the signal since only smaller weights are put on in succession.

(2) If the conversion is adequately fast but the data has unwanted noise [Fig. 9.21(b)], the A/D will sample whatever is present when the encode com-

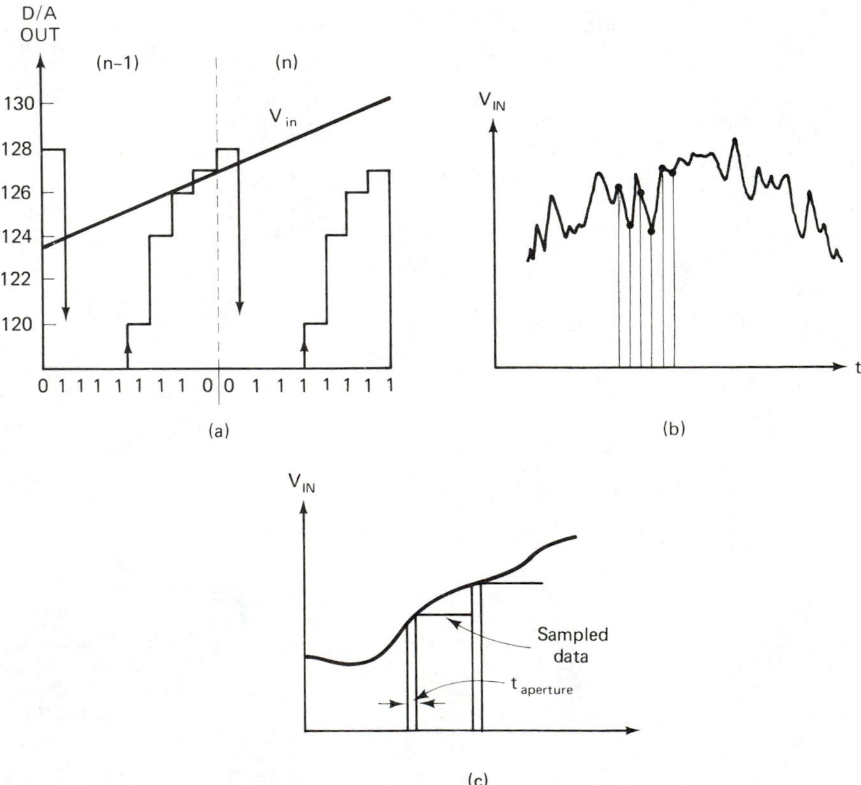

Figure 9.21 Example of data acquisition subtleties using the successive approximation method. (a) Errors due to the input signal changing during the sampling interval. (b) Rapid encoding of a noisy signal will "accurately" digitize the noise, too. (c) Use of sample and hold results in a more precise snapshot of the signal at times t_n.

mand is received. The signal-to-noise ratio is far less than the potential resolution of the measurement. Intelligent use of the successive approximation method requires analog filtering of the data to eliminate as much noise as possible.

(3) The settling time of the D/A limits the encode speed: the data throughput rate of an N-bit A/D can be no faster than $(N\tau_{D/A})^{-1}$ where $\tau_{D/A}$ is the D/A settling time.

Problem (1) is minimized by adding a *sample-and-hold* stage to the input. A sample and hold circuit uses an FET switch to selectively connect the input to the A/D, storing a sample of the input voltage level on a capacitor (Chapter 20). The sample is latched rapidly at precise time intervals, and held while the longer Encode is done [Fig. 9.21(c)]. As a result, a stable input is presented to the D/A. Error due to time variation of the signal is avoided by latching the sample in an aperture time t_{ap} which is quite fast. No loss of resolution occurs if the maximum rate of change of the input signal is:

$$(dV_{in}/dt)_{max} = 2^{-n}\, V_{FS}/t_{ap} \tag{9.12}$$

where V_{FS} is the full-scale input range and n is the number of bits in the D/A. For example, for a 10-bit conversion with $V_{FS} = 10$ V, an aperture time $t_{ap} = 100$ ns is adequate for input changes up to 10^7 V/s.

An accurate *sampled* picture of a waveform can thus be formed even if the signal is changing faster than the D/A settling rate, since the sampling time is known within t_{ap}, which is much shorter than $\tau_{D/A}$.

9.7 MISCELLANEOUS A/D METHODS

Although the dual slope and successive approximation methods can handle most A/D applications, four other methods find uses in special situations. These are:

(1) The *tracking* A/D, for ultra-high-resolution measurements if the signal does not change suddenly from one data point to the next

(2) The *comparator ladder* or *analog level detector*, for ultra-high-speed measurements at low resolution

(3) The *voltage-frequency converter*, for remote encoding with digital data transmission over a single wire

(4) The *companding* A/D, for measurements which require the highest dynamic range

A fifth method, *pulse modulation*, combines features of the tracking and voltage-to-frequency methods for high-speed serial transmission of digitized data. Although important in communications and digital audio recording, this method is little used in scientific or computing applications, and will not be explained in this book.

9.7.1 The Tracking A/D Converter

Suppose one needs to detect very small changes in a signal which is not changing very rapidly. Examples include a stable dc voltage reference signal or a battery backup power supply for a computer, where an early warning that the voltage was falling could avert a disastrous loss of data. The successive approximation method wastes time by putting all the weights back on the scale for each measurement. The *tracking* A/D keeps most of the weights on the scale at all times, changing only less significant bits as needed to follow signal variations. As a result, it can actually be faster than the successive approximation method for the same resolution, since only those bits which are changing need to be remeasured. Like the successive approximation method, the tracking A/D uses a D/A and comparator in a feedback loop (Fig. 9.22), and resembles a digital closed loop servo system. The shift register and storage latch of the successive approximation A/D are replaced by a high-resolution up/down counter. The counter is directed to count up or down by a comparator which senses whether the present D/A output is smaller or larger than the input signal. The first conversion is quite slow, requiring 2^N clock cycles to balance an N-bit signal. Compare this with the N clock cycles of the successive approximation A/D. However, once balanced, the circuit quickly tracks changes in the input signal. For example, to follow millivolt-level drift in a 10.0 V reference signal, the total resolution must be 14 bits (1 in 16384). Only a few clock cycles are needed to track the 1 in 10^4 voltage fluctuation, compared with 14 clock cycles in the successive approximation method.

The tracking A/D speed limit is characterized by its *slew rate*, the rate at which the closed loop can follow input signal changes:

$$dV/dt \leqslant 2^{-n} V_{FS} f_c$$

where f_c is the clock frequency. Thus, a 10-bit A/D with a 1-MHz clock can follow input signals changing at up to 10^4 V/s (assuming $V_{FS} = 10$ V).

Test your understanding. What is the limiting slew rate for a 14-bit A/D? Suppose the input signal is a 1-V sine wave. How high a frequency could be tracked for 10- and 14-bit examples?

Figure 9.22 Tracking A/D converter.

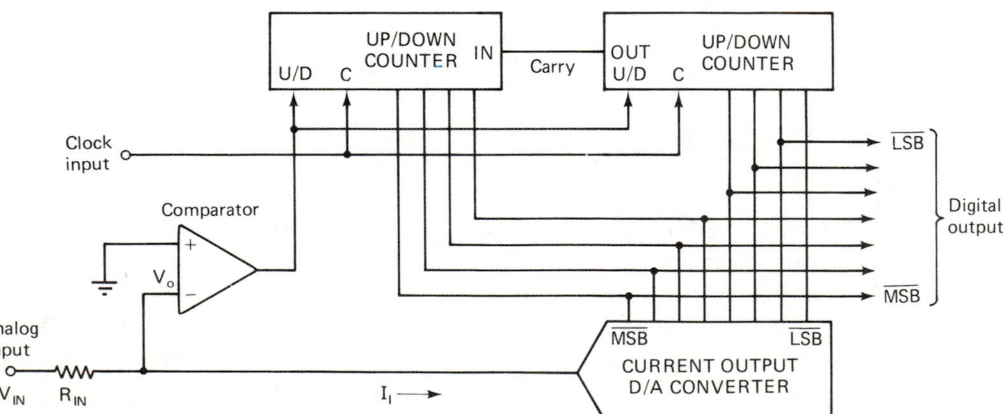

9.7.2 Comparator Ladder or Analog Level Detector

The comparator ladder A/D [Fig. 9.23(a)] resembles the classic analog comparison or *potentiometric* method. A reference voltage V_{ref} is sampled at intervals along a string of resistors and fed to a comparator array. The analog input voltage is applied to the second terminal of each comparator. The output of any compara-

Figure 9.23 Comparator ladder A/D. (a) Circuit and truth table. (b) High/low alarm circuit. (*Courtesy* Texas Instruments Incorporated)

Function table

Nominal input A (mV)*	Outputs				
	Q1	Q2	Q3	Q4	Q5
< 200 mV	H	H	H	H	H
200 mV	L	H	H	H	H
400 mV	L	L	H	H	H
600 mV	L	L	L	H	H
800 mV	L	L	L	L	H
1000 mV	L	L	L	L	L

H = high level L = low level
*Threshold variations are proportional

TL489

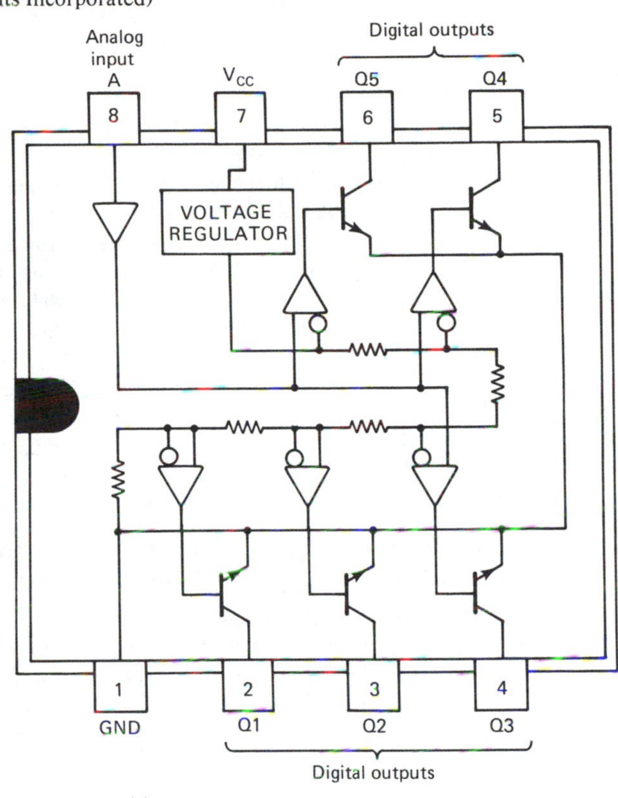

(a)

(b)

TABLE 9.7 ANALOG LEVEL DETECTOR A/D SPECIFICATIONS[a]

Device	Input	Steps	Display Type
TL489	Linear	5	LED
TL487	Logarithmic	5	LED
TL490	Linear	10	LED
TL480	Logarithmic	10	LED
TL491	Linear	10	VF
TL481	Logarithmic	10	VF

Input impedance: 100 KΩ (480 series) or 10 KΩ (490 series)

Input voltage 1.0 V FS (5-step units) or 2.0 V FS (10-step units)

200 mV per step (linear units) or
3 dB per step (5-step logarithmic units)
2 dB per step (10-step logarithmic units)

LED = open collector (40 mA current sink) outputs to drive LED's directly

VF = open emitter (25 mA current source, up to 35 V) to drive vacuum fluorescent displays directly

Supply voltage range 10 V to 18 V

[a]*Source:* Texas Instruments Incorporated.

tor will be low if the input voltage is lower than the corresponding point along the resistor ladder. The point where comparator outputs change from high to low is a measure of the input voltage. This method may sound crude, but it has some advantages. For example, connecting the outputs to a string of LED's (now available as an LED bar generator) gives a visual output of the signal's magnitude with no other electronics. The output is digital, though not binary. Ten output wires correspond to a 1-in-10 resolution, equivalent to less than 4 binary bits of information. The conversion is very fast, since the only limiting factor is comparator switching speed. The digital output is available immediately.

Inexpensive IC versions of the comparator ladder A/D are useful in level-detector and alarm circuits because of the simplicity of circuit construction. Examples of the inexpensive analog level detector IC's, the TL400 series, are described in Table 9.7. A simple example circuit is shown in Fig. 9.23(b). Outputs Q_n corresponding to voltages smaller than the input signal go low (connected to ground through a switching transistor), while outputs at higher tap points float high. The number of LED's lit is a measure of the input signal magnitude.

9.7.3 Video Encoders and The Digital Oscilloscope

Fast versions of the comparator ladder are used in high-speed A/D conversion called *video encoding.* A demonstration example is a crude digital oscilloscope (Fig. 9.24). Flat panel display units are presently under intensive development because of their potential to replace the cathode-ray tube (CRT), the last survivor

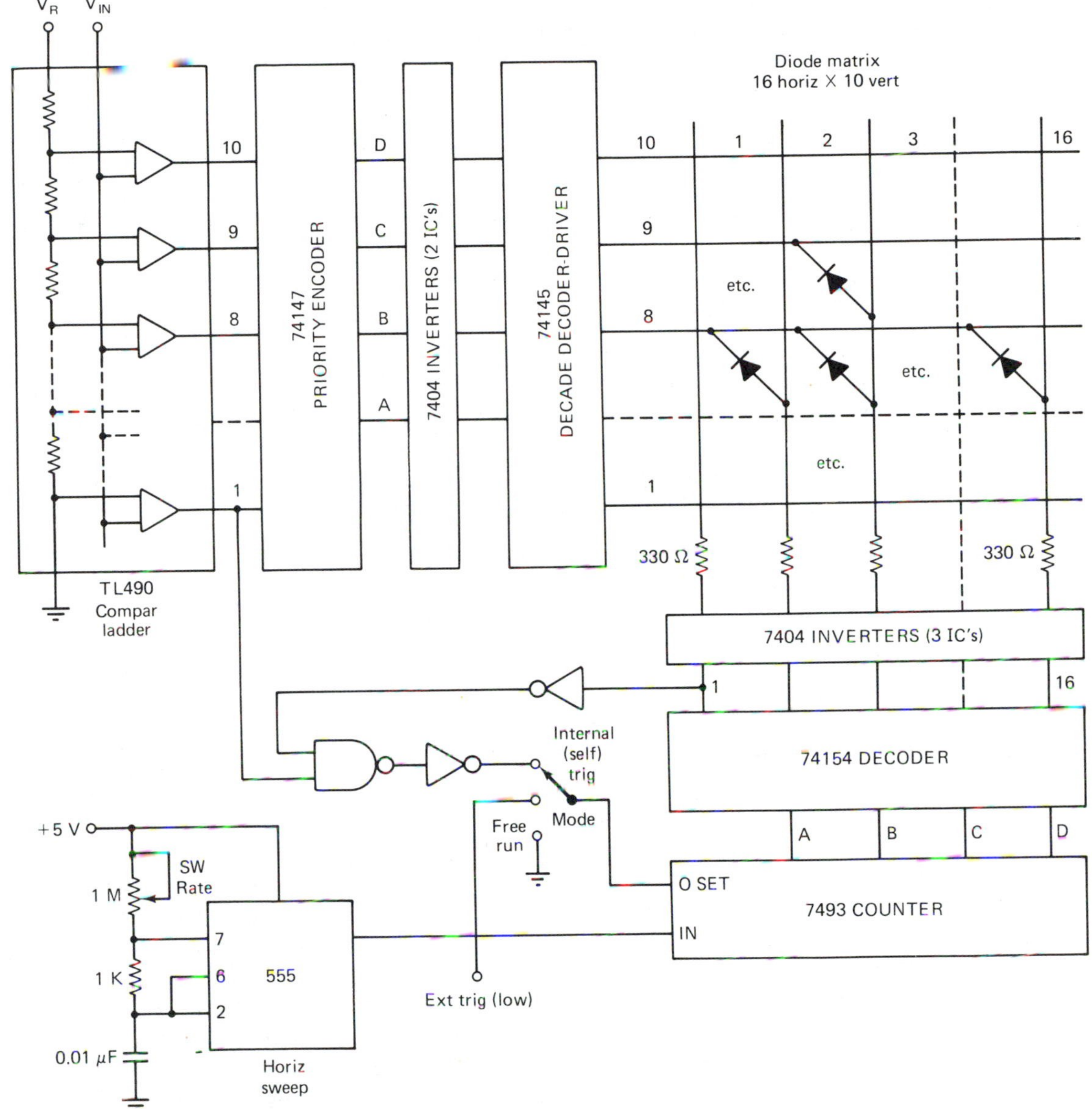

Figure 9.24 Digital oscilliscope demonstration using a flat panel LED or LCD matrix.

of vacuum-tube technology. CRT's are expensive and delicate. A flat panel display composed of LED's or *liquid crystal displays* (LCD) would be a cost-saving and rugged improvement. The digital oscilloscope is a good demonstration example, since it takes fewer points to display a curve on a scope than a full TV picture. A comparator array (e.g., TL339) or complete ladder comparator (e.g., TL490) is connected to the input signal (Fig. 9.24). The display panel is a matrix of LED's in this demonstration example, although lower power LCD's are a more likely commercial alternative if their switching speed limitations can be overcome. At any instant only one column line is connected to ground and only one row line is high, so only one LED will light up. The ladder outputs are applied to a *priority encoder* whose output is a binary word corresponding to its largest high input. This converts the ladder network output to binary form. The binary output is then fed to a decade decoder-driver (74175) whose selected output goes low, providing a current sink to ground for one row of the diode matrix. The *X*-axis circuit generates the horizontal sweep of the oscilloscope by bringing one column at a time high, at a sweep rate set by the 555 timer. The display panel approximates a crude but effective flat panel display of the original analog waveform.

Question. Since both the input to the 74147 priority encoder and the outputs from the 74145 decoder are decimal lines, why was the priority encoder necessary?

Comparator ladder A/D's such as the National 3914 have the 1-and-only-1 decoding done internally and available as a jumper-wired option.

9.7.4 Voltage-to-Frequency Converter A/D

The voltage-to-frequency (V/F) method uses a voltage-controlled oscillator (Chapter 17) whose output frequency is a linear function of the input voltage. Since the output is a pulse train, a digitized signal may be sent over a single wire, facilitating remote data loggers. The digitized signal can also be transmitted on a telephone line. The single wire output also has advantages in medical electronics where it is important to isolate the patient from dangerous voltages, using an optoisolator.

The V/F converter (Fig. 9.25) is an analog/digital hybrid. The input signal is compared with the voltage across a discharging RC combination. The comparator output fires a one-shot when it crosses a certain voltage threshold. The pulse is fed back around to add some charge to the capacitor. The larger the input voltage, the more often a pulse will be sent back to keep the capacitor charged to match the input voltage. The output pulse frequency is a linear measure of the input voltage amplitude (see Prob. 9.9). Modern IC V/F converters include a square-wave output rather than a string of pulses, since narrow pulses might not survive a long transmission path.

Relevant V/F specifications include:

(Non)linearity. Specified as a percentage of full-scale frequency, the linearity of IC V/F's is typically better than 0.1%, comparable to that of dual slope IC's in a similar cost range.

Frequency range. The greater the frequency range, the greater the resolution for a given counting period. The V/F pulse train is converted to a binary reading using a frequency counter. The longer one waits, the higher the

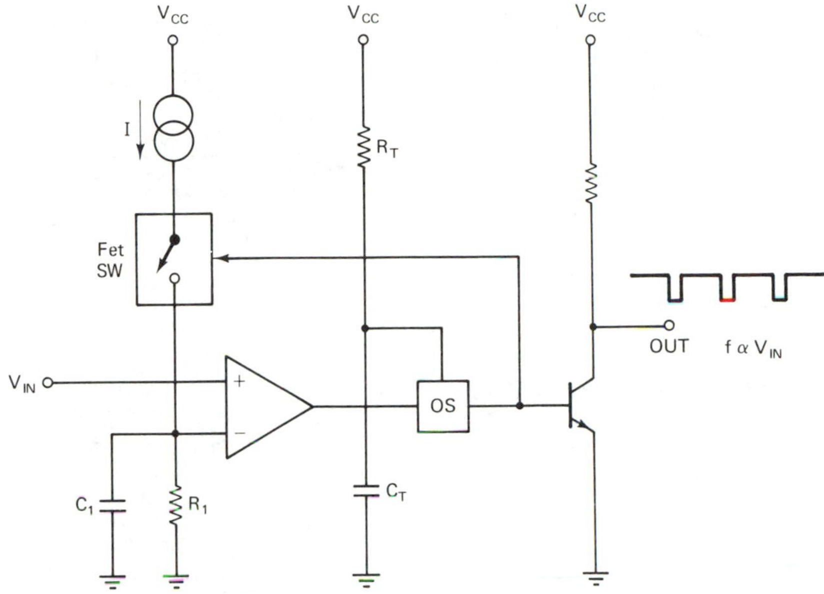

Figure 9.25 Voltage-to-frequency (V/F) A/D converter.

resolution. For example, if the highest V/F output frequency is 100 KHz, the measurement has 5-digit resolution if the counting interval is 1 s.

Dynamic Range. What is the ratio of largest to smallest signals giving meaningful results? Some IC V/F's maintain linearity down to 0.001 Hz. If the upper limit frequency is 10^5 Hz, the range of measurable voltages covers 8 decades. This dynamic range is superior to any of the A/D methods previously discussed.

Examples of inexpensive IC V/F converters are the Teledyne 9400 and the National LM 331. These have an upper frequency limit of 100 kHz, linearity approaching 0.01%, and lower frequency limits of about 0.1 Hz.

9.7.5 Companding A/D and D/A

The *companding* method squeezes many bits of resolution out of fewer bits of binary data by compressing and then expanding the signal's dynamic range. An audio example illustrates the problem. If the largest excursions in the groove of a phonograph record are 1000 times larger than the smallest excursions above background noise or mechanical reproduction limits of the material, no more than 10 bits of information may be encoded meaningfully. This is referred to in audio terminology by saying that the recording has a 60-dB dynamic range (20 dB = a factor of 10). If the information is compressed when recorded, with low-level quiet passages amplified to make a bigger impression, and then the process is reversed on playback by attenuating low-level passages on the output side, the dynamic range is effectively expanded. Such a process can be carried out digitally using a

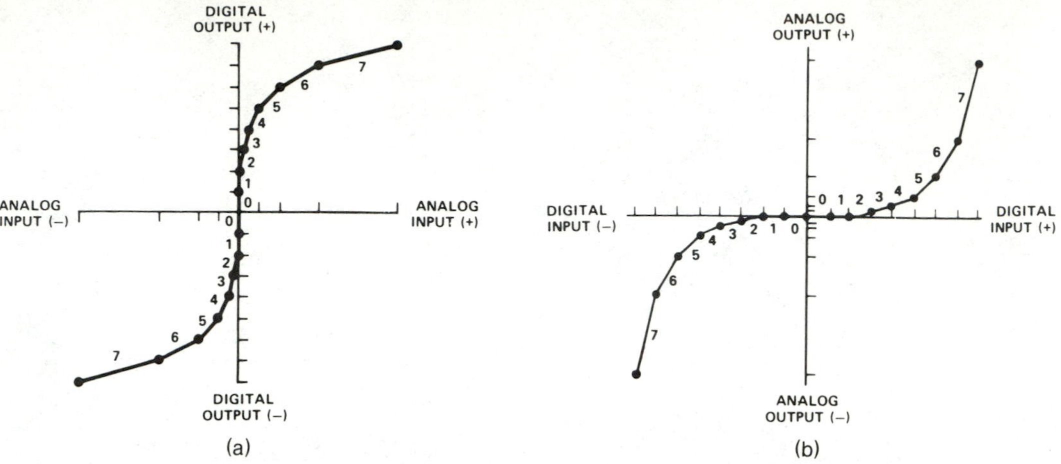

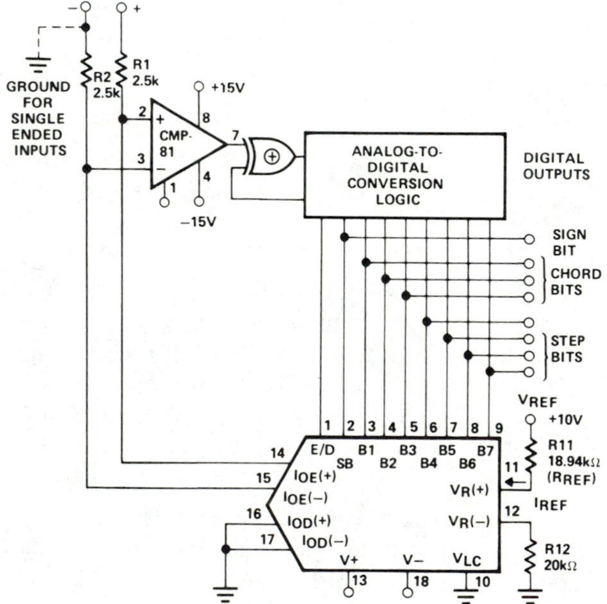

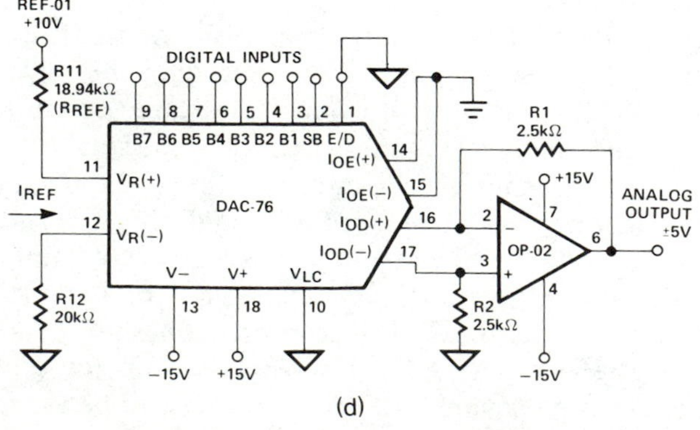

Figure 9.26 Companding transfer functions for (a) encoding and (b) decoding. Companding circuits for (c) A/D encoding and (d) D/A decoding. (*Courtesy* Precision Monolithics Inc.)

companding D/A–A/D converter combination, giving a 72-dB dynamic range with only 8-bit inputs. The 8-bits of information used directly would give only a 48-dB (= 20 log 256) dynamic range.

The companding principle [Fig. 9.26(a)] uses a nonlinear transfer function for both A/D and D/A. The *encode* transfer function's nonlinearity is shaped so that relatively small input signals result in relatively large digital outputs. The digitized signal is played back through a nonlinear D/A converter [Fig. 9.26(b)] whose transfer characteristic removes the nonlinearity: the product of the two transfer functions is a straight line. The *companding* or compression/expansion process in this example uses 3 bits to select one of 8 nonlinear "coarse" outputs and 4 additional bits to select one of 16 linear steps within the coarse intervals. The eighth bit represents the sign. The coarse steps are related to one another as powers of 2. Steps at the lower end of the range give output signal changes with effectively much higher resolution than the number of bits indicate. For example, the smallest step of an 8-bit device, corresponding to 0.025% full scale, has an equivalent resolution of 12 bits (72 dB). The step sizes and resolution and specifications of a commercial IC, the DAC-76, are summarized in Table 9.8.

In the compression mode [Fig. 9.26(c)], the D/A is used with an external successive approximation register and comparator to make a piecewise-linear A/D converter [Fig. 9.26(c)]. The nonlinear digital signal must be played back to analog form through the same device [Fig. 9.26(d)] to provide the expansion half of the cycle.

PROBLEMS

9.1. Analyze the logic of Fig. 9.10(b) which steers clock pulses to the count-up or count-down inputs. Show using waveforms that the correct clock signal is received each time a borrow or carry pulse is generated. You will need to look up the characteristics of the 7470 clocking, and the characteristics of the 74193 borrow and carry pulses. How and when does gate C function?

TABLE 9.8 COMPANDING D/A CONVERTER: RESOLUTION OF COARSE NONLINEAR STEPS[a]

Chord number	Step size normalized to full scale	Step size as percent of full scale	Resolution and accuracy of equivalent binary D/A
0	2	0.025	Sign + 12 bits
1	4	0.05	Sign + 11 bits
2	8	0.1	Sign + 10 bits
3	16	0.2	Sign + 9 bits
4	32	0.4	Sign + 8 bits
5	64	0.8	Sign + 7 bits
6	128	1.6	Sign + 6 bits
7	256	3.2	Sign + 5 bits

DAC-76

COMDAC® COMPANDING D/A CONVERTER

MONOLITHIC LOGARITHMIC DAC

FEATURES

- **Sign Plus 12-Bit Range with Sign Plus 7-Bit Coding**
- **12-Bit Accuracy and Resolution Around Zero**
- **Sign Plus 72dB Dynamic Range**
- **True Current Outputs: −5V to +18V Compliance**
- **Tight Full Scale Tolerance Eliminates Calibration**
- **Low Full Scale Drift Over Temperature**
- **Conforms with Bell System μ-255 Companding Law**
- **Multiplying Reference Inputs**
- **Low Power Consumption and Low Cost**
- **Ideal for PCM, Audio, and 8-Bit μP Applications**
- **Outputs Multiplexed for Time Shared Applications**

COMDAC® TRANSFER CHARACTERISTIC

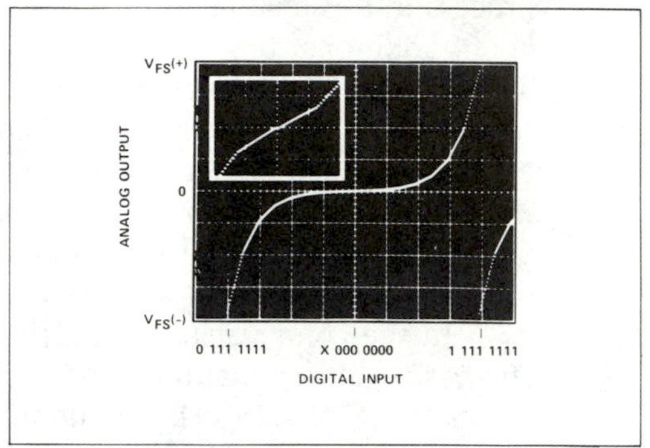

GENERAL DESCRIPTION

The DAC-76 monolithic COMDAC™ D/A Converter provides the dynamic range of a sign +12-bit DAC in a sign +7-bit format. A companding (compression/expansion) transfer function is implemented by using three bits to select one of eight binarily-related chords (or segments) and four bits to select one of sixteen linearly-related steps within each chord. Accuracy is assured by specifying chord end point values, chord nonlinearity, and monotonicity over the full operating temperature range.

The 8-bit format with a sign +72dB dynamic range is especially useful in control systems using 8-bit microprocessors, RAMs and ROMs. Low distortion multiplying capability and conformance with the Bell System μ-255 logarithmic law for PCM transmission make the DAC-76 ideal for use in audio applications. Other applications include servo controls, stress and vibration analysis, digital recording and speech synthesis. Additional applications are listed on the last page.

EQUIVALENT CIRCUIT

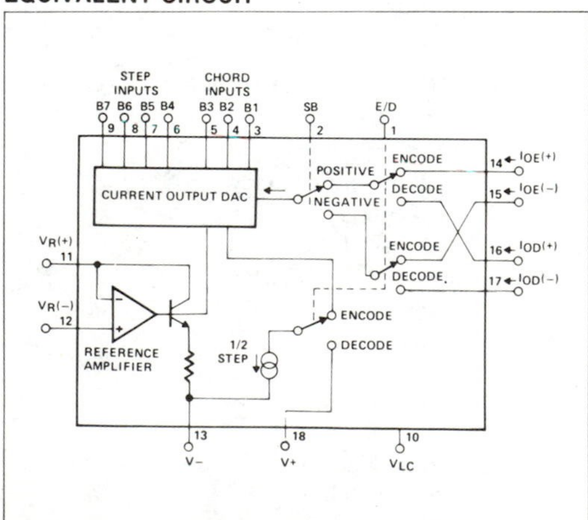

ORDERING INFORMATION & PIN CONNECTIONS

ENCODE/DECODE SELECT: 1 = ENCODE	1 E/D	V+ 18	POSITIVE POWER SUPPLY
SIGN BIT INPUT: 1 = POSITIVE	2 SB	IOD(−) 17	DECODE OUT: E/D SB = 00
MOST SIGNIFICANT CHORD BIT INPUT	3 B1	IOD(+) 16	DECODE OUT: E/D SB = 01
SECOND CHORD BIT INPUT	4 B2	IOE(−) 15	ENCODE OUT: E/D SB = 10
LEAST SIGNIFICANT CHORD BIT INPUT	5 B3	IOE(+) 14	ENCODE OUT: E/D SB = 11
MOST SIGNIFICANT STEP BIT INPUT	6 B4	V− 13	NEGATIVE POWER SUPPLY
SECOND STEP BIT INPUT	7 B5	VR(−) 12	NEGATIVE REFERENCE INPUT
THIRD STEP BIT INPUT	8 B6	VR(+) 11	POSITIVE REFERENCE INPUT
LEAST SIGNIFICANT STEP BIT INPUT	9 B7	VLC 10	THRESHOLD CONTROL

18-PIN HERMETIC DUAL-IN-LINE (X-Suffix)

MODEL	TEMP RANGE	ACCURACY
DAC-76BX	−55°/+125°C	±1/2 STEP
DAC-76X	−55°/+125°C	±1 STEP
DAC-76EX	0°/+70°C	±1/2 STEP
DAC-76CX	0°/+70°C	±1 STEP
DAC-76DX	0°/+70°C	±1 1/2 STEP

Military Temperature Range Devices with MIL-STD-883 Class B Processing

DAC76BX/883
DAC76X/883

[a]*Source:* Precision Monolithics Inc.

9.2. Design a symmetric triangular wave generator using a CMOS up/down counter such as the 4516. Show all external gating required to control the up/down counting.

9.3. How could the preset inputs be altered to make the triangular waveform of Fig. 9.10(b) contain a different number of steps? How does the wiring of *carry* to *load* and *borrow* to *clear* make this possible?

9.4. What is the relationship between speed and precision of the dual slope A/D converter of Fig. 9.14? Suppose the clock frequency $f_c = 1$ MHz. How many measurements could be made per second with 3-digit precision [$N = 10^3$ in Eq. (9.10)]? How many digits of resolution are available if the sampling speed is made 10 times faster?

9.5. Refer to dual slope A/D block diagram (Fig. 9.14) and detailed circuit diagram (Fig. 9.17).

(a) What functions are performed by the control logic (Fig. 9.14) at each point in the cycle?

(b) How is this executed for the circuit of Fig. 9.17?

(c) How is the fixed time T_1 generated in this circuit?

9.6. Sketch on a single page on a common time axis the waveforms you expect to see at the following points in the circuit of Fig. 9.17: (a) Integrator input; (b) Integrator output; (c) Comparator output; (d) Reset clock; (e) Counter output (analog graph of digital reading vs. time); (f) Ramp up control.

9.7. How could you change the 3-digit circuit of Fig. 9.17 to add another digit of resolution? Assume the clock frequency is unchanged. Precisely how much will this slow the voltage measurement, in points per second?

9.8. Design a circuit to accurately measure the amplitude of a chosen point of a waveform displayed on an oscilloscope, using a successive approximation A/D such as Fig. 9.19. Derive the cycle start pulse from the scope's sweep trigger pulse, but include a 555 one-shot adjustable delay so the measured point may be set anywhere on the displayed waveform. The selected point can be made brighter than the rest of the display, using the cycle start pulse to drive the scope's *z*-axis brightness control. Assume for simplicity that all scope signals are TTL-compatible.

9.9. Show that the output frequency of the V/F converter (Fig. 9.25) is linearly proportional to input voltage. (**Hint**: Differentiate the expression $V = V_o \exp(-t/RC)$ for the RC pair. Find the time interval Δt required for the voltage to fall by a fixed amount ΔV from the input voltage. ΔV is the fixed hysteresis of the comparator.

REFERENCES

More complete bibliographic information for the books listed below appears in the annotated bibliography at the end of the book.

Analog Devices, *Analog-Digital Conversion Handbook*

BRACEWELL, *The Fourier Transform and its Applications*

HENRY, *Electronic Systems and Instrumentation*

HIGGINS, *Experiments with Integrated Circuits*, Experiments 14, 15, and 16

MALMSTADT et al., *Electronic Measurements for Scientists*

STEARNS, *Digital Signal Analysis*

10

Differential Amplifiers and the Magic of Feedback

10.1 INTRODUCTION

In 1927 Black discovered that feeding back a portion of the output signal to the input could result in a substantial improvement in amplifier performance (Fig. 10.1). The problem originated in vacuum tube amplifier "repeaters" to boost the strength of telephone conversations over long distances. The cumulative distortion of a number of repeaters in series became excessively large. By taking some of the amplifier's output and feeding it back to the input, where the two could be electronically compared for fidelity, the distortion of the output signal could be corrected. Without this discovery, cross-country telephone calls would be impossible. This simple idea, which at first seems to defeat the amplifier's purpose of amplifying signals, has a variety of suprisingly powerful consequences.

The reader who is unfamiliar with discrete electronic circuits and their analysis will need to review Chapter 1. However, the details of Chapter 10 are not essential for the balance of the text. Many readers may be content with the results of Section 10.4. Chapter 10 concerns itself with *why* an operational amplifier behaves as it does.

10.2 THE BASIC DIFFERENTIAL AMPLIFIER

Most op amps look similar to Fig. 10.2(a). The first stage [Q_1 and Q_2, Fig. 10.2(b)] is called the *long-tailed pair*. Details of Fig. 10.2(a) which influence nonideal op amp operation will be described in Chapter 12. In Chapter 10, we will analyze *ideal* behavior, and in particular the negative feedback resulting from the

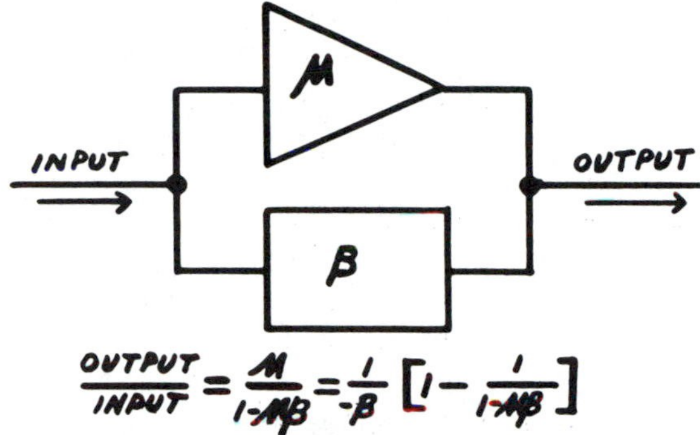

Figure 10.1 Copy of the original sketch of the negative feedback idea as conceived by Harold Black on the Hoboken ferryboat, August 2, 1927, enroute to work at the Bell Laboratories. (*Courtesy* Harold Black and Bell Laboratories)

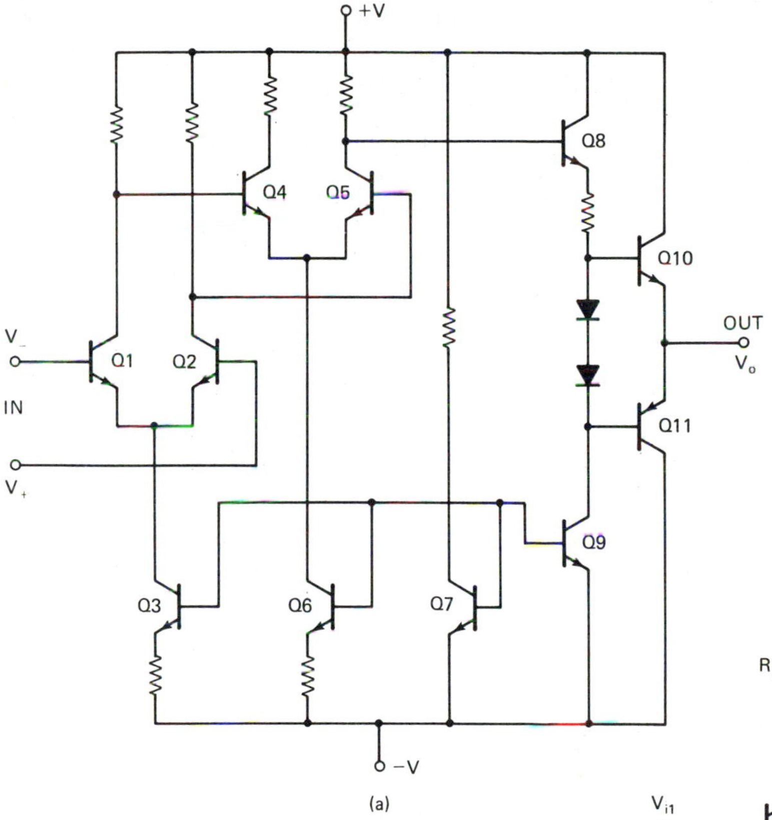

(a)

Figure 10.2 (a) The basic operational amplifier circuit and (b) first stage "long tailed pair."

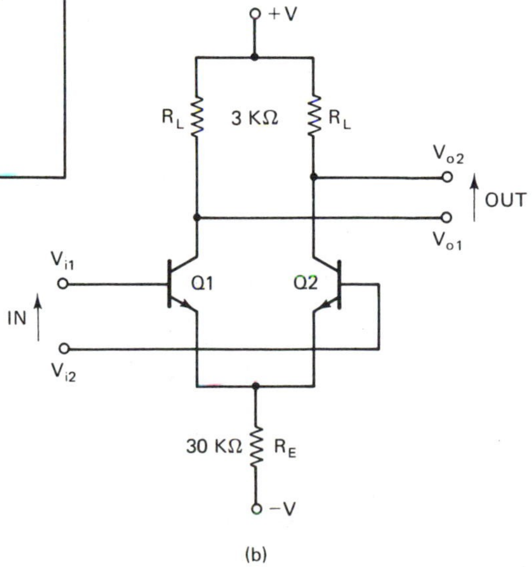

(b)

emitter resistor R_E in Fig. 10.2(b). Although R_E resembles the emitter resistor used in bias stabilization of a typical ac amplifier, its function is quite different.

10.3 A NEGATIVE FEEDBACK EXAMPLE: UNBYPASSED EMITTER RESISTOR

It is useful to analyze a simple negative feedback example. Consider a single stage transistor amplifier [Fig. 10.3(a)] with an emitter resistor R_E unbypassed by the capacitor usually found in an ac transistor amplifier circuit. R_E substantially lowers the amplifier gain, as may be seen intuitively. Suppose a signal voltage turns on the transistor. More emitter current flows through R_E, raising the voltage drop across it. This drives the emitter voltage more positive with respect to the base, tending to turn off the transistor, so the emitter current falls. This sequence is called *negative feedback:* the circuit opposes the amplification of signals.

This qualitative argument may be readily extended quantitatively to show that the gain is substantially reduced and the input impedance is substantially increased by R_E. The transistor circuit of Fig. 10.3(a) has the *small signal equivalent*

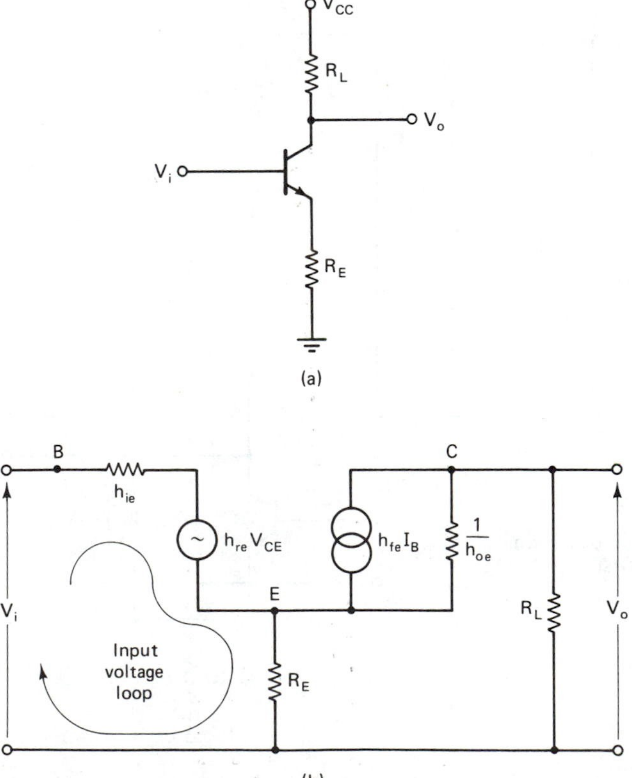

Figure 10.3 Effect of an unbypassed emitter resistor on a single stage transistor amplifier. (a) Transistor circuit. (b) Hybrid equivalent circuit.

circuit shown in Fig. 10.3(b). Small signal* or *hybrid parameter* equivalent circuits, a linear approximation model for calculating circuit gains and impedances, were introduced in Chapter 1.

10.3.1 Gain Calculation

The output voltage is determined by the collector current I_C through R_L.

$$V_o = I_C R_L \tag{10.1}$$

The collector current is determined by the base current I_B and the current gain h_{fe} of the transistor.

$$I_C \simeq h_{fe} I_B \tag{10.2}$$

The base current is determined by Ohm's law and a Kirchoff voltage loop at the input [Fig. 10.3(b)]. This differs from the input voltage loop in an ordinary transistor amplifier due to the presence of R_E. Since $h_{fe} \simeq 10^2$, the voltage drop across R_E is dominated not by I_B but by I_C. The voltages which must balance around this loop are

$$V_i = h_{ie} I_B + R_E(I_E + I_B) = h_{ie} I_B + R_E I_C \tag{10.3}$$

Note the cross-coupling from input to output due to the position of R_E [Fig. 10.3(b)]. The value of I_B determines I_C, but I_C in turn influences I_B via the *feedback* term ($I_C R_E$). Solving Eqs. (10.1) through (10.3) leads to the voltage gain

$$V_o / V_i = R_L h_{fe} / (h_{ie} + R_E h_{fe}) \tag{10.4}$$

Comparing this with the voltage gain $R_L h_{fe} / h_{ie}$ of the ordinary amplifier where R_E is either absent or bypassed, we see that R_E reduces the gain by a large amount ($h_{fe} >> 1$). The current gain of the transistor increases the effect of R_E so it becomes the dominant term in the denominator for typical circuit values (h_{ie} = $2K\Omega$, $R_L = 3K\Omega$; $h_{fe} = 50$).

$$V_o / V_i \simeq R_L / R_E \tag{10.5}$$

Surprisingly, negative feedback makes the gain of the *circuit* independent of the gain of the active *device* used to produce the amplification! This feature is characteristic of negative feedback; it does not mean that one can simply unplug the transistor.

10.3.2 Input Impedance Calculation

The input impedance is raised substantially as a consequence of negative feedback. Input impedance is defined as the ratio of input voltage to input current:

$$R_i = V_i / I_i = V_i / I_B \tag{10.6}$$

* The distinction between capital V for bias voltages and lower case v for small signal voltages will not be made here, since only signal voltages will be considered.

The increase in R_i may be seen intuitively by noting that current flow in R_E raises the voltage at point E [Fig. 10.3(b)]. Less voltage drop appears across h_{ie} for a fixed V_i, and the base current I_B is reduced, raising R_i. The analysis follows by using the Kirchoff voltage loop [Eq. (10.3)], rewritten to eliminate I_C:

$$R_i = h_{ie} I_B + R_E(h_{fe} I_B) \qquad (10.7)$$

Dividing through by I_B yields the result:

$$R_i = h_{ie} + h_{fe} R_E \qquad (10.8)$$

The input impedance of a normal transistor amplifier is just h_{ie}. Since h_{fe} is large, feedback substantially increases R_i. For example, with $h_{ie} = 2\ \mathrm{K\Omega}$, $R_E = 30\ \mathrm{K\Omega}$, $h_{fe} = 50$, then $R_i = 1.5\ \mathrm{M\Omega}$.

10.4 DESIRED FEATURES OF A DIFFERENTIAL AMPLIFIER

There are many situations where the voltage *difference* across two points in a circuit must be measured (Fig. 10.4). For example, grounding either point may alter the system's behavior. An oscilloscope with a *single-ended* input (normal scope probe) may not be used, and one must turn instead to an oscilloscope with a *differential* input (two probes, neither one grounded). Similarly, in the case of a voltmeter, one would ensure that the voltmeter was floating (neither input grounded). The differential measuring instrument (Fig. 10.5) has a linear ampli-

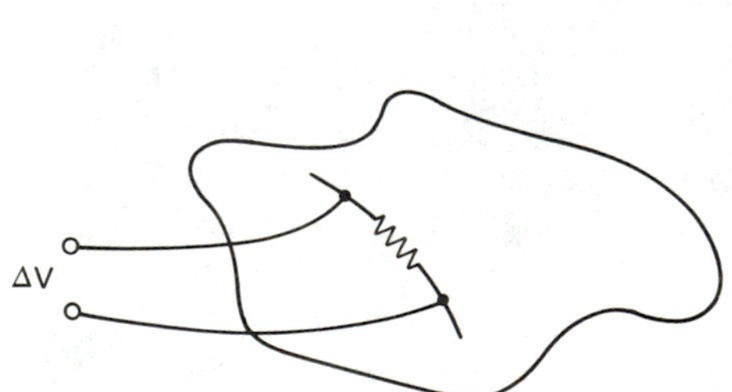

Figure 10.4 The problem of measuring a potential *difference* between two points in a system.

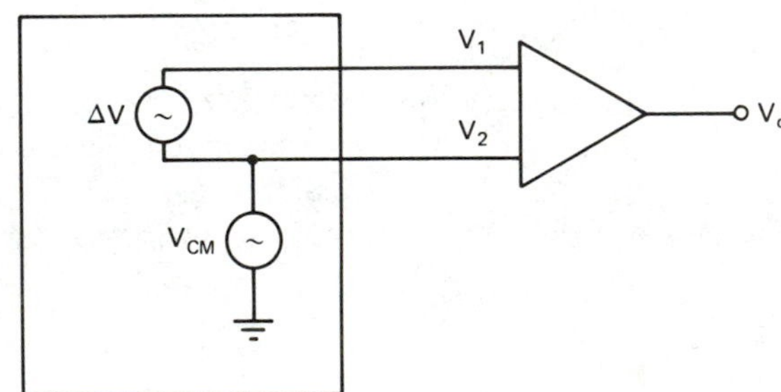

Figure 10.5 Generalized signal source and generalized differential amplifier.

fier, represented by the triangle, which provides gain for voltage differences appearing across its inputs. The generalized input V_i, shown to the left, may be broken down into two independent voltage sources: one drives a voltage difference across the amplifier's input, and the other drives both terminals up and down in common. The first, the *differential* input ΔV, is the signal of interest. The second, the *common mode* signal V_{cm} is undesired. The amplifier's response to this generalized input may be described by

$$V_o = A \Delta V + g V_{cm} \qquad (10.9)$$

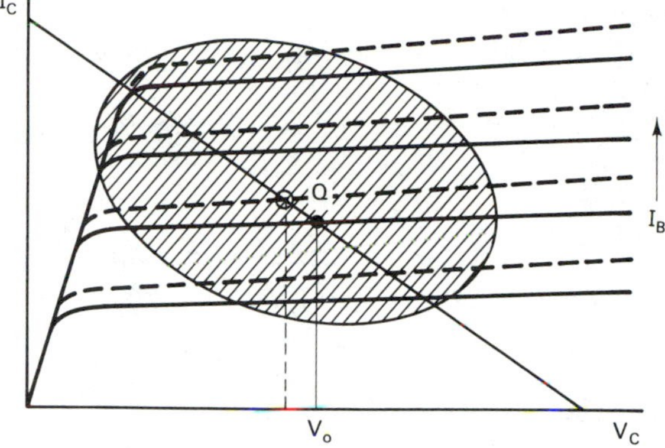

Figure 10.6 Problems of the single transistor dc amplifier. The operating point yields zero offset and thermal drift of characteristic curves (dashed lines) shifts the output voltage.

The desired amplifier has the differential gain A much larger than the common mode gain g. A figure of merit is the ratio:

$$A/g = \text{CMRR} \tag{10.10}$$

defined as the *common mode rejection ratio* (CMRR).

The differential amplifier must respond to signals as low in frequency as dc, and yet be insensitive to drift in the operating point of the transistors. Consider the dc characteristic curves of a transistor amplifier (Fig. 10.6). Whatever operating point Q is selected, the output voltage V_o is nonzero even when the input signal voltage V_i is zero, because one must bias the circuit into a linear region. This dc bias is removed by a coupling capacitor in an ac amplifier, which is not possible here. Even if the dc bias is subtracted, V_o will drift due to temperature fluctuations in the operating environment. As temperature increases, the transistor characteristic curves shift upwards. The operating point shifts, and the output voltage falls, even in the absence of an input signal change. Such drift can overwhelm millivolt level signals.

10.5 THE SOLUTION: BALANCED DIFFERENTIAL AMPLIFIER

The solution to these problems is the two-transistor amplifier of Fig. 10.2(b). The transistors are matched so their gains are as close as possible. The problem of a dc output in the absence of an input signal is automatically solved by taking the output as a voltage *difference* between the two collectors. If the operating points are identical, the output voltage difference is zero in the absence of an input voltage difference. The problem of output voltage fluctuations due to thermal drift is also solved (to the extent that the two transistors are matched), since shifts in the output voltage of one transistor are matched by corresponding shifts of the second transistor, with both transistors in the same thermal environment. Deviations from this ideal are responsible for nonideal op amp behavior (Chapter 12).

The desired differential and common mode behavior of Eq. (10.9) are also achieved by this configuration. The essential results are:

Differential Amplifier Properties

(1) Differential gain equal to the gain of a single transistor amplifier *without* decrease due to R_E.

(2) Common mode gain equal to the single transistor amplifier with unbypassed R_E, i.e., less than 1.

As a result, the differential gain is large ($A \geqslant 10^2$), but the common mode gain is significantly less than 1. The details involve hybrid parameter analysis which we will ignore.

10.5.1 Intuitive Analysis of the Differential Amplifier

Common mode gain. For a common mode signal, $V_{i1} = V_{i2}$. R_E makes the gain fall for either Q_1 or Q_2 (Section 10.3), so the output voltage of either transistor is *reduced* [Eq. (10.5)]. With typical circuit values, the common mode gain is considerably less than 1. This is further reduced because the output signal is the difference ($V_{O2} - V_{O1}$). These output voltages are *equal* for a common mode input, to the extent that the two transistors are matched, so the difference voltage and common mode gain are 0.

The alert reader may have noticed the subterfuge in this argument. We ignored the fact that an input voltage on transistor 1 develops a feedback signal across R_E which is seen by transistor 2.

Differential mode gain. Here, the input voltages are equal and opposite

$$V_{i1} = -V_{i2} \tag{10.11}$$

As a consequence, equal and opposite currents flow through R_E. The feedback voltage across R_E due to a differential signal is 0, so R_E can have no effect on the gain. The gain is just that of an ordinary transistor amplifier, with R_E absent.

The desired result has been achieved: the differential mode gain is large, and the common mode gain is small.

10.6 CURRENT SOURCE BIASING AND THE ORIGIN OF OP AMP BIAS CURRENT

As shown in Eq. (10.5) the common mode gain becomes smaller as R_E is made larger. This is why in standard op amp designs, the emitter resistor of the long-tailed pair [Fig. 10.7(a)] is replaced by a transistor driven as a constant current source [Fig. 10.7(b)], since a current source acts as a nearly infinite resistance (slope of I-V curve $= 1/R \simeq 0$). In many designs developed for integrated circuits, large resistors of discrete circuits are often replaced by transistors. It is easier to make yet another transistor on the IC (since the Si slab is there anyhow) than it is to make a large resistor.

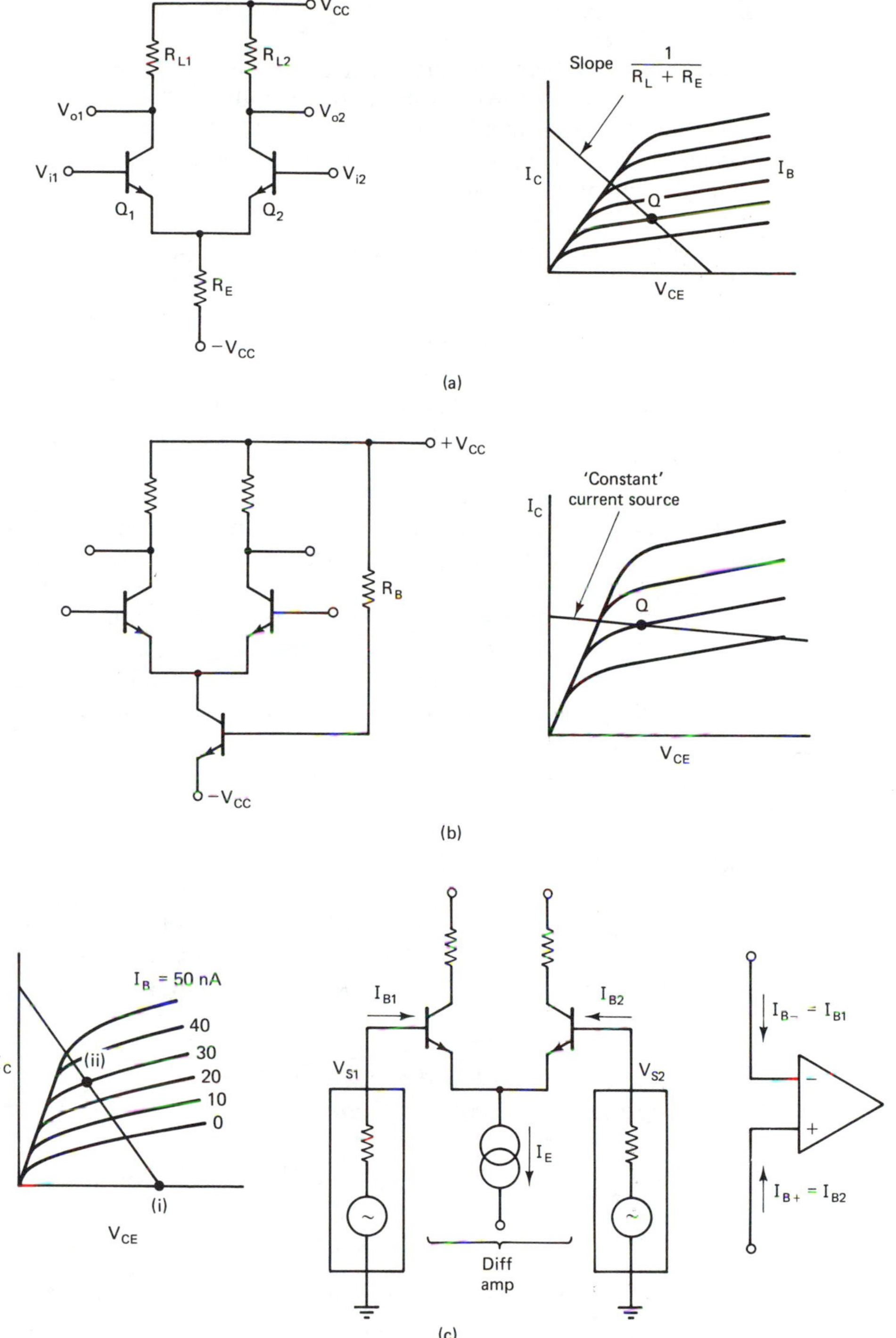

Figure 10.7 Use of a current source to set the operating point, and origin of op amp bias current. (a) Conventional long tailed pair, with load line determined by $(R_E + R_L)$. (b) Current source replacing R_V flattens the load line, reducing common mode gain. (c) Op amp base bias current is drawn from the sources.

Question. Where does the base bias current for the long-tailed pair come from? No base biasing network is shown for Q_1 and Q_2. Clearly the circuit will not function as a *differential* amplifier if $I_B = 0$ (Fig 10.7(c), operating point *i*), since only one sign of input signal will drive the transistor out of cutoff into the linear amplifying region of the characteristic curves.

Answer. Although shown floating, in any application the bases in the long-tailed pair are of course connected to a source [Fig. 10.7(c)]. The current source biasing circuit sets the value of I_E, and a corresponding value of I_B ($= I_E/h_{fe}$) is drawn in from the input signal sources, setting the operating point [*ii* in Fig. 10.7(c)]. As a consequence, any op amp has certain input current *demands* which, although neglected in ideal analysis, have effects which must often be estimated or controlled. This is the origin of the *bias current* specification of an op amp (Chapter 12).

Question. If the differential amplifier is biased with a constant current source, how can it ever amplify a signal, since the current cannot change?

Answer. It amplifies only *differential* signals. With equal and opposite input signals, the transistor collector currents change in equal and opposite directions. As long as one goes up and the other goes down, the net current flow I_E through the biasing source need not change. Yet changes do appear in the voltage at the collectors of Q_1 and Q_2.

10.7 THE MAGIC OF FEEDBACK

Negative feedback has a variety of surprising and powerful properties, summarized in Table 10.1. The techniques developed here are useful either in the analysis of discrete amplifier circuits with feedback or in the analysis of operational amplifier circuits.

10.7.1 Negative Feedback Makes Circuit Gain Independent of Amplifier Gain

Consider the ideal amplifier shown in Fig. 10.8(a). The only property which matters, ideally, is the voltage gain *a*.

$$V_o = aV_s \tag{10.12}$$

This is called the *open loop* gain, to distinguish it from the *circuit* gain with feedback, which may be very different. Consider the feedback example shown in Fig. 10.8(b). A voltage divider at the output feeds back to the input a fraction β of the output voltage, determined by the ratio $R_2/(R_1 + R_2)$. The input voltage V_s applied to the amplifier is therefore the sum of the feedback voltage and input from the outside world, V_i. V_s is referred to as the summing-point voltage. The output

TABLE 10.1 NEGATIVE FEEDBACK PROPERTIES

(1) Circuit gain is independent of amplifier gain. Amplifier gain stability becomes less important. Detailed circuit analysis becomes unimportant (black box approach).
(2) Amplifier DC drift and ac noise are reduced.
(3) Amplifier distortion is reduced.
(4) Frequency response is extended.
(5) Input impedance is increased.
(6) Output impedance is decreased.
(7) Potential problem of stability with respect to oscillation. Techniques for analysis: gain-phase plot; Nyquist diagram.
(8) Shape the circuit gain vs. frequency to do desired operations: add, subtract, differentiate, integrate, bandpass filter, oscillate.
(9) A closed-loop system of feedback amplifiers can simulate a differential equation representing a physical system.
(10) Closed-loop feedback control: voltage and current regulation, regulation of physical variables (position, temperature....).
(11) Nonlinear mathematical operations: multiply, divide, square, square root.

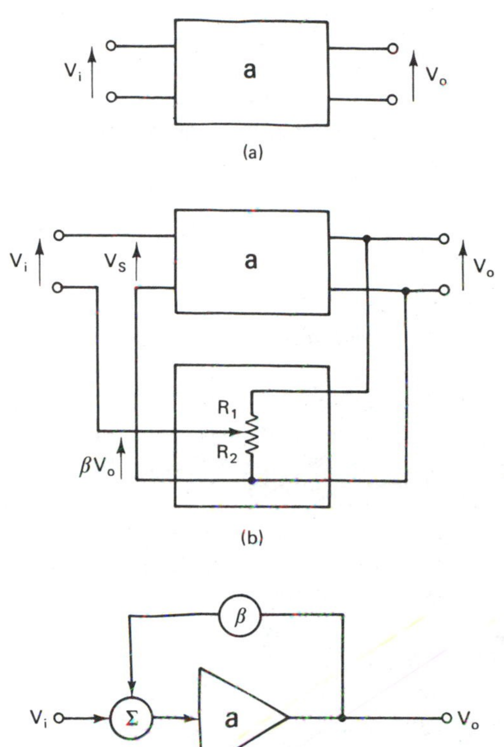

Figure 10.8 The feedback amplifier. (a) Amplifier block. (b) Feedback connection. (c) Generalized form for feedback block diagrams.

voltage is determined by V_s, which in turn is determined by the output voltage and the input voltage via a Kirchoff voltage loop at the input.

$$V_s = V_i + \beta V_o \tag{10.13}$$

Eliminating V_s yields the closed loop circuit gain.

Golden Rule

$$\frac{V_o}{V_i} = \frac{a}{1-\beta a} \qquad (10.14)$$

If $|\beta a|$ is large compared to 1, the gain is substantially smaller than the open loop gain a. In fact, if $|\beta a| >> 1$, then the denominator is approximately equal to βa, and a cancels.

$$V_o/V_i = -1/\beta \qquad (10.15)$$

The circuit gain is independent of the amplifier open loop gain. This result makes circuits with negative feedback relatively insensitive to shifts or drift in the properties of the active devices (see Prob. 10.3). This result also makes the *black box* approach possible. In most of the text we find it possible to disregard the inside of the box, characterizing it by certain properties such as the value of the gain a. To facilitate this, a more general approach to systems modeling [Fig. 10.8(c)] is preferred over the specific *voltage feedback* example just considered. The resistive voltage divider becomes a multiplicative element β, and the voltage summing wiring becomes an abstract summing element Σ.

The sign of the term βa matters and determines whether the feedback will be negative, leading to stability, or positive, leading to oscillations.

$\beta a < 0$	Negative feedback	Control	Feedback opposes amplifier
$\beta a > 0$	Positive feedback	Oscillation or switching	Feedback reinforces amplifier

The instability possible when βa is positive is seen in Eq. (10.14). The denominator will vanish if βa happens to reach $+1$ for some circuit condition or some frequency. The circuit gain can become infinite, which is another way of saying that there can be an output even in the absence of an input. Of course the circuit gain does not really become infinite. Rather, the gain limits itself, since the value of a falls under large signal conditions, and the circuit output either latches up at the value of the power supply voltage or oscillates back and forth at the frequency at which βa happens to equal 1. The first case leads to a two-state switch, and the second case leads to an oscillator. Both situations are useful and will be described in later sections. Positive feedback can also occur when it is not intended, and op amp circuits designed without adequate attention to the possibility of positive feedback may oscillate. We will now consider several of the negative feedback properties listed in Table 10.1, assuming negative gain $-a$.

10.7.2 Negative Feedback Reduces Drift and Noise

Consider a nonideal amplifier (Fig. 10.9) which has an output D in the absence of an input voltage. D may be due, for example, to nonideal amplifier properties (Chapter 11) such as offset, drift, and noise.

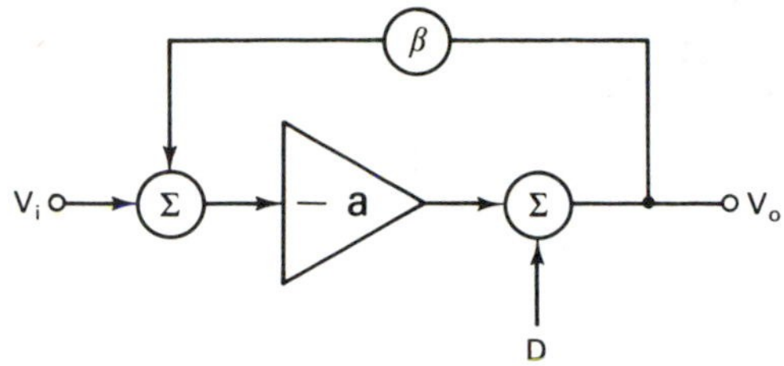

Figure 10.9 Distortion source D introduced into the feedback loop.

$$V_o = -aV_s + D \tag{10.16}$$

The analysis follows that of the ideal amplifier, evaluating the output voltage in terms of the summing-point voltage. *This sequence of steps is basic in the analysis of nearly any op amp or negative feedback problem and should be mastered.*

$$V_s = V_i + \beta V_o \qquad \text{(summing – point voltage)} \tag{10.17}$$

$$V_o = -aV_i - \beta a V_o + D \qquad \text{(output voltage)} \tag{10.18}$$

Solving for the output voltage by eliminating V_s yields

$$V_o = -aV_i/(1 + \beta a) + D/(1 + \beta a)$$

$$\simeq -V_i/\beta + D/(+\beta a) \qquad (if\,|\beta a| >> 1) \tag{10.19}$$

The drift or noise D has been reduced by $|\beta a|$, which is large compared to 1. Negative feedback thus has large error-correcting power, as long as the error signal originates on the output side of the op amp (See Prob. 10.5).

10.7.3 Negative Feedback Reduces Distortion

In many cases, the amplifying element is not ideally linear. For example, the push-pull output stage of an audio power amplifier can exhibit distortion at low signal levels. The nonlinearities of device characteristic curves can lead to harmonic distortion at large signal levels. Modern designs use negative feedback to reduce the problem of distortion by many orders of magnitude.

It is simplest to model the distorting amplifier (Fig. 10.10) as an ideal op amp followed by a nonlinear stage T. This is representative of realistic situations, since most distortion comes in the final output power stages. The analysis is straightforward following the procedures used in arriving at Eq. (10.14).

$$V_B = TV_A$$

$$V_s = V_i + \beta V_o \tag{10.20}$$

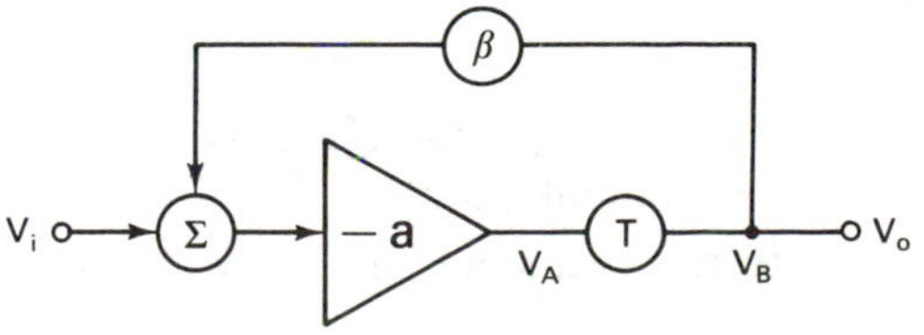

Figure 10.10 Element T with a nonlinear transfer function introduced into the feedback loop.

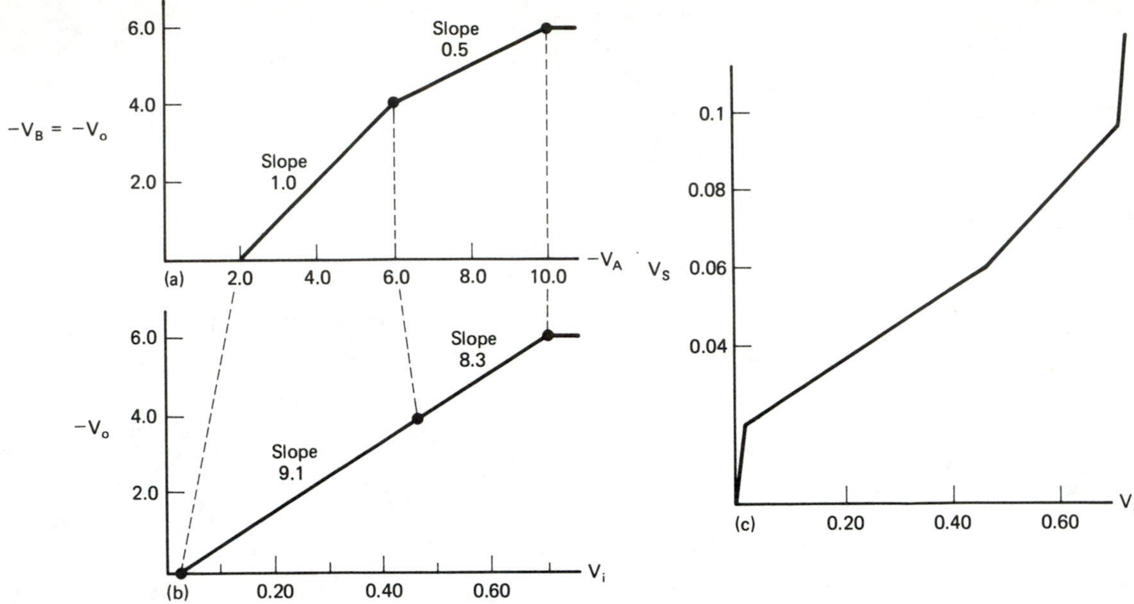

Figure 10.11 The effect of a nonlinear transfer function (a) is much reduced (b) by negative feedback which *compensates* (c) for the distortion.

But $V_A = -aV_s$, leading to

$$V_o/V_i = -aT/(1 + \beta aT) \tag{10.21}$$

With $aT >> 1$, $V_o/V_i \simeq -1/\beta$, independent of distortion T. But T is not really a linear function. Doing the analysis honestly is complicated and not very illuminating. It is preferable to consider a graphical example which illustrates the idea. Suppose T introduces nonlinearities as shown in Fig. 10.11. There is initially no response to an input, then there is a linear region whose slope then falls off, and finally a cutoff region where increasing input results in no further increase in output. Graphical analysis of the input-output response in the presence of negative feedback may be done by evaluating Eq. (10.21) at the breakpoints of the nonlinear device. Suppose $a = 10^2$, $\beta = 0.1$. The gain has been chosen low enough to magnify the problem, yet still illustrate the improvement which feedback brings. The relevant voltages at the breakpoints of Fig. 10.11(a) are given in Table 10.2, using Eqs. (10.20) and (10.21). The resulting input-output relationship [Fig. 10.11(b)] is surprisingly linear. The initial inactive threshold is compressed, and the shift in slope when the gain falls is scarcely noticeable. Cutoff remains cutoff, since the output voltage range cannot be increased by feedback. The explanation for this improvement is as follows:

(1) Because a is very large, only a very tiny input signal is needed to sweep past the initial inactive threshold (first break point).

(2) As long as $|\beta aT| >> 1$, a decrease in slope of $T = V_B(V_A)$ is unimportant (second break point).

TABLE 10.2 EXAMPLE OF REDUCED DISTORTION WITH NEGATIVE FEEDBACK

$-V_o =$ $-V_B$	$-V_A$	$V_S =$ $-V_A/a$	$-\beta V_o$	$V_i =$ $V_S - \beta V_o$
0	0	0	0	0
0	<2.0	<0.02	0	<0.02
0	2.0	0.02	0	0.02
4.0	6.0	0.06	0.40	0.46
6.0	10.0	0.10	0.60	0.70
6.0	>10.0	>0.10	0.60	>0.70

(3) Negative feedback tries to keep the difference between output and input sensed at the summing point as small as possible. In doing so, a feedback signal is generated whose distortion *compensates* for amplifier distortion by distorting V_s in the *opposite* direction [Fig. 10.11(c)].

10.7.4 Negative Feedback Extends Frequency Response

Consider an amplifier whose gain is a function of frequency (Fig. 10.12). Any amplifier's gain falls off above some cutoff frequency f_c due to capacitance in the circuit, either inherent or added for stability. Since the capacitance appears across the output of a Thevenin equivalent source resistance, it creates a low-pass filter. In the presence of negative feedback, the low-frequency gain is $1/\beta$ rather than a. This is true regardless of the precise value of a as long as $\beta a >> 1$. As a result, the circuit gain will remain $1/\beta$ even after a is above its cutoff frequency and starts to fall. The circuit gain does not start to fall until a is low enough so that βa is comparable to 1. The closed-loop frequency response has therefore been considerably extended by negative feedback, as shown in Fig. 10.12.

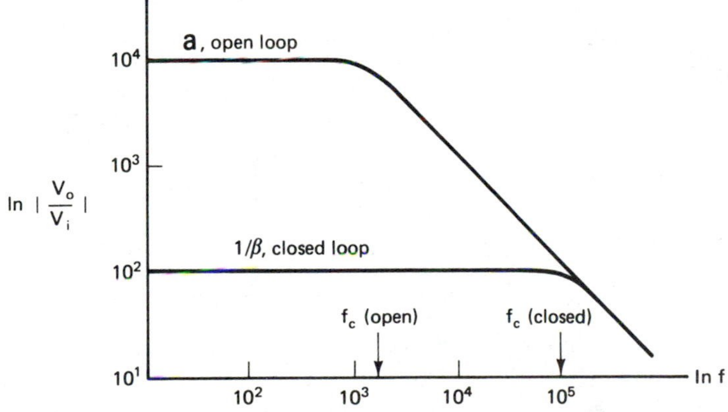

Figure 10.12 The frequency response of an amplifier is extended with closed loop negative feedback.

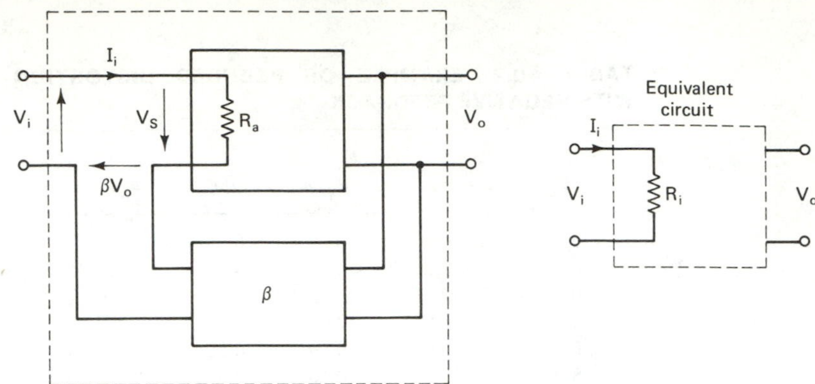

Figure 10.13 The input impedance of an amplifier is increased by negative feedback.

10.7.5 Negative Feedback Increases Input Impedance

Input impedance in the presence of negative feedback may be analyzed by a black box approach, as shown in Fig. 10.13. Suppose that the physical amplifier (with gain a and amplifier input resistance R_a) plus negative feedback β may be replaced by an equivalent circuit whose effective input resistance is R_i, defined as the ratio V_i/I_i. Physically, the feedback voltage βV_o balances out nearly all of V_i. As a result, a smaller input voltage V_s is applied to the amplifier. The input current is therefore reduced by feedback, so the effective input resistance is increased. The details follow straightforwardly. Referring to Eq. (10.14), the feedback voltage is

$$\beta V_o = [\beta a/(1 - \beta a)] V_i \tag{10.22}$$

This is indeed very close to V_i as long as $|\beta a| >> 1$. The input current, by Ohm's law, is the difference voltage $(V_i - \beta V_o) = V_s$, divided by the actual amplifier input resistance,

$$I_i = \frac{V_s}{R_a} = \frac{1}{R_a}\left(1 + \frac{\beta a}{1 - \beta a}\right) V_i = \frac{V_i}{(1 - \beta a)R_a} \tag{10.23}$$

The effective input resistance (impedance) is the ratio

$$|R_i| = |V_i/I_i| = |(1 - \beta a)|R_a >> R_a \tag{10.24}$$

This is $|\beta a|$ times larger than the physical amplifier's input impedance.

10.8 FEEDBACK METHODS

10.8.1 Current and Voltage Feedback

It is useful to recognize several standard feedback methods encountered in a discrete amplifier circuit. The example shown in Fig. 10.14(a) has already been analyzed in Section 10.2 as a precursor to differential amplifier feedback. Viewing the transistor as a black box amplifier, we see that what is fed back to the input is a signal proportional to the output current. This form of feedback is therefore called

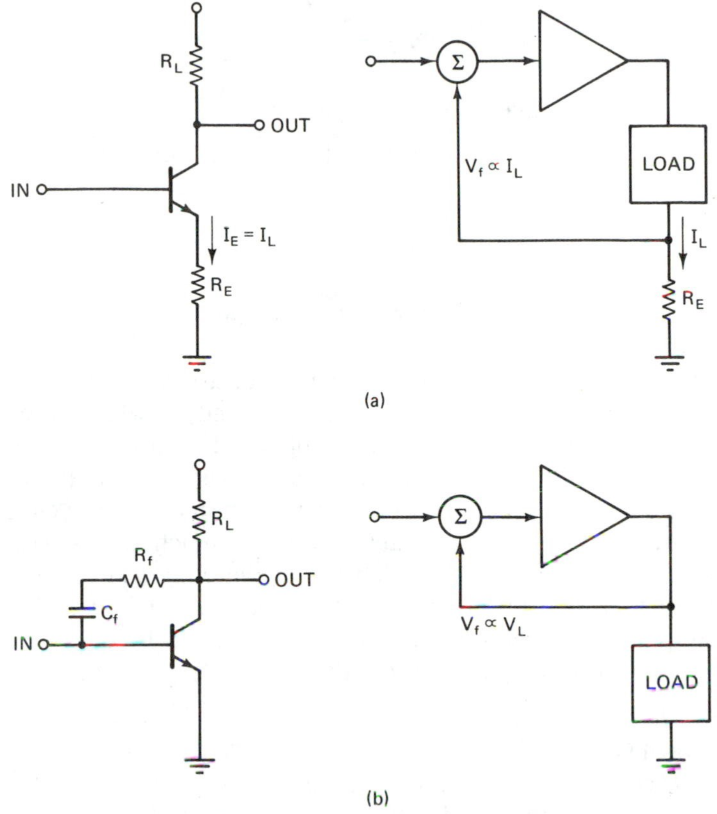

Figure 10.14 Feedback methods. (a) Current feedback. (b) Voltage feedback.

current feedback. Its use will be seen again in Chapter 13 in current-regulated power supplies. A second form of feedback is shown in Fig. 10.14(b). This form of feedback feeds a sample of the output voltage back to the input, and is called *voltage* feedback (already analyzed in Section 10.7). Although it is useful to be able to recognize these forms of feedback, we will not analyze them in detail, since in this book we restrict ourselves for the most part to black box circuit analysis. The power of the black box method is that in most cases it is not necessary to do detailed circuit analysis.

10.8.2 Operational Amplifier Feedback

An ideal operational amplifier (Fig. 10.15) is a differential amplifier whose output is proportional to the voltage difference applied to its inputs.

$$V_o = -aV_s \quad (a = \text{open loop gain}) \tag{10.25}$$

This *single-ended* output voltage (only a single wire is shown) is generally measured with respect to a circuit ground, as shown in Fig. 10.15, although this ground

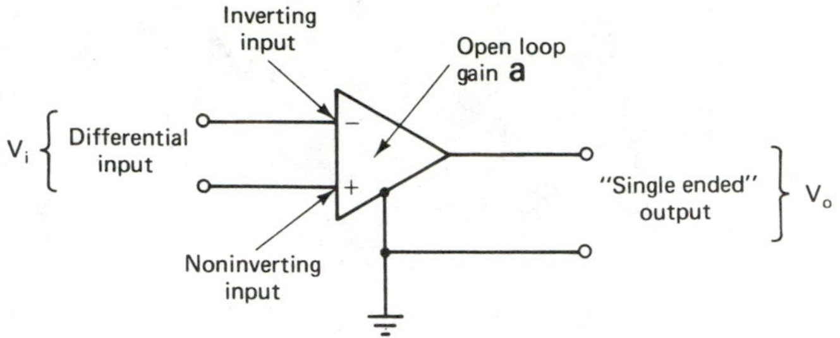

Figure 10.15 Conventional nomenclature for an operational amplifier (op amp).

connection is generally left off in circuit diagrams. In many applications, the circuit may not be connected to an earth ground, in which case the output is referenced to the point halfway between plus and minus power supply voltages. This can be useful in referencing the "ground" point in a circuit to some particular place in a measuring device, or in construction of op amp circuits with only a single battery. The input terminal with the (−) attached to it is called the *inverting input*, and the other (+) is called the *noninverting input*. This convention is a reminder that the gain of an op amp is negative [Eq. (10.25)] to facilitate negative feedback in the standard op amp connection, to be discussed. With an ideal op amp, two conventions are followed:

Ideal Op Amp

(1) No current flows into either the (+) or (−) input.
(2) There is no output voltage in the absence of an input voltage.

In practice, modern operational amplifiers are extremely close to this ideal.

The standard op amp feedback connection is shown in Fig. 10.16. The noninverting terminal is connected to the system ground; if this connection is left off in a circuit diagram, it is implicitly assumed. The relation between output voltage and input voltage follows from two basic assumptions: current is conserved, and no current flows into the op amp. The first is of course always true, and the second is an assumption of the ideal op amp. As a result, the current flow in R_i and R_f may be equated. Each is related to the voltage difference across each resistor, by Ohm's law.

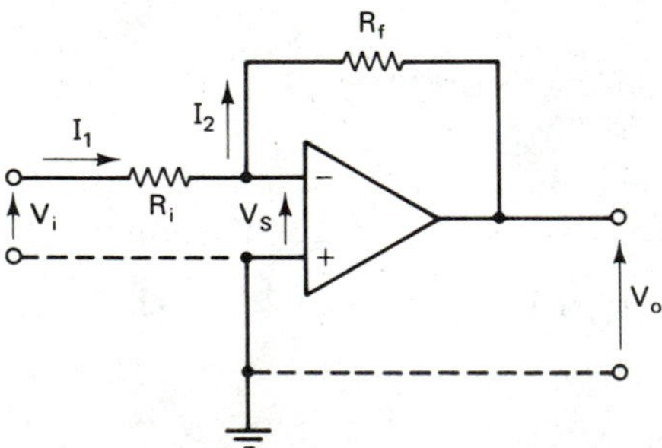

Figure 10.16 Operational amplifier feedback.

$$I_2 = I_2$$

$$(V_i - V_s)/R_i = (V_s - V_o)/R_f \tag{10.26}$$

Neither V_s nor V_o is yet known, but they are related via the gain of the op amp under *open loop* conditions [Eq. (10.25)]. Using this to eliminate V_s leads to

$$V_o = \frac{-aV_i}{1 + (R_i/R_f)\,(1 + a)} \tag{10.27}$$

Under typical circuit conditions, $a \simeq 10^5$, $R_i/R_f \geqslant 0.01$, and the 1's in the denominator are negligible, so the closed loop gain is

$$V_o \simeq -(R_f/R_i)\,V_i \tag{10.28}$$

This is the fundamental op amp feedback relationship. Note that the gain of the circuit is independent of the amplifier gain a. As a consequence, the circuit gain may be set merely by varying passive components R_i and R_f. In the derivation above, the op amp *input* is assumed to play a very passive role, merely sampling the voltage appearing at V_s. This is not the same as saying that the op amp may simply be unplugged without changing the circuit function! We will now show that it is the op amp output functioning through Eq. (10.25) that drives the right amount of negative feedback current I_2 to keep V_s as close as possible to 0. V_s may be obtained from Eqs. (10.25) and (10.27),

$$V_s = -V_o/a = [aR_i/R_f]^{-1} \tag{10.29}$$

In typical circuit applications, R_i/R_f is in the range of 10^{-2} to 10^2. The open loop gain a may range from 10^4 to 10^5. As a result, V_s is only 10^{-2} to 10^{-7} of V_i. This is the sense in which op amp feedback is negative. Since a is very large, a tiny signal at V_s would drive V_o to the full power supply voltage. With the negative feedback connected (Fig. 10.15), the output voltage provides feedback to keep this from happening. Negative feedback drives V_s nearly to 0. For this reason, V_s is often called a *virtual ground*, and the inverting terminal is often called the *summing point* of the op amp.

10.8.3 Simplified Op Amp Analysis

It is sufficient in most cases to believe that V_s will be 0 once negative feedback is connected. This will considerably simplify the analysis. In the above example, assuming V_s is 0, equating the currents in R_i and R_f, and using Ohm's law leads to

$$V_s = 0$$

$$I_1 = I_2$$

$$(V_i - 0)/R_i = (0 - V_o)/R_f$$

or

$$V_o/V_i = -R_f/R_i \tag{10.30}$$

The result is the same as the basic op amp relationship [Eq. (10.28)]. Both these approaches to op amp feedback should be mastered, and both will be used again and again. If in doubt, or if nonideal op amp behavior is to be analyzed, the more general approach will always work. But in most cases, the simplified approach of assuming a virtual ground at the summing point will work satisfactorily. Note that the quantity R_f/R_i appears to play the role of $1/\beta$ in the voltage feedback example of Section 10.3.

10.9 STABILITY PROBLEMS

Referring back to the voltage feedback example, Eq. (10.14), the right-hand side will blow up if the product βa happens to reach 1. This may happen even if the dc value of $\beta a < 0$, since both β and a may include a frequency-dependent phase shift. The feedback network may include components other than resistors, or the amplifier may have unavoidable capacitance with phase lag in a as a consequence. If the combined phase shift reaches 180°, the sign in Eq. (10.15) will flip so that negative feedback becomes positive, βa may reach $+1$ at some frequency, and oscillations may occur. In this section, some terminology useful in understanding op amp specifications (Chapter 12) will be introduced by an example.

> **Stability Criterion.** As the frequency increases, the magnitude of βa must fall below 1 before its phase shift reaches 180°.

Example: Two-Stage Amplifier

The gain of a two-stage dc amplifier falls off above some frequency, as shown in Fig. 10.17(a). This is due to capacitive effects which short out the circuit's amplifying properties at high frequencies. The capacitance may be either inherent, as in the junction capacitance of a transistor, or explicitly added for stability purposes. The rate of falloff above the cutoff frequency amounts to 20 db per stage per decade of frequency, or a total of 40 db/decade. Also shown in Fig. 10.17(a) is the phase shift accompanying this roll-off in gain. Each stage acts as an RC filter with a maximum 90° phase shift, or a total maximum phase shift of 180° for the two-stage amplifier. Instability in this circuit may be predicted by checking the value of $|\beta a|$ when the phase shift approaches 180°.

Instability may be seen more directly, however, by presenting the same information in a *Nyquist diagram.* The magnitude and phase of βa may be plotted together as a vector, shown in Fig. 10.17(b). As the frequency increases from dc, the vector swings around from $-180°$ (negative feedback) and also shrinks in size. Since the maximum phase shift of this amplifier is 180°, the trajectory executes a portion of a spiral in the negative half plane ($y < 0$). We plot this in the complex plane, with the real part the horizontal component and the imaginary part the vertical component. The point $\beta a = +1$ at which oscillations or instability would occur is shown as a cross in Fig. 10.17(b). This danger is approached (for a fixed a) as the feedback fraction β is increased. But since the maximum phase shift is limited to 180°, the trajectory may only *approach* the danger point and never actually cross it. As a result, the two-stage amplifier is unconditionally stable.

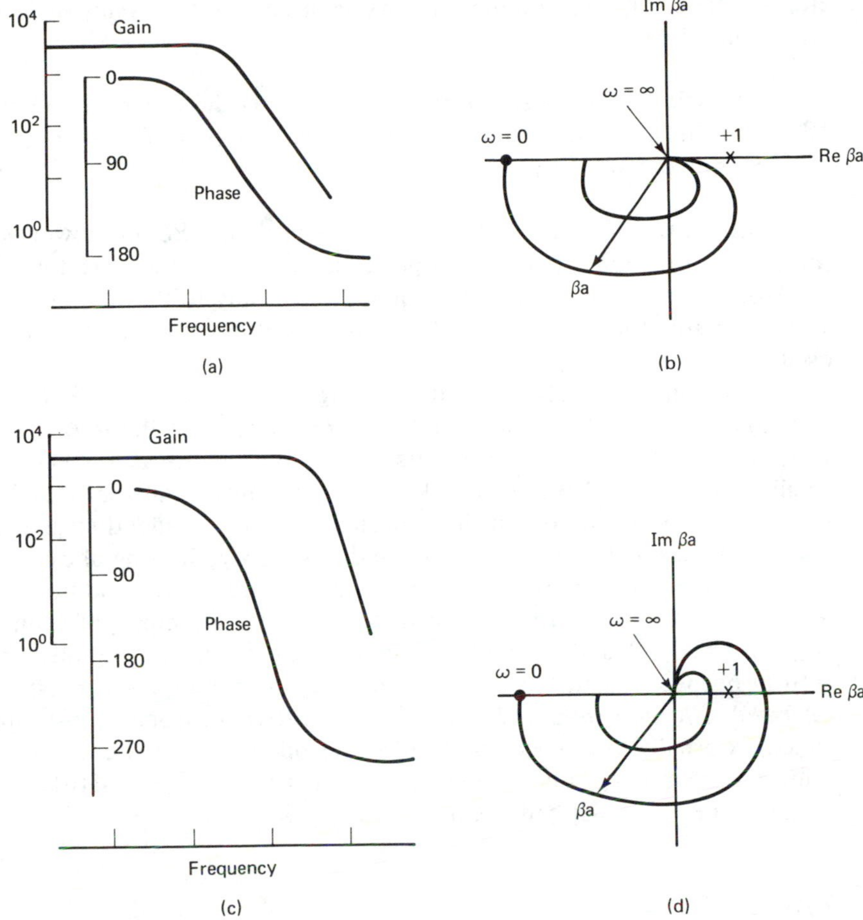

Figure 10.17 (Left side) Gain and phase plots for (a) two- and (c) three-stage amplifiers. (Right side) A plot of the product βa in the complex plane (magnitude and phase) leads to a stability criterion for (b) two- and (d) three-stage amplifiers.

Example: Three-Stage Amplifier

The above example is not typical of op amps, since the two-stage amplifier gain is positive (each transistor yielding a negative voltage gain). To achieve the negative gain required for op amp operation, three stages may be employed. This leads to a larger roll-off in the magnitude of the gain, Fig. 10.17(c), and a maximum phase shift of 270°. Viewed as a vector plot or Nyquist diagram [Fig. 10.17(d)], the trajectory followed by the vector βa as frequency increases can wrap around the danger point $\beta a = +1$. Keeping the open loop gain constant and varying β, it is possible to select β such that the instability is crossed by the trajectory, and the circuit will oscillate. This is a standard method of producing an oscillator using feedback. Equally often, however, an op amp circuit which is not intended to oscillate will be found to do so inadvertently. To avoid such situations, one must alter the trajectory of βa in the complex plane so that it does not enclose the instability point $\beta a = +1$.

Question. Suppose the trajectory of βa is the outer curve shown in Fig. 10.17(d). Since this does not touch the instability point $\beta a = +1$, is there any problem due to instability?

Answer. Very likely there will be a problem. Before power is applied to the circuit, the gain a is 0. When power is turned on, the trajectory βa expands out from 0 and cuts the instability point on its way to its final destination. It is very likely that the circuit will lock into the instability point and either latch up or oscillate.

Although this analysis was done using the voltage feedback idea, it also applies to conventional op amp circuits, where Z_i/Z_f plays the role of β. Instability with large feedback therefore occurs in op amp circuits at low gain, i.e., Z_f/Z_i small. This is why there are two kinds of op amps: *compensated* and *uncompensated.* A *compensated* op amp has internal capacitance added so that as the frequency increases, the open loop gain a drops slowly, starting at a low frequency. As a result, the compensated op amp is stable under any conditions of negative feedback, but also has a poor high-frequency gain. Examples of compensated op amps are the 741 and CA3140. Sometimes the fast response inherent in an op amp is needed and the low cutoff frequency of a compensated type cannot be tolerated. An *uncompensated* op amp has no internal capacitor, but capacitance is added externally as needed to stabilize the operation for the actual circuit gain selected. Examples of uncompensated op amps are the 709, LM101, and CA3130. Real op amps are described in Chapter 12.

PROBLEMS

10.1. Draw the hybrid equivalent circuit of the transistor differential amplifier shown in Fig. 10.2. Explain the function of all resistors. Explain why (in words and sketches rather than mathematics) the gain is the same as a single transistor amplifier, even though the emitter resistor is not bypassed. Do this by considering the ac signal across R_E resulting from a *differential* input across terminals 1 and 2, i.e., when $V_1 = -V_2$. Compare this with the situation when the input is single-ended, i.e., signal put in at 1, and terminal 2 grounded.

10.2. Consider the basic negative feedback amplifier [Fig. 10.8(c) and Eq. (10.14)]. Suppose $a = 10^5$, and $\beta = 10^{-2}$. What is the circuit gain? Now suppose the amplifier gain shifts to 5×10^4 due to some drift in transistor properties. By what percentage does the circuit gain change as a result of this 100% change in amplifier gain?

10.3. Consider an amplifier whose open-loop gain a is 10^5 but starts to fall above a cutoff frequency of 10^4 Hz by a factor of 10 in gain per decade of frequency. Suppose the feedback network of Fig. 10.8(b) has $R_2/R_1 + R_{2)} = 0.01$. At what frequency will a circuit gain fall by a factor of 2 from its dc value f? **Suggestion:** Analyze the problem graphically.

10.4. The improvement in performance when a nonideal error signal is introduced into a negative feedback circuit depends upon where it is introduced in the circuit. Consider the nonideal amplifier of Fig. 10.9, but now put the drift or noise signal in *before* the op amp rather than after. Show that negative feedback results in no improvement in performance and no reduction in the output error *relative to the amplified input signal.* An error of this kind is discussed in Chapter 12 under Input Voltage Offset.

10.5. What is the "feedback fraction" β for the op amp circuit of Fig. 10.16? **Hint:** What fraction of V_o finds its way back to the inverting terminal? The answer is not quite R_i/R_f as implied in the text, but should be consistent with rewriting Eq. (10.27) as

$$\frac{V_o}{V_i} = \frac{-(aR_f)(R_i + R_f)}{1 + [(aR_i)/(R_i + R_f)]}$$

10.6. Suppose that the signal source V_i applied to the input of the op amp circuit of Fig. 10.16 has a source resistance R_s. Show that the *effective* source impedance, seen looking back from the output V_o, is reduced to

$$R_o = \frac{R_s + R_i}{[R_i/(R_i + R_f)]a}$$

What size is the effective source impedance, for $R_s = 10^3\ \Omega$, $R_i = 10^4\ \Omega$, $R_f = 10^6\ \Omega$, and $a = 10^5$?

REFERENCES

More complete bibliographic information for the books listed below appears in the annotated bibliography at the end of the book.

BENEDICT, *Electronics for Scientists and Engineers*

BROPHY, *Basic Electronics for Scientists*

HIGGINS, *Experiments with Integrated Circuits*, Experiment 17

MELEN & GARLAND, *Understanding IC Operational Amplifiers*

11

Basic Op Amp Applications

Op amps can represent a precise electronic analog of mathematical operations. Signal voltages may be added, subtracted, integrated, or differentiated just as if they were mathematical functions. The output response to a given input may be tailored for an application by shaping the *transfer function* as a function of frequency. In this chapter we will assume an ideal op amp with these properties:

Ideal Operational Amplifier:
(1) Infinite gain
(2) Completely stable under any conditions of feedback
(3) Unlimited frequency response
(4) Infinite input impedance
(5) Zero output impedance
(6) Zero voltage drift
(7) Zero current offset and drift
(8) Perfect common mode rejection

The real world is not far from this ideal with modern op amps, so the idealized analysis of this chapter is sufficient in most applications. In the derivations below, and in any application where ideal op amp assumptions are made, it can be assumed that negative feedback keeps the voltage across the summing point equal to zero. The derivations are similar to those leading to Eq. (10.30). The more general approach leading to Eq. (10.28) may always be used, however, and it is suggested that this approach be tried for several of the examples below. The more general approach *must* be used where nonideal behavior of op amps is important; examples will be shown in Chapter 12.

11.1 WEIGHTED ADDER

A set of independent signals may be summed using the circuit of Fig. 11.1. Any number of signals are connected to the summing point through input resistors $R_1...R_n$. The output is the weighted sum of the inputs, with the weighting set independently by the ratios $R_f/R_1...R_f/R_n$.

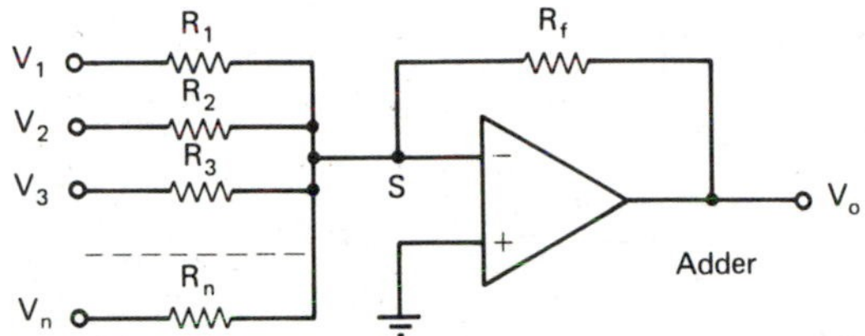

Figure 11.1 Summing amplifier circuit.

Analysis

Assumption 1. Virtual ground at the summing points.
Assumption 2. No current flows into the op amp.

Since current is conserved, the sum of the input currents flowing through $R_1...R_n$ must equal the current flowing through R_f (assumption 2). Ohm's law and assumption 1 (virtual ground) lead to

$$(0 - V_o)/R_f = (V_1 - 0)/R_1 + (V_2 - 0)/R_2 + \cdots + (V_n - 0)/R_n \quad (11.1)$$

Solving for V_o gives

$$V_o = -R_f(V_1/R_1 + V_2/R_2 + \cdots + V_n/R_n) \quad (11.2)$$

The inputs are decoupled from one another: no voltage appears at input 1 due to the voltage applied at any of the other inputs.

Question. Why does no voltage appear at terminal 2 as a result of the voltage input at terminal 1?

Answer. Because the point *s* is a virtual ground. A signal applied to terminal 1 does not alter the potential at point *s*, so no change in potential is seen at terminal 2. See Probs 11.1 and 11.2.

This feature is very useful in avoiding interaction of signal sources. A large signal applied to the *output* of a source can sometimes alter its behavior, shifting the operating point or introducing harmonic distortion or mixing. Such problems are completely avoided when signals are added using this op amp circuit.

11.2 SUBTRACTOR

The output of this circuit (Fig. 11.2) is the difference of the two input voltages V_1 and V_2. This is useful in mathematical operations, or as a difference amplifier when neither point in the signal source is grounded. A subtractor may be used to feed the output of an analog system (A) to the input of a digital system (D), where the grounds of the two systems are independent or contain considerable noise relative to one another (Fig. 11.3). Connecting the ground system A to input terminal V_2 and its signal output to terminal V_1 has the effect of bringing the signal into system D referenced to its own ground, without connecting the two system grounds together.

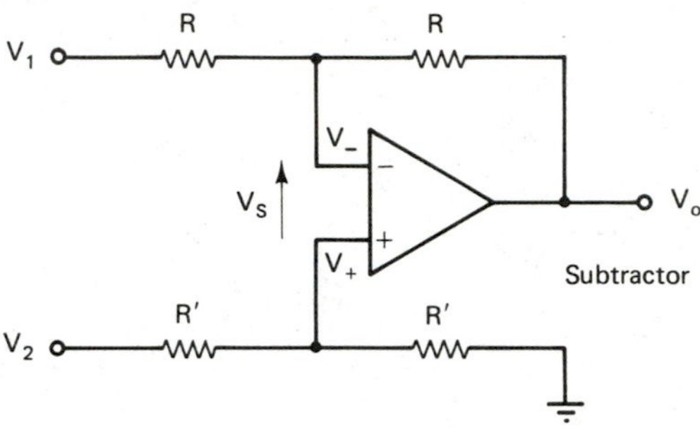

Figure 11.2 Subtractor circuit. Each pair (R, R) or (R′, R′) must be precisely matched, but it is not necessary that R = R′.

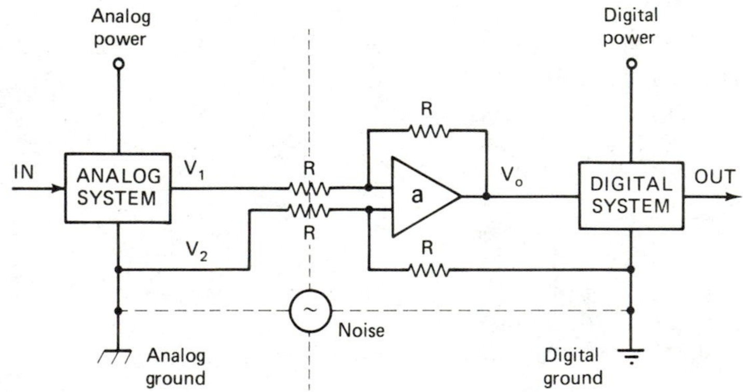

Figure 11.3 Subtractor application for ground isolation. The op amp output is referenced to the ground to the right, and (not shown) gets its power from the digital system. The signal coming from the left has its own ground reference, with common mode noise between the two grounds.

Subtractor Analysis

Assumption 1. Virtual ground is not appropriate, since the op amp's V_+ input is not grounded. As a result, the summing point s will not be 0 V. Negative feedback will try to make the voltage *difference* V_s between the inverting terminal V_- and noninverting terminal V_+ equal to 0.

Assumption 2. No input current flows into the op amp.

Since the input terminals of the op amp are passive observers, their voltages are determined by the voltage dividers formed by the two resistor pairs. The voltage V_+ is half of V_2, and the voltage at V_- is the average of the voltages appearing at V_1 and V_o. Assuming that negative feedback tries to keep V_+ equal to V_- leads to

$$V_s = V_- - V_+ = 0 = (V_o + V_1)/2 - V_2/2 \tag{11.3}$$

The two resistor values R' and R need not be equal; their values do not appear in the result

$$V_0 = V_2 - V_1 \tag{11.4}$$

However, to keep the subtraction precise, each *pair* must balance within the precision needed for the subtraction (see Prob. 11.3).

11.3 INTEGRATOR

Sometimes one needs to measure the sum of what has gone on over past times. A simple op amp circuit shown in Fig. 11.4 can provide a precise voltage analog of this mathematical operation of integration.

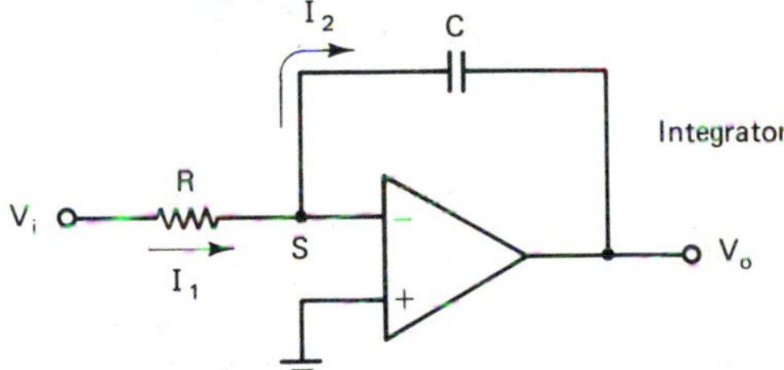

Figure 11.4 Integrator circuit.

11.3.1 Integrator Analysis

Assume $V_s = 0$, $I_1 = I_2$. These two assumptions are the basic ritual in analyzing most op amp circuits, and need not be commented on further. The relationship between current and voltage through the feedback element in this case is not given by Ohm's law, because current flow in a capacitor is proportional to the *rate of change* of voltage across it. This leads to

$$V_o = -(1/RC)\int V_i dt \tag{11.5}$$

It is useful to compare this circuit with a passive low pass RC filter, which also can integrate under appropriate conditions (Section 11.6). Even with the same R and C values, the active integrator does not suffer from the frequency limitations of the passive integrator, which only approximates integration for signals whose frequency is small compared to $1/RC$. However, nonideal op amp properties show up most visibly in an integrator, as output drift in the absence of an input. One must provide a reset switch across C to discharge its capacitor when needed. Integrator drift is covered in detail in Chapter 12.

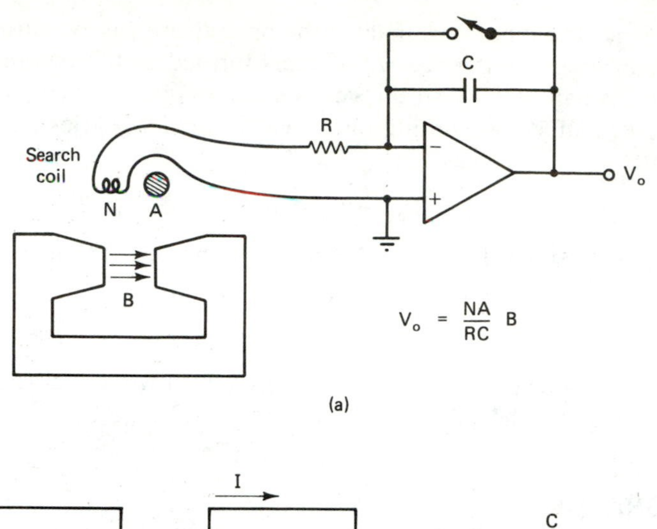

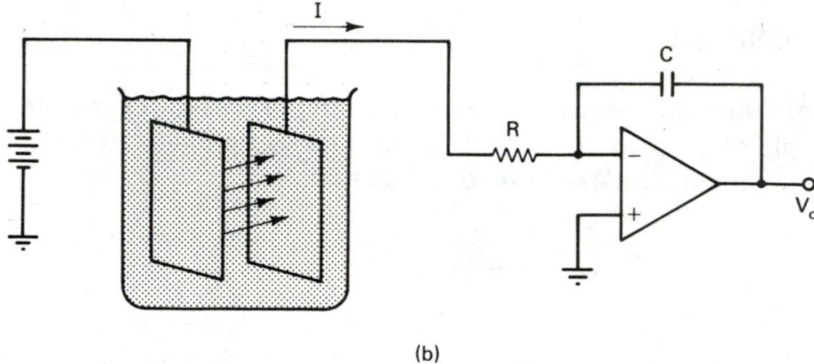

Figure 11.5 Integrator applications. (a) Magnetic field measurement; V_o is a measure of B when the search coil is moved from far away ($B = 0$) into the magnet. (b) Coulometric measurement of total charge or number of moles of material electrodeposited.

11.3.2 Integrator Applications

An integrator provides an easy way to measure a magnetic field [Fig. 11.5(a)]. A search coil (N turns of wire of area A) is connected to an op amp integrator circuit. The magnetic field B is measured by inserting the coil into the magnet from far away ($B = 0$). The change in integrator output voltage at the end of this process is a precise measure of B at the final position of the coil. This follows from Faraday's law: since the voltage induced in the coil is equal to the rate of change of magnetic flux Φ within it, the integral of the voltage during the insertion process measures the magnetic field at the final position.

$$V_{in} = \frac{d\Phi}{dt} = -NA \; dB/dt$$

$$V_{\text{out}}(RC)^{-1} \int_{\text{outside}}^{\text{inside}} V_{in}\,dt = (\text{NA/RC})\text{B} \tag{11.6}$$

Since the parameters N, A, R, and C are readily measurable, Eq. (11.7) provides a relatively precise (<1%) method of absolute magnetic field calibration. The method is particularly useful in high resolution (<0.01%) measurements of spatial field uniformity. To map the field, connect the output to the y-axis of an xy recorder, whose x-axis is connected to a pot which measures coil position.

Another application of integrators is in electrochemistry, in the coulometric measurement of the amount of material electroplated on a surface [Fig. 11.5(b)]. The integral of the current equals the charge, which by another Faraday's law is precisely related to the number of moles N which has moved across the cell.

$$V_{\text{out}} = -(RC)^{-1} \int V_{in}\, dt = (1/C) \int I\, dt = Q/C = (F/C)N \quad (11.7)$$

where F is Faraday's constant, $F = 9.65 \times 10^4$ coulombs/mole. Note that the value of resistor R drops out in charge measurements (see Prob. 11.6).

11.4 DIFFERENTIATOR

Reversing the role of R and C in the integrator results in the differentiator (Fig. 11.6), which helps to resolve fast signal changes in the presence of undesired low frequency noise. The analysis follows the basic ritual, and will be left as an exercise (see Prob. 11.7). The result is

$$V_o = -RC\frac{dV_i}{dt} \quad (11.8)$$

The problem with an ideal differentiator is that the magnitude of the output increases linearly with input frequency indefinitely, since

$$\left|\frac{d}{dt} \sin \omega t\right| = |\omega \cos \omega t| = \omega \quad (11.9)$$

A signal passed through a differentiator may be overwhelmed by radio frequency noise from a nearby station or a local noise source such as a SCR power controller. Practical differentiators are always modified from the ideal, limiting the linear range to the signal range of interest (see Prob. 11.9).

Examples of differentiator applications in signal processing are shown in Fig. 11.7. In case (a), the sine wave signal of interest is obscured by lower frequency background noise. Feeding this signal through a differentiator results in an output signal whose amplitude is increased relative to the noise by the ratio of their two frequencies [see Prob. 11.8(a)]. In case (b), the signal is an edge or pulse riding

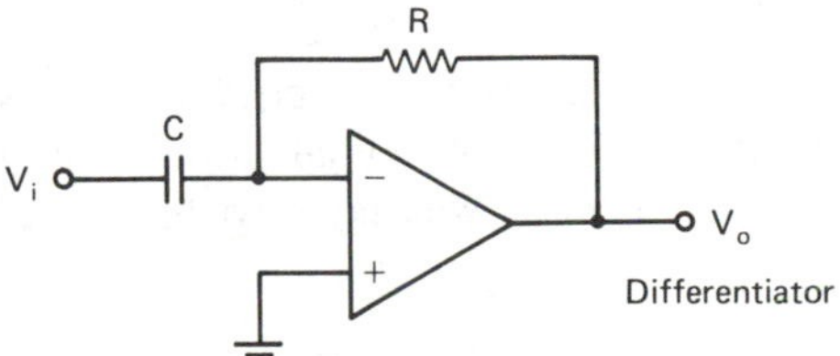

Figure 11.6 Differentiator circuit.

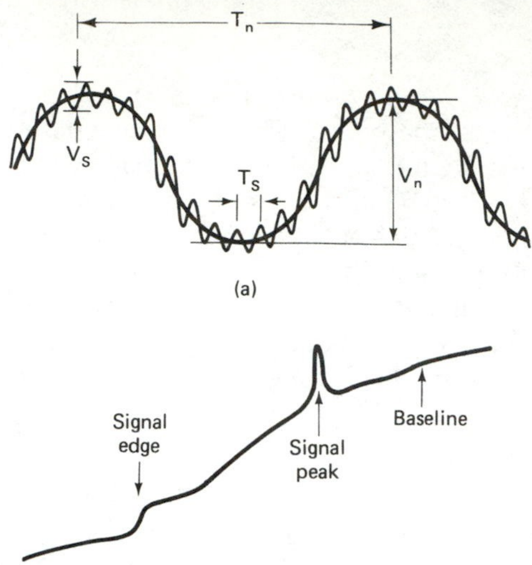

Figure 11.7 Example of inputs where a differentiator will extract signals either from (a) lower frequency noise, or (b) a slowly varying baseline.

on top of a much larger slowly varying baseline. Since the baseline slope is not zero, subtraction (= zero offset) will not help. Differentiation will bring out the signal relative to the baseline [see Prob. 11.8.(b)]. The relative enhancement is proportional to the ratio of slopes of signal and baseline. The shape of the signal is altered, since it too is differentiated, but the change is readily related to the true signal shape.

11.5 THE TRANSFER FUNCTION CONCEPT

These diverse approaches to analyzing op amp circuits may be unified by the concept of the transfer function (Fig. 11.8). Consider for the moment a sinusoidal input signal. It will be convenient to use exponential notation (see Appendix 3).

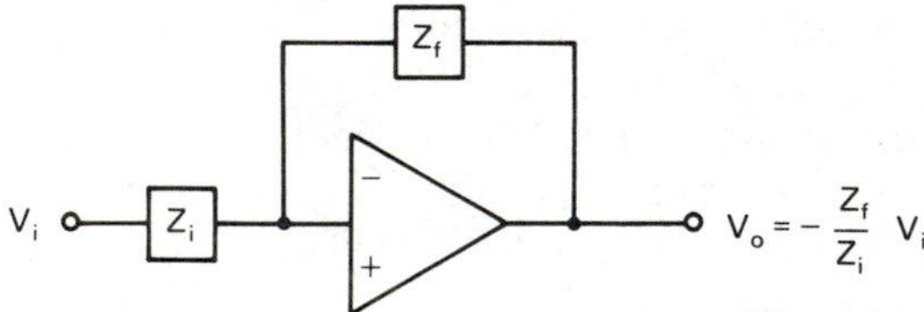

Figure 11.8 Generalized transfer function circuit.

11.5.1 Integrator Transfer Function

When a sine wave signal is run through an analog integrator (Fig. 11.4), the result is

$$V_o = -(1/RC)\int V_i(t)\,dt = -V_i/RC\int e^{j\omega t}dt$$
$$= -V_i(t)/(j\omega RC) \tag{11.10}$$

The output scales in amplitude inversely with the frequency. This is a characteristic signature of integrator response to a sinusoidal input [Fig. 11.9(a)]. Noting that the impedance of a capacitor is $1/j\omega C$, we may rewrite Eq. (11.10) as

$$V_o = -\frac{Z_f}{Z_i}V_i \tag{11.11}$$

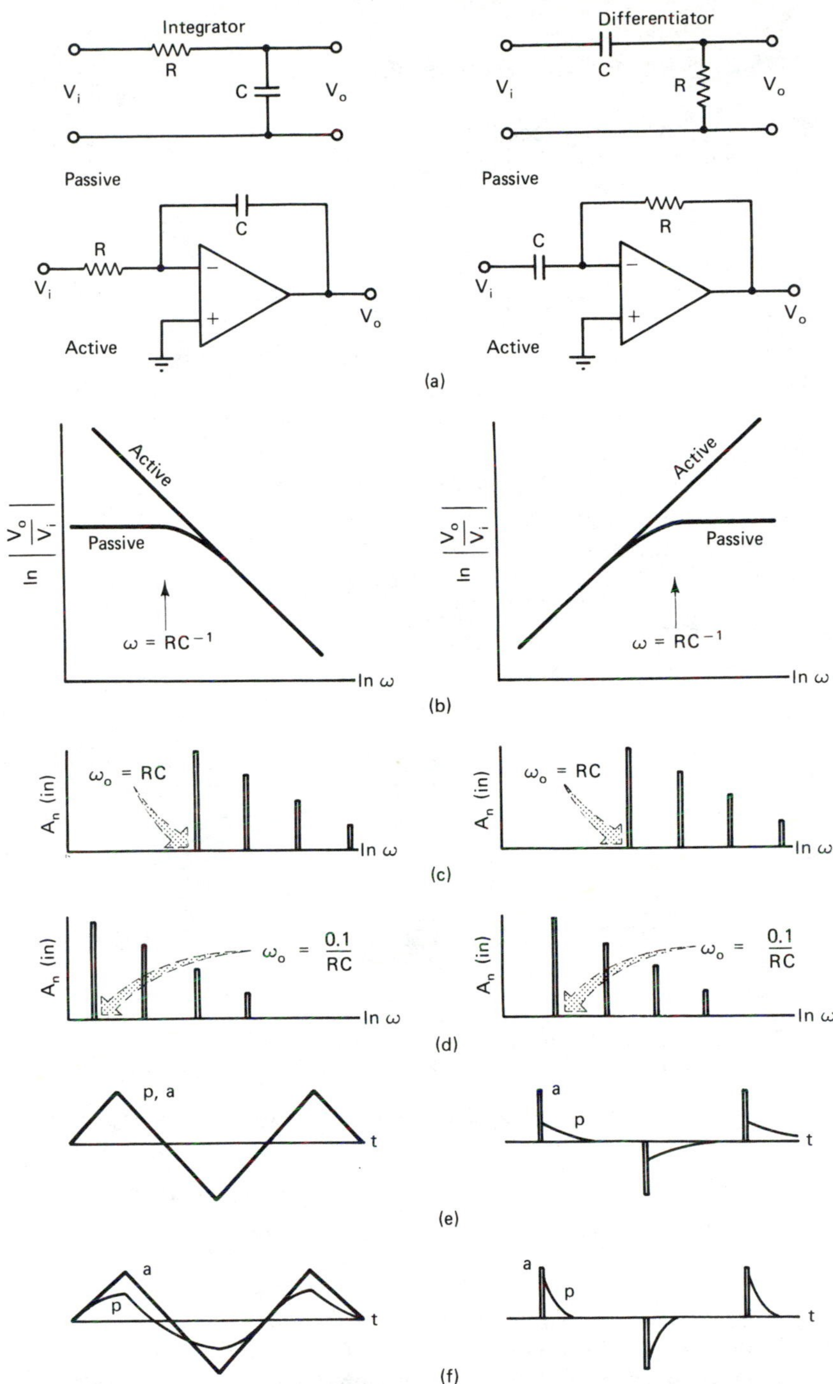

Figure 11.9 Comparison of passive and active integrator and differentiator circuits. (a) Circuits. (b) Transfer functions. (c) Spectrum of a square wave of frequency w_o = (RC of the filter)$^{-1}$; the corresponding output signal is shown in (e). (d) Spectrum of a square wave of frequency $w_o RC = 0.1$; the output signal is (f).

11.5.2 Differentiator Transfer Function

Following the same procedure, the mathematical or electronic analog result of differentiating a sinusoidal input signal is

$$V_o = -RC\frac{dV_i(t)}{dt} = -j\omega RC\,V_i \tag{11.12}$$

The output amplitude scales linearly with frequency, which is the characteristic signature of differentiator response to a sinusoidal input signal [Fig. 11.9(b)]. Identifying the impedances of input and feedback components $Z_f = R$ and $Z_i = 1/j\omega C$, Eq. (11.9) can be written in precisely the same *form* as Eq. (11.8) for the integrator. In both examples, the result looks like an ordinary negative feedback amplifier [Fig. 10.16 and Eq. (10.28)], except that resistors R_i and R_f are replaced by more general impedances. We postulate that when the input and feedback elements are any combination of linear passive feedback elements with two terminals (Fig. 11.8), the response of the circuit is

Golden Rule

$$V_o/V_i = -Z_f/Z_i = \text{transfer function} \tag{11.13}$$

This result is the "Golden Rule" for analysis of op amp circuits. Although shown here only for sinusoidal inputs, it can be shown to be more generally true, although the proof requires more advanced mathematics.

11.6 APPLICATIONS OF THE TRANSFER FUNCTION CONCEPT

Exercise

Compare the response of passive and active differentiators and integrators for a square wave input signal of period $T = RC$. Repeat for period $T = 10\ RC$.

Solution The transfer function for the passive filters shown in Fig. 11.9 is obtained by considering the components as a voltage divider. For the low-pass filter,

$$V_o/V_i = Z_c/(R + Z_c) = 1/(1 + j\omega RC) \tag{11.14}$$

and for the high-pass filter,

$$V_o/V_i = R/(R + Z_c) = j\omega RC/(1 + j\omega RC) \tag{11.15}$$

The low-pass filter acts approximately as an integrator for $\omega RC >> 1$, since in that regime $V_o/V_i \simeq 1/j\omega RC$. Similarly, the high-pass filter acts approximately as a differentiator for $\omega RC << 1$, since in that region, $V_o/V_i \simeq j\omega RC$. The passive filter will work adequately as long as the signal is a sine wave with frequency in the region where these approximations are valid. For signals more complex than a sine wave, significant waveshape distortion will occur if the attenuation of harmonics differs significantly from the ideal differentiator or integrator. Placing the same R and C into *active* op amp circuits [Fig. 11.9(a)] results in transfer functions without these limitations. The waveform examples shown in Fig. 11.9(c) to (f) illustrate the difference. For an input square wave with $\omega_o RC = 1$ (where $\omega_o = 2\pi/T$), both the passive and active integrator produce essentially indistinguishable outputs, the expected triangular wave. But for lower signal frequencies (e.g., $\omega_o RC = 0.1$), the passive integrator misses the lower harmonics, producing a rounded waveform. When the same square wave is applied to a differentiator, the passive version works poorly for $\omega_o \mathrm{RC} = 1$ because the high frequency components are attenuated. Even for $\omega_o RC = 0.1$, the passive differentiator works rather poorly, though the first few harmonics fall within the ideal differentiator regime. Reproducing a pulse-like shape requires many higher harmonics, which are attenuated by the passive differentiator transfer function.

Exercise

Analyze and plot the transfer function of the active differentiator (Fig. 11.10), which has been modified to limit its response to noise at frequencies above the signal range.

Solution This is a straightforward application of the transfer function concept. The output/input ratio is the ratio of feedback to input impedances [Eq. 11.11]. In this case,

$$Z_i = 1/(j\omega C_1)$$

$$Z_f = RZ_c/(R + Z_c) = R/(1 + j\omega RC_2) \tag{11.16}$$

and the resulting transfer function is

$$V_o/V_i = j\omega RC_1/(1 + j\omega RC_2) \tag{11.17}$$

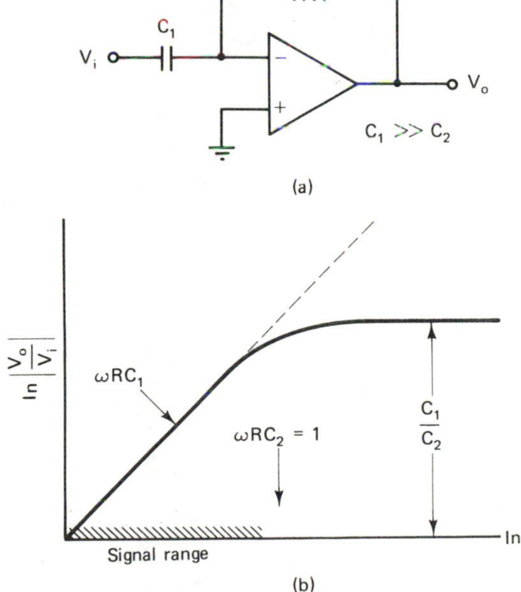

Figure 11.10 (a) Nonideal differentiator circuit. The extra capacitor C_2 limits the gain at high frequencies, as shown by (b) the transfer function.

The transfer function [Fig. 11.10(b)] has an initial region of differentiator action, rolling off to a constant at frequencies $\omega RC_2 >> 1$. Capacitor C_2 is selected to make the cutoff occur above the signal range of interest. This transfer function is similar to that of a passive differentiator (see Fig. 11.9) but has the advantage that the slope and the cutoff may be adjusted independently, and, unlike the passive version, gain is possible ($C_1/C_2 >> 1$, usually).

11.7 APPLICATIONS OF THE NONINVERTING CONNECTION

11.7.1 Source Loading and the Voltage Follower

Although the standard op amp connection (Fig. 10.16) is the most common, it suffers from the disadvantage of low input impedance, $Z_i = R_i$. Source loading effects will set a lower limit to R_i, yet limitations in real op amps (Chapter 12) prevent R_i from being larger than 1 MΩ or so. A second class of op amp circuits brings the input directly to the noninverting terminal. This is very useful in isolating or buffering a high-impedance source. These circuits present to the source the full input impedance (10 MΩ or greater) of the op amp. The feedback connection is returned to the inverting terminal, so feedback is still negative. The two connections are compared in Fig. 11.11. Rather than adjusting the feedback current so the inverting terminal (summing point) voltage is kept zero [Fig 11.11(a)], the feedback voltage is kept equal to the input voltage (Fig 11.11(b)]. In (simplified) analysis of such circuits, current conservation still is useful, but the virtual ground concept is replaced by the faith that negative feedback will find a way to keep $V_+ = V_-$.

The simplest noninverting circuit is the voltage follower (Fig. 11.12). If the above rule is true, any change in V_i must be accompanied by a change in $V_f = V_o$. The circuit therefore has a gain of +1. Note the lack of a minus sign, common in all circuits where the input goes to V_+. The circuit properties are *not* the

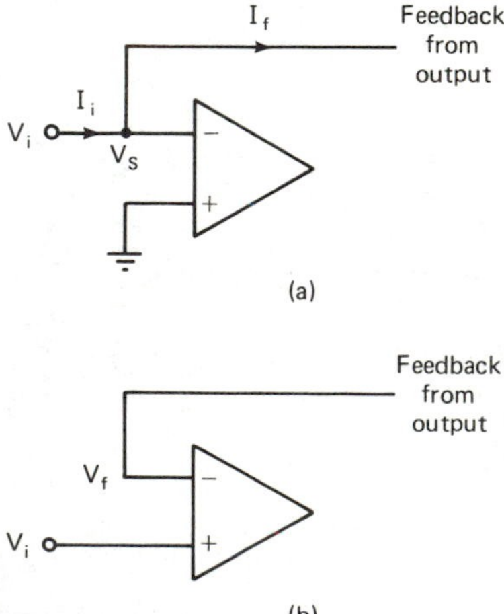

Figure 11.11 Comparison of input and feedback in (a) inverting connection and (b) noninverting connection.

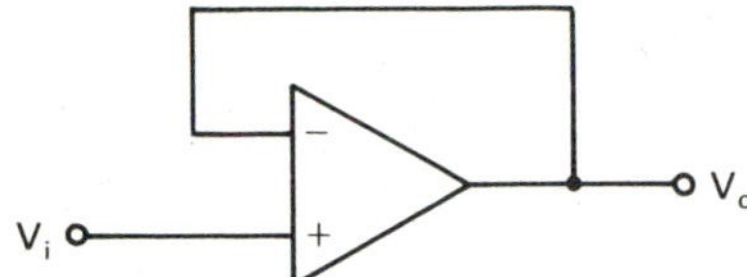

Figure 11.12 Unity gain amplifier or voltage follower.

same as a wire connected from V_i to V_o (with no op amp at all), since the source impedance presented to the next stages is the low source impedance of the op amp itself ($<100\ \Omega$ or even lower), thus effectively isolating the original source from loading.

In spite of obvious advantages in source isolation, noninverting circuits are seen relatively less often than their inverting counterpart. This is largely a result of history. Noninverting circuits have a potential error resulting from imperfect cancellation within the op amp of common mode input signals, as measured by the common mode rejection ratio (CMRR). Modern op amps have a high enough CMRR so that this limitation, discussed in Section 11.6, is usually not serious once understood.

11.7.2 Follower with Gain

This circuit (Fig. 11.13) not only isolates the source but provides gain. It will be analyzed here assuming finite op amp gain, to facilitate explicit comparison with the original idea of voltage feedback (Section 10.8). Analysis proceeds by the usual two-step process. The output is determined by the voltage difference across the two inputs

$$V_o = a(V_+ - V_-) = a(V_i - V_-) \tag{11.18}$$

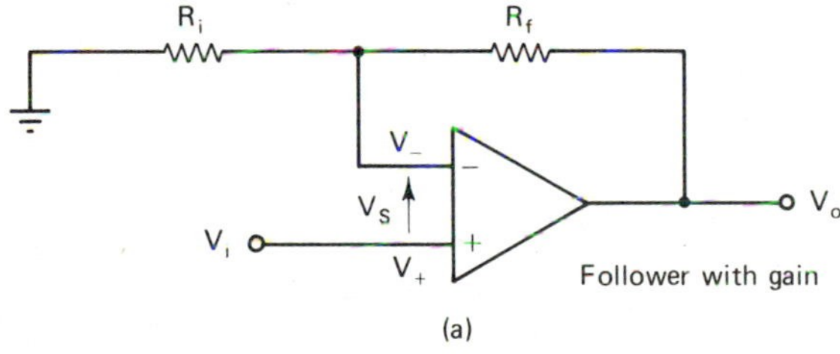

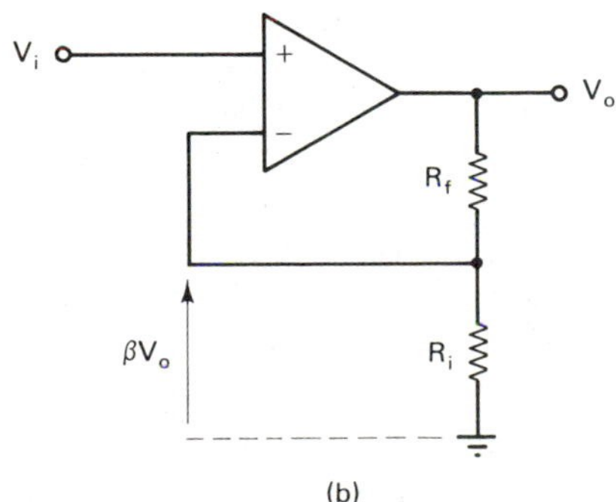

Figure 11.13 Noninverting amplifier or follower with gain. (a) Basic circuit. (b) Circuit redrawn to show the voltage feedback.

Assuming no current flow into the op amp,

$$(0 - V_-)/R_i = (V_- - V_o)/R_f$$

$$V_o/R_f = V_-(1/R_i + 1/R_f) = V_-(R_i + R_f)/R_iR_f \tag{11.19}$$

Combining these two with a little algebra to eliminate V_- gives

$$V_o = [V_i(R_i + R_f)/R_i][1 + (R_i + R_f)/(aR_i)]^{-1}$$

$$\simeq V_i[(R_i + R_f)/R_i] \qquad \text{for } 1/a << 1 \tag{11.20}$$

As long as $a >> 1$, the gain is approximately $(R_i + R_f)/R_i$. It is useful to compare this with the inverting amplifier.

COMPARISON OF TWO OP AMP CONNECTIONS

	Inverting	Noninverting
Gain	$-R_f/R_i$	$+(R_i + R_f)/R_i$
Gain range	(0 to ∞)	(1 to ∞)
Input impedance	R_i	R(op amp) $>> R_i$

In the limit of $R_f = 0$, the noninverting amplifier is equivalent to the voltage follower.

The connection between the follower with gain and the original idea of voltage feedback (Section 10.8) may be made by redrawing Fig. 11.13(a) as shown in Fig. 11.13(b). The feedback fraction β is generated by the voltage divider formed by R_i and R_f.

$$\beta = R_i/(R_i + R_f) \tag{11.21}$$

Rewriting Eq. (11.20) in this form,

$$V_o = \frac{V_i}{\beta}\frac{1}{1 + (\beta a)^{-1}} = \frac{aV_i}{1 + \beta a} \tag{11.22}$$

The form is identical with Eq. (10.14), except for the sign.

Once these comparisons with earlier circuits are clear, it is usually sufficient to analyze noninverting circuits by the simpler procedure of assuming infinite open-loop gain, as shown in the following example.

Exercise

Analyze the noninverting op amp amplifier (Fig. 11.13). Assume an ideal op amp of infinite gain, and calculate the transfer function.

Assumption 1. Negative feedback tries to keep the voltage difference $V_+ - V_- = 0$. In noninverting circuits, neither of these voltages will be zero, so the virtual ground concept is inappropriate.

Assumption 2. Whatever V_o is, voltage V_- will be the fraction of V_o set by the voltage divider R_i and R_f. The op amp input terminals are passive onlookers.

Combining these two assumptions

$$V_- = V_+ = V_i$$

$$V_- = V_o R_i/(R_i + R_f)$$

with the result

$$V_o = (R_i + R_f)/R_i \tag{11.23}$$

This result could also have been obtained using current conservation instead of the voltage divider idea. The essential point in analyzing any noninverting circuit is to imagine that the op amp with negative feedback will do whatever it needs to do to make the voltage at V_- equal the input voltage applied to V_+.

PROBLEMS

11.1. The statement is made following Eq. 11.2 that in a summing circuit (Fig. 11.1), the inputs are decoupled from one another due to the virtual ground at s. How accurate is this statement?

(a) Calculate the voltage appearing at the summing point s due to a 1-V input signal at terminal 1. Assume open-loop gain $|a| = 10^4$, $R_1 = R_2 = 1\ \text{K}\Omega$, $R_f = 1000\ \text{K}\Omega$. Assume that V_2 has been set to zero when doing the numerical calculation.

(b) How much current would flow back into terminal 2 (grounded) as a result of this summing-point voltage? This is a measure of how much will flow into a signal source at terminal 2.

11.2. A passive summing circuit is shown in Fig. 11.14.

(a) Calculate the output signal V_s resulting from a 1-V input signal at terminal 1. Assume V_2 has been set to zero.

(b) How much current would flow back into terminal 2 (grounded) as a result of input signal 1? Assume $R = 1\ \text{K}\Omega$.

(c) Compare this with the result of Prob. 11.1.

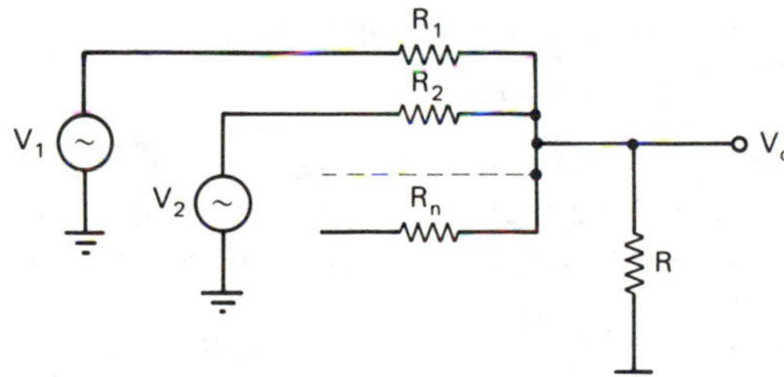

Figure 11.14 Passive summing circuit. V_o is proportional to the sum $V_1 - V_n$, but there is no virtual ground to isolate the sources from one another.

11.3. Suppose the resistors in a subtractor circuit (Fig. 11.2) are not perfectly matched. Let the feedback resistor be $R + q$, where $q/R << 1$, and let the resistor connecting V_+ to ground be R' and r, where $r/R' << 1$. Calculate the output V_o as a function of inputs V_1 and V_2. Assume $a = \infty$, i.e., solve by setting $V_+ = V_-$. How large an error will be made if $q/R = +0.01$ and $r/R' = -0.01$?

11.4. The ground isolation provided by a subtractor circuit (Fig. 11.3) deteriorates when the resistors are not balanced. Calculate how the common mode leakage depends upon the imbalance. Use the notation and results of Prob. 11.3, but set $q = 0$ for

$$V_1 = -V_d/2 + V_{cm}$$

$$V_2 = +V_d/2 + V_{cm}$$

and calculate V_o. What fraction of the common mode signal leaks through if $r/R' = .01$?

11.5. If an integrator is to be used for waveshaping an ac signal, errors due to dc drift (Chapter 12) can be eliminated if the response is cut off below the signal frequency range. Calculate the transfer function for such a modified integrator (Fig. 11.15). Suppose $R_1 = 10^6 \Omega$ and $C = 1\ \mu$F.

(a) What value of R_2 should be selected if the circuit is to accurately integrate signals of frequency 10 Hz or greater?

(b) What will the dc gain of circuit (a) be?

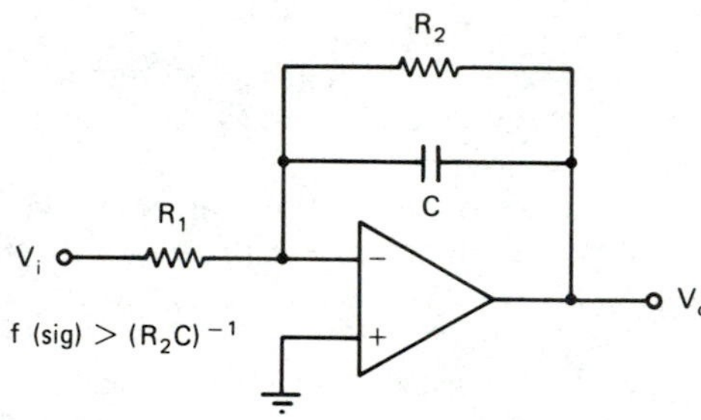

Figure 11.15 Nonideal integrator, modified to cut off the response to low frequency signals.

11.6. Electrometer or current follower.

(a) Show that an inverting amplifier op amp circuit with the input resistor left out [$R_i = 0$; see Fig. 11.16(a)] functions as a current-to-voltage converter, with

$$V_o = -I_{\text{in}} R_f$$

Estimate the lower limit to the currents measurable with this technique. (**Answer**: $< 10^{-15}$ A with general-purpose op amps; see Chapter 12.)

(b) Show that an integrator with the input resistor left out [$R_i = 0$; see Fig. 11.16(b)] functions as a charge-to-voltage converter, with

$$\Delta V_o = -\ Q/C$$

Here, ΔQ is the integral of the current I since the last reset. Estimate the lower limit to the charge measurable with this technique. (**Answer**: 10^{-14} coulombs or 10^5 electrons with general purpose op amps; see Chapter 12).

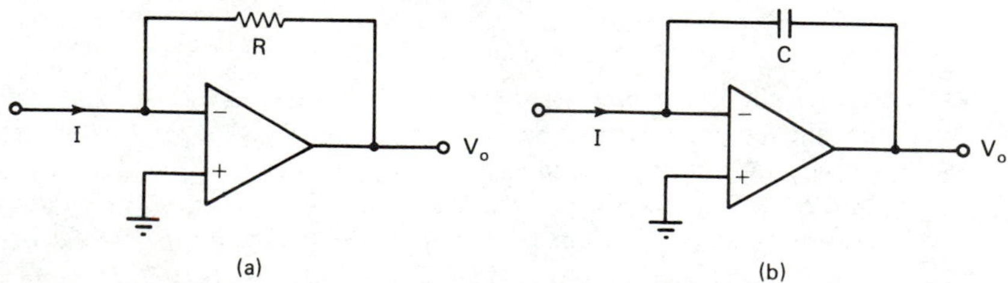

Figure 11.16 (a) Electrometer or current follower. (b) Charge to voltage converter.

11.7. Analyze the differentiator circuit (Fig. 11.6) and derive Eq. (11.8). Assume an ideal op amp ($a = \infty$) and virtual ground.

11.8. Sketch the output signal when each of the two input signals (a) and (b) shown in Fig. 11.7 is fed through a differentiator (ideal). Include an estimate of the enhancement of signal relative to noise or baseline.

(a) Assume $\omega_n \simeq \omega_s/20$ and $V_s = V_n/5$ (input signal-to-noise ratio 1:5). The answer can be quite precise, including output signal-to-noise ratio.

(b) The answer when the "signal" is an edge (or shoulder) or a peak riding on top of a slowly drifting base line will only be an estimate, based on graphically estimating the relative *slopes* of the various portions of the curve.

11.9. If the limiting of high-frequency gain in the nonideal differentiator (Fig. 11.10) is not sufficient, a further modification (Fig. 11.17) is possible. Calculate the transfer function and sketch it, assuming $R_1 = 10^6 \Omega$ and $C_1 = 1\,\mu$F. Select the values of R_2 and C_2 needed to make the differentiator cut off at frequencies above 10^4 Hz.

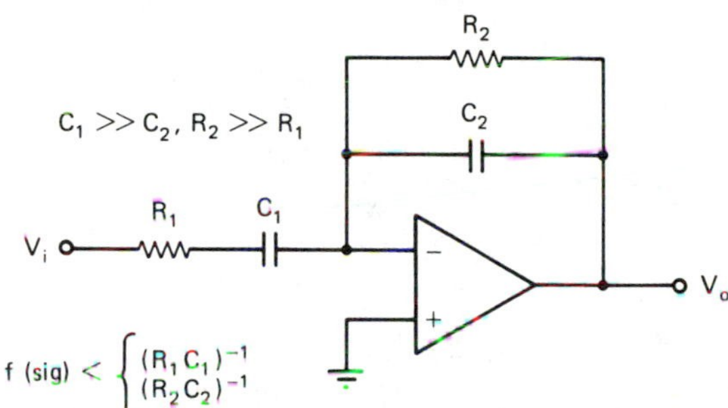

Figure 11.17 Nonideal differentiator with gain which falls off at high frequencies.

11.10. The modified differentiator circuit (Fig. 11.17) is a useful rate-of-change indicator to monitor the charging or discharging of a storage battery. Its output could be used to drive an indicator to warn when the battery voltage starts to fall near the end of life of a discharge cycle. Select component values to achieve the following goals:

(a) Output voltage of at least 1 V for an input rate of 10^{-3} V/s.

(b) About 10^{-3} V of 60-Hz noise at the input should result in less than 0.1 V noise at the output.

11.11. A circuit which is capable of extremely high gain with a single op amp is shown in Fig. 11.18.

(a) Explain its operation and calculate the gain.

(b) Compare with other ways (e.g., inverting or noninverting basic amplifier circuits) of getting the same gain. Consider limits on component ratios. Other limitations involve op amp realities (Chapter 12).

Figure 11.18 Ultra high gain amplifier. The circuit has a gain of 1000 with the component values shown.

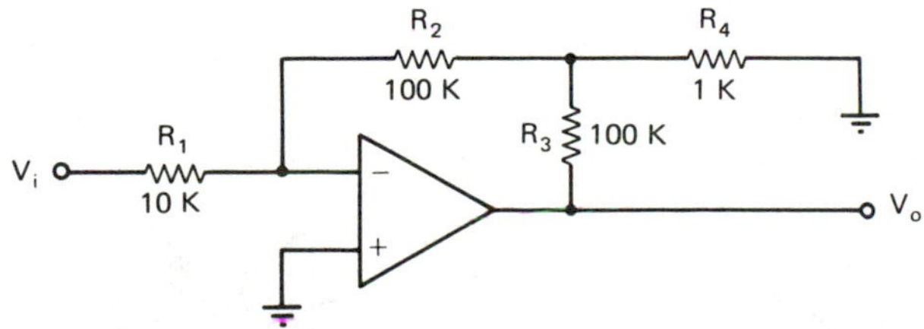

11.12. **(a)** Show that the feedback fraction β for the inverting amplifier (Fig. 10.16) is identical with that of the noninverting amplifier (Fig. 11.13), $\beta = R_i/(R_i + R_f)$. The trick is to disconnect the op amp output R_2 and determine what voltage appears at V_s when a signal of value V_o is applied to the output side of R_2. The input V_i is set to zero here (superposition principle).

(b) Although it is clear that the noninverting amplifier has a block diagram identical with that of Fig. 10.8(c), it is less clear how to draw that block diagram for the inverting amplifier. By opening the feedback loop at the op amp output and grounding terminal V_o (superposition), show that the fraction of the input V_i which gets to the summing point is $R_f/(R_i + R_f) = R_2/(R_1 + R_2)$. Show that this is equivalent to placing an attenuator block to the left of the block Σ in Fig. 10.8(c).

(c) Show that (a) and (b) together result in a block diagram which makes the general feedback equations [Eq. (10.14)] and the specific inverting amplifier equations [Eq. (10.27)] equivalent.

REFERENCES

More complete bibliographic information for the books listed below appears in the annotated bibliography at the end of the book.

GRAEME, *Designing with Operational Amplifiers*

HIGGINS, *Experiments with Integrated Circuits*, Experiment 18

HOENIG and PAYNE, *How to Build and Use Electronic Devices Without Frustration, Pain, etc.*

JUNG, *IC Op Amp Cookbook*

MELEN & GARLAND, *Understanding IC Operational Amplifiers*

National Semiconductor, *Linear Applications Handbook*

PHILBRICK, *Applications Manual*

12

Operational Amplifier Realities: The Nonideal Op Amp

In this chapter, the terms which characterize nonideal op amp behavior will be introduced. A brief look will be taken into several standard op amps to see how they work and how nonideal behavior originates. Since our interest is in op amp applications rather than design, the look inside will be very brief and focused on understanding those characteristics which must be carefully specified or corrected for in order to select and use an op amp wisely in a given application.

12.1 INTERPRETING OP AMP SPECIFICATIONS

A summary of the specifications important in selecting an op amp is given in Table 12.1. An *equivalent circuit* for the nonideal operational amplifier is given in Fig. 12.1. The components and sources shown are not meant to represent what is actually inside, but are the simplest way of adequately modeling real op amp performance. The various parameters will now be discussed briefly.

The first few parameters describe the op amp's ability to supply what negative feedback demands of it. *Open-loop gain a* has already been used extensively and needs no further comment. *Rated output* V_{max}, I_{max} limits the op amp as an ideal source. The maximum output voltage V_{max} is limited by the power supply, and is typically near that value. Maximum output current I_{max} is limited by the output transistors. In early designs, self-destruction was possible due to overloading, but most present designs incorporate a protection circuit which allows even an output short circuit to be tolerated. *Output resistance* R_o is the effective resistance in a Thevenin equivalent; it sets the maximum current which the op amp

TABLE 12.1 SHORT GLOSSARY OF OP AMP SPECIFICATIONS

Symbol	Definition
a	**Open Loop Voltage Gain** The ratio of output signal voltage to differential input signal voltage
V_{max}, I_{max}	**Rated Output** The peak values of output voltage and currrent which can be simultaneously supplied by the amplifier
R_o	**Output Resistance** The effective output source resistance of the amplifier when operated without feedback (open loop)
R_i	**Differential Input Impedance** The effective input impedance measured between the two input terminals when the amplifier is operated open loop
f_u	**Unity-Gain Bandwidth** The frequency at which the small-signal open-loop gain crosses the unity gain level
$(dV_o/dt)_{max}$	**Slew Rate** The maximum rate of change of output voltage for which the amplifier can maintain linear operation
f_p	**Full Power Response** The maximum sine wave frequency at which the amplifier can operate linearly while supplying rated output voltage and current to a resistive load
t_s	**Settling Time** The time interval required, following the application of a step input signal, for the output voltage to rise and settle to within a specified percentage of its full value
t_{ol}	**Overload Recovery Time** The time interval required for the amplifier to return to linear operation from saturation following removal of an overdrive signal
V_{os}	**Input Offset Voltage** The differential dc input voltage required to provide zero output voltage when no other signal or common mode voltage is present and one input is connected to signal common
V_d	**Input Voltage Drift** The average rate of change of input offset voltage with temperature, power supply voltage, or time
I_b	**Input Bias Current** The dc biasing current which must flow into or out of each input of the amplifier when both input signal and input offset voltage are zero
I_d	**Input Bias Current Drift** The average rate of change of input bias current with temperature, power supply voltage, or time
I_{os}	**Input Offset Current** The difference between the two input bias currents
$(I_{os})_d$	**Input Offset Current Drift** The average rate of change of input offset current with temperature, power supply voltage, or time
V_n	**Input Noise Voltage** The equivalent differential input noise voltage which produces the observed output noise when the amplifier inputs are terminated in zero source impedance
I_n	**Input Noise Current** The equivalent input noise current (at either input) which would produce the observed output noise when that input is terminated in a large source impedance
R_{cm}	**Common Mode Impedance** The impedance measured between either input and common
V_{cm}	**Maximum Common Mode Voltage** The peak value of input voltage which may be applied simultaneously to both inputs, while retaining linear operation of the amplifier
CMRR	**Common Mode Rejection Ratio** The ratio of differential voltage gain to the common mode voltage gain

Adapted from the Burr Brown Applications Manual. The symbols correspond to the equivalent circuit of Fig. 12.1.

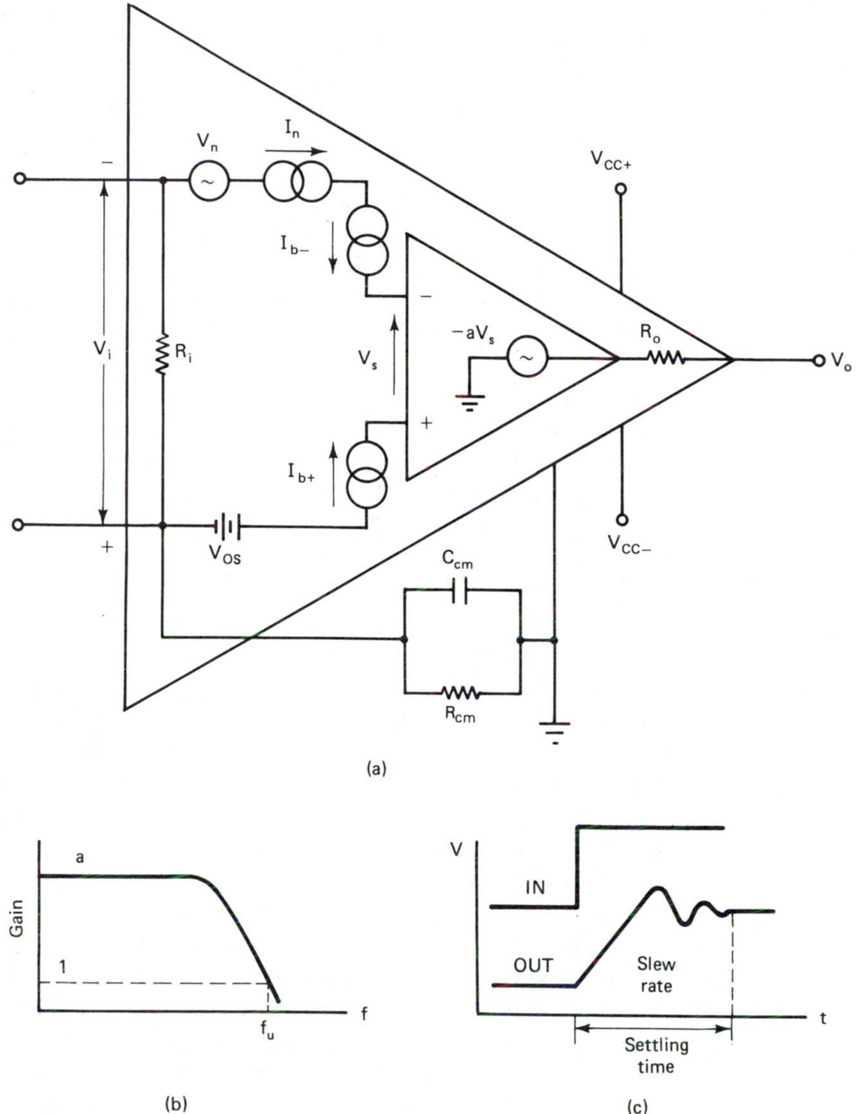

Figure 12.1 (a) Equivalent circuit for a nonideal op amp. Refer to Table 12.1 for definitions. (b) Open loop gain as a function of frequency, showing unity gain crossover. (c) Slew rate $(dV_o/dt)_{\max}$ and settling time t_s in response to a step input.

can deliver, and may be estimated (when not given) as $V_{\max}/I_{\max}$. *Differential input impedance* R_i is usually large enough to be ignored, except when the input signals have a high source impedance, or contain high common mode noise. In high-frequency applications, an additional capacitance C_i is effectively in parallel with R_i.

The next several specifications concern the op amp's frequency response in the frequency domain or the time domain. They are of course related, but used in different applications (sine wave input vs. step input), so it is simpler to define them separately. *Unity gain bandwidth* f_u [Fig. 12.1(b)] specifies the frequency below which the op amp gain is unimportant in the closed loop circuit gain. *Slew rate* $(dV_o/dt)_{\max}$ establishes how fast the op amp can respond to a sudden change of input. Slew rate is important, for example, in comparators and function generators. *Full-power response* f_p is a result of the rate limiting set by the slew rate. This frequency applies to large-signal operation, whereas f_u applies to small-signal operation. The two frequencies may differ by several orders of magnitude (f_u being larger), and two op amps with identical unity gain bandwidth may have very different full-power response frequencies (see for example Table 12.2). *Settling time* t_s, [Figure 12.1(c)] is the time required to settle to within a specified percentage of the final value for a step input. The *recovery time* is much longer since it refers to a situation where the op amp has been driven out of its linear operating region; the various transistors have been either saturated or cut off. To avoid having the op amp go into a nonrecovering (*latched up*) state, it is advisable in switching applications to limit output excursions to less than the supply voltage, avoiding overloads entirely.

The next several specifications concern unwanted signals seen at the output which have nothing to do with the input signal. These specifications are usually "referred to the input," which means that in an equivalent circuit model (Fig. 12.1) they appear on the input side, and in any calculation of *output* errors the parameter value must be multiplied by the *circuit* gain. A comparison with input signals is therefore independent of closed-loop gain. These errors can be discussed one at a time, with the input signal set to zero. The superposition principle guarantees that the result under operating conditions is the sum. *Input offset voltage* V_{os} and *input voltage drift* V_d are output voltage in the absence of an input voltage, resulting from unbalanced components in the first differential amplifier. The input offset voltage is the less serious of the two, since it can be nulled to an arbitrarily small value using an external trim potentiometer. What matters more is how stable the null is as time goes on. The dominant factor is the temperature of the op amp, so the most relevant specification is drift V_d with respect to temperature. In routine applications, the total drift is estimated by assuming reasonable fluctuations in room temperature ($\pm$ 10°C maximum). In more demanding applications, one may need to take special precautions to thermally stabilize the input stages of a low-level circuit. *Input bias current* I_b is the current needed to bias the bases of a bipolar transistor input stage into a linear operating point, and is perhaps the most dominant nonideal parameter for a bipolar op amp. On the other hand, with the advent of inexpensive FET input op amps with negligible bias current (leakage current in the picoamp range), this parameter is becoming less of a problem. With bias current either balanced out (bipolar; see Section 12.4) or small (FET), what becomes important is the *difference* in bias currents between the two inputs, called the *offset current* I_{os}. Here too, the relevant question for stability is the drift in these quantities with respect to time or temperature.

Input noise voltage V_n and *input noise current* I_n are relevant only in low-level (microvolt or nanoamp) applications. Noise sources are inherent in semiconductor device operation, originating in statistical fluctuations of current flow within semiconducting materials (Chapter 18). The *output* noise voltage contains two terms: one which is independent of input source resistance, and a second which scales linearly with input source resistance. For this reason, it is necessary in Fig. 12.1 to assume two independent noise sources, a voltage source V_n and a current source I_n. Only the noise voltage source is important as long as the signal's source resistance is less than about 10^7 Ω. Noise level is specified in $\text{nV}/(\text{Hz})^{1/2}$. Since noise occurs at all frequencies, the total root-mean-square (rms) value depends upon the bandwidth Δf of the circuit. The laws of random variables give the square root dependence (see Chapter 18). To estimate the total rms noise, multiply the noise level specification of the op amp by $(\Delta f)^{1/2}$.

The final group of specifications arise because the op amp is not ideally differential. Although the input stage is ideally floating, in fact it is connected to the power supply common through a very large resistance, the *common mode impedance* R_{cm}, which in high-frequency applications includes an additional parallel capacitance C_{cm}. R_{cm} sets the upper limit to the input impedance of the noninverting or voltage follower connection but is unimportant in noninverting connections, since other resistances are smaller. It also becomes important in low-level signal applications from a high-impedance source (see Chapter 18). The *common mode rejection ratio* (CMRR) originates in slightly different gains of the input transistors. CMRR values of 80 dB (factor of 10^4) are typical of general-purpose commercial op amps, and 120 dB (factor of 10^6) or larger is possible in premium designs.

12.2 INTERNAL CIRCUITRY OF GENERAL-PURPOSE IC OP AMPS

To see where nonideal properties come from, it is useful to look inside several industry standard devices. The only type to be discussed is the monolithic IC (total integrated circuit on a silicon chip), which now dominates over the older hybrid type (integrated circuit transistors plus thick-film resistors and capacitors sharing a ceramic substrate) and discrete type (discrete transistors, capacitors, resistors, etc., potted within an epoxy block).

In the 1970's, monolithic IC op amps were used in high volume commercial applications where the low cost offset the poor performance (high bias currents, low frequency response) and justified designing around some self-destruct problems. Discrete or hybrid devices were standard for industrial, scientific, and military applications, with a typical cost per op amp of \$10 to \$100. The situation shifted spectacularly with the introduction of cheap and reliable monolithic devices. The 741 and the 301, for example, are prototypes of general-purpose device families, and cost now only about \$0.25 apiece (a factor of 100 decrease!). One key distinction is whether a device is (internally) *compensated* (741) or *uncompensated* (301). This refers to the presence or absence of an internal capacitor to make the device stable against oscillations when feedback is applied. The large-area capacitor "plate" is clearly visible in a photomicrograph [Fig 12.2(a)] of the 741.

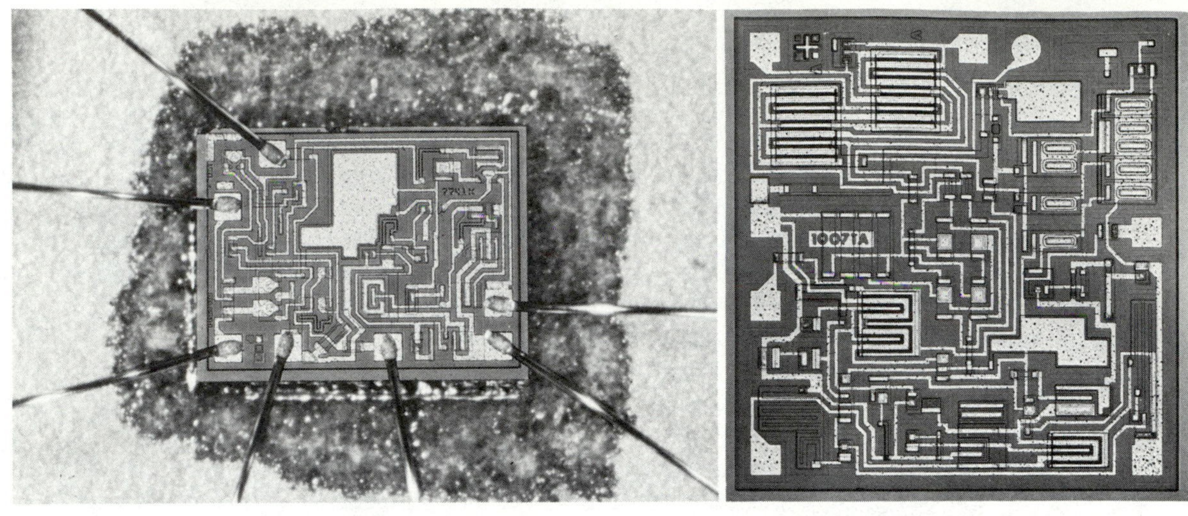

(a) (b)

Figure 12.2 (a) Photograph of the 741 op amp IC chip. Note the on-chip capacitor. (*Courtesy* Fairchild Camera and Instrument Corporation) (b) Photograph of 3140 op amp IC chip. (*Courtesy* RCA Solid State Division)

The distinction is made between high volume manufacturers of general-purpose "industry standard" devices (National, RCA, Motorola, Signetics...) and manufacturers of lower volume "premium" devices (Analog Devices, Philbrick, Burr Brown, Precision Monolithics...). The distinction is becoming less firm as premium manufacturers have entered into large volume monolithic production, and general-purpose manufacturers have introduced unusually high performance devices. But in demanding applications (microvolts, nanoamps, ultrahigh speed, ultralow noise, wide temperature range), the premium manufacturer offers a more tightly specified product, at a consequently higher cost.

The discussion of internal workings will begin with the 741, the prototype of the compensated op amp. The 741 had a revolutionary impact in replacing discrete component circuits and in making the op amp revolution available to everyone. The 741 is still a widely used general-purpose type, and is adequate for most large-signal audio frequency circuits. Next we discuss *uncompensated* op amps for high-frequency applications, with the 301 as the prototype. Finally, we examine higher performance op amps, with two new families, the BiMOS and the BiFET, plus improved bipolar devices. These are the industry standards of the next decade. BiMOS and BiFET op amps achieve bias currents 1000 times smaller than the 741 using an FET input stage. For example, the MOSFET input of a 3140 is seen as the symmetrical "maze" in the upperleft corner of Fig 12.2(b). Improved bipolar devices have higher bias currents than FET input devices, but achieve much lower offset voltage and noise. Condensed specifications for a selection of industry standard devices are given in Table 12.2.

One of the basic tricks in understanding the function of various transistors inside an IC is to separately identify groups which function as signal amplifiers and groups which function only as bias current sources or level shifters. A transistor makes a better current source than a resistor, and also takes a much smaller area

TABLE 12.2 COMPARATIVE SPECIFICATIONS OF INDUSTRY STANDARD IC OPERATIONAL AMPLIFIERS

	Bipolar			BiMOS		BiFET	
Device	741C	301A	LM-11C	3140	3130	LF351	LFT356
Input offset voltage V_{os}, mV (max)	6	7	0.6	15	15	10	0.5
Input voltage drift $(dV_{os}/dT)_{max}$, μV/°C	15	30	5	6	?	10	3
Input bias current I_b, nA (max)	500	250	0.11	0.05	0.05	0.2	0.05
Input offset current I_{os}, nA (max)	200	50	0.01	0.03	0.03	0.1	0.01
Input noise voltage V_n, nV/(Hz)$^{1/2}$	20–30	20	150	40	~40	16	12
Input impedance R_i, MΩ	1	4	10^5	1.5×10^6	1.5×10^6	10^6	10^6
Open loop voltage gain a, V(out)/mV(min)	50	50	50	50	50	25	50
Common mode rejec. ratio CMRR, dB (typ)	90	90	110	90	90	100	100
Unity gain bandwidth f_u, MHz	1	>4	5	4.5	15	4	4
Slew rate $(dV_o/dt)_{max}$, V/μs	0.5	1.5	0.3	0.9	30	13	12
Output swing using ± 15V $V_{(max)}$, V (typ)	10	12	12	13	13	12	12
Output current $(I_o)_{max}$, mA	25	20	15	20	20	20	25
Power dissipation P_{max}, mW (typ)	500	500	500	600	600	500	570
Compensation	int.	ext.	ext.	int.	ext.	int.	int.
Approximate price, (dollars)	0.25	0.25	6.00	1.25	1.40	1.25	5.00

on an IC. Level-shifting devices shift output voltage levels from stage to stage in a multistage dc amplifier, which has no decoupling capacitors to isolate the dc biases. Transistors whose function is only biasing or level shifting are readily identified, because their base currents are fixed, generated by connection to (fixed) power supply voltages or to their own collector.

Another trick used in integrated circuits makes possible not only biasing without large resistors, but also bias current independent of the transistor current gain. The evolution of this idea is shown in Fig. 12.3. The emitter resistor R_E in the long-tailed pair is replaced by a current source transistor Q [Fig. 12.3(b)], resulting in improved common mode rejection (Section 10.4) and larger voltage gain. But generating the small base current for Q requires very large value of R_B (nanoamps would take megohms). This is avoided by the *current mirror* configuration [Fig. 12.3(c)]. Two transistors are connected base to base. If these are matched, a given base-emitter voltage will result in identical emitter currents in both transistors. Setting the level of current in Q_2 forces the same level of current

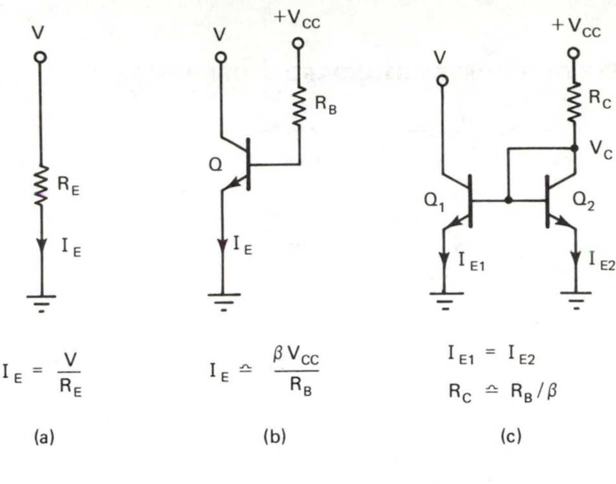

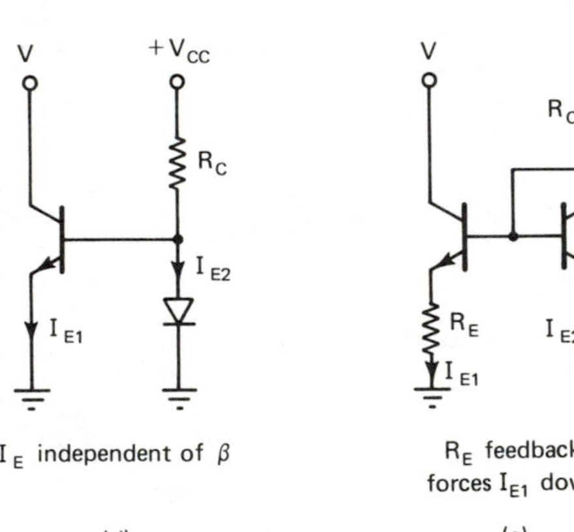

Figure 12.3 Evolution of IC bias circuit sources. (a) Simple emitter resistor, replaced by (b) transistor current source. (c) Mirror pair lowers the value of resistance needed, using (d) a diode-connected transistor. Further reductions in resistance values result from (e) emitter feedback.

to flow in Q_1. Since R_C is in a collector lead rather than a base lead, its value is two orders of magnitude smaller than R_B, for $\beta \simeq 10^2$. Transistor Q_2, with its collector shorted to the base, is called a *diode-connected* transistor, equivalent to Fig. 12.3(d). The current supplied is

$$I_{E2} = (V_{CC} - V_{BE2})/R_C \simeq V_{CC}/R_C \tag{12.1}$$

The second step follows as long as the power supply voltage is much larger than the diode forward drop. The current is independent of β, and current adjustment depends only on the value of a passive component. A further modification allows adjustment of bias current over a very wide range, using still smaller resistor values. The trick combines negative feedback of an emitter resistor [Fig. 12.3(e)] with the fundamental relationship that the voltage across a base-emitter junction varies as the logarithm of the collector current. For Q_2, this is set simply by the bias resistor R_C

$$\begin{aligned} V_{BE2} &= (kT/e)\ \ln(I_{C2}/I_{C0}) \\ &= (kT/e)\ \ln(V_{CC}/R_C I_{C0}) \end{aligned} \tag{12.2}$$

where I_{C0} is the reverse bias leakage current ($\simeq 10^{-7}$ A) of the basic diode I-V relationship, $I_C \simeq I_{C0} \exp(+eV/kT)$. The base bias of Q_1 is lowered by an additional feedback voltage V_{FB}:

$$V_{BE1} = (kT/e)\ \ln(I_{C1}/I_{C0}) - V_{FB} \tag{12.3}$$

which depends upon that transistor's collector current

$$V_{FB} = I_{C1}\ R_E \tag{12.4}$$

Since the base-emitter voltages are equal, Eqs. (12.2) and (12.3) can be equated, giving

$$\ln(I_{C1}/I_{C2}) = (e/kT)\ V_{FB} = V_{FB}(mV)/(26\ mV) \tag{12.5}$$

where $kT/e = 26$ mV at room temperature. For every 26 mV appearing across the feedback resistor R_E, the ratio I_{C1}/I_{C2} is increased by a *factor* of $\ln(2) = 2.7$. (Test your understanding with Prob. 12.1.) As a result, desirably low bias current (submicroamps) in Q_1 is supplied by a current in Q_2 of nearly 1 mA, resulting in comfortably low resistor values ($R \simeq 20$ K Ω).

12.2.1 Compensated Op Amps: The 741 Family Tree

The 741 Op Amp (Table 12.2) has adequately high open-loop gain, low offset current, and high input resistance for many applications. The 741 is internally compensated; no external capacitance is needed for stability under any conditions of resistive feedback. The tradeoff is limited bandwidth (Fig. 12.4). The open-loop gain of the 741 starts rolling off at about 10 Hz. This is because one cannot sharply cut off the frequency response and still maintain stability. Unconditionally stable behavior requires the roll-off shown of 20 dB/decade, or a factor of 10 in gain per factor of 10 in frequency. As a consequence, the 741 and most other compensated op amps are limited to operation in the audio range (see Prob. 12.2). Since the precise value of the gain does not matter as long as it is large (10^2 or

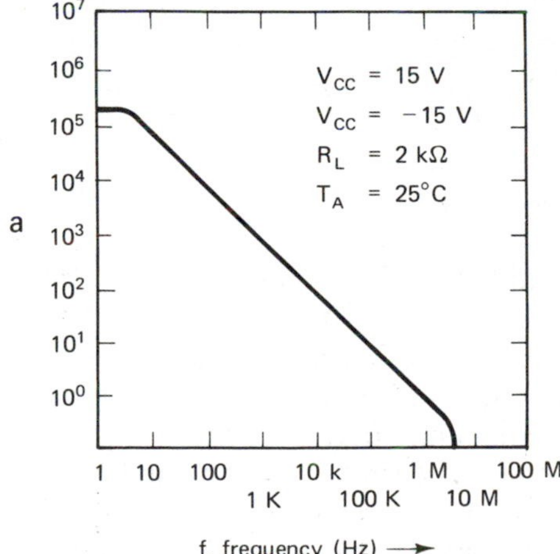

Figure 12.4 Open-loop gain as a function of frequency for the 741 op amp.

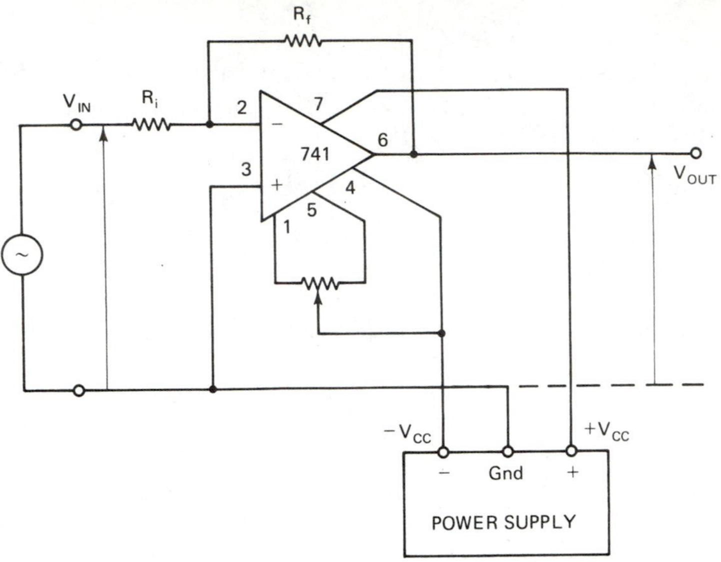

Figure 12.5 Offset nulling circuit for the 741 op amp.

greater), the 741 is used in very many applications where high frequency response is not needed and internal frequency compensation is convenient.

The 741 provides a simple means of offset nulling, as shown in Fig. 12.5. Internally generated offsets due to mismatched components are readily corrected by a single variable resistor connected to the minus power supply. In this and most other general-purpose IC op amps, there is no explicit ground connection, even though the output is single-ended and therefore implicitly referenced to the power supply. The process of voltage offset balancing brings the internal common or "local" ground into adjustment with a point halfway between positive and negative power supply voltages. This arrangement allows optional operation from a single power supply, with the common floating at half the power supply voltage.

The 741 is virtually indestructible, with internal protection against short-circuiting the output. Finally, the 741 will deliver a large output current (25 mA). Because of its reliability, relatively good performance, and extremely low cost ($0.25), the 741 has long been an industry standard for most nondemanding applications.

12.2.2 741 Circuit Explanation

The schematic circuit diagram for the 741 is shown in Fig. 12.6. In understanding an op amp circuit as complex as this (20 transistors), first separately identify the stages which provide signal gain. The 741 is in effect a two-stage amplifier, which reduces the complexity of frequency compensation. Transistors Q_1 through Q_4 form the input-stage differential amplifier. Q_{16} and Q_{17} provide the second-stage voltage gain, with additional output current gain provided by emitter followers Q_{14} and Q_{20}. Other transistors accomplish biasing or level shifting. The 741 introduced new transistor functions: the active load and protection switches to prevent self-

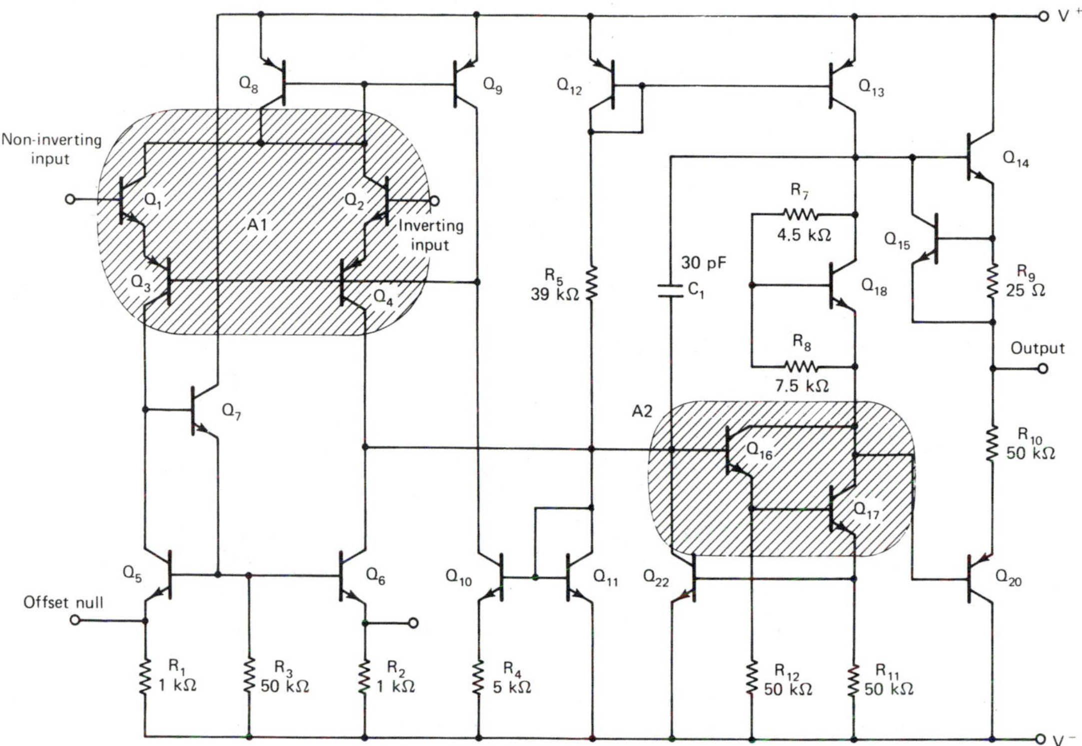

Figure 12.6 Schematic circuit diagram of the 741 op amp.

destruction. A simplified block diagram (Fig. 12.7) shows the function of all transistors.

Input stage. Transistors $Q_1 + Q_3$ and $Q_2 + Q_4$ form the differential input stage. Two transistors in series (Q_1 feeds Q_3, etc.) provide the high gain per stage needed to achieve adequate open-loop gain in a two-stage amplifier. Q_1 and Q_2 are emitter followers (output taken at the emitter), and Q_3 and Q_4 are common base amplifiers (input at the emitter). Q_5 and Q_6 have a dual function: they form an equivalent high resistance or *active load* for Q_3 and Q_4, resulting in increased voltage gain for the common base amplifier, and they function as a differential amplifier for the external offset nulling signal. Q_7 provides the bias to turn on Q_3 and Q_4. Bias currents for the input stage are provided by a complicated arrangement of mirror pairs. Q_{12} generates a current in Q_{11}, which is reflected over to Q_{10} (though reduced because of the emitter feedback due to R_4), which in turn generates a series current in Q_9 which is reflected across another mirror pair to Q_8. The bias current of Q_3 and Q_4 is effectively driven by the mirror pair Q_{10} and Q_{11}.

Question. An emitter current source is usually shared by a differential pair so that one transistor's current decreases as the other transistor's current increases, keeping the total supplied current constant. But here, Q_5 and Q_6 have their bases connected together and are *separately* connected to the differential

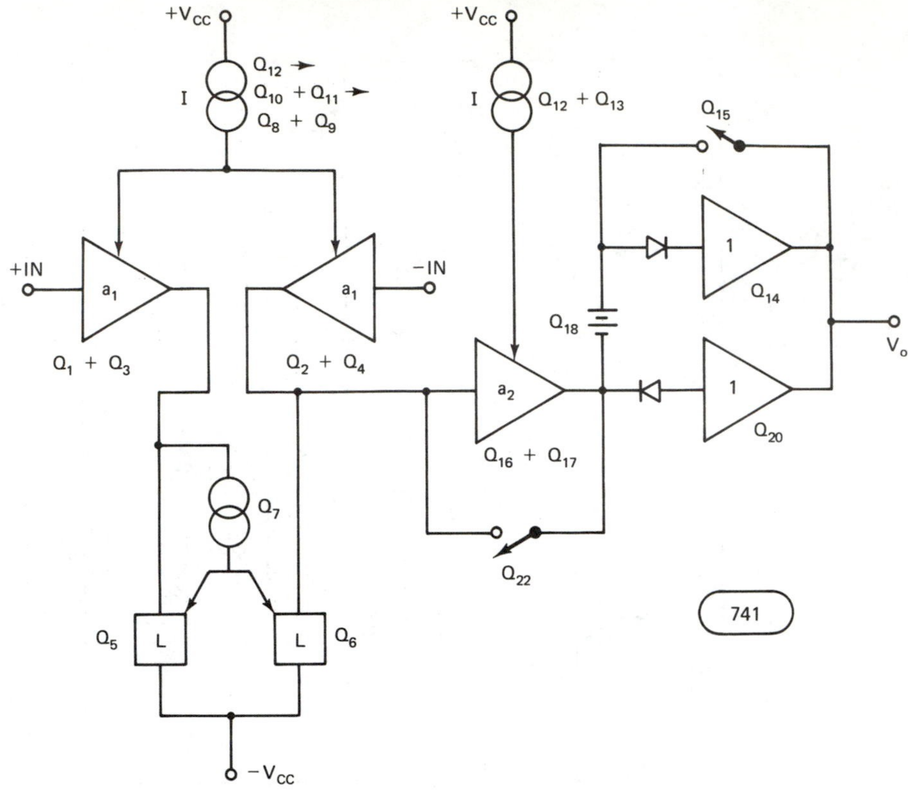

Figure 12.7 Simplified block diagram of the 741 internal operation.

pair. How can the current in either the Q_1–Q_3 or the Q_2–Q_4 leg change separately in response to a differential input signal?

Answer. This clever feature of the 741 is one reason for the very high gain in the two-stage design. Since the bases of Q_5 and Q_6 are driven in common, their currents must indeed stay identical. The output of this first stage, however, is taken as a single-ended connection from the collector of Q_4 to the base of Q_{16}. The identical emitter currents of Q_5 and Q_6 change in response to the bias by Q_7, whose base is driven by a signal connection to Q_3 rather than by a constant-current source. A differential input signal will cause I_{C3} to decrease by ΔI and I_{C4} to increase by ΔI. Current I_{E5} decreases by ΔI, since it is in series with Q_3. The current I_{E6} must also decrease by ΔI, since its base is driven in common with Q_5. But the only way that the current in Q_4 can *decrease* by ΔI while that in Q_6 *increases* by the same amount is if a current $2\Delta I$ flows into the base of the next stage Q_{16}. This efficiently brings the amplified current out of the first stage.

Second stage. Q_{16} and Q_{17} form a high-gain voltage amplifier, because of the Darlington connection which *multiplies* the two gains. The current output of this stage is further amplified by Q_{14} and Q_{20}, a pnp–npn complementary symmetry pair functioning as emitter followers. Q_{12} and Q_{13} form a current mirror bias source for the Q_{16}–Q_{17} amplifier. Q_{18} is a fixed voltage level shifter, lifting the voltage output of Q_{17} by a fixed amount on its way to the input of Q_{14}. The output pair operates so that, depending upon the sign of the output, only one of Q_{14} or Q_{20} is conducting at any time. With no input signal, both devices are turned off, resulting in low quiescent current drain in the output stage. Protection against self-destruction is provided by two transistors. Q_{15} acts as a switch (Fig. 12.7) which turns *on* and shorts out Q_{14} whenever Q_{14} begins to *source* an excessive amount of current. Q_{20} is protected against excessive current *sinking* by Q_{22}, which turns on and shorts out amplifier Q_{16}–Q_{17} whenever excessive current flows in R_{11}. Finally, the internal 30-pF capacitor in the output stage provides the high frequency roll-off (Fig. 12.4) which stabilizes the device. A photograph of a 741 IC chip [Fig. 12.2(a)] shows this on-chip capacitor.

Since the 741 was introduced in 1968, it has sold more units than the total of all other types combined since the op amp was invented. Allthough made technically obsolete by improved devices, the 741 is still popular.

12.2.3 Uncompensated Op Amps: The 301 Family Tree

Referring to Table 12.2, the 301 op amp has the low cost of the 741, but with a frequency response extending to more than 1 MHz. An external capacitor stabilizes operation under the specific closed-loop gain of a particular application. The circuit for the 301 (Fig. 12.8) resembles the 741 but with some more complicated

Figure 12.8 Schematic circuit diagram of the 301A op amp, a standard general purpose uncompensated type.

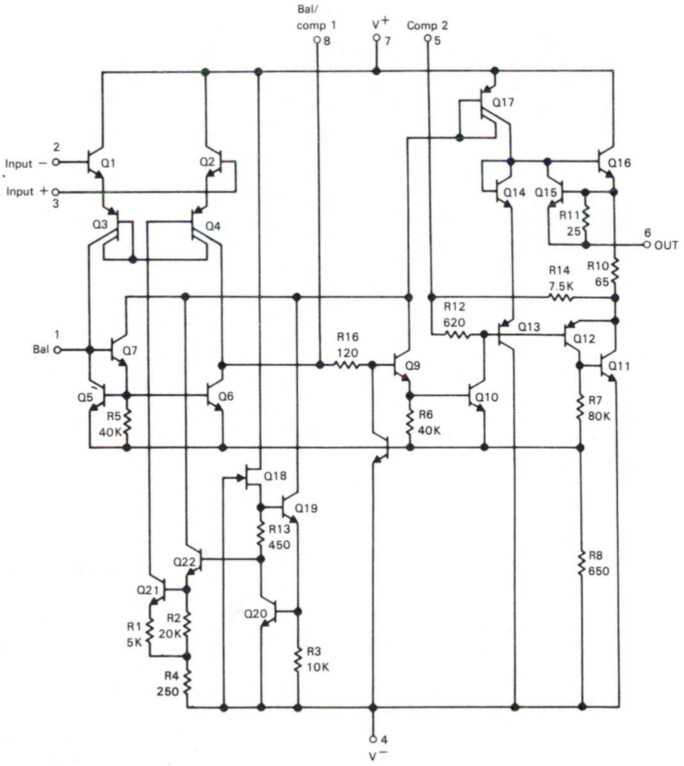

design details (note the multiple collector transistors) beyond the scope of this book. It uses two stages of gain, requiring only one external capacitor for compensation (stabilization). To achieve high gain per stage, each amplifier is rather complicated. Improved biasing circuits reduce bias and offset currents significantly below the 741. The connection point for external compensation is shared with one of the offset voltage terminals. Trimming offset voltage of the 301 is therefore awkward; this op amp is ordinarily used only for ac applications where the dc offset can be ignored.

Frequency compensation. The connections for frequency compensation of the 301 are shown in Fig. 12.9. Examples of gain as a function of frequency in compensated operation are shown in Fig. 12.10. The capacitance values have been selected to keep the *closed-loop* voltage gain flat to above 100 KHz. This requires larger capacitance for smaller closed-loop gain, decreasing the cutoff fre-

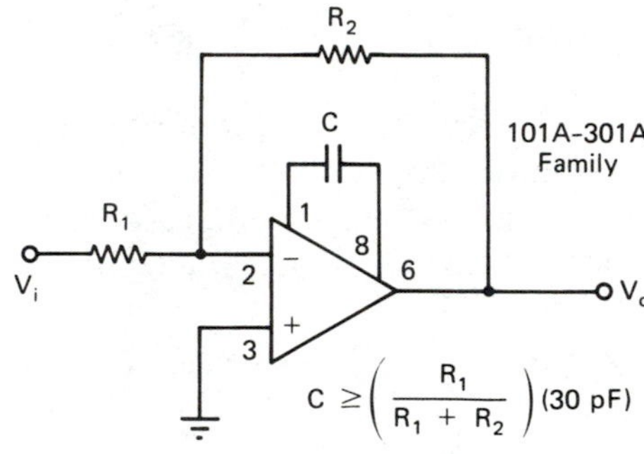

Figure 12.9 Frequency compensation circuit of the 301 op amp.

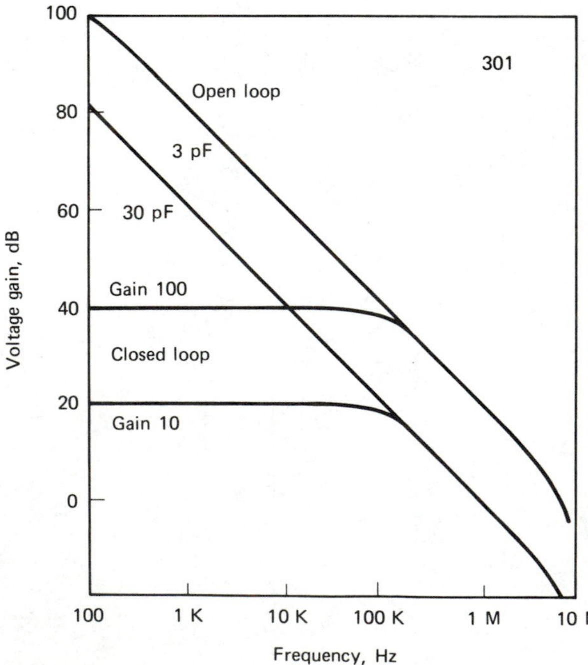

Figure 12.10 Gain as a function of frequency for the 301. (a) Compensation capacitor values which shift the open loop cutoff frequency downwards have been selected to keep the closed loop gain cutoff frequency at 100 KHz.

quency of the *open-loop* gain as the closed-loop gain decreases. This strategy avoids βa ever becoming equal to $+1$ (Section 10.9). Since smaller closed-loop gains correspond to larger values of β, extra capacitance is needed so the vector βa does not encircle the point $+1$ at high frequencies (see Sec. 10.9).

Although technically superseded by improved uncompensated op amps, the 301 series remains a popular choice in economy applications requiring higher frequency response than the 741.

12.3 HIGH PERFORMANCE IC OP AMPS

12.3.1 Bi-MOS Devices: The 3140 Family

A spectacular improvement in general-purpose op amp performance came with RCA's introduction of the CA3140 op amp, which is pin-compatible with the 741, but has bias and offset currents more than 1000 times smaller and input resistances about 10^6 times higher (Table 12.2). This is due to a metal-oxide-semiconductor (MOS) FET input stage [Fig. 12.2(b)]. The layer of oxide isolating the inputs draws only negligible leakage current from the input source, compared to a bipolar base junction with its bias current. In applications where these improved specifications are crucial, the BiMOS family, combining bipolar and MOS technology, has become an industry standard, replacing not only the 741 series, but also many much more expensive premium op amps. In addition to the compensated 3140, Table 12.2 includes the uncompensated 3130, a 301 family replacement with frequency response extending above 15 MHz.

A BiMOS device, however, is not nearly as rugged as junction transistor (bipolar) or junction FET (JFET) devices. Although the MOSFET input stage is diode-protected against various dangers, it can still be destroyed fairly easily, by oxide breakdown due to the buildup of static charges (which cannot leak off through the high resistance input), or from excessive voltages or self-heating. BiMOS devices are best stored in conductive plastic, and require grounding tools during installation. They can also be destroyed by changing connections with the power supply still turned on.

Other weak points of BiMOS devices include:

(1) Offset voltage significantly higher than other general-purpose but bipolar devices (Table 12.2) and much higher than that of premium types.

(2) Significantly higher noise than bipolar types. The 3140 noise level of 40 $\mathrm{nV/(Hz)^{1/2}}$ is larger than that of a 741, and significantly higher than the better bipolar types. Higher noise is typical of FET op amps.

For this reason, a MOSFET input op amp is best reserved for very low current amplifiers, high-impedance signal sources, or applications such as integrators and sample-and-hold amplifiers in which the very high input impedance and very low bias current provide a dominant advantage.

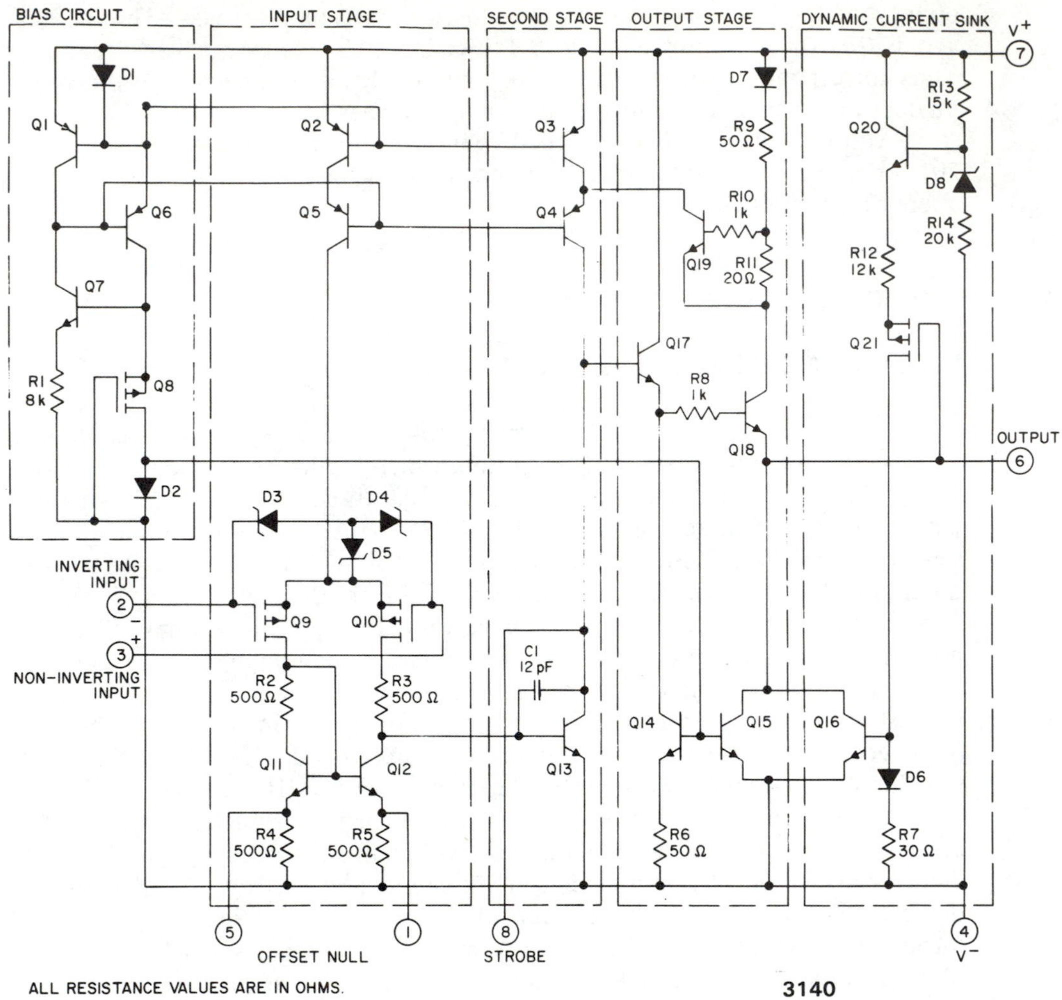

Figure 12.11 Schematic circuit diagram of the CA 3140 op amp. (*Courtesy* RCA Solid State Division)

12.3.2 3140 Circuit Explanation

The schematic circuit diagram of the 3140 (Fig. 12.11) has a simplified block diagram shown in Fig. 12.12.

Input stage. The input stage in some ways resembles the 301 or 741 series, except that the input differential pair Q_9 and Q_{10} are MOSFET's. Transistors Q_2 and Q_5 provide a bias current source, driven from a bias current circuit including transistors Q_1 and Q_6–Q_8. Q_{11} and Q_{12} form active loads for Q_9 and Q_{10}. Since their bases are connected together, their currents must change in common, and *differential* currents created between Q_9 and Q_{10} are forced to the next stage, Q_{13} (the same trick was used for the 741).

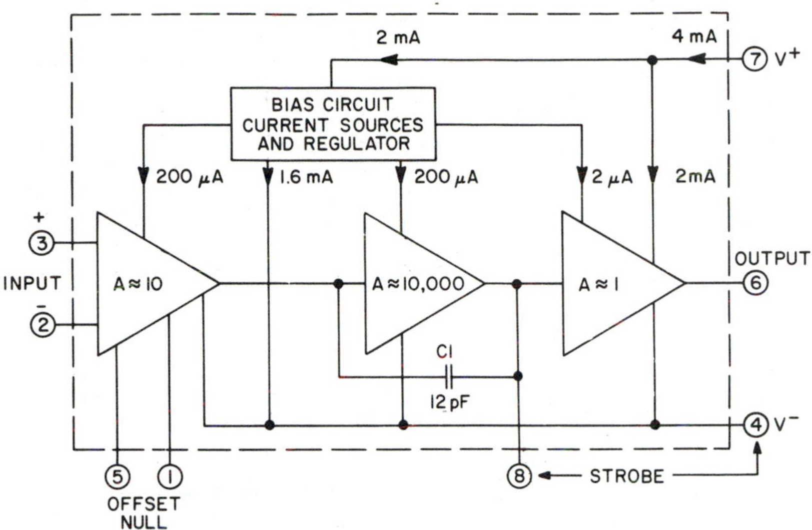

Figure 12.12 Block diagram of the 3140.

Second stage. The second stage is a superhigh-gain ($a = 10^4$) voltage amplifier Q_{13} with bias current provided by Q_3 and Q_4. Note the internal compensating capacitor in this stage.

Output stage. The output stage begins with an emitter follower Q_{17}, which feeds output current amplifier Q_{18}. These are biased by the current mirror pair Q_{14} and Q_{15}. R_6 reduces the bias current of Q_{17} to a lower value than that of Q_{18} [see Eq. (12.5)]. The 3140 includes internal protection against an output overload. When excessive current is drawn from Q_{18}, the voltage drop across R_{11} turns on switch Q_{19}, which shorts out the Q_{17}–Q_{18} amplifier. Operation of the output stage in the opposite polarity is quite different from previous designs. Q_{16} forms the output current *sink*, and is biased into operation by Q_{20} and Q_{21}. When output voltage levels fall to the point where Q_{18} begins to turn off, the same output voltage level applied to the gate of Q_{21} has the effect of turning on Q_{16}, providing the output current sink.

12.3.3 BiFET Op Amps

The BiFET family, which incorporates both FET and bipolar technology, has the best advantages of BiMOS but is considerably more rugged, since junction rather than MOS FET's are used. Both the JFET's and the bipolar transistors are created on the same IC (Fig. 12.13), using ion implantation. BiFET's have about an order of magnitude higher bias current than BiMOS devices (Table 12.2), but are otherwise very similar in their specifications. BiFET's such as the 351, 353, and 357 are pin-compatible replacements for single, dual, and quad versions of the 741, 747, 748 series. The TL080, 070, 060 series (not listed in Table 12.2) from Texas Instruments are a BiFET family with specifications comparable to the 350 series introduced by National Semiconductor. In spite of the stability against oscillations

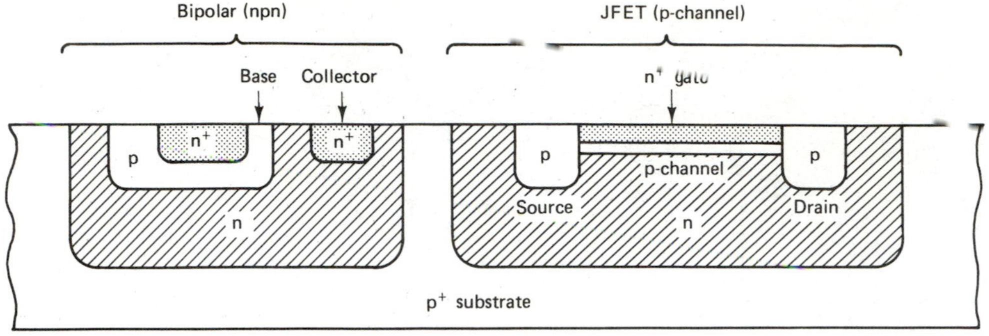

Figure 12.13 Cross-section of BiFET IC fabrication. (Copyright 1980 National Semiconductor Corporation).

provided by their internal compensation, BiFET's as well as BiMOS devices have high frequency response, extending to well beyond 1 MHz.

12.3.4 BiFET vs BiMOS vs Bipolar

Although less rugged than a BiFET, BiMOS devices retain an advantage: the output voltage can swing to within a volt of the power supply "rails." When operated from a single-ended power supply, the BiMOS family interfaces readily between analog circuits and digital logic. The supply voltage can be as low as $\pm$ 2 V, so in single-ended operation the 5 V digital standard is adequate. When larger values are used on the analog side, the *strobe* terminal of the 3140 can be wired to a Zener diode to clamp the output voltage at TTL-compatible levels (Fig. 12.14). The BiMOS amp can tolerate a much larger common mode voltage than a BiFET, which also facilitates analog–digital interfacing. In single power supply operation the op amp sets its internal "ground" at $V_{cc}/2$, so an analog input signal centered at 0 V has an apparent common mode voltage $V_{cc}/2$.

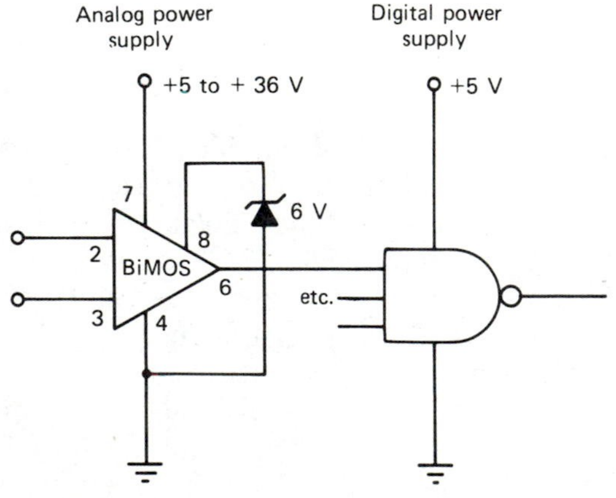

Figure 12.14 Analog to digital signal interface using single end TTL-compatible BiMOS op amp. Pin connections refer to the 3140 (Fig. 17.11).

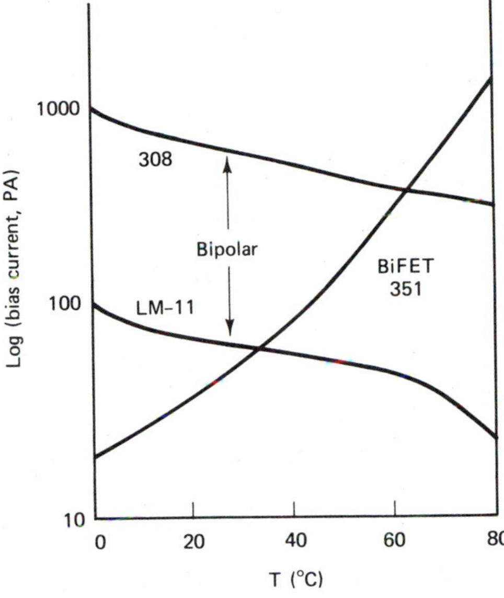

Figure 12.15 Bias current comparison vs. temperature for a BiFET and several bipolar op amps.

Negative aspects of both FET input families include:

(1) Poor offset voltage and drift, comparable to the 741. The few exceptions (e.g., LFT356 listed in Table 12.2) are high-priced premium devices.

(2) The bias current of a BiFET, though low, varies strongly with temperature, increasing by a factor of 2 for every 10°C increase in temperature. If the circuit will experience a wide temperature range, the nominally low bias current of a BiFET is inferior to that of a better bipolar device (Fig. 12.15). By 100°C, the BiFET 351 is even worse than the inexpensive bipolar 308.

(3) Though excellent as low-current amplifiers or as integrators, general-purpose FET amplifiers are not a good solution for low-noise *voltage* amplification, because of offset voltages in millivolts and drifts of nearly 10 μV/°C. One turns then to a premium op amp. Two examples listed in Table 12.2 are the LM-11 (Bipolar) and LFT356 (BiFET). Premium op amps are discussed in Section 12.6 after the consequences of op amp limitations have been made more clear.

12.4 ERRORS DUE TO INPUT VOLTAGE OFFSET AND BIAS CURRENT

12.4.1 Input Voltage Offset Error

V_{os} is an output voltage in the absence of an input, due predominantly to the imbalance of transistors in the first stage. V_{os} is independent of input bias current. Since the circuit is linear, one can treat independent error signals separately, and then sum the results. The equivalent circuit (Fig. 12.1) shows V_{os} as a (hypotheti-

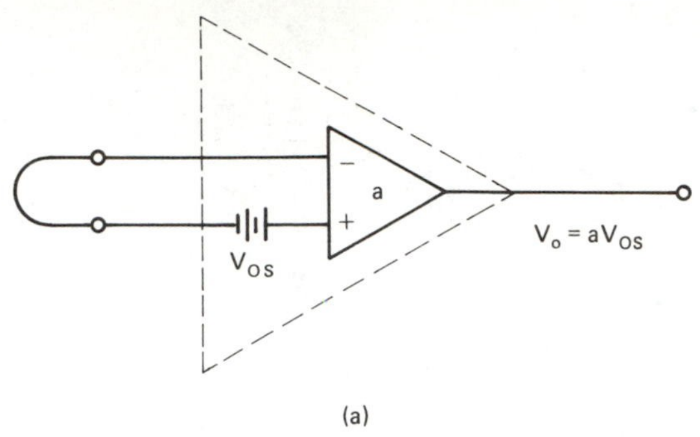

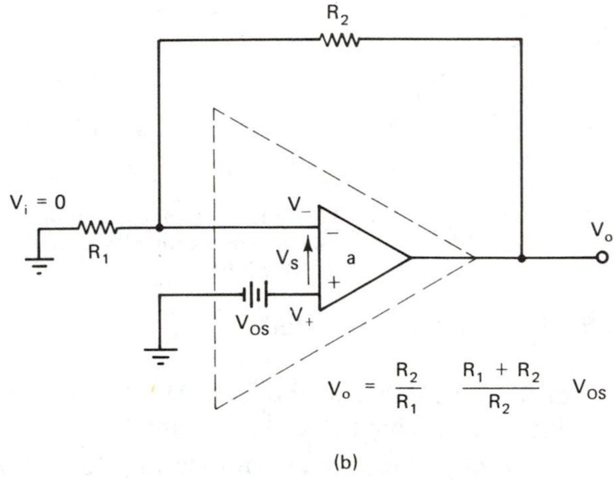

Figure 12.16 Equivalent circuit model of input voltage offset. (a) Open loop, input shorted. (b) Closed loop, input shorted.

cal) battery connected between the external input and the input to the (hypothetical) ideal op amp. This portion of the circuit is shown separately in Fig. 12.16(a). The voltage appearing at the output when the inputs are shorted is represented as an equivalent battery in series with the input. Even though offset voltages are relatively small (~1 mv at the input), the consequence is a huge voltage at the output, (V_o ~100 V!) because of the large open-loop gain (~10^5). Of course this cannot happen with a power supply voltage ~15 volts, since the output latches when it reaches the power supply voltage. The problem is not so severe with closed-loop negative feedback.

In normal closed-loop operation [Fig. 12.16(b)], the offset voltage is considerably reduced. The proof follows the basic procedure used to solve nearly any op amp problem. Suppose that the negative input is grounded, and that the offset voltage appears as an equivalent battery in series with the noninverting terminal, also grounded. Negative feedback drives summing point V_- to a voltage nearly equal to V_{os}, in order to make the voltage difference between plus and minus input terminals zero. Neglecting input bias current, the current flow in R_1 equals that in R_2:

$$I = (V_{os} - V_-)/R_1 = (V_- - V_o)/R_2 \tag{12.6}$$

The op amp amplifies by a whatever signal appears across its input terminals

$$V_o = -a(V_- - V_{os}) \tag{12.7}$$

Eliminating V_- between the two previous equations,

$$V_o = \frac{-aV_{os}}{1 + a[R_1/(R_1 + R_2)]} \tag{12.8}$$

And for normal conditions $a >> 1$,

$$V_o = \left[\frac{(R_1 + R_2)}{R_1}\right] V_{os} = \frac{R_2}{R_1}\left[\frac{(R_1 + R_2)}{R_2}\right] V_{os} \tag{12.9}$$

Thus negative feedback has greatly reduced the effect of offset voltage. On the other hand, if the gain R_2/R_1 is large, $(R_1 + R_2)/R_2 \simeq 1$, and the output offset scales very nearly with the closed-loop gain for signal voltages. So although negative feedback reduces the offset voltage, it does not improve the *ratio* of signal to offset, which is usually the quantity of most concern. When the error signal occurs *before* the amplification, negative feedback provides no *net* improvement.

The cure is *offset nulling*, which introduces a bias voltage to cancel V_{os}. Although V_{os} varies from one op amp to another, it is relatively constant for a specific device, so simple nulling circuits (e.g., Fig. 12.5) can adjust it to zero. The important factor is then not the size of the offset but the amount it *drifts* with time or temperature. This drift is sufficiently indeterminate for inexpensive op amps like the 741 that it is not usually even specified. For critical applications where voltage offset errors of 100 μV or less are serious, one turns to higher-quality (premium) op amps whose offset drift is clearly specified, and may be as low as 1 μV/°C (see Table 12.3).

TABLE 12.3 PREMIUM OP AMP SELECTION

	Bias current nA (max)	Offset voltage mV (max)	$(dV_{os}/dT$ μV/°C	Slew rate V/μs	Approximate price (dollars)
General Purpose IC					
AD 301 A	250	7.5	30	1–10	1.50
AD 301 AL	30	0.5	5	1–10	10.00
AD 741 C	500	6.0	no spec	0.5	1.50
AD 741 J	200	3.0	20	0.5	2.00
FET Input IC					
AD 506 J	0.015	3.5	75	3	15.00
AD 515 J	3×10^{-4}	3	50	0.3	18.00
AD 542 J (Trifet)	0.05	2	20	3	4.00
Low voltage drift IC					
AD 504 J	200	2.5	5	0.1–2.5	14.00
AD 510 J	25	0.1	3	0.1	10.00
AD 517 J	5	0.15	3	0.1	6.00
AD OP-07 A	3	0.075	1.3	0.17	7.50

Courtesy Analog Devices, Inc.

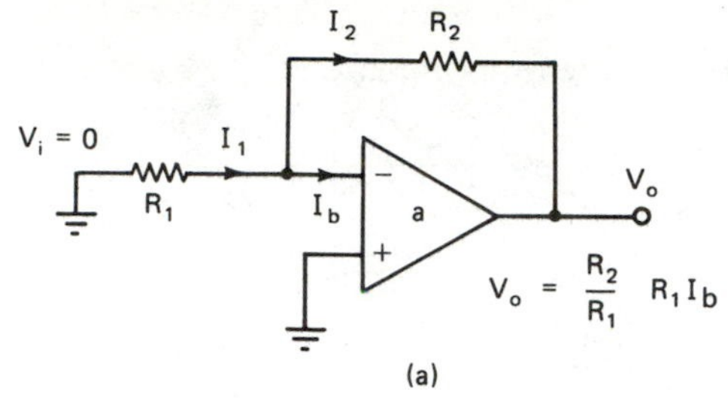

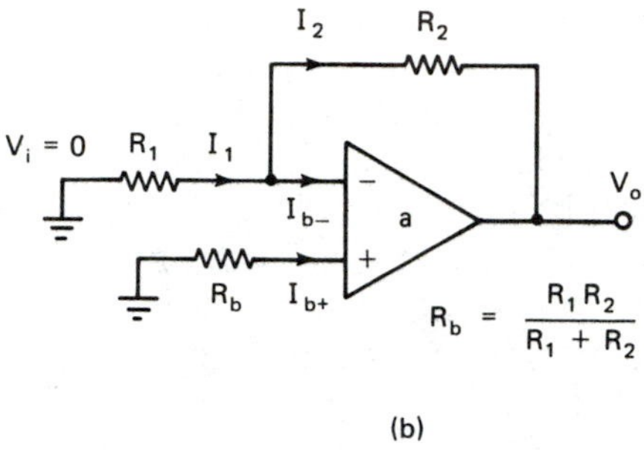

Figure 12.17 (a) Equivalent circuit model of output voltage error due to input bias current. (b) Reduction of bias current error using an additional resistor in the noninverting input.

12.4.2 Bias Current Error

Input voltage offset will be ignored here, by the superposition principle, so the error due to input bias current I_b may be considered by itself. The bias current portion of the equivalent circuit (Fig. 12.1) is simplified in Fig. 12.17(a). With the input grounded, an error is caused by the Ohmic voltage drop due to I_b.

Current flowing into and out of the summing point node is conserved, except that now an additional term is present.

$$(0 - V_-)/R_1 = I_b + (V_- - V_o)/R_2 \tag{12.10}$$

Together with $V_o = -a(V_- - V_+)$ this yields.

$$V_o\left[\frac{1}{a}\left(\frac{1}{R_1} + \frac{1}{R_2}\right) + \frac{1}{R_2}\right] = I_b \tag{12.11}$$

The factor containing a^{-1} is negligibly small. The approximate result is

$$V_o = I_bR_2 = (R_2/R_1)I_bR_1 \tag{12.12}$$

The output voltage error due to input bias current scales as the product of the bias current I_b and the feedback resistor R_2. This sets an upper limit on the value of the feedback resistor used in an op amp circuit. For example, with $I_b = 10^{-7}$ A (741), a 10-M Ω feedback resistor yields a sizable output error of 1 V. R_2 for a 741 is usually limited to about 1 M Ω (see Prob. 12.3). The second version of Eq. (12.12) shows that for a given gain (R_2/R_1), the bias current error (relative to the signal) scales with R_1. Since R_1 also sets the input impedance, there is a conflict between obtaining high input impedance and obtaining low bias current error for the inverting connection. This conflict is not present for the noninverting (follower with gain) connection (see Prob. 12.10).

Bias current errors are inherent with bipolar op amps, since the transistors in the input stage require a bias current to drive them into their linear operating region. For a given op amp, the output error may be reduced by inserting a resistor R_b between the positive input and ground, causing an ohmic voltage drop. By proper choice of the resistor, the voltage drops appearing at V_- and V_+ may be made equal, resulting in a zero output error voltage. In more demanding applications, a much more satisfactory solution is to select an FET op amp, with tolerable magnitudes of bias current and tolerable offset current errors.

12.4.3 Bias Current Balancing

The output error voltage due to bias current is minimized by choosing

$$R_b = \frac{R_1 R_2}{R_1 + R_2} \tag{12.13}$$

This follows because, intuitively, when R_b is properly chosen and V_o is brought to zero, R_1 and R_2 are effectively in parallel (both connected to ground), and the bias current I_{b-} is drawn though the parallel combination. V_o can be zero only when the voltages V_+ and V_- are equal, or

$$R_b\, I_{b+} = [R_1 R_2/(R_1 + R_2)]I_{b-} \tag{12.14}$$

If the bias currents are equal at both terminals (offset current negligible), the optimum value of R_b is therefore given by Eq. (12.13). With the use of this current-balancing technique, the dominant error is not bias current, but the bias current difference or *offset* current. With a bias current balancing resistor R_b [Eq. (12.13)] in place, an upper limit to the output voltage error is then $I_{os} R_2$. If necessary, R_b may be replaced by a variable resistor in an attempt to eliminate the offset currents. However, since bias and offset current may vary with temperature, the better solution in more demanding applications is always the selection of a better op amp.

Some may find the above "bootstrap" proof objectionable, since it assumes that the problem is solved ($V_o = 0$) in order to solve for the answer. A more detailed and rigorous approach (below) yields the same result.

Exercise: Optimum Bias Current Balancing Resistor

Assume an op amp with finite open loop gain a. The analysis follows earlier procedures except for new bias current terms. Current conservation and Ohm's law applied at the inverting terminal lead to

$$V_- = -R' I_{b-} + V_o(R'/R_2) \tag{12.15}$$

where

$$R' = R_1 R_2/(R_1 + R_2)$$

The voltage at the noninverting terminal is

$$V_+ = -R_b I_{b+} \tag{12.16}$$

Solving for the output voltage (after a little algebra) gives

$$V_o = a\,[R_b I_{b+} - R' I_{b-}][1 - aR'/R_2]^{-1} \tag{12.17}$$

The goal is to adjust the passive components such that $V_o \simeq 0$. As long as the bias currents are equal, this is achieved by making $R_b = R'$, proving the desired result, Eq. (12.13).

12.5 INTEGRATOR DRIFT DUE TO NONIDEAL OP AMP BEHAVIOR; THE PREMIUM OP AMP

A spectacular example of nonideal behavior can be observed by constructing an integrator from a standard 741 op amp. Even with the input grounded, the output voltage drifts rapidly until the output saturates at the power supply voltage, usually within seconds. The necessary bias current charges the feedback capacitor, resulting in this output voltage variation. We now explore the magnitude of this error and its connection with nonideal op amp properties and passive component values.

12.5.1 Integrator Drift Due to Op Amp Bias Current

A simplified equivalent circuit is shown in Fig. 12.18(a). As usual, we write the equation for current conservation at the summing-point node. The relationship between current and voltage in the capacitor involves a time derivative:

$$I_2 = C \frac{d}{dt} (V_- - V_o) \tag{12.18}$$

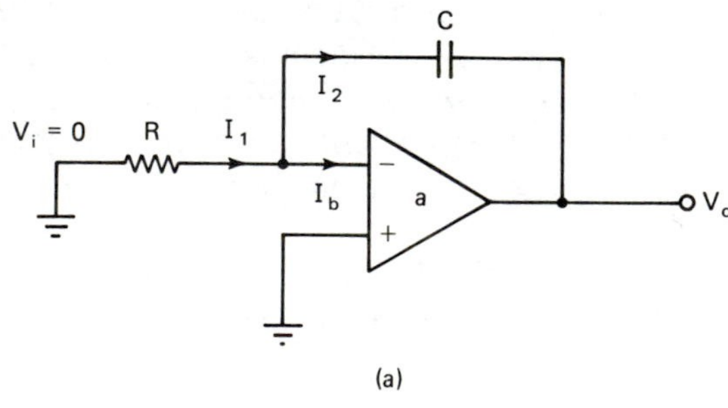

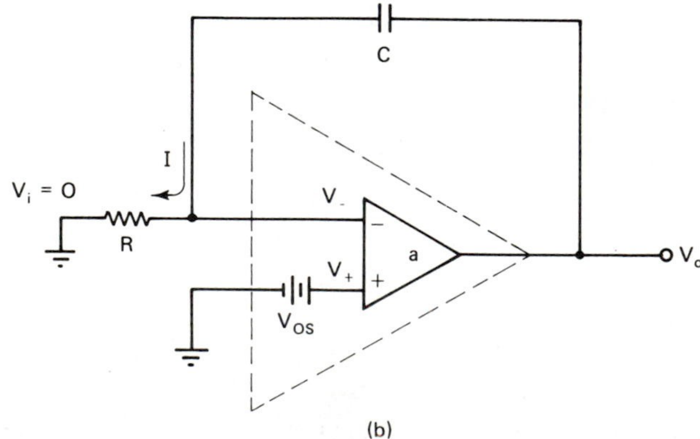

Figure 12.18 Integrator drift due to nonideal op amp behavior. (a) Effect of input bias current. (b) Effect of input voltage offset.

As a result, the expression for current conservation will be a differential equation:

$$(0 - V_-)/R = I_b + C \frac{d}{dt} (V_- - V_o) \tag{12.19}$$

Inserting $V_o = -a(V_- - V_+) \simeq -aV_-$ and eliminating V_-,

$$V_o/aR - C(1 + 1/a)(dV_o/dt) = I_b \tag{12.20}$$

The general behavior may be seen without solving the differential equation. Eq. (12.20) may be simplified, since $1/a << 1$. To a first approximation,

$$dV_o/dt = I_b/C \tag{12.21}$$

This corresponds to a uniformly increasing voltage ramp. The rate of change of output scales proportionally to I_b and inversely as the value of C. As a result, the output drift of an integrator may be reduced by making C as large as possible. Capacitor values of order 1 μF are typical for integrators which are stable over a period of seconds (see also Prob. 12.6).

12.5.2 Integrator Drift Due to Op Amp Voltage Offset

A simplified equivalent circuit for this situation is shown in Fig. 12.18(b). The input is grounded. Current conservation at the summing-point node leads to a differential equation

$$I = V_-/R = C \frac{d}{dt} (V_- - V_o)$$

The output voltage is proportional to the voltage difference across the inputs:

$$V_o = -a(V_- - V_{os})$$

Eliminating V_- leads to

$$(V_o/a + V_{os})/R = C \frac{d}{dt} [V_o(1 + 1/a) + V_{os}] \tag{12.22}$$

The behavior occurring right after the capacitor has been shorted may be determined without solving the differential equation. On the right-hand side, V_{os} is assumed constant, and hence has no time derivative: $dV_{os}/dt = 0$. The term $1/a$ may be neglected compared to 1. On the left, assuming V_o small when the capacitor is shorted, $V_o/a << V_{os}$. This leads to

$$dV_o/dt = V_{os}/RC \tag{12.23}$$

The output voltage ramps up linearly in time with a rate proportional to V_{os} and $1/RC$. Therefore, to make the rate of change due to this error small, the time constant RC should be made as large as possible. There are, of course, practical limits to this procedure, since making RC large also reduces the magnitude of the integrator's response to an input signal (see Prob. 12.6).

12.5.3 Integrator Drift Example

We complete this discussion by illustrating the magnitude of integrator drift for a general-purpose op amp such as the 741, comparing the relative effects of input bias current and offset voltage. Suppose an integrator circuit is constructed with R = 100 KΩ and $C = 1\ \mu$F, so that $RC = T = 10^{-1}$ s. Using the results of the previous section, the initial output voltage drift is:

Bias current	Offset voltage
$I_b \simeq 2 \times 10^{-7}$ A	$V_{os} \simeq 10^{-3}$ V
$dV_o/dt = I_b/C \simeq 0.2$ V/s	$dV_o/dt = V_{os}/RC \simeq 0.01$ V/s

It is clear that drift is a serious problem when attempting to make a dc integrator with a 741. The integrator will drift off scale in a matter of seconds. With this (typical) choice of component values, bias current is the more serious problem. It is *not* possible to use a bias current–balancing resistor [Eq. (12.13)] in an integrator circuit (see Section 14.5). The magnitude of both errors is reduced by increasing *C*, but practical component values limit this approach, and the ratio of signal to error is not improved. For these reasons, a 741 or similar general-purpose bipolar op amp is not used as a dc integrator, although it may be used as a nonideal integrator for signal processing in the audio range by placing a resistor (R~ 1 MΩ) in parallel with *C*.

12.5.4 How to Select a Premium Op Amp

The cost of an op amp is a relatively small fraction of the cost, for example, of a NASA experiment to land on Mars. The more conservative premium approach is also useful in many applications less demanding than a Mars landing. In research applications, the requirements demand higher performance than a general-purpose device, and the extra cost is justified by the more tightly specified performance.

Example

Suppose a given application requires a dc integrator which will be stable for several minutes. The nature of the signal and its frequency range results in the selection of $R = 10^6\ \Omega$ and $C = 1\ \mu$F. As a design criterion, limit the output voltage drift to 0.1 V over 1 min. Specify the maximum tolerable input current offset and input voltage offset. Then select an op amp which will achieve this goal reliably at minimum cost from Tables 12.2 and 12.3.

Solution

Tolerable values of offset voltage and offset current are found by following the procedure of the previous numerical exercise.

Voltage Offset

$$dV_o/dt = V_{os}/RC < (0.1/60)\text{V/s}$$

Therefore,

$$V_{os} < RC\ dV_o/dt = 1.6 \times 10^{-3}\ \text{V} = 1.6\ \text{mV}$$

Bias Current

$$dV_o/dt = I_b/C < 1.6 \times ^{-3}\ \text{V/s}$$

$$I_B < C\ dV_o/dt = 10^{-6}\ F \cdot 1.6 \times ^{-3}\ \text{V/s}$$

$$= 1.6 \times ^{-9}\ A = 1.6\ \text{nA}$$

Referring first to the industry standard IC op amps (Table 12.2), the general-purpose 741/301 bipolar devices will of course not satisfy the 1.6-nA bias current requirement. The high-performance MOSFET's and BiFET's have bias currents which are at least an order of magnitude below the required limit, but their offset voltage does not meet the requirement. Even if offset voltage is nulled, the stability over the operating temperature range may be inadequate. There are two alternative solutions: use a selected higher quality device within the same family, or use a premium op amp.

Selected higher quality units are designated by slightly different part numbers. For example, going from an LF351 to an LF351A or from a 3140 to a 3140B reduces the offset voltage to 2 mV, tolerably close to the design limit. It also increases the price by a factor of 10. The LFT356 is a similar quality device, at a similar high price. Before ordering such a selected part, ask what the delivery time will be. Six to eight weeks is not uncommon.

The other approach is to use a premium op amp from a manufacturer specializing in such devices. Table 12.3 lists a subset of devices available from one such manufacturer, Analog Devices. The prices listed are for comparison only, and are subject to wide variation. Op amp prices, as with other IC's, tend to decrease as sales push production up along the "learning curve."

In Table 12.2, op amps are grouped into categories for ease in selection. The category General Purpose includes bipolar op amps such as the 741. The price is about 5 times what a 741 costs from a discount house, but the product is somewhat more tightly specified. The FET category includes op amps with input bias currents in the desired range for this exercise, though with rather large offset voltage drift as a function of temperature. When this is a problem, one turns to the *High Accuracy Low Voltage Drift* grouping with offset voltage drifts 10 to 100 times lower. The nanoamp-range bias currents of this group are, however, several orders of magnitude larger than picoamp bias FET versions. Chopper-stabilized low drift op amps (not listed) have $dE_{os}/dT < 0.5\ \mu\text{V/°C}$. Other important groups include *Wide Band Width*, with slew rates up to 500 V/μs; *Electrometers*, with bias currents of 10^{-14} A, used in measuring ultra-small currents; *High Ouput* op amps, with 100 mA output current for power-driving applications; and *Isolated* op amps for biomedical applications with optical isolators to decouple input sources from the power supply.

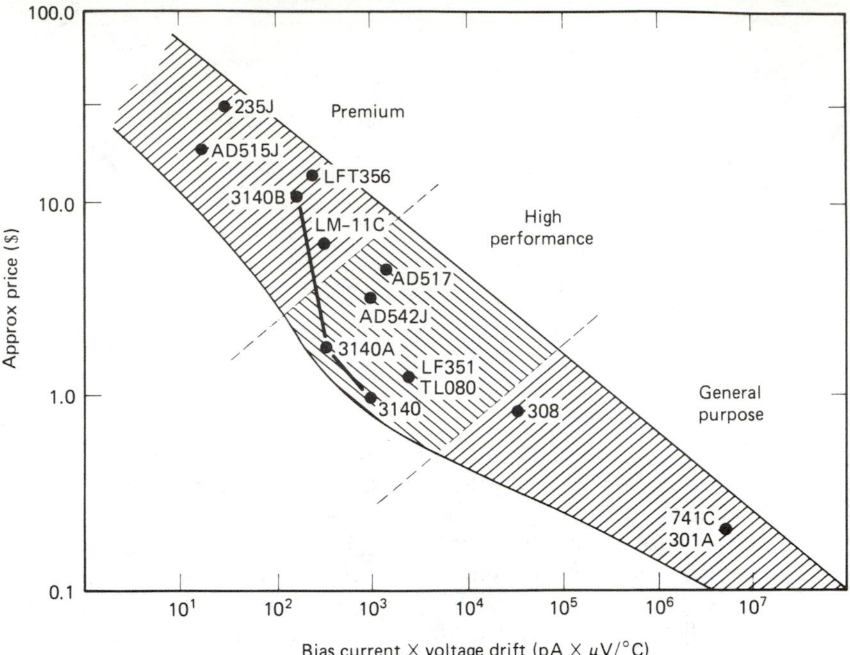

Figure 12.19 A figure of merit for measuring price vs. performance of op amps is the product of bias current and rate of change of offset voltage with temperature.

For our integrator application, the FET input group has bias current low enough to meet the design criterion. The offset voltages of 2 mV to 3.5 mV are marginally acceptable. Although V_{os} can be nulled to zero, a large shift as the device warms up might cause unacceptable errors. In this example, selecting an AD 542 for the integrator leads to offset voltage drift of only a fraction of a millivolt over a typical temperature range of 10°C , well within the design range of 1.6 mV. The AD542 also costs a factor of 4 less than the premium AD506 or AD515. The AD506 is an older device; improved technology is usually cheaper. (Note the similar price decrease from old standby to newer model in the sequence AD504–AD510–AD517.) The AD515 is a superlow bias current device, better than this application demands. The AD542 is thus a "best buy," at a cost of under $5.00. This is also a factor of 2 less than better versions of the LF351 or 3140. The lower-cost products from a premium manufacturer are less expensive than the more tightly specified products of a volume manufacturer, and also more available.

Selecting an op amp is thus a compromise between specifications, price, and availability. Although the specifications will vary with the application, a useful figure of merit is the product of bias current and offset voltage drift. Offset voltage can often be trimmed to zero, leaving the drift. Bias current is not usually balanced since FET op amps became available. A plot of op amp price as a function of the product $I_b \times (dV_{os}/dT)$ displays (Fig. 12.19) a suprising consistency which spans all op amp types (bipolar, BiMOS, BiFET) and grades (general-purpose, high-performance, premium).

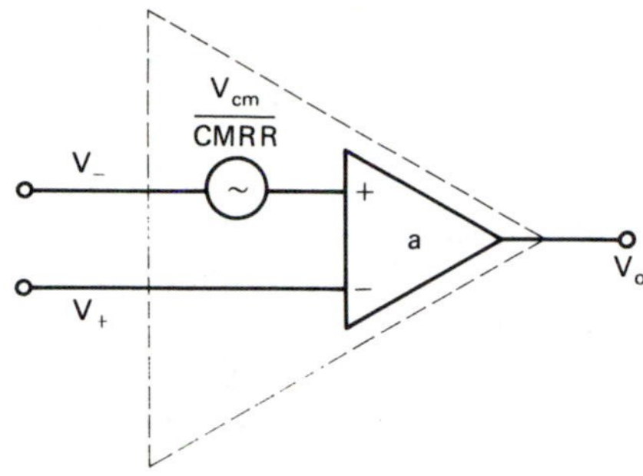

Figure 12.20 Equivalent circuit model of an op amp with finite CMRR. The ideal op amp has an additional input source V_{cm}/CMRR, where V_{cm} is the average of V_+ and V_-.

12.6 COMMON MODE ERRORS

The op amp's response to differential and common mode inputs is characterized by two different gains, a and g, respectively [Eq. (10.9) and Fig. (10.5)]. The finite common mode rejection is modeled as an additional input source V_{cm}/CMRR in series with the input signal, connected to an ideal op amp whose CMRR is infinite (Fig. 12.20). The common mode signal is defined as the average of the signals appearing at the two inputs of the real op amp:

$$V_{cm} = (V_- + V_+)/2 \tag{12.24}$$

For a single-ended input, common mode limitations only give a small gain error for the circuit. For a differential input circuit or for a subtractor, common mode errors limit the signal to noise improvement resulting from the differential connection.

12.6.1 Common Mode Errors for a Voltage Follower

In this case (Fig. 12.21), since $V_o \simeq V_i$ and $V_- = V_o$, the common mode signal is approximately

$$V_{cm} = (V_o + V_i)/2 \simeq V_i$$

The op amp amplifies whatever voltage difference appears across its inputs

$$V_o = a[V_i - (V_o + V_i/\text{CMRR})]$$

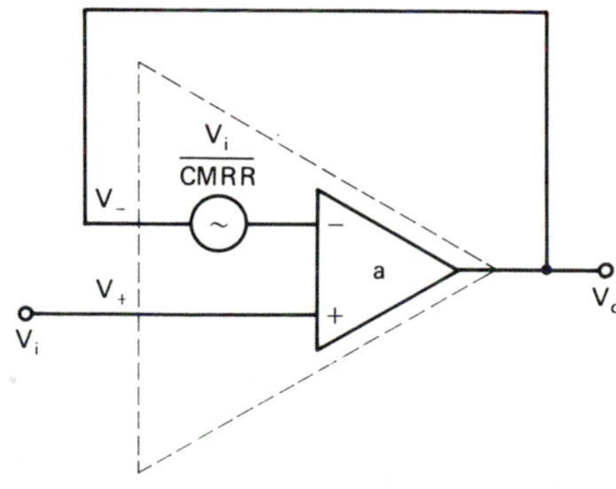

Figure 12.21 Common mode rejection of a voltage follower.

Solving for V_o,

$$\frac{V_o}{V_i} = \frac{1 - 1/\text{CMRR}}{1 - 1/a} \tag{12.25}$$

There is an additional error over and above the error due to finite op amp gain. Since CMRR is usually less than a, common mode errors are usually the dominant gain error in a follower circuit. For example, for a 741, CMRR = 80 dB = 10^4 but $a = 10^5$, so V_o/V_i is in error by 1 part in 10^4. This gain error is usually not serious. But if the CMRR is a function of signal strength, distortion can be introduced. Note the bootstrap method used to solve this problem. The ideal circuit is used to estimate V_{cm}, which is then used in the nonideal circuit model of Fig. 12.21.

12.6.2 Common Mode Error for an Inverting Amplifier

An inverting amplifier with a hypothetical common mode error source is shown in Fig. 12.22(a). Common mode errors for this circuit turn out to be very small. The plus input is grounded, and feedback reduces the minus input to a virtual ground, so neither input ever gets a large voltage, and V_{cm} is very small as a result. For

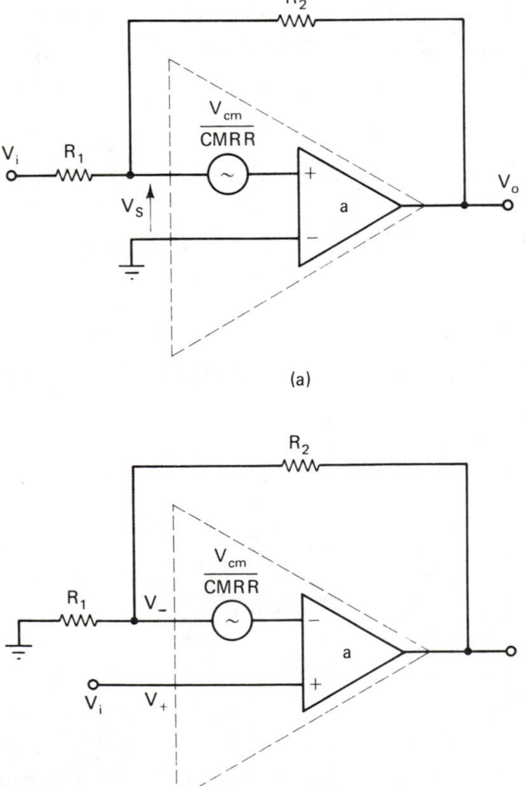

Figure 12.22 Common mode rejection of (a) an inverting amplifier; (b) a noninverting amplifier.

this reason, the inverting connection was used almost exclusively when finite CMRR was a serious problem.

The magnitude of the error may be estimated as follows. V_{cm} is the average of V_- and 0:

$$V_{cm} = (V_- + 0)/2 = V_-/2$$

The op amp amplifies whatever appears across its inputs:

$$V_o = -a(V_s + V_{cm}/\text{CMRR}) = -aV_-[1 + (2\text{CMRR})^{-1}] \qquad (12.26)$$

Balancing the input and feedback currents in the usual way leads to an expression for V_-:

$$V_- = V_i[(R_1 + R_2)/R_2] + V_o[(R_1 + R_2)/R_1] \qquad (12.27)$$

This may be used to eliminate V_- in Eq. (12.26).

$$\frac{V_o}{V_i} = \frac{-(R_2/R_1)[1 + (2\text{CMRR})^{-1}]}{1 + (2\text{CMRR})^{-1} + (1/a)[R_1/(R_1 + R_2)]} \qquad (12.28)$$

If $1/a$ is truly negligible, the gain errors in numerator and denominator *cancel*, leaving no net common mode error for this circuit. The virtual ground results in a small common mode voltage for the inverting amplifier circuit.

12.6.3 Common Mode Error for a Noninverting Amplifier

A follower with gain with an additional common mode error source is shown in Fig. 12.22(b). Since neither input is grounded, there is a chance to generate a large common mode signal. The common mode signal is the average

$$V_{cm} = (V_- + V_+)/2 \simeq 2V_i/2 = V_i$$

The op amp amplifies whatever voltage difference appears across its inputs:

$$\begin{aligned} V_o &= a[V_+ - (V_- + V_{cm}/\text{CMRR})] \\ &= a[(1 + 1/\text{CMRR})V_i - aV_-] \end{aligned}$$

Balancing currents through R_1 and R_2 leads to an expression for V_-

$$V_- = [R_1/(R_1 + R_2)]V_o$$

which can be used to eliminate V_-, resulting in

$$\frac{V_o}{V_i} = \frac{[(R_1 + R_2)/R_1](1 + 1/\text{CMRR})}{1 + (1/a)[(R_1 + R_2)/R_1]} \qquad (12.29)$$

Thus, there is gain error of the order of 1/CMRR. This is why inverting connections, with their lower common-mode error, are most often found in op amp recipes. Nevertheless, the high input impedance makes the noninverting connection extremely valuable. If one uses premium op amps with CMRR as high as 120 dB (10^6), common mode gain errors will be negligible.

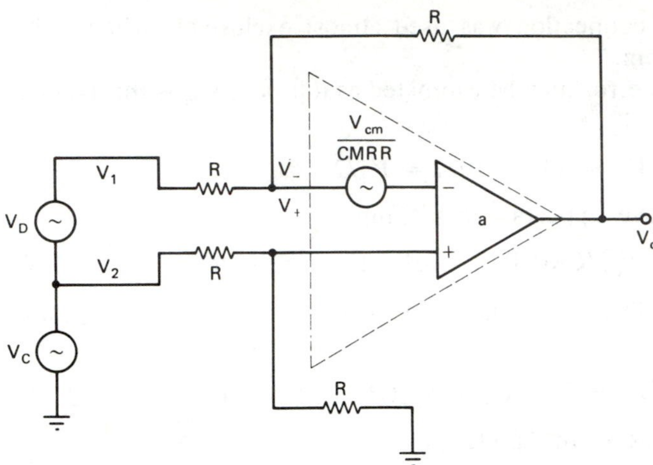

Figure 12.23 Subtractor circuit in the presence of a common mode error signal V_C.

12.6.4 Real World Common Mode Problems

In real measurements, common mode sources have more subtle origins than the ones shown here. Common mode signals can originate in differences in grounding of op amp and signal source, and differences in source resistances between input connections. These real-life complications will be discussed in Chapter 19 in connection with the measurement of low-level signals. The results there tend to *reverse* the above picture, showing that the inverting connection is inferior to the noninverting connection (used cleverly) in differential amplifier circuits when the common mode source is *external* to the circuit. To see this, consider a subtractor circuit (Fig. 12.23) with an outside world common mode error signal V_C. Op amp terminal V_+ has a voltage

$$V_+ = V_2/2 = V_C/2$$

The voltage on terminal V_- is

$$\begin{aligned} V_- &= (V_1 - V_o)/2 \\ &= (V_D + V_C - V_o)/2 \end{aligned}$$

Ideally, the error signal V_C cancels. Balancing $V_+ = V_-$ leads to

$$V_o = V_C - V_C + V_D$$

But there appears at each op amp input terminal an error voltage $V_C/2$ in addition to V_D.

$$V_{cm} = V_C/2$$

The gain for this signal is 1/CMRR, so the overall circuit gain is

$$V_o = V_D + V_C/(2\ \text{CMRR}) \tag{12.30}$$

Thus, if the common mode error noise V_C is 10^3 times larger than the differential signal V_D, an appreciable noise level will get through the subtractor.

PROBLEMS

12.1. The 741 has a typical input bias curent of 200 nA (2×10^{-7} A). Refer to Fig. 12.6 and answer the following questions.

(a) What must be the current supplied to the emitters of Q_1 and Q_2 by Q_{18}? Assume $\beta = 100$ for Q_1 and Q_2.

(b) How much feedback voltage is subtracted by R_4?

(c) What current must flow in the collector of Q_{10}?

(d) *Estimate* the collector current in Q_{10} caused by its biasing network. Compare this answer to (c). The current in R_5 is set by the supply voltage (assume 15 V) but with about 1.0 V subtracted by each of the *on* transistors Q_{13} and Q_{11}. An answer agreeing within a factor of 4 is good enough for this estimate.

(e) Suppose that the simpler biasing scheme of Fig. 12.3(b) were used. What value of R_B would have to be fabricated on the chip?

12.2. At what frequency will the 741 no longer act ideally as a gain of 10 ($R_f/R_i = 10$) amplifier?

12.3. For a given circuit gain, an inverting amplifier can be constructed many ways; the actual choice of component values is a compromise involving nonideal op amp properties.

(a) For a gain-of-100 inverting amplifier, evaluate the output errors due to offset voltage (assume $V_{os} = 1$ mV) and bias current (assume $I_b = 10^{-7}$ A) for values of (R_2/R_1) ranging from (100 KΩ/1 KΩ) to (100 MΩ/1 MΩ). Which error dominates in each case?

(b) Which error dominates for high values of circuit gain? Which error dominates for high values of circuit resistance?

(c) For gain-of-10 amplifier, what is the highest input impedance circuit which can be constructed before the error due to an input bias current of 10^{-8} A is larger than a 1-mV input signal?

12.4. As with the inverting amplifier, the output errors of an integrator are more useful when expressed in a form where the gain has been extracted [as in Eqs. (12.9) and (12.12)] so the error can be compared to the input signal. The difference is that here the comparison is between drift *rate* and signal *frequency.*

(a) Write an expression for the total output error signal due to V_{os} and I_b, factoring the "gain" $1/RC$.

(b) For an input sine wave signal of 1-V magnitude but variable frequency, is the error *relative to the input signal* more serious at high or low frequencies? Explain.

(c) Discuss a strategy for picking component values to reduce the drift problem relative to the input signal.

12.5. Design an integrator which will be stable to 1.0 V over several minutes. Suppose the signal requires $RC = 1$ s, but the largest available capacitor with low leakage is 1 μF.

(a) What is the upper limit for op amp offset voltage and bias current?

(b) Referring to the op amp selection Tables 12.2 and 12.3, decide if any of the general-purpose op amps will do. If not, select the proper family of premium op amps and find the lowest-price solution in this family.

(c) If offset voltage can be nulled, the problem of offset voltage drift with temperature remains. Suppose that the integrator will be in an environment where temperature fluctuates over a 20°C range. Will your solution be adequately stable?

12.6. Select an op amp capable of measuring a microvolt input in a gain-of-100 amplifier without being overwhelmed by offset voltage and ofset current errors. What is the lowest-cost solution?

12.7. Select an op amp which, when connected as an electrometer, will be able to measure a picoamp (10^{-12}) level input current, without bias current and offset voltages getting in the way. What is a minimum-cost solution?

12.8. Suppose an op amp has an open-loop source resistance (seen from the output) of 100 Ω. What is its source resistance when connected as a gain-of-100 amplifier with $R_f = 1\ \text{M}\Omega$, $R_i = 10\ \text{K}\Omega$? Does the output source resistance depend upon the choice of an inverting or noninverting connection?

12.9. Compare the input impedance of the inverting and noninverting amplifier connections. Show that the conflict bewtween high-input impedance and low bias current error is not present in the noninverting connection.

12.10. Show that for the noninverting amplifier (follower with gain), the signal and error gains in the presence of offset voltages and bias currents are given by

$$V_o = [(R_2 + R_1)/R_1](V_i + V_{os} + I_b R_s)$$

under conditions of high source impedance, $R_s >> R_1$ or R_2. Refer to Fig. 12.24(a).

12.11. Although the noninverting amplifier ideally does not load the source, bias current requirements can lead to substantial errors when the source impedance is large. Show that the output voltage error due to input bias current can be reduced to 0 (for $I_{os} = 0$) or at least reduced by the ratio I_{os}/I_b by the additional resistor shown in Fig. 12.24(b), whose value is:

$$R_c = R_s - R_1 R_2/(R_1 + R_2)$$

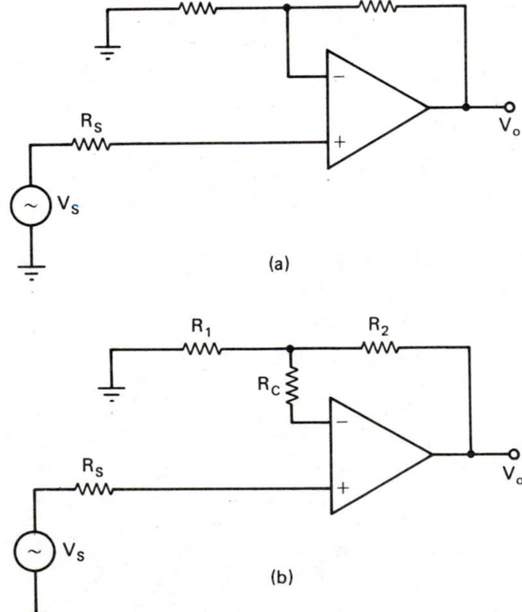

Figure 12.24 Bias current balancing of non-inverting amplifier (see Problem 12.11).

12.12. What limits the practical gain of a single op amp stage? Consider a gain-of-1000 amplifier. Compare the limitations due to source loading, bias current, and offset voltage of the following three alternatives:

(a) Inverting amplifier

(b) Noninverting amplifier

(c) Supergain circuit of Fig. 10.18

12.13. The derivations of errors due to offset voltage and bias current are "honest" derivations assuming finite op amp gain. In practice, the virtual ground method could be used instead. Provide simplified analysis of the output error in the following situations.

(a) Input voltage offset and an inverting amplifier [Fig. 12.16(b) and Eq. (12.9)]. Negative feedback makes $V_+ = V_-$. **Hint**: Note how Eq. (12.9) resembles a voltage divider.

(b) Input bias current and an inverting amplifier [Fig. 12.17(a) and Eq. (12.12)]. Assume that negative feedback creates a virtual ground. **Hint**: How can it be, as *apparently* implied by Eq. (12.13), that all of the bias current is drawn through R_2 and none through R_1?

Now repeat the above simplified analysis for an integrator circuit, for

(c) Input voltage offset [Fig. 12.16 and Eq. (12.23)]

(d) Input bias current [Fig. 12.17 and Eq. (12.21)]

REFERENCES

More complete bibliographic information for the books listed below appears in the annotated bibiliography at the end of the book.

Analog Devices, *Data Acquisition Products Catalog*

GRAEME, *Designing with Operational Amplifiers*

HIGGINS, *Experiments with Integrated Circuits*, Experiment 18

JUNG, *IC Op Amp Cookbook*

MELEN & GARLAND, *Understanding IC Operational Amplifiers*

National Semiconductor, *Linear Databook*

National Semiconductor, *Linear Applications Handbook*

ROBERGE, *Operational Amplifiers*

TOBEY, GRAEME, & HEULSMAN, *Operational Amplifiers*

13

Op Amp Applications in Regulation and Control

Among the most useful applications of op amps is the control of variables in the outside world. The value of the variable is sensed and fed back to the op amp input to provide an error signal. The error signal is amplified and fed back in such a way that the controlled variable is made to equal the reference signal. The controlled variable need not be a voltage, but could be a temperature, a light intensity, an angular or linear position, or nearly any physical or chemical variable. In this chapter, several examples of op amp regulators will be analyzed. The main goal will be to see how the negative feedback loop is closed for each case, and to understand quantities such as the open-loop gain when the feedback is not entirely electrical.

13.1 VOLTAGE REGULATOR

Perhaps the most familiar example of op amp regulation is found in a voltage-regulated dc power supply. The basic circuit of the voltage regulator is shown in Fig. 13.1(a), drawn to illustrate the important sections. Power is delivered to the load through a power transistor, which acts as a valve controlling the flow of current from a power source. This source is typically rectified and filtered, but has, in addition to its dc component, considerable ac ripple which must be removed. The value of the input voltage may also fluctuate considerably, and this fluctuation must be removed from the output. The output voltage is sensed across a voltage divider (*output sensing*) with a variable fraction of the output fed to an op

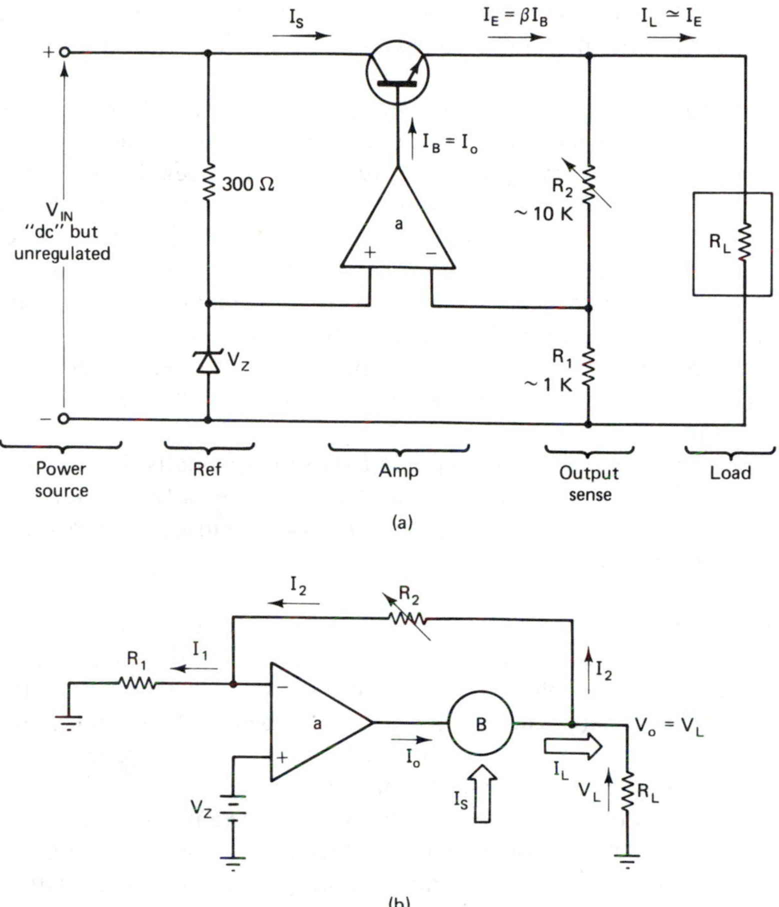

Figure 13.1 Fig. 12.1(a) Basic voltage regulator circuit. (b) Voltage regulator redrawn as a conventional op amp circuit.

amp. Also input to the op amp is a reference voltage, provided in this case by a Zener diode. A resistor biases the Zener onto the constant-voltage portion of its characteristic curve, so the Zener acts as a battery, providing a source of constant voltage for comparison. Since the output voltage sample and the reference voltage are fed to opposing terminals of the op amp, the difference or error signal appears at the summing point and is amplified greatly. The op amp output is fed to the base of the power transistor, whose emitter current is then fed to the load. Although a 741 can deliver about 20 mA, regulators must often regulate currents of amps. The transistor has the function of a current booster, providing a current gain of $I_E/I_B = B$ where $B \simeq 10^2$. (The symbol B is used in this chapter for transistor current gain to avoid confusion with the feedback fraction β.) The transistor may also be viewed as an emitter follower with voltage gain = 1. The output voltage at R_L follows the output voltage of the op amp, regardless of the current required, i.e., regardless of variations of the value of R_L (*load* regulation).

The usual feedback connection, with an element Z_f between the op amp output and inverting input, does not appear. Nonetheless, the feedback is present.

Question. *Is the feedback loop closed and is the feedback negative?* Before doing a precise analysis of this circuit, it is useful to see qualitatively how regulation occurs. Assuming that the circuit is initially regulating, suppose that the output voltage fluctuates upward for some reason. The voltage applied to the inverting terminal of the op amp also fluctuates upwards. Since V_Z remains constant, the error voltage applied between the op amp input terminals increases. This increase in the error signal causes a decrease in op amp output voltage, because $a < 0$. The smaller signal at the base of the transistor is translated into a decrease in the emitter current; hence the voltage across the load drops. This is opposite to the initial fluctuation. The feedback loop is closed so as to prevent such fluctuations from occurring.

This is a useful technique in exploring qualitatively the stability with respect to changes in a feedback system. One supposes a fluctuation to occur, and then follows it around to see if the feedback loop is closed in such a way as to to oppose the fluctuation.

Detailed Analysis. It is useful in analyzing the circuit to redraw it more similarly to a standard op amp circuit [Fig. 13.1(b)]. The circuit resembles a standard inverting amplifier, except that the output voltage is attached to a load resistor R_L, and there is an extra stage of gain before the feedback loop is closed. The function of the extra circle (labeled B) may be voltage gain (see Prob. 13.1) or current gain (the present case, since the transistor is in an emitter follower configuration). In this case, the circle B has voltage gain 1 and current gain B. Since the Zener diode presents a constant voltage input, it is replaced by its equivalent circuit, a battery. We explore how the output voltage varies with V_Z and with the value of variable resistor R_2, and calculate the value of the open-loop gain A and the feedback fraction β.

The feedback loop is closed, so the summing-point voltage difference will be made equal to zero. The analysis is straightforward:

$$I_1 = I_2$$

$$V_-/R_1 = (V_o - V_-)/R_2, \quad \text{or}$$

$$V_o/R_2 = V_-(1/R_1 + 1/R_2) = V_-(R_1 + R_2)/(R_1 R_2) \tag{13.1}$$

The voltage delivered to the load is related to the load current I_L

$$V_o = I_L R_L \tag{13.2}$$

This current results from the boosted op amp output current I_o

$$I_L + I_2 = BI_o \simeq I_L \qquad (R_2 >> R_L) \tag{13.3}$$

where the last step follows since $I_L >> I_2$. The output voltage is bigger than the summing point voltage by the gain a.

$$V_o = -a(V_- - V_Z) \tag{13.4}$$

Eliminating V_- leads to

$$V_o = \frac{aV_Z}{1 + aR_1/(R_1 + R_2)} = V_Z\left(\frac{1}{a} + \frac{R_1}{R_1 + R_2}\right)^{-1} \tag{13.5}$$

The regulated output is proportional to V_Z with a scale factor that depends upon the values of R_1 and R_2. The $1/a$ term is negligibly small ($a >> 1$). The current gain B appears nowhere in the result. This follows from the fact that the feedback signal is voltage-sensing; if R_L were to change, I_L would be automatically compensated to keep V_o constant. As a result, it is not necessary to know the detailed properties of the transistor (e.g., its characteristic curves or small-signal properties). Comparing the form of Eq. (13.5) with the usual result $V_o/V_1 = -1/\beta$, the quantity which acts as the feedback fraction β is the voltage divider $R_1/(R_1 + R_2)$. Some nonideal properties are also traceable to Eq. (13.5). *Line regulation* is a measure of how constant V_Z remains as the input voltage varies. To improve line regulation, the 300-Ω biasing resistor may be connected to the output of the circuit rather than to its input. The *load regulation* depends on how constant a remains as the output load R_L fluctuates. Variation of B with load is relatively unimportant in closed-loop operation, since the voltage sensing feedback will automatically adjust to compensate for it. However, the circuit will fail to regulate when the load resistance falls to the point where B cannot deliver the current required by Eq. (13.2). Finally, the circuit provides a variable output voltage by varying the ratio $R_1/(R_1 + R_2)$.

In situations where the output *voltage* is beyond the op amp range, a voltage booster (transistor amplifier) stage may be used for the circle B. The analysis is very similar (Prob. 13.1).

13.2 CURRENT REGULATORS

Some situations require a constant current rather than a constant voltage. For example, in the measurement of electrical resistivity, a constant current is passed through the material, and the voltage across the sample is measured. A constant current is necessary if the changes in resistivity as a function of temperature or time are to be measured. The design of a current regulator follows the same principles as a voltage regulator, except that the output *current* is sensed rather than the output voltage. Since most op amps amplify voltage, the output current is converted to a voltage by passing the signal through a small current-sampling resistor r [Fig. 13.2(a)].

Qualitative examination of current regulator operation is straightforward. Negative feedback adjusts the current so that no voltage difference appears across the op amp inputs: $I_L r = V_Z$. A constant current is therefore maintained. The balance is independent of R_L, so the current is constant no matter how the load resistance varies. Of course there are practical limitations. For increasing R_L, the voltage across the load increases until it reaches the supply voltage V_{in}. Any

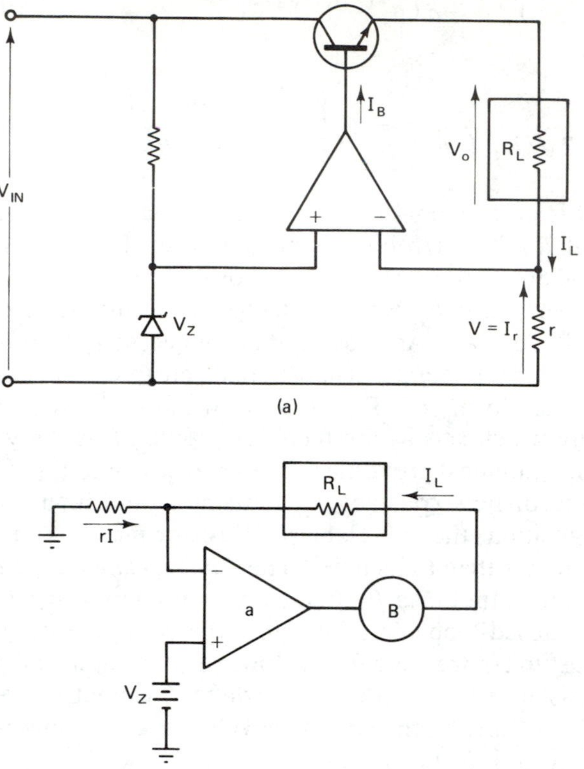

Figure 13.2 (a) Basic current regulator circuit. (b) Current regulator redrawn as a conventional op amp circuit.

further increases will eliminate the voltage drop across the transistor, so the transistor will cut off, and the circuit will no longer be able to regulate. Commercial current-regulated power supplies include a variable-voltage *compliance* as a safety feature to limit the output voltage and guard against burning out the test circuit or damaging the power supply (see Section 13.3).

Question. Is the feedback loop closed so it provides *negative* feedback?

Answer. An increase in I_L will cause an increase in the current in r, raising the voltage at the minus terminal of the op amp. The op amp output voltage falls, as does the transistor base current I_B, causing a fall in I_L, which opposes the original change. Therefore the feedback loop is closed in such a way as to provide a negative feedback path.

Detailed Analysis. It is useful to redraw the circuit of Fig. 13.2(a) to resemble the conventional op amp inverting amplifier, as shown in Fig. 13.2(b). The resemblance to a follower-with-gain indicates why the circuit functions as a constant-current device. The summing-point voltage $(V_- - V_+)$ can only be made zero if the voltage at the inverting terminal is made equal to V_Z. This requires a current in the sampling resistor r

$$I_L = V_Z/r \tag{13.6}$$

I_L can only come through the feedback loop, and in doing so flows through the load R_L, providing the desired current source. If R_L varies, the output voltage V_o automatically adjusts up or down, keeping the current in the feedback path constant. The detailed analysis is given in Prob. 13.2.

The basic current-regulator circuit has a number of disadvantages. All the load current flows through the sensing resistor r, whose value must remain constant. To avoid changes in r with self-heating, this resistor is usually a special high-current shunt. These are not usually variable, so the current from this circuit is not readily varied. In addition, with the transistor emitter follower (Fig. 13.2), the *sign* of the output current is fixed. In some applications, the current must follow an input signal whose sign could change. Finally, with the circuit of Fig. 13.2, neither end of R_L can be grounded. This limits the circuit to situations where the load can remain floating, and may be a source of noise. All but the last restriction is removed with the circuit shown in Fig. 13.3. The reference signal V_R is connected to an inverting amplifier, rather than to a high-impedance terminal as in Fig. 13.2. This will result in current drain from the reference supply, which is kept small by the use of a current-scaling trick. By the use of a *complementary symmetry* pnp–npn output transistor pair, one transistor or the other can turn on to follow a reference signal of either sign, driving current through R_L in either direction. This circuit has potential *crossover distortion*, as one transistor turns off and the other turns on. But since the transistor booster is within the feedback loop, its precise gain is irrelevant as long as the loop gain is large, and the nonlinearities of crossover distortion are reduced to zero. The circuit operates by forcing the voltage V_3 to equal the reference voltage V_R. Since the junction of R_1 and R_2 is a virtual ground, currents I_3 and I_f from point V_3 through R_2 and R_3 are scaled by roughly the ratio of the resistances. Since I_3 is driven by the transistors through

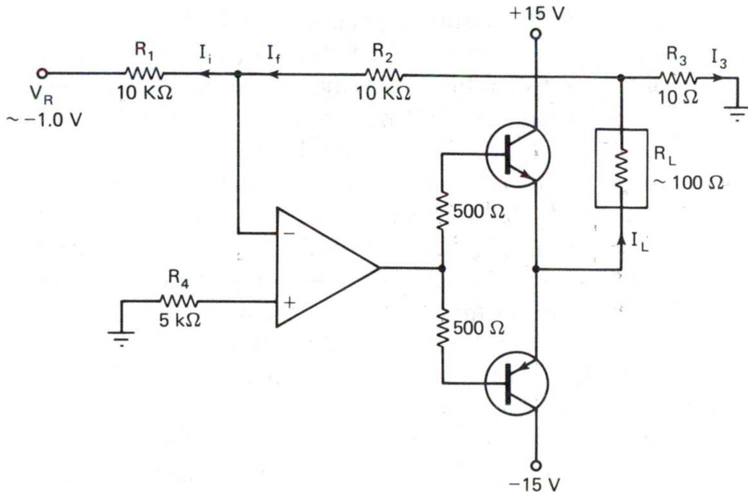

Figure 13.3 Dual polarity current source which employs current scaling.

R_L, a current source results: I_L is independent of the value of R_L. The detailed analysis is left for Prob. 13.3. The result is

$$I_L = V_{\text{ref}}(1/R_2 + 1/R_3) = V_{ref}/R_3 = I_i(R_1/R_3) \tag{13.7}$$

The current has been scaled up from the reference current by the ratio of resistances, 1000 to 1 in this example. The current can be made adjustable by choosing a variable resistor for R_1.

13.3 COMMERCIAL REGULATED POWER SUPPLIES

In applications where the utmost flexibility and the highest specifications are desired, commercial power supplies provide the simplest and cheapest solution. In selecting a power supply, one must specify: whether it is to operate as a voltage source or a current source; stability with respect to variations in the load, input line voltage, or temperature; stability over long periods of time; and whether it is desirable to control the regulated variable by an external voltage or resistance. These specifications are defined in detail in Table 13.1. Specifications for one example are given in Table 13.2. A power supply such as this is easily adjustable and may operate either in a constant-current or a constant-voltage mode, with automatic crossover from one to the other. When operating in the constant-voltage mode, the constant-current threshold setting provides a convenient means of limiting the maximum current I_S supplied to the external load. Similarly, the voltage adjustment V_S may be used to limit the voltage applied to an external circuit when operating in the constant-current mode. These provide useful limits protecting both power supply and the external circuit against unexpected changes. The operating characteristics of a dual-mode power supply are shown in Fig. 13.4. The circuit may operate as a constant-voltage (CV) source [Fig. 13.4(a)] or as a constant-current (CC) source [Fig. 13.4(b)]. It may also operate in a CV/CC mode [Fig. 13.4(c)]. With no load attached, the output is $V_o = V_S$, set by a front panel control. When a load resistance is attached, the output current increases, while the output voltage remains constant. The operating point is the intersection of a line of slope R_L with the rectangular characteristic curve of the power supply. As R_L decreases, I_o increases, with essentially no change in V_o until the output current reaches I_S, which has been set by the front panel control setting. At this point, the supply automatically changes into a constant-current mode. Further decrease in R_L results in a drop in V_o, with no change in the output current I_o. The operating point in this mode traverses the nearly vertical line of a current source. This circuit provides full protection against any overload condition, since both V_S and I_S are under operator control and may be set to ensure protection for both the load and the power supply.

The basic schematic circuit of a constant-voltage–constant-current power supply is shown in Fig. 13.5. Two error-sensing amplifiers control output voltage and current. The constant-voltage amplifier acts as a zero-output impedance source, varying the output *current* whenever the load resistance changes, while

TABLE 13.1 POWER SUPPLY SPECIFICATIONS[a]

Automatic Tracking A master–slave connection of two or more power supplies, each of which has one of its output terminals in common with *one* of the output terminals of *all* the other power supplies. Auto-tracking allows simultaneous proportional output from all power supplies with a single control knob.

Constant Current (CC) Power Supply A power supply that acts to maintain its output current constant in spite of changes in load, line, temperature, etc., as shown in Fig. 13.4(a). Thus, for a change in load resistance, the output current remains constant while the output voltage changes by whatever amount necessary to accomplish this.

Constant Voltage (CV) Power Supply A power supply that acts to maintain its output voltage constant in spite of changes in load, line, temperature, etc., as shown in Fig. 13.4(b). Thus, for a change in load resistance, the output voltage remains constant while the output current changes by whatever amount necessary.

Constant Voltage/Constant Current (CVCC) Power Supply A power supply that operates as a constant-voltage source or a constant-current source, depending on load conditions. It acts as a CV source for comparatively large values of load resistance R_L and as a CC source for comparatively small values of R_L. The crossover or transition between these two modes of operation occurs automatically as shown in Fig. 13.6(c) at a critical value $R_L = R_C = V_S/I_S$, where V_S and I_S are values of stabilized output voltage and current set by the operator.

Constant Voltage/Current Limiting (CV/CL) Power Supply A power supply similar to the CV/CC above, except that above the current threshold the output current is limited but not stabilized, i.e., it will not serve as a good current source.

Crowbar See **Overvoltage**.

Drift Also referred to as *stability*, but really an instability. The maximum change of output voltage or current during a specified period of time following the warm-up time, with all other variables held constant. Drift is essentially the dc component of *noise*, and is related to the internal temperature rise of the power supply.

Line Regulation Also known as *source effect*, the change in the steady-state value of the stabilized output voltage (CV) or current (CC) resulting from a specified change in the source voltage, with all other variables remaining constant.

Load Regulation Also known as *load effect*, the change in the steady-state value of the stabilized output voltage (CV) or current (CC) resulting from a specified change in load resistance, with all other variables remaining constant. The specified change in the CV mode is usually from open circuit to a value which results in the maximum rated current. In the CC mode, the specified change is usually from a short circuit to the value which results in the maximum rated output voltage.

Load Transient Recovery Time The time interval between a specified step change in the load current (CV) or load voltage (CC) and the time when the stabilized output returns to within a specified percentage of its previous output.

Output Impedance The slope $\Delta V/\Delta I$ of the output voltage–current characteristic curve. The output impedance of an ideal CV power supply would be zero and that of an ideal CC power supply would be infinite at all frequencies. For a CV supply, the output impedance as a function of frequency of load disturbance is approximately a resistance in series with an inductance.

Overvoltage Protection Protection of the power supply and/or connected equipment against excessive output voltage. It is a separate circuit which monitors the output voltage and rapidly (order of microseconds) places a low-resistance ("crowbar," usually an SCR switch) across the output terminals whenever a preset voltage limit is exceeded. Such a supply must also be protected by means of limiting the output current.

Remote Programming Also known as *remote control*. The setting of the power supply output by an external control variable, such as a variable resistance, voltage, or digital signal.

Remote Sensing A means by which a CV power supply monitors the stabilized voltage directly at the load, using extra sensing leads. Since no current flows in these sense leads, this prevents errors due to voltage drops in the current-carrying leads to the load.

TABLE 13.1 CONTINUED

Ripple and Noise The residual ac component superimposed on the dc output. It may be specified in terms of its rms or peak-to-peak value. Measurement of ripple and noise with an instrument of insufficient bandwidth may conceal high-frequency spikes (from SCR switching, for example) detrimental to the load.

Temperature Coefficient The maximum change in a power supply's stabilized output voltage or current per degree Celsius following a change in the ambient temperature within specified limits, with all other variables held constant.

[a]Adapted from Hewlett-Packard's *Power Supply Catalog and Handbook.*

TABLE 13.2 SPECIFICATIONS OF A COMMERCIAL POWER SUPPLY[a]

Specification		Value
DC Output: Voltage and current spans indicate range over which output may be varied using front panel controls.	Volts	0–10V
	Amps	0–1A
Load Effect (Load Regulation): Voltage load effect is given for a load current change equal to the current rating of the supply. Current load effect is given for a load voltage change equal to the voltage rating of the supply.	Voltage	4 mV
	Current	500 μA
Source Effect (Line Regulation): Given for a change in line voltage between 104 and 127 V ac.	Voltage	4 mV
	Current	750 μA
Ripple and Noise: (rms/p-p, 20 Hz to 200 MHz)	Voltage	200 μV/1 mV
	Current	150 μA/50 μA
Temperature Coefficient: Output change per degree Celsius change in ambient following 30-min warm-up.	Voltage	0.02% plus 1 mV
	Current	6 mA
Drift (Stability): Total drift in output (dc to 20 Hz) over 8-h interval under constant line, load, and ambient temperature.	Voltage	0.1% plus 5 mV
	Current	15 mA
Resolution: Minimum output voltage or current change that can be obtained using front panel controls.	Voltage	5 mV
	Current	75 μA
Output Impedance (Typical): Approximated by a resistance in series with an inductance.		5 MΩ, 1 μH
Load Effect Transient Recovery (Load Transient Recovery): Time required for output voltage recovery to within the specified level of the nominal output voltage following a change in output current equal to one-half the current rating of the supply.	Time	50 μs
	Level	15 mV
Output Mode: Constant-voltage/constant-current, or constant-voltage/current-limited.		CV/CC
DC Output Isolation: Supply may be floated at up to the given level above ground.		300 V

[a]Courtesy Hewlett-Packard. **Note:** the supply is a CV/CC type, so the specifications are listed separately for both modes.

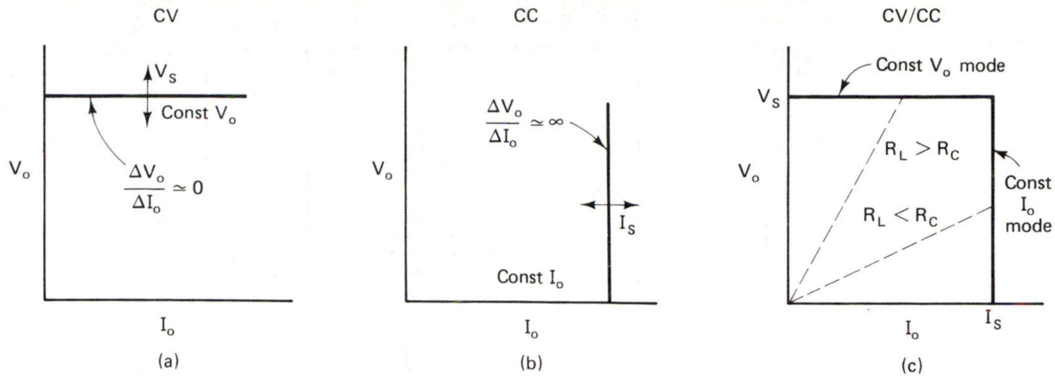

Figure 13.4 Three modes of power supply operation. (a) Constant voltage. (b) Constant current. (c) Constant voltage/constant current with automatic crossover.

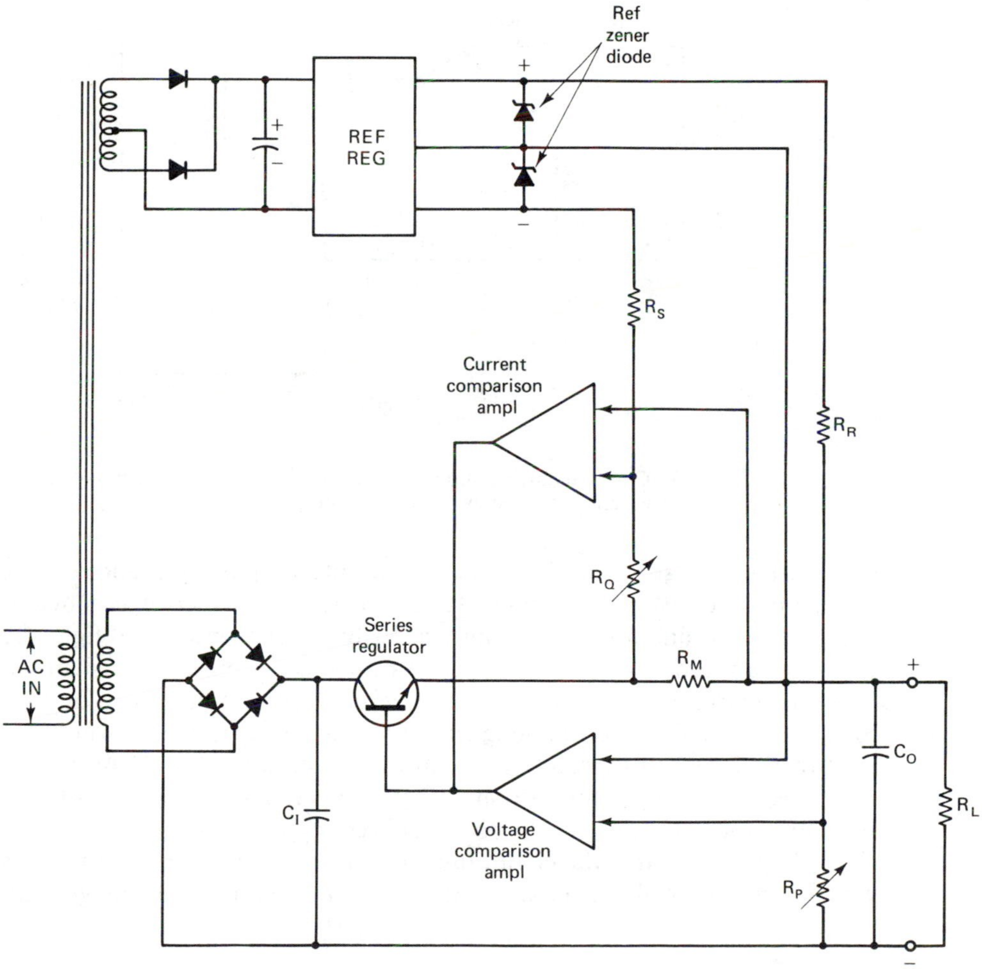

Figure 13.5 Basic circuit of a commercial constant-voltage/constant-current power supply.

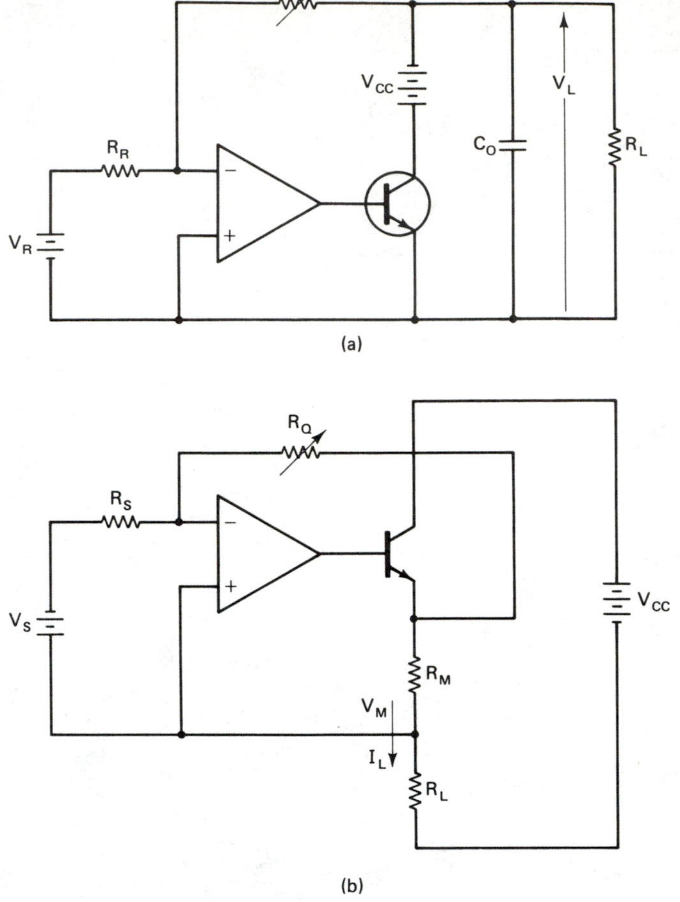

Figure 13.6 Constant-current mode of the commercial power supply of Fig. 13.5. (a) Actual circuit. (b) Redrawn as a conventional op amp circuit.

the constant-current amplifier acts as an infinite-output impedance source, varying the output *voltage* in response to any load resistance change. Obviously, the two amplifiers cannot operate simultaneously, or else one would fight the other. For a given value of R_L, the power supply must act as either a constant voltage or a constant current supply. Transfer between these two modes is accomplished automatically by suitable decoupling circuitry (not shown), which switches between the two modes when R_L reaches the ratio V_S/I_S shown in Fig. 13.4(c).

The voltage regulator portion of Fig. 13.5 is redrawn in Fig. 13.6(a). Viewing the transistor and unregulated power source V_{CC} as a gain block B [as in Fig. 13.1(b)], it is clear that the circuit functions as a variable gain-inverting amplifier, which adjusts the voltage across the transistor to keep the load voltage at a value

$$V_L = -(R_P/R_R)\,V_R \tag{13.8}$$

In a general-purpose power supply such as this, both output terminals are floating, and may be grounded on either side as desired. This is possible because both the unregulated power source and the reference source are taken from the secondary coil of a transformer (Fig. 13.5).

The circuit for the current source portion has an advantage over those previously discussed. Not only is the current readily adjustable, but also the load may be grounded. This is possible because the reference supply can be floated in this design, acting as a battery. The current source portion of Fig. 13.5 is redrawn in Fig. 13.6(b). Again, an inverting amplifier circuit can be identified. The op amp adjusts the voltage across the transistor so that the voltage V_o across R_M is held constant. The current in R_M (and hence in R_L; see Prob. 13.4) is then

$$I_L = (R_Q V_S)/(R_S R_M) \qquad (13.9)$$

The examples of Fig. 13.6 illustrate a useful procedure. When an unfamiliar feedback circuit is encountered (such as Fig. 13.5), see if it can be redrawn to resemble a familiar circuit, such as the inverting amplifier.

13.4 IC VOLTAGE REGULATORS

In instrumentation applications where a dedicated power supply is desired, the best approach is to use an IC voltage regulator. Multiple printed-circuit (pc) board systems frequently include inexpensive regulator IC's on each card. This approach has several advantages. It provides decoupling between various parts of the circuit, which is particularly useful in avoiding noise in a low-level analog portion of an instrument, introduced by switching transients of digital circuitry in the same system. In addition, if a component failure occurs on one portion of the system, the rest of the system is isolated and protected from destruction.

The typical IC voltage regulator includes the reference voltage, op amp, and power transistors functionally equivalent to the voltage regulator circuit of Fig. 13.1. The actual circuits are a good deal more complicated, however, due in part to additional protection circuitry.

The voltage regulator IC is employed as a three-terminal black box [Fig. 13.7(a)]. The only external component required is an input capacitor, to provide filtering for the rectified input voltage. It is standard practice to put this on the individual pc board, to decouple the board from external fluctuations. The output capacitor is not necessary, but improves the transient response of the circuit. The same three-terminal regulator can also be used as a current source [Fig. 13.7(b)]. The regulated current is set by the value of the current-sampling resistor R, which provides the voltage for the error-sensing amplifier. An adjustable output voltage larger than the nominal fixed voltage of the IC regulator can also be obtained [Fig. 13.7(c)]. The performance of the device is somewhat deteriorated, and more demanding applications instead use another kind of IC regulator such as the LM 723 (Table 13.3) whose output voltage is internally adjustable. Finally, in applications where the current demands of the load exceed the current capability of the IC, a current-sharing trick [Fig. 13.7(d)] may solve the problem. The transistor supplies the majority of the current to the load, with a small fraction supplied by

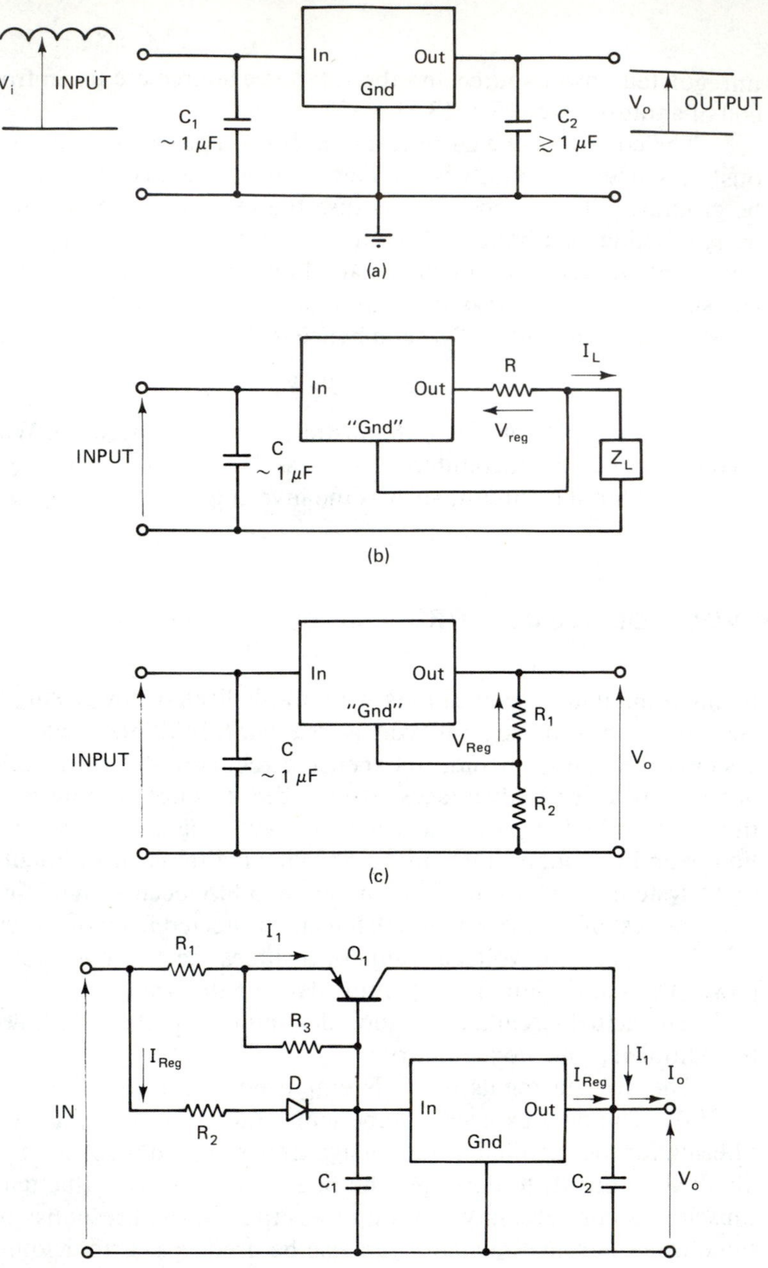

Figure 13.7 IC voltage regulator applications. (a) Fixed voltage regulator. (b) Current regulator; I is set by the value of R. (c) Adjustable voltage regulator. (d) Boosted current voltage regulator.

the regulator. The line regulation and load regulation are not deteriorated in this application; rather, the IC is made to respond to whatever fluctuations occur, since the transistor has a fixed bias and does not function as an error-correcting amplifier.

TABLE 13.3 SPECIFICATIONS OF TYPICAL IC VOLTAGE REGULATORS

Parameter	LM109	LM320 Series	LM123	LM340 Series	LM723	Units
Output voltage	5.0	Fixed (5 to 24)	5.0	Fixed (5 to 24)	Adjustable 2 to 37	V
Line regulation	4	10	5	$\leqslant$100	0.01% V_{out}	mV
Load regulation	50	50	25	$\leqslant$200	0.03% V_{out}	mV
Noise	40	200	40	80	20	μV
Input voltage range	7 to 35	(V + 1.0) to 35	6.5 to 20	(V + 2.0) to 35	40	V
Quiescent current	5	2	12	7	1.3	mA
Maximum output current	>1.0	1.5	3.0	>1.0	0.15	A
Power dissipation	Internally limited	Internally limited	30	Internally limited	0.8	w
Approximate cost (dollars)	1	<2	5	2	<1	
Other features						
Current limit	Yes	Yes	Yes	Yes	Yes	
Thermal shutdown	Yes	Yes	Yes	Yes	No	
Terminals	Three	Three	Three	Three	Eleven	
External components	1 capacitor	1 capacitor	1 capacitor	none	Several	
Input/output minimum voltage differential	>2	>1	>1.5	>2	>5	V
Other					General regulator or controller	

Specifications for some typical IC voltage regulators are given in Table 13.3. The low cost makes it clear why it is now unnecessary to spend time in voltage regulator design. Most of the specifications have already been defined in this chapter and correspond to those of commercial power supplies. Most of the IC's feature internal current limiting and a thermal shutdown circuit which removes the device from the circuit if its temperature exceeds a certain value due to excessive power dissipation.. The user must provide an input voltage large enough to allow the device to turn on, but keep the input voltage below a certain upper limit. The lower limit below which the device will not operate is called the *dropout voltage.* Larger than necessary input voltages can result in the dissipation of a substantial amount of heat, which should be kept to a minimum. To provide adequate heat sinks, these devices typically have a fin to facilitate bolting the device down to a heat-sink plate. Among the device families listed in Table 13.3 are fixed 5.0-V regulators for digital logic applications and several series of regulators with fixed-output voltages available in one of many useful values. In addition, one type (the LM 723) allows internal adjustment of the output voltage, and can be used as the heart of a general-purpose high-quality regulator or controller.

13.5 VOLTAGE-CONTROLLED RESISTOR

Another important example of op amp control is an electronically determined resistance. This extremely useful function, which acts as an electronic gain control, can be accomplished with a field-effect transistor as a voltage-variable resistor. With two such FET's arranged in a mirror configuration, the output can be made a precise function of the control voltage, regardless of the characteristic curves of the FET. Examples are shown in Fig. 13.8. The trick is to put one FET in the feedback loop of an op amp circuit. For example, in Fig. 13.8(a) the op amp output will provide the correct voltage to the FET's gate to establish a virtual ground at the op amp's minus input. The sign of V_C must be chosen so that the FET gate is back-biased. The input current is set by the control voltage

$$I_i = V_C/R_C \tag{13.10}$$

The feedback current depends upon the current driven between the FET source and drain. It is convenient to view the FET in this mode as a voltage-variable resistance R_{Q1}. The source-drain current acts as the negative feedback current I_f.

$$I_f = V_{\text{ref}}/R_{Q1} \tag{13.11}$$

The two currents I_f and I_i must balance. Since the two transistors have their gates connected in common, the second FET follows the first. Its resistance, and therefore the resistance seen looking from the output, is

$$R_{Q2} = R_{Q1} \tag{13.12}$$

Combining the above,

$$R_{Q2} = (V_{\text{ref}}/V_C)R_C \tag{13.13}$$

A voltage-variable resistance has been achieved. Its value varies *inversely* as the control voltage and is independent of the characteristic curves of the FET device.

Variations upon this basic idea may be used to construct a voltage-variable resistance *proportional* to the control voltage, rather than to its inverse. Examples are shown in Fig. 13.8(b) and (c). In the first example, the output resistance is only approximately linear in V_c.

$$R_{Q2} = V_C R_C/(V_{\text{ref}} - V_C) \tag{13.14}$$

This is close to linear as long as $V_{\text{ref}} >> V_C$. If this is not adequate, a variation utilizing a current source [Fig. 13.8(c)] may be used. The analysis (Prob. 13.5) shows a precisely linear relationship between control voltage and output resistance.

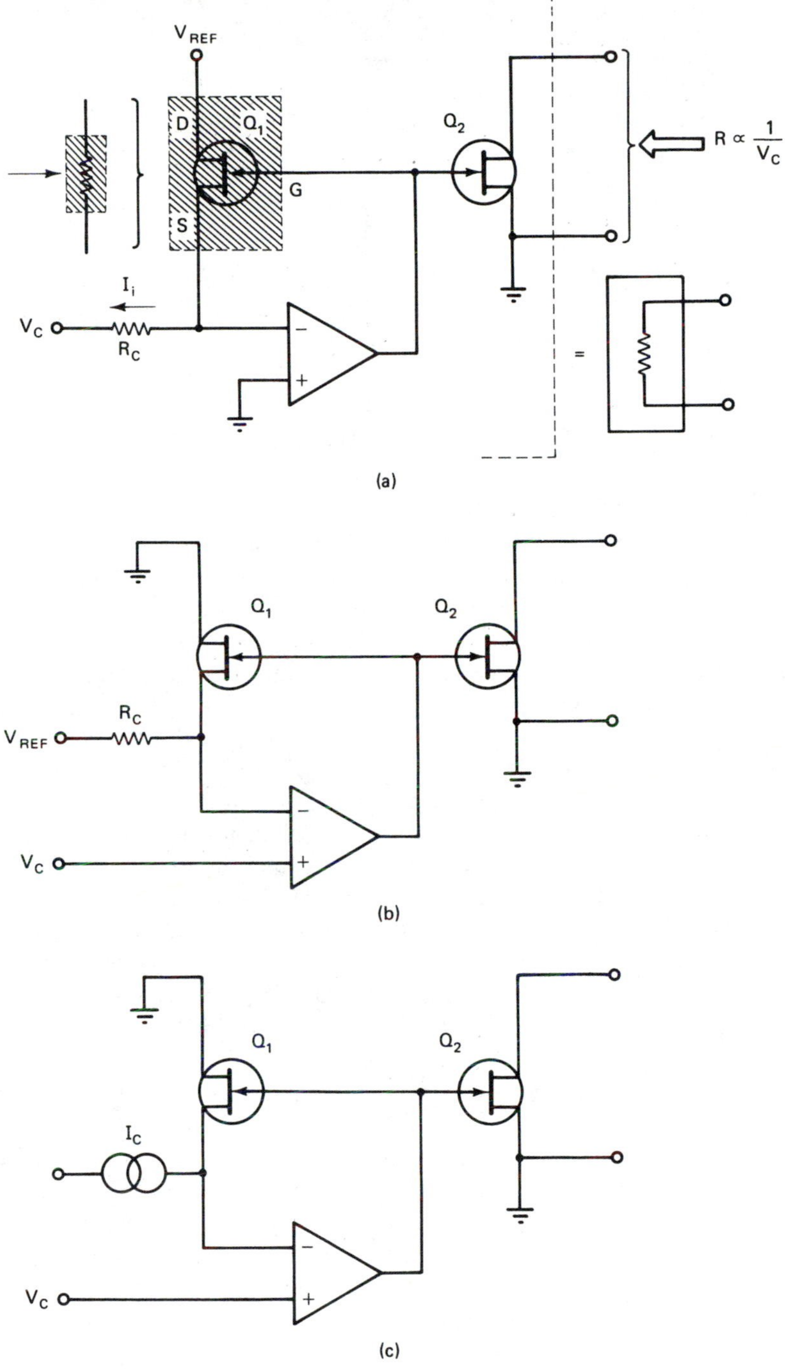

Figure 13.8 Voltage-controlled resistor circuits. (a) Two-terminal resistance whose value is precisely proportional to the inverse of the control voltage. Other versions have resistances approximately (b) or precisely (c) linear in the control voltage. The characteristics of the FET's do not matter, as long as the two are matched.

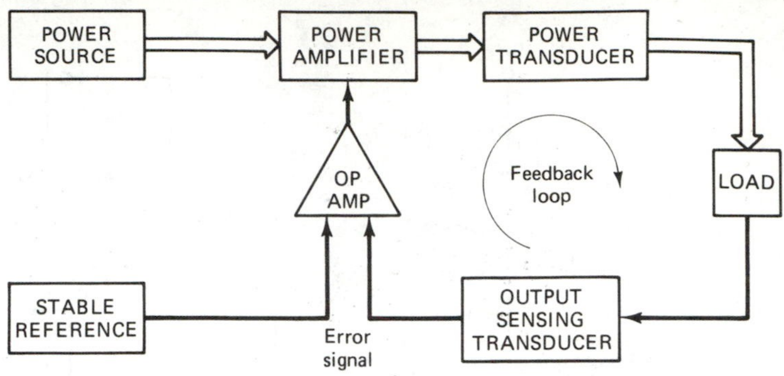

Figure 13.9 Generalized feedback control circuit.

13.6 THE REGULATION OF PHYSICAL VARIABLES

Negative feedback may also be used to regulate physical or chemical variables which are not electrical but are electrically *controllable*, as shown in Fig. 13.9. The output (or controlled variable) is sensed by a transducer which converts the variable into a voltage, for use as an error signal. Design of such regulators requires the conversion from electrical signals to physical power in the load, with conversion back into electrical signals in the output-sensing transducer. An understanding of transducer characteristic curves (transfer function) is generally adequate to close the feedback loop properly, and to estimate the open-loop gain of the system. Stable operation requires that the feedback loop be closed so that the feedback is negative. Otherwise, the device may explode when it is first turned on! The question of stability is more subtle when the feedback loop is closed in the physical world, because there may be inertia or lag in the system. This can cause a delay between the application of an electrical signal to the load and the transmission of a feedback signal from the transducer, which can readily lead to unstable oscillations in the system. We consider here an example in which the output system's response is fast enough that instability is not ordinarily a problem, but can be introduced in a systematic way to illustrate the idea.

13.6.1 Exercise: Light Regulator

Design a light regulator which controls a light source to keep the overall intensity of the light falling on a surface constant, regardless of the variation in other sources of light in the vicinity. A photoconductor, whose resistance R_p decreases as the light intensity increases, is available as a sensor. Assume that R_p is 100 KΩ at the operating point. A solution is shown in Fig. 13.10. The analysis requires answering several questions.

(1) How is the feedback loop closed?

(2) Is the connection to the op amp correct for negative feedback? That is, will the circuit regulate at a stable operating point?

(3) Why is the top end of the output sensing circuit wired back to V_{in} rather than to the output, as in the voltage regulator?

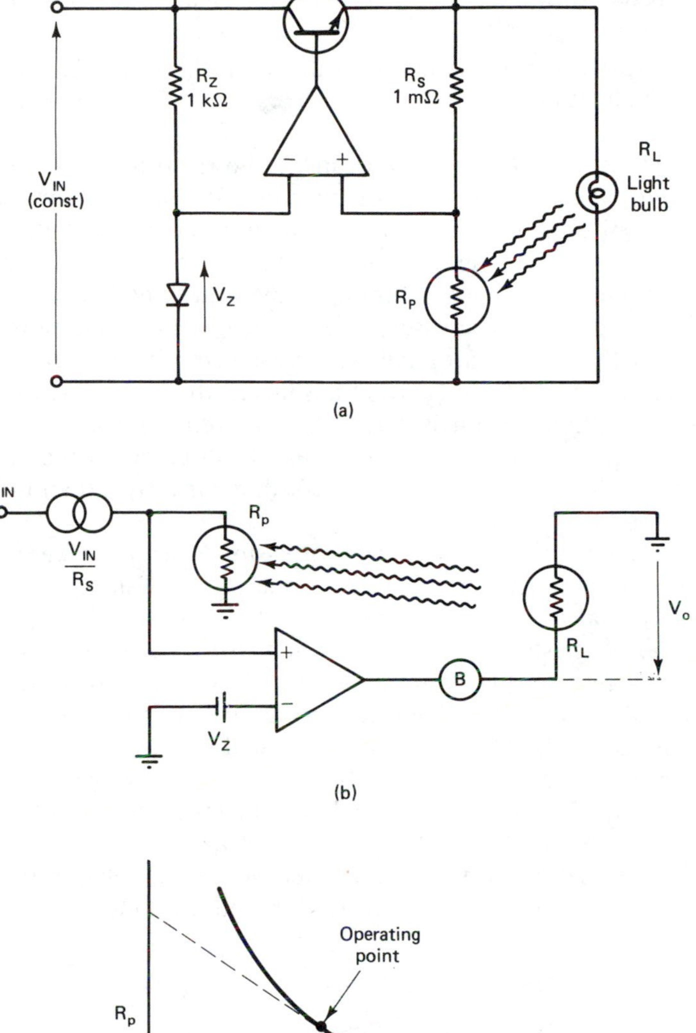

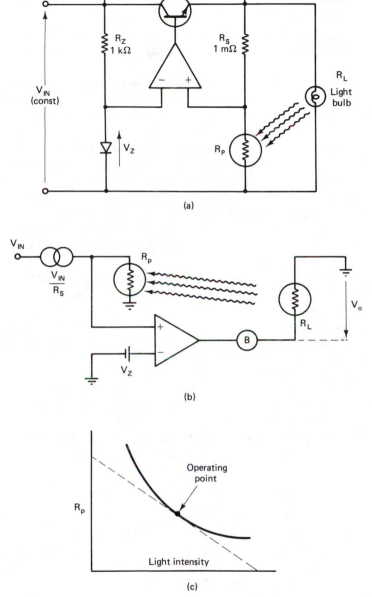

Figure 13.10 (a) Light regulator circuit, which keeps the total light intensity falling on a photoresistor constant. (b) The light regulator circuit redrawn to resemble a conventional op amp circuit. Note that the feedback loop is closed by an optical path. (c) Characteristic curve of the transducer (photoresistor).

(4) Why must $R_S >> R_p$ for proper operation?

(5) Analyze the operation and show that a stable operating point is possible. Show how R_S, V_Z, and R_p (and its dependence upon light intensity) enter in.

(6) It looks as if the op amp is connected as a comparator, with no feedback resistor. Show that this is not so. Calculate the open-loop gain A and the feedback fraction β.

(7) Explore the possibility of instability if there is a time lag in the response of the light bulb.

The feedback connection can be shown to be correct for stable operation, following the procedure used earlier in this chapter. Suppose that the reference voltage V_Z is increased. Follow the circuit response through to see if the error signal applied to the op amp V_+ terminal is in the right direction to match the original change, keeping the summing-point voltage difference zero. If V_Z increases, the op amp output decreases, as does the output of the transistor. The power applied to the light source falls, decreasing the light intensity. This reduces the amount of light received at the transducer, which *increases* its electrical resistance. Since the photoresistor is driven by a current source ($R_S >> R_p$), the voltage supplied to V_+ will increase, matching the direction of the original change of V_Z. If the op amp terminals were reversed, a runaway situation (positive feedback) would result.

Where is the feedback path? Resistor R_Z is not a feedback resistor but supplies a bias current for the Zener diode to establish V_Z. Although the circuit of Fig. 13.10(a) is drawn to resemble a voltage regulator [compare Fig. 13.1(a)], the top end of resistor R_p is connected not to R_L but to a constant voltage V_{in}. The feedback loop in this circuit is closed not by wires but by an optical path. This is clearer if the circuit is redrawn [Fig. 13.10(b)] to resemble the standard op amp circuit. The only feedback path is provided by a light signal from the output light source to the input light transducer. In the analysis which follows, a number of gross simplifications will be made to illustrate the method of approach when a portion of the feedback path is nonelectrical.

Suppose the nonlinear characteristic curve, relating the transducer's resistance R_p to light intensity J received by the transducer, is a straight line over the operating region:

$$R_p = R_{po}(1 - \alpha J) \tag{13.15}$$

Although this limits the validity of the solution, it serves to illustrate how electrical-to-physical and physical-to-electrical variables are introduced. The output light intensity J is proportional to the electrical power input, and hence approximately to V^2 or I_o^2. It is sufficient to consider small excursions about an operating point, with a linear relationship:

$$J = \gamma V_o \tag{13.16}$$

Here, γ includes the resistance of the load and other geometric factors such as the fraction of light received by the transducer. V_o depends upon the voltage difference applied to the op amp inputs

$$V_o = -aB(V_z - V_+)$$

$$J = \gamma\alpha B\ (V_+ - V_z) \tag{13.17}$$

Assume that $R_s >> R_p$, i.e., R_p is driven by a current source. Then

$$V_+ = R_p I_{\text{ref}} = I_{\text{ref}} R_{po}(1 - \alpha J) \tag{13.18}$$

Combining Eqs. (13.17) and (13.18)

$$J = \gamma a B[-V_z + I_{\text{ref}} R_{po}(1 - \alpha J)] \tag{13.19}$$

The output light intensity is

$$J = \frac{\gamma a B(I_{\text{ref}} R_{po} - V_Z)}{1 + \gamma a B I_{\text{ref}} R_{po} \alpha} = \frac{I_{\text{ref}} R_{po} - V_Z}{I_{\text{ref}} R_{po} \alpha} \tag{13.20}$$

where the second quality follows for $aB >> 1$.

In the approximate solution, the gain aB drops out. The circuit therefore has closed-loop negative feedback. Comparison with the usual result $V_o/V_{\text{in}} = a/(1 - \beta a) \simeq 1/\beta$ shows that the quantity which plays the role of the open-loop gain βa must be the combination $[\gamma a B I_{\text{ref}} R_{po} \alpha]$. This combination is dimensionless, as required:

$$[\gamma]\ [aB]\ [I_{\text{ref}} R_{po}]\ [\alpha] = \left[\frac{\text{light}}{\text{voltage}}\right] [1] \left[\frac{\text{voltage}}{\text{light}}\right] \tag{13.21}$$

Here, *light* is short for light intensity J. By inspection, the feedback fraction is $\beta = I_{\text{ref}} R_{po} \alpha$. It is illuminating to approach the problem from the simplified point of view of seeing what the op amp needs to do to keep the summing-point voltage zero (Prob. 13.6).

This circuit provides an excellent test system for trying out basic control systems ideas. Circuit values are given on Fig. 13.10 to facilitate its use as a demonstration. What happens when an additional reflecting surface, such as a piece of paper, is brought close to the light bulb and photoresistor? What happens when the distance between light bulb and photoresistor is changed? What happens when the ambient light intensity in the room is changed? What happens when a capacitor is placed in parallel with the light bulb or with the photoresistor? Pick C such that $RC \sim 1$ s. This last example illustrates visually how lag in a system's response can lead to oscillations in closed-loop control.

13.6.2 The Control of Position with Servomotors

Linear or angular position is controlled using a servomotor, a low-voltage dc (hence bidirectional) motor. The most familiar example is found in an X–Y recorder, as shown in Fig. 13.11. In order to keep the voltage difference at the op amp terminals zero, the error signal generates an output which adjusts the servomotor–reference voltage combination until

$$X = V_{\text{in}}/V_{\text{ref}}$$

Although a variable resistor (called a slidewire) is shown, most modern designs lower the long-term noise using a noncontact reference element such as a capacitor. Note that the feedback path is closed only by the mechanical link; there is no

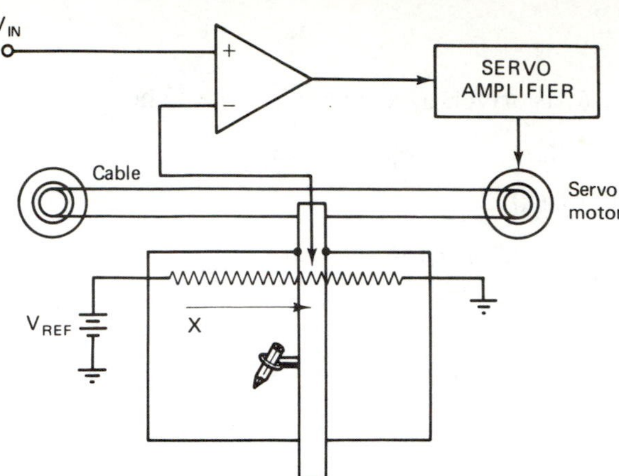

Figure 13.11 XY recorder servomechanism feedback circuit.

gain-adjusting feedback resistor. This is called *potentiometric* recording, after the null-seeking design of the classical potentiometer.

It is possible to reconnect the components of Fig. 13.11 to make a circuit in which the pen position is *inversely* proportional to the input voltage (see Prob. 13.7). This is useful in recording physical measurements in which the parameter varies as the *inverse* of an independent variable ($1/T$ in semiconductor measurements, for example) but the transducers available measure the independent variable directly (e.g., measuring temperature T). In another modification, the same components can be reconnected to measure the ratio of two voltages A/B (see Prob. 13.8). A typical application of such *ratiometric* recording is the measurement of optical absorption vs. wavelength. Since the ratio A/B is the fraction transmitted, direct ratiometric recording saves a troublesome normalization calculation. However, this mechanical system suffers from sluggishness at one end of the range and instability at the other, and ratiometric measurements in which the ratio is evaluated electronically (Chapter 15) are to be preferred.

Since op amps can not supply the current (0.1 A or greater) needed to drive a servomotor, power transistor current booster circuits are used. An example, shown in Fig. 13.12, uses a pnp–npn complementary symmetry pair acting as emitter followers, where only one transistor is *on* for either sign of op amp output. If sluggishness of servomotor response (due to a *deadband*, in which a small error signal is not sufficient to start the motor moving) is a problem, an op amp *limiter* circuit is used. The error signal is amplified with very high gain in the region of the motor's deadband, and then limited (using diodes; see Chapter 15) at a value providing reasonable motor speed (*slewing* speed) while approaching a null.

Another positioning problem occurs when it is desired to keep a solar collector pointed to maximize the energy input. Since the solar flux is not constant in magnitude, the previous method of comparing against a fixed reference signal will not work. One can sweep a transducer mounted with the collector and look for a maximum influx. This may be connected into an automatic negative feedback positioning circuit, either by time-differentiating the output signal (the error signal will change sign on either side of the peak influx), or by superimposing an oscillatory motion on top of the linear motion, using ac lock-in techniques (Chapter 18). However, a simpler concept, adequate in many cases, is to use two transducers

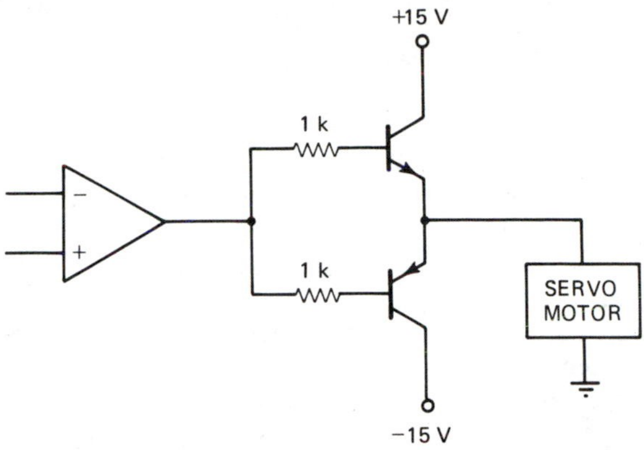

Figure 13.12 Servodriver amplifier (op amp current booster) circuit.

mounted so that the light falling on them is equal only when the collector is correctly positioned [Fig. 13.13(a)]. The signal generated when the collector is off the optimum position generates an error signal of the correct sign to drive the collector towards the peak. Examples using two photoresistors and a subtractor circuit, or two photovoltaic cells and an inverting amplifier, are shown in Fig. 13.13(b) and (c).

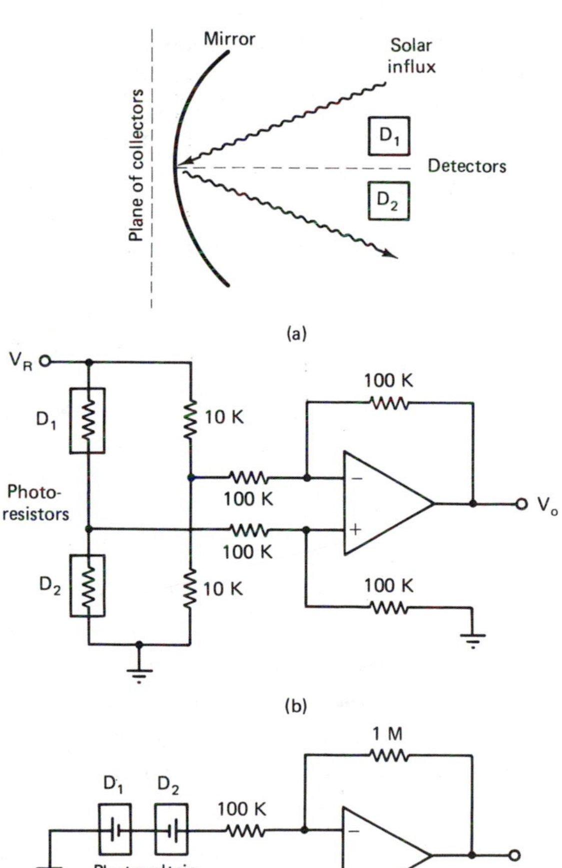

Figure 13.13 Solar collector positioning example. (a) Double detectors with equal output at solar influx peak angle. (b) Photoresistor bridge error signal generating circuit. (c) Photovoltaic cell error signal generating circuit.

PROBLEMS

13.1. Suppose the block B in Fig. 13.1 is not a unity gain emitter follower but a voltage amplifier (voltage gain B) to boost the op amp output voltage range. Calculate the output voltage as a function of circuit parameters as in Section 13.1. Show that the role of the open-loop gain is played by the product aB. What sets the output voltage under high-gain conditions?

13.2. Analyze the current source of Fig. 13.2 by the methods of Section 13.1. Assume an op amp gain a and current booster gain B. Find the open-loop gain and determine how I varies as a function of circuit parameters.

13.3. Show that the load current I_L in the scaled current source circuit of Fig. 13.3 is given by Eq. (13.7). By comparing I_i and I_f, calculate the voltage V_3. Show that to create this voltage, the transistors must drive the required current through the load R_L, regardless of the value of R_L.

13.4. Verify that the constant-current portion of the CVCC power supply of Fig. 13.5 may be redrawn as shown in Fig. 13.6(b). For simplicity, draw the Zener reference source as a battery. Show that the load is not in the feedback loop in this circuit, and that the load current is given by Eq. (13.8).

13.5. Analyze the voltage-controlled resistor of Fig. 13.8(c) and show that it gives a two-terminal resistance precisely linear in V_c.

13.6. Consider the light regulator circuit of Fig. 13.10(b) from a simplified point of view. See what the op amp needs to do to keep the summing-point voltage zero. Set up expressions for the voltages V_+ and V_-, equate the two, and solve for the resulting light intensity J as a function of circuit parameters. The gain aB should appear nowhere in your solution. Show that if the reference voltage V_Z is reduced, the resulting increase in output light intensity results in a change in V_+ of the correct sign to keep the two input signals balanced.

13.7. How can the components of Fig. 13.11 be reconnected to record the *inverse* of the input voltage, i.e., $X = 1/V_{\text{in}}$?

13.8. How can the components of Fig. 13.11 be reconnected to make a ratiometric recorder, whose position X is proportional to the ratio of two independent voltages, i.e., $X = A/B$?

REFERENCES

More complete bibliographic information for the books listed below appears in the annotated bibliography at the end of the book.

Hewlett-Packard, *DC Power Supply Handbook* (Hewlett-Packard Corp.)

HIGGINS, *Experiments with Integrated Circuits*, Experiment 19; see also Project Lab: *Light regulator*

HOENIG & PAYNE, *How to Build and Use Electronic Devices without Frustration, Pain, etc.*

JUNG, *IC Op Amp Cookbook*

National Semiconductor, *Voltage Regulators Handbook*

PHILBRICK, *Applications Manual*

14

Operational Notation and Linear Simulation

14.1 OPERATIONAL NOTATION

In the basic feedback configuration of Fig. 11.8, as long as Z_f and Z_i are resistors R_2 and R_1, the transfer function is simply the ratio R_2/R_1. But when the input or feedback impedance contains a reactive element, calculating the transfer function requires complex numbers for sine wave inputs, working in the frequency domain, or differential equations for more general inputs such as pulses, working in the time domain. Examples of these approaches have been given in Sections 11.5 and 12.5. Operational notation is a powerful method of avoiding complex algebra in the frequency domain or differential equations in the time domain. It is also not restricted to sine wave inputs or, indeed, inputs of any standard form. The method makes it possible to understand the stability of systems such as negative resistance devices (Fig. 14.1).

Operational notation consists in replacing a derivative with the algebraic operation *s*, and of replacing an integration with the algebraic operation $1/s$.

Operational Notation

$$d/dt \to s$$

$$\int dt \to 1/s \qquad (14.1)$$

This replacement allows manipulation of algebraic expressions rather than derivatives or integrals. Those familiar with operators in mathematics may

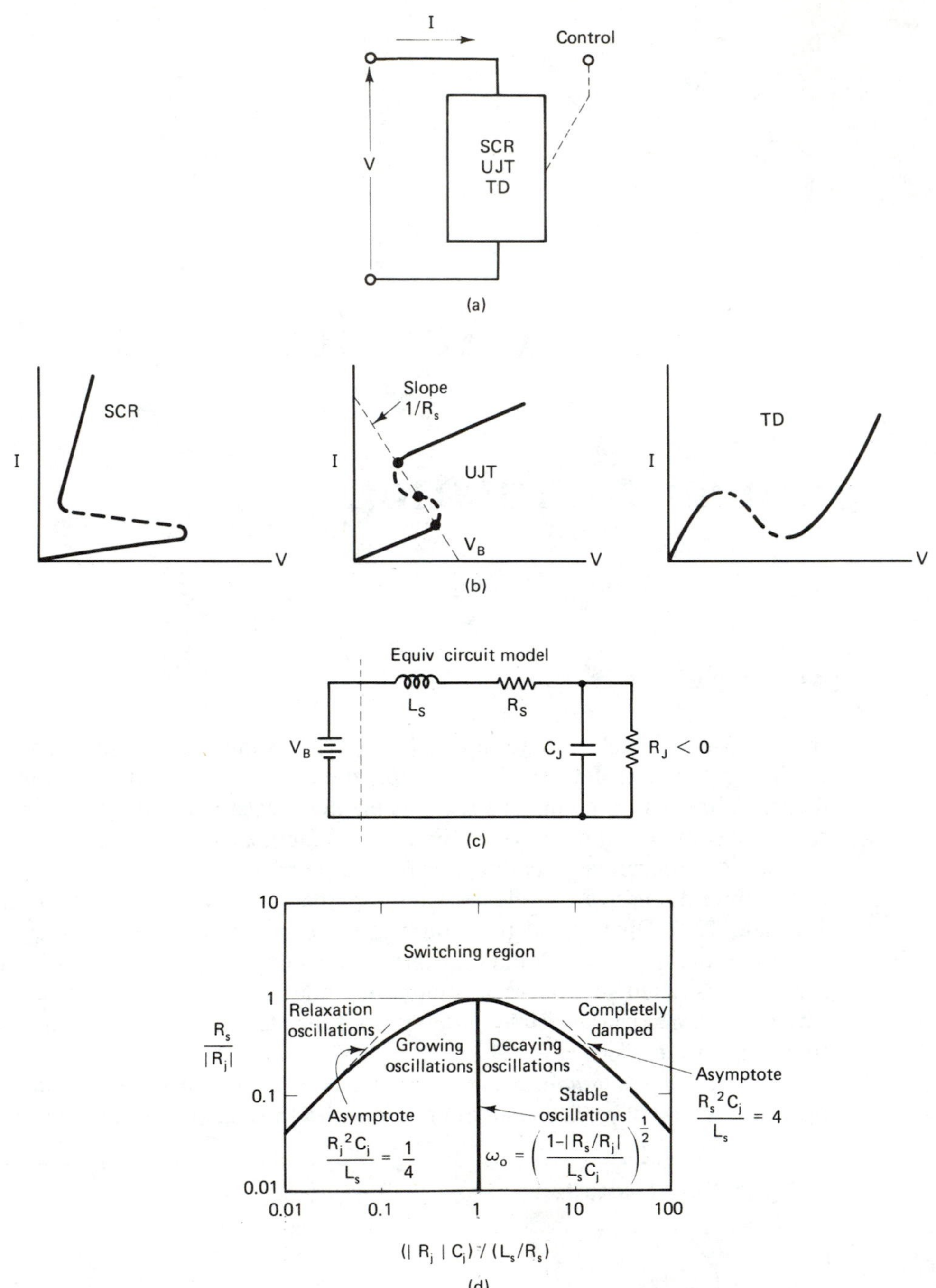

Figure 14.1 So-called negative resistance devices have an equivalent circuit model whose stability or instability can be studied using the methods of this chapter.

wonder if the meaning of differentiation and integration has been lost by this replacement. The method has a rigorous foundation, but the treatment in this chapter will be limited to a simple procedure for *using* the method.

14.2 APPLICATION EXAMPLE: INTEGRATOR

Consider the simple integrator (Fig. 11.4). In operational notation, the integrator's transfer function is:

$$V_o = -1/(RC)\int V_i\, dt = -V_i/(RCs) \qquad (14.2)$$

With a sine wave input, the integral relationship between output and input becomes algebraic [Section 11.5, Eq. (11.10)]:

$$V_o = -V_i/(RCj\omega) \qquad (14.3)$$

Comparing Eqs. (14.2) and (14.3), s may be identified with the quantity $j\omega$ when the input is a sine wave. However, operational or *s-plane* notation is much more general than this, and is not restricted to sine wave inputs. The proof involves Laplace transforms. The example of sine wave inputs allows a ready interpretation of the results in the familiar frequency domain.

To extend this approach to more general circuits, we work out an s-plane notation for the impedance of circuit components. An *impedance* is defined as the ratio of voltage to current. The integral and derivative I(V) relationship for capacitors and inductors leads to

$V_L = L\, dI/dt$	$Z_L = V(s)/I(s) = Ls$	[14.4(a)]
$V_c = (1/C)\int I\, dt$	$Z_C = V(s)/I(s) = 1/(Cs)$	[14.4(b)]

We now make an intuitive leap and hypothesize a generalized transfer function to specify the system response to any input. In Fig. 11.8, all feedback components are combined in box Z_f, and input components are in box Z_i. It is assumed for now that each of these boxes has only two terminals and contain only linear circuit elements. As long as the boxes contain only R, L, and C elements, the generalized transfer function is:

$$V_o/V_i = -Z_f(s)/Z_i(s) \qquad (14.5)$$

In this chapter, Eq. (14.5) will be assumed true, and its consequences will be explored. Several applications of operational notation will now be given.

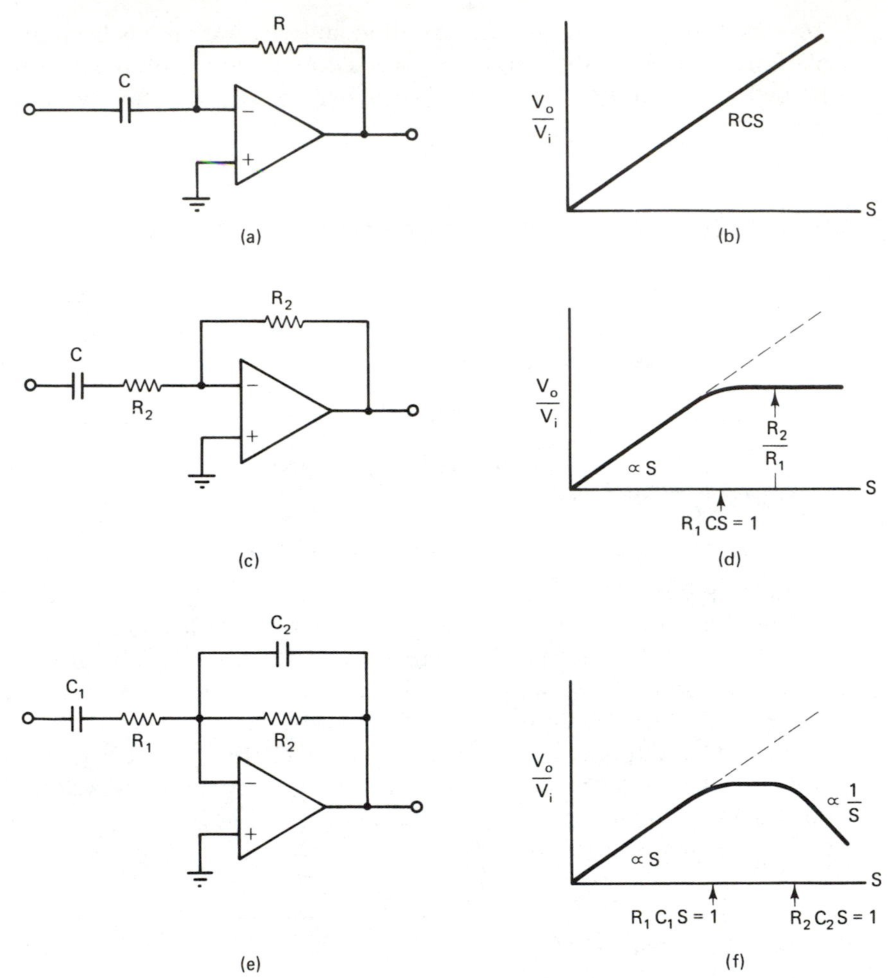

Figure 14.2 Differentiators and their transfer functions. (a) Ideal. (b) Modified with cutoff. (c) Modified with stronger cutoff.

14.3 EXAMPLE OF OPERATIONAL NOTATION: DIFFERENTIATOR WITH A CUTOFF

The ideal differentiator [Fig. 14.2(a)] has a disadvantage: the gain increases linearly with frequency. As a result, this circuit is susceptible to noise and pickup of radio frequencies. In practice, one always picks an upper limit frequency for a differentiator, adding other components to produce a cutoff above that frequency.

Consider the circuit shown in Fig. 14.2(b), whose transfer function may be readily analyzed using operational notation. The transfer function is (Prob. 14.1)

$$-Z_f/Z_i = -R_2/[R_1 + (sC)^{-1}] = -R_2C_1s/(1 + R_1C_1s) \qquad (14.6)$$

A plot of the transfer function shows that the desired cutoff occurs when $|R_1Cs| > 1$. Other components may be added to make the rate of cutoff even sharper [Fig. 14.2(c)]. The transfer function here is (see Prob. 14.1):

$$-Z_f/Z_i = -R_2C_1s/[(1 + R_2C_2s)(1 + R_1C_1s)] \tag{14.7}$$

The result has the linear response of a differentiator for small s, but flattens out when the first time constant of extra components reaches $|RCs| = 1$. At large s, the transfer function falls off as $1/s$. The precise shape of the curve may be adjusted by manipulating the independent time constants R_1C_1 and R_2C_2.

14.4 THE FREQUENCY DOMAIN AND COMPLEX ALGEBRA

Fig. 14.2 demonstrates how operational notation is useful in quickly plotting transfer functions of complicated networks. Most readers will (correctly) feel uneasy, knowing that complex phase information lies hidden. A rigorous result for the transfer function in the frequency domain is obtained by the replacement $s = j\omega$, and the magnitude and phase of the result may be calculated for a sine wave input. The result for the differentiator with a single cutoff is

$$|Z_f/Z_i| = RC\omega/[1 + (\omega RC)^2]^{1/2}$$

$$\tan\phi = -1/(\omega R_1C) \tag{14.8}$$

Qualitatively, the result $|Z_f/Z_i|$ is similar to the curve in Fig. 14.2(d), even though it corresponds to a somewhat different equation [Eq. (14.6) vs. Eq. (14.8)]. In nearly all situations, a plot of the transfer function as a simple algebraic function of s is adequate to determine the asymptotic behavior at low and high frequencies and to explore cutoff frequencies. At the very least, s-plane notation provides a way of avoiding much complex algebra until the very end of the calculation. This is its principal benefit in the case of sine wave inputs, where the more standard impedance method also works reasonably well. The advantage of operational notation is somewhat greater in the time domain, as demonstrated in the following example.

14.5 APPLICATION OF OPERATIONAL NOTATION: INTEGRATOR DRIFT

The analysis of nonideal integrator operation due to input bias current and voltage offset can be carried out using the simple equivalent circuit model of Chapter 12, but now without differential equations, by applying operational notation.

Consider the simplified equivalent circuit of a real op amp with general feedback networks Z_f and Z_i shown in Fig. 14.3. The input voltage at the plus terminal is V_{os}. With a closed feedback loop, the voltage at the minus input terminal will be made to equal this value. As a result, current conservation at the summing point node takes the form of

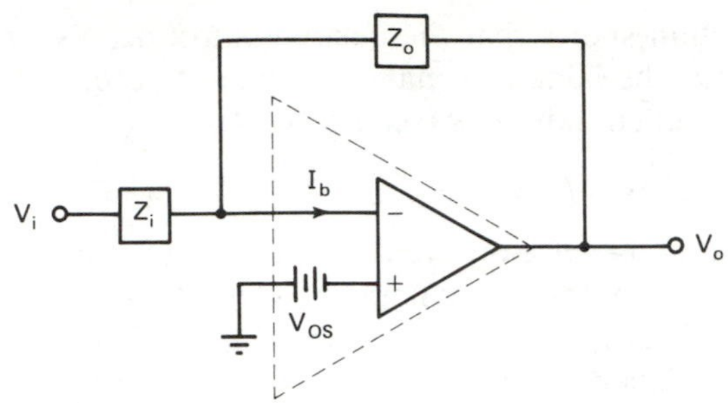

Figure 14.3 Generalized "golden rule" block diagram for a real op amp.

$$-V_{os}/Z_i = (V_{os} - V_o)/Z_f + I_b$$

Solving for the output V_o leads to

$$V_o = [(Z_i + Z_f)/Z_i]\,V_{os} + Z_f I_b \qquad (14.9)$$

This general result may be used to calculate the effects of nonideal op amp behavior with any feedback network which can be drawn in the form shown in Fig. 14.3.

Consider the integrator (Fig. 11.4). Since $Z_i = R$ and $Z_f = 1/sC$, the output voltage is of the form

$$V_o = \{[R + 1/(sC)]/R\}\,V_{os} + I_b/(sC) \qquad (14.10)$$

The resulting behavior is best seen graphically (Fig. 14.4). Since the interest is in the dc drift behavior of the circuit, attention may be confined to the low-frequency end ($RCs << 1$), where both terms in Eq. (14.10) vary inversely with s. The significance may be seen by simplifying Eq. (14.10):

$$sV_o = V_{os}/(RC) + I_b/C \qquad [s << 1/(RC)] \qquad (14.11)$$

Since s corresponds to a time derivative, Eq. (14.11) indicates that the *rate of change* of output contains two terms, the first proportional to V_{os} and varying inversely with RC and the second proportional to I_b and inversely proportional to C. Therefore, the integrator output will drift linearly in time. This is identical to the previous result [Eqs. (12.22) and (12.24)] which was derived using differential equations. The analysis has been simplified by using the s-plane or operational notation.

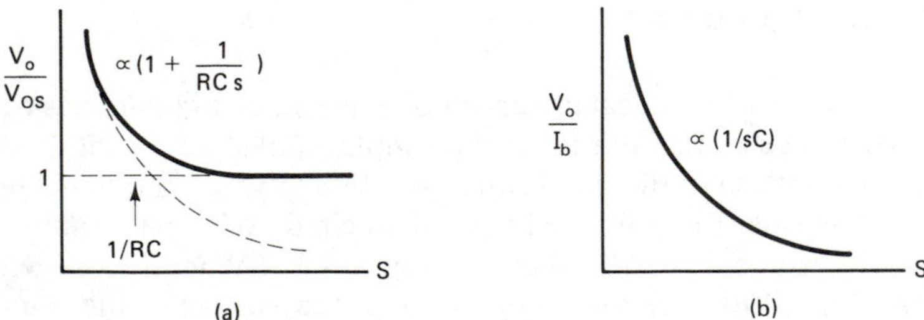

Figure 14.4 Contributions to the output voltage drift of an integrator.

Note the procedure: the problem is solved algebraically with operational notation, and then translated back at the end to the time domain. In this simple example, the final step was done intuitively. In more complicated problems, the Laplace transform is necessary, although in many situations, inspection of the transfer function in the s-plane is sufficient.

In summary, operational notation is a way to solve many op amp transfer function problems algebraically, either changing back at the end to the time domain (having avoided differential equations) or changing back to the frequency domain (having avoided most complex algebra).

Exercise: Optimum Bias Current Balancing for an Integrator using Operational Notation

How should R_b be chosen to minimize the output voltage due to input bias current for an integrator? Does there exist a value of R_b which will *zero* the dc output voltage? **Suggestion**: Use operational notation.

Solution Follow the derivation for an inverting amplifier with additional resistor R_b on the plus terminal [Fig. 12.17(b)], with the substitutions appropriate for an integrator: $Z_f = 1/(Cs)$ and $Z_i = R$. As in the inverting amplifier case (Eq. 12.18), the output voltage is

$$V_o[1 - aZ'/Z_f] = a[-V_{os} + I_{b+}R_b - Z'I_b] \qquad (14.12)$$

where

$$1/Z' = 1/Z_f + 1/Z_i$$

$$Z' = R/(1 + RCs) \qquad (14.13)$$

If V_{os} has been nulled, an output voltage null is achieved only if

$$I_{b+}R_b = Z'I_{b-} \qquad (14.14)$$

And in the event that $I_{b+} = I_{b-}$ (offset current small)

$$R_b = Z' \qquad (14.15)$$

Since Z' [Eq. (14.13)] involves a $1/(1 + s)$ term, there is no purely resistive value of R_b which will balance the drift due to offset current.

14.6 OTHER EXAMPLES OF OPERATIONAL NOTATION

The operational method may be extended to generate many useful transfer functions which would be complicated to obtain by the method of complex impedance. A variety of useful networks is shown in Table 14.1. The *transfer impedance* for the circuits is defined as the ratio V/I, where I is the current flowing out of the right-hand side of the network when a voltage V is applied to the left with the right-hand side grounded. Both two- and three-terminal networks are used. Inductive circuits have been omitted; they are *rarely* used in op amp circuits, since their function can be synthesized actively (Chapter 16).

As an example, consider the double integrator shown in Fig. 14.5. Applica-

TABLE 14.1 TRANSFER IMPEDANCES IN *S*-PLANE NOTATION FOR A SELECTION OF PASSIVE NETWORKS. TRANSFER IMPEDANCE IS THE RATIO OF VOLTAGE APPLIED ACROSS THE TERMINALS TO CURRENT FLOW THROUGH THE TERMINALS

Network	Transfer impedance	Special relations
Passive networks I		
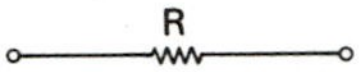 R	R	
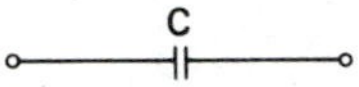 C	C	
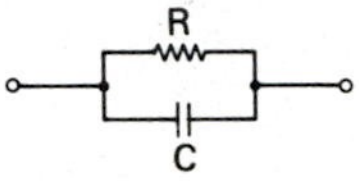 R, C	$\frac{R}{1 + sT}$	$T = RC$
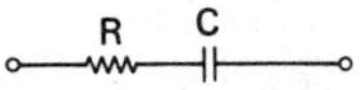 R C	$\frac{1}{sC} \quad (1 + sT)$	$T = RC$
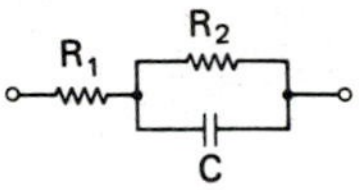 R$_1$, R$_2$, C	$(R_1 + R_2) \quad \frac{(1 + sR_eT)}{(1 + sT)}$ $R_e < 1$	$T = R_2C$ $R_e = \frac{R_1}{R_1 + R_2}$
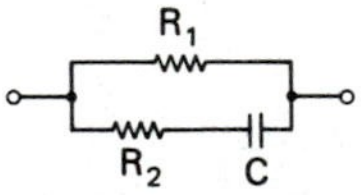 R$_1$, R$_2$ C	$R_1 \quad \frac{(1+sR_eT)}{(1 + sT)}$ $R_e < 1$	$T = (R_1 + R_2)C;$ $R_e = \frac{R_2}{R_1 + R_2}$
Passive networks II		
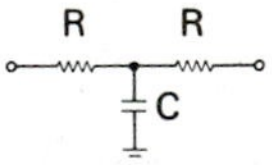 R R, C	$2R\,(1 + sT)$	$T = \frac{RC}{2}$
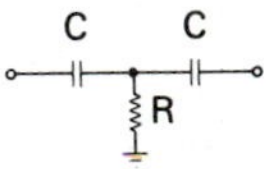 C C, R	$\frac{2}{sC} \quad \frac{(1 + sT)}{(sT)}$	$T = 2RC$
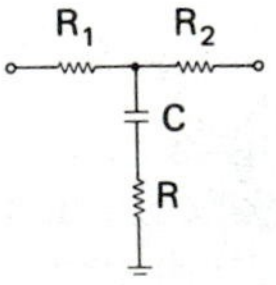 R$_1$ R$_2$, C, R	$2R_1 \quad \frac{(1 + sT)}{(1 + sR_eT)}$ $R_e < 1$	$T = (R_2 + \frac{R_1}{2})C$ $R_e = \frac{2R_2}{2R_2 + R_1}$
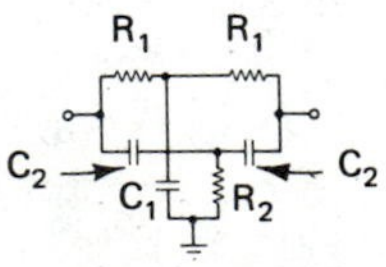 R$_1$ R$_1$, C$_2$, C$_2$, C$_1$, R$_2$	$2R_1 \quad \frac{(1 + sT_1)}{(1 + s^2T_1T_2)}$	$T_1 = \frac{R_1C_1}{2}$ $T_2 = R_1C_2$ $R_1C_1 = 4R_2C_2$

TABLE 14.1 CONTINUED

Network	Transfer impedance	Special relations
Passive networks III		
	$\dfrac{C_1 + 2C_2}{sC_1C_2}\left[\dfrac{(1+sT_1)(1+sT_2)}{s^2T_1T_2}\right]$ $T_1 < T_2$	$T_1 = RC_1$ $T_2 = R(C_1 + 2C_2)$
	$(2R_1 + R_2)(1 + sT_1)(1 + sT_2)$ $T_1 < T_2$	$T_1 = \dfrac{R_1R_2C}{(2R_1 + R_2)}$ $T_2 = R_1C$
	$(2R_1 + \dfrac{R_1^2}{R_2})\ \dfrac{1 + sT_2}{(1 + sT_1)(1 + sT_3)}$ $T_2 \leqslant T_1 \leqslant T_3$	$T_1 = R_1C_1$ $T_2 = \dfrac{R_1R_2}{R_1 + 2R_2}\ (C_1 + C_2)$ $T_3 = R_1C_2$

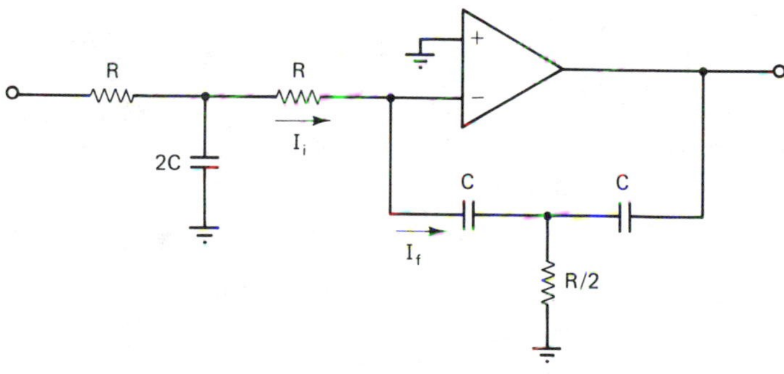

Figure 14.5 Double integrator with only one op amp.

tion of the "golden rule" $V_o/V_i = -Z_f/Z_i$ and the transfer impedances given in Table 14.1 yields

$$V_o/V_i = G(s) = -1/(RCs)^2$$

for the indicated component values. Since s^{-1} corresponds to integration, the s^{-2} term must correspond to a double integration, performing the same operation as two op amp integrators in series. Many such transfer functions can be generated using Table 14.1 as a look-up table, but it is best to have a clear understanding of the actual principles involved. The double integrator is as complicated an example as will be encountered, and is analyzed using fundamental principles in Prob. 14.3.

14.7 MATHEMATICAL AND ELECTRONIC ANALOGS

Earlier chapters have shown that op amp circuits such as integrators and differentiators can simulate with voltages $V(t)$ the mathematical operations of addition, subtraction, integration, and differentiation. The dynamics of physical, chemical, and biological systems are governed by mathematical relationships. For example, Newton's second law, $F = m(d^2x/dt^2)$, has an electronic analog representation in which F and x are voltages related by differentiation or integration. The mathematical response of physical systems to an external force is given by the solution of the differential equation describing its dynamical laws of motion. These may be Newton's laws for mechanical systems, the laws of electricity and magnetism for an electronic component, or the laws of fluid mechanics for wave motion in liquids. This situation is summarized in Table 14.2, and an example of two analogous systems is shown in Fig. 14.6. The common mathematical behavior of systems makes it possible to simulate dynamics with electronic analogs, using op amp circuits. Physical variables and parameters (such as mass, restoring force, viscosity) can easily be changed by altering the corresponding electrical circuit elements (resistors, capacitors). As a result, a simulation experiment can proceed much faster than the corresponding physical experiment. Although this approach

TABLE 14.2 ELECTRONIC ANALOGS ORIGINATE THROUGH A COMMON MATHEMATICAL PICTURE OF SYSTEM DYNAMICS IN WHICH ANALOGOUS PHYSICAL PARAMETERS APPEAR

Parameter	Mechanical	Thermal	Electrical
Applied force	Force F	Temperature gradient dT/dx	Voltage V
Displacement	Displacement x		Charge Q
Velocity	$dx/dt = v$	Heat flow Q	Current $I = dQ/dt$
Acceleration	$d^2x/dt^2 = a$		$dI/dt = d^2Q/dt^2$
Inertia	Mass m		Inductance L
Elasticity	Compliance $1/K$	$\left(\text{Heat capacity } C\right)^{-1}$	$\left(\text{Capacitance } C\right)^{-1}$
Frictional dissipation constant	Friction constant b	Thermal resistance $1/K$	Electrical resistance R
Momentum	$m\ dx/dt$		LI
Kinetic energy	$(m/2)(dx/dt)^2$		$LI^2/2$
Potential energy	$Kx^2/2$		$CV^2/2$
Power	Fv		VI
Dissipation force	$F = bv$	$dT/dx = Q/K$	$V = IR$
Elastic force	$F = Kx$	$\Delta T = (1/C)\int Qdt$	$V = 1/C\int Idt$
Kinetic force	$F = m(dv/dt)$		$V = L(dI/dt)$

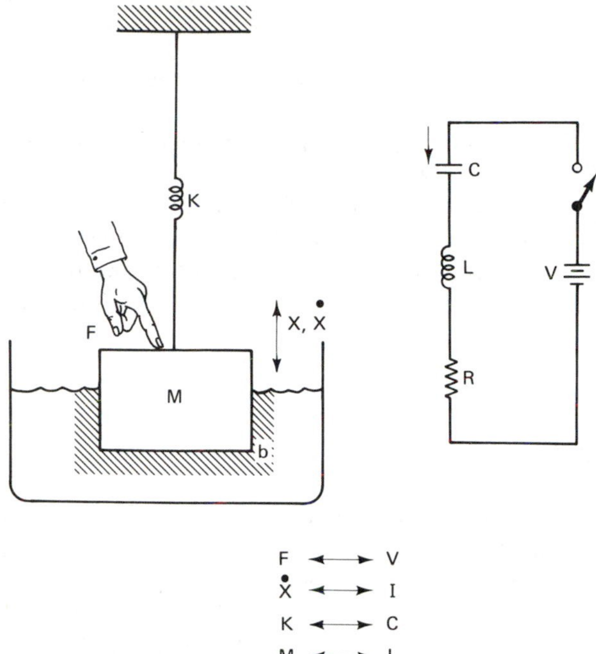

Figure 14.6 Example of a physical system: vibrating spring and mass, with damping. The analogous series RLC electrical system is also shown.

uses a *model* of the physical world, it can include all the complexity of the best mathematical model available. It can provide solutions which are very difficult to obtain by other means. For example, the mathematical or digital computer solution of *nonlinear* systems is formidable. However, nonlinear systems are readily set up in analog simulation, using accurate analog multipliers. This approach is not limited to classical systems; quantum mechanical problems lend themselves equally well to analog simulation. Variation of the parameters demonstrates in an intuitive way why only certain energies are eigenvalues of the problem (i.e., Why do quantum numbers exist?).

Although analog computation has largely been overwhelmed by digital computers with their high capability at low cost, analog IC's of high quality and low cost are now available. A multiplier which formerly cost $1000 now costs less than $10. An op amp good enough to build a stable integrator once cost $100 but now is under $1. As a result, a simulation problem which formerly required an expensive general-purpose analog computer now can be set up with a handful of parts costing less than $100. A resurgence of analog simulation is overdue.

14.8 ANALOG SOLUTION OF THE DAMPED HARMONIC OSCILLATOR

14.8.1 Mathematical Representation

It will be useful to introduce the subject by means of a specific example, the vibrating spring and mass system, called in physics the *damped harmonic oscillator*. Physicists often find it useful to view the real world by looking for mechanical, thermal, hydraulic, or electrical analogies with this second-order system (Table 14.2).

The system, shown in Fig. 14.6, has a mass m, suspended by a spring with spring-constant k, suspended in a fluid whose damping coefficient (force/ velocity) has the value b. The system is excited by a force $F(t)$ from the outside world. The response is described by Newton's law, $F_{tot} = \Sigma F_i = ma$, where the summation includes the external force, the spring restoring force, and the viscous damping force. This leads to a second-order linear differential equation with constant coefficients:

$$m(d^2x/dt^2) = \sum F_i = -b(dx/dt) - kx + F(t) \qquad (14.16)$$

14.8.2 Setting Up the Op Amp Simulation

In the analog simulation of this system, the position, velocity, and acceleration d^2x/dt^2 become voltages; the physical parameters (mass, spring constant, damping) become combinations of resistors and capacitors. The differential equation is constructed using the Kirchoff voltage law around a closed loop of op amps.

To design the circuit, consider that if the various forces are fed to a summing amplifier [Fig. 14.7(a)], its output will be the acceleration $m(d^2x/dt^2)$, by Newton's law [Eq. (14.16)]. The two force terms proportional to velocity and to displacement can be obtained by integrating the acceleration twice in succession [Fig. 14.7(b)]. These may then be fed back (together with one inversion of sign) as inputs to the original summing amplifier. The complete circuit is shown in Fig. 14.8.

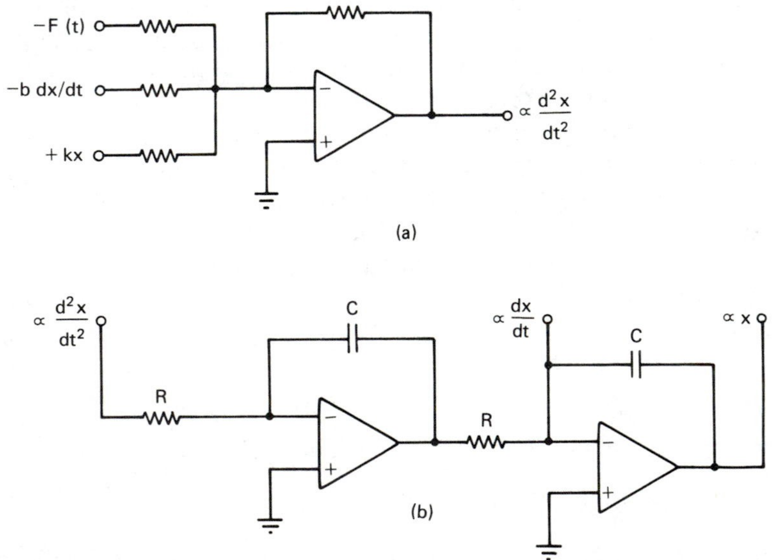

Figure 14.7 (a) The sum of the forces equals the product of mass m and acceleration a. (b) Obtaining dx/dt and x by integration.

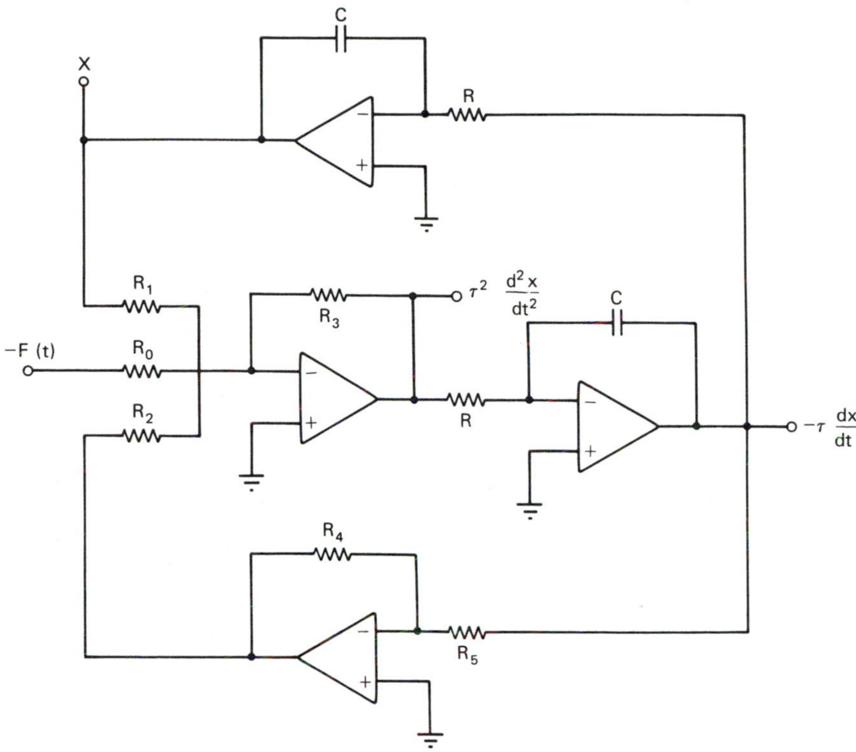

Figure 14.8 Final form for electronic analog simulation of the damped harmonic oscillator. Initial condition circuitry is not shown.

The system's behavior will depend upon the nature of the stimulus provided from the outside world by the force $F(t)$. If $F(t)$ is a step function, the system will relax to a new equilibrium position corresponding to additional stretching of the spring. If $F(t)$ is an impulse, the system will oscillate, as if kicked, for a period of time set by the damping constant b. If $F(t)$ is an oscillation from a sine wave oscillator, the system will show resonant response, with natural frequency and Q set by circuit parameters.

Initial conditions must be specified. Since this is a second-order differential equation, there are two initial conditions: the initial velocity $(dx(0)/dt)$, and the initial position $x(0)$. Since both x and dx/dt appear as voltages at the output of integrators, initial conditions are set by charging the integrator capacitors to the appropriate voltage values prior to running the simulation. An example of an initializing circuit is shown in Fig. 14.9. This is called a *three-mode integrator*, the modes being: *set* (put in the initial condition), *hold* (sit there and don't let it go), and *run* (apply the input to the integrator and let the system take off).

Examples of actual recorded data from this system are shown in Fig. 14.10. A step of force (square-wave generator), corresponding to suddenly pushing on the spring, causes oscillations which decay, and the voltage settles to a new value corresponding to a more stretched spring. If the system is given a kick (pulse), there is initially no change in position, but the velocity suddenly changes due to

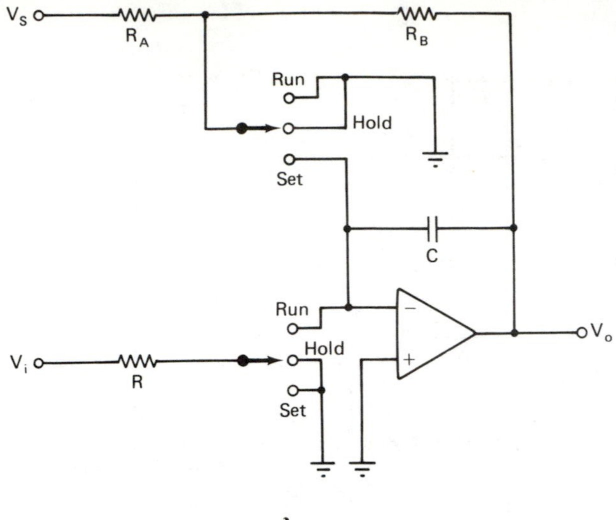

$$V_o = -\frac{1}{RC}\int_0^t V_i\,dt + V_{I.C.}$$

$$V_{I.C.} = -(R_B/R_A)\,V_S$$

Figure 14.9 Setting initial conditions using a three-mode integrator switch.

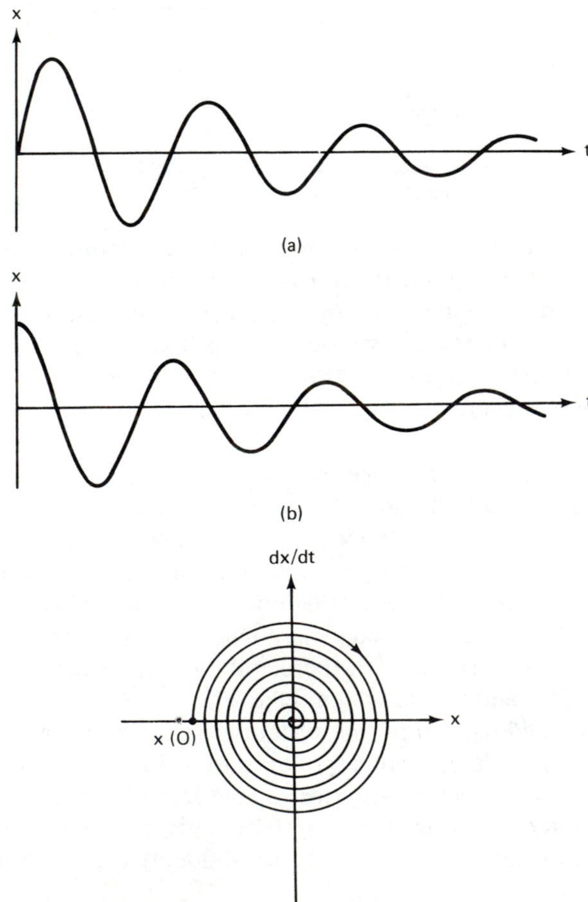

Figure 14.10 Examples of actual voltage-time recordings of the response of the system of Fig. 14.7 to a variety of initial disturbances. (a) Step response: $F(t)$ is a step. (b) Impulse response: $F(t)$ is a pulse. (c) A useful picture of system response formed by recording dx/dt as a function of x.

the impulse; both x and dx/dt eventually decay back to their original positions. A useful way of plotting these results together, called a *Lissajous figure*, is shown in Fig. 14.10(c). The decay of position and velocity form a spiral, due to the fact that x and dx/dt are 90° out of phase with one another. This general shape is characteristic of an oscillating linear system. Nonlinear responses result in shapes more complicated than a spiral, with distortion from harmonics in the response. The values of natural frequency and damping are set by resistor values

$$\omega_o = (1/\tau)(R_3/R_1)^{1/2}$$
$$b/m = (1/\tau)(R_3R_4/R_2R_5) \tag{14.17}$$

where $\tau = RC$. The proof is left for a problem (Prob. 14.5)

Question. What sets the time scale of the system? What determines the R and C values of the integrators?

Answer. For an electronic integrator,

$$V_o = -\int V_i(t)(dt/\tau) \tag{14.18}$$

The quantity $\tau = RC$ acts as a scale factor, speeding up or slowing down the simulation of the real system. Thus, events which in nature happen in picoseconds or in years may be simulated in the laboratory in a more convenient time scale merely by picking the time-scaling variable appropriately.

Question. How does integrator drift affect circuit operation?

Answer. Surprisingly, integrator drift is not a serious problem, even over long times, once the loop is closed. This is true even for relatively poor op amps. The output of a given integrator in response to the input bias current is automatically fed back around the closed loop, and in this case has the right sense to cancel integrator drift over long time periods. However, large bias currents can cause erratic initial response and will complicate the setting of initial conditions. A good rule of thumb is to choose op amps so that integrator drift *before* the feedback loop is closed is small over a time interval τ compared to the oscillation amplitude *after* the feedback loop is closed. For this reason, a 741 should be used only for simulations in the audio frequency range, and a better (FET) op amp should be used for simulations whose time constant is about 0.1 s or slower.

Question. Since an integrator transfer function contains a scale factor RC, what limits the integrator outputs from being so large as to go off scale, or so small as to be buried in the noise?

Answer. To see the problem, suppose that the feedback loop is opened and a sine wave signal of frequency ω_o is applied somewhere in the circuit. The output of the next integrator will be proportional to the time scale of the system.

$$\left|\int A\sin(\omega_o t)(dt/\tau)\right| = A/\omega_o\tau \tag{14.19}$$

If $\omega_0 \tau << 1$, the output will overload. The integrator climbs to the power supply voltage during ½ cycle of signal. In the opposite case, $\omega_0 \tau >> 1$, the integrator does not have time to charge much before the signal changes its polarity. The output will be much smaller than the input and may be smaller than the noise level of the system. Reliable behavior in analog simulation requires careful control of the magnitudes of voltages appearing at various points in the circuit. This is called *scaling*, and books on analog computation deal extensively with the subject. For our purposes it is sufficient to know that scaling problems can be avoided by matching the natural frequency integrator and time constant such that $\omega_0 \tau \simeq 1$.

14.9 STABILITY

Analog simulation has the advantage that one can readily vary the parameters characteristic of the system, and explore the response under a variety of circumstances more easily than in the real world. The simulation approach has been applied to diverse systems such as population dynamics, ecology, and sociology. Critics of this approach point out that a predicted catastrophe or instability may be only the result of using too simple a model. When used with care, however, analog simulation remains a powerful method of graphically understanding or predicting the instability of electrical, mechanical, chemical, physical, biological, and social systems.

The stability of the damped harmonic oscillator will be explored in detail. Since this is the prototype of a general second-order linear system, the behavior of many physical devices may be explored by analogy. It is therefore useful to explore all regions of behavior, even conditions that are not physical for the spring and mass, such as negative damping, since those conditions do occur in other systems.

14.9.1 Stability of the Damped Harmonic Oscillator

Consider the equation of motion for the spring and mass system, Eq. (14.16). For the moment, suppose there are no external forces, i.e., $F(t) = 0$. Solutions will be of the form

$$x = A\ e^{j\omega t} + B \tag{14.20}$$

A and B will be chosen to meet initial conditions. If this trial solution satisfies the differential equation, it must be *the* solution (uniqueness theorem). Substituting the trial solution into the differential equation results in

$$-\omega^2 x = -(b/m)j\omega x - (k/m)x \tag{14.21}$$

$x(t)$ cancels, leaving an algebraic equation for ω, which may be solved to give the natural frequencies ω_o of the system. In the most general situation, the solutions may be complex.

$$\omega_o = +j(b/2m) \pm (k/m - b^2/4m^2)^{1/2} \tag{14.22}$$

As a check, with no damping ($b = 0$) the result is $\omega_o = (k/m)^{1/2}$, the familiar result for the spring and mass system. With damping, however, ω has both a real and an imaginary part. Insertion of ω_o into the trial solution shows the damping behavior:

$$x = A\exp(-\omega_2 t)\exp(j\omega_1 t) \tag{14.23}$$

where

$$\omega_1 = (k/m - b^2/4m^2)^{1/2}$$

$$\omega_2 = b/2m$$

The imaginary part Im of ω_o leads to a negative exponential term $\exp(-\omega_2 t)$, which corresponds to the decaying amplitude of the damped harmonic oscillator. The real part (Re) of ω_o leads to the oscillatory term $\exp(j\omega_1 t)$. This solution exhibits the expected critical damping: oscillatory behavior vanishes when $Re(\omega_o) = 0$, or $k/m = (b/2m)^2$.

Stability Criterion. The parameters of this analog system can take on any values, even those which are unrealistic for a real spring and mass. It is clear that for a stable system, the amplitude of the solution must be damped or constant, since a positive real exponential will grow out of bounds. Under what conditions will this system be stable?

(1) To make the exponential term in Eq. (14.23) damped, it is necessary that $\omega_2 = b/m > 0$. This obvious result corresponds to friction or viscous damping.

(2) A positive real exponential can also result if ω_1 should become imaginary. As a result, stability requires that ω_1 be real, or $k/m > b^2/4m^2$. This condition corresponds to the spring opposing the displacement ($k/m > 0$), plus damping terms that are not large enough to overdamp the system.

The various limiting situations are mapped in Fig. 14.11, in a parameter space whose axes are $(b/2m)$ and (k/m). In the figure, only the top right quadrant is physically accessible for the spring and mass, with the regions A and B corresponding to damped and overdamped oscillations, respectively. A physical system with the other regions accessible can become either a negative resistance oscillator or a negative resistance switch (Fig. 14.1 and Prob. 14.10).

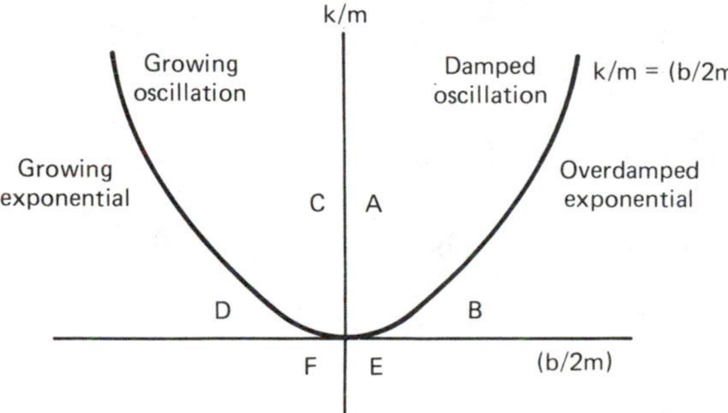

Figure 14.11 Physically distinct domains of behavior of the generalized second-order system, plotted as a function of the parameters $(k/m)^{1/2}$ and $b/2m$ of the damped spring and mass system.

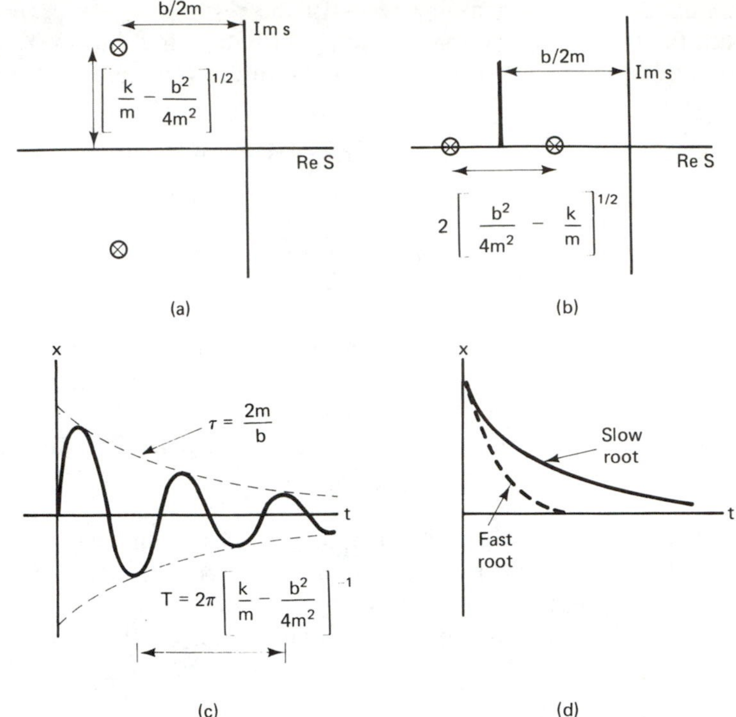

Figure 14.12 Complex s-plane view of the stability of a second-order system. (a) Complex zeros in the characteristic equation. (b) Corresponding ringing response in time to a step input. (c) Zeros both on the negative real axis. (d) Corresponding damped exponential response.

14.10 THE COMPLEX S-PLANE VIEW AND THE CHARACTERISTIC EQUATION OF THE SYSTEM

A very general way of viewing stability and instability of physical systems is in the complex s-plane. We make the identification $s = j\omega$, as in operational notation, in a trial solution such as Eq. (14.20). The solution of Eq. (14.21) then becomes

$$s_o = -b/2m \pm j(k/m - b^2/4m^2)^{1/2} \tag{14.24}$$

This is called the *characteristic equation* of the system. System stability depends upon where the roots lie in the *complex s-plane.* For example, consider region A in Fig. 14.11, where $b/2m > 0$ and $(k/m)^{1/2} > b/2m$. The roots of the characteristic equation are shown plotted in the s-plane in Fig. 14.12(a). The real part is negative, corresponding to a decaying exponential, and there is an imaginary part, which determines the oscillatory frequency. The corresponding time dependence is a damped oscillation [Fig. 14.12(b)]. The details of the behavior at $t = 0$ depend upon initial conditions. The roots corresponding to region B, where

TABLE 14.3 PHYSICALLY DISTINCT REGIONS OF PARAMETER SPACE FOR THE SECOND-ORDER ELASTIC SYSTEM[a]

Region	b/m	k/m	Other	Roots	Solution
A	>0	>0	$k/m > b^2/4m^2$	$-a \pm j\beta$	Damped oscillation
B	>0	>0	$k/m < b^2/4m^2$	$-a \pm \beta$	Damped exponent
C	<0	>0	$k/m > b^2/4m^2$	$a \pm j\beta$	Growing oscillation
D	<0	>0	$k/m < b^2/4m^2$		
E	>0	<0			
F	<0	<0			

[a]The missing portions are left for a problem

$b/2m > 0$ and $(k/m)^{1/2} < b/2m$, are shown in Fig. 14.12(c). The system is overdamped and now both roots are on the real axis. There are two solutions with different rates of decay, with time dependence shown in Fig. 14.12(d). The amplitude of the fast root s_2 quickly drops to zero, and most of the time dependence is determined by the slow root s_1. The *magnitude* of the response is also dominated by the root which is closest to the origin. All the other regions of behavior in Fig. 14.11 correspond to well-defined roots in the *s*-plane with well-defined time-dependent solutions. The relationship between circuit parameters and domains of behavior is summarized in Table 14.3.

A general stability criterion may be induced from this particular example. Table 14.3 shows that if a root of the characteristic equation has a positive real component, the solution will have an exponentially growing amplitude.

> **Stability Criterion:** For stability in the absence of an outside excitation, there can be no roots of the characteristic equation lying in the right half of the *s*-plane.

PROBLEMS

14.1. Analyze and plot the transfer function of the differentiator with a cutoff, Fig. 14.2(b) and (c). Show which component values may be adjusted to set the *two* independent cutoff frequencies in the transfer function of Fig. 14.2(c).

14.2. Show that for a nonideal differentiator to work properly, one should pick $R_f C_f << R_f C_i$ and $R_i C_i << R_f C_i$. R_f and C_i specify the time constant of the ideal differentiator, and R_i and C_f are the input resistor and feedback capacitor added [Fig. 14.2(d)] to limit the high frequency gain. **Hint**: the value of $R_f C_i$ is selected so the highest signal frequency does not lead to an output voltage overload for typical ($\sim$ 1 V) input voltage amplitudes.

14.3. Analyze the double integrator shown in Fig. 14.5 and show that the transfer function is given by

$$V_o/V_i = G(s) = -1/(RCs)^2$$

Method: Virtual ground and current conservation. Compute I_i as a consequence of V_i and I_f as a consequence of V_o, and equate. An equivalent parallel RC results in both the input and feedback networks due to the "virtual" ground.

14.4. The open-loop gain and phase shift as a function of frequency are shown in Fig. 14.13 for a hi-fi amplifier operating without feedback. Suppose that a negative feedback loop is added with feedback fraction $\beta = 0.01$ (frequency independent). Calculate and plot on the same scale as Fig. 14.13 the magnitude of closed-loop gain. Over what frequency range does the result display a gain flat within 3 dB? Do this two ways. (a) Ignore the phase shift (this approach is incorrect). (b) Include the phase shift. Some algebra is required for calculating numerical values from complex data.

$$|V_o/V_i| = |a/(1-\beta a)|$$

$$a = |a|\exp(j\phi)$$

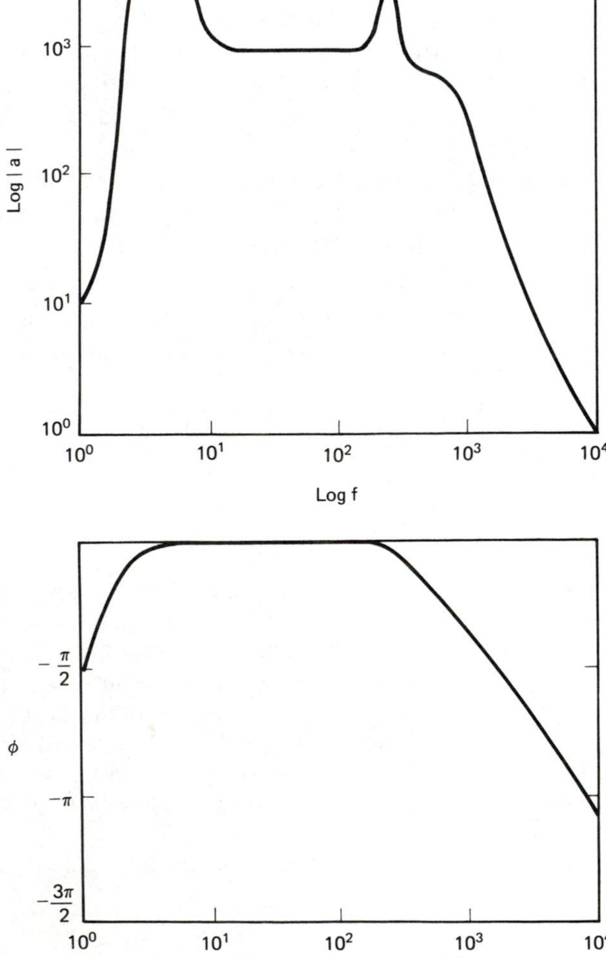

Figure 14.13 Gain and phase as a function of frequency for a hypothetical hi-fi amplifier without negative feedback.

14.5. Verify the relationship between system and circuit parameters [Eq. (14.17)] for the circuit shown in Fig. 14.8. (**Procedure**: Write the differential equation for the electronic circuit. Group the terms to resemble Eq. (14.16) and make the term-by-term equivalence.)

14.6. Complete the empty portions of Table 14.3. Sketch where the roots of the characteristic equation lie in the complex s-plane for these regions, and sketch the expected form of the time-dependent solutions. Regions with $k/m < 0$ are also unstable, since the condition for instability requires that one of the roots be negative imaginary.

14.7. Show that when the integrator of Fig. 14.9 is in the *Set* mode, the output voltage (and hence the initial condition) is

$$V_o = -(R_B/R_A)\,V_s$$

14.8. When the analog simulation of Fig. 14.8 is excited by an oscillatory force, it simulates a series-tuned RLC circuit with easily variable Q. Obtain a second-order differential equation for the current in the series-tuned RLC circuit, and, by comparison with Eqs. (14.16) and (14.17), relate $Q = \omega_o L/R$ to the damping coefficient b/m.

14.9. Referring to Table 14.2, why is it that the dynamics of the "potential element" in the mechanical system involves a simple algebraic relationship while the corresponding element in the thermal, hydraulic, and electrical analogs involves an integral relationship? **Hint**: No, analogous systems do not fail. Carry through the procedure by which the electrical potential element relationship was obtained, but substitute the corresponding entry in the mechanical system column.

14.10. Explore the instability of a negative resistance device using the equivalent circuit model [Fig. 14.1(c)]. R_J can be *negative* over a portion of the device's characteristic curve: $dR_J/dV < 0$. Write an expression for the impedance Z of the equivalent circuit. Find the natural frequencies ω where $Z = 0$; ω will be complex, and the sign of the real and imaginary parts determines the behavior shown in Fig. 14.1(d). Verify all of the critical parameter values for the transitions between types of behavior. **Warning**: This problem requires a good understanding of complex numbers.

14.11. (For the physicists) Set up the mathematics, then the analog computer solution for a falling body with friction proportional to velocity. Go into the lab and try it. What parameters determine the terminal velocity?

14.12. (For the chemists) Set up the mathematics, then the analog computer solution for a first-order chemical rate process. Go into the lab and try it.

REFERENCES

More complete bibliographic information for the books listed below appears in the annotated bibliography at the end of the book.

DAVIS, *Feedback and Control Systems*

HIGGINS, *Experiments with Integrated Circuits*, Experiment 20

ITT, *Reference Data for Radio Engineers*

JACKSON, *Analog Computation*

PHILBRICK, *Applications Manual*

ROBERGE, *Operational Amplifiers*, for more on stability

15

Nonlinear Analog Circuits

15.1 DIODES AS SWITCHES: PRECISION AC VOLTMETERS

In an ac voltmeter, diodes are used to rectify the ac signal, and the result is fed to a conventional dc meter movement. Since the diode turn-on voltage is about 0.5 V, this simple design does not allow measurement of ac voltages below a few volts. An op amp can provide enough gain so the rectifying diode may be fully conducting even for input signals smaller than a millivolt. The concept is illustrated in Fig. 15.1. For positive inputs, this circuit functions as a voltage follower, since a forward-biased diode is essentially a short circuit. The diode is connected on the output side of the op amp, whose large gain makes it possible for the diode to be fully turned on with a very small input voltage. For example, if $a = 10^5$, a voltage across the op amp input terminals of only $(0.5\text{ V})/10^5 = 5\ \mu\text{V}$ will suffice to turn on the diode. The usual golden rules force $V_- = V_i$ for $V_i > 0$, since this polarity forward-biases the diode. A measure of the ac voltage magnitude may be then obtained in either of two ways. The current in the feedback loop, V_i/R, may be measured by a conventional meter (Example 1 below). Alternatively, the voltage V_-, which will equal V_i during the positive half-cycles, may be filtered and measured by various means (Example 2 below). By itself, Fig. 15.1 is not a practical circuit, since the opposite polarity signal reverse-biases the diode, leaving the feedback loop open and the behavior unpredictable.

Question. Why doesn't the diode forward voltage drop introduce error into the measurement?

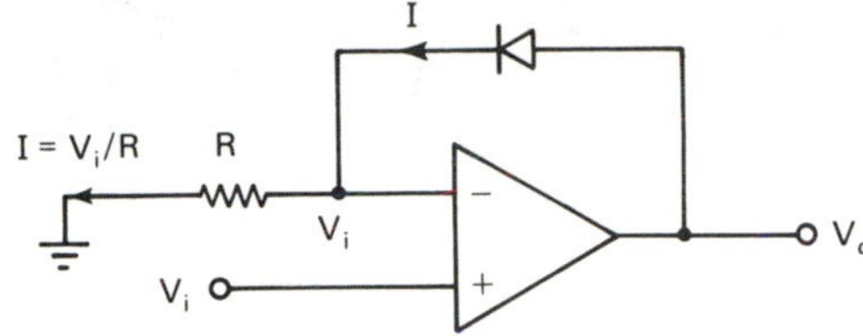

Figure 15.1 A diode in the feedback loop will rectify even very small input voltages, due to the large open loop gain of the op amp.

Answer. The diode forward drop appears only at the op amp *output* voltage. Since the op amp makes $V_- = V_i$, the output voltage will reach $V_i + V_d$. However, the input voltage is detected as feedback current or as voltage at the inverting input rather than as op amp output voltage, so the diode forward drop causes no error in the *measured* quantity.

Example 1: Precision ac Millivoltmeter. A full-wave rectifying meter circuit with millivolt sensitivity is shown in Fig. 15.2. The circuit provides a current path for input voltages of either sign, and functions by measuring the current in the feedback loop. A current path is provided through the feedback loop for both positive and negative input signals; the circuit functions as a follower in either case. The current $I_f = V_i/R_1$. The current is routed through the meter in the same direction for either sign of input voltage, even though the direction of the current in R_1 alternates with the sign of the input voltage. As in the example of Fig. 15.1, the diode forward drops are irrelevant; the op amp open-loop gain dominates the diode nonlinearity. Negative feedback keeps $V_s = V_1$, so the magnitude of the current is set by the input voltage and is independent of the diode drops. Capacitor C in parallel with meter resistance R functions as a filter. The *dc average* value of the half-wave rectified V_i appears across the meter, as in an RC filtered power supply. This provides a dc input for the moving coil-meter, independently of frequency.

This circuit is convenient only when it is acceptable to measure a current, and was designed for a conventional moving-coil meter. In situations (such as digitized measurements) where it is more desirable to detect a voltage rather than a current, the next circuit is preferable.

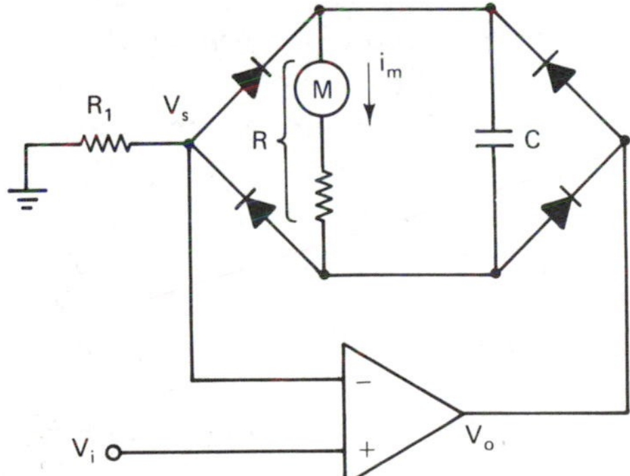

Figure 15.2 Precision AC millivoltmeter, used principally when the desired output is a conventional moving coil analog meter.

Example 2: Precision Rectifier with Variable Gain. A clever precision rectifier is shown in Fig. 15.3. The circuit functions as an absolute value circuit, presenting at its output a full-wave rectified version of the input signal. In addition, it has a variable and potentially very large gain. Circuit operation, which

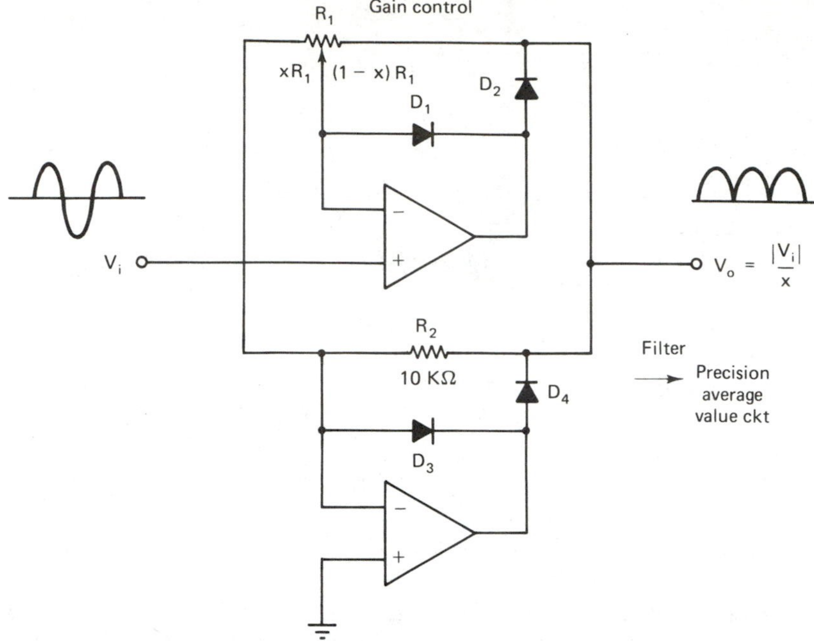

Figure 15.3 Improved precision AC millivoltmeter with variable gain.

involves voltage followers with diodes in the feedback loop, is most easily understood by following the current paths for positive and negative inputs separately. The circuit of Fig. 15.3 is redrawn in Fig. 15.4 separately for each input polarity, replacing On diodes with short circuits and Off diodes with open circuits. When V_i is positive, [Fig. 15.4(a)], amplifier A_1 functions as a follower with gain (see Section 4.7). The circuit output is V_i/x, where x represents the setting of the variable resistor $(0 < x < 1)$. The voltage at point P is held at 0, because A_2 is functioning as a voltage follower whose positive input is a ground. A_2 serves no further function for positive inputs.

For negative inputs, the equivalent circuit looks very different [Fig. 15.4(b)]. Here, A_1 functions as a simple follower, presenting a voltage V_i to the pot tap. A_2 is an inverting amplifier with gain. The input and feedback resistors for this amplifier are xR and R, so the gain is $-1/x$. As a result, the output signal during negative half-cycles is also equal to $|V_i/x|$. The circuit functions therefore as an absolute-value circuit, presenting an output voltage proportional to the instantaneous magnitude of V_i. The gain of the circuit is variable by a single pot, even though the signal flow paths are very different for positive and negative signals.

Question. What is the effect of diode forward drops on the output of this circuit?

Answer. The diodes are inside feedback loops, and as a result the diode drops become irrelevant. For example, a more accurate representation of amplifier A_1's operation for positive inputs is shown in Fig 15.4(c). This is a follower

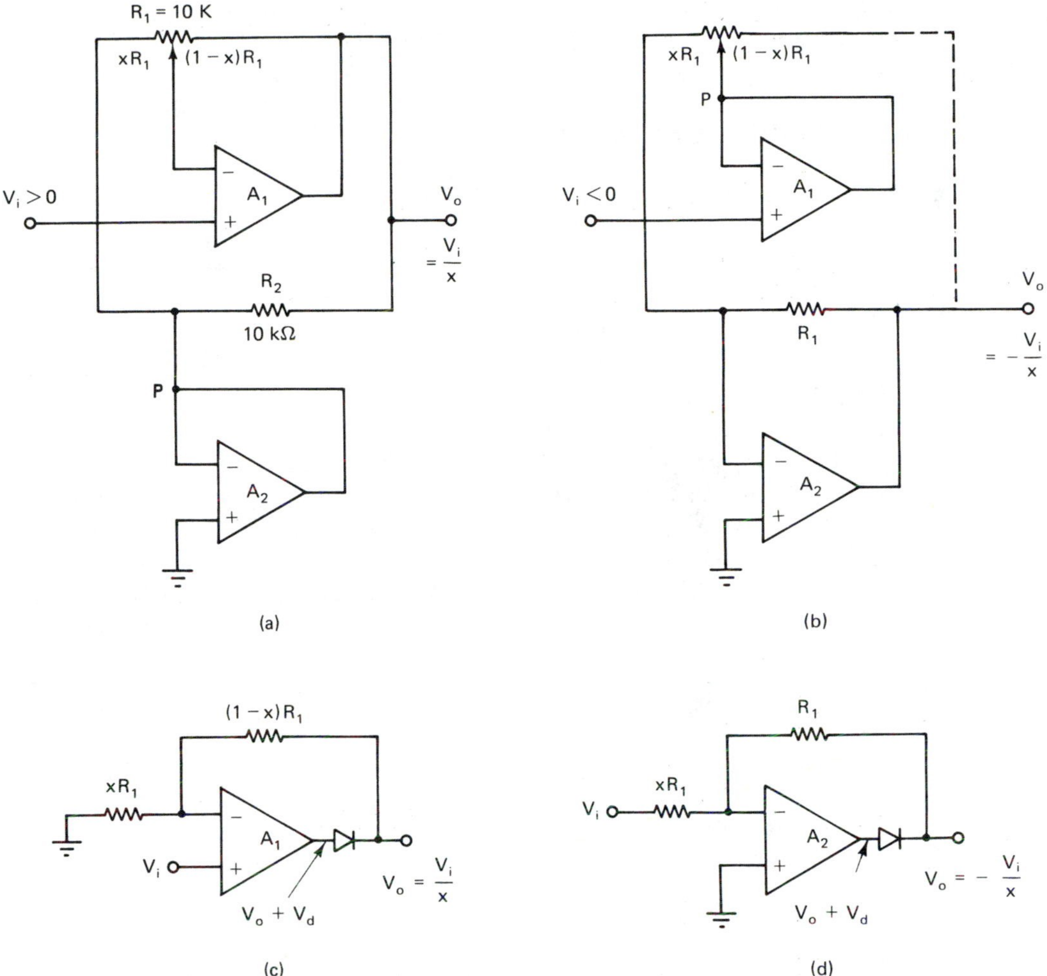

Figure 15.4 Current flow paths and equivalent circuits for the AC millivoltmeter of Figure 15.3. The diodes may be viewed as either open or short circuits, for either positive (a) or negative (b) input half-cycles. A more careful consideration of the diode drops for positive (c) and negative (d) half-cycles demonstrates that the diode drops are indeed irrelevant, since the diodes are inside the feedback loop.

with gain. The feedback loop forces $V_s = V_i$. But V_s is generated by V_o through the voltage divider xR and $(1 - x)R$. As a result,

$$V_o = \frac{(1 - x)R + xR}{xR} V_i = V_i/x \tag{15.1}$$

The diode is *inside* this feedback loop, so the output of the op amp will need to be larger than V_o by the diode forward drop V_d to provide the balance condition. But since the op amp output voltage is never measured or used, the diode forward drop is irrelevant. In a similar way, the forward drop of the feedback diodes of A_2

do not affect circuit operation, since only the virtual ground at V_- is being used. The analysis of negative input operation is left for a problem (Prob. 15.1).

15.2 DIODE FUNCTION GENERATORS

15.2.1 Limiter or Bound Circuit

It is often useful to produce nonlinear transfer functions. A simple example is a limiter which keeps the op amp output within its linear range (i.e., less than the power supply voltage) when the input signal would otherwise drive it out of range. This prevents "latching up" and resultant recovery delays. The standard limiter circuit (Fig. 15.5) uses back-to-back Zener diodes in the feedback loop in parallel with resistor R_2. The circuit is a normal amplifier until V_o exceeds either Zener diode threshold voltage. Above this, the Zener presents a low resistance, putting a *bound* to further increase in V_o. Other circuits discussed below introduce *breakpoints* or sudden changes in slope. Another common application of diodes is nonlinear function generation to convert a triangular wave input signal into a reasonable approximation of a sine wave. We reserve for Chapter 19 a discussion of a circuit which generates *backlash*, also called *hysteresis* or *deadband.*

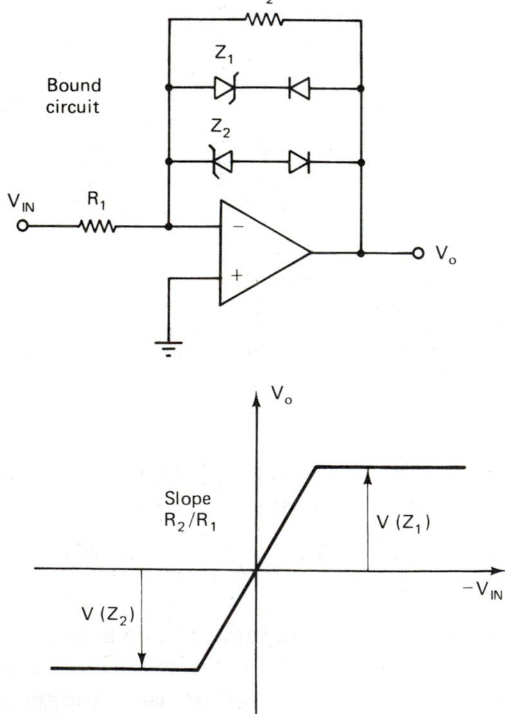

Figure 15.5 Bound or limiter circuit. The output voltage is clamped at the Zener diode breakdown voltages.

15.2.2 Diode Breakpoint Circuits

A summary of the four standard circuits providing concave upwards or concave downwards breakpoints is shown in Fig. 15.6. The analysis assumes that the diode is either an open circuit or short circuit, depending on its bias. The diode drop V_d may need to be taken into account in actual circuit application. The analysis of the circuit of Fig. 15.5(d) is given here for illustration. The circuit provides a concave downwards breakpoint with adjustable breakpoint and adjustable slope.

Below the breakpoint, the diode and the connection to the reference voltage V_R are irrelevant, so the circuit functions as a simple voltage divider with output slope $R_L/(R_S + R_L)$. The breakpoint occurs when the diode becomes forward-biased, i.e., whenever V_o begins to exceed V_R

$$V_i = V \text{ (breakpoint)} = V_R(R_S + R_L)/R_L \tag{15.2}$$

Above the breakpoint, the diode is short circuit. The output voltage is determined by the current in the load I_L, which is the sum of currents I_1 in R_S and I_2 in R_R. Each of these depends upon voltage differences across the respective resistors.

$$V_o = I_L R_L = (I_1 + I_2)R_L = (V_i - V_o)(R_L/R_S) + (V_R - V_o)(R_L/R_R)$$

which after some algebra becomes

$$V_o = \frac{V_i R_L R_R}{R_S R_R + R_L R_S + R_L R_R} + \frac{V_R R_L R_S}{R_S R_R + R_L R_S + R_L R_R} \tag{15.3}$$

This expression describes the slope and y-intercept of Fig. 15.5(d) above the breakpoint.

15.2.3 Sine Waves from Triangular Waves

A natural extension of diode breakpoint circuits is sine wave generation using a triangular wave and a string of diodes with their breakpoints at different voltage values [Fig. 15.7(a)]. The two variable resistors provide adjustment of both the slope and the breakpoint of the curve. The output of this circuit is similar to a triangle but with the tops rounded as shown. Connecting a string of such circuits, with provision for both negative and positive input signals, results in an approximation to a sine wave [Fig. 15.7(b)]. For any given value of input voltage, each diode is either on or off, and and the output may be computed by straightforward (albeit messy) passive circuit analysis. Such devices are available commercially to generate fixed or variable functions.

Although this class of circuit is adequate for routine applications, the resulting sine wave (actually a series of line segments rather than a continuous curve) has limited harmonic quality. Sine waves produced in this manner rarely have harmonic distortion below 2%. The *derivative* of this signal has very large harmonic distortion, because the derivative of a series of straight-line segments has steps. In applications requiring either low harmonic distortion or derivatives, more conventional sine wave oscillators are preferred.

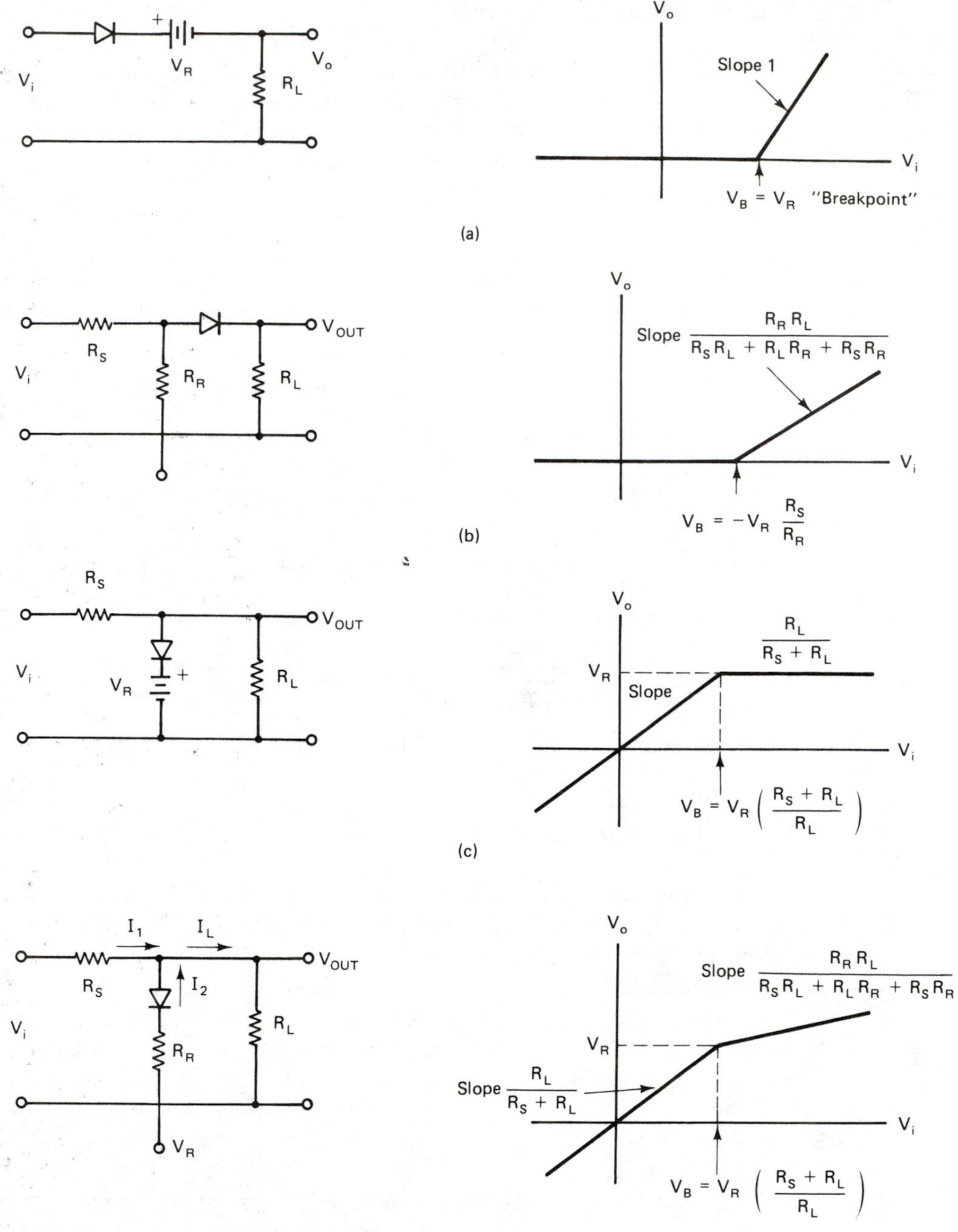

Figure 15.6 Diode breakpoint circuits for function generation. Concave upwards, with (a) unit slope, and (b) with independently adjustable breakpoint and slope above the breakpoint. Concave downwards, with (c) independently adjustable breakpoint and slope below the breakpoint, and (d) independently adjustable breakpoint and slopes above and below the breakpoint.

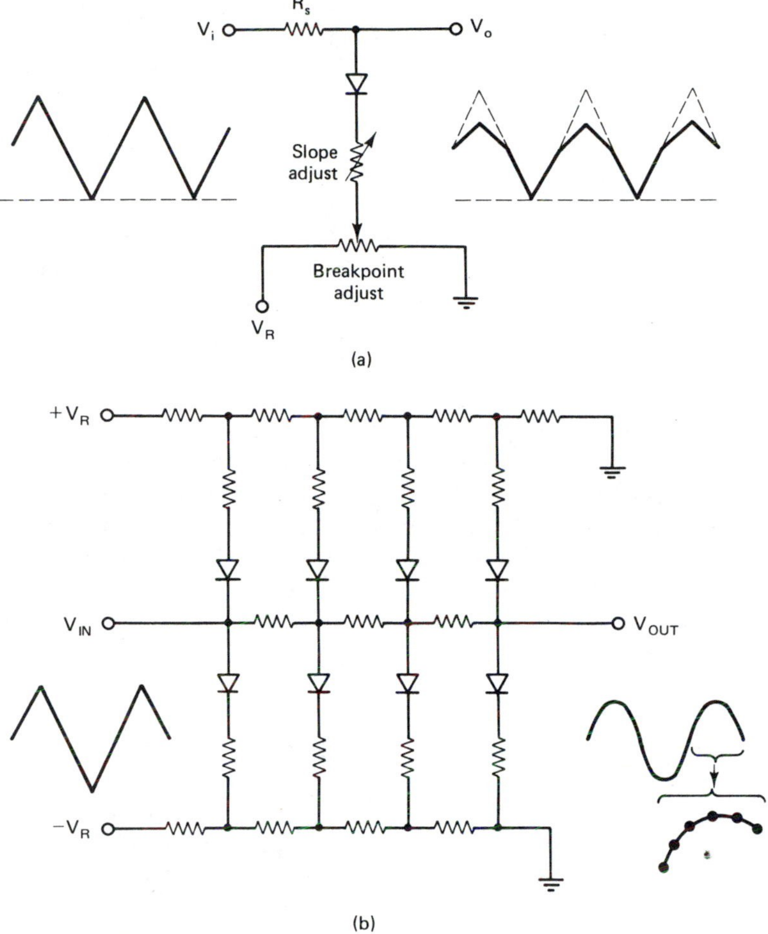

Figure 15.7 Diode function generator circuits. (a) Single breakpoint rounds the sharp peaks of a triangular wave. (b) A set of such breakpoints can be adjusted to transform a triangular wave into a good approximation to a sine wave.

15.3 LOGARITHMIC AMPLIFIERS

Many measurements produce a wide dynamic range of input signals.

(1) A vacuum system may require a range of at least 6 to 8 decades of pressure, and viewing the pump-down progress involves looking over that full range. A logarithmic amplifier can make a circuit with the required wide dynamic range, with an output which covers a certain *fraction* of its range for each *decade* of input pressure.

(2) In spectroscopic applications with light, x-rays, or nuclear radiation, it may be necessary to normalize the intensity received from the sample at a given

wavelength. The log amp provides a simple way to do this, by taking the difference of the logarithm of sample intensity J and the logarithm of incoming source intensity J_o, which is equivalent to the logarithm of the ratio J/J_o.

(3) In many systems, the physical or chemical variable is inherently an exponential function of some independent variable. For example, electrical conductivity of semiconductors and chemical rate processes depend exponentially on $1/T$. By passing the measured signal through a log amp, the exponential variation is made linear, and the experimental result (activating energy of a semiconductor or rate constant in a chemical reaction) is put in a form to be recorded graphically as a straight line, so the desired parameters can be read off with a ruler.

This section shows how a diode or transistor in the feedback loop of an op amp results in a logarithmic amplifier with a dynamic range of at least 6 decades. Nonideal behavior due to the high effective impedance of the feedback element occurs at low values of feedback current, since the op amp has certain input bias current demands. This problem may be largely eliminated by the use of an op amp with an FET input stage. A second problem is thermal drift, due largely to the inherent temperature dependence of the diode or transistor's characteristic curve. In demanding applications thermal drift can be largely eliminated by cancellation circuits.

15.3.1 Diode and Transistor Log Elements

The circuit shown in Fig. 15.8(a) has an approximate logarithmic response $V_o(V_i)$. Since the feedback element is nonlinear (does not follow Ohm's law), the transfer function is not simply $V_o/V_i = Z_f/Z_i$. On the other hand, given a

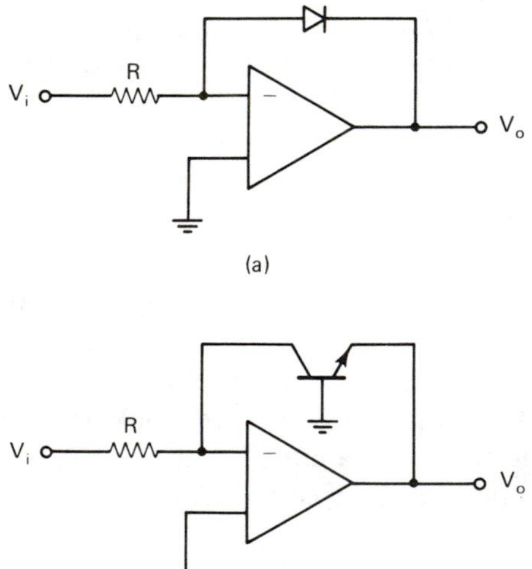

Figure 15.8 (a) A diode in the feedback loops of an inverting amplifier produces a logarithmic transfer function. (b) A more accurate log amp results if a transistor is used in the feedback loop.

chance, the op amp will still try to make the summing point a virtual ground. Log circuits demand an input signal of a single polarity ($V_i > 0$ in this example) so the diode is forward-biased and current can flow. The diode current (see characteristic curve, e.g., Fig. 1.13) is, to a very good approximation, exponentially related to the voltage across it.

$$I = I_0[\exp(eV/kT) - 1] \tag{15.4}$$

where

I_0 = reverse bias current in diode

e = charge of electron = 1.6×10^{-19} coulombs,

k = Boltzmann's constant = 1.4×10^{-23} joules/K,

T = absolute temperature (K)

$e/kT = 1/(25.6\ m\text{V})$ at 300 K

The voltage drop V across the diode terminals in this circuit is $V_- - V_o \simeq -V_o$. For a forward bias greater than about 100 mV, the 1 may be neglected. This current must be balanced against the input current I_i through the resistor, V_i/R:

$$V_i/R = I_0 \exp(-\alpha V_o) \tag{15.5}$$

or

$$V_o = -\alpha^{-1}\ln(V_i/I_0R) \tag{15.6}$$

where $\alpha = e/kT$. The output is quite accurately the logarithm of the input, as long as the input current is large compared to I_0. Since I_0 is $\sim 10^{-9}$ A, this condition is easily obtained.

15.3.2 Practical Log Amp Circuit

It is conventional to use a silicon npn transistor rather than a diode in the feedback loop, since its I–V curve is more accurately exponential than that of a typical diode. The transistor is in a grounded base configuration [see Fig. 15.8(b)]. Since the collector is tied to a virtual ground at the summing point, the collector and base are effectively shorted together, and the transistor acts as a diode. For forward-biased current flow, V_i must be positive, and V_{out} will be negative. The behavior is identical to that of Eq. (15.4). Typical characteristics for a circuit similar to Fig. 15.8(b) are shown in Fig. 15.9. A range of about 6 decades of input signal current is compressed into an output voltage range of 0.2 V to 0.8 V, the forward-biased turn-on range of the base emitter junction. The circuit takes an accurate logarithm over about 5 decades. Nonideal or nonlinear behavior comes in at the high-current end due to ohmic losses in the semiconductor material. The nonlinear curvature at the low-current end comes when the current approaches I_0, i.e., $\exp(eV/kT) \sim 1$ in Eq. (15.4). Errors at low current may also result from op amp input bias current.

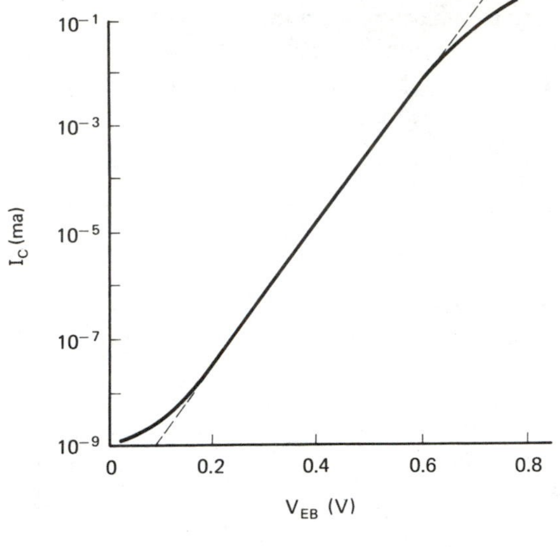

Figure 15.9 Measured transfer function for the log amp circuit of Fig. 15.8(b).

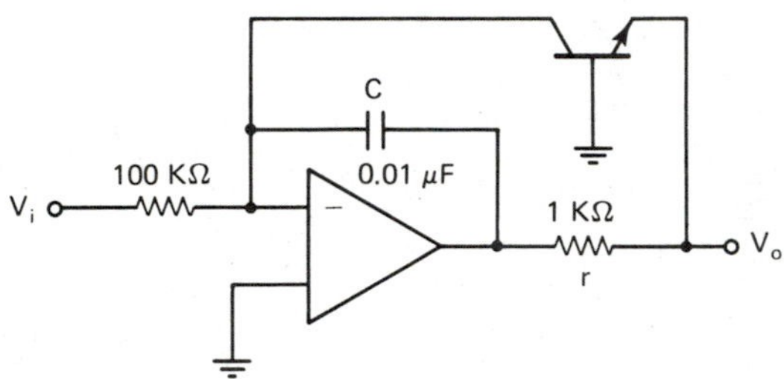

Figure 15.10 Practical-stabilized log amp circuit.

15.3.3 Correcting for Nonideal Behavior

A practical log circuit is shown in Fig. 15.10. It is essential to use a high-quality low bias current op amp, since the output voltage error due to input bias current is $V_{out} = I_b R_2$. In the log amp circuit, the effective resistance is huge at low current levels: $R_f \sim 0.1\ \mathrm{V}/10^{-9}\ \mathrm{A} = 10^8\ \Omega$. As a result, a bipolar op amp will not do, since bias currents of nanoamps will force the output to drift tremendously or latch up at the power supply voltage. This problem is easily removed by using an FET input op amp. For instance, with a CA3140, input bias current $I_b = 2 \times 10^{-12}$ amps causes a negligibly small offset, $V_{out} = (2 \times 10^{-12}\ \mathrm{A}) \times (10^8\ \Omega) = 2 \times 10^{-4}$ V.

A feedback capacitor is added to avoid instability, and to reduce high-frequency noise. Because of the junction capacitance in the transistor, the possibility of an additional 90° phase shift exists, and the instability points in the op amp may be shifted to a dangerous position. In addition, since the gain is huge at the low-current end, the circuit is sensitive to noise. The capacitor across the op amp, together with a small (~ 1 KΩ) resistor in series with the transistor emitter, limits the high-frequency gain. It may also be necessary to put a very large resistor (100 MΩ) across C to prevent long-term integrator-like drift.

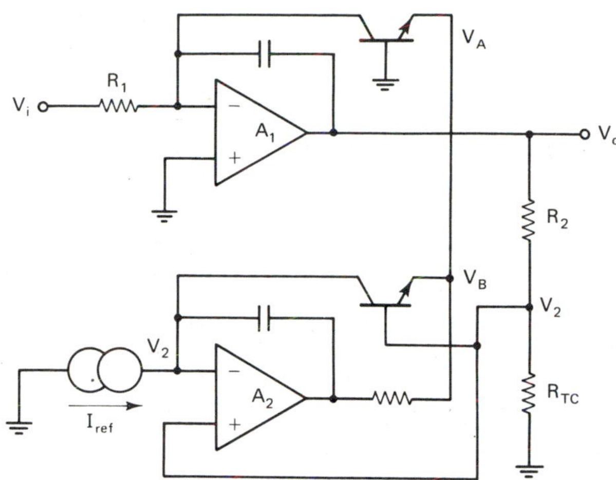

Figure 15.11 Temperature-compensated log amp circuit.

Because in Eq. (15.6) the factor α contains T, this circuit is inherently unstable with changes in temperature. Since it is the *absolute* temperature which appears in these expressions, fluctuations of 10° K amount to an error of only a third of a percent. In demanding applications where this would be a problem, cancellation circuits are used. An example is shown in Fig. 15.11 (see Prob. 15.2).

15.3.4 Log Amp Applications: Log Ratio Meter for Spectrophotometer

A spectrophotometer measures the transmission of light through a sample as a function of wavelength λ. The resulting spectrum has peaks and valleys indicating characteristic optical absorption, which is uniquely related to the chemical composition, molecular composition, or defect structure in the material. However, the input intensity J_o may be a function of wavelength, or may drift with time. Although one may compensate for this by a separate measurement and normalization, it is an unnecessary nuisance. Suppose one wishes to follow changes (for example, a chemical reaction) as a function of time; any drift or change in the input would result in an erroneous measurement. The situation may be improved by normalizing the results automatically (Fig. 15.12). The conventional transmission method is altered by measuring both the transmitted intensity J and a sample of the incident intensity J_o. These two measurements are then fed to a log amp circuit. By using two transistors in a single package, much of the thermal error is cancelled when the subtraction is taken. The output is

$$V_o = (e/kT)\ [\ln \mu\, J_o - \ln d - (\ln J_o - \ln d)] \tag{15.7}$$

where $\mu = J(\text{transmitted})/J_o$, and $d = I_{C0}R_1$. Terms related to I_{C0} cancel in this expression, which may be simplified.

$$V_o = (e/kT)\ \ln[\mu\, J_o(\lambda)/J_o(\lambda)] = (e/kT)\ \ln(\mu) \tag{15.8}$$

Since the magnitude of $J_o(\lambda)$ cancels, drift or wavelength dependence of source

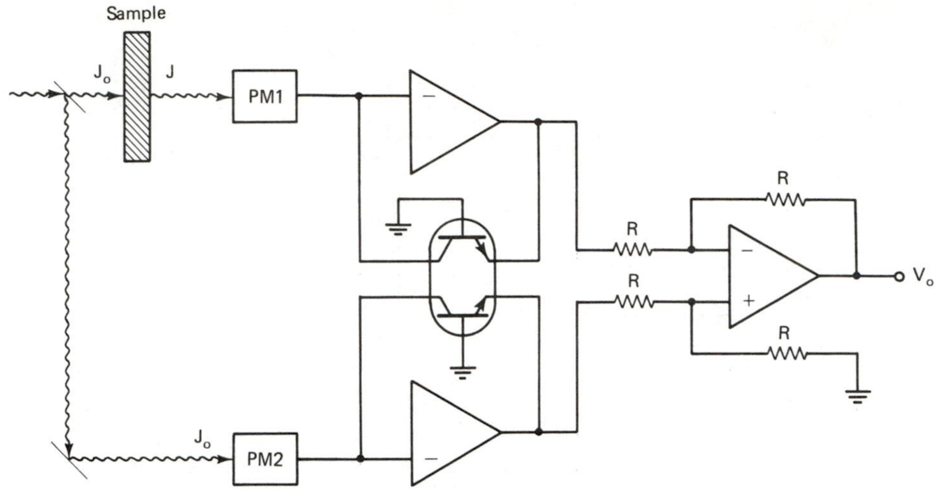

Figure 15.12 Log ratio circuit. The output is proportional to J/J_o, where J is a light intensity fed to the photomultiplier.

intensity has been removed. This arrangement is also convenient in measurement of the absorption coefficient α in the sample, since the fraction μ of light transmitted through a sample of thickness d is

$$\mu = \exp(-\alpha d) \tag{15.9}$$

The log ratio circuit reads out the absorption coefficient directly.

15.3.5 Linear Thermometer

Although in most cases the presence of the term kT/e causes an undesired thermal drift in log amp circuits, it can be made use of to provide a thermometer precisely linear in T, calibrated in the absolute Kelvin temperature scale. The same circuit can also measure the ratio of two fundamental constants: (Boltzmann's constant k)/(charge of the electron e), although there are other methods with somewhat greater accuracy. With a constant input voltage reference signal, the output is then an accurately linear measure of temperature. In practice, one uses the IC temperature transducer discussed in Chapter 1 (Section 1.15), which makes use of the base-emitter voltage containing the kT/e factor and no other temperature-dependent terms.

15.4 MULTIPLIERS AND THEIR APPLICATIONS

Multipliers play an important role in instrumentation, with applications in control and signal processing. For example, a sine wave may modulate a physical variable, with the magnitude of the sine wave dictated by a second signal, a dc voltage from another physical variable or from a computer. Here, a multiplier acts as a voltage-controlled attenuator or volume control. In signal processing, one uses multipliers to shape a test signal: a tone burst for pulsed nuclear magnetic resonance or for noise-response testing, or the envelope of a musical tone in an analog electronic music system. Other instrumentation examples include:

(1) *Spectrophotometer* In a measurement of the absorption of light by a material, the transmitted light must be normalized by the intensity of the incoming light to eliminate fluctuations. A multiplier can provide this normalization.

(2) *Precision rms converter* Most ac meter circuits rectify the signal and measure the average magnitude $<|V|>$. This is not the same as the rms value, $<V^2>^{1/2}$ except for simple sine wave. "True rms" meters traditionally used joule heating, which is proportional to $<V^2>$, with transducers like thermocouples. A calibrated rms voltmeter of this type was very expensive. A measure of a true rms value can be obtained more easily in an analog circuit, using multipliers to provide both the squaring and the square root function.

(3) *Vector voltage* When a network contains elements equivalent to capacitance or inductance, its transfer function is not a simple numerical quantity, but has both magnitude and phase (measured with respect to the source) or two vector components. Multipliers may be used either to project out the in-phase and out-of-phase components of the signal or to compute the true magnitude and phase of the signal.

(4) *Analog Simulation* Many of the more interesting systems one would like to solve cannot be done with the linear methods of Chapter 14. In electromagnetic or in quantum mechanical problems, there is usually a potential energy term to be multiplied by the function.

$$H\,\psi(r) = [T + V(r)]\,\psi(r) = E\,\psi(r) \tag{15.10}$$

Or, two or more linear equations may be coupled; again, a multiplication is required. An analog multiplier opens up this whole realm of nonlinear differential equations.

Multiplier waveforms. Multipliers can generate unusual and useful signals. Examples of multiplier waveforms are given in Fig. 15.13. It is suggested that the output waveform be compared against the expected results, either in the time domain or by analyzing the frequency spectrum for each of the input signal combinations (Prob. 15.3). Example (a) is a pure beat pattern, equivalent to suppressed carrier modulation in communications. Example (b), a *tone burst*, is useful in testing circuits or physical systems, and in the physics of music. (What will a short tone burst sound like? See Prob. 15.4.) Example (c) finds application in oscilloscopes which "chop" waveforms in order to display several at a time. When used in this manner, the multiplier may be viewed as a switch. In example (d), the two input signals are close together in frequency, and the resulting difference term in the spectrum (see Prob. 15.5) shows up as a low-frequency component. (Can you also see the high frequency harmonic?)

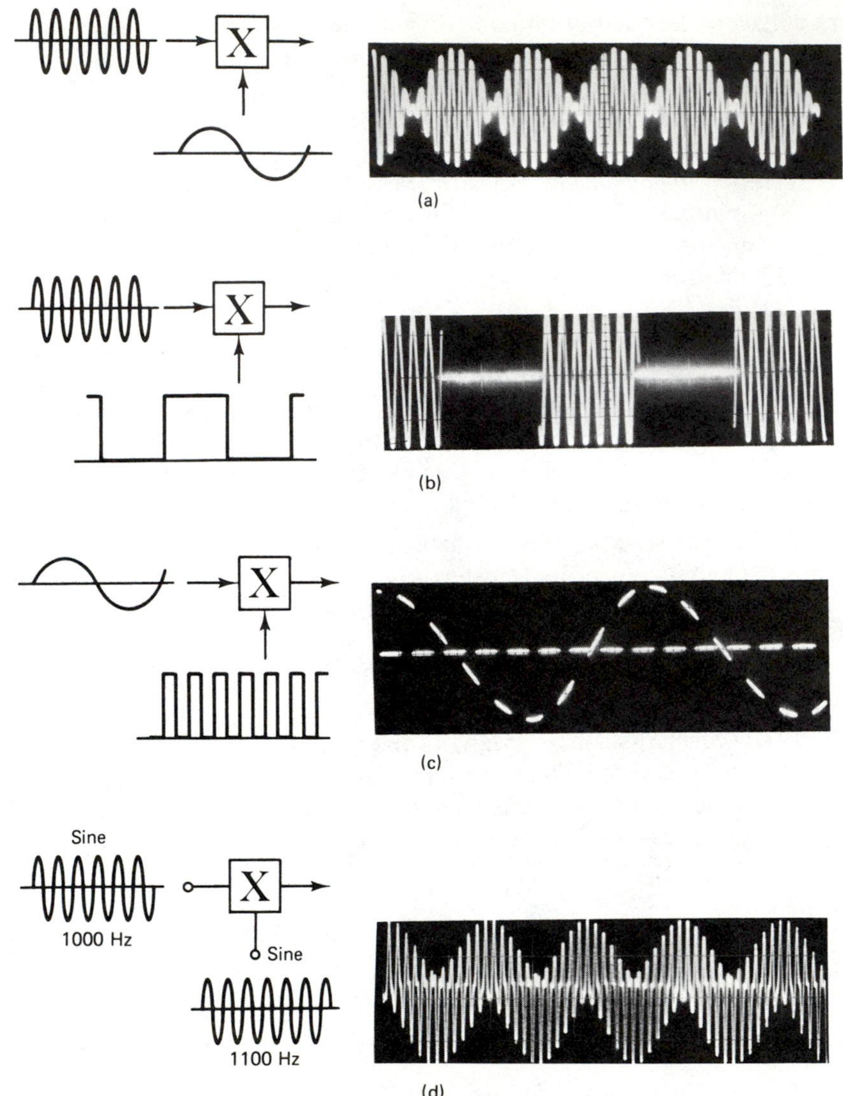

Figure 15.13 Examples of multiplier waveforms (an audio presentation is even more interesting!). (a) Product of two sine waves producing a beat. (b) A sine times a slower square produces a tone burst. (c) A sine times a higher frequency square produces a chopped waveform. (d) Sine times nearby sine. Note the low frequency difference tone.

15.5 SUMMARY OF MULTIPLIER METHODS

A summary of multiplier methods is given in Table 15.1. Each of these methods and its particular advantages or disadvantages will be explained briefly in Section 15.7. A log multiplier has the best dynamic range. The highest accuracy is pro-

TABLE 15.1 MULTIPLIER METHODS

Method	Comments: Advantages	Comments: Disadvantages
Quarter square (diode function generator)	4-quadrant	Slope discontinuities, poor dynamic range, expensive (~$100); *obsolete*
Log/antilog	Wide dynamic range (6 orders of magnitude), good divider; lowest feedthrough (~1m V)	Only one quadrant; drift
Pulse width	High accuracy (⩽ 0.1%), 4-quadrant	Expensive (⩾ $100)
Transconductance	Inexpensive (~$5), easy to use, 4-quadrant, widest possible bandwidth (1 MHz); *most widely used method*	Only moderate accuracy (~0.5%)

vided by a pulse-width multiplier. The transconductance multiplier is now the accepted standard for most applications, since the accuracy (⩽ 1%) is adequate for many situations and the cost is very low. A comparison of specifications for one example of the three principal types is given in Table 15.2. These specifications are explained in the following section.

Multipliers have benefited from monolithic IC technology both in performance and in lowered cost. For example, a multiplier with an accuracy of 0.1% cost about $1000 in the 60's but now can be obtained for about $100. The cost

TABLE 15.2 MULTIPLIER SPECIFICATION EXAMPLES[a]

Technique	Transconductance	Pulse-width	Log-antilog	Units
Specification				
Model	AD 533 K	427 K	434 A[b]	
Accuracy	1.0	0.20	0.5	% Fullscale
Output offset voltage	Trimmable to 0	5	2	mv
Nonlinearity	0.5	0.04	0.2	%
Feedthrough				
Untrimmed	100	20	2	mv p-p
External trim	<1	4		mv p-p
Bandwidth				
Small signal	1.0	0.1	0.1	Mhz
Full power	0.75	0.03	0.03	Mhz
Output noise (5 Hz to 10 kHz)	600	50	300	μv rms
Input resistance	$>10^3$	10	100	KΩ
Input bias current	3000	3000	10	nA
Price	15	200	75	Dollars

[a] *Courtesy* Analog Devices., Inc.

[b] No longer available; included for comparison purposes only.

reduction has been more spectacular for multipliers in the 1% range of accuracy, which cost more than $100 in the 60's, and are now available for less than $10. If two-quadrant multiplication with about 2% accuracy is adequate, an Operational Transconductance Amplifier is available for only about $1.

15.6 A GUIDE TO MULTIPLIER TERMINOLOGY

Important specifications for any multiplier are shown in Fig. 15.14.

Number of Quadrants The terms four-quadrant, two-quadrant, or one-quadrant refer to the allowed sign of the input variables [see Fig. 15.14(a)]. A four-quadrant multiplier allows any sign of input. With a one-quadrant

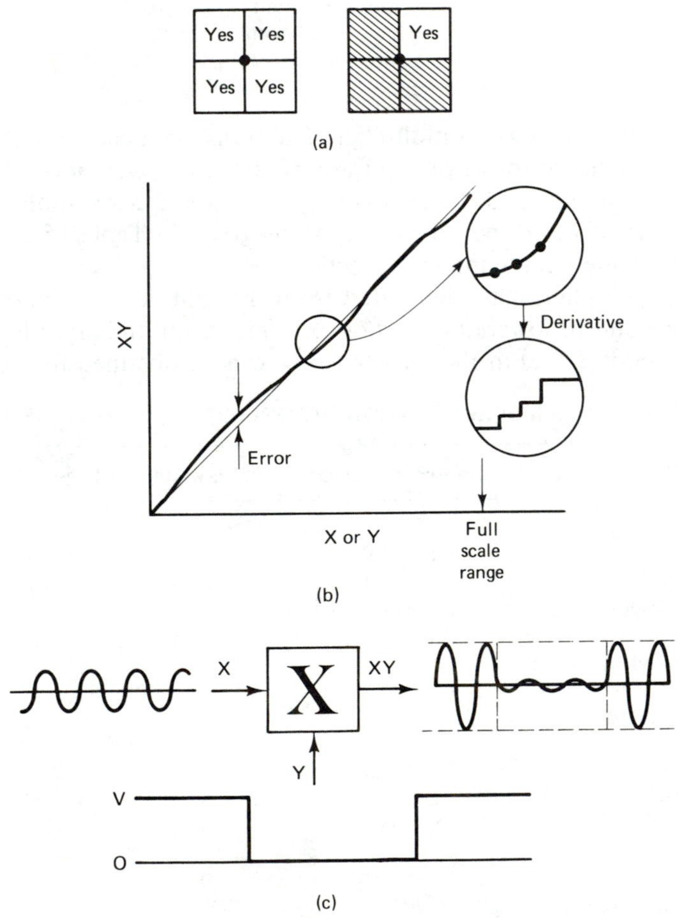

Figure 15.14 Multiplier specifications. (a) 4-quadrant vs. 1-quadrant. (b) Nonlinearity errors; the inset shows the effect of a derivative on breakpoints such as in a diode function generator quarter square multiplier. (c) Feedthrough, an output signal which is nonzero even when one input is zero.

multiplier, both inputs must be positive. A log amplifier made into a multiplier is a single-quadrant device since the transistor in the feedback loop must be forward-biased. Barring unusual circumstances, a four-quadrant multiplier circuit is the best choice.

Linearity The linearity of a multiplier is usually specified as a percentage error over the full-scale x or y range. If a particular application does not use the full scale, the nonlinearity as a percentage of the range used may be much worse [Fig. 15.14(b)]. A quarter-square multiplier can cause a problem when a wide dynamic range is required or when a derivative is taken, since its breakpoints may be amplified and introduce significant errors in the output. Errors may be amplified when a multiplier is made into a divider, since small errors at the low end of the y-range are amplified in $1/y$. A multiplier which is used as a divider must therefore have a wide *dynamic range* in order to have relatively constant errors over the range of the *output* variable. Since manufacturers generally do not publish the percentage error as a function of input signal level, some understanding of circuit operation is needed to predict linearity for a specific application. These problems have become relatively less important with the increasing acceptance of the transconductance multiplier, which has an inherently wide dynamic range and does not contain breakpoints or other gross nonlinearities.

Feedthrough Feedthrough is leakage of one input signal through to the output when the other input is set to zero. This is shown schematically in Fig. 15.14(c). Many transconductance multipliers include a provision for external trimming of feedthrough, so that it may be made as small as needed for a given application.

15.7 MULTIPLIER METHODS

15.7.1 Quarter-Square Multiplier

A quarter-square multiplier (Fig. 15.15) uses two squaring circuits (typically diode function generators) and three op amps. An adder and a subtractor generate combinations of X and Y, and a final subtractor is used to eliminate all but the cross term XY. Although the concept is elegant (see Prob. 15.7), it will not be developed here. The practical realization uses diode function generators, whose relatively high cost and objectionable breakpoints have made the quarter-square multiplier obsolete after the transconductance multiplier was introduced.

15.7.2 Log Multiplier-Divider

As shown in Fig. 15.16, a product can be obtained by taking the log of both inputs and then the antilog of their sum. Conventional log amps with transistors in the feedback loop (see Fig. 15.8) are used in the log stage; an op amp with a transistor in the input stage then provides the antilog (see Prob. 15.8). The practical realiza-

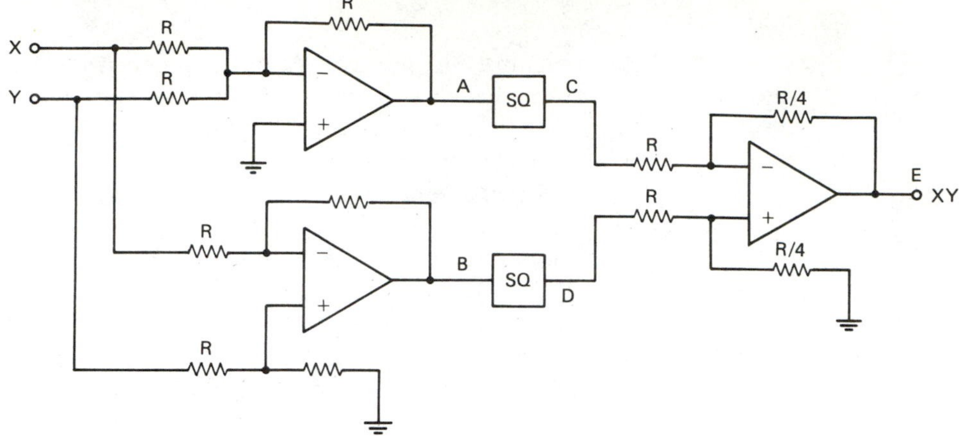

Figure 15.15 Quarter-square multiplier circuit. Now of historical interest only.

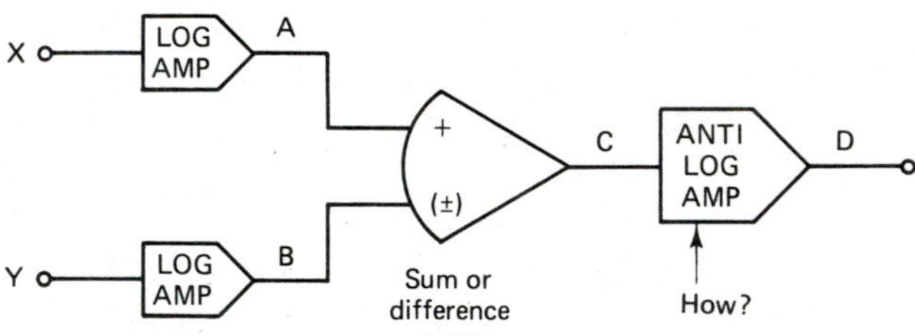

Figure 15.16 Log-antilog multiplier circuit, for signals with widest dynamic range.

tion of this circuit is more complicated, due to the necessity to eliminate temperature drifts in the log circuits (Fig. 15.11). Because of the relatively high cost of stable log circuits, this technique is selected only when extremely high dynamic range is essential.

15.7.3 Pulse-Width Modulation Multiplier

The basic idea of a pulse-width modulation multiplier, shown in Fig. 15.17(a), is that the average value of a train of pulses is proportional to the product of the pulse amplitude and duration (actually the *duty cycle* or fraction of time high). If one sets the duration of a pulse train with one of the inputs, V_Y, and the amplitude with the other, V_X, the product may be obtained by taking the average value of the resulting pulse train [Fig. 15.17(b)]. The voltage-to-time conversion can be accomplished with an integrator arranged to connect a switch to V_X for a time which is proportional to the value of the input signal V_Y. The relatively complex circuitry of the pulse-width multiplier limits its applicability to situations where highest accuracy is absolutely essential.

15.7.4 Transconductance Multiplier

Because the transconductance multiplier is now the industry standard method, it will be analyzed thoroughly. For further details, see the *Nonlinear Circuits Handbook* in the references. For a pn-junction diode or for a transistor with the collec-

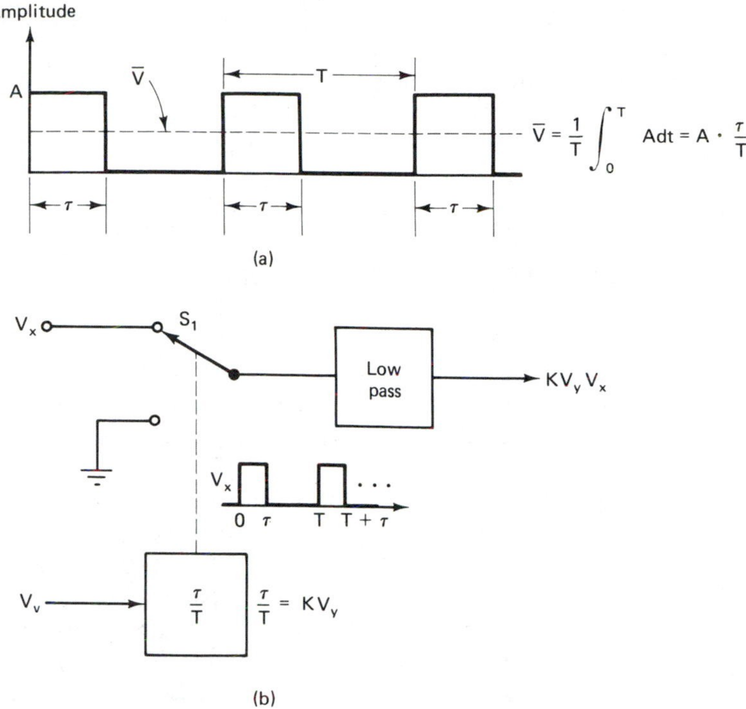

Figure 15.17 Pulse width multiplier circuit. (a) Basic principle: the average value of a pulse train is the product of the height times the fraction of time high (duty cycle). (b) Block diagram.

tor and base shorted together (transdiode), the relationship between collector current and base-emitter voltage is exponential.

$$I_C = I_{C0}[\exp(eV_{BE}/kT) - 1] \tag{15.11}$$

Here, I_{C0} = collector leakage current = 10^{-12} A or less at 300° K, and the other quantities are as defined for the diode [Eq. (15.4)].

When a relationship between two variables is exponential, an infinitesimal change in the output variable is proportional to the variable itself times the infinitesimal change in the input variable.

$$dI_C = (e/kT)I_C dV_{BE} \tag{15.12}$$

The term *transconductance* follows because this derivative has the units of ohms^{-1}, called *mhos*. Multiplier action results if one of the inputs sets the magnitude of I_C and the other input sets the change dV_{BE}. The output, taken as the *change* in collector current, is proportional to the product of inputs. It is assumed that $I_C >> I_{C0}$ so the 1 is negligible in Eq. (15.11), but also that I_C is small enough that additional terms due to the resistance of the semiconducting material can be neglected. This allows a wide range, since $I_{C0} \sim 10^{-12}$ A, and $I_C(\max) = 10^{-4}$ A for 1% error due to resistive voltage drop. At this point, the transconductance $(e/kT)I_C$ reaches about 1/260 mho, and parasitic resistances of a few ohms are a 1% error.

A simple multiplier based on this principle is shown in Fig. 15.18(a). The Y input provides the bias current of a differential pair. The X input is applied to the base of one transistor, unbalancing the pair, and resulting in difference in current $I_{C1} - I_{C2}$ proportional to the product $V_X V_Y$. This circuit has serious limitations. The Y input voltage is far from being the current source required in the transconductance principle. This is corrected in more complicated circuits by a voltage-to-current conversion. Furthermore, the Y input must have the appropriate sign to drive the transistors in the forward direction, which limits the input to a single sign of input. Limitations on the X input are primarily due to the nonlinear $I-V$ relationship [Eq. (15.11)], so that linear multiplication [Eq. (15.12)] results only for very small inputs dV_{BE}. For this reason a voltage divider is placed at the X input in Fig. 15.18. Finally, the scaling drifts with temperature because of the term e/kT.

A simple and elegant improvement (due to B. Gilbert) is shown in Fig. 15.18(b). The X input is a difference *current* which is converted to a voltage by diodes D_1 and D_2. Because of the diode $I-V$ curve, the conversion is logarithmic. This voltage then becomes the input for the pair Q_1 and Q_2, whose transfer function is exponential. The combination of a logarithmic conversion and a following exponential conversion leads to a cancellation of both the nonlinearity and the temperature dependence. (This analysis assumes base currents small compared to the diode currents.) More rigorously, the X input signal is converted to a difference current ΔX, which when added to bias current I_X becomes the current in diodes D_1 and D_2. The resulting voltage across these diodes is proportional to the logarithm of the currents

$$V_1 = (kT/e)\ \ln(I_{D1}/I_{D01}) \qquad (15.13)$$
$$V_2 = (kT/e)\ \ln(I_{D2}/I_{D02})$$

Figure 15.18 (a) Transconductance multiplier principle. (b) Method of improving transconductance multiplier linearity. (Due to B. Gilbert; *courtesy* Analog Devices, Inc.)

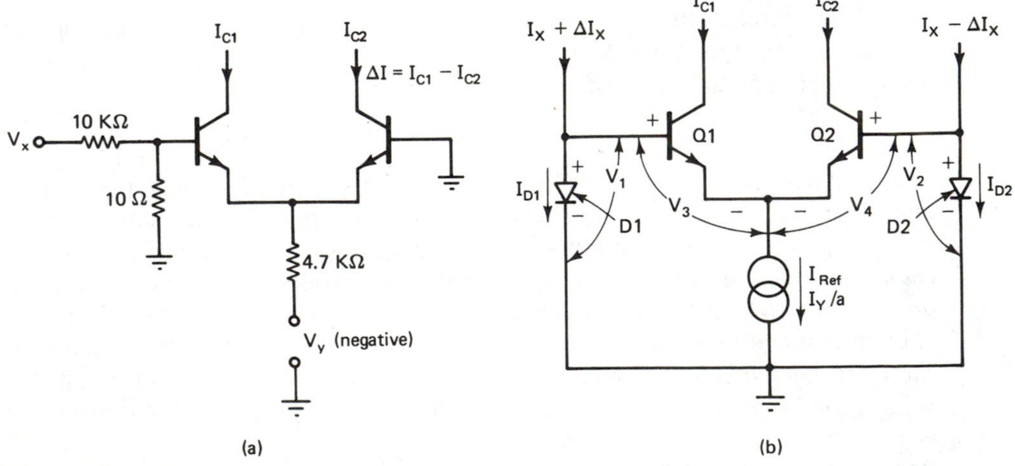

The collector currents in Q_1 and Q_2 are related to the base-emitter voltages by

$$V_{BE1} = V_3 = (kT/e)\ \ln(I_{C1}/I_{C02}) \tag{15.14}$$

$$V_{BE2} = V_4 = (kT/e)\ \ln(I_{C2}/I_{C02})$$

These four voltages form a Kirchoff voltage loop:

$$V_1 - V_3 + V_4 - V_2 = 0 \tag{15.15}$$

Inserting Eqs. (15.13) and (15.14) leads to

$$\ln\,(I_{D1}/I_{D2}) = \ln(I_{C1}/I_{C2}) \tag{15.16}$$

The kT/e term has cancelled, so this circuit will be temperature-independent. It is assumed that the transistors and diodes are perfectly matched, so their leakage currents cancel in Eq. (15.16). If the logs of the ratios are equal, then the ratios must be equal:

$$I_{D1}/I_{D2} = I_{C1}/I_{C2} \tag{15.17}$$

This result provides the multiplier action, since the ratio of the output currents is proportional to the ratio of input currents. Introducing the X input as a difference current ΔI_X and the output as a difference current ΔI_C, leads to

$$\frac{I_X + \Delta I_X}{I_X - \Delta I_X} = \frac{I_Y/2 + \Delta I_C}{I_Y/2 - \Delta I_C} \tag{15.18}$$

Solving for ΔI_C,

$$\Delta I_C = (\Delta I_X I_Y)/2I_X \tag{15.19}$$

The output difference current is proportional to the product of the X input difference current ΔI_X and the Y input current. The output is inversely proportional to the bias current I_X, which is used to set the scale factor for the device. This circuit is a two-quadrant multiplier, since V_X may have either sign but V_Y may have only a positive sign as required to turn on the transistors.

Four-quadrant multiplier. A four-quadrant multiplier circuit which is the present industry standard multiplier is shown in Fig. 15.19. Qualitative operation is as follows. Transdiodes Q_1 and Q_4 cancel a nonlinearity for two identical transconductance pairs formed by Q_2–Q_3 and Q_5–Q_6, which are connected in parallel. The X input voltage generates a difference current by means of the pair Q_7 and Q_8. An identical configuration formed by Q_9 and Q_{10} provides for differential Y input of either sign. The Y input voltage thus determines how much of the bias current I_B goes into the pair Q_2–Q_3 and how much flows into the pair Q_5–Q_6. These output currents are then combined in the output subtractor in such a way that the only term left is proportional to the product XY. Quantitative analysis provides the final result.

$$V_o = (V_X V_Y)/(I_A R) \tag{15.20}$$

The output is a product of input voltages, with the bias current I_A serving as a scale factor, a controllable gain, or a third Z input.

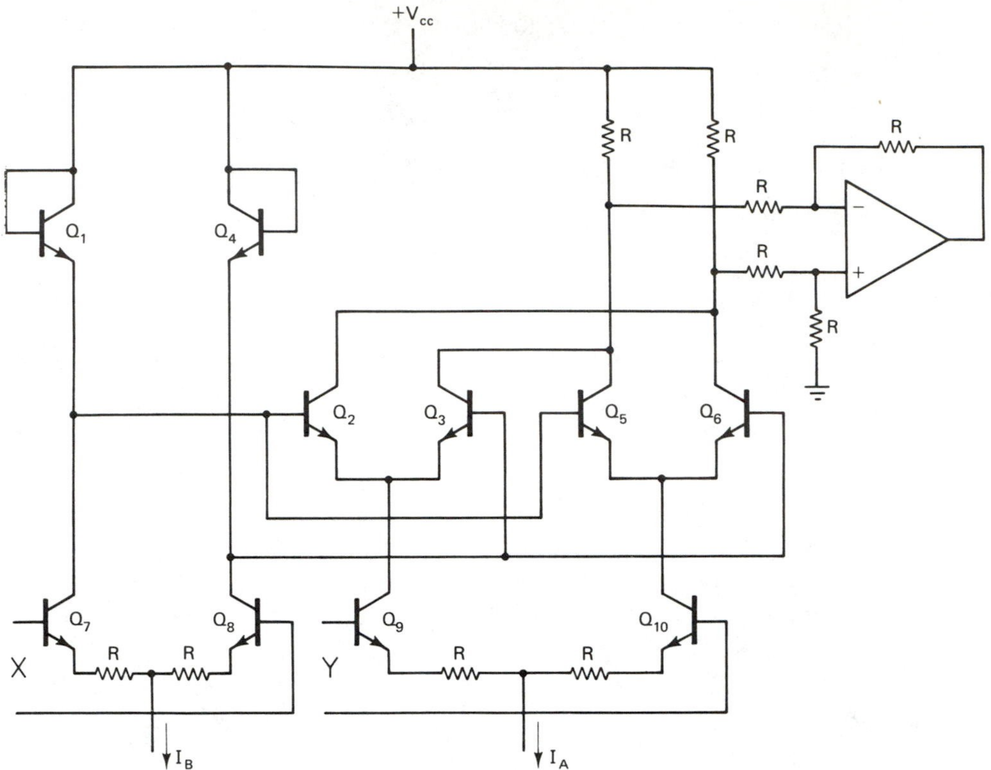

Figure 15.19 Extension to a 4-quadrant transconductance multiplier. (*Courtesy* Analog Devices, Inc.)

15.7.5 Operational Transconductance Amplifier

Closely related to the transconductance multiplier is the *operational transconductance amplifier* (OTA). An example, the RCA CA 3040, is shown in Fig. 15.20. The circuit's principle of operation is virtually identical to a two-quadrant multiplier. An equivalent block diagram is shown in Fig. 15.20(b). The transconductance principle enables the amplifier bias current I_B to function as a gain control. This class of amplifier differs from the normal op amp or normal multiplier in having an output which is a current source ($R_s = 15\ \text{M}\Omega$ for the CA 3080). OTA applications utilize this "voltage-controlled current source" feature extensively. The gain is set not by the usual pair of feedback resistors, but by the value of I_B. The bias current adjusts the transconductance g_m of the device (Fig. 15.21) over a 4-decade dynamic range. The output current is given by

$$I_o = g_m V_i \tag{15.21}$$

and therefore the output voltage across the load R_L is

$$V_o = g_m R_L V_i \tag{15.22}$$

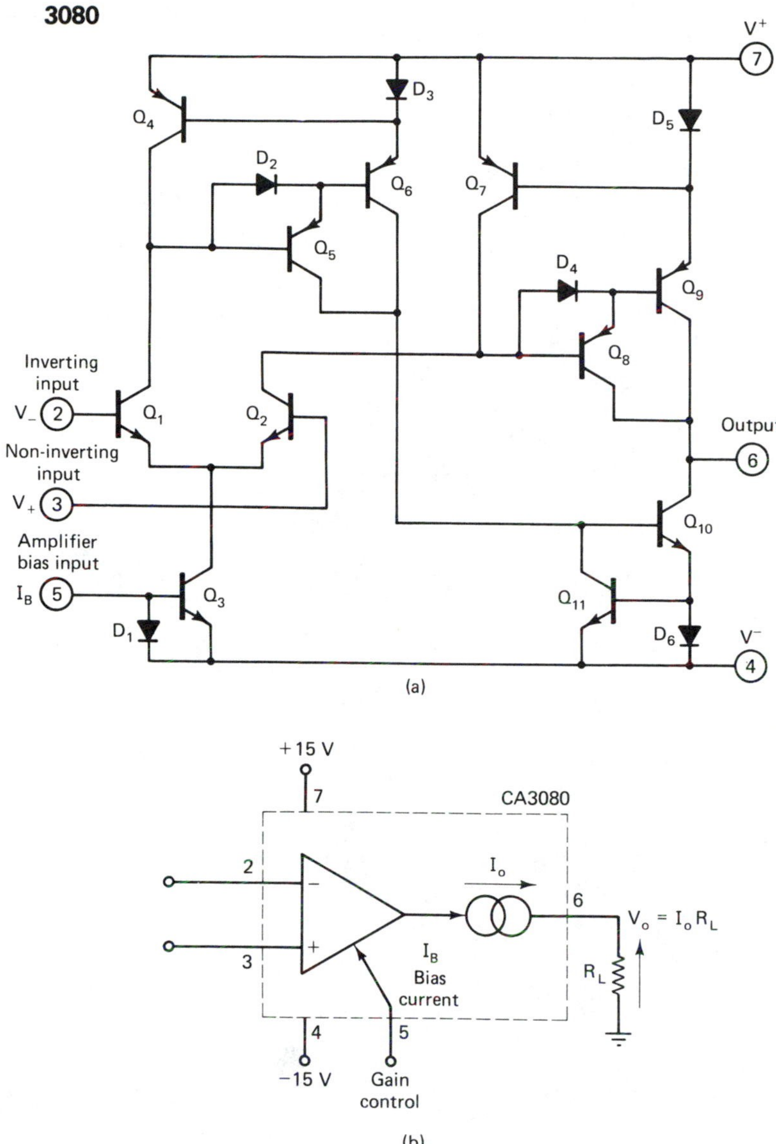

Figure 15.20 Operational transconductance multiplier. (a) CA 3080 circuit. (b) Equivalent circuit. (*Courtesy* RCA Solid State Division)

Note that the circuit gain depends upon the value of the load resistance R_L. In applications where it is inconvenient to provide an input bias *current* source to adjust g_m, a controlling input *voltage* in combination with a large resistor may be used instead. However, this results in very nonlinear response at low input voltage levels (due to the high input resistance of the base-emitter junction of Q_3 at low voltage). A much better solution is to use a voltage-to-current transducer; a complete circuit example is shown in Fig. 15.22.

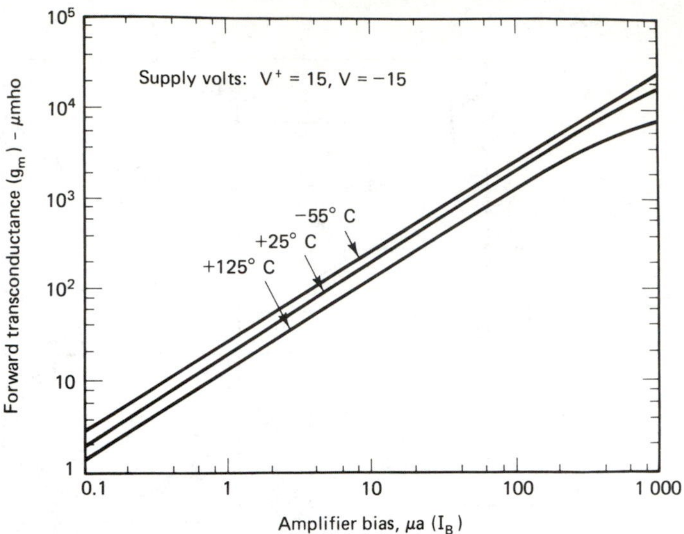

Figure 15.21 Characteristic curves of CA 3080 transconductance multiplier. (*Courtesy* RCA Solid State Division)

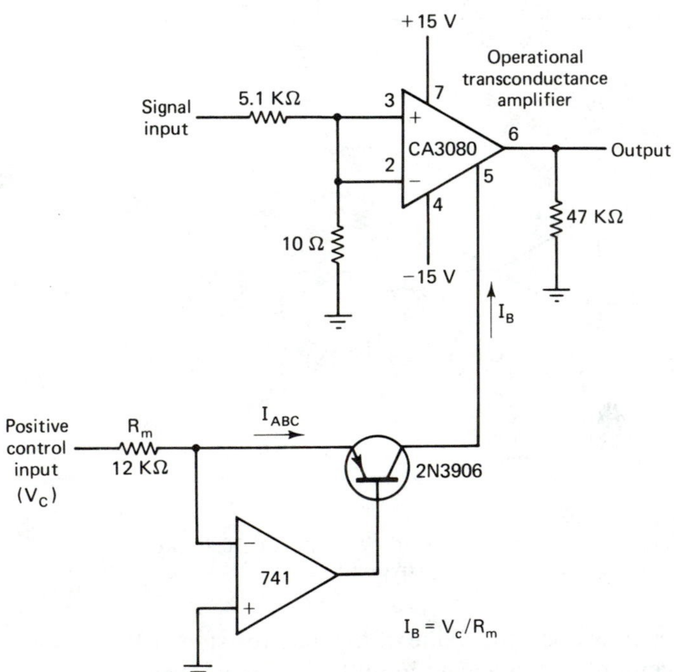

Figure 15.22 Linearization of the operational transconductance multiplier's gain as a function of input voltage, using a voltage-to-current converter (see Fig. 12.2). The values of R_2 and R_1 are adjusted to keep V_o from clipping at the power supply level for actual values of V_c. (*Adapted from* Jacob Moskowitz, *Electronics.* (March 17, 1977): 96.)

Micropower applications of OTA's utilize the fact that the total supply current and power dissipation vary over 4 decades as the input bias current I_B is varied. Thus, one can apply an OTA as a conventional op amp amplifier with closed-loop gain R_f/R_i, but with the open-loop gain adjusted to be only as large as necessary to keep ($\beta a > 1$), maintaining device power dissipation at the minimum possible. Some situations utilize this capability to switch to standby low power while maintaining operating high power when needed.

This class of amplifiers is widely used in applications where a low-cost (~$1) voltage-controlled gain function is needed, since the cost is a factor of 5 to 10 lower than that of a quality linear transconductance multiplier. The examples listed below will be discussed in greater detail in the appropriate later chapters.

The fast slew rate of an OTA output (50 V/s for the CA 3080) make it especially useful in switching applications such as analog multiplexing and sample-and-hold (Chapter 19). Linear applications include active filters, utilizing the current source output. Although the OTA is inferior in linearity to general-purpose multipliers and is limited to two quadrants, the low cost and reasonable ($\gtrsim$ 1%) linearity recommend it for modulation of one signal by another, envelope shaping, and voltage-controlled function generators.

15.8 MULTIPLIER GYMNASTICS

Multipliers have an extraordinary variety of applications, many based upon feedback tricks in which the multiplier appears to "lift itself up by its own bootstraps." A few will be illustrated here.

15.8.1 Divider

A divider circuit using a multiplier is shown in Fig. 15.23(a). One input is applied to the multiplier; the other input is applied to the noninverting terminal of an op amp. The op amp's feedback loop is closed through the second multiplier input, so the multiplier output is the input *A* times the op amp output *C*. Negative feed-

Figure 15.23 Multiplier gymnastics. (a) Using a multiplier to make a divider (bootstrap technique). (b) Squaring circuit. (c) Using a squaring circuit to generate a square root bootstrap technique).

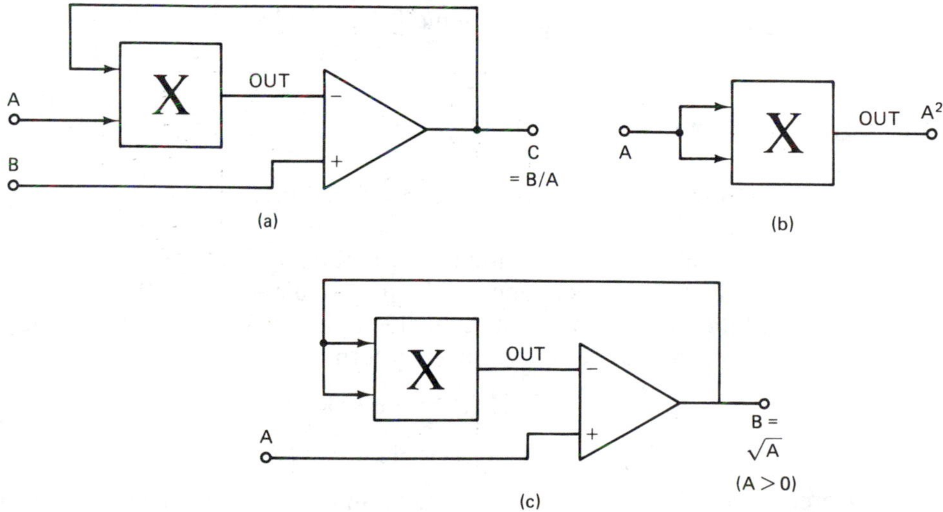

back drives the voltage at the op amp inverting input to match the noninverting input, giving the result

$$B = AC, \text{ or} \qquad (15.23)$$

$$C = +B/A$$

15.8.2 Square and Square Root Circuits

The *squaring circuit* is shown in Fig. 15.23(b). The operation is obvious; both inputs are tied together resulting in the output proportional to the square. A squaring circuit may be used in a closed-loop feedback configuration similar to the divider to form the square root circuit shown in Fig. 15.23(c). Negative feedback alters output B to make the summing-point voltage equal to zero. This can only happen if

$$B^2 = A, \text{ or} \qquad (15.24)$$

$$B = [A]^{1/2}$$

15.8.3 Other Multiplier Applications

Other examples of multiplier applications are shown in the following figures. The analysis is left for problems.

Vector summation. The output is proportional to the square root of the sum of squares of inputs (Fig. 15.24). Note that only one multiplier is used to perform three nonlinear operations (Prob. 15.8).

RMS or DC converter. The output is proportional to the square root of the time average of V^2_{in} (Fig. 15.25 and Prob. 15.10).

Automatic Gain Control. The output is a faithful image of the input ac sig… waveshape, but has a magnitude which is constant and independent of in… signal magnitude (Fig. 15.26 and Prob. 15.11).

15.9 MULTIPLIERS IN SIMULATION

15.9.1 Coupled Equations: Foxes and Rabbits

The classical host–parasite of foxes and rabbits problem of ecology us… plier to simulate the dependence of the fox population on both the pre… foxes and the number of rabbits available to eat (Fig. 15.27). The m… ples two first-order systems by a product term

$$dY/dt = -aY + bXY$$

$$dX/dt = +cX - dYX$$

Here, a and c are death rates of foxes and birth rates of ra…

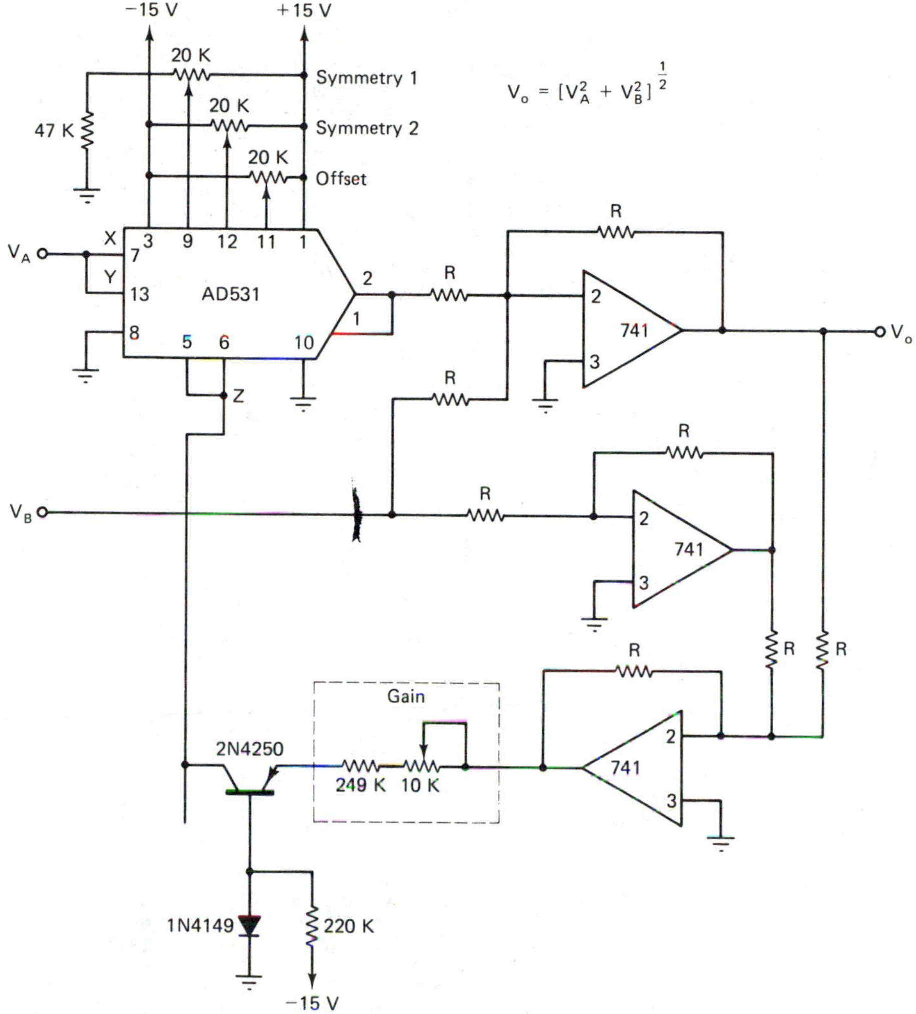

Figure 15.24 Practical vector summation or magnitude circuit, using a multiplier whose transfer function is XY/X, and a clever bootstrap technique (see Prob. 16.9) to generate $V_o = (V_x^2 + V_y^2)^{1/2}$. (*Courtesy* Analog Devices, Inc.)

Without rabbits, the foxes would die out (note the minus sign), while rabbits would multiply without limit (note the plus sign). Foxes breed at a rate which depends upon not only how many foxes there are, but also how many rabbits there are to eat (coefficient b). Rabbits have an extra death rate which depends upon how many foxes there are (coefficient $-d$). The coupled equation exhibits oscillations in both populations [Fig. 15.27(b)] whose character depends upon initial conditions. Did the rabbits have a chance to get ahead because there weren't many foxes at first? All those rabbits will make a later surge in the fox population. More complicated simulations with more interdependent populations exhibit less wild oscillations; diversity brings stability.

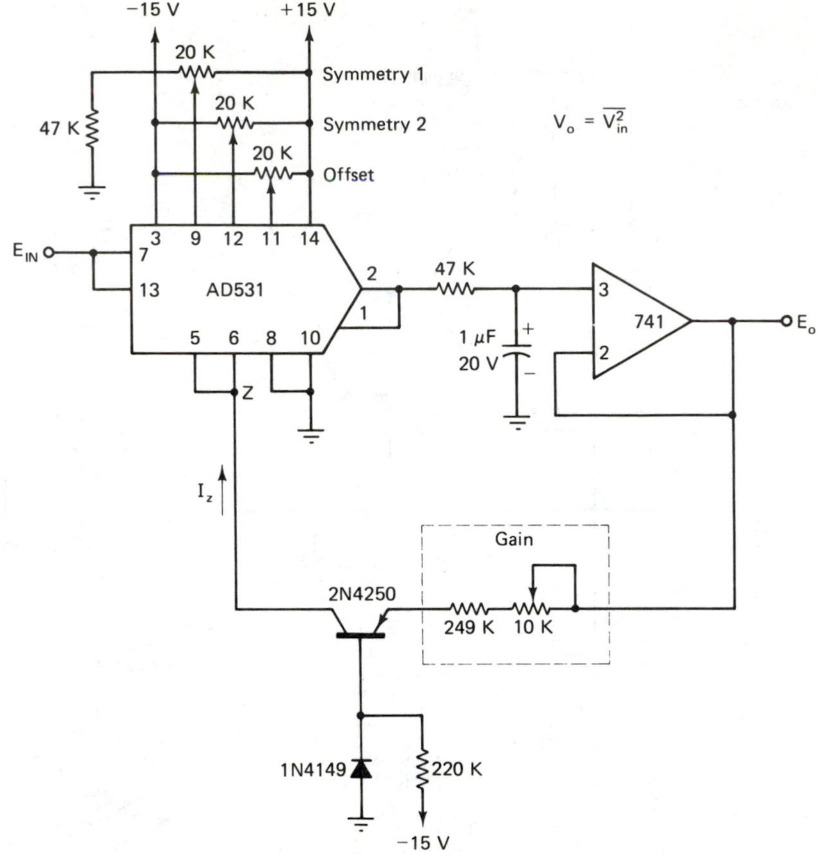

Figure 15.25 RMS to DC converter. Note the use of the bootstrap technique (see Prob. 16.10) as in Fig. 15.25 to generate the squaring and square root simultaneously with a single multiplier. (*Courtesy* of Analog Devices, Inc.)

15.9.2 The Analog Solution of Quantum Mechanics Problems

This approach provides visually convincing graphic solutions. A wave function ψ in a potential well $V(r)$ is bounded only for certain quantized energy levels which are solutions of Schrödinger's equation (Fig. 15.10). The probability amplitude ψ is represented as a voltage. The r-space variable is replaced by time, and the potential $V(r)$ becomes $V(t)$, provided by a function generator. The kinetic energy operator $T = d^2/dx^2$ becomes a second time derivative d^2/dt^2. A multiplier is needed to multiply the Hamiltonian $(T + V)$ by the wave function in this simulation. Fig. 15.28 shows an example of the use of a multiplier in the problem of a particle in a box. The Schrödinger equation for the problem is then

$$d^2\psi/dt^2 + (V(t)/K)d\psi/dt = (E/K)\psi(t) \qquad (15.26)$$

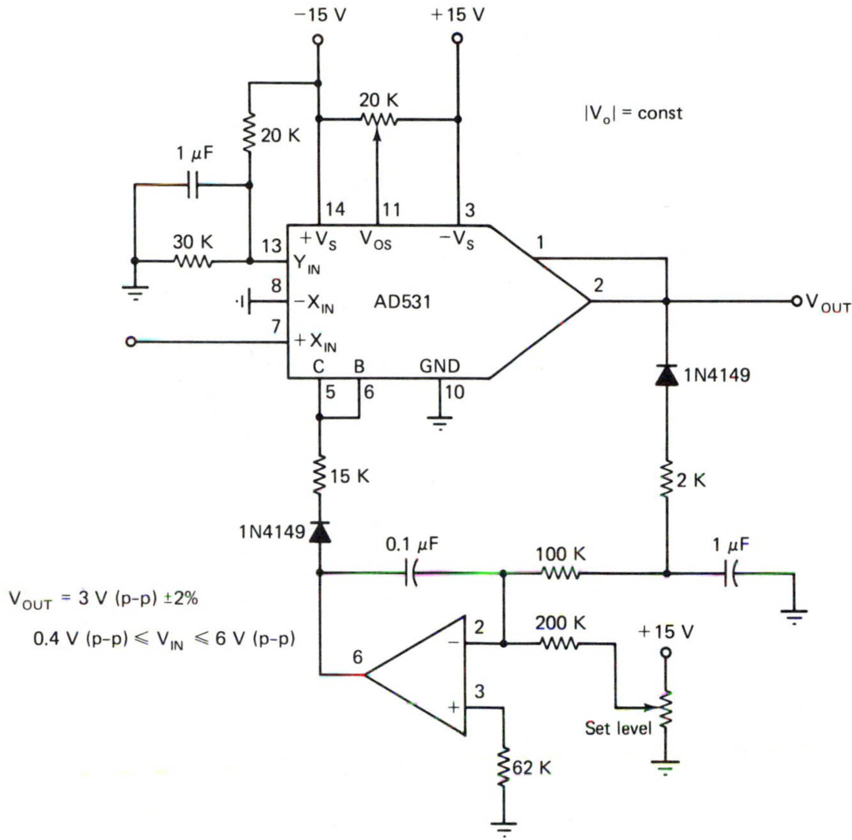

Figure 15.26 Automatic gain control circuit. The multiplier gain (Z input) is adjusted to keep the output signal amplitude a constant. (*Courtesy* Analog Devices, Inc.)

The box potential is generated by a timing circuit which switches V initially to ground, switching the integrators to the *run* mode to start the clock. When the particle reaches the edge of the box, the switch brings V to V_m, the height of the box. The wave function is observed to diverge except at very carefully adjusted energy values, the eigenvalues E_n [Fig. 15.28(b)]. Energy values above V_m lead to free-particle wave-like behavior whose period and amplitude change at the edge of the box.

Although these applications are not new, good but inexpensive multipliers have only recently become available. Multipliers are now freely used in signal generation, signal processing, and simulation. The simulation examples of Figs. 15.27 and 15.28, which would formerly have required the resources of an analog computer, can now be set up on a single breadboard for a cost of about \$100.

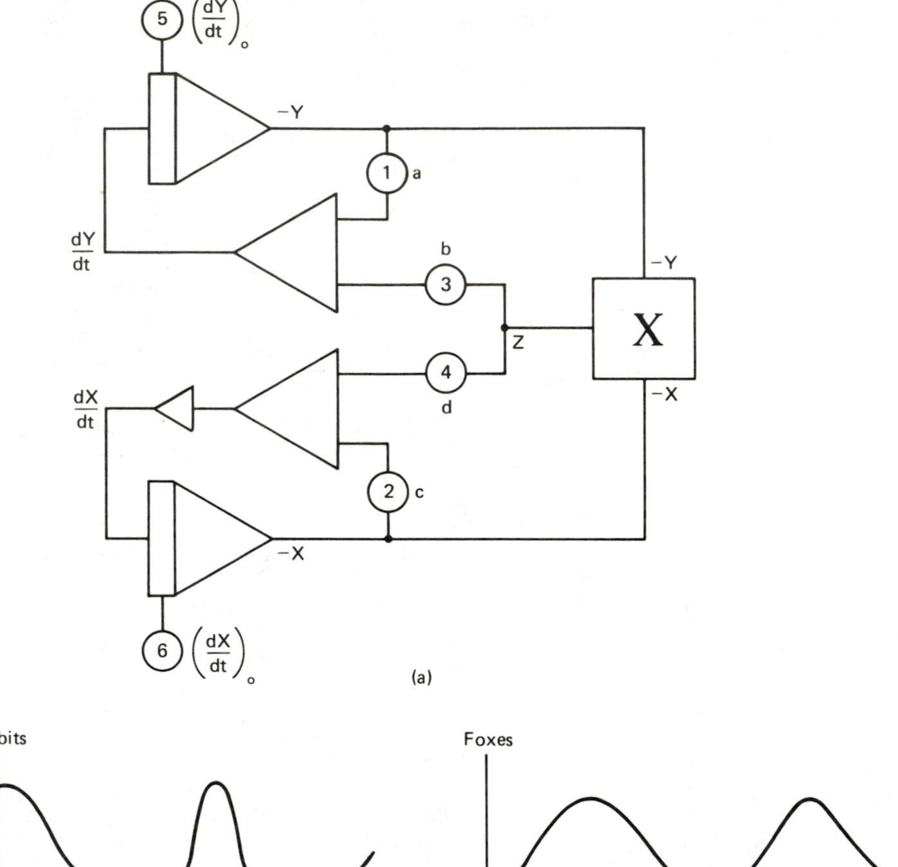

Figure 15.27 Simulation of a host-parasite problem (foxes and rabbits). (a) Circuit. The simplified convention of analog computation has been adopted. Amplifiers with boxes on the input are integrators, with initial conditions set by coefficients (circles). Other circles are pots or variable gain amplifiers which set other coefficients. All triangles are inverting amplifiers. (b) Output of a simulation example, showing oscillations in both populations.

PROBLEMS

15.1. **(a)** Describe how the circuit of Fig. 15.3 reduces to the circuit of Fig. 15.4(b) for $V_i < 0$. In each case, list which diodes are on (short circuit) and off (open circuit).

(b) A key feature of the precision millivoltmeter of Fig. 15.3 is the reversal of the sign of the gain when the input polarity reverses, with no change in the mag-

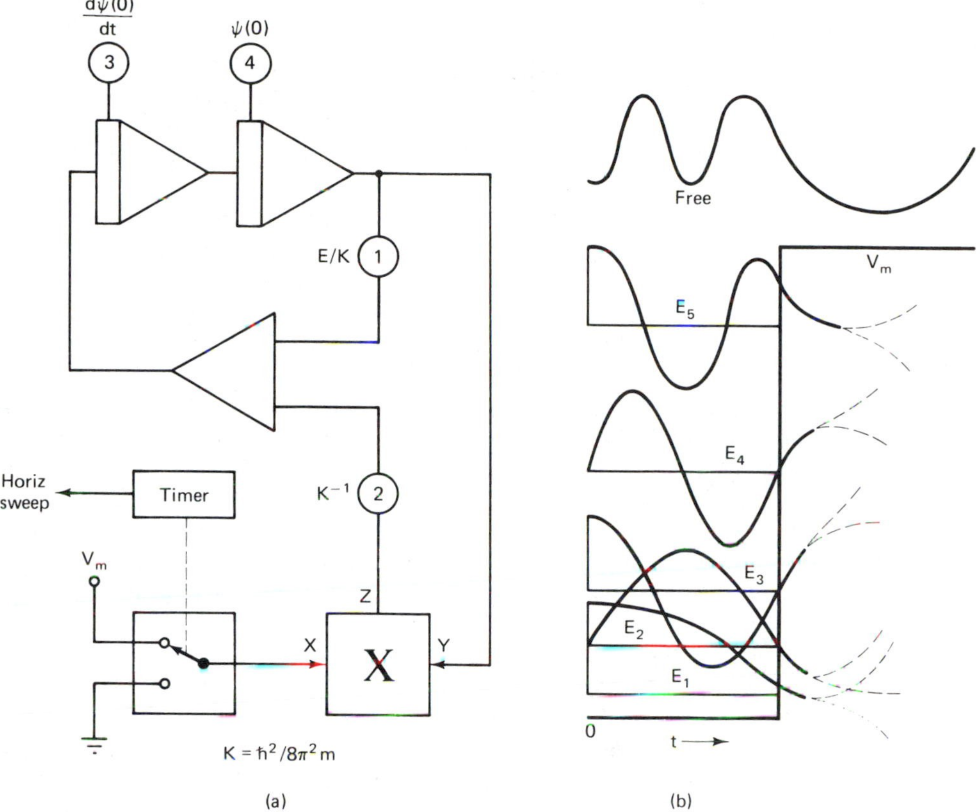

Figure 15.28 Simulation of the quantum-mechanical particle in a box. (a) Circuit. (b) Observed behavior of the wave function analog. The wave function is well behaved for $(E < V_m)$ only at energy values $E/V_m = 0.036, 0.128, 0.289, 0.532$, and 0.835. There are only 5 bound states for this particular box. Boundary conditions are established as initial conditions at $t = 0$, the *center* of the box, because the behavior is symmetric.

nitude of the gain. Show that the transfer functions of Fig. 15.4(c) and (d) are $V_o = +V_i/x$ and $V_o = -V_i/x$, respectively.

15.2. Show that the transfer function of the temperature-compensated log amp of Fig. 15.11 is

$$V_o = (1 + R_2/R_{TC})\ (kT/e)\ \ln \frac{V_i\ I_{C02}}{R_1 I_{\text{ref}} I_{C01}}$$

where I_{C01} and I_{C02} are the prefactors in the $I-V$ curve of Eq. (15.4) for transistors 1 and 2. The effective subtraction provided by the lower "bootstrap follower" cancels these prefactors $(I_{C02} \simeq I_{C01})$, and judicious selection of the temperature-compensating resistor R_{TC} can also cause the temperature-dependent kT/q to cancel. **Hint:** A_2 acts as a voltage follower whose input is $V_o\ R_{TC}/(R_2 + R_{TC})$. Solve for voltages V_A and V_B in terms of $\ln(V_i/R_1)$, and equate the two. Note, however, that both the collector and the base of the lower transistor are offset by an amount V_2.

15.3. Explain the waveforms shown at the multiplier outputs of Fig. 15.13. First explain the outputs viewed as a function of time. Then, for examples (a) and (d), explain how the observed output frequency spectrum compares with the frequency spectrum to be expected from the multiplication of the inputs given. (**Hint to avoid explicit Fourier analysis:** A product of two sine functions may be broken up into the sum of two sine functions at the sum and difference frequencies.)

15.4. Explore the frequency spectrum of Fig. 15.13(b) as the on-time at the pulse is made increasingly short. How well can the pitch be defined? Show this in the frequency domain. (**Hint:** The Fourier transform of a pulse has the form $(\sin x)/x$. What is the connection between the width of the pulse in the time domain and the width of its Fourier transform in the frequency domain?)

15.5. Referring to Fig. 15.13(d), when two signals close in frequency are multiplied, what causes the low-frequency term in the output? A higher frequency can also be seen as a distortion of the waveshape in Fig. 15.13(d). What is its origin? **Procedure:** Fourier analysis by trigonometry as in Prob. 15.3. Sketch the spectrum.

15.6. If in Fig. 15.13(d) one of the sine inputs is replaced by a triangular wave, the low-frequency component in the waveform also looks like a triangular wave. Explain in terms of the frequency spectrum of a triangular wave and the mixing which occurs in the multiplier.

15.7. Show how the quarter-square multiplier circuit (Fig. 15.15) generates the product XY. Evaluate the signals at points A, B, C, D, and E. The square boxes are squaring circuits: output = $(\text{input})^2$.

15.8. **(a)** Verify that reversing the position of the transistor and resistor in Fig. 15.8(b) results in an *antilog* amplifier, whose output is exponentially related to the input.

(b) Derive the transfer function for the log multiplier/divider of Fig. 15.16. Specify the voltage at each of points A, B, C, and D in terms of input voltages X and Y.

15.9. Show that the circuit of Fig. 15.24 generates $[V_a^2 + V_b^2]^{1/2}$ using only a single multiplier. The multiplier has the transfer function XY/Z. In this case Z is the sum of two signals $(-V_o + V_B)$. Write an expression for the output V_o in terms of this transfer function, noting the bootstrap action bringing V_o back as a Z input. Simple algebra then leads to the desired result. Ignore the details of the transistor, which function to generate a control current from this signal.

15.10. Explain the operation of the rms to DC converter (Fig. 15.25). The multiplier has a transfer function $V_o = XY/Z$. Show that the bootstrap feedback of the average value (note the low-pass filter) of the multiplier output performs the squaring and the square root function needed for the rms value $V_o = (V_{\text{in}}^2)^{1/2}$ using only a single multiplier. Ignore the function of the transistor which provides voltage-to-current conversion.

15.11. Explain the operation of the automatic gain control circuit of Fig. 15.26. The multiplier has the transfer function $V_o = XY/Z$. Note the rectifier filter between the output ac signal and the reference comparison subtractor circuit. The left-hand diode functions only as a protection circuit.

15.12. How would an automatic gain control circuit be constructed using a CA 3080 operational transconductance amplifier? Sketch the circuit.

15.13. For the linearized OTA of Fig. 15.22, what values of R_i should be chosen (keeping all other values unchanged) to keep V_o below 10 V when $V_i = 10$ V and $I_B = 1$ ma?

REFERENCES

More complete bibliographic information for the books listed below appears in the annotated bibliography at the end of the book.

Analog Devices, *Nonlinear Circuits Handbook*

GRAEME, *Designing with Operational Amplifiers*

HIGGINS, *Experiments with Integrated Circuits*, Experiments 21 and 22

PHILBRICK, *Applications Manual*

TOBEY, GRAEME, & HEULSMAN, *Operational Amplifiers*

16

Active Filters

16.1 INTRODUCTION

An *active filter* uses active devices (transistors, op amps) to simulate and extend the characteristics of older designs with passive (R,L,C) components only. Active filters now predominate over passive filters in most applications below 1 MHz, because they do not use expensive and bulky inductors, they can easily be tuned, and their performance corresponds closely with the ideal. A comparison between active and passive filter features is given in Table 16.1.

Size, weight, cost. A passive filter which includes a bandpass function must have an inductor. Active filters do not require inductors, consequently saving size, weight, and cost. The savings are substantial, since a high-Q inductor suitable for the audio range costs several dollars and is bulky and heavy. The op amp which replaces it costs 10 times less and has one-fiftieth the size and weight.

Tuning. The cutoff or center frequency of a passive filter is adjusted by varying an inductance or capacitance. Variable capacitors for the audio range do not exist, and variable inductors are expensive. The adjustment is further complicated when the filter design involves several L or C values in matched ratios. By contrast, tuning an active filter is generally done using a single resistor which is cheap and easily adjustable.

Bandwidth. The bandwidth $\Delta\omega$ of a bandpass filter is measured by a figure of merit called Q, where $Q = \omega_o/\Delta\omega$, and ω_o is the center frequency.

TABLE 16.1 ACTIVE AND PASSIVE FILTER COMPARISON

Criterion	Passive	Active
Size, weight, cost	Limited by inductors	No inductors
Tuning	Must adjust L or C	Adjustable by a resistor
Bandwidth	Limited by inductor Q	$Q \geqslant 100$ easily achieved
Performance over frequency range		
sub-audio (< 10 Hz)	Impossible	Excellent
audio	Inconvenient	Excellent
RF (> 100 KHz)	Good	Good
Gain	Insertion loss	Gain > 1
Impedance match at input and output	Must match	Not necessary
Multiple stages	Successive loading	No problem
Transfer function manipulation	Subtle; nonideal inductors and interacting stages	Straightforward and powerful

Bandwidth of passive filters is limited by the Q of the inductor, due to unavoidable series resistance in the windings or to hysteresis losses in the magnetic core. By contrast, with an active filter, the Q is readily adjusted with a resistor to achieve the desired bandwidth, and Q values in excess of 100 are easily achieved.

Performance over frequency range. Because of the above, the active filter is preferred over the passive equivalent in the audio range. In the *subaudio* range (less than 10 Hz), passive LC filters become impossibly bulky and expensive, since $\omega_o = 1/(LC)^{1/2}$. Since the cutoff or center frequency of active filters can be extended merely by shifting R and C values, active filters have had a great impact in extending the benefits of signal processing to low-frequency phenomena. Applications in geophysics, for example, extend down in frequency to 0.001 Hz. On the other hand, in frequency ranges above 10^5 Hz, passive filters become relatively good performers, since air core or other inexpensive RF inductors can be used. They become superior in frequency ranges above 10 MHz, where op amp frequency limitations become severe, but high-Q passive filters based on crystal resonance become possible. Passive filters are still commonly used in the RF range, when the center frequency and Q are fixed.

Gain. Since passive filters attenuate a signal, careful passive filter design must pay attention to maximizing the power transfer from input to output, measured by the *insertion loss* of a filter. But with an active filter, the op amp can provide gain, which can usually be adjusted independently of other circuit properties.

Impedance match. Since any filter will be connected to a source and load, care must be taken to avoid signal power loss due to impedance mismatch-

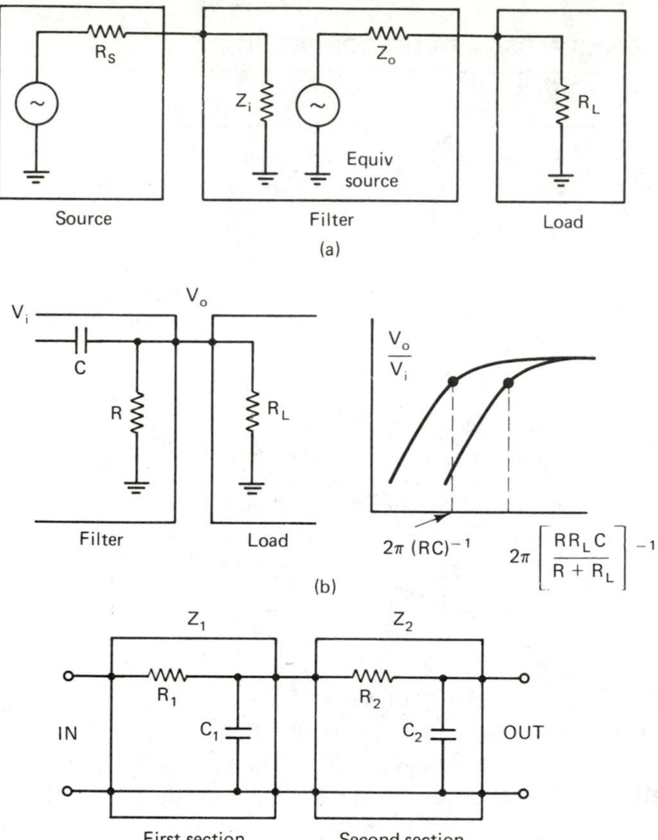

Figure 16.1 (a) Generalized filter, with equivalent input impedance Z_i and source impedance Z_o. (b) Loading example for a high pass filter. (c) Cascading two low-pass filters in series.

ing. With passive filters, power loss is minimized by matching the filter's input impedance to the impedance of the source, and output impedance to the impedance of the load, as shown in Fig. 16.1(a). This is inconvenient, since source and load impedances are not always known in advance. In addition, the shape of the response curve (transfer function) will vary with source and load impedances. An example is shown in Fig. 16.1(b). By contrast, because an active filter presents a high imput impedance and low output impedance, matching between source and load is usually unnecessary.

Multiple stages. To obtain a sharper cutoff or narrower bandwidth than a single stage, several filter stages are cascaded in series. With passive filters, this leads to degradation in the signal strength due to the insertion loss of each stage, and to changes in the actual transfer function due to the interaction of successive stages. As an example, the cutoff frequency of the high-pass filter shown in Fig. 16.1(b) is shifted when a load is attached, unless $R_L >> R$.

$$f_c^{-1} = \tau = RC \qquad \text{ideal}$$

$$f_c^{-1} = \tau = [RR_L/(R + R_L)]C \qquad \text{actual} \tag{16.1}$$

By contrast, active filters in series do not interact, due to their high input impedance and low output impedance.

Transfer function manipulation. The generation of an arbitrary transfer function using passive inductors is complicated by the limited Q of real inductors and the interaction of successive stages. But with active filter design techniques, the synthesis of an arbitrary transfer function is straightforward, because the real circuit differs little from the ideal calculation.

16.2 LOW-PASS ACTIVE FILTERS

A low-pass filter extracts a low-frequency signal of interest from high-frequency noise which obscures it (Fig. 16.2). Desired characteristics and best solutions will be considered in detail for low-pass filters. Designing a high-pass filter proceeds similarly, but with the location of R and C reversed.

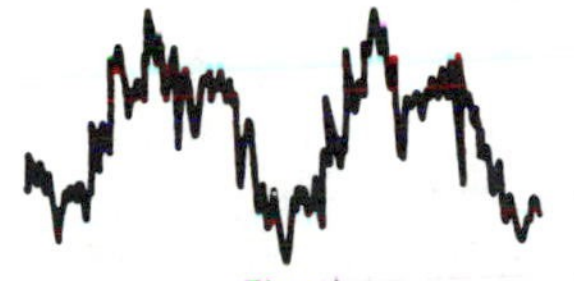

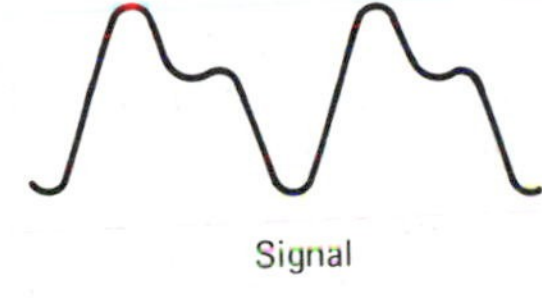

Figure 16.2 Example of data for which a low-pass filter is useful in improving the signal-to-noise ratio.

16.2.1 First-Order Low-Pass Filters

The simplest passive and active low-pass filters are shown in Fig. 16.3(a) and (b). These filters are called first-order because the differential equation governing their behavior (charging of an RC network) is first-order, and the corresponding transfer functions are first-order in *s*.

$$V_o/V_i = 1/(1 + j\omega RC) = 1/(1 + sRC) \qquad \text{passive} \qquad (16.2)$$

$$V_o/V_i = (-R_2/R_1)/(1 + sR_2C) \qquad \text{active} \qquad (16.3)$$

The transfer functions look very similar [Fig. 16.3(c) and 16.3(d)]. Both roll off above the cutoff frequency at 6 dB per octave, or 20 dB per decade of frequency. The transfer functions differ in that the low-frequency gain of the active filter is adjustable by the ratio R_2/R_1, whereas the low-frequency gain of the passive stage is fixed at 1. In addition, the gain and cutoff frequency of the active stage can be adjusted independently, since the cutoff frequency $(R_2C)^{-1}$ is independent of the value of R_1.

16.2.2 Second-Order Low-Pass Filters

Since the relatively slow rate of roll-off of the first-order low pass does not provide much attenuation of noise which is near the signal frequency range, a sharper cutoff is often needed (see Prob. 16.1) With passive filters, the simplest solution is to

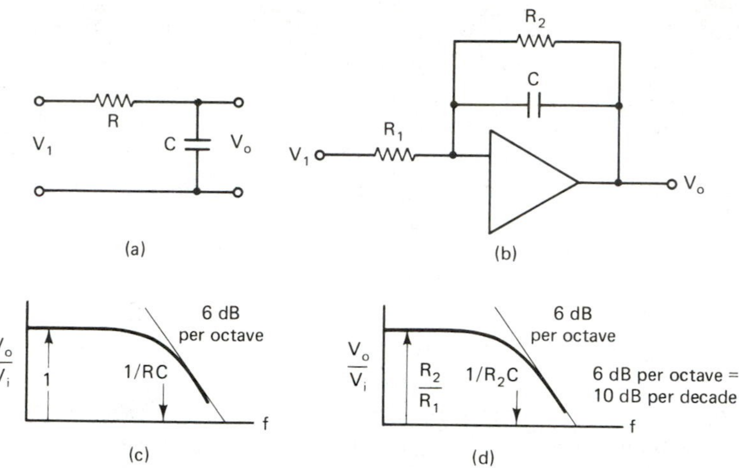

Figure 16.3 First-order low-pass filters. (a) Passive RC circuit. (b) Active version. (c) Transfer function for passive filter. (d) Transfer function for active filter.

cascade several sections in series, as shown in Fig. 16.1(c). The problem is that section 2 loads section 1, degrading the signal amplitude and altering the transfer function. The cutoff frequency and rate of roll-off will differ from that expected. As long as $Z_2 >> 1/(j\omega C_1)$, each stage may be analyzed as a voltage divider, and the overall transfer function is the product of two terms similar in form to Eq. (16.2). This yields a faster roll-off (proportional to $1/s^2$). However, if the impedance levels of the two sections are comparable, current flow into section 2 alters the current flow in section 1. The problem then requires the solution of Kirchoff equations of the network. This problem may be avoided by making $Z_2 >> 1/(j\omega C_1)$, but this leads to a more serious insertion loss if Z_2 approaches R_L [see Fig. 16.1(a)]. Cascading passive filters improves the signal-to-noise ratio, but reduces the amplitude of the signal.

Although the first-order active filter section of Fig. 16.3(b) could be cascaded indefinitely without this problem, a more elegant solution using only one op amp per second-order stage is shown in Fig. 16.4. The transfer function is

$$V_o/V_i = -[1 + b(RCs) + (RCs)^2]^{-1} \tag{16.4}$$

This has the desired second-order shape [Fig. 16.4(b)], with a 12 dB per octave roll-off above the cutoff frequency. The parameter b, which is adjusted by the ratio $C_2/C_1 = b^2/9$, allows the step response of the filter to be adjusted from overdamped to underdamped. There will be no ringing or overshoot as long a b $>$ 2. This follows (as in Chapter 14) by looking for the location in the complex plane of the natural frequencies s_o. Solving for the zeros in the denominator of Eq. (16.4)

$$S_o = (2RC)^{-1}[-b + (b^2 - 4)^{1/2}] \tag{16.5}$$

For $b > 2$, there are two real and negative roots [Fig. 16.5(a)], and the step response will be an exponential whose time constant is $\tau = 1/RC$ for $b = 2$. For

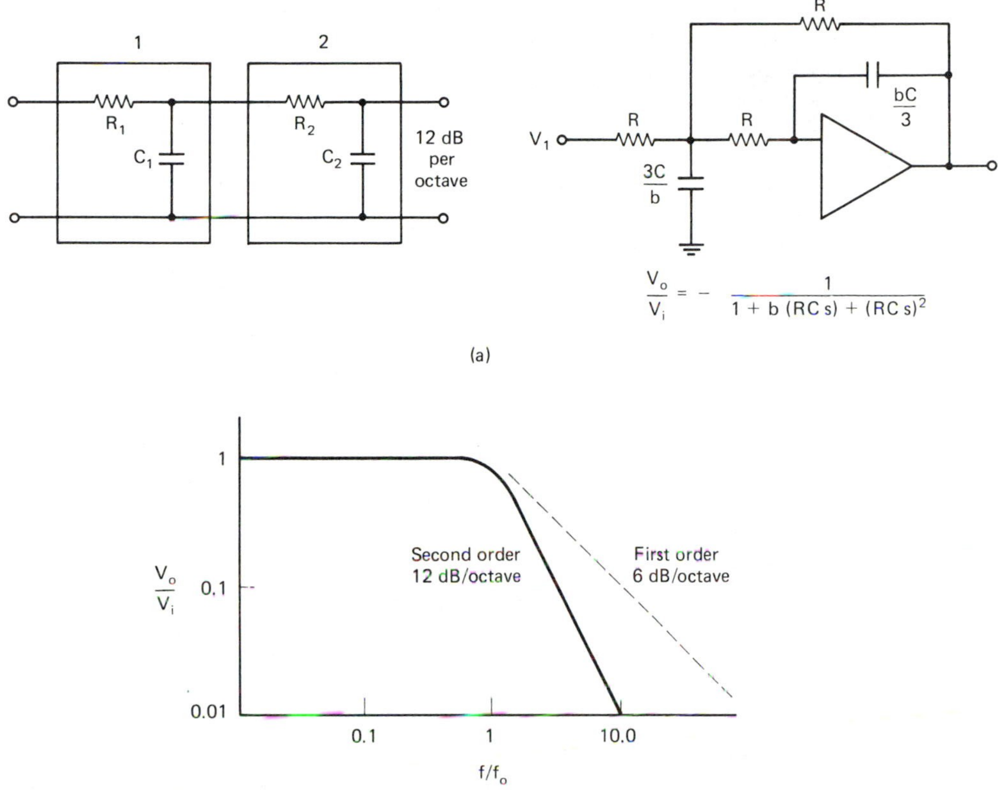

Figure 16.4 (a) Second-order low-pass active filter on right, with improved performance over the passive version shown on left. (b) Transfer function of the active second order low pass.

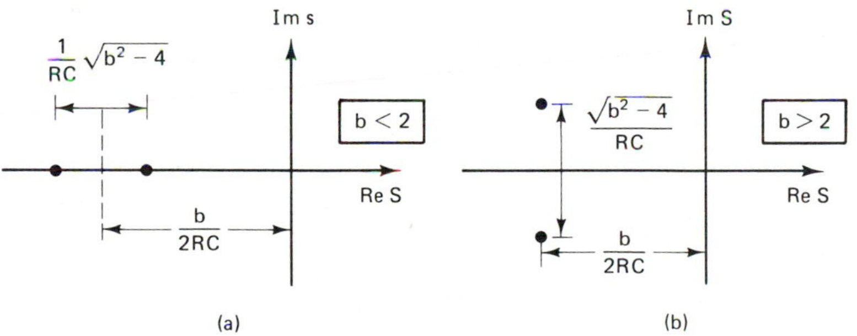

Figure 16.5 Natural frequencies of the second-order low-pass active filter. (a) Overdamped. (b) Underdamped.

$b < 2$, the two roots move off the real axis [Fig. 16.5(b)], and the response to a step will therefore have ringing or overshoot. This may actually be desired when the output must settle to a final value as quickly as possible.

The reader may see the essential results intuitively by analogy with a series RLC circuit. The characteristic equation for the natural frequencies of the series RLC circuit is of identical form with the denominator of Eq. (16.4). A criterion for overdamped (no ringing) response, $(R/L)^2 > 4/LC$, corresponds exactly to the criterion that $b > 2$ in the second-order low pass (see Prob. 16.2).

Exercise: Analysis of the Second-Order Active Filter

The analysis of the second-order low pass is our first example of a multiple feedback path. A circuit diagram for a general second-order section is given in Fig. 16.6. Since this configuration is applicable to a low-pass, high-pass, or bandpass filter, depending upon where the resistors and capacitors are located, it will be analyzed with gen-

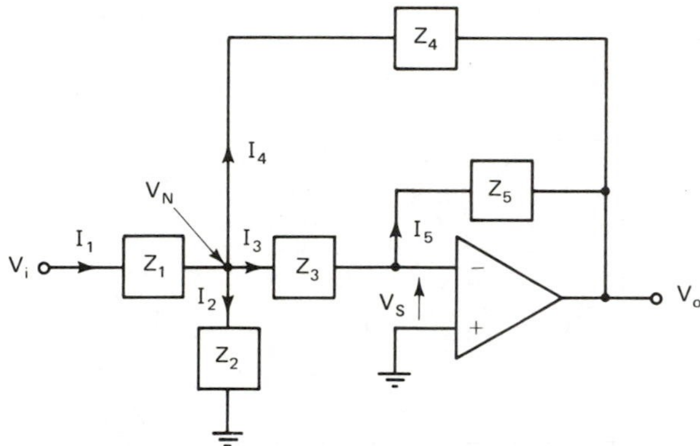

Figure 16.6 Generalized multiple feedback circuit. Any second-order filter section may be synthesized by insertion of appropriate components.

eral impedances Z_n. The desired filter may by then be synthesized by replacing a given Z with a resistor or capacitor. The analysis follows the usual negative feedback procedure. Current conservation at node N results in

$$I_1 = I_2 + I_3 + I_4 = (V_i - V_N)/Z_1 \tag{16.6}$$

The individual current flows in elements Z_1 to Z_5 are given by

$$\begin{aligned} I_2 &= V_N/Z_2 \qquad & I_3 &= (V_N - V_s)/Z_3 = V_N/Z_3 \\ I_4 &= (V_N - V_o)/Z_4 \qquad & I_5 &= (V_s - V_o)/Z_5 = -V_o/Z_5 \end{aligned} \tag{16.7}$$

Since there is no current flow into the op amp, $I_5 = I_3$. The algebra becomes simple if, instead of impedances Z_i, we use admittances $Y_i = (Z_i)^{-1}$. Substituting Eq. (16.7) into Eq. (16.6),

$$Y_1 V_1 = V_N(Y_1 + Y_2 + Y_3 + Y_4) - Y_4 V_o \tag{16.8}$$

Eliminating V_N and some algebra results in the transfer function

$$V_o/V_i = (-Y_1 Y_3)[Y_5(Y_1 + Y_2 + Y_3 + Y_4) + Y_3 Y_4]^{-1} \tag{16.9}$$

To synthesize a low-pass filter, let $Y_1 = Y_3 = Y_4 = R^{-1}$; $Y_2 = sC_1$; $Y_5 = sC_2$. Substituting these values into Eq. (16.9) yields

$$V_o/V_i = -[(R^2 C_1 C_2)s^2 + 3RC_2 s + 1]^{-1} \tag{16.10}$$

This is the expected form for a second-order low-pass filter provided that C_1 and C_2 are related by the parameter b:

$$C_1 = 3C/b; \qquad C_2 = bC/3$$

The cutoff frequency is $\omega_c = RC$.

16.3 HIGH-PASS ACTIVE FILTERS

High-pass active filters may be synthesized from low-pass filters by reversing the components R and C. An example of a *first*-order high-pass filter is shown in Fig. 16.7. This is the familiar differentiator circuit of Chapter 11, with additional components added to cut off the circuit response above the signal frequency range. Fig 16.8 shows a *second*-order high-pass filter using the multiple feedback approach of the previous section.

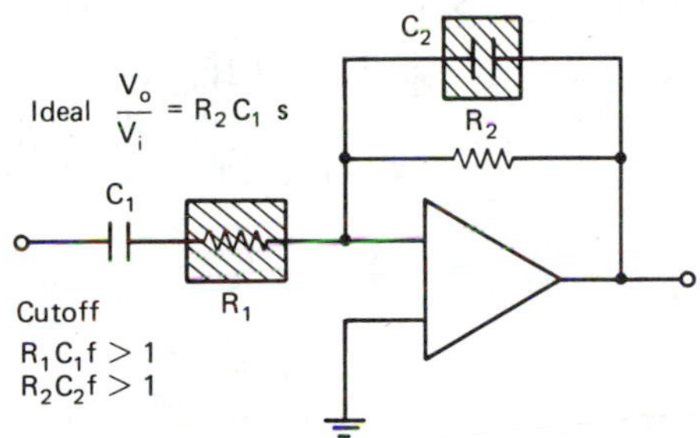

Figure 16.7 First-order high-pass active filter (differentiator). The shaded components are nonideal additions to stabilize high frequency behavior.

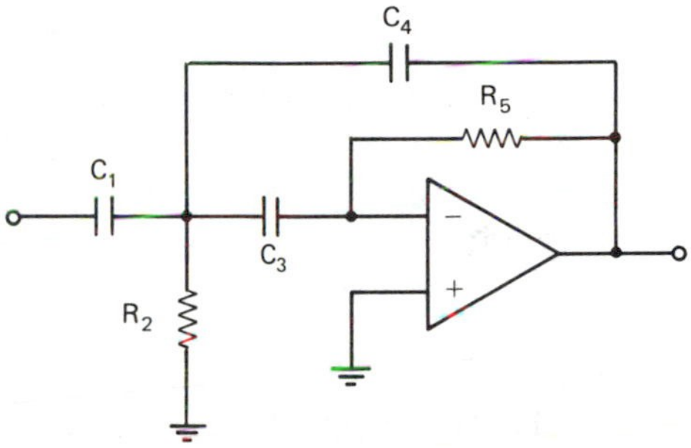

Figure 16.8 Second-order high-pass active filter, using multiple feedback design.

16.4 BANDPASS ACTIVE FILTERS

The strongest advantages of active filters come into play in bandpass filter design. Although passive bandpass filters utilize the resonance of an LC combination or its equivalent (e.g., quartz crystal), no inductor is necessary in any of the active methods to be shown. The general features desired in a bandpass filter are shown in Fig. 16.9. The principal characteristics are a sharp peak at the frequency ω_o where enhancement of the signal is desired, and the bandwidth, the distance $\Delta\omega$ over which the voltage amplitude falls by a factor of $2^{1/2}$ from its magnitude at the peak. This is conventionally expressed as the 3-dB point (or half-power point, since power $\propto V^2$). The bandwidth of a filter is conventionally expressed in terms of Q.

$$Q = \omega_o/\Delta\omega \qquad (16.11)$$

The most desirable design would allow *independent* variation of the center frequency and the bandwidth, each adjusted by a single component.

Most active bandpass filters emulate the shape of *the universal resonance curve*

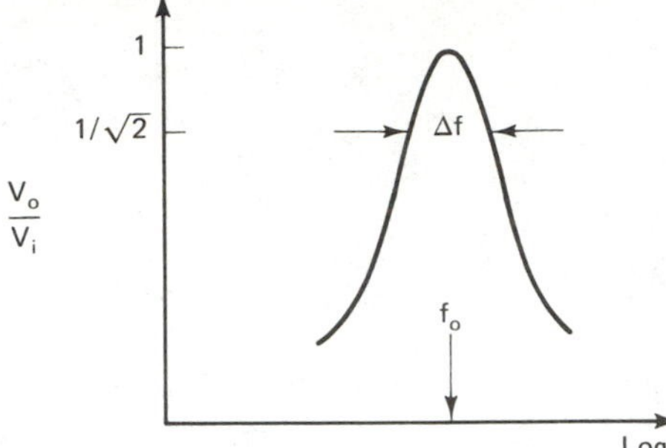

Figure 16.9 Desired shape of a bandpass filter transfer function.

$$V_o/V_i = G[1 + jQ(\omega/\omega_o - \omega_o/\omega)]^{-1} \tag{16.12}$$

where G is the gain at $\omega = \omega_o$. The magnitude has the form:

$$|V_o/V_i| = G[1 + Q^2(\omega/\omega_o - \omega_o/\omega)^2]^{-1/2} \tag{16.13}$$

Filters designed to resemble resonant behavior will have a transfer function similar to Eq. (16.13), and are called the *second-order section.* More complex bandpass filters with differently shaped transfer functions can be synthesized by combining such basic building blocks.

Four methods of synthesizing an active bandpass filter will be discussed here:

(1) *Multiple feedback method* A version of a general second-order filter of Fig. 16.6

(2) *The notch method* Infinite feedback impedance at a certain frequency, using RC networks whose output voltage falls to zero at a single frequency

(3) *Analog simulation* Like using an analog computer to simulate a passive RLC network whose differential equation is related to Eq. (16.12)

(4) *Biquad Bandpass* Simulating an active inductor (called the *gyrator* method)

Of these four most common methods, the multiple feedback method is in practice the most troublesome. The notch method requires only one op amp and provides a sharply peaked bandpass, but frequency adjustment requires the simultaneous variation of three components whose values must be closely matched. The bandwidth is not easily adjustable, and the shape of the resonance curve differs significantly from that of the universal filter [Eq. (16.13)]. The notch method is restricted to situations with a single permanently fixed frequency. Both analog simulation and biquad bandpass methods use three op amps, but this is no great disadvantage, since four general-purpose op amps are available within a single IC package. For example, LM 324 or 348 = four 741's; LF 347 = four 351's. These methods have the advantage that the center frequency and Q are independently adjustable. In addition, both filters have multiple outputs available simultaneously. Both provide a low-pass output, and the analog simulation filter also has a high-pass output. The biquad filter is preferred for its design simplicity whenever the bandpass plus low-pass output are sufficient. In cases where the component values of the biquad filter become too high for convenience or result in inadequate temperature stability, the analog simulation filter is preferred, since its parameters are adjusted by component *ratios.*

16.4.1 Multiple Feedback Bandpass Filter

An appropriate selection of components in the general multiple feedback section, (Fig. 16.6) will provide a bandpass function (Fig. 16.10). Inserting the specific components into the admittance formula [Eq. (16.9)],

$$\frac{V_o(s)}{V_{in}} = \frac{-s(R_1C_4)^{-1}}{s^2 + R_5{}^{-1}(C_3{}^{-1} + C_4{}^{-1})s + (R_5C_3C_4)^{-1}(R_1{}^{-1} + R_2{}^{-1})} \tag{16.14}$$

The natural frequency is, neglecting a small shift due to damping

$$\omega_o = [(C_3C_4R_5)^{-1}(R_1{}^{-1} + R_2{}^{-1})]^{1/2} \tag{16.15}$$

and the Q is

$$Q = [R_5(R_1{}^{-1} + R_2{}^{-1})]^{1/2}[(C_3/C_4)^{1/2} + (C_4/C_3)^{1/2}]^{-1} \tag{16.16}$$

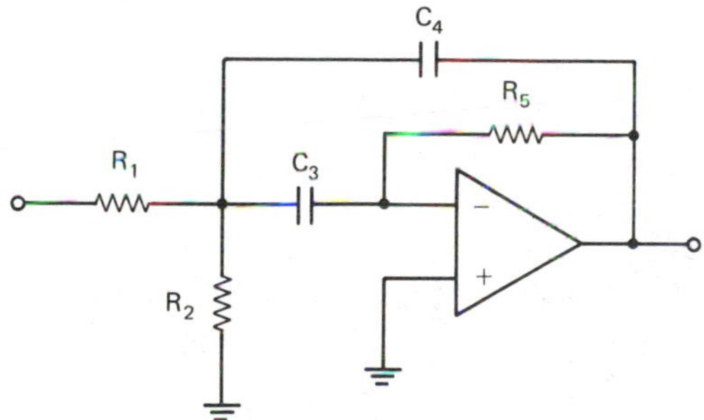

Figure 16.10 Second-order bandpass active filter using multiple feedback design.

Selecting component values to adjust ω_o, Q, and gain of this filter is tricky, because many components appear in more than one parameter. Assume that R_2 is large enough to be neglected compared to R_1. Since both Q and the gain depend only upon component *ratios*, ω_o may be adjusted by selecting the magnitude of C and R values, without changing Q or the gain, as long as the *ratios* C_4/C_3 and R_5/R_1 are kept unchanged. The value of Q may then be adjusted by selecting the ratio R_5/R_1. Since the same ratio appears in the gain, adjusting the Q also shifts the gain. A more serious disadvantage of this filter is that Q depends upon the square root of component ratios, so that a high-Q (~ 100) requires extremely high component ratios ($\sim 10^4$).

16.4.2 Bandpass Filter Using a Notch in the Feedback Loop

Perhaps the oldest method of creating a bandpass filter is to use a *notch* filter in the feedback loop. This creates a peak in Z_f at a certain frequency due to the null in the notch filter's output/input response. For example, the twin-T filter [Fig.

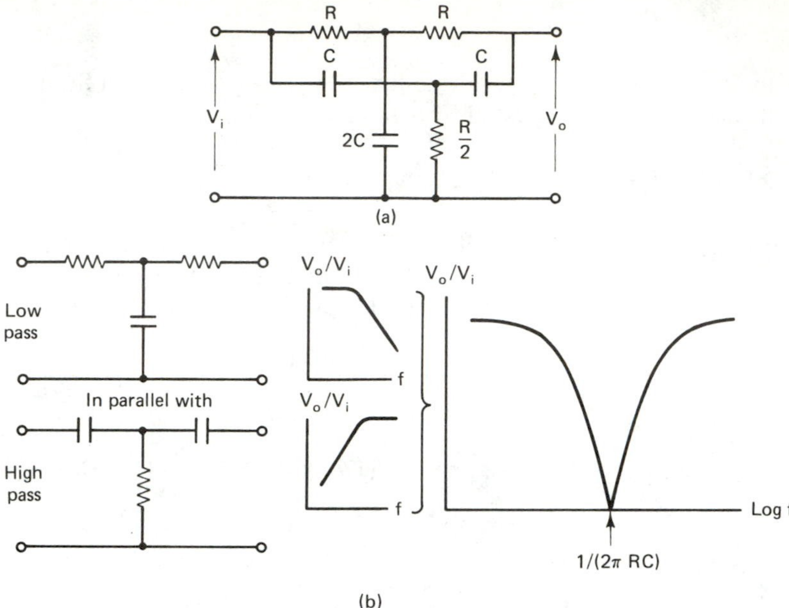

Figure 16.11 (a) Twin-T filter circuit. (b) The twin-T's transfer function viewed as the net effect of a high pass and a low pass filter in parallel.

16.11(a)] may be viewed as a low-pass filter in parallel with a high-pass filter [Fig. 16.11(b)]. Since the low-pass output falls off as frequency increases, and the high-pass output falls off as the frequency decreases, there will be a frequency in the middle where the output is a minimum. In fact, with the correct choice of component values [Fig. 16.11(a)] the output has a true null at a particular frequency [Fig. 16.11(b)]. The minimum is a true *zero* at the frequency at which the high-pass and low-pass contributions to the output are precisely equal and 180° out of phase. (A more precise description of the twin-T filter requires network analysis. See the references at the end of the chapter.) For the circuit shown in Fig. 16.11(a), the null frequency is

$$f_o = 1/(2\pi RC) \tag{16.17}$$

To see why a notch filter can act as an infinite impedance, consider a three-terminal network (Fig. 16.12). At the null, $I_1 = 0$, although there is still a current I_2 to ground. The transfer impedance $(V_2 - V_1)/I_1$ is infinite when I_1 falls to 0.

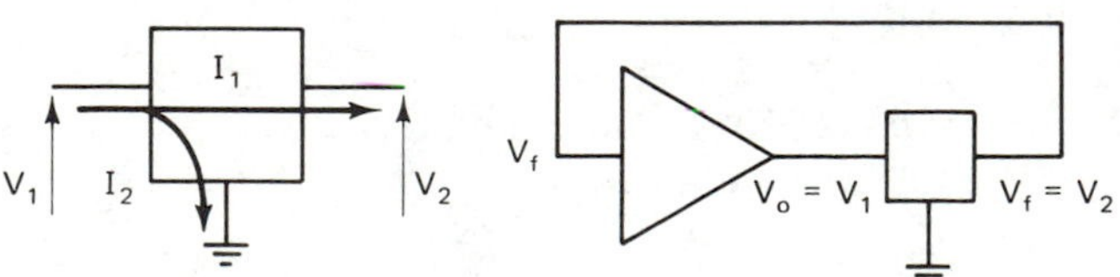

Figure 16.12 Three terminal network used as a feedback component. The third terminal can significantly alter the feedback current.

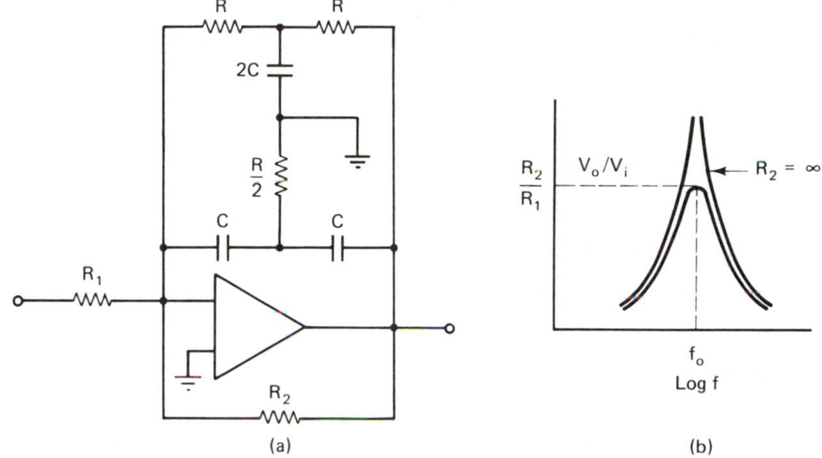

Figure 16.13 Twin-T active filter bandpass. (a) Circuit. (b) Transfer function.

A bandpass filter utilizing the twin-T notch is shown in Fig. 16.13(a). Resistor R_2 is added to adjust the gain at f_o. For reasons mentioned above, the twin-T circuit is only useful as a fixed-frequency bandpass. The peak can be a good deal sharper than the peak of a second-order section, however, giving excellent frequency selectivity.

16.4.3 Bandpass Using Analog Simulation

The same combination of op amp circuits that simulated the spring-and-mass mechanical system (Chapter 14) can simulate an electrical RLC system, since they share the same differential equation. An example is shown in Fig. 16.14. This filter uses one less op amp than the circuit of Fig. 14.8. A subtractor circuit replaces the summing amplifier, which eliminates the inverter stage.

The solution is found by considering how V_{HP} generates V_{BP} and V_{LP}

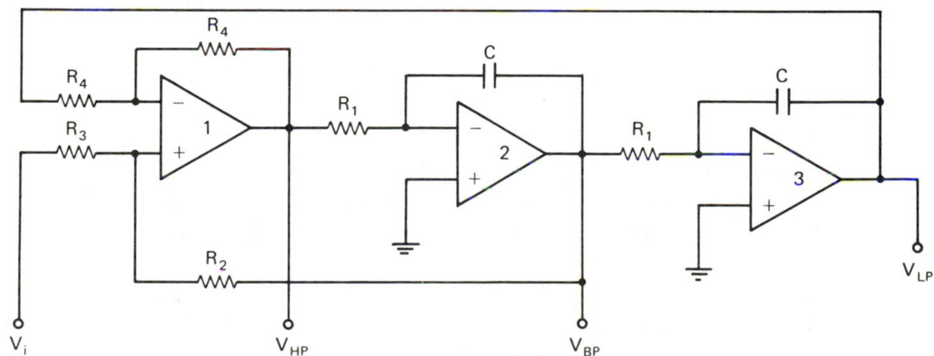

Figure 16.14 Analog simulation bandpass active filter (also called the state variable filter).

through op amps 2 and 3, and then how V_{HP} is generated by the same voltages through op amp 1 and the input V_i. From Fig. 16.14,

$$V_{BP} = -R_1 Cs V_{LP} \tag{16.18}$$

$$V_{HP} = -R_1 Cs V_{BP} = (R_1 Cs)^2 V_{LP} \tag{16.19}$$

V_{HP} is generated by the subtractor circuit of op amp 1. Its transfer function is found by equating $V_+ = V_-$, each being the weighted average of voltages at the ends of the resistor pairs (R_3, R_2) and (R_4, R_4). The result is

$$V_{HP} = -V_{LP} + 2V_{BP}(1 + R_2/R_3)^{-1} + 2V_i(1 + R_3/R_2)^{-1} \tag{16.20}$$

The output at each of the three points of interest may now be solved using Eqs. (16.18) and (16.19)

$$\frac{V_{LP}}{V_i} = \frac{2(1 + R_3/R_2)^{-1}(R_1 C)^{-2}}{s^2 + 2(1 + R_2/R_3)^{-1}(RC)^{-1}s + (R_1 C)^{-2}} \tag{16.21}$$

$$\frac{V_{BP}}{V_i} = \frac{-2(1 + R_3/R_2)^{-1}(R_1 C)^{-1}s}{s^2 + 2(1 + R_2/R_3)^{-1}(R_1 C)^{-1}s + (R_1 C)^{-2}} \tag{16.22}$$

$$\frac{V_{HP}}{V_i} = \frac{+2(1 + R_3/R_2)^{-1}s^2}{s^2 + 2(1 + R_2/R_3)^{-1}s + (R_1 C)^{-2}} \tag{16.23}$$

The relationship between this result and the bandpass filter parameters ω_o, Q, and the gain G may be seen by rewriting Eq. (16.12) in s-plane notation.

$$G(s) = G_o \alpha \omega_o s \; [s^2 + \alpha\omega_o s + \omega_o^2]^{-1} \tag{16.24}$$

Compare Eq. (16.24) with Eq. (16.22) and equate equal powers of s to determine the natural frequency and Q:

$$\omega_o = (R_1 C)^{-1} \tag{16.25}$$

$$\alpha = 1/Q = 2(1 + R_2/R_3)^{-1} \tag{16.26}$$

The tuning procedure for this filter is particulary simple, due to the independence of filter parameters. However, since the value of Q depends upon the ratio of two resistance values, the upper limit on Q with this filter is about 100.

16.4.4 Biquad Bandpass Using an Active Inductor

An equally useful bandpass filter (Fig. 16.15) is similar to the previous circuit, except that one of the integrator capacitors has a resistor in parallel. The combination formed by op amps 2 and 3 is in effect an *active inductor.* The transfer function inside the dashed box of 2 is identical with that of an inductor, even though what is inside the box is very different.

We calculate the current flow I_f out of terminal A in response to a voltage applied at terminal B. Terminal A, connected to the virtual ground of op amp 1, will be treated as a true ground. Current flow in resistor R_f depends upon voltage V_{o2}, which is determined by voltage V_B. The combination of op amps 2 and 3 forms a noninverting integrator.

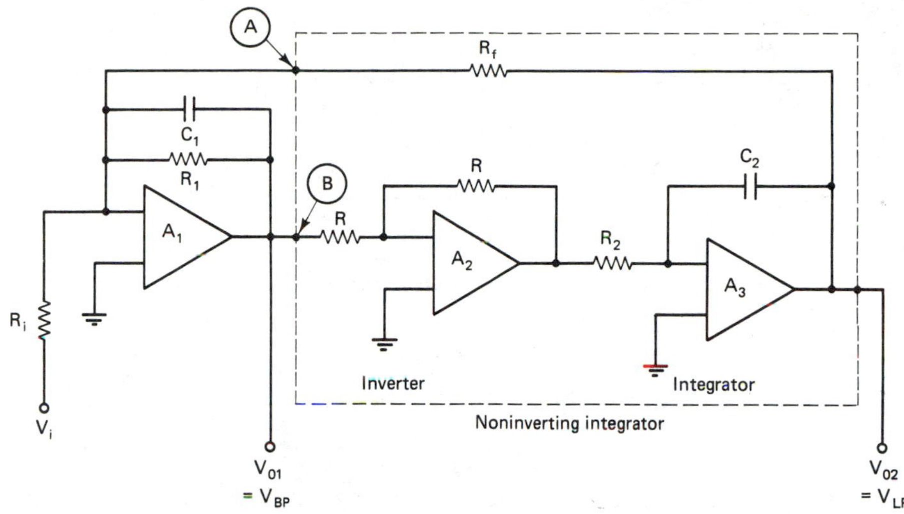

Figure 16.15 Biquad bandpass active filter. The dashed box acts as an active inductor viewed between terminals A and B.

$$V_{o2} = \frac{+V_{o1}}{R_2C_2s} = \frac{V_B}{R_2C_2s} \tag{16.27}$$

Since V_a is a virtual ground, the current in feedback resistor R_f is

$$I_f = \frac{V_{o2}}{R_f} = \frac{V_B}{R_fR_2C_2s} \tag{16.28}$$

This is identical in form with the I–V relationship of an inductor; $I(s) = V(s)/(sL)$. The *effective* inductance of this active inductor (sometimes called a *gyrator*) is determined by the values of two resistors and a capacitor:

$$L_{eff} = R_fR_2C_2 \tag{16.29}$$

The complete circuit is equivalent to a parallel RLC resonant filter in the feedback loop of op amp A1. The transfer function may then be evaluated using the golden rule.

$$V_o/V_i = Z_f/R_i \tag{16.30}$$

where Z_f is the parallel resonant RLC combination. The feedback admittance of the parallel RLC combination is

$$\begin{aligned} Z_f^{\ -1} &= R_1^{\ -1} + sC_1 + (sL)^{-1} \\ &= [s^2 + s(R_1C_1)^{-1} + (LC_1)^{-1}](sC_1^{\ -1}) \end{aligned} \tag{16.31}$$

If we introduce expressions for the resonant frequency ω_o and Q

$$Q = R_1/(\omega_oL); \quad \omega_o^2 = 1/(LC_1) \tag{16.32}$$

$$\omega_o/Q = \omega_o\alpha = \omega_o^2L/R_1 = 1/(R_1C)$$

The feedback admittance becomes

$$1/Z_f = (C_1/s)\ (s^2 + \alpha\omega_os + \omega_o^2) \tag{16.33}$$

and the transfer function takes on a useful standard form

$$V_o/V_i = G(s) = -Z_f/Z_c$$

$$= -\frac{s}{R_i C_1} \frac{1}{(s^2 + \alpha\omega_o s + \omega_o^2)} \tag{16.34}$$

The form, with a denominator quadratic in s and a numerator linear in s, is identical to the general form for the transfer function of a second-order bandpass filter. The relationship between filter parameters and component values is given by

$$\omega_o^2 = 1/LC_1 = (R_f R_2 C_1 C_2)^{-1} \tag{16.35}$$

$$Q = R_1/(\omega_o L) = \omega_o R_1 C_1 = R_1 C_1/(R_f R_2 C_1 C_2)^{1/2}$$

The gain at resonance is determined by the ratio R_1/R_i, because the inductive and capacitive components of the parallel impedance cancel at resonance. Component values appear more than once in Eq. (16.35), so filter parameters are interdependent. A noninteracting adjustment procedure can be found, if the adjustment is done in the proper order. The adjustment procedure is:

(1) First the natural frequency is chosen using C_1, C_2, and R_2. R_f may be used as a convenient frequency trim.

(2) Then Q is adjusted by varying R_1. Since R_1 does not appear in the expression for ω_o, the natural frequency is unchanged.

(3) Then adjust the gain by selecting the value of R_i, which does not appear in the expressions for Q and ω_o and will therefore not affect these parameters.

16.5 NOTCH ACTIVE FILTERS

The design of notch active filters which reject a specific frequency is very similar to that of bandpass filters, and will not be discussed in detail.

The most common notch filter utilizes a twin-T network in the input side [Fig. 16.16(a)]. The twin-T makes $I_1 = 0$ at a certain frequency by shorting the

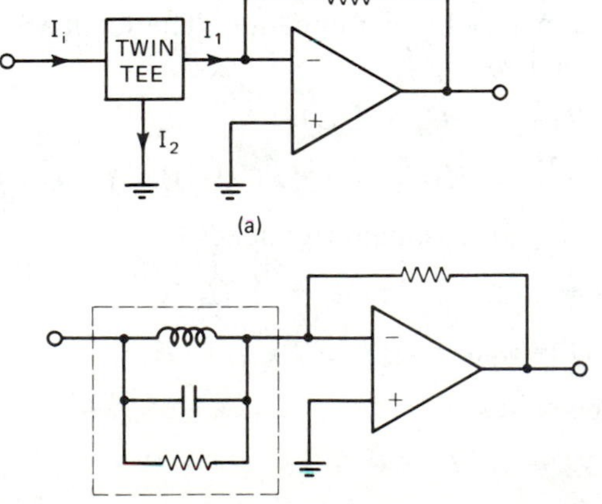

Figure 16.16 Notch filter designs. (a) Twin-T. (b) Resonant circuit. In both cases, the input impedance Z_i in effect go to infinity at the notch frequency.

input current to ground through I_2. Z_i is then effectively infinite, and the gain falls to zero.

A notch filter may also be designed with a parallel resonant circuit [Fig. 16.16(b)] in the input side. The inductor L shown within dashed lines may be replaced by an active inductor.

PROBLEMS

16.1. Suppose a signal contains noise of roughly equal amplitude (signal-to-noise ratio $S/N = 1$) but of 10 times higher frequency. Calculate S/N at the output of first- and second-order low-pass filters whose cutoff frequency has been selected to optimize the S/N enhancement. Sketch the waveforms at input and output.

16.2. Explore the analogy between the criterion $(R/L)^2 > 4/LC$ for no ringing of a series RLC circuit when excited by a step to the criterion $b < 2$ for the second-order low pass, Fig. 16.4 and Eq. (16.4). Look for 0's in Z of the series RLC and solve for the (complex) natural frequencies. Compare with the text result for the 0's in the denominator of Eq. (16.4).

16.3. Show that the transfer function of the general second-order section [Eq. (16.9)] becomes that of the second-order low pass [Eq. (16.4)] when the components of Fig 16.6 are used.

16.4. Consider the possibility of synthesizing a band reject filter which has a notch ($V_o/V_i = 0$) at some frequency ω_o, using the multiple feedback second-order filter of Fig. 16.6. Consider what criterion is necessary, and select what kind of components should go into the Z_n. If possible, make each lump a single component.

16.5. Solve for the transfer function of an analog simulation filter shown in Fig. 16.14, by analogy with the method of poles in the denominator (e.g., Eq. 16.4). Solve for the bandpass, low-pass, and high-pass transfer functions. Show that

$$\omega_o = (R_1 C)^{-1}$$

$$\alpha = (R_4/R_2)$$

$$G_{BO} = -(R_2/R_3)$$

REFERENCES

More complete bibliographic information for the books listed below appears in the annotated bibliography at the end of the book.

GRAEME, *Designing with Operational Amplifiers*

HIGGINS, *Experiments with Integrated Circuits*, Experiment 23

HOENIG & PAYNE, *How to Build and Use Electronic Devices without Frustration, Pain, etc.*

JUNG, *IC Op Amp Cookbook*

PHILBRICK, *Applications Manual*
ROBERGE, *Operational Amplifiers*
TOBEY, GRAEME, & HEULSMAN, *Operational Amplifiers*

17

Oscillators and Function Generators

17.1 OSCILLATOR DESIGN IDEAS AND METHODS

Oscillators are used to generate test signals to probe the response of electronic, physical, or chemical systems. There are many excellent op amp oscillator circuits, some of which will be explored in this chapter. Since oscillators are a practical application of the stability ideas discussed in Chapter 14, it is useful to begin with a review of those ideas. An overview will then be given of various classes of oscillators and function generators. The balance of this section describes the design and analysis of specific oscillator types.

17.1.1 Basic Oscillator Ideas

What makes an oscillator? The origin of oscillation is related to the basic feedback circuit shown in Fig. 17.1(a), whose transfer function is

$$V_o/V_i = a/(1 - \beta a) \qquad \text{closed-loop gain} \qquad (17.1)$$

In most negative feedback situations, one avoids instability by making sure that βa never reaches 1. To make an oscillator, however, one ensures that βa *does* reach 1 at some frequency, where the divergence in Eq. (17.1) provides an output even in the absence of an input. It is useful to open the feedback loop [Fig. 17.1(b)] and calculate the *open-loop transmission* from the voltage $V_f = \beta V_o$ fed back.

$$V_f/V_i = \beta a \qquad \text{open-loop transmission} \qquad (17.2)$$

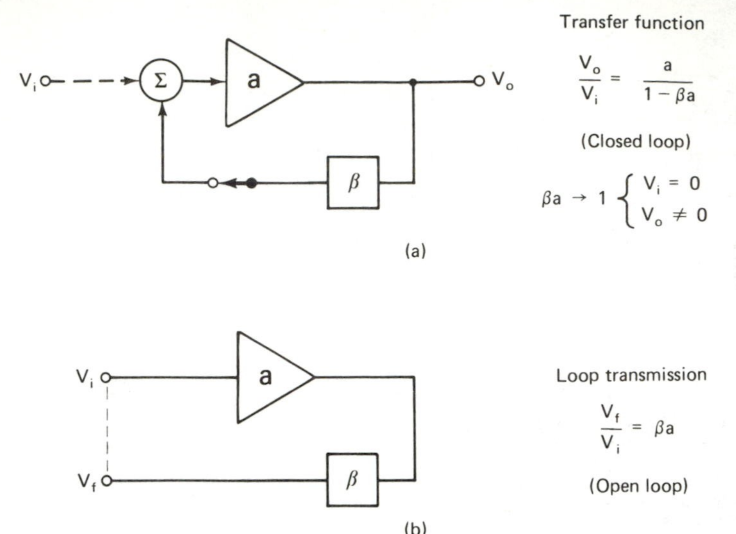

Figure 17.1 (a) Closed loop structure of a feedback oscillator. (b) Open loop structure and the "loop transmission" factor.

The necessary and sufficient condition for oscillation is that the open-loop transmission βa reach 1, so the denominator of the transfer function reaches 0.

What determines the frequency? Suppose that a is independent of frequency and has a negative sign. An oscillator may be constructed by selecting a feedback network where the phase of β varies rapidly and crosses π at some value of ω_o. This will select the frequency at which the circuit *could* oscillate. The criterion for oscillation can also be met by using a noninverting amplifier. In this case, the circuit will oscillate when the phase of β passes through 0. An example of an oscillator's Nyquist diagram is shown in Fig 17.2(a). As the frequency increases, the vector βa follows the contour shown, and crosses 1 at some ω_o which

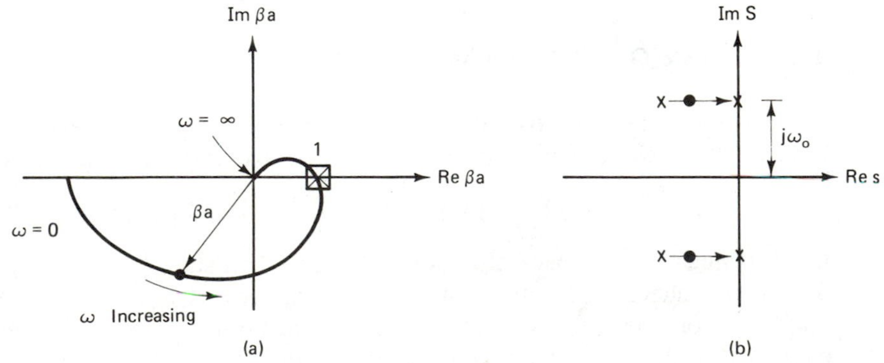

Figure 17.2 (a) Example of contour traced by βa as a function of frequency in an oscillator. The contour crosses 1 and hence would oscillate ($\beta a = 1, \beta a - 1 = 0$). (b) Example of the location in the s-plane of the complex pole pair of an oscillator. When $\beta a = 1$, the poles lie on the imaginary axis.

determines the frequency of the oscillator. This corresponds to poles in the complex transfer function crossing the imaginary axis, shown in Fig. 17.2(b). The criterion for oscillation may be summarized as

$$\begin{aligned} \beta(\omega_o)a = 1 &\longrightarrow \text{oscillation at } \omega_o \\ \text{if } a < 0, \quad \Phi_\beta(\omega_o) &= \pi \\ \text{if } a > 0, \quad \Phi_\beta(\omega_o) &= 0 \end{aligned} \tag{17.3}$$

where $\Phi_\beta(\omega_o)$ is the phase of β at frequency ω.

Will it oscillate? The circuit will oscillate only if the amplifier gain is large enough so that $|\beta a|$ can be 1 when Φ_β crosses the critical value. Therefore,

$$a \geqslant |1/\beta(\omega_o)| \tag{17.4}$$

In addition, a stable oscillator frequency will result if the phase Φ_β passes through π or 0 rapidly (curve 2 in Fig. 17.3) since this provides a very narrow range of frequency where oscillation may occur. Otherwise, when Φ_β passes slowly through the critical value (curve 1), a variation of a or of $|\beta(\omega)|$ with temperature can shift the operating frequency.

What starts the oscillation, and what sets the amplitude? The oscillation is triggered by any fluctuation; any unstable circuit may suddenly oscillate when a transient or other noise occurs. The fluctuation in circuit parameters caused by turning on the power is enough to start most oscillators. For example, it is sufficient if the curve of [Fig. 17.2(a)] merely encircles the point 1, since the instability point will necessarily be crossed as $|a|$ increases from 0 when power is turned on. The amplitude of oscillation generally involves both the feedback amplitude and some nonlinear response in the circuit at large signal levels. In a

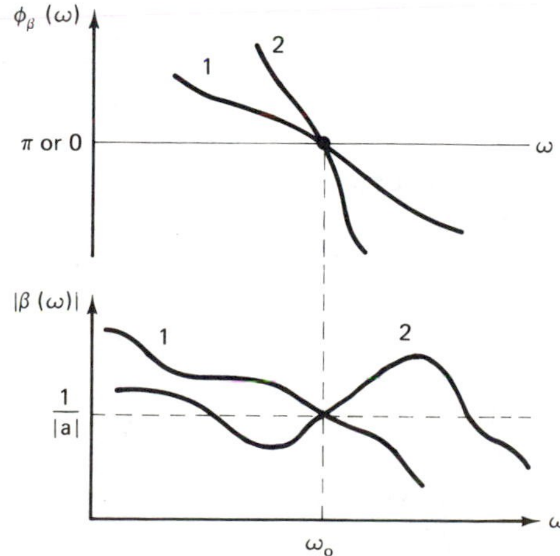

Figure 17.3 Example of a Bode plot for an oscillator: the phase (a) and magnitude (b) of βa as a function of frequency. In case 2, phase varies more rapidly through the critical value than in case 1, corresponding to a more stable frequency. However, the magnitude of $\beta(\omega_o)$ in case 2 is not quite large enough for sustained oscillations, although it is in case 1.

discrete-component oscillator, with values of $|a| \leqslant 100$, the amplitude rises until the output voltage swings outside the linear portion of the transistor characteristics and gets clipped. The design value of $|a|$ is chosen just larger than the critical value [Eq. 17.4], and the amplitude rises until $|a|$ is reduced by nonlinear effects to the critical value for sustained oscillation. One would think that this mechanism would result in severely distorted sine waves. However, as long as the phase varies rapidly near ω_o, sustained oscillation will occur only at one frequency. The harmonics of the clipped wave are fed back at the wrong phase to reinforce themselves, and negative feedback restores the output to nearly sinusoidal form.

With op amp oscillators where $|a|$ is much larger, or in any situation where the distortion must be kept low, amplitude-limiting techniques are preferable to reliance upon amplifier nonlinearity. Examples are discussed below.

17.1.2 Classes of Oscillators

RC feedback oscillators. This class of oscillators uses an RC network to achieve a phase shift of π ($a < 0$). Examples include phase shift oscillator, twin-T oscillator, Wien bridge oscillator, and multiple feedback oscillator.

The multiple feedback oscillator and similar analog simulation designs use several op amps, and are principally used where ultralow distortion levels or several signals in precise 90° phase relationships to one another are required. This oscillator is analyzed in Prob. 17.4.

Resonant circuit oscillators. A high-Q resonant circuit provides the input to an amplifier, whose output is fed back to the resonant circuit in the right phase to sustain the oscillation. The resonant circuit may involve RLC components, a quartz crystal, or active simulation (simulated inductor).

Switching oscillators. In contrast to linear oscillators, switching oscillators use a nonlinear two-state circuit element (transistor switch or an op amp comparator) to switch at a certain voltage level. An RC charging network feeds the switch, whose output is fed back to the RC network. Examples include sweep circuits and function generators.

Negative resistance oscillators. An instability is generated by a device whose characteristic curves include a negative resistance ($dI/dV < 0$) region. This can be used to drive either a nonlinear switching oscillator (*relaxation* oscillator) or a linear resonant circuit oscillator. Examples of devices with negative resistance regions include tunnel diode, diac, unijunction transistor, and neon lamp (see Fig. 14.1 and Prob. 14.10).

Negative resistance oscillators are particularly useful in power supply circuits and for ultrahigh-frequency applications. However, since they typically employ discrete devices rather than integrated circuits, their analysis falls outside the thrust of this book.

TABLE 17.1 COMPARISON OF OSILLATOR FEATURES

	Frequency stability	Pure sine wave	Easy change of frequency	Many wave shapes	Can be triggered
RC feedback					
phase shift	Fair	Fair	Poor	No	No
Wien bridge	Good	Exc	Good	No	No
twin T	Good	Good	Poor	No	No
Resonant Circuit					
RLC	Good	Exc	Poor	No	No
active	Good	Exc	Fair	No	No
crystal	Exc	Exc	No	No	No
Switching					
sweep	Fair	No	Exc	Good	Exc
function generator	Fair	Poor	Exc	Exc	Good
Negative resistance	Fair	Fair	Exc	No	No

Each class of oscillators has advantages and disadvantages which determine which is preferred in a given application. A summary is shown in Table 17.1. For applications where maximum frequency stability is desired but the frequency does not need to be changed, a crystal oscillator is chosen. For pure sine waves, either a tuned circuit oscillator or, if the frequency must be readily variable, a Wien bridge oscillator is chosen. The widest range of frequency variation is available for a variety of waveforms in the function generator.

17.2 RC FEEDBACK OSCILLATORS

17.2.1 Phase Shift Oscillator

Although conceptually simple, the phase shift oscillator is not used as a general-purpose device, because a frequency change involves varying three components simultaneously. The basic circuit is shown in Fig. 17.4. Each RC stage can provide a maximum phase shift of 90°, so the three-stage RC network can easily provide the 180° phase shift needed for positive feedback. (Two RC stages would not unless stray capacitance were also present.) The feedback network is driven by an amplifier of finite gain A; the amplifier input comes from feedback network. The combination provides (given the proper phase shift) the necessary reinforcement or *regeneration* to create an oscillatory signal.

A rough analysis of the phase shift oscillator begins by assuming that each stage produces a phase shift of 60°. The transfer function of each stage is

$$\frac{V_o}{V_i} = \frac{j\omega RC}{1 + j\omega Rc} = \frac{(\omega RC)^2 + j\omega RC}{1 + (\omega RC)^2} = X + jY \qquad (17.5)$$

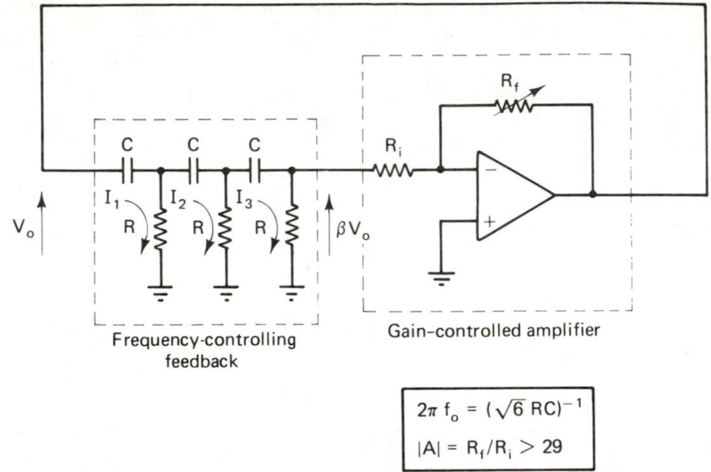

Figure 17.4 Circuit diagram for the basic RC phase shift oscillator.

The necessary 60° per stage comes when

$$\tan^{-1}\frac{Y}{X} = \tan^{-1}\frac{1}{\omega RC} = 60°$$

$$\omega_o = \frac{0.58}{RC} \tag{17.6}$$

At that frequency, the attenuation is

$$\frac{V_o}{V_i} \simeq 0.5 \tag{17.7}$$

so that the total attenuation of three stages is roughly

$$\beta = \frac{V_f}{V_i} = (0.5)^3 = \frac{1}{8} = \frac{1}{A(\omega_o)} \tag{17.8}$$

In practice, the stages are not independent and a more accurate solution is required. Simple network theory shows the natural frequency to be

$$\omega_o = (6^{1/2}RC)^{-1} = \frac{0.408}{RC} \tag{17.9}$$

and to sustain oscillations, the amplifier gain must be at least

$$|A| > \left|\frac{1}{\beta}(\omega_o)\right| = 29 \tag{17.10}$$

Exercise

Show that the natural frequency and minimum amplifier gain for the phase shift oscillator are as given in Eqs. (17.9) and (17.10).

Solution The simplest approach is by the *mesh method* of network analysis. The feedback voltage $\beta V_o = I_3R$, where I_3 is the current flowing in the right-hand mesh of the network (see Fig. 17.4). It is assumed that there are no loading effects between the amplifier and the network. The solution to the network problem is

$$\beta = I_3R = \{ [1 - 5(\omega RC)^{-2}] + j [(\omega RC)^{-3} - 6(\omega RC)^{-1}] \}^{-1} \qquad (17.11)$$

Since the amplifier has a negative gain, the criterion for oscillation is $\Phi_\beta = \pi$. This can only happen if β lies along the real axis or $Im(\beta) = 0$. This requirement determines the natural frequency

$$\omega_o = (6^{1/2}RC)^{-1} \qquad (17.12)$$

Although this ensures a frequency value at which Φ_β will cross π, it does not ensure that a contour of βa crosses through 1, as in Fig. 17.2(a). The circuit will oscillate at the natural frequency only if the gain is large enough so that

$$|1/A| < Re[\beta(\omega_o)] = \left[1 - \frac{5}{(\omega_o RC)^2}\right]^{-1} = \left[1 - \frac{5}{1/6}\right]^{-1} = \frac{1}{29} \qquad (17.13)$$

This process may be understood either as shifting the contour plot of βa to cross 1 [Fig. 17.2(a)] or as shifting the poles in the transfer function to intersect the imaginary axis [Fig. 17.2(b)].

The quadrature oscillator (Fig. 17.5) is similar in principle to the phase shift oscillator but has an additional output 90° out of phase. The analysis is simpler than for the phase shift oscillator. If all three time constants R_iC_i are made equal, the natural frequency is simply $(RC)^{-1}$. The analysis is a useful exercise in transfer functions (note the noninverting integrator) and stability analysis (see Prob. 17.2).

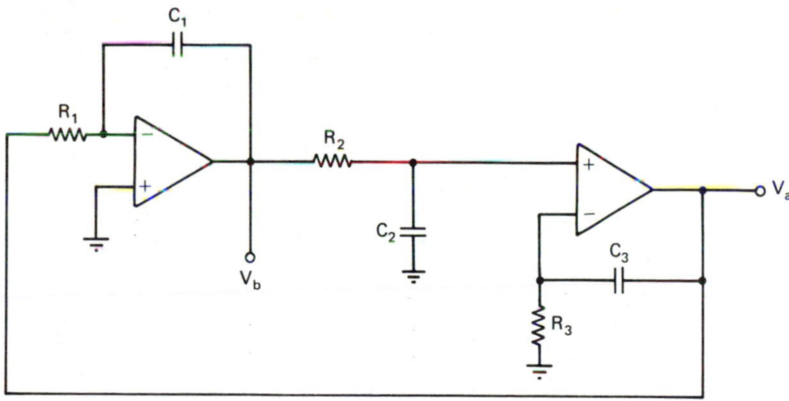

Figure 17.5 Circuit diagram for the quadrature oscillator.

17.2.2 Twin-T Oscillator

This circuit uses negative feedback and a notch filter in the feedback loop to eliminate feedback at a certain frequency. A working circuit is shown in Fig. 17.6. With an ideally balanced twin-T, no current is passed at a frequency ω_o where the currents in the upper and lower portions are equal and opposite. The effective impedance of the feedback network (and hence the circuit gain) is therefore very large. In addition, the T is unbalanced to give a net 180° phase shift to the feed-

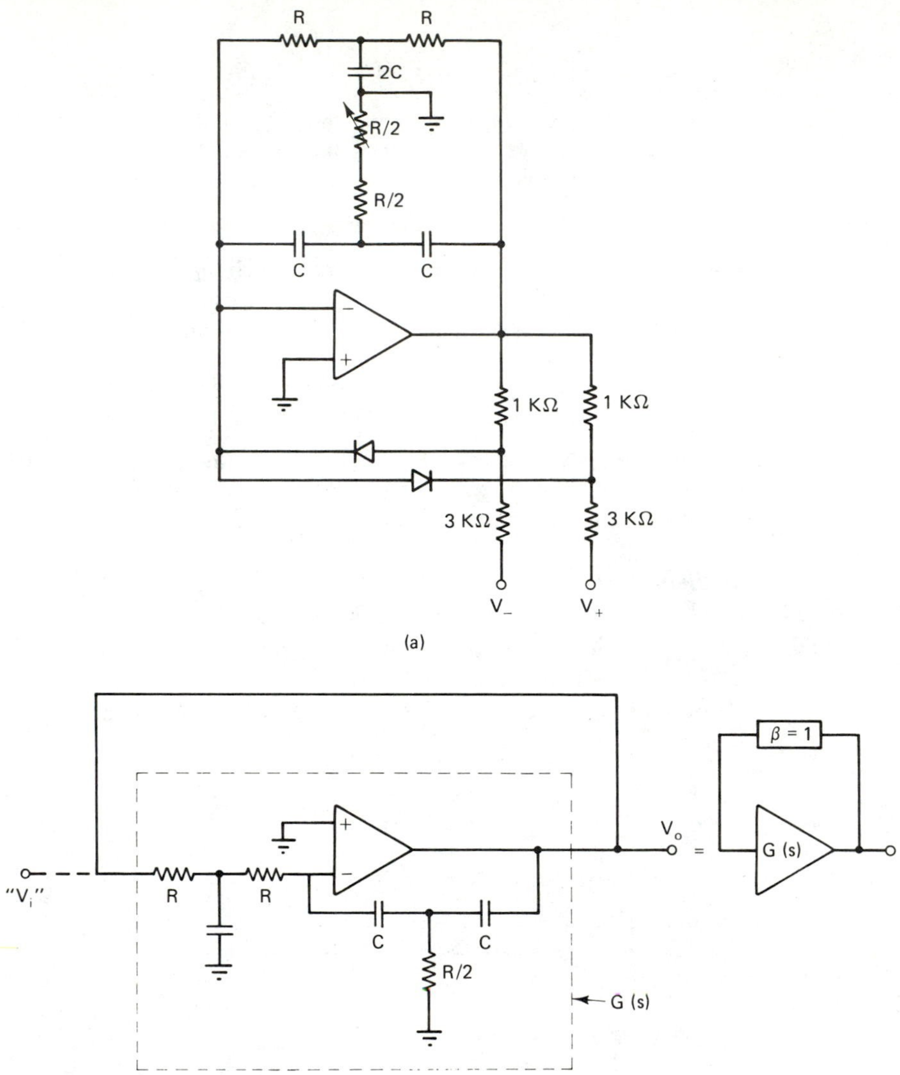

Figure 17.6 (a) Working circuit of twin-T oscillator, with diode limiting of amplitude. (b) Twin-T oscillator redrawn as a double integrator.

back signal at ω_o. The amplitude of the oscillation is limited by the second (negative) feedback path including a diode clamp for gain limiting (see below).

Although this circuit can be analyzed using methods similar to the phase-shift oscillator, it is more illuminating to use an *s*-plane view because it introduces a new way of making an oscillator. The twin-T oscillator circuit, redrawn in Fig. 17.6(b), is identical with the double integrator (Problem 14.3) whose transfer function is

$$\frac{V_o}{V_i} = \frac{1}{(RCs)^2} = G(s)_{\text{open loop}} \tag{17.14}$$

The wire closing the feedback loop in Fig. 17.6(b) is equivalent to setting the feedback fraction $\beta = 1$. This gives the closed loop transfer function

$$\frac{V_o}{V_1} = \frac{G(s)}{1 - \beta G(s)} = \frac{-(RCs)^{-2}}{1 + (RCs)^{-2}} \tag{17.15}$$

The circuit will oscillate at the value of s which makes the denominator equal to 0, or

$$s_o = \pm\, j/RC$$

$$\omega_o = 1/RC \tag{17.16}$$

Another way of making an oscillator, then, is to find an amplifier with a frequency dependent transfer function whose value passes through -1 at some frequency, and connect the amplifier in a unity gain follower configuration.

Although in theory the poles are pinned to the imaginary axis of the s-plane, in practice component values are not ideally matched, and the poles may shift. The twin-T is therefore *detuned* with a variable resistor [Fig. 17.6(a)]. The poles are shifted to the right to ensure *instability*, and a limiter network is added for amplitude control.

17.2.3 Wien Bridge Oscillator

This is perhaps the most common sine wave oscillator design used in commercial signal generators, because of the ease of tuning over a wide range, the relatively good frequency stability, and the low harmonic distortion. The circuit [Fig. 17.7(a)] contains a four-terminal bridge as a feedback network. Note the presence of both a negative feedback path to the inverting terminal and a positive feedback path to the noninverting terminal. The differential signal fed to the op amp is

$$\frac{V_+ - V_-}{V_o} = \left[3 + \left(j\frac{\omega}{\omega_o} - \frac{\omega_o}{\omega}\right)\right]^{-1} - \left[1 + \frac{R_2}{R_1}\right]^{-1} \tag{17.17}$$

where $\omega_o = 1/(RC)$. The goal is to provide a single frequency at which the output of the bridge is 0, which eliminates the stabilizing feedback, and hence results in oscillations. This requires that both the real and the imaginary parts of Eq. 17.17 vanish. The imaginary part vanishes at ω_o, which sets the natural frequency, and the real part vanishes at the same frequency provided that $R_2/R_1 = 2$.

Another way of drawing the same circuit makes the feedback paths look simpler and makes the instability easier to understand [Fig. 17.7(b)]. One of the basic blocks is an op amp follower with gain. The input to this block is provided by the series parallel RC network which sets the feedback fraction β. The generalized block diagram is given in Fig. 17.7(c), where the original circuit, with both negative and positive feedback paths, has been reduced to the familiar negative feedback connection. Instability results when βA reaches 1, where A is the gain of the amplifier block.

$$A = 1 + \frac{R_2}{R_1} \qquad \beta = \frac{Z_p}{Z_s + Z_p} \tag{17.18}$$

where Z_p and Z_s are the impedances of the parallel series RC networks.

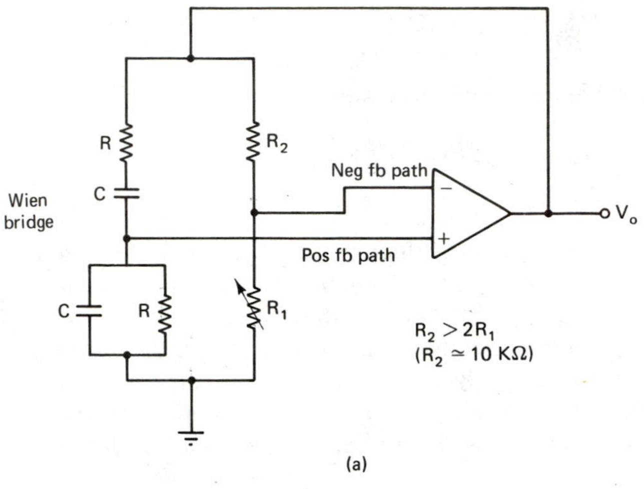

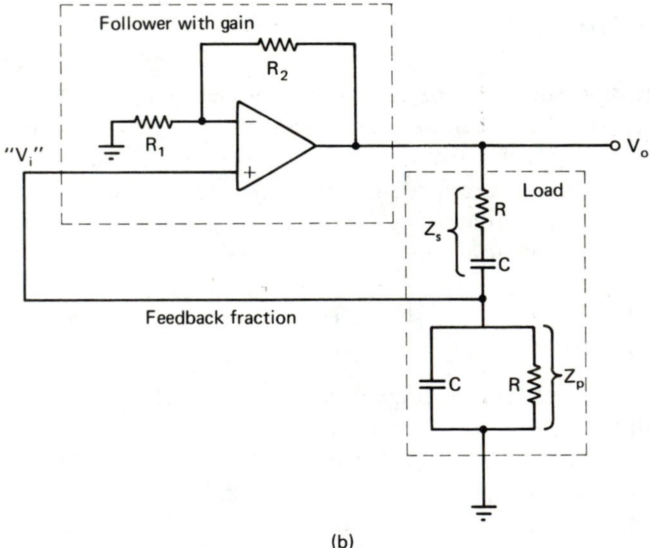

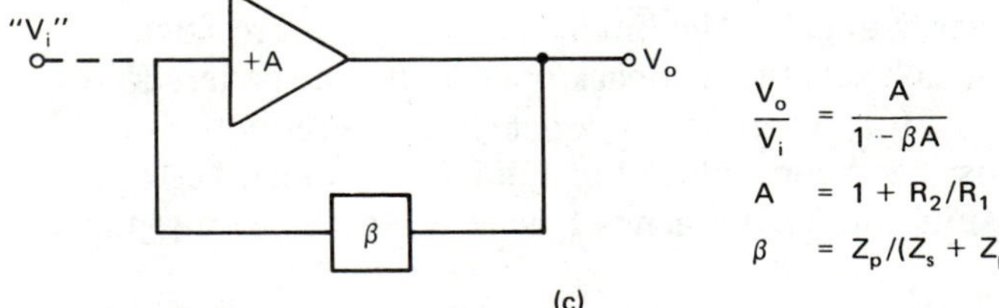

$$\frac{V_o}{V_i} = \frac{A}{1 - \beta A}$$

$$A = 1 + R_2/R_1$$

$$\beta = Z_p/(Z_s + Z_p)$$

Figure 17.7 Wien bridge oscillator circuit. (a) Basic circuit viewed as a bridge. (b) The same circuit redrawn as a follower with gain, fed by a series-parallel RC network. (c) Equivalent circuit for (b).

$$Z_p = \frac{R}{1 + RCs} \qquad Z_s = \frac{1 + RCs}{Cs} \tag{17.19}$$

This reduces to

$$\beta = \frac{RCs}{(RCs)^2 + 3RCs + 1} \tag{17.20}$$

The conditions for oscillation are

$$0 = 1 - \beta A = 1 - \frac{RCs(1 + R_2/R_1)}{(RCs)^2 + 3RCs + 1}$$

$$0 = (RC)^2 s^2 + RC\left[3 - 1 + \frac{R_2}{R_1}\right]s + 1 \tag{17.21}$$

For sustained oscillations rather than a damped resonance, the damping (term linear in s) must be 0 or negative. This is equivalent to shifting the poles across the imaginary axis as in Fig. 17.2(b). This requires that

$$1 + \frac{R_2}{R_1} \geqslant 3;$$

$$R_2 \geqslant 2R_1 \tag{17.22}$$

The inequality ensures that the poles lie in the right half-plane. The location on the imaginary axis determines the natural frequency of oscillations, which, for $R_2 = 2R_1$, is given by

$$0 = (RC)^2 s^2 + 1$$

$$s_o = \pm\, j/(RC) \quad \text{or} \quad \omega_o = 1/(RC) \tag{17.23}$$

Circuits such as the Wien bridge have enhanced frequency stability compared to the phase shift oscillator (Fig. 17.8). The phase changes *rapidly* near the frequency of oscillation, so a shift in phase angle due to changes in component values will shift the natural frequency only slightly. As a result, the value of ω_o is tightly defined.

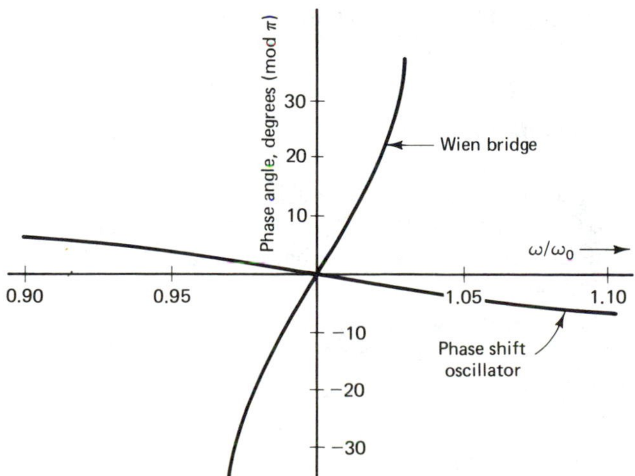

Figure 17.8 The frequency stability of an oscillator is connected with the rate at which the phase passes through the condition of instability. Here, the Wien bridge oscillator is far more stable than the phase shift oscillator, as component values change and the phase angle shifts.

17.2.4 Gain-Limiting Circuits for Oscillators

To ensure continued stable oscillations in spite of component value variations, the gain of the amplifier is normally made higher than the critical value, as shown by the inequality in Eq. (17.22). This shifts the poles of the transfer function slightly to the right of the imaginary axis in the s-plane, into the region of unstable growing oscillations. To avoid the distortion which occurs if the amplitude grows to the value of the power supply voltage, some means of automatic limiting is needed. Modern limiter circuits use active devices in the feedback loop. One example [Fig. 17.9(a)] uses diodes which turn on whenever the amplitude of the signal exceeds the desired output. The resistor network is a pair of voltage dividers whose center taps are the weighted averages of the output signal and a reference voltage. Whenever the voltage across a diode exceeds the forward turn-on voltage, a new path for negative feedback appears, and the gain is lowered because a much smaller 1-KΩ resistor is connected across the feedback loop. This circuit introduces some harmonic distortion, which, although not visible on an oscilloscope, is still appreciable ($\geqslant 1\%$). A more sophisticated circuit [Fig. 17.9(b)] uses an FET as a voltage-controlled resistor at the amplifier input. The transistor is driven by a dc signal generated from the output sine wave. As the output signal amplitude rises, the FET channel resistance increases, which decreases the amplifier gain and reduces the output amplitude. The dc correction signal is generated by a rectifier-integrator combination which incorporates a very clever feedback idea. The integrator circuit will ramp continuously up or down depending upon whether the current from the (negative) reference signal or the (positive) output sample is larger. Negative feedback adjusts the output signal level to keep the average current flow to the integrator zero, and in doing so automatically biases the FET into its correct region of operation.

17.3 TUNED CIRCUIT OSCILLATORS

For applications where a stable frequency is needed, a tuned circuit oscillator is preferred. The characteristic frequency may be set by an LC resonant circuit, with a real or simulated inductor, or by a quartz crystal.

17.3.1 LC Resonant Circuit Oscillator

The basic idea of all resonant tuned circuit oscillators is shown in Fig. 17.10, which is based upon the classic Colpitts oscillator. Suppose there is a current oscillating in the LC circuit. The current creates a feedback signal which has the right phase to keep the oscillation going. With respect to the ground point between the two capacitors, the LC resonator causes a phase difference of 180° between the op amp input and output so that the oscillations are reinforced by negative feedback. The frequency is $\omega_o = [L(C_1 + C_2)]^{-1/2}$.

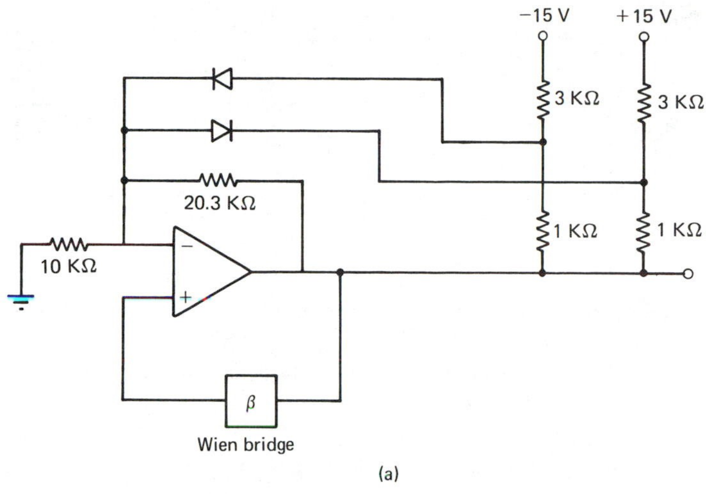

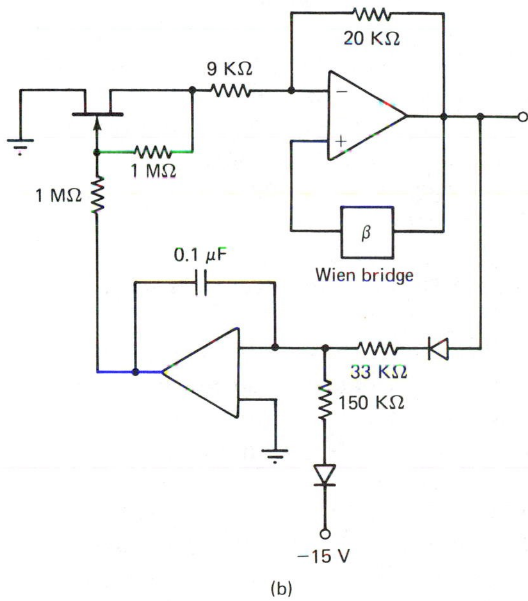

Figure 17.9 Amplitude limiter circuits using the Wien bridge oscillator as an example. (a) Diode limiter. (b) FET (voltage controlled resistor) limiter.

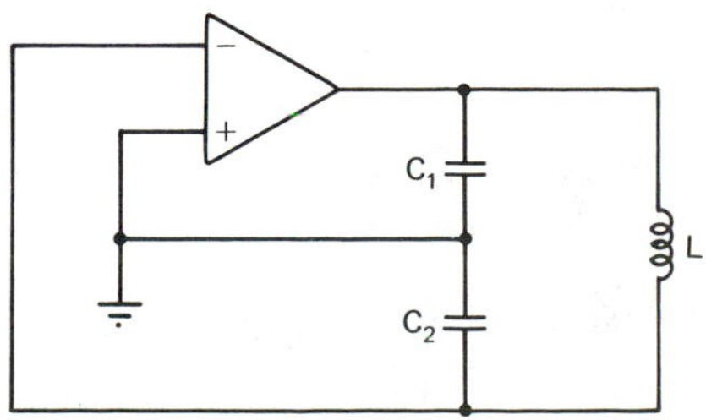

Figure 17.10 LC resonant circuit or Colpitts oscillator.

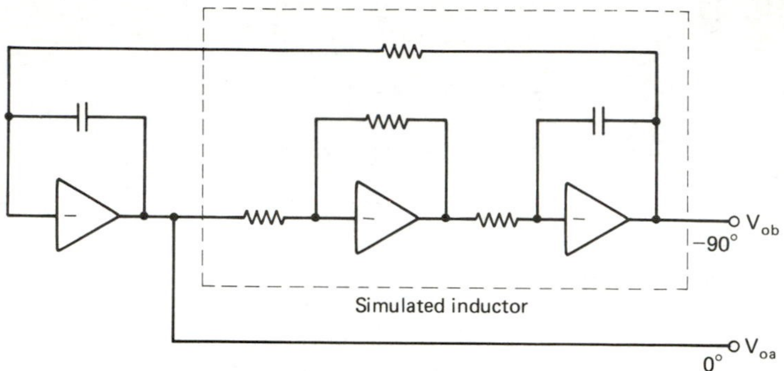

Figure 17.11 Biquad oscillator with simulated inductor.

17.3.2 Simulated Inductor (Biquad) Oscillator

As with active filters, the bulky and expensive inductor may be replaced by a simulated inductor. An oscillator based on this principle may be made simply by removing the damping resistor from the biquad bandpass filter (Chapter 16), as shown in Fig. 17.11. This is functionally identical with an RLC bandpass with infinite "gain," i.e., an output in the absence of an input. This oscillator has a second output 90° out of phase with the first. The circuit can be shown (Prob. 17.6) to be closely related to the analog simulation oscillator (Fig. 17.22) or to the quadrature oscillator (Fig. 17.5).

17.3.3 Quartz Crystal Oscillator or Clock

In applications requiring highest frequency stability, a quartz crystal replaces the LC resonant circuit. High stability results because the low thermal expansion of quartz makes the mechanical resonance relatively insensitive to temperature. The electrical coupling to the mechanical resonance is a consequence of the piezoelectric effect, with the equivalent circuit shown in Fig. 17.12(a). The capacitor plated onto the quartz crystal couples the mechanical resonance to the electrical circuit, but also adds a parallel capacitance C' to the series LC equivalent of the crystal resonance. The combination has two resonances [Fig. 17.12(b)], which are quite close together, since $C' >> C$. A working oscillator can lock on to either one. Two practical crystal clock circuits are shown in Fig. 17.12(c).

17.4 SWEEP GENERATOR

The following sections will explore sweep generators used in oscilloscopes, basic function generators and voltage-controlled oscillators used in many test instruments, and common IC function generators.

A sweep generator is the basis for the horizontal deflection circuit of an oscilloscope. It is also useful in situations where an experimental parameter is to be

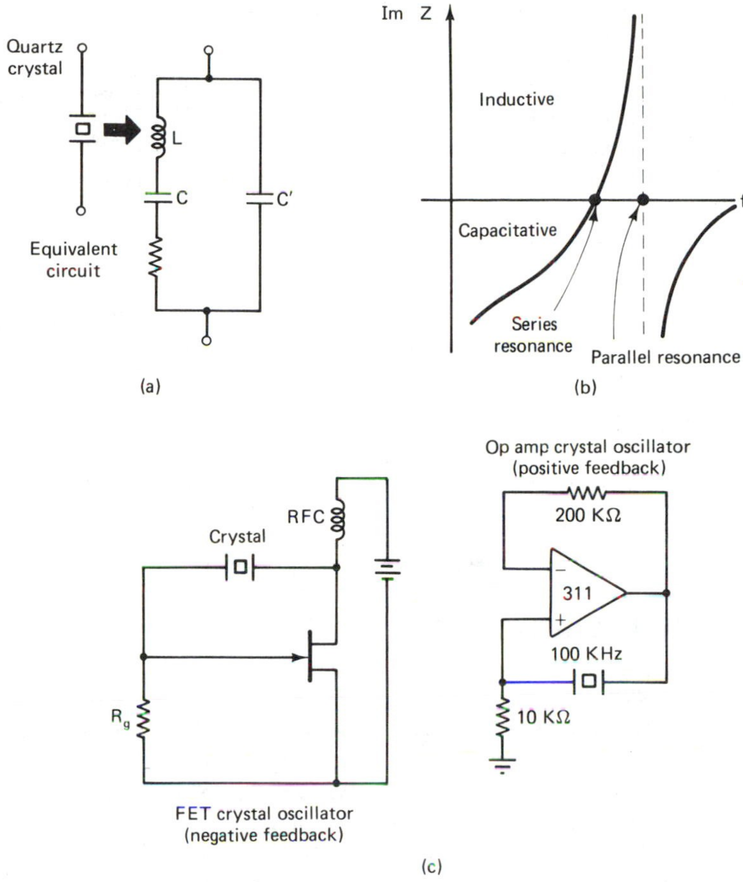

Figure 17.12 Quartz crystal oscillator or clock. (a) Electrical analog equivalent circuit of the crystal. (b) Characteristic impedance of (a). (c) Practical crystal clock circuits. The component labeled RFC (*radio frequency choke*) is an inductor which provides a dc bias path for the *FET* while blocking ac current flow.

varied linearly in time. The basic circuit of a sweep generator is shown in Fig. 17.13. An integrator, driven by a current source, generates a linear ramp signal. The ramp may be reset to zero by an electronic switch. The sweep will repeat itself indefinitely if the switch is driven by a comparator whose input compares the integrator output with a fixed reference voltage. The integrator is therefore reset whenever its output voltage exceeds a certain threshold. The sweep rate is governed by the integrator time constant RC; the reset time is dictated by the value of rC. In an oscilloscope, the beam is switched off during the reset time so the retrace is not seen. General-purpose op amps do not usually have adequate slew rates for this application, and a special-purpose comparator op amp is preferred (Chapter 19). The integrator op amp is selected for the frequency range of interest. For low-speed (1-Hz) applications, an op amp with low bias current must be used, and for high-speed (10^6-Hz) applications, a high-speed op amp is required.

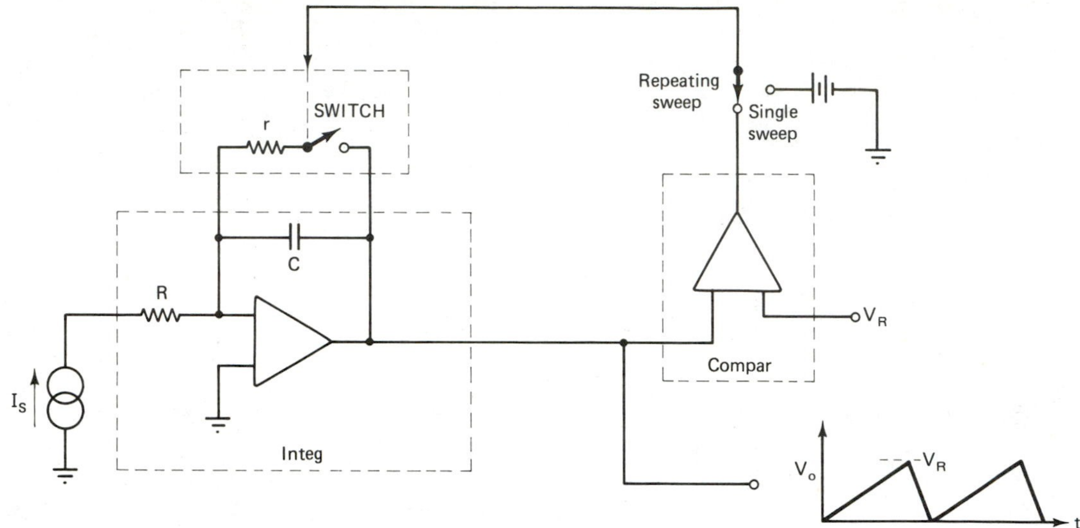

Figure 17.13 Sweep circuit block diagram.

17.5 FUNCTION GENERATORS

Function generators, which are based on switching principles, are the most useful way of generating a wide variety of waveforms. The basic design allows variations of frequency over an extremely wide range (up to 9 decades in a commercial instrument) by varying only one or two components. *Voltage control* of frequency allows for the automatic plotting of network transfer functions. *Gated* oscillators make possible a "tone burst" with precise control of the starting phase. Function generators are intermediate between analog (linear) and digital (two-state) devices, since they typically involve an integrator (analog), a switch (digital), and a comparator (analog input and digital output).

17.5.1 Basic Function Generator Circuit

The basic function generator involves both analog integration and a digital comparator. A basic two–op-amp function generator is shown in Fig. 17.14. Zener diodes clamp the output of the comparator at a value less than the power supply voltage, to avoid latch-up. The positive feedback path makes this a comparator with hysteresis (see Sec.19.1). The closed-loop operation is shown graphically in Fig. 17.14(b). When the comparator output is high, the voltage at point A decreases until it reaches $-V_Z$. The comparator then changes to the low state and the voltage at point A increases until it again reaches a change-of-state threshold at $+V_Z$. The feedback provides the correct sign to result in an oscillator (as opposed to a latch) with symmetric square and triangular wave outputs. The period is determined by

$$T = \frac{2\Delta V}{dV/dt} = \frac{4V_Z}{V_Z/RC} = 4RC \tag{17.24}$$

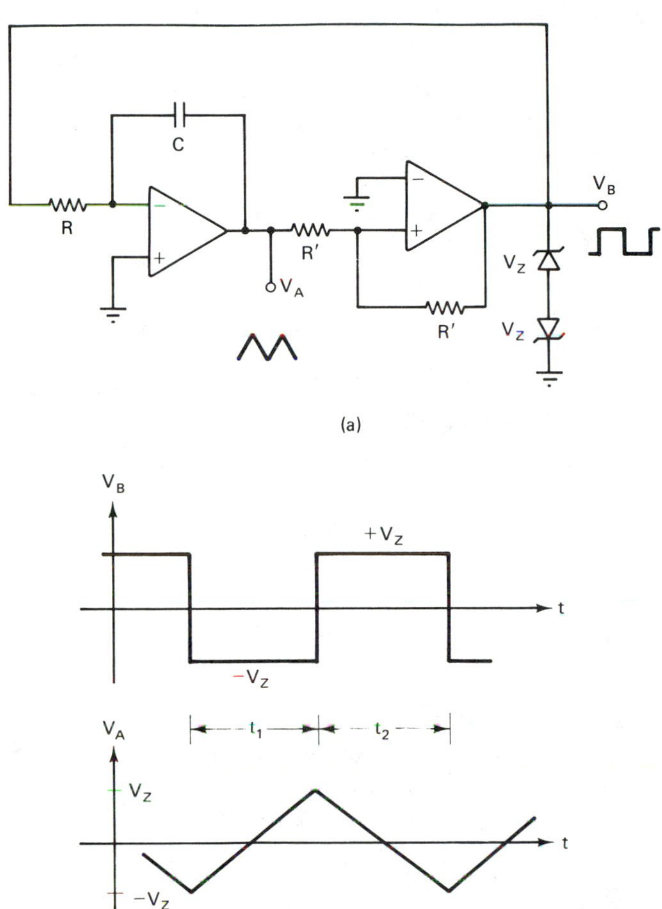

Figure 17.14 Basic function generator. (a) Circuit. (b) Waveforms.

The frequency is determined by two passive components and is independent of V_Z (and incidentally is independent of power supply voltages as well). This results in a stable frequency.

17.5.2 Voltage-Controlled Oscillators (VCO's)

The frequency may be controlled by using a variable external voltage for the integrator input. The comparator then functions only as a switch. Two idealized VCO's are shown in Fig. 17.15. Either an analog multiplier or an analog switch may be used. Although it would seem that the multiplier or switch could be avoided by summing in a control voltage to the integrator of Fig. 17.15, this is in fact not possible (see Prob. 17.9).

Commercial function generator instruments and integrated circuits use inexpensive current switching techniques. Two examples are shown in Fig. 17.16. The circuit shown in Fig. 17.16(a) is similar to the basic function generator of Fig. 17.15, except that a comparator is used to change the current paths into the integrator. The magnitude of the current is set by the external voltage V_C. The cir-

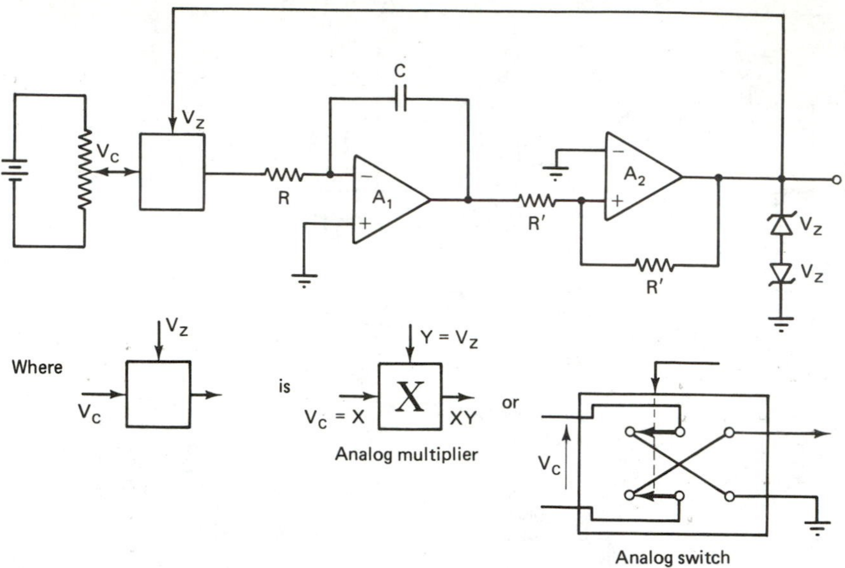

Figure 17.15 Voltage controlled function generator methods.

cuit shown in Fig. 17.16(b) uses a *current differencing* or *Norton* op amp, whose output voltage is proportional to the difference between input currents. The comparator switches the current flow paths so that currents of $\pm I$ charge and discharge the capacitor on alternate half-cycles. Since I is proportional to V_C, the period scales with V_C.

17.6 COMMERCIAL FUNCTION GENERATORS

The development of circuits such as those in the preceding sections has led to the availability of general-purpose function generator test instruments. A list of important function generator specifications is given below. Function generators having all of these features are available for about $500. More modest function generators which are very useful as voltage-controlled oscillators are available for only about $100. An example is shown in Fig. 17.17(a). A selection of the output waveforms in various modes is shown in Fig. 17.17(b).

Waveforms. Unlike the sine wave oscillator, most function generators have at least three basic waveforms available. Due to the switching mechanism used to generate the basic oscillation, both a square wave and a triangular wave are usually available simultaneously. A sine wave can be created from the triangular wave by diode function generation, though the distortion is generally a factor of 10 to 100 larger than in a conventional sine wave oscillator. In addition, a pulse-like wave form is often available by distortion of the square wave duty cycle, and a sawtooth or ramp waveform by distortion of the triangular wave.

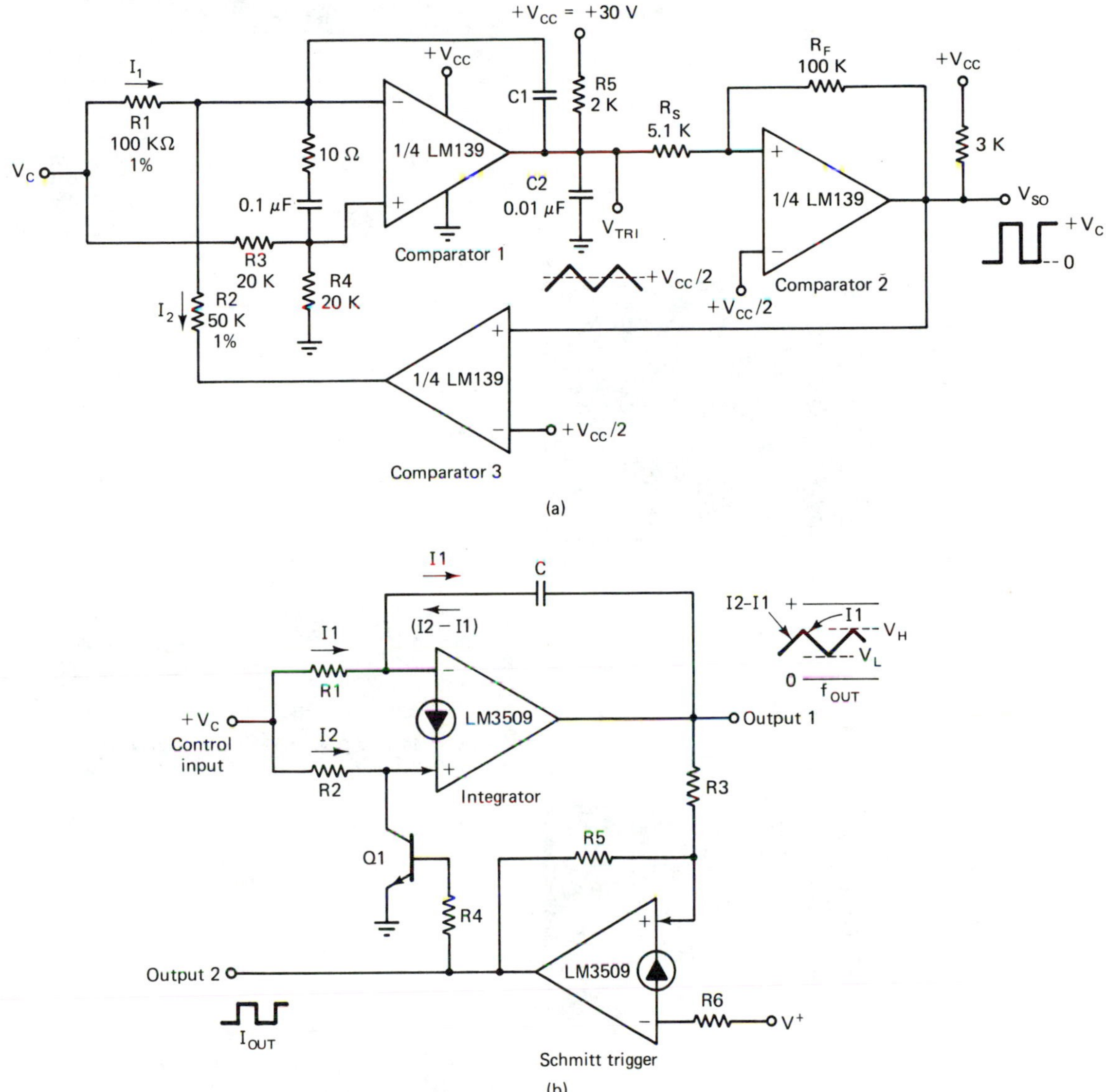

Figure 17.16 Practical voltage-controlled function generator circuits. (a) Comparator switching of integrator current. (b) Circuit employing a Norton current-differencing op amp. (Copyright 1980 National Semiconductor Corporation)

Frequency range. Since the frequency of the basic oscillator is set simply by an *RC* value, the frequency range of such devices is significantly wider than that of sinusoidal oscillators. A typical range is from 10^{-2} Hz to 10 MHz.

Voltage programming. If the frequency of the oscillator may be varied by an external voltage, the instrument is called *voltage-programmable* or *voltage-*

Figure 17.17 (a) A commercial low-cost function generator is a versatile test instrument. (*Courtesy* Wavetek)
(b) to (f) Function generator-voltage controlled oscillator waveforms.

(a)

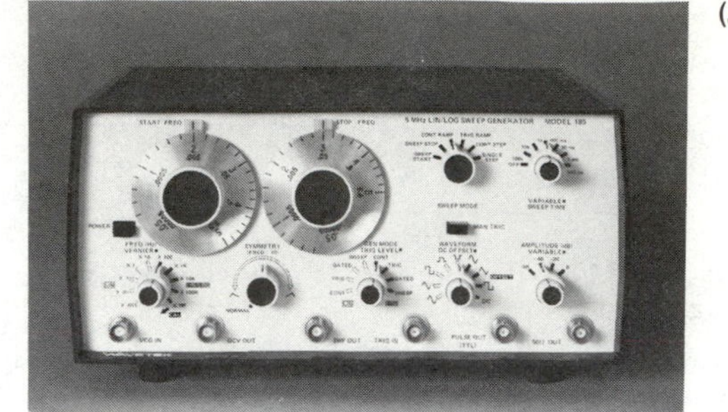

(b)

Triangle

Sine

Square

(c)

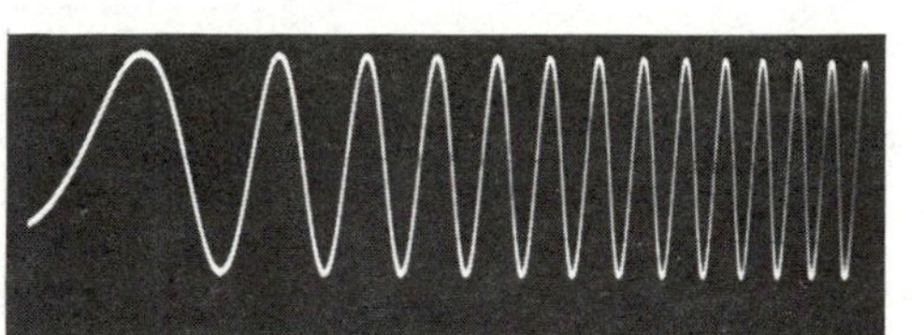

Frequency sweep

(d)

Amplitude modulation

(e)

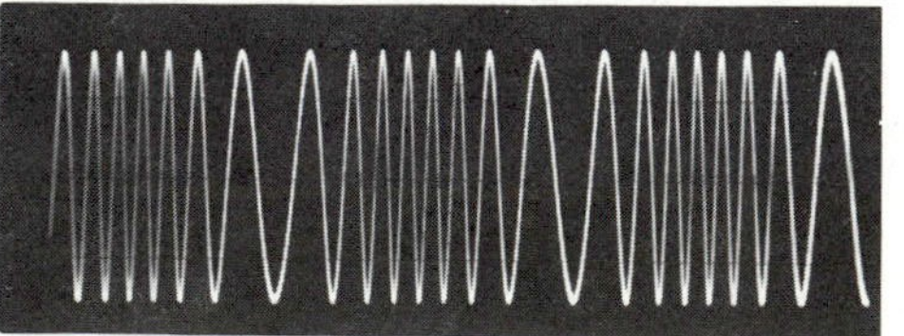

Frequency modulation

(f)

Pulse

Tone burst

controlled. The range over which frequency may be swept varies from 10 to 1 to as much as 1000 to 1 in the better instruments. This feature makes function generators useful as test instruments in automatic recording of frequency response in electronic, physical, or chemical systems.

Internal sweep. Often, a second independent low-frequency function generator is built in to a function generator to internally generate the sweep voltage. If a signal proportional to this internal ramp is available at the front panel, it may be connected to the horizontal axis of an *xy* plotter. Thus, a response or transfer function $V_o(f)$ may be generated and plotted automatically.

Gated (tone burst) operation. It is often possible to operate a function generator in synchronization with another oscillator via the *sync* terminal. In addition, many units can start and stop oscillation on external command. This *gated* signal may be used to generate a tone burst to record the impulse response of a system.

Duty cycle. The fraction of time in which the periodic square wave output is high or low may be varied over a wide range. The *duty cycle* is defined as the percentage of the period during which the square wave is high; a symmetric square wave has a 50% duty cycle. A signal with a small (large) duty cycle is often referred to as a positive (negative) pulse. Pulses are useful as probes of impulse response in analog systems, as trigger signals, and in digital systems as pulsers or clocks.

17.7 INTEGRATED CIRCUIT FUNCTION GENERATORS

In recent years a number of medium-scale integration (MSI) IC's have become available which are useful as clock circuits and function generators. A simple square wave timer may be constructed with acitve circuits costing less than $1, and a multifunction (square, triangle, sine) function generator with voltage-controlled frequency entails an active-circuit-parts cost of only $5.

Examples of IC function generator chips (Table 17.2) are compared in this section.

17.7.1 555 Timer

The current industry standard for all but the most demanding applications is the 555 timer. Its prominent features are an extremely wide range of frequencies and the ability to drive a large output current. On the other hand, it has only one output waveform and does not allow voltage control of frequency. The operation of the circuit was presented in Chapter 6 (see Fig. 6.10 and 6.11). The 555's frequency of operation is given by

$$f = 1.46[(R_1 + R_2)C]^{-1} \tag{17.25}$$

TABLE 17.2 COMPARISON OF IC FUNCTION GENERATORS

IC	555	566	XR-205	8038
Originator	Signetics	Signetics	Exar	Intersil
Waveforms				
square	Yes	Yes	Yes	Yes
triangle	No	Yes	Yes	Yes
sine @ distortion	No	No	Yes @ 2.5%	Yes @ 1%
pulse	No	No	Yes (limited)	Yes
ramp or sawtooth	No	No	Yes	Yes
Frequency range	0.1 Hz to 1 MHz	1 Hz to 1 MHz	0.1 Hz to 5 MHz	10^{-3} Hz to 1.0 MHz
Voltage programming range	None	10 to 1	10 to 1	1000 to 1
Duty cycle range	Fixed 50%	Fixed 50%	20 to 80%	2% to 98%
Frequency modulation	No	No	Yes	Yes
Amplitude modulation	No	No	Yes	No
Gated (tone burst)	No	No	Yes	No
Frequency stability				
vs. temp, ppm/degree C	50	100	300	50
vs. power supply, %/V	0.1	2	0.2	0.05
Rise time (square wave)	100 ns	20 ns		100 ns
Other functions	One shot		Gated toneburst	
Other features	200 ma output			
Cost (dollars)	0.50	2.00	8.00	5.00

Since both the charging voltage and the reference voltage for the comparators come from the same source, the frequency is independent of the supply voltage. The circuit has a wide range of period, ranging from microseconds to hours! The 555 has been called a "ubiquitous" IC. Using resistive or capacitive transducers, many measurements which formerly used analog voltage methods are now done using time or frequency.

17.7.2 566 and XR-205 Function Generators

The 556 IC adds to the 555 an additional buffered triangular wave output and a moderate (10 to 1) range of voltage programming. The block diagram for the 566 function generator is shown in Fig. 17.18(a). The feedback loop is similar to that of the basic function generator (Fig. 17.14), but the analog integrator is replaced by a capacitor driven by a current source, providing an output voltage which increases linearly in time. The principal advantage of this change is the ease with which a voltage-controlled current can be generated for voltage control of frequency. The internal equivalent circuit of this IC is shown in Fig. 17.18(b). An analysis of its operation is left for Prob. 17.8.

The XR-205 is typical of a family of more complex IC function generators. It adds to the functions of the 566 a sine wave output, although the distortion is rather substantial. The most unusual features of this IC are an internal multiplier, which facilitates gating and tone burst signal generation, and amplitude and frequency modulation of the output signal.

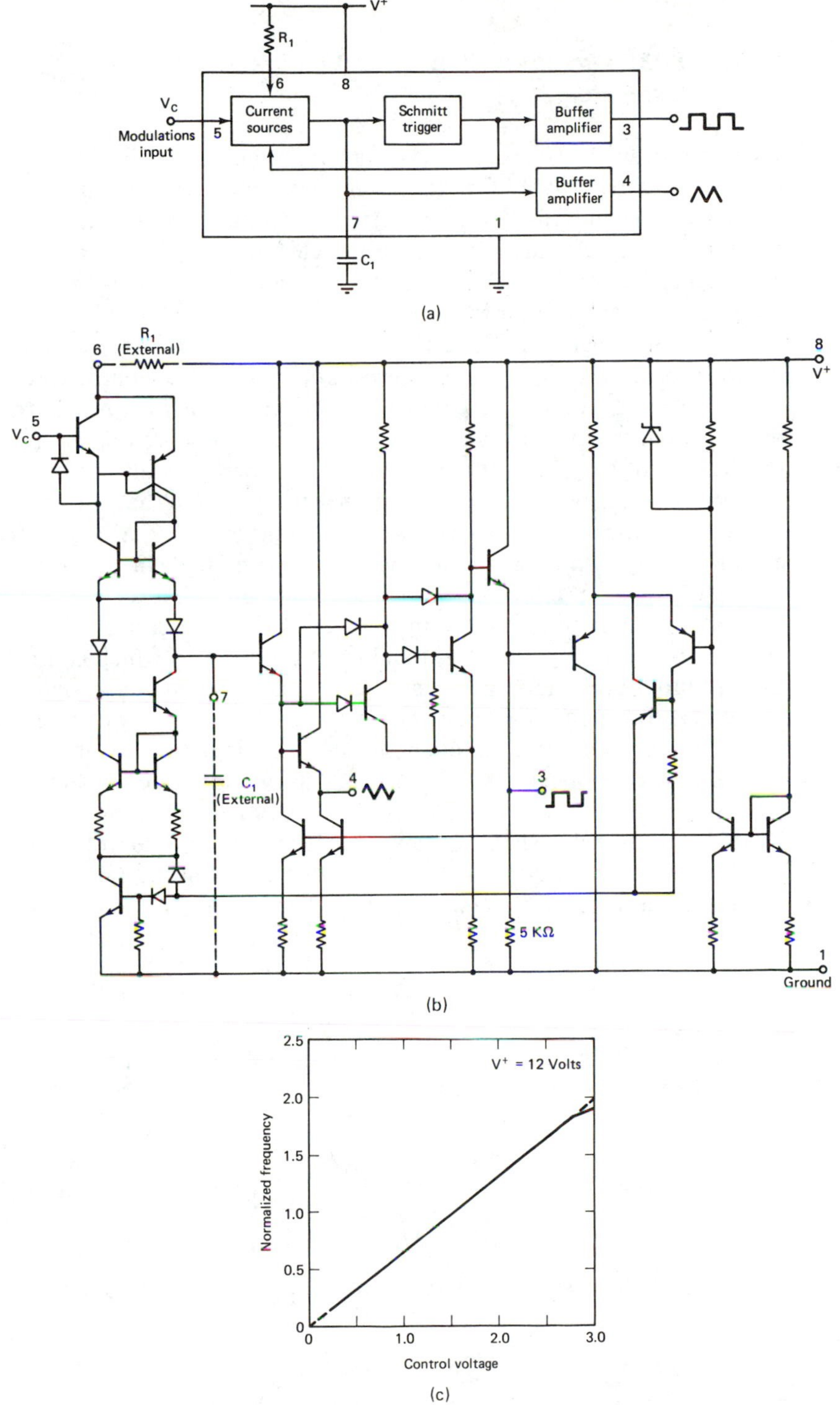

Figure 17.18 566 function generator. (a) Block diagram. (b) Equivalent circuit. (c) Voltage control of frequency. (Permission to reprint granted by Signetics Corporation, a subsidiary of U.S. Philips Corp., 811 E. Arques Avenue, Sunnyvale, CA 94086.)

17.7.3 8038 Waveform Generator/Voltage-Controlled Oscillator

This extraordinary IC has three waveforms available simultaneously, including a sine wave of reasonably low harmonic distortion. Its frequency spans the range from microseconds to hours. The frequency may be voltage-controlled over a 1000 to 1 range. A wide range of duty cycles makes it functional as a true pulser. The capabilities of this \$5 IC compare with commercial instruments which only a few years ago cost 100 times more.

A block diagram of the 8038 waveform generator is shown in Fig. 17.19(a). As with the 566, a capacitor driven by a current source generates the ramp function. The arrangement of comparators and flip flop is identical to that of the 555 timer. Two current sources are used, and in square wave operation one has twice the current of the other. When the switch is open, current flows into the capacitor, which charges at a rate proportional to $I_1 = I$. When the switch is closed, however, a current $I_2 = 2I$ is drawn by current source 2. Since source 1 provides current I, an additional I must be drawn from the charge stored on C. This discharges the capacitor at a linear rate precisely identical to the charging rate, generating a triangular waveform. Two comparators are used, with high and low voltage comparison points to set and reset the flip flop which controls the switch. Buffer amplifiers isolate the square wave and triangular wave signals from external loads. Additional circuitry converts the triangular wave to sinusoidal form, providing a total of three simultaneous output waveforms.

External connections of timing resistors and capacitor and the connection for external frequency modulation or frequency sweeping are shown in Fig. 17.19(b). External resistor R_A and R_B set the magnitude of current I_1 and I_2. The comparators are set to switch when the ramp voltage reaches ⅓ and ⅔ of the voltage difference between pins 8 and 11. The charge time is given by

$$t_1 = CV_{c+}/I_1 = \frac{CV_{cc}/3}{V_{cc}/5R_A} = 5R_AC/3 \tag{17.26}$$

The discharge time depends upon the difference current when the switch is closed, hence on both R_B and R_A

$$t_2 = \frac{CV_{c-}}{I_1} = \frac{CV_{cc}/3}{\left[\frac{2}{5}\frac{V_{cc}}{R_B} - \frac{1}{5}\frac{V_{cc}}{R_A}\right]} = \frac{5}{3}\left[\frac{R_AR_BC}{2R_A - R_B}\right] \tag{17.27}$$

Thus, a 50% duty cycle is achieved when $R_A = R_B$. The frequency of operation is given by

$$f = (0.3/R_AC)(2R_A - R_B)/R_A = 0.3/RC \qquad \text{if } R_A \doteq R_B \tag{17.28}$$

The frequency is independent of supply voltages, since *both* the magnitude of the current sources and the comparator thresholds depend linearly upon the supply voltage. Good frequency stability results.

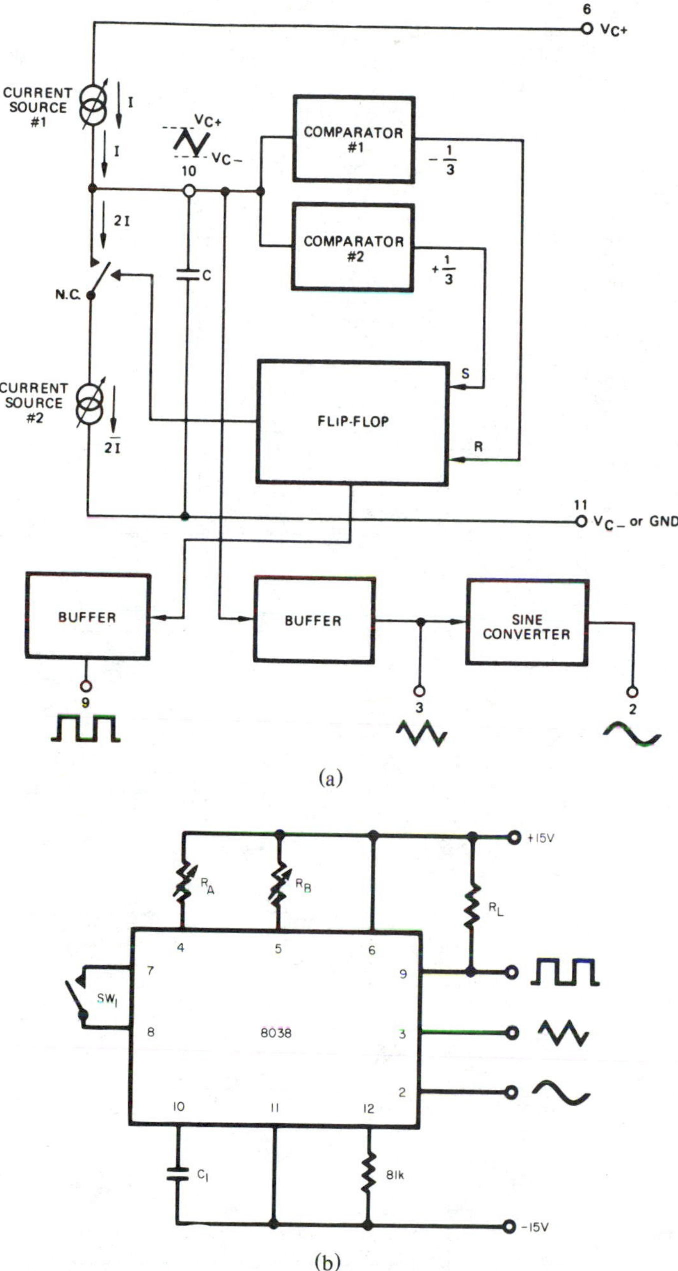

Figure 17.19 (a) Block diagram of 8038 waveform generator. (b) Connections of the 8038 showing external timing components and external frequency modulation or sweeping. Power supply voltages V_{C+} and V_{c-} are assumed in Eq. 17.26 and 17.27 to have magnitude V_{cc}, with $V_{c+} = -V_{c-}$. (*Courtesy* Intersil Inc.)

An internal schematic circuit of the 8038 is shown in Fig. 17.20. Since most of its components resemble blocks described earlier, they will not be discussed in detail. Of special note, however, is the clever triangle-to-sine converter on the right-hand side, which provides a surprisingly accurate sine wave. This circuit

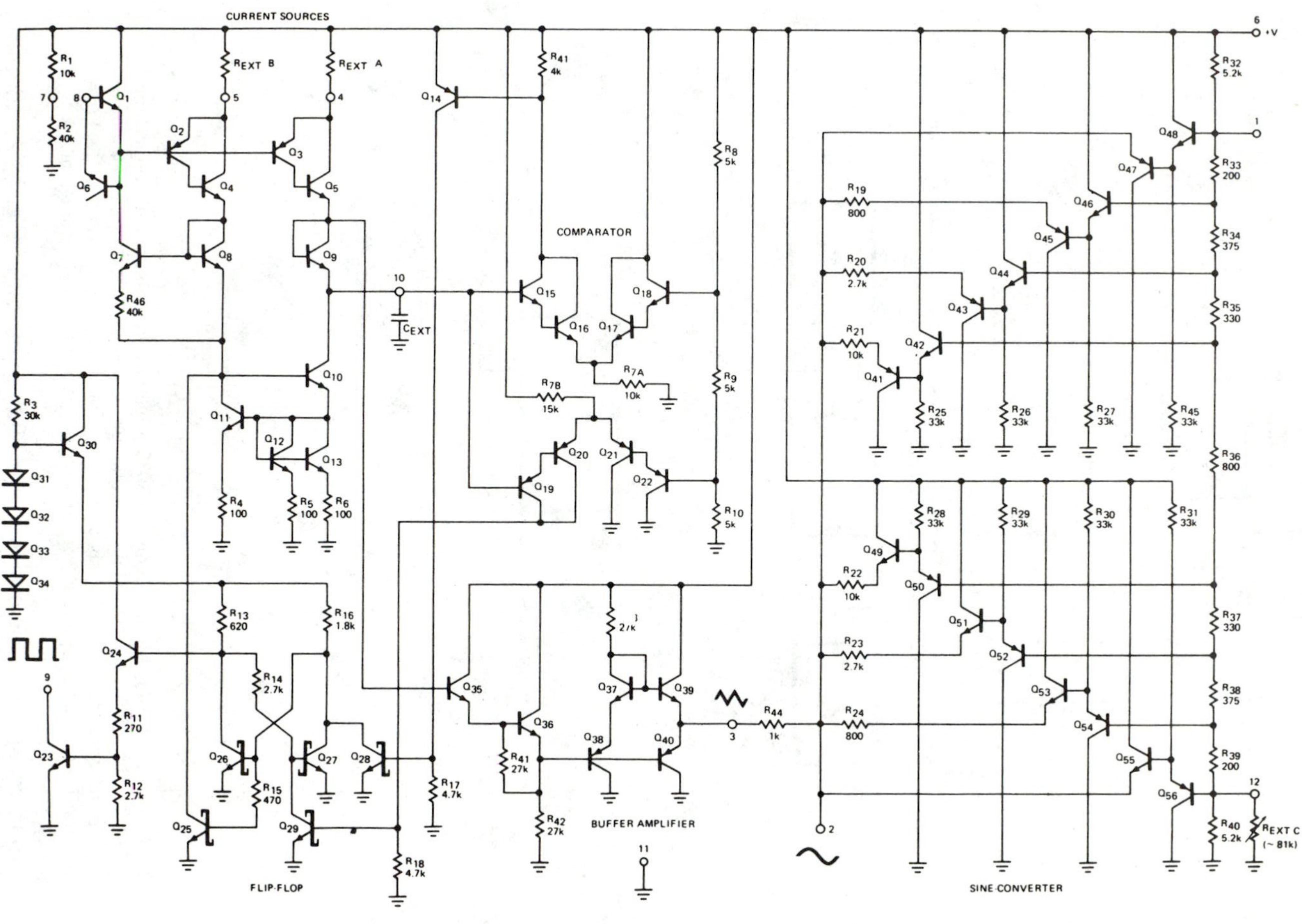

Figure 17.20 Equivalent schematic circuit diagram of 8038 function generator-VCO. (*Courtesy* Intersil Inc.)

operates by loading the triangular wave (with its 1-KΩ source resistor) by an amount that changes with the size of the signal. Resistive loads (10 KΩ, 2.7 KΩ, 800 Ω, and 0 Ω) are switched in to generate the increasingly flatter portions of the sine wave as it approaches its peak. Viewed from output terminal 2, the sine wave source looks like a Thevenin equivalent whose voltage value is set by an amplitude-switched voltage divider. Operation of the circuit is illustrated in Fig. 17.21. The base of transistor Q_1 is attached to a fixed point on a voltage divider ladder network. The base current I_1 is constant, resulting in constant collector current and a constant voltage V_2, which provides a back bias for transistor switch Q_2. When the triangular-wave voltage is sufficiently negative, the voltage at point V_3 is high enough to forward-bias Q_2, which turns on and connects another resistor in parallel with the voltage divider to ground. Since the slope V_o/V_{in} is set by $R_1/(R_1 + R_2)$, the output waveform flattens out as shown in Fig. 17.21(c).

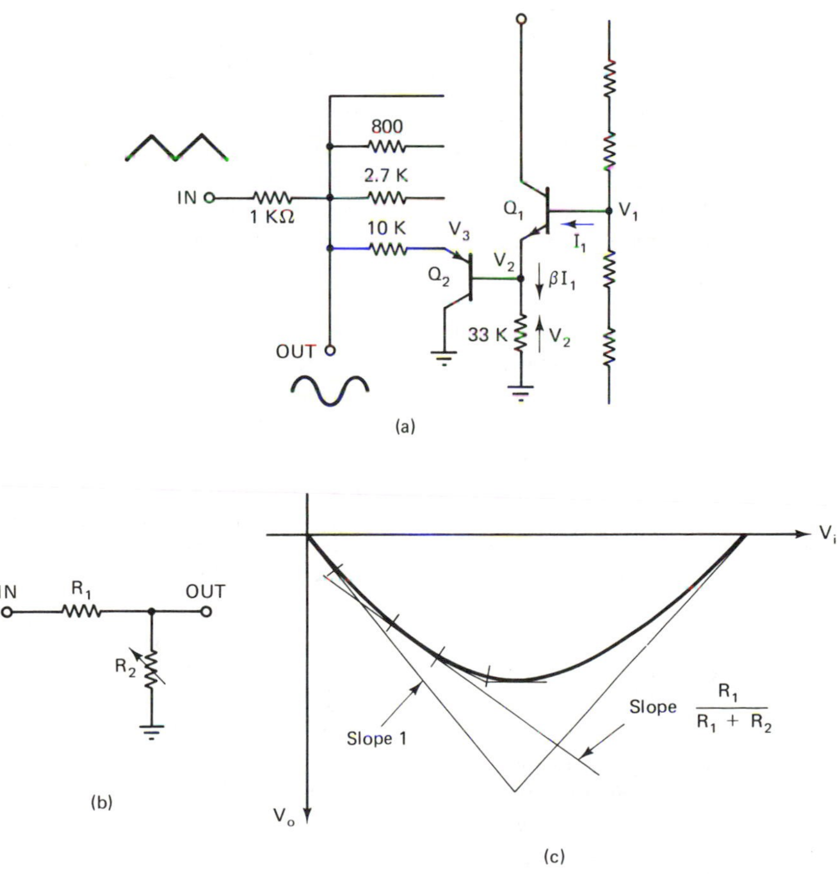

Figure 17.21 Operation of the triangle to sine converter. A portion of the 8038 circuit (a) functions as a voltage-switched voltage divider, (b) generating a piecewise linear approximation to a sine wave (c).

As the input amplitude falls further, other breakpoints are reached and successively smaller resistors are switched in parallel, flattening the curve, until the horizontal portion of the sine wave peak is reached by switching in a short circuit. An identical circuit with complementary transistors provides similar waveshaping for the positive half-cycle. In spite of the small number of breakpoints, the curve is surprisingly smooth (1% harmonic distortion is achievable), due partly to the rounding which occurs because the breakpoint circuits turn on gradually rather than sharply.

PROBLEMS

17.1. Show that the criterion for oscillation of the phase shift oscillator [Eqs. (17.9) and (17.10) corresponds to shifting the poles of the transfer function [Eq. (17.1)] to the imaginary axis. Use Eq. (17.11) and the correspondence $s \longrightarrow j\omega$.

17.2. **(a)** Show that the open-loop transmission of the quadrature oscillator (Fig. 17.5) is

$$\frac{V_f}{V_i} = \frac{-1}{R_1C_1s}\left[\frac{R_3C_3s + 1}{(R_2C_2s + 1)R_3C_3s}\right]$$

where the left and right bracket terms are the closed-loop transfer functions of the left and right op amp circuits, respectively. Note the noninverting integrator.

(b) For the case where all three time constants are equal, show that poles in the closed-loop transfer function occur on the imaginary axis at

$$s_o = j\omega_o = \pm 1/RC$$

17.3. How does a twin-T oscillator (Fig. 17.6) differ from a twin-T bandpass filter (Fig. 16.13)? This network has a phase shift which rapidly shifts from π to $-\pi$ across the natural frequency. Show that by unbalancing the T to give a net positive feedback (hence oscillation) corresponds to shifting the locations of the complex poles in the transfer function to the right of the imaginary axis.

17.4. Show that the circuit of the analog simulation oscillator (Fig. 17.22; adapted from W. C. Jung, *Electronics*, Feb. 5, 1976, p. 90) has the required transfer function for an oscillator, i.e., a complex pole pair in the right half-plane. **Procedure:** Break the circuit at X and find the conditions under which the open-loop transmission crosses 1.

17.5. In the analog simulation oscillator (Fig. 17.22), show how one op amp can be eliminated using a noninverting integrator. This leads to the quadrature oscillator (Fig. 17.5). The extra RC element is necessary to shift the transfer function of the noninverting integrator back to the standard form $1/(RCs)$.

17.6. Show by redrawing the circuit of Fig. 17.22 that the biquad oscillator of Fig. 17.11 is equivalent to the simplest form of the analog simulation oscillator (with growth feedback removed).

17.7. Refer to the block diagram of the voltage-controlled function generator (Fig. 17.15). Show that the use of a voltage-controlled current source provides an output frequency proportional to the control voltage.

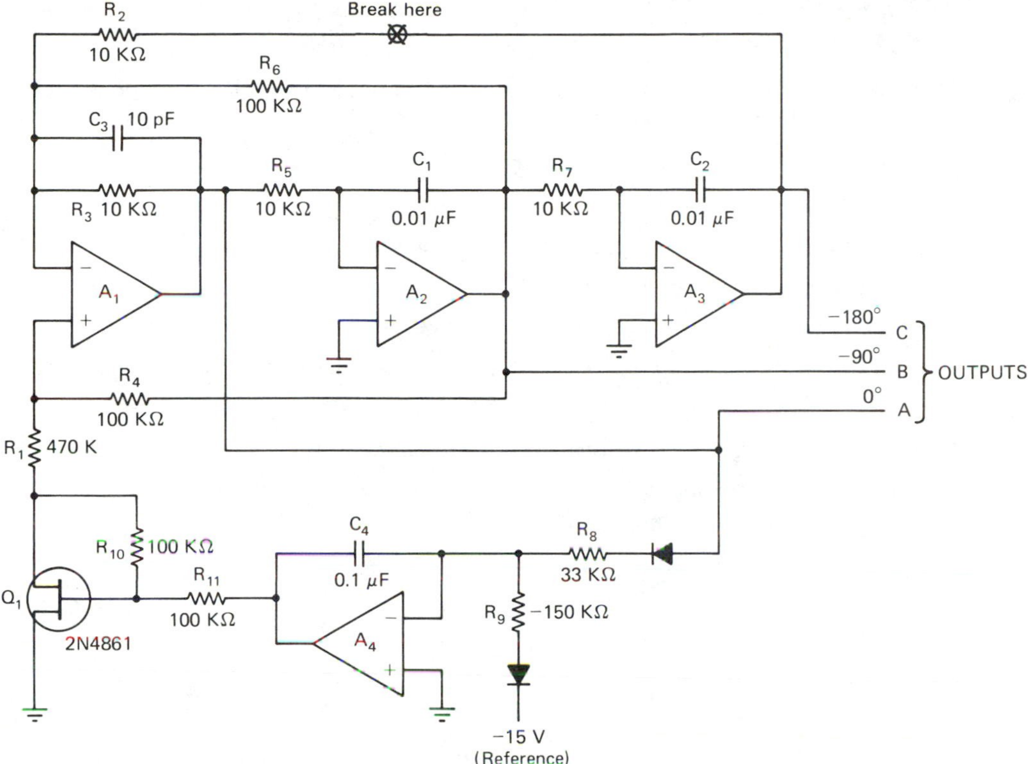

Figure 17.22 Analog simulation oscillator. Op amp A_4 and the associated diode and FET circuitry plays no part in the basic oscillator loop but functions as a low distortion automatic gain control as in Fig. 17.9(b). As expected from its resemblance to the damped harmonic oscillator simulation (Fig. 14.8), three simultaneous outputs at 90° phase intervals are possible. (*From* W.C. Jung *Electronics*. (Feb. 5, 1976): 90.)

17.8. Refer to the equivalent circuit of the 566 function generator [Fig. 17.18(b)]. Identify the function of each group of transistors, and outline the general blocks. Useful categories are current mirror bias source, switch, current source, Darlington pair, emitter follower (buffer), and Schmitt trigger.

17.9. Consider constructing a voltage-controlled function generator by summing into the integrator of Fig. 17.14 a control voltage V_c instead of using a multiplier or switch to alter the ramp rate. Show that this will alter the duty cycle but not the overall period, hence failing as a VCO technique.

17.10. Sketch a block diagram for an instrumentation circuit to plot automatically the transfer function V_o/V_i of a system as a function of frequency. Available building blocks: function generator (voltage-controlled frequency with dc *sweep output* proportional to Δf); precision rectifier (ac/dc converter); *xy* recorder.

REFERENCES

More complete bibliographic information for the books listed below appears in the annotated bibliography at the end of the book.

BENEDICT, *Electronics for Scientists and Engineers*

BROPHY, *Basic Electronics for Scientists*

DIEFENDERFER, *Principles of Electronic Instrumentation*

GRAEME, *Designing with Operational Amplifiers*

HIGGINS, *Experiments with Integrated Circuits*, Experiments 24 and 25

HOENIG and PAYNE, *How to Build and Use Electronic Devices without Frustration, Pain, etc.*

JUNG, *IC Op Amp Cookbook*

MALMSTADT et al., *Electronic Measurements for Scientists*

PHILBRICK, *Applications Manual*

ROBERGE, *Operational Amplifiers*

TOBEY, GRAEME, & HEULSMAN, *Operational Amplifiers*

18

The Measurement of Small Signals

The brain wave signal shown in Fig. 18.1 can be successfully measured using present-day instrumentation techniques. The signal is small enough ($\sim 10^{-6}$ V) that ac power lines and radio frequency interference can be serious noise sources. The source impedance is high enough ($\sim 10^{-5}$ Ω) to cause problems due to grounding, common mode noise, and common mode limitations of the amplifier. Nonetheless, available integrated circuits and techniques can overcome all these problems once their origin is understood. It is now possible to measure easily voltages and currents as small as a microvolt (10^{-6} V) or a nanoamp (10^{-9} A), a nanovolt (10^{-9} V) or a picoamp (10^{-12} A) with care, using low-noise instrumentation, or a picovolt (10^{-12} V) or femtoamp (10^{-15} A) using superconducting instrumentation. To see the significance of these numbers, note that 10^{-15} A corresponds to the passage of only a few electrons per millisecond.

This chapter will first deal with noise, occurring either as a fundamental limit, or as a result of nonideal amplifier behavior, or as a result of external noise sources. We then discuss low-noise op amps and special low-noise circuits for broad-band preamplifiers. Next comes a special class of narrow-band detection techniques called *lock-in* or coherent detection. Finally, we deal with a special class of frequency-tracking circuits called *phase-locked loops.*

18.1 NOISE AND ITS ORIGINS

It is useful to distinguish between three distinct kinds of noise (Fig. 18.2): fundamental noise sources; amplifier noise sources; and external noise sources. *Fundamental noise sources* originate in basic thermodynamics and can therefore not be

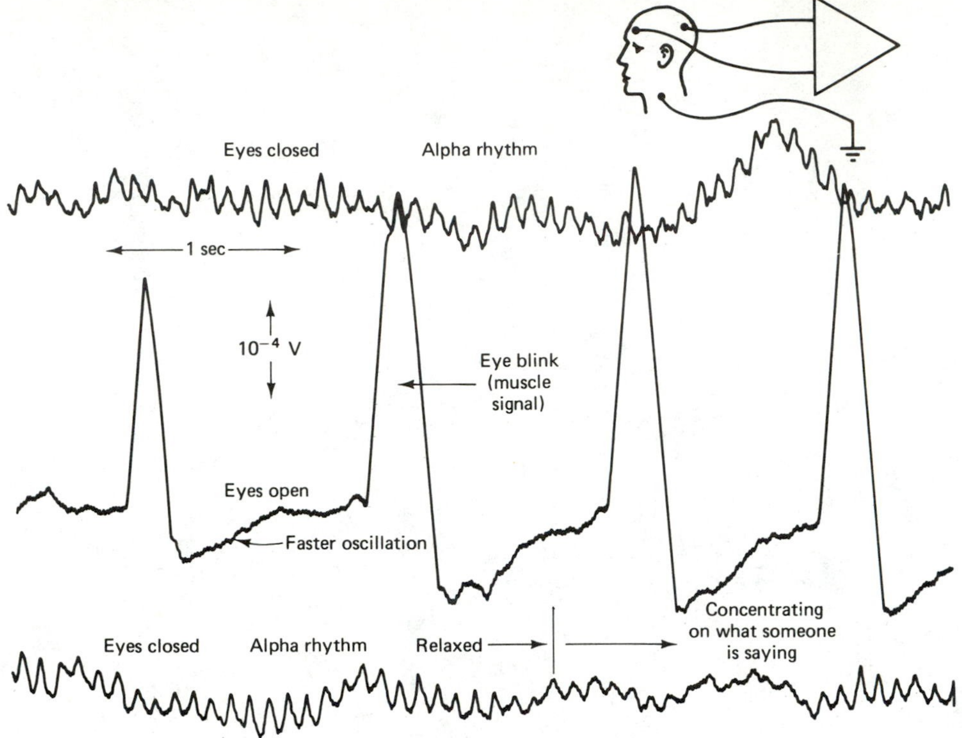

Figure 18.1 Actual lab record of brainwaves, using inexpensive instrumentation amplifier constructed from op amps.

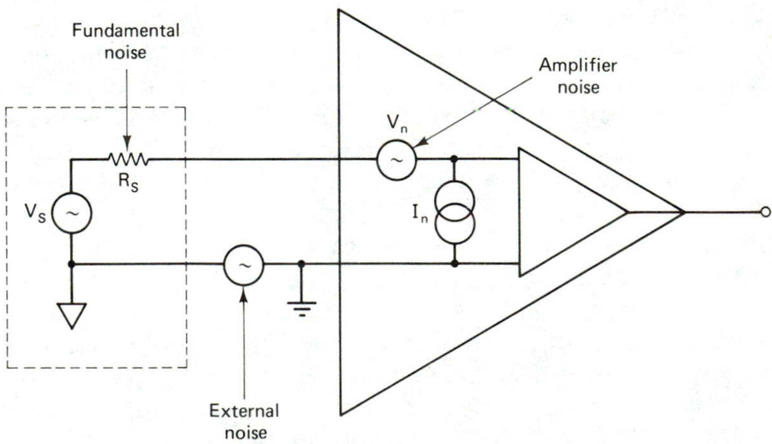

Figure 18.2 Noise sources.

eliminated. However, their effect can be minimized by controlling the source resistance, amplifier bandwidth, or temperature. *Amplifier noise sources* are a result of noise added by the amplifier to the amplified signal which exceeds the fundamental noise level. Amplifier noise can be greatly reduced by careful device selection and adjustment of the measurement frequency and bandwidth. Examples of *external noise sources* are pickup of ac line voltage, radio frequency interference from nearby radio or TV stations, and switching noise from instruments such as temperature controllers. Minimizing external noise problems involves careful grounding and shielding and special circuitry to compensate for common mode limitations of amplifiers. All of these techniques will be described in the following sections.

18.1.1 Fundamental Noise Sources

Johnson noise is due to Brownian motion in a resistive element R. Fluctuations in the thermal motion of charge carriers can lead to a momentary imbalance in the current flow [Fig. 18.3(a)]. This translates into a momentary voltage across the resistor. The size of Johnson noise is

$$V_n^2 = < V^2 >_{\text{rms}} = 4kTR\,\Delta f \qquad (18.1)$$

Here, k is Boltzmann's constant (1.4×10^{-23} joules/°K), T is the *absolute* temperature, R is the source resistance, and Δf is the bandwidth of the amplifier. The

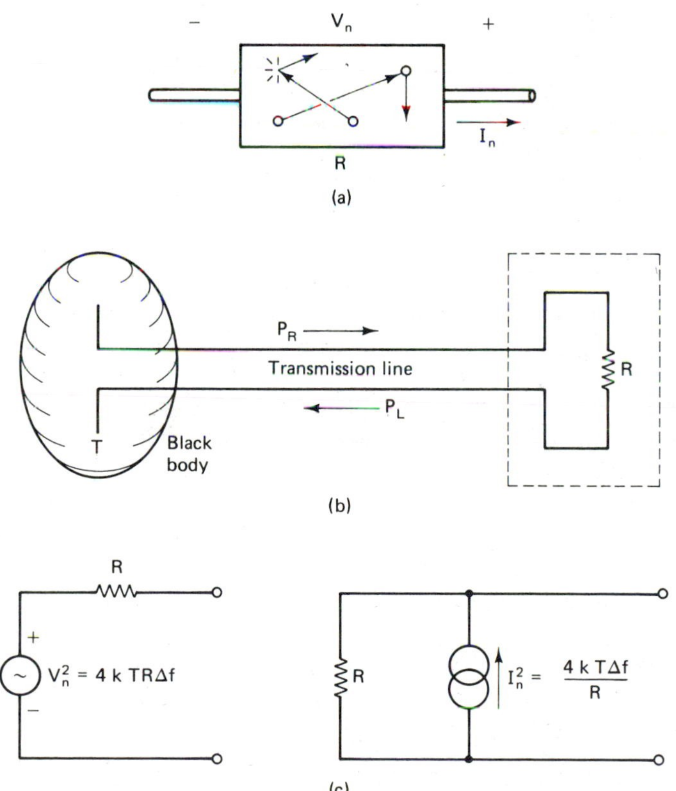

Figure 18.3 Johnson noise. (a) Origin in current fluctuations. (b) Black body power balance explanation of Nyquist. (c) Equivalent circuit models.

Johnson noise *voltage* varies as $(TR\Delta f)^{1/2}$. As a numerical example, suppose $T = 300°$ K, $R = 100$ KΩ, and $f = 10^4$ Hz.

$$V_n = [(1.7 \times 10^{-20}) \times 10^5 \times 10^4]^{1/2}$$
$$= 4 \times 10^{-6} \text{ V}$$

This is certainly not negligible in low-level measurements. Therefore, the source resistance is kept as low as possible, and the amplifier bandwidth is made no wider than demanded by the measurement. Finally, some detectors are operated in a cooled mode, to lessen the importance of temperature.

Johnson noise is thermodynamic in origin, as illustrated in Fig. 18.3(b). A black body at temperature T is coupled to a resistor R by a dipole antenna and a transmission line. If the system is in steady state, the average power transmitted to the right (P_R) from the black body must equal the average power (P_L) originating in the resistor and moving to the left. The original explanation of Eq. (18.1) by Nyquist followed the theory of black body radiation in physics. The temperature T term comes from the equipartition theorem, the dependence on bandwidth Δf comes from the number of modes of motion per unit frequency range along the transmission line (like the modes in a guitar string), and the factor R enters because the black body noise power $4kT\Delta f$ must balance the resistor's noise power V^2/R. The same result can also be obtained microscopically by looking inside the resistor. From this point of view, the T term originates from the mean square velocity of the moving charges (hence mean square current), the term R originates in converting this current to a voltage, and the term Δf is there because all modes of motion are equally likely.

The equivalent circuit for Johnson noise is an ideal noise-free resistor in series with a voltage noise source or in parallel with a current noise source [Fig. 18.3(c)]. The effect of Johnson noise may be reduced by controlling each of the factors in Eq. (18.1). R depends upon the signal source and may not be controllable. Since considerable improvement can be made by lowering T, some measuring circuits such as detectors of infrared radiation are cooled to temperatures of 1°K or below. The biggest improvement comes from keeping the bandwidth Δf of the measuring device as narrow as possible. Filters should be used to limit the bandwidth of the amplifier to no larger than the bandwidth of the signal. The extremely narrow bandwidth capability of lock-in amplifiers therefore favors this technique when it is applicable.

18.1.2 Amplifier Noise

A plot of amplifier noise vs. frequency is shown in Fig. 18.4(a). *Shot noise* is wideband (*white*) noise originating in the quantization of charge in transistors. The charge densities in semiconductors are low enough that statistical fluctuations occur (over and above Johnson noise). An additional and typically much more important component varies as $1/f$ at low frequencies, and is called *flicker noise* or *pink noise*. Pink noise is observed in a wide variety of phenomena related to a random walk: where you get to next is not quite random (white) but depends upon

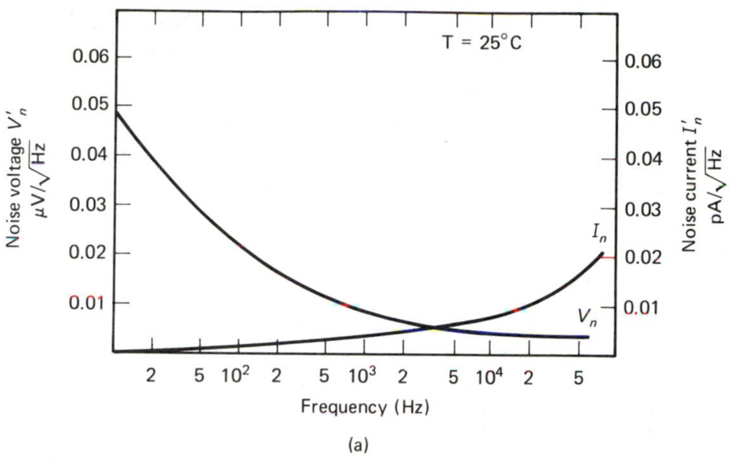

(a)

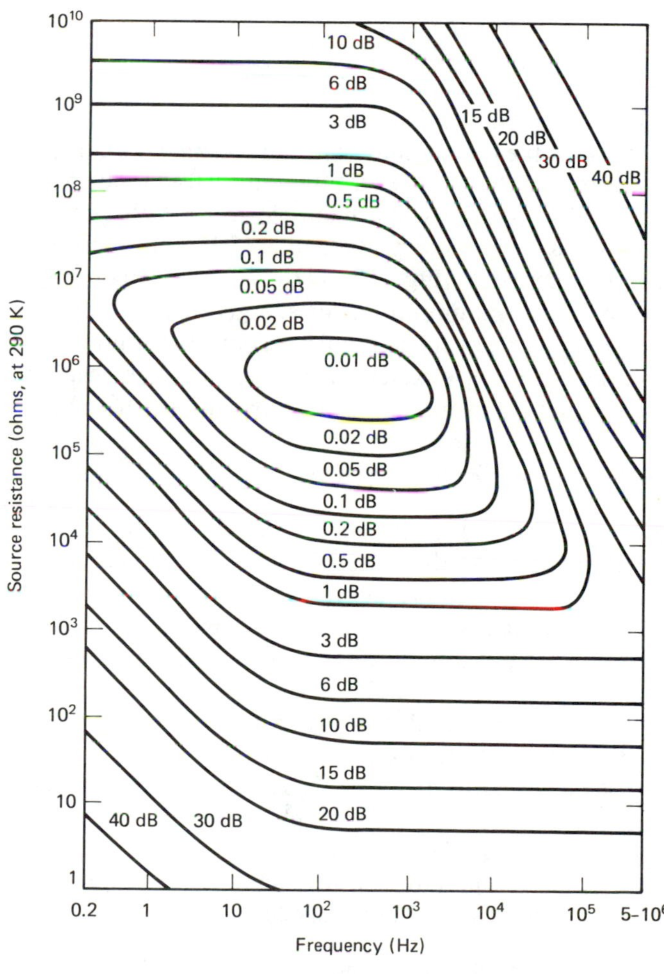

(b)

Figure 18.4 (a) Source of amplifier noise. (b) Noise figure contours for a low noise preamp. (*Courtesy* EG&G Princeton Applied Research)

where you've been recently (pink). The dc offset drift of an op amp is the low-frequency limit of $1/f$ noise. This noise source is the principal limit to low-frequency measurements, and is best avoided by shifting the signal frequency upwards by modulation techniques (Section 18.3). In addition to this voltage noise V'_n, a noise component is observed at high frequencies which acts as a current source I'_n.* A real amplifier therefore has as its equivalent circuit an ideal (noise-free) amplifier with two additional noise sources (Fig. 18.2). Because noise source I'_n drives current back through the signal source, there will be a term in the amplifier noise proportional to the external source resistance.

A convenient way of summarizing the quality of amplifier performance is to use noise figure contours. The *noise figure* (NF) is a measure of the extra noise added by the amplifier over and above the noise in the source.

$$\begin{aligned} \mathrm{NF} &= 20 \log\left[\frac{(S/N)_{\text{input}}}{(S/N)_{\text{output}}}\right] \qquad (18.2) \\ &= 10 \log \frac{V_s{}^2(4kTR_s f_n)^{-1}}{V_s{}^2\{[4kTR_s + V'_n{}^2 + (I'_n R_s)^2]f_n\}^{-1}} \end{aligned}$$

Here, S/N is the signal-to-noise ratio measured at amplifier input and output, f_n is the frequency at which the noise is measured, and the other terms are defined in Fig. 18.2. Noise figure measurements plotted as a function of frequency and source resistance generate *noise figure contours.* An example for a high-quality low-noise preamp is shown in Fig. 18.4(b). Since voltage noise falls with increasing frequency while current noise increases [Fig. 18.4(a)], a "best" operating frequency exists where the noise figure is a minimum for any given value of R_s. Above about $10^8\ \Omega$ in this example, current noise dominates and the minimum disappears. The total noise and the extra noise added by the amplifier can readily be estimated from noise figure contours. The total noise (referred to the input) is, from Eq. (18.2),

$$V_n(\text{total}) = (4kTR_s f_n)^{1/2} \quad 10^{(\mathrm{NF}/20)} \qquad (18.3)$$

For example, if $R_s = 10^2\ \Omega$ and $f_n = 100$ Hz, the NF read from Fig. 18.4(b) is 20 dB. The noise is a factor of 10 higher than the Johnson noise limit, making V_n (total) $= 0.13\ \mu$V. Intelligent use of noise figure curves can improve the situation. The frequency could be shifted to the optimum region (10^2 to 10^3 Hz in this example), and the source resistance might also be altered with a transformer to shift the source impedance level.

Noise figure contours are not usually available for op amps. However, an estimate may be made from noise voltage specifications, usually either given over a particular bandwidth, or expressed as noise in nV/(Hz)$^{1/2}$. To convert a specification given in the latter form into a noise estimate, multiply the specification by the *square* of the signal bandwidth [see Eq. (18.3)]. The relevant specification at dc is the offset voltage as a function of temperature, which is the limiting value of $1/f$ noise as f approaches zero. A comparison of noise specifications for selected

*The prime reminds us that V'_n and I'_n are like densities measured in units of [volts/Hz$^{1/2}$] and [amps/Hz$^{1/2}$], respectively. The *net* noise is an integral over frequency.

TABLE 18.1 A COMPARISON OF LOW-NOISE OP AMPS[a]

Type	Premium 741	AD510 IC	AD235 chopper	NE5532 low noise	AD520 instrument
Offset voltage (μV/°C)	>20	1	0.1	>20	5
Input noise					
(μV, broadband; 0.1 Hz–10 KHz)	>10	1	10	0.6	2
(nV/(Hz)$^{1/2}$	~50	10	500	5	Unspecified
Approximate price (dollars)	1	15	60	15	30

[a]AD indicates Analog Devices; NE indicates National Semiconductor.

IC op amps is given in Table 18.1. Even a "premium" 741 is not suitable for microvolt measurements. Inexpensive IC op omps such as the AD 510 have specifications comparable to much more expensive chopper-stabilized op amps which formerly dominated in microvolt-level applications. A generation of IC op amps such as the NE5532 is optimized for low noise ac amplification, offering noise on the order of 5 nV/(Hz)$^{1/2}$, although not recommended as a dc amplifier because of its poor V_{os} drift. Thus, one selects an op amp to fit the specific situation, and the choice depends upon whether the dominant problem is dc drift or ac noise.

18.1.3 External Noise

Noise is picked up from various sources outside the circuit, coupled capacitatively or electromagnetically. An example of an external noise spectrum is shown in Fig. 18.5. The dominant components are harmonics of the ac line frequency (predom-

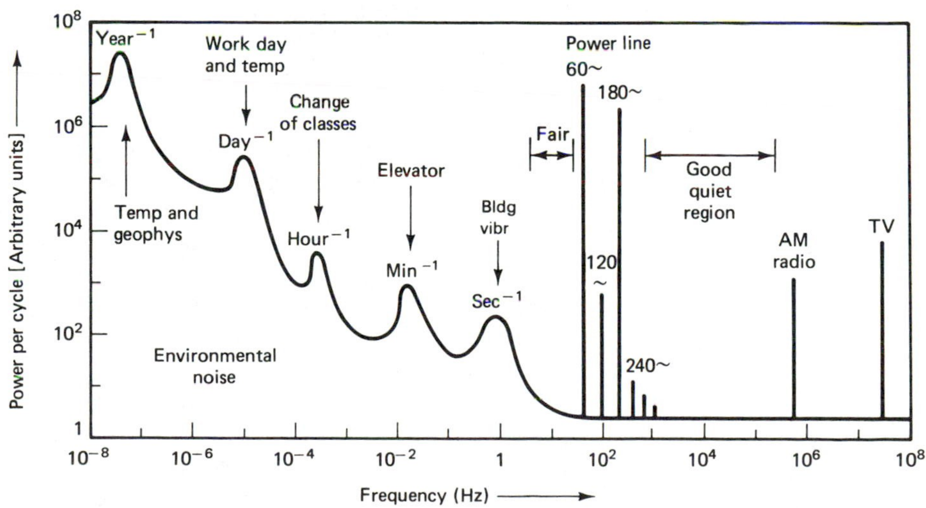

Figure 18.5 Sources of external noise. (*Adapted from* PAR application notes, *courtesy* EG&G Princeton Applied Research)

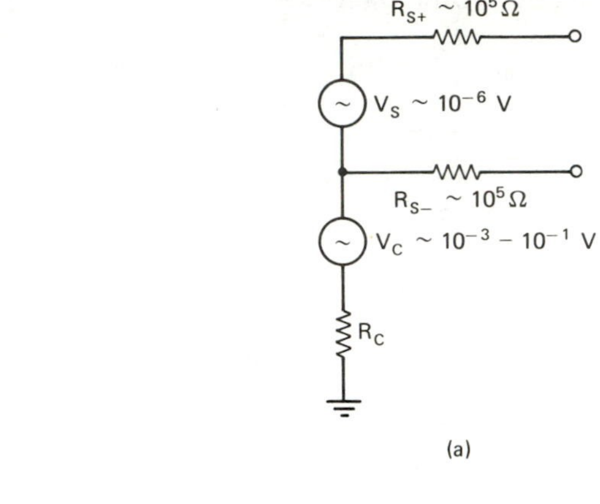

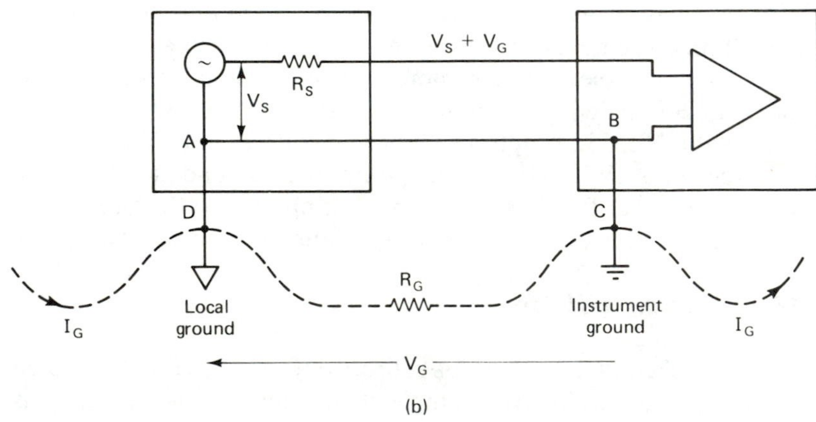

Figure 18.6 (a) Examples of a real signal V_s with common mode noise plus possible error sources due to source resistance and common mode resistance R_c. (b) Origin of ground loop problems. The triangular symbol designates a local ground whose potential may differ from the instrument ground.

inantly the fundamental and third harmonic) and local radio and TV stations. There are also peaks in the audio range due to mechanical vibrations; building vibrations at $\sim$1 Hz; electrical noise peaks at daily intervals, due to fluctuations in power usage (the start and end of the work-day are worst); and even lower-frequency fluctuations due to climate or geophysical events.

Whenever possible, one uses ac modulation techniques to shift signals up to a quiet region of the spectrum (Section 18.3). In addition, careful attention must be paid to grounding, and to shielding the amplifier, source, and interconnections. When shielding of the source is difficult, there will be *common mode* noise due to the source floating up and down in response to external noise [Fig. 18.6(a)]. Even if the signal is grounded at the source, large external noise, together with appreciable resistance R_G between source and ground, can lead to a common mode signal of the order of millivolts [Fig. 18.6(b)]. This is called a *ground loop* problem. The potential V_G at point D may be far from that of the instrument ground point C. If this "grounding" is done via the third wire of an ac power line, large and indeterminate currents I_G will flow through R_G, causing a large error voltage V_G, predominantly at the line frequency and its harmonics. In addition, points ABCD form a loop which, if large in area, will pick up electromagnetically large induced error signals, again principally at ac line frequencies. The rules for avoiding ground loops are:

(1) Return chassis and shield common terminals to a "true" ground at a *single* point.
(2) Make only one ground connection in a circuit such as Fig. 18.6(b). Cut connection A–D if possible. If floating the source leads to noise pickup of other kinds, cut connection B–C.

The latter solution is most often convenient, and requires a carefully balanced *differential* amplifier such as the one discussed below.

18.2 INSTRUMENTATION AMPLIFIERS

Compare the effectiveness of standard op amp configurations (Fig. 18.7), assuming a source [Fig. 18.6(a)] with signals V_s in the microvolt range and source resistances (R_s, R_c) as large as $10^5\ \Omega$. This is typical of the brain wave signal (Fig. 18.1). A gain of about 10^4 will be needed to bring the signal up to the level of a volt. The usual inverting connection [Fig. 18.7(a)] can provide enough gain, but the input impedance of this connection (10 KΩ in this example) is so low that little signal voltage will reach the amplifier *input.* A follower-with-gain [Fig. 18.7(b)] does offer high input impedance with high gain, but amplifies common mode error signals equally well. The common mode and differential mode gains are equal, or CMRR = 1. The subtractor configuration [Fig. 18.7(c)] offers differential amplification but loads the source. In addition, perfect rejection of common mode signals is achieved only when the input and feedback resistors are precisely matched ($R_2/R_1 = R_4/R_3$). But when the subtractor is connected to the source, the source resistances add to R_1 and R_3. If source resistances R_{s+} and R_{s-} are unequal, as is often the case, the common mode gain is no longer zero.

The *instrumentation amplifier* configuration [Fig. 18.7(d)] solves all these problems. Two op amps connected as followers-with-gain prevent the source from being loaded. A third op amp connected as a subtractor amplifies the difference between follower outputs. Since all the gain is in the first stage, only op amps A_1 and A_2 must be premium quality and carefully nulled. If further gain is needed and the signal is ac, a low-pass filter is put at the output to prevent offset voltages in the low-level end from overloading later circuits. The first stage has a perfect CMRR, *independent* of component matching. This surprising result comes from the fact that the same current [I in Fig. 18.7(d)] flows through the feedback loops of both A_1 and A_2 and through the coupling resistor R_1.

$$I = (V_2 - V_{s+})/R_2 = (V_{s+} - V_{s-})/R_1 = (V_{s-} - V_3)/R_3 \qquad (18.4)$$

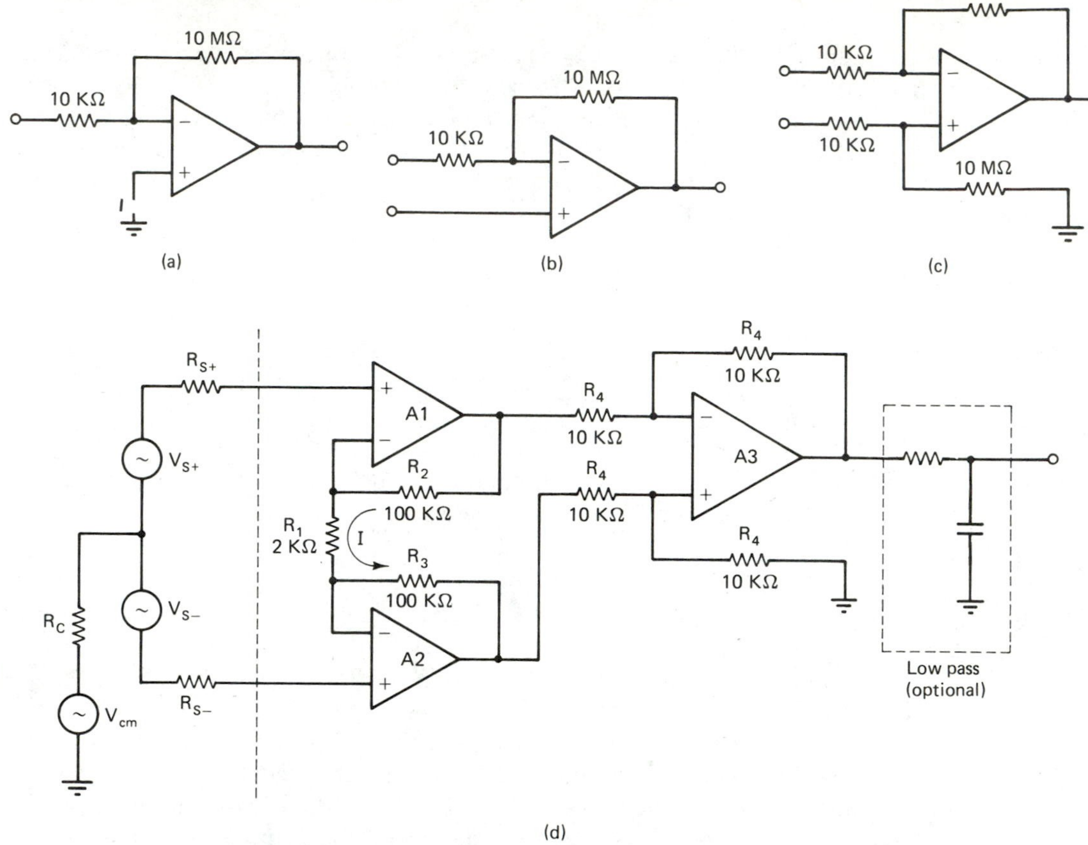

Figure 18.7 Possible ways to measure a small signal. (a) Inverting amplifier. (b) Follower with gain. (c) Subtractor. (d) The instrumentation amplifier circuit; A_1, A_2 are low noise, low offset, low bias current premium op amps.

Here, we have split the source V_s in half (V_{s+}, V_{s-}) to emphasize differential symmetry, and have assumed that negative feedback finds a way to make $V_+ = V_-$ at any op amp input. The gain of each follower is:

$$V_2 = (1 + R_2/R_1)\,V_{s+} - (R_2/R_1)\,V_{s-} + V_{cm} \qquad (18.5)$$
$$V_3 = (1 + R_3/R_1)\,V_{s-} - (R_3/R_1)\,V_{s+} + V_{cm}$$

Each amplifier acts as a follower-with-gain for the signal presented to its plus terminal, an inverting amplifier for the other signal, and a unity gain follower for the common-mode signal. The low common-mode gain follows by symmetry. Since the common-mode signal gives an identical voltage at both ends of R_1, no common-mode voltage appears across it. R_1 is effectively a short circuit for V_{cm}, and A_1 and A_2 act independently as unity gain followers.

The output of the subtractor is

$$\begin{aligned} V_o &= V_3 - V_2 \\ &= \{1 + [(R_2 + R_3)/R_1]\}(V_{s-} - V_{s+}) + V_{cm} - V_{cm} \qquad (18.6) \end{aligned}$$

Athough the resistors in the subtractor must be carefully matched in pairs to reduce the common-mode gain below 1, the cancellation is independent of component matching in the first stage. The first-stage gain is purely differential, with no error if R_2 is not exactly equal to R_3. This would *not* be true for two followers-with-gain. The cross-coupling provided by R_1 presents a measure of the lower input to be amplified by the upper amplifier, and vice versa, so component imbalance cancels.

The instrumentation amplifer is now available in an IC package from several manufacturers. The user adds one or two resistors to set the gain. The op amps are closely matched, and noise pickup from external circuit wiring is minimized. An example is included in Table 18.1.

18.2.1 Common-Mode Input Impedance

This op amp specification becomes important in a differential amplifier. Consider the equivalent circuit shown in Fig. 18.8. Although it is the differential input impedance r_i which determines whether the amplifier will load the source, it is the common-mode input impedance r_{cm} which determines how seriously the common-mode rejection will be degraded by an imbalance in source impedance. Suppose r_i is large enough that it does not load the source. The rejection of the common-mode signal by the differential configuration depends upon keeping equal the common-mode signals appearing at V_+ and V_-. Ignoring r_c (which does not affect the result) and setting $V_s = 0$ (by superposition)

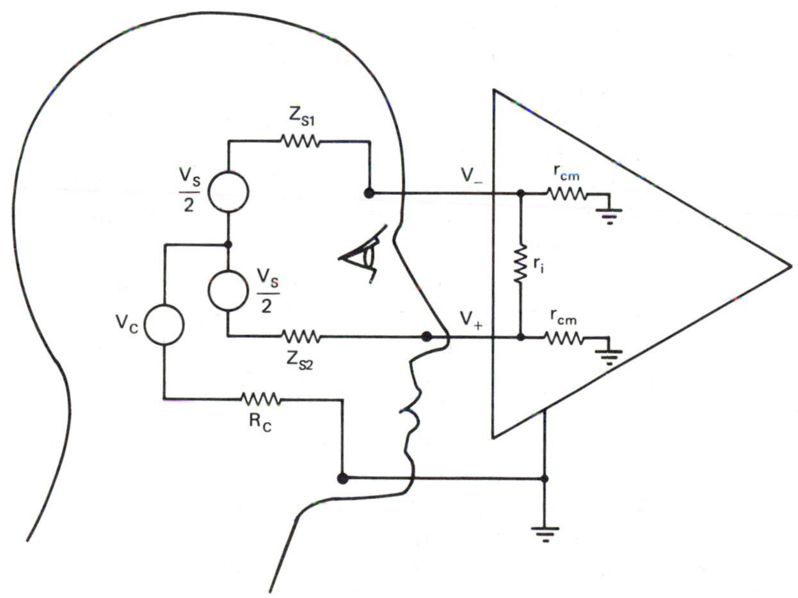

Figure 18.8 Equivalent circuit for analyzing common mode error due to an unbalanced source.

$$V_- = r_{cm} V_{cm} / (Z_{s1} + r_{cm}) \tag{18.7}$$
$$V_+ = r_{cm} V_{cm} / (Z_{s2} + r_{cm})$$

As long as $r_{cm} >> Z_s$

$$V_- - V_+ = (Z_{s1} - Z_{s2}) V_{cm} / r_{cm} \tag{18.8}$$

The importance of large r_{cm} is clear. If the imbalance in source impedance approaches $10^5\ \Omega$, a common-mode impedance r_{cm} of even $10^6\ \Omega$ will attenuate V_{cm} by only a factor of 10. Since V_{cm} may be several orders of magnitude larger than V_s, the signal will be swamped by the common-mode noise. With an op amp optimized for high CMRR, $r_{cm} \simeq 100\ \text{M}\Omega$ or larger, so a millivolt-level common-mode signal is attenuated to the microvolt level at the differential amplifier input.

18.3 COHERENT DETECTION METHODS: CHOPPER-STABILIZED AND LOCK-IN AMPLIFIERS

The lock-in amplifier or phase-sensitive detector detects signals coherent in frequency with an ac reference signal at frequency f_R. It rejects all incoherent frequency components, even noise many times larger than the coherent signal. The phase-sensitive detector may be thought of as a switch driven by a reference voltage $V_R(f_R)$. The switch closes for half of each cycle, storing whatever is present on an RC low-pass filter. If the time constant $\tau = RC >> f_R{}^{-1}$, a dc signal is stored which is proportional to the coherent part of the signal. The lock-in acts as a narrow-band amplifier, with bandwidth $\Delta f = 1/RC$. Since RC may be of the order of seconds, and f_R may be as high as 10^5 Hz, the equivalent Q of the circuit can easily reach 100,000!

There are two modes of coherent detection. A dc or slowly varying signal V_{in} can be fed through a *chopper*, a switch driven by V_R, creating a pulsating signal whose amplitude is proportional to V_{in}. This signal is then ac amplified and phase-detected [Fig. 18.9(a)]. If the measured quantity depends on some variable within the experimenter's control, that *variable* can be modulated, giving V_{in} an ac component which is phase-detected at the modulation frequency [Fig. 18.9(b)]. The chopper method attenuates random noise with the low pass and suppresses errors due to dc drift of the amplifier. However, a dc error in V_{in} will get through. The modulation method shares these advantages but in addition rejects nonrandom signals which are incoherent with V_R. Suppose one is measuring a small resistance, $R \sim 10^{-4}\ \Omega$. The chopper method works only if care is taken to eliminate dc offsets from thermoelectric voltages in the connecting leads. But with the modulation method, the measuring current is made ac, and thermoelectric error signals are suppressed. Either mode is useful in measuring signals below the dc offset and drift limits of even premium op amps.

Because the chopper method uses ac amplification, one can measure *nanovolt*-level dc signals, 1000 times smaller than the amplifier's dc offset. The signal frequency is translated upwards past the noisy $1/f$ region [Fig. 18.4(a)], so

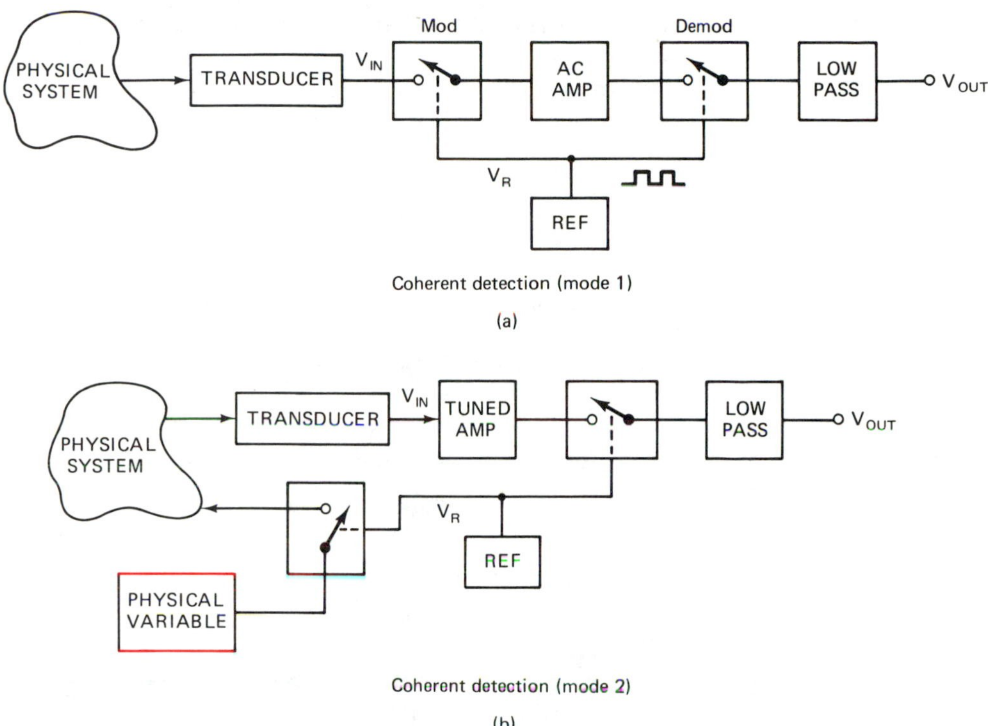

Figure 18.9 Coherent detection. (a) Chopper-stabilized mode. (b) Lock-in mode.

one can reach the Johnson noise limit of the source. The Johnson noise can be minimized with the modulation method because the amplifier bandwidth Δf can be narrowly centered about f_R [see Eq. (18.1)]. Since the reference frequency is under one's control, noisy spectral regions can be avoided (Fig. 18.5). Shifting a physical event with its own characteristic frequency domain to a quiet region of frequency can retrieve a signal otherwise buried.

18.3.1 How Chopper-Stabilized Amplifiers Work

Suppose one is measuring a low-level voltage which cannot be chopped or modulated at the source. The definition of "low level" is changing, as monolithic op amps with microvolt capability become available which cost ($10 or so) less than the cheapest chopper amplifier (at least $50). Nonetheless, the chopper-stabilized op amp remains important for long term nanovolt-level drift and submicrovolt noise levels. Specifications for a typical commercial device are given in Table 18.2.

The chopper-stabilized circuit (Fig. 18.10) has two amplifiers. Amplifier A_1 chops the low-frequency input, amplifies it with a high-gain ac amplifier to eliminate dc offset problems, and then demodulates back to the original signal frequency. The tradeoff is poor high-frequency response in A_1, since the time constant of the demodulator must be slower than the period of the chopper. The gain

TABLE 18.2 CHOPPER-STABILIZED OP AMP EXAMPLE[a]

Noise level	1 Hz bandwidth	0.5 μV
	10 Hz bandwidth	1 μV
	10 Hz to 10 KHz	5 μV
Offset voltage	vs. temperature	0.1 μV/°C
	vs. time	5 μV/year
Bias current	Initial	100 pA
	vs. temperature	1 pA/°C
Open-loop gain		10^7 V/V
Input impedance		300 KΩ

[a]The model 235. *Courtesy* Analog Devices.

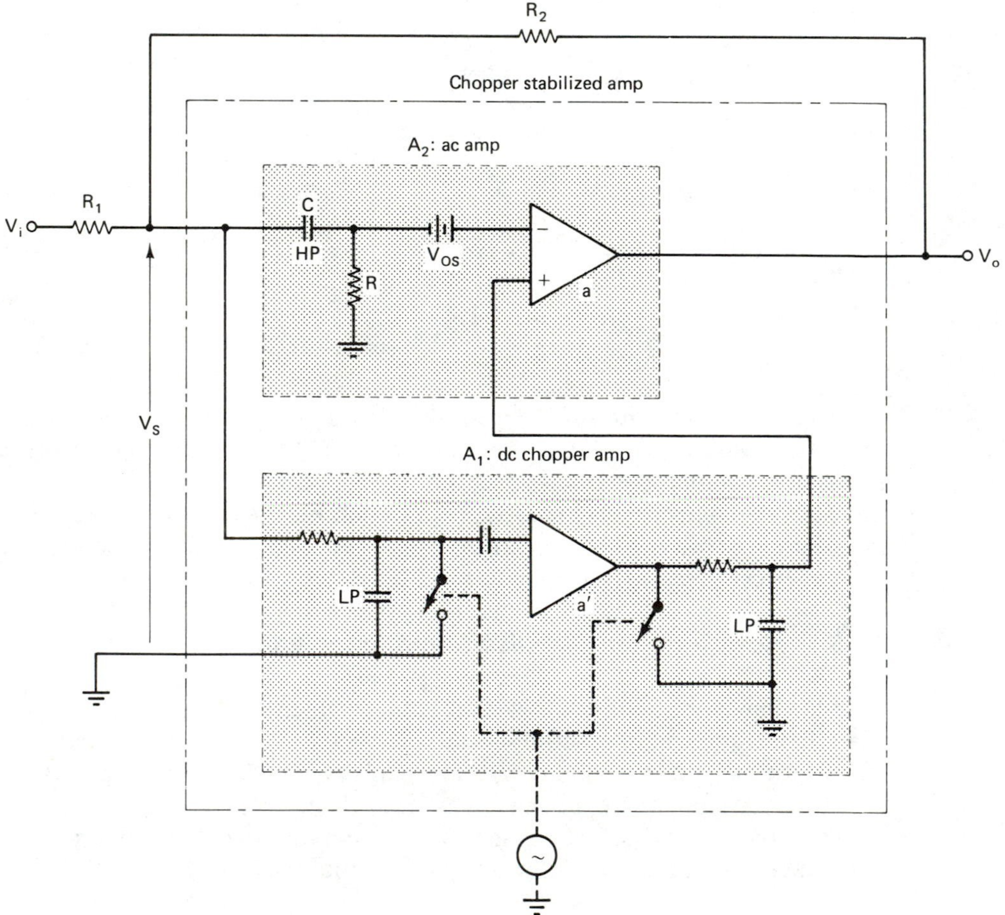

Figure 18.10 Chopper-stabilized op amp circuit.

of A_1 is a' at dc but falls off as $a'/(1 + R'C's)$ at higher frequencies. Amplifier A_2 is a conventional op amp with offset voltage V_{os}, which is ac coupled to the input. A_2 provides gain at higher frequencies where the gain of A_1 falls. The interconnection of the two amplifiers, called the *feedforward technique*, reduces the offset voltage error of A_2 by a large factor a'. Normally, V_{os} prevents V_s from reaching zero. But here the dc path is broken by capacitor C. Output offset error fed back to V_s is instead amplified by a' and fed to the *noninverting* terminal of A_2 to cancel V_{os}. Following the usual negative feedback assumptions,

$$V_o = V_s(1 + R_f/R_i) - (R_f/R_i)V_1 \tag{18.9}$$

The output of A_2 is related to what appears across its inputs:

$$V_o = -aV_i = -a\frac{V_sRCs}{1 + RCs} + V_{os} - a'V_s \tag{18.10}$$

Combine the above two equations with a bit of algebra, eliminating V_s and dropping terms of order $1/a$.

$$V_o = -(R_f/R_i)V_1 + \frac{(1 + R_f/R_i)V_{os}}{a' - [RCs/(1 + RCs)]} \tag{18.11}$$

If the extra amplifier a' were not present, the denominator term $\{a' - [RCs/(1 + RCs)]\}$ would have been absent. Feedforward reduces the dc offset voltage V_{os} error by a large factor a'.

18.3.2 The Isolation Amplifier

The ac portion of the chopper amplifier (Fig. 18.9) provides an opportunity to decouple input from output via an isolation transformer. Op amps with this feature, called *isolation amplifiers*, have the highest common-mode rejection (CMRR $>$ 120 dB). Isolation amplifiers find uses in medical instrumentation since they isolate the subject from any possible danger of dc voltages, even if a circuit fails. The isolation amplifier is also valuable in harsh environments where high noise levels or stray voltages must be prevented from getting into the measuring system.

18.3.3 How Lock-In Amplifiers Work

The operation of the lock-in circuit [Fig. 18.9(b)] can be understood graphically. The chopper, also called a *phase detector* or *mixer*, effectively multiplies the signal V_{in} by $+1$ or -1, coherent with the reference signal V_R [Fig. 18.11(a)]. The phase detector output V_{PD} has a dc component, proportional to the amplitude of V_{in}, which is extracted by the low-pass filter. When a noise signal incommensurate with V_R is chopped [Fig. 18.11(b)], the phase detector output has no net dc component, so the low-pass output is zero. This is how the lock-in rejects noise. The size of the dc output voltage will depend upon the relative phase between signal and reference [Fig 18.11(c)]. When signal and reference are 90° out of phase, the output is zero, even when both are coherent. Commercial lock-in instruments have a variable phase adjustment for tuning up purposes or for measurement of the phase angle.

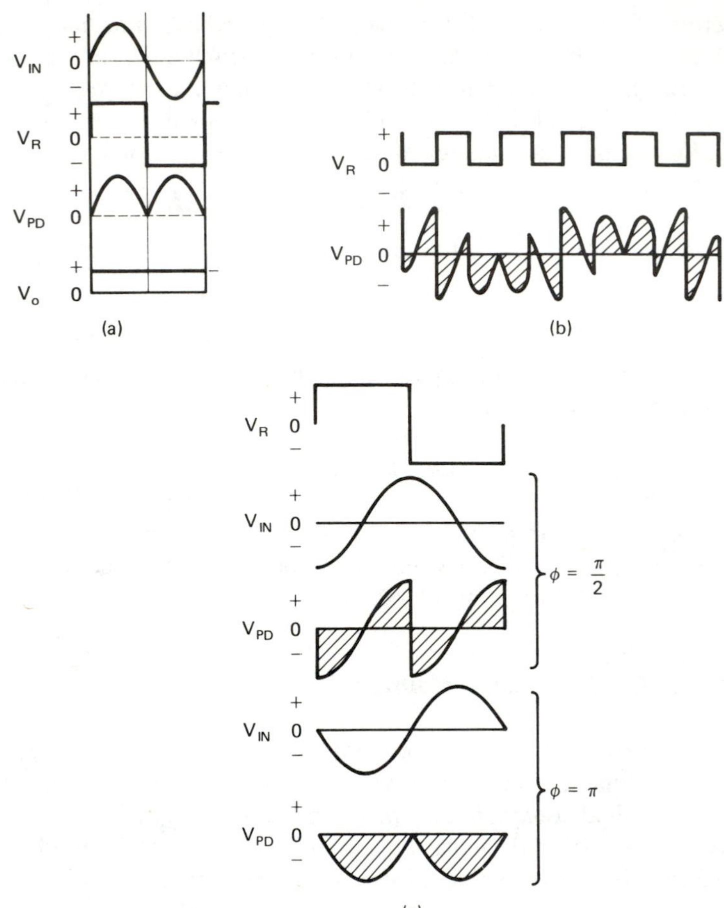

Figure 18.11 Lock-in amplifier waveforms. (a) Signal, reference, phase detector output, and averaged output V_o. (b) Cancellation of phase-detected in coherent noise. (c) The dc output (shaded area) varies with phase difference between signal and reference.

Some applications are shown in Fig. 18.12. In measuring optical absorption [Fig. 18.12(a)], the light source is chopped by a rotating wheel, which also generates V_R. The detector is presented with pulses of light which have passed through the sample, and also stray light (noise). Since the noise is not coherent with V_R, it is eliminated by the lock-in. Recovery of signals which would otherwise be buried in noise is possible [Fig. 18.12(b)]. Sometimes phase information is also important. For example [Fig. 18.12(c)], the speed of sound in a gas or a material may be measured by varying the distance x between source and detector. The dc output of the lock-in will vary as sin x because of the phase shift with distance. When the dc output goes through one cycle, x has changed by one wavelength λ. The sound velocity, v, may be calculated, given the reference frequency ($v = f\lambda$).

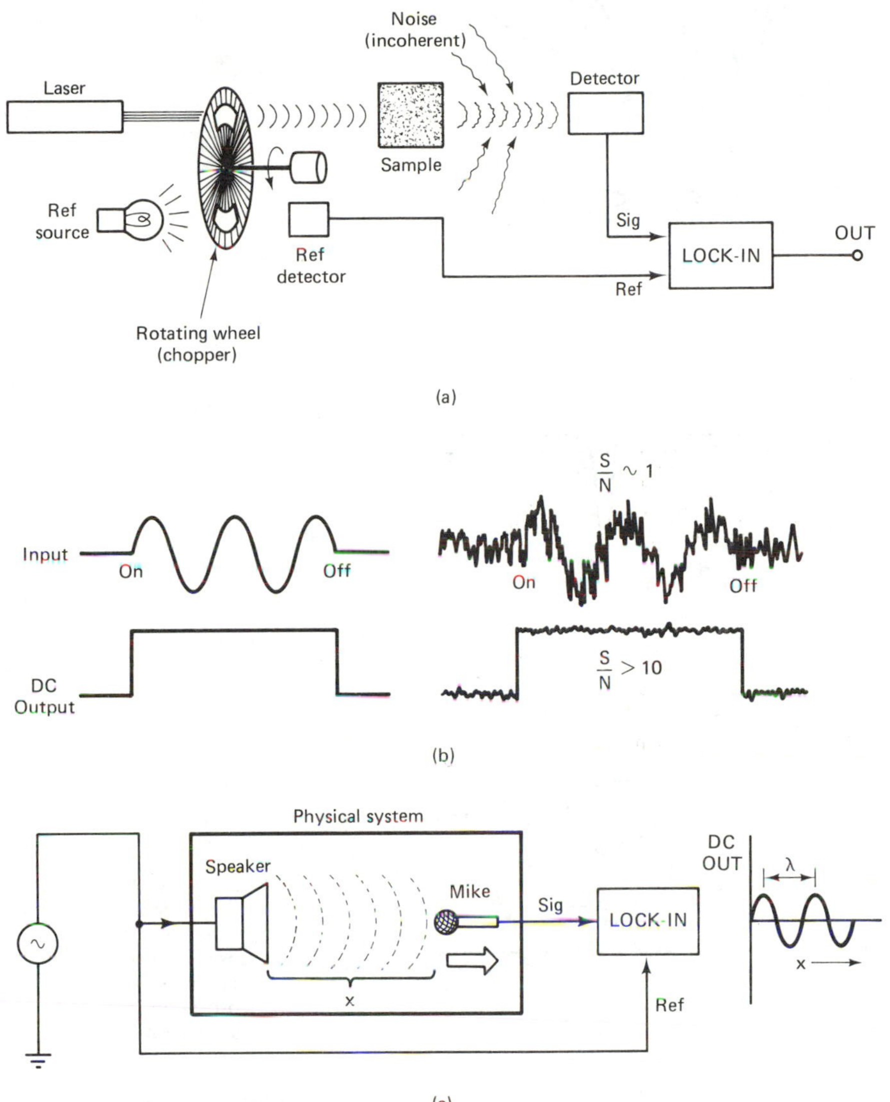

Figure 18.12 Lock-in applications. (a) Optical absorption. (b) Signal to noise enhancement. (c) Velocity from phase measurement.

When plotted as a function of the modulating variable, a graph of lock-in output is equivalent to a derivative with respect to the modulating variable (see Prob. 18.4). A given modulation amplitude V_R causes variations in the ac signal amplitude proportional to the *slope* [Fig. 18.13(a)] of the function. An example is optical absorption as a function of wavelength λ [Fig. 18.13(b)]. The diffraction grating is rotated by a small angle at frequency f_R, modulating the wavelength of transmitted light intensity J passing through the entrance slit of the detector. The

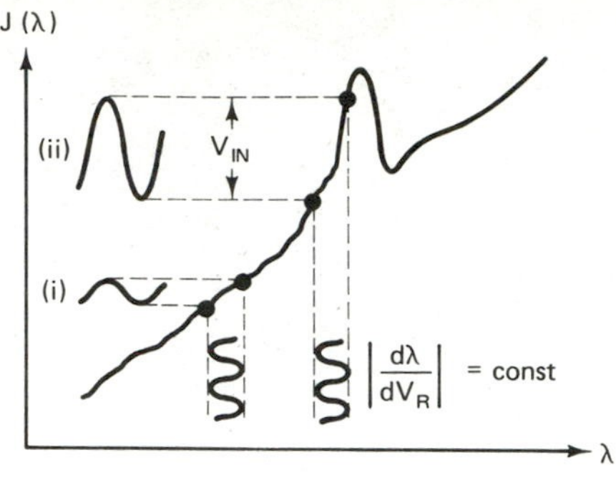

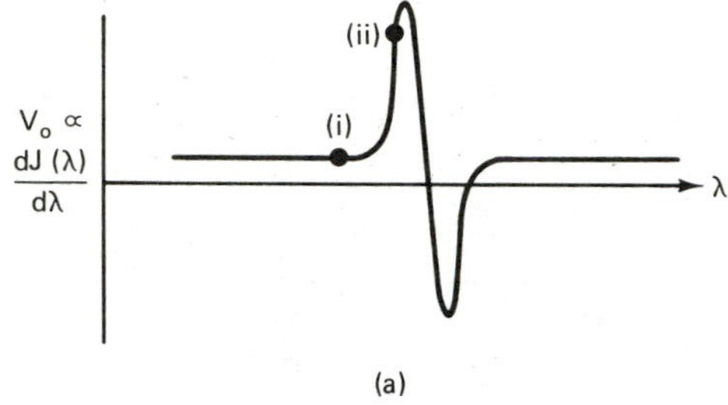

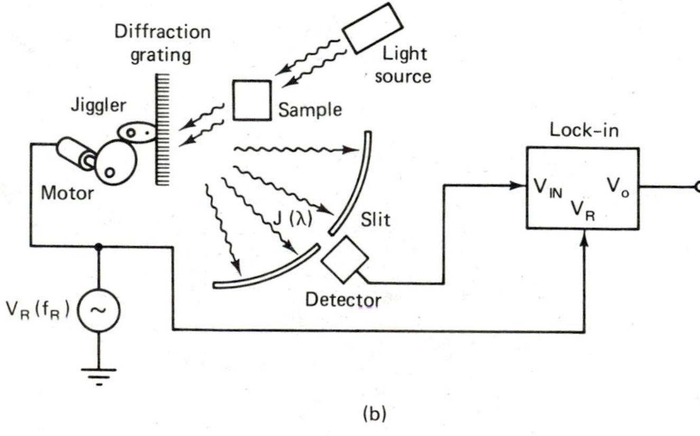

Figure 18.13 Modulation of an independent variable takes a derivative of the signal *J.* (a) Structure in the absorption spectrum is enhanced in the derivative. (b) Application to optical absorption. The entrance slot selects a specific wavelength λ.

lock-in output is proportional to $dJ/d\lambda$. This method is particularly valuable when features of interest are superimposed upon a large but slowly varying baseline. The baseline is suppressed, since the derivative of a nearly constant term is small.

Lock-in circuits. The simplest phase detector is an FET switch driven by the reference signal [Fig. 18.14(a)]. An inverting amplifier is included to recover both halves of the ac signal, giving the full-wave rectified appearance of Fig. 18.11. A *reed relay* (a hermetically sealed and mechanically actuated switch) or a *photochopper* (light-emitting diode driven by V_R, modulating the gain of a phototransistor amplifier) might be found in instruments designed for low-level currents or voltages. But as their cost decreases (under $1 apiece) and performance increases, IC MOS switches have rapidly replaced mechanical or photochoppers in nearly all applications.

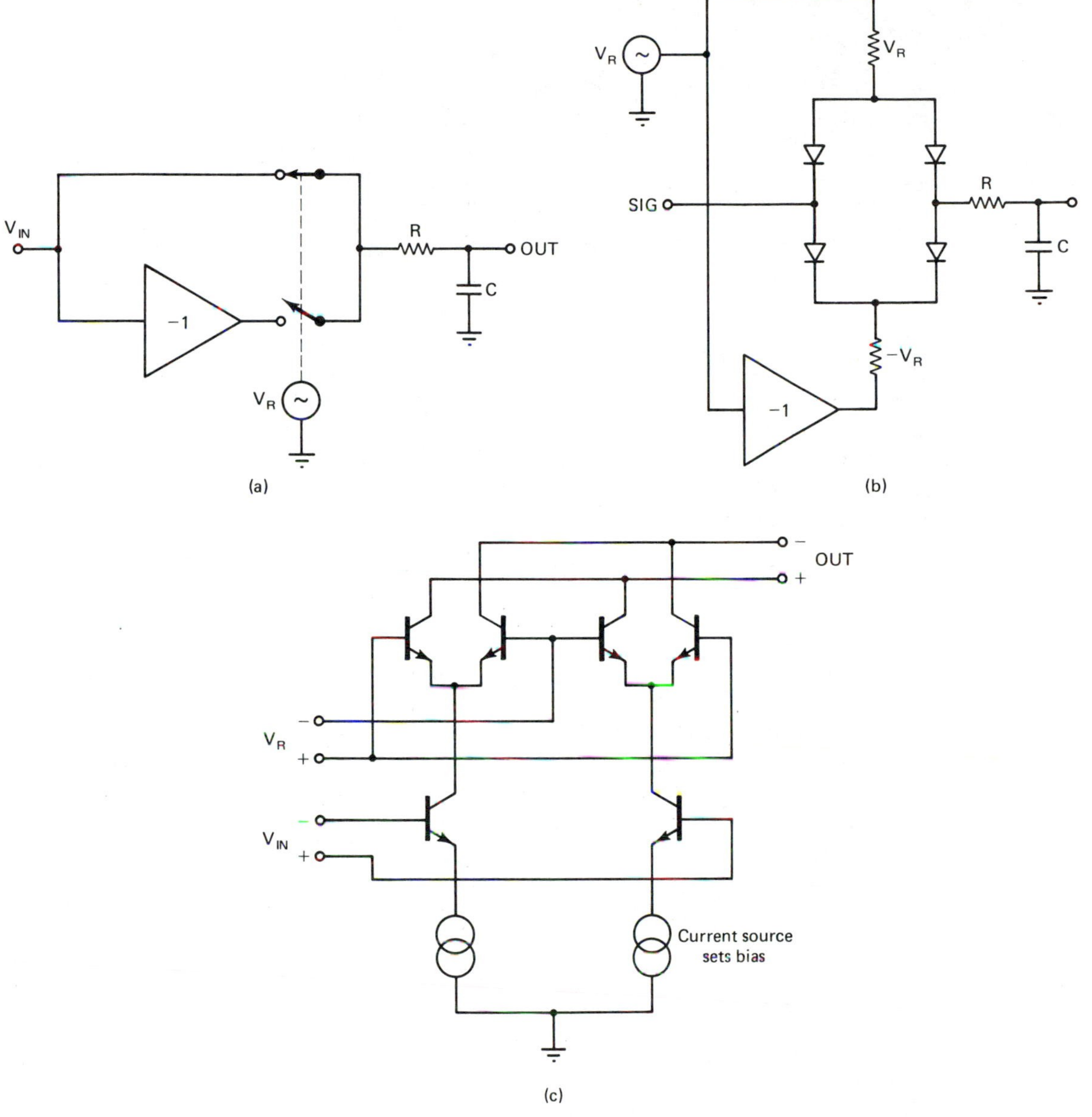

Figure 18.14 Lock-in circuits. (a) FET switch chopper. (b) Ring modulator. (c) Balanced modulator, equivalent to MC 1496 IC.

An older solid-state switching method uses the large resistance change of diodes between their on-state and off-state. The reference signal turns the diodes on and off, altering the signal flow path. The *ring modulator* [Fig. 18.14(b)] uses diode switches. When the diodes are driven *on* by V_R, the signal finds a low-resistance path to the RC integrator. In the other half-cycle, the signal is blocked. The symmetry of the ring configuration prevents the reference signal from appearing at the integrator and providing a false output. But the signal voltage must

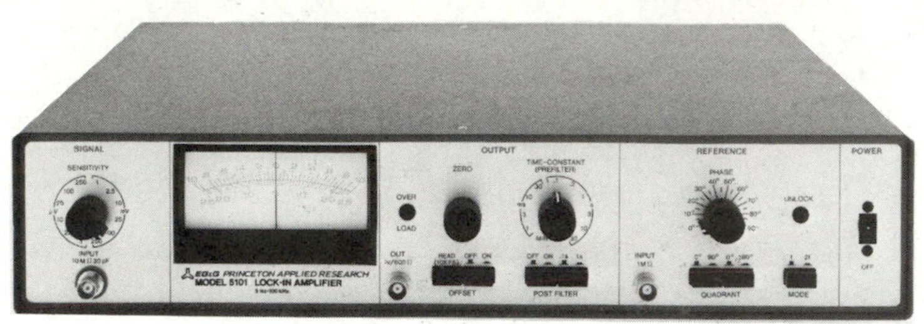

Figure 18.15 Example of a commercial lock-in amplifier. (*Courtesy* EG&G Princeton Applied Research)

be small compared to V_R so it is the reference voltage which turns the diodes on and off.

The limitations of diode switches are eliminated in the transistor switch shown in Fig. 18.14(c). The lower pair of transistors is operated as a linear differential amplifier. The upper two pairs are two-state switches, driven by V_R. One transistor of each pair is always on and the other off, forming a pathway to the output which reverses when V_R changes sign. This circuit is now available as an IC called a *balanced modulator-demodulator.*

Research applications of the lock-in technique generally use high-quality commercial instruments available from manufacturers such as Princeton Applied Research, Princeton, N.J., and Ithaco Inc., Ithaca, N.Y. An example is shown in Fig. 18.15 and Table 18.3. A high-quality but inexpensive CMOS analog switch-type lock-in is described in the article by Temple listed in the References. In this circuit, the reference signal and the phase shifter are generated digitally, representative of methods used in commercial instruments.

A simple yet convincing example of a lock-in's ability to recover a coherent signal from larger incoherent background is shown in Fig. 18.16. The circuit uses a clever integrator subtraction technique (explained in Chapter 19) to remove ambient light from the photodiode signal. The ambient light could otherwise introduce a dc offset large enough to overload the input op amp A_1. Any dc signal in the output of A_1 ramps integrator A_2 up or down until its output cancels the dc input signal.

TABLE 18.3 SPECIFICATIONS OF A LOW-COST COMMERCIAL LOCK-IN AMPLIFIER

Frequency range of reference signal	5 Hz to 150 KHz
Sensitivity (direct)	1 μV
Sensitivity (with preamp):	
Direct-coupled	100 nV
Transformer-coupled	1 nV
Common-mode rejection	85 dB
Noise figure (with impedances matched)	<0.1 dB
Output stability	<0.1% °C
Equivalent noise bandwidth (30 s time constant)	As low as 0.03 Hz
Dynamic reserve (asynchronous signals)	Up to 1000 times Full scale

[a]The Par model 5101. *Courtesy* Princeton Applied Research

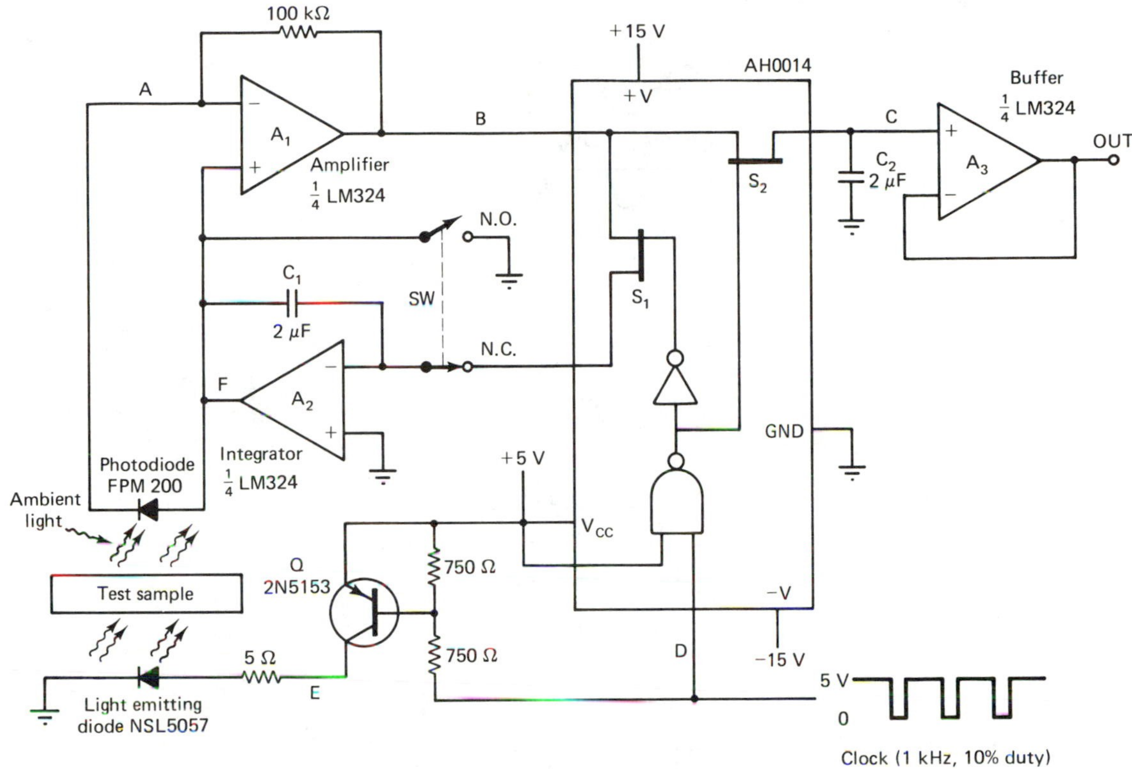

Figure 18.16 Lock-in with cancellation of dc offset signals (ambient light in this example). Circuit operation may be understood by examination of waveforms at points A to F. The integrator action which subtracts the ambient light may be defeated temporarily using switch SW. (*From* H. H. Martensen, *Electronics.* (June 15, 1975): 124.)

18.4 PHASE-LOCKED LOOPS

The phase-locked loop (PLL) is a powerful signal processing technique which can track a signal's *frequency* coherently and recover it from noise. The PLL resembles the lock-in amplifier, except that the reference signal is generated internally. Although the PLL idea is not new, it has come into wide application only in recent years, with the development of inexpensive (<$5) IC phase-locked loops. PLL's are now used widely in communications electronics, and will be used increasingly in measuring instruments.

18.4.1 Basic Phase-Locked Loop Operation

Basic PLL operation involves two building blocks: the coherent detector and the voltage-controlled oscillator. The coherent detector [Fig. 18.17(a)] is a multiplier followed by a low-pass filter. Although analyzed in Section 18.3 in terms of switching, coherent detection may be also accomplished with linear circuits. The product of two sine wave signals contains sum and difference frequencies,

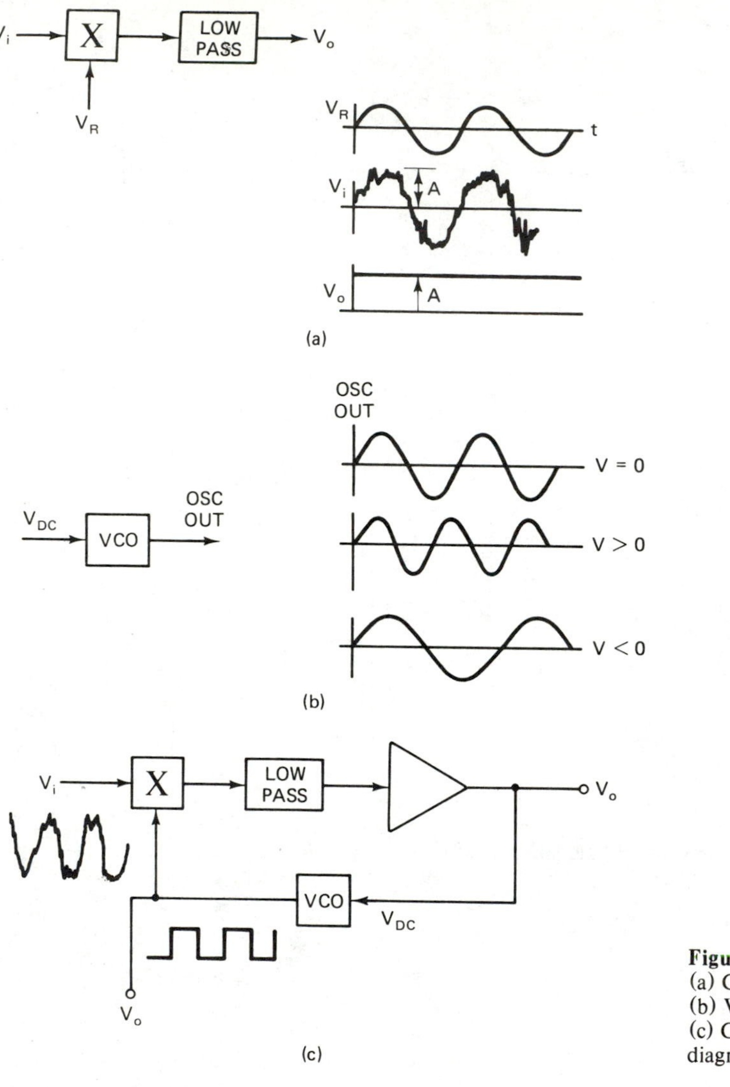

Figure 18.17 Phase-locked loop. (a) Coherent detector block. (b) Voltage-controlled oscillator block. (c) Complete phase-locked loop block diagram.

$$V_{\text{in}} = A_{\text{in}} \sin (\omega_{\text{in}} t) + \text{noise} \tag{18.12}$$

$$V_R = A_R \sin (\omega_R t + \phi) t$$

$$V_{\text{in}} V_R = (A_{\text{in}} A_R/2)\{ \cos [(\omega_s - \omega_r)t + \phi] - \cos [(\omega_{\text{in}} + \omega_R)t + \phi]\} + (\text{noise})(A_R \sin \omega_R t)$$

The low-pass filter eliminates both the sum frequency term and the noise term, since $\omega_r \tau_{\text{LP}} >> 1$, $\omega_s \tau_{\text{LP}} >> 1$. However, if the signal includes a component coherent with the reference signal ($\omega_{\text{in}} = \omega_R$), there will be a dc output:

$$V_{\text{out}} = A_{\text{in}} A_R \cos \phi \tag{18.13}$$

The second building block, the voltage-controlled oscillator (VCO) [Fig. 18.17(b)], has a natural frequency which can be shifted up or down by a dc voltage.

The phase-locked loop circuit connects these two building blocks together in a closed feedback loop [Fig. 18.17(c)]. When the PLL is "locked," the dc voltage generated at the low-pass filter output is just the right amplitude to drive the VCO frequency coherently with the signal. If the signal frequency changes, the VCO frequency changes to follow it. This makes a PLL an FM detector, since the low-pass output is a measure of the frequency modulation of the input signal. Because of the time constant of the low pass, the PLL is relatively insensitive to random noise, retaining phase memory of the input signal, and improving the signal-to-noise ratio.

18.4.2 Typical Phase-Locked Loop Specifications

Practical PLL specifications include (Table 18.4):

> *Free-Running Frequency* The natural frequency of the VCO in the absence of an input signal (or when running "out of lock"). This frequency is set by an external capacitor/resistor combination.
>
> *Lock Range* The range of input frequencies over which the VCO will remain locked.
>
> *Capture Range* The range of input frequencies over which the PLL can go from the unlocked to the locked condition when an input is applied.

The capture range is narrower than the lock range [Fig. 18.18(a)] because of the dynamics of the "capture transient" [Fig. 18.18(b)]. The dc signal which drives the loop into lock comes from the output of the low-pass filter. Initially, the signal and reference frequency are unrelated. During the capture process, the difference frequency term in Eq. (18.12) is small in amplitude at first, because it is attenuated by the filter. As its average value drives the VCO towards the signal frequency, this difference in frequency becomes smaller, and a larger dc component is passed by the filter, shifting the VCO frequency further. The process

TABLE 18.4 CONDENSED SPECIFICATIONS OF THE NE565 PHASE-LOCKED LOOP

Supply voltage	± 5 to ± 12 V @ 10 mA
Input impedance	10 KΩ
Minimum input level for tracking	1 mV rms
Maximum VCO frequency	500 kHz
Lock range	$\sim \pm$ 100% f_o (max)
VCO output ($V_{cc} = \pm 6$ V)	+5.2 V (logic 1); −0.2V (logic 0)
VCO rise time	20 ns; fall time 50 ns
VCO output current	1 mA (sink); 10 mA (source)
Output voltage ($V_{cc} = \pm 6$ V)	4.5 V, with full-scale swing $\pm$ 2 V

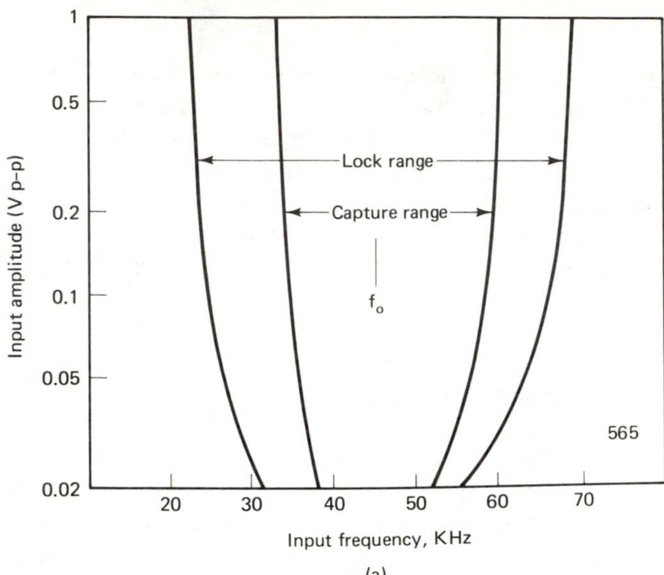

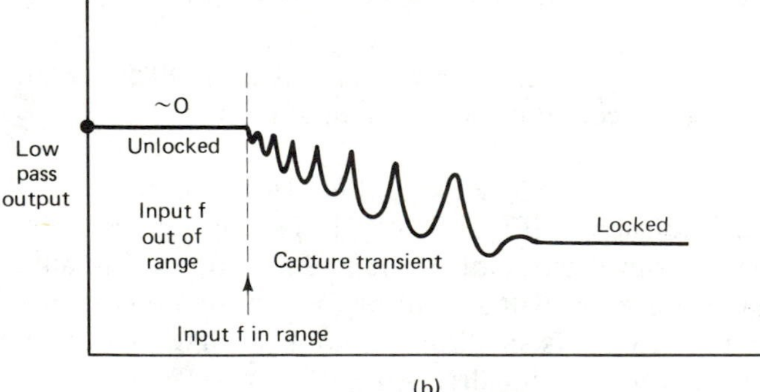

Figure 18.18 Phase-locked loop behavior. (a) Capture range and lock range. (b) The capture transient.

continues until the VCO locks on to the signal, and the difference frequency is dc. The lock range is larger than the capture range because only when locked will there be a pure dc signal to drive the VCO. The difference in lock range and capture range is dependent on the low-pass filter cutoff frequency, which is under the user's control. These ranges become narrower as the input signal amplitude decreases. This can be understood from Eq. (18.13). The feedback loop selects the value of V_{out} necessary to drive the VCO frequency to the input frequency by adjusting ϕ (since A_R is constant). But since the value of cos ϕ is bounded, the useful phase range is limited to $\pm\ \pi/2$. As the input signal amplitude A_s decreases, the range of V_{out} decreases proportionally, decreasing the frequency tracking capability of the loop.

18.4.3 Phase-Locked Loop IC's

The 565 is a useful PLL IC. Although limited in frequency range to 0.5 MHz, the 565 has a very wide lock range ($\Delta f = 1.2 f_o$, where f_o is the natural frequency of the VCO) and a supply voltage range (± 5 to ± 12 V) compatible with digital logic. All internal points in the block diagram [Fig. 18.17(c)] are externally accessible, facilitating the most general applications.

The phase detector section is a multiplier or balanced modulator, similar to Fig. 18.14(c). The VCO resembles the 566 function generator of Fig. 17.18. Consult the Signetics PLL references for further circuit details.

A block diagram of the 565 with pin connections is shown in Fig. 18.19(a). Three external components are required. The VCO free-running frequency f_o is set by R_1C_1:

$$f_o = 1.2/(4R_1C_1) \text{ Hz} \tag{18.14}$$

The time constant of the low pass is set by an internal resistor and C_2:

$$\tau = RC_2 \; (R = 3.6 \text{ K}\Omega) \tag{18.15}$$

The lock range is proportional to f_o but also depends on the supply voltage:

$$f_1 = 8f_o/V_{CC} \text{ Hz} \tag{18.16}$$

The capture range depends on the lock range and varies inversely with the low-pass time constant:

$$f_c = (1/2\pi) \quad [2\pi \; f_L/(RC_2)]^{1/2} \tag{18.17}$$

A wide variety of supply voltages are possible. A split ($\pm V$) supply is not essential, and a single-ended (0 to $+V$) supply can be used if attention is paid to input biasing. Both input and output are differential, to accommodate power supply flexibility and to ease matching of signal levels to other devices.

Single-ended power supply. Both inputs (pins 2 and 3) must be dc-biased above ground. *The bias on both must be identical,* and may range from 0 V to 4 V. A typical biasing scheme is shown in Fig. 18.19(b). The output dc level at pin 7 will be floating high (near V_{CC}) and must be isolated from later circuits. If an output level referenced to ground is needed, use a comparator between pins 6 and 7 [Fig. 18.19(c)]. With a single-ended supply, both the VCO input and output will float above ground. This may cause problems in circuits with TTL logic levels. Although it is possible to capacitively decouple and resistively clamp the correct levels, interfacing the VCO and phase detector (pins 4 and 5) to TTL levels is not recommended when a single-ended power supply is used.

Split power supply. The input may be single-ended. If the source has a specific impedance (e.g., 600 Ω), pin 3 must be grounded through a resistor of the same value. Normally, pin 3 is grounded directly. The output will be floating near $+V$ and a comparator will normally be used (pins 6 and 7) to reference the digital output to the next stage [Fig. 18.19(c)]. Alternatively, an analog subtractor may be connected across pins 6 and 7 if a linear output is desired. The 565 (unlike other members of the 560 series) does not have internal Zener diodes at the power supply leads. The split supply must be carefully matched (~1%), or else circuit operation may be erratic.

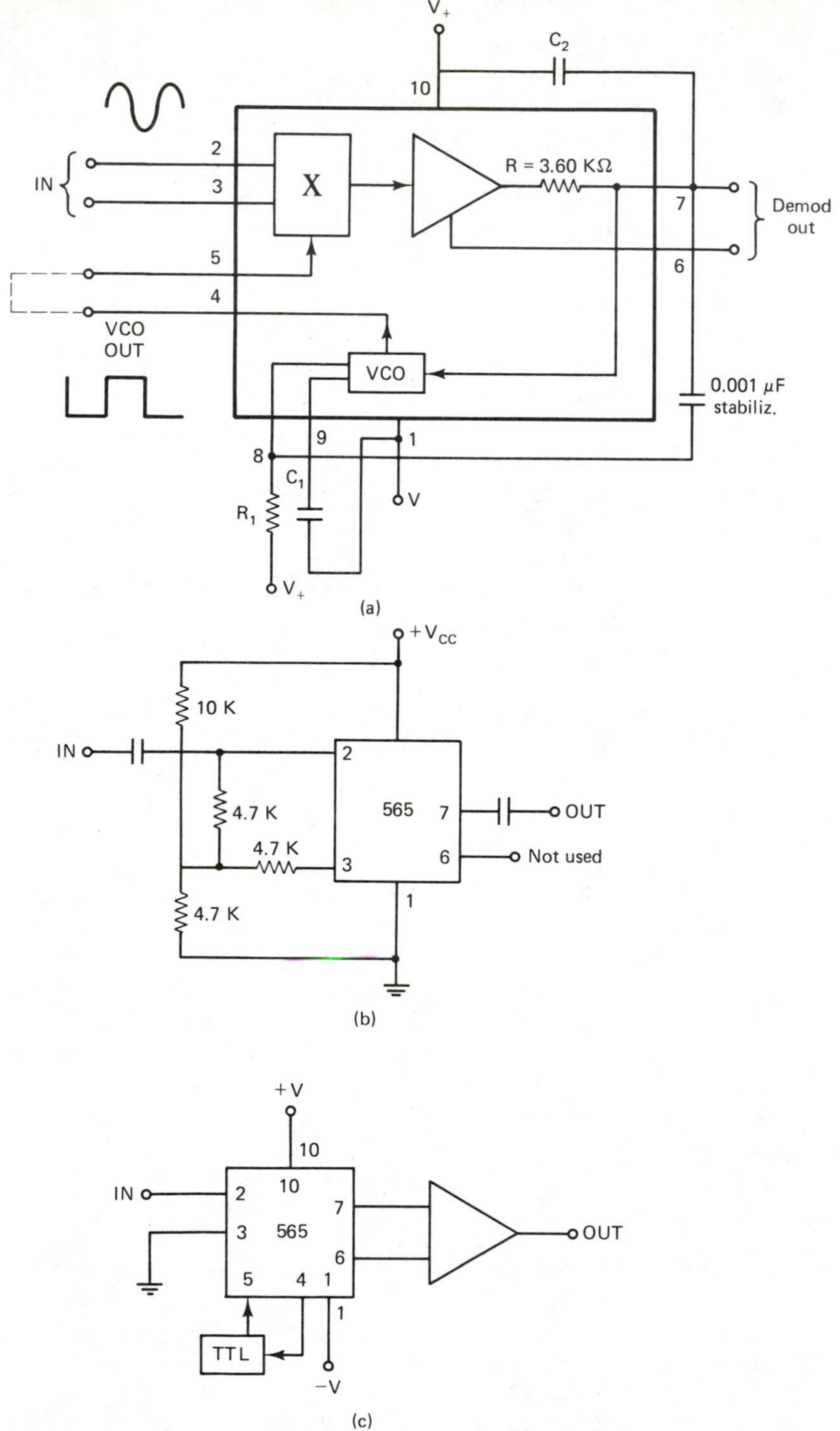

Figure 18.19 (a) Practical working circuit for 565 phase-locked loop. Block diagram with pin connections and external components. (b) Single-ended biasing. (c) Split power supply.

18.4.4 Phase-Locked Loop Applications

A few of the applications most relevant to scientific instrumentation will be sketched here. Many other applications are given in the Signetics literature (see References). The PLL is a hybrid; it detects an analog signal, though the information detected is not the amplitude but the frequency. Any information which can be represented as a frequency can be measured with a PLL.

Frequency-to-voltage converter. The basic PLL circuit generates a dc output proportional to the input frequency *difference* from the free-running frequency of the VCO.

FM demodulator. The basic PLL circuit is already an FM demodulator, since the VCO output is proportional to input signal frequency excursions. The passband of the low-pass filter must extend high enough to cover the bandwidth of the FM signal.

Lock-in amplifier. If the feedback loop is broken at V_o in Fig. 18.17(c), the result is a coherent detector or lock-in amplifier [Fig. 18.20(a)]. However,

Figure 18.20 Phase-locked loop applications. (a) Lock-in amplifier. (b) AM demodulator.

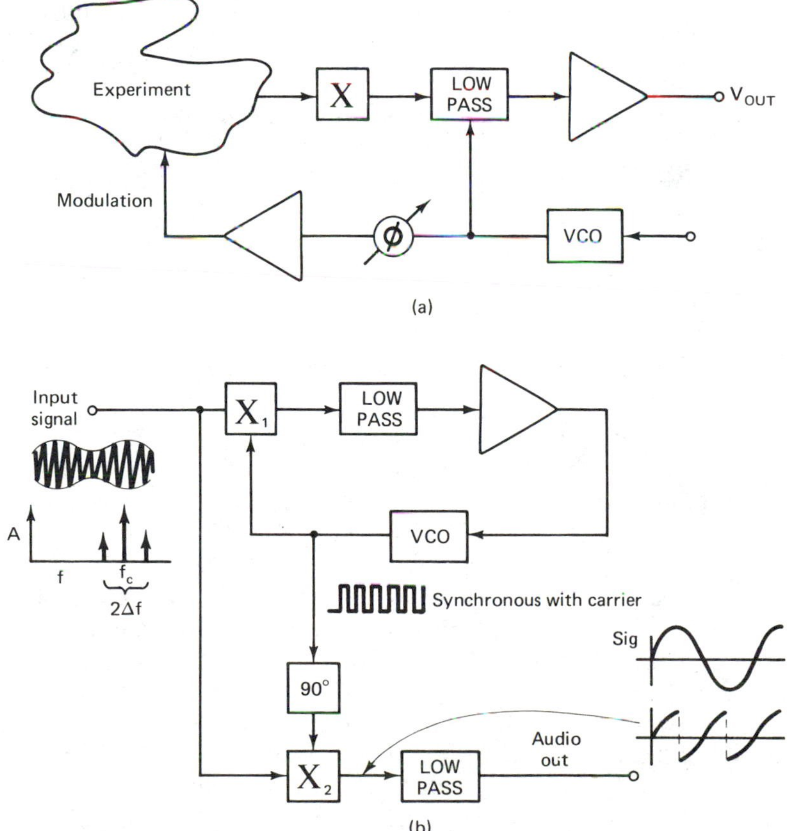

since the detected signal contains no amplitude information, this lock-in is only useful if the measured quantity *frequency modulates* the reference signal.

AM demodulator. An AM demodulator may be made by adding to the PLL a 90° phase shifter, plus another multiplier and low pass [Fig. 18.20(b)]. The 561 and 567 PLL's have this additional circuitry built in. The input to the second multiplier is a square wave synchronous with the carrier f_c but 90° out of phase. The extra 90° makes the output zero when the input is of constant amplitude (carrier only).

Harmonic generation and frequency synthesis. By breaking the loop between VCO and multiplier, elements may be inserted to cause tracking at precise multiples of the input frequency. Suppose the extra element is a frequency counter or digital divider [Fig. 18.21(a)]. The VCO finds it necessary to generate a signal N times higher than V_{in} for the frequency of V_R to match the input. When the divider is a string of flip flops, the output frequency is a power of 2 times the input [Fig. 18.21(b)]. The factor N need not be a power of 2, since standard MSI counters (e.g., 7493 or 8281) can be wired to make N an arbitrary integer. Since the PLL by itself can lock onto harmonics, it is possible to synthesize frequencies related to the input by an arbitrary rational fraction M/N [Fig. 18.21(c)]. A single stable frequency reference (e.g., crystal clock) can thus be used to generate a series of equally stable frequencies.

Although the principal applications of this technique have been in communications, there are also applications in instrumentation. For example, a single PLL with a divider of large N (say 32 or 64) can precisely time-slice a repetitive waveform for encoding by a computer. This is far better than *incommensurate* sampling for digital signal processing.

There are some practical limitations to the divide-by-N technique.

(1) The lock range is limited to a fixed fraction Δf of the VCO output. Viewed from the input, the lock range will appear N times narrower, since

$$\Delta f_{VCO} = N\ \Delta f_{in}$$

(2) Since the VCO *phase* is related to the input signal amplitude, the phase of the multiplied output will vary if the input *amplitude* varies.

(3) The multiplied-by-N output may exhibit incidental frequency modulation, due to some frequency components leaking through the low-pass filter. Extra filtering in the low pass is necessary, with reduced capture range and underdamped transient response.

Data transmission or recording via PLL frequency shift keying. The PLL's capability to convert frequencies to voltages and the reverse makes possible inexpensive data transmission and recording schemes for both analog and digital signals.

Low-frequency (dc to 100 Hz) data is below the range of a standard audio tape recorder, but can be brought to the recorder's optimum audio frequency

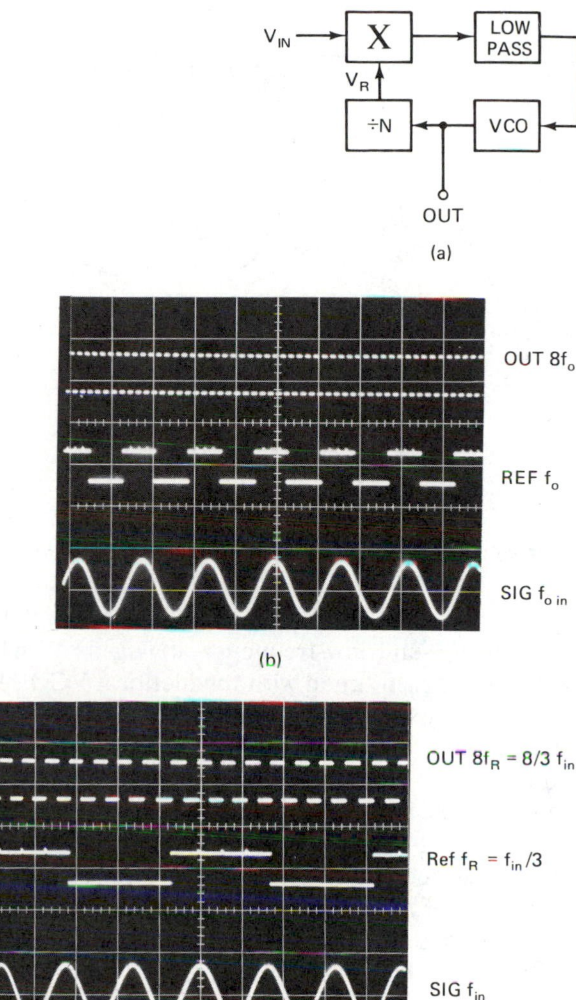

Figure 18.21 Frequency multiplication. (a) Circuit. (b) $N = 8$ multiplication. (c) M/N multiplication with $f_o = (8/3)f_{in}$.

range with a PLL (Fig. 18.22). In the *record* mode, the data modulates the frequency of the VCO. In the *playback* mode, the entire PLL is used. Analog data can be recorded with inexpensive portable equipment at remote locations, returning the tape for analysis with more sophisticated and less portable instruments. The result is equivalent to an FM instrumentation tape recorder costing several thousand dollars, but costs less than $100, including cassette tape recorder. One practical problem is dc offset caused by drift of the VCO center frequency between the time the record is made and the time it is played back. This can be eliminated by choosing the VCO center frequency to be a multiple of the line frequency, and adding a *calibrate* mode. The VCO frequency is tuned to zero the dc offset when the input is a 60-Hz square wave.

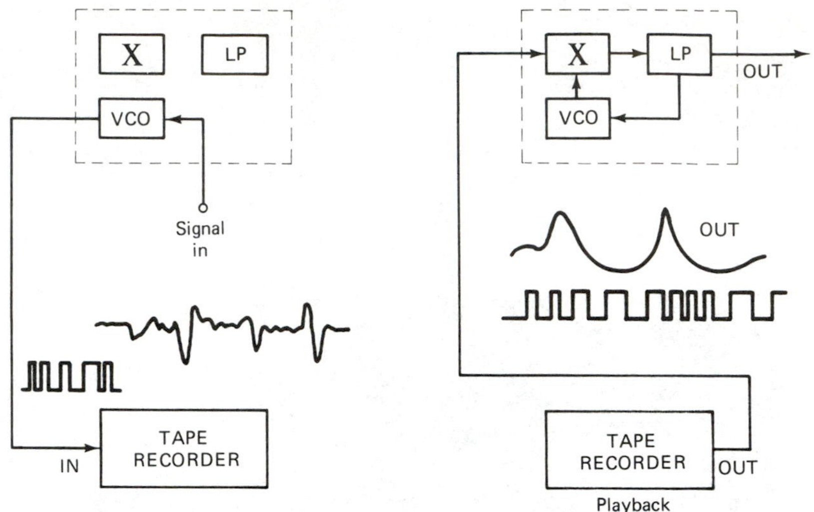

Figure 18.22 Analog data recording using a PLL and tape recorder. (a) Record circuit. (b) Playback circuit.

This concept can also be used to send low-frequency analog data on telephone lines, using a PLL on either end, each tuned with the identical VCO center frequency. This makes possible inexpensive coincidence or correlation experiments in geophysics, for example.

PLL's also find application in *digital* data recording and transmission, for situations where moderate speed (~1000 bits or ~100 characters per second) is adequate. The circuit can transmit digital data over telephone lines, enabling access to a remote computer. The basic technique originated in digital telephone transmission and is called *frequency shift keying* (FSK). Long-distance telephone transmission of voice signals is done digitally, because noise is reduced when information is sent in digital form. In the FSK technique, the two binary voltage levels (logic 0 and logic 1) are encoded as two distinct frequencies.

We describe an FSK application to an inexpensive digital tape recorder. Digital phone line communications is similar but, with the generator and detector separated, careful clock frequency adjustment is required. The block diagram (Fig. 18.23) is a two-state (binary) version of the analog recorder previously

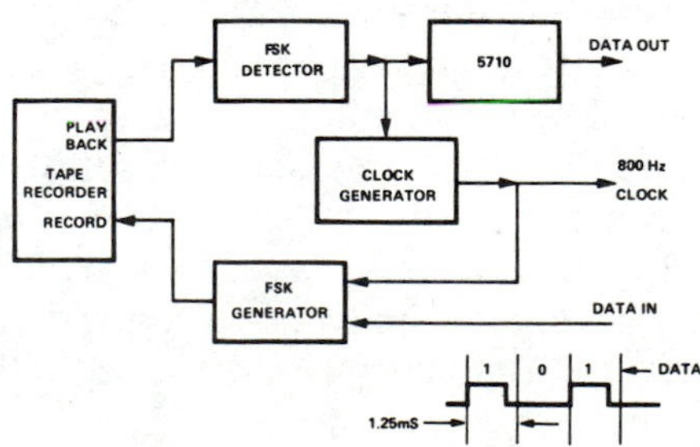

Figure 18.23 Digital tape recorder block diagram.

described. The principal difference is the addition of a clock. An 800-Hz bit rate is selected to drive the signal source and for synchronization of frequency shift keying (FSK) generator and detector. The 800-Hz rate is about eight times faster than teletype speed, but the two binary tones (at 6 × 800 Hz and 8 × 800 Hz) are still well within the frequency range of conventional tape recorders. The FSK generator uses the VCO of a PLL to alter the frequency recorded as binary data goes from 1's to 0's. The FSK detector decodes the ac signal on the tape back to binary levels. Further details of this application are given in the Signetics references.

PROBLEMS

18.1. For the source shown in Fig. 18.6(a), calculate the output signal as a consequence of both the differential and the common-mode input signals for each of the amplifier configurations shown in Fig. 18.7.

18.2. Analyze the instrumentation amplifier of Fig. 18.7 and show that the gains are as given in Eq. (18.6).

18.3. What would be the differential and common-mode gains in the instrumentation amplifier of Fig. 18.7 if the cross-coupling resistor R_1 were replaced by two separate resistors ($R_1/2$ each) to ground, giving two independent followers with gain? What component matching would then be necessary to achieve performance equivalent to Eq. (18.6)?

18.4. When a dc signal V_s is chopped, amplified, and phase-detected, the dc output is proportional to V_s. If V_s is a function of some variable X, show that when X is modulated ($X = X_o + \delta X \cos \omega_o t$), the signal will have an ac component whose phase-detected output will be proportional not to V_s but to dV_s/dX. **Hint**: Expand $V(X)$ in a power series about X_o.

18.5. Show that if the detector of Prob. 4 is driven at $2\omega_o$, the dc output is proportional to d^2V/dX^2. This technique is often used to enhance fine structure in $V(X)$, for example in tunneling measurements in superconductors. Suggest a method of generating the second-harmonic reference signal using a diode.

18.6. Using a reference frequency of 10^3 Hz and a time constant of 1 s, what is the effective bandwidth of the lock-in amplifier? In an ordinary tuned amplifier, how high a Q would be necessary to achieve this bandwidth?

18.7. Would you choose a chopper or the modulation method for the following measurements? Explain your choice. If you choose modulation, what do you modulate? Draw a block diagram for each case.

(a) Measuring slow intensity variations in a double-star system in the presence of atmospheric light scattering

(b) Measuring the resistivity of a material, on the order of a micro-ohm; application; galvanomagnetic measurements in metals

(c) Measuring a weak magnetic susceptibility $\chi(H)$ with a large constant term χ_o plus a small fine structure term $\chi_f(H)$ which has a strong field dependence. You wish to measure only $\chi_f(H)$ as a function of magnetic field H.

18.8. Sketch the waveforms you expect to see at the output of the 565 phase detector [Fig. 18.17(c)] under the conditions listed below. Recall that the phase detector is not a multiplier but a switch. Show also the resultant dc level at the output of the low-pass filter. Assume the VCO center frequency to be 50 kHz.

(a) Signal at 60 kHz; VCO at 50 kHz (unlocked condition)

(b) Both VCO and signal locked at 50 kHz. What about the relative *phase* of the two signals? (**Hint**: What dc input voltage is required to drive the VCO to 50 kHz?)

(c) Now shift the signal to 55 kHz and assume the PLL tracks. What about the relative phase now? What will happen to the dc output?

(d) Now make the input signal considerably smaller in amplitude, leaving the frequency unchanged. Assuming the PLL stays locked, what will happen to the dc output of the low-pass filter? What will happen to the relative phases of signal and VCO?

18.9. To generate multiple harmonics, more than one PLL is necessary, but not as many as the number of frequencies, since one can divide by a large N whose factors contain several of the desired harmonics. Design a circuit to produce harmonics 2, 4, 6, 8, 10, 12 with a minimum number of PLL's. A PLL can lock on quite high harmonics ($N \sim 100$ is possible). Represent the dividers symbolically ($\div N$ will do) and assume any integer value is possible.

18.10. Complete the analysis that lies between Eq. (18.12) and Eq. (18.13). In particular, it is useful in understanding the capture process to look at the difference frequency component when ω_s is not exactly equal to ω_r.

18.11. Complete the analysis of the PLL AM demodulator [Fig. 18.20(b)]. In particular, show that for an input signal of the form

$$A_c \sin (2\pi f_c t) [1 + A_s \sin (2\pi f_s t)]$$

the output signal will be proportional to

$$A_s \sin (2\pi f_s t)$$

REFERENCES

More complete bibliographic information for the books listed below appears in the annotated bibliography at the end of the book.

BROPHY, *Basic Electronics for Scientists*

DIEFENDERFER, *Principles of Electronic Instrumentation*

GARDNER, *Phaselock Techniques*

HENRY, *Electronic Systems and Instrumentation*

HIGGINS, *Experiments with Integrated Circuits*, Experiments 26 and 27

Manufacturers' Application Notes: see, especially, Analog Devices *Analog Dialog*; Hewlett-Packard *Application Notes*; Princeton Applied Research *Technical Notes*

MORRISON, *Grounding and Shielding Techniques in Instrumentation*

Nyquist theorem. The original discovery by J. B. Johnson, *Physical Review*, 32 (1928) p. 97; the explanation by H. Nyquist, *ibid.* 32, (1928) 110.

SHEINGOLD, *Transducer Interfacing Handbook*

Signetics, *Analog Applications Manual*

TEMPLE, *American Journal of Physics*, 43 (1975) 801. A do-it-yourself lab quality lock-in.

19

Analog Switching and Digital Filtering

Analog switching is another bridge between analog and digital worlds, just like D/A's and A/D's. An analog switch is a hybrid device, since digital signals control and steer analog voltages. Analog switches and comparators have been used in many circuits described earlier, notably function generators, timing devices, and digital voltmeters. Now that the background has been developed, we will take a look inside both kinds of devices (Sections 19.1 and 19.3). Some analog circuits utilize nonlinear switching operation and result in level or slope-sensitive behavior such as the slew-rate limiting filter (Section 19.2). Analog switches and multiplexers make possible digital control of almost any of the analog circuits described earlier (Section 19.4), or the sampling of analog voltages for digital or analog processing (Section 19.5). A class of LSI sample-and-hold devices called the *charge-coupled device* makes possible complex signal processing which uses discrete time sampling and digital filter analysis but operates on linear signals (Section 19.6). This brings us to a natural conclusion to this volume on electronics, since the extension to purely digital signal processing is best done not with electronic hardware but with software inside a computer.

19.1 COMPARATORS

Comparators are an interface between analog and digital domains, converting a continuous linear analog signal into a two-state digital signal. The simplest comparator [Fig. 19.1(a)] is an open-loop op amp, whose output V_o is high or low depending upon whether the input voltage V_{in} is smaller or larger than a reference voltage V_R. Comparator applications include:

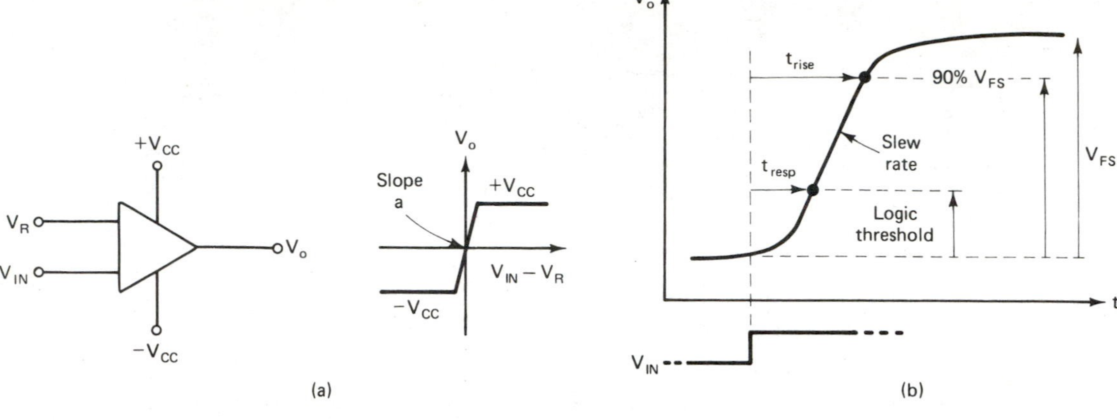

Figure 19.1 The comparator. (a) Schematic symbol. (b) Transfer function. (c) Time response to a step input.

Threshold detectors: Does the input voltage exceed a set threshold? In pulse applications, a comparator circuit called a discriminator is used to pass only those pulses significantly above a given noise level.

Squaring circuits: A sine wave is converted to a square wave for precise timing or phase measurement. Or, a noisy pulse is "cleaned up" into a pulse suitable for digital processing.

Timing circuits: The threshold-detecting ability times *when* a voltage crosses a set voltage level, and performs a switching function at that moment. Examples include function generators and the logic circuits inside dual slope digital voltmeters.

19.1.1 Comparator Op Amps and Specifications

Many traditional linear op amp specifications are irrelevant in a comparator. What matters most is speed of response. Although one could use a 741 or other standard op amp as a comparator in the audio range, internal compensation seriously limits the response speed. Since the application is usually open-loop, frequency compensation for stability under negative feedback is irrelevant. One either uses a faster (uncompensated) op amp such as a 301 or a special-purpose comparator op amp such as a 311. If a conventional op amp is used, the most important specification is speed, measured by the *rise time* and *slew rate.*

Rise Time (t_{rise}). The time it takes to reach a given percentage (typically 90%) of the full-scale output.

Slew Rate (SR). The maximum rate of change of voltage during one transition.

These specifications are illustrated in Fig. 19.1(b). For example, a 101–301 op amp has a rise time of about 1 μs and a slew rate of about 20 V/μs.

Specialized *comparator* op amps are significantly faster than any conventional op amp, and introduce several new specifications, the most important being *response time.*

Response Time (t_{resp}). The time interval between the application of an input step function and the time when the output crosses the logic threshold voltage.

The term *logic threshold* is used because comparators usually drive digital logic circuits which make a transition when an input crosses a certain voltage value [Fig. 19.1(b)]. A logic threshold value of 2.5 V is normally adopted for a 0-V to 5-V output transition.

Response times for two typical comparators are shown below.

Comparator	111,211,311	160,260,360
Response time	200 ns	16 ns

The 160–360 series are examples of very high speed comparators (once referred to whimsically in manufacturer's literature as "damn fast"). The 111–311 comparator is an industry standard, and plays the role for comparators which the 741 plays for conventional op amps. Specifications for the 311 are given in Table 19.1.

The offset voltage and bias current specifications resemble the 741. The response time varies from the value given by a factor of 2, depending on how far the output transistor is brought out of saturation, because of charge stored at pn junctions. The common mode limits restrict input signals from voltages near either power supply value, where operation becomes unpredictable. This presents no limitation with analog ($\pm V_{cc}$) power [Fig. 19.2(a)], since neither input will ordinarily approach the supply voltages. When operated from a single power supply [Fig. 19.2(b)] neither input may reach ground. Although adequate for inputs from a TTL low ~0.4 V) state, an input signal which must truly pass through ground requires a comparator op amp such as the 139–339 series, whose common-mode range includes ground even when operated from a single power supply.

TABLE 19.1 SPECIFICATIONS OF THE 311 COMPARATOR

Input offset voltage (max)	7.5	mV
Input bias current (max)	250	nA
Voltage gain (min)	40	V/mV
Response time (typ)	200	ns
Common-mode limits below V_+	1.0	V
above V_-	0.25	V

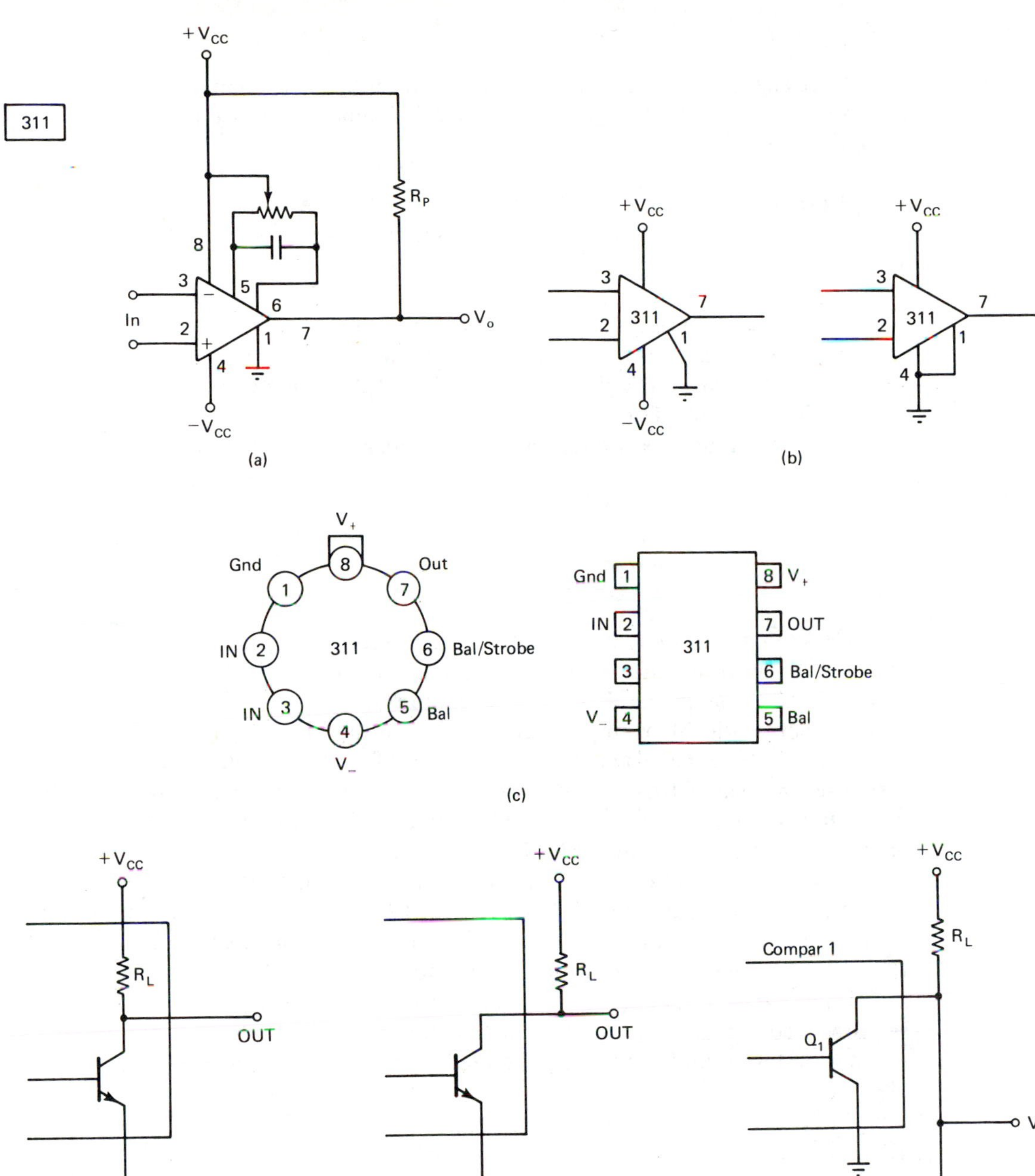

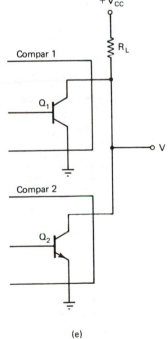

Figure 19.2 (a) Circuit connections for a typical comparator, the 311. (b) Comparison of connection for dual power supply and for single power supply operation. (c) Pin diagrams for a typical comparator, the 311. (d) Open collector outputs require an external heat load resistor and (e) allow several comparator outputs to be wired together.

Pin connections [Fig.19.2(c)] are unlike standard 741-compatible op amps. The comparator pin labeled *ground* (pin 1 for the 311) supplies a ground reference for the output transistor. In single-supply operation of the 311, pins 1 and 4 are tied together. But when V_- is tied to a negative supply for analog operation, pin 4 is tied to ground, completing the level shifting interface from analog to digital circuits. The resistor in Fig. 19.2(a) supplies offset voltage balancing. The capacitor compensates the comparator against bursting into oscillations near threshold. This can occur when the input signal is a voltage ramp or slow sine wave or has a high source impedance ($> 1\ \text{K}\Omega$).

Comparators typically require an external load resistor. This *open-collector output* [Fig. 19.2(d)] allows increased flexibility in interfacing, and allows several comparators to be tied together at the output [Fig. 19.2(e)]. If two *conventional* op amps are tied together at their output, the output state will be indeterminate, and an unprotected op amp may be destroyed.

19.1.2 Comparator Circuits

The basic comparator circuit of Fig. 19.1 suffers from several limitations. If the input signal crosses V_R slowly, the output voltage will make a slow transition of little use in precise timing. If the input is noisy, the comparator will make several transitions or even oscillate before settling into its final state. This is unacceptable, since digital circuits can detect even nanosecond-speed settling noise.

Both of these problems are eliminated by the *comparator with hysteresis*, also called the analog Schmitt trigger (Fig.19.3). The circuit of Fig. 19.3(c) is intended for analog inputs which may go either positive or negative. Hysteresis is introduced because the point at which the device switches depends upon the previous output state. This follows from the positive feedback connection between output and input. Although comparators are predominantly two-state devices, linear op amp analysis works during the nearly vertical transitions [Fig 19.1(a)]. To locate the thresholds on the circuit of Fig. 19.3(a), calculate the voltage V_A on the weighted voltage divider formed by R_1 and R_2.

$$V_A = V_R R_2/(R_1 + R_2) + V_o R_1/(R_1 + R_2) \tag{19.1}$$

The values of input voltage which cause a change of state are

$$V_{\text{in}} = V_A = V_R R_2/(R_1 + R_2) \pm V_{cc} R_1/(R_1 + R_2) \tag{19.2}$$

The threshold is a fraction of the reference voltage V_R, plus or minus a hysteresis voltage [Fig. 19.3(b)]. The amount of hysteresis is adjusted by the ratio $R_1/(R_1 + R_2)$, and can vary from 0 to $2V_{cc}$.

For the circuit of Fig. 19.3(c), the thresholds are determined by

$$V_A = V_{\text{in}} R_2/(R_1 + R_2) \pm V_{cc} R_1/(R_1 + R_2) = 0 \tag{19.3}$$

This results in switching at

$$V_{\text{in}} = \pm V_{cc} R_1/R_2 \tag{19.4}$$

This comparator's switching is centered around 0 V, with hysteresis adjusted by

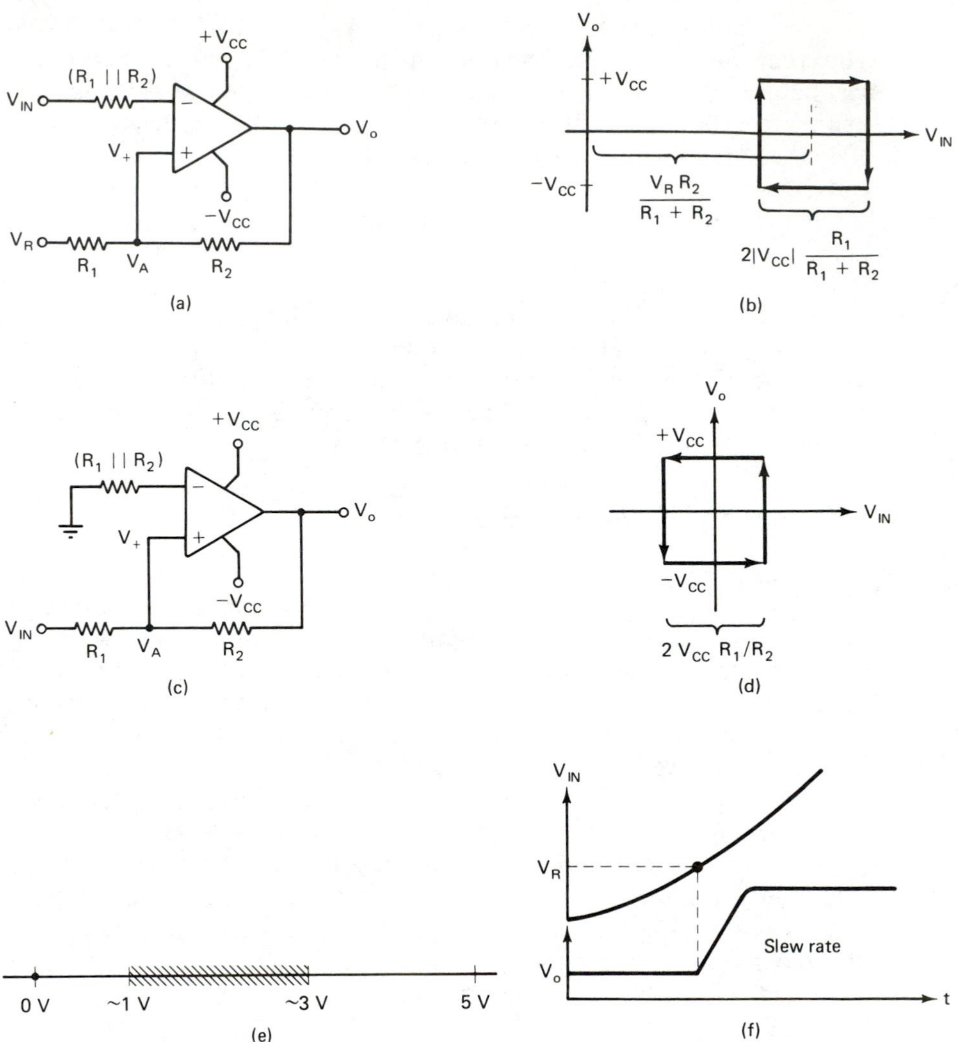

Figure 19.3 (a) Circuit and (b) transfer function for comparator with hysteresis with independent threshold and hysteresis control. (c) Circuit and (d) transfer function for comparator with adjustable hysteresis about 0 V.

ratio R_1/R_2. Because of the differing input connections in these two circuits, the sense of rotation [arrows in Fig. 19.3(b) and (d)] is opposite. If V_o is high, a transition to low is effected by an input high state for Fig. 19.3(a), and an input low state for Fig. 14.4(c).

Choice between these circuits and selection of hysteresis range depends upon the application. For example, in digital interfacing, hysteresis is chosen to create a "dead zone" [Fig. 19.3(e)] in the indeterminate region between the 0 and 1 state of logic circuits. The positive feedback of these comparator circuits has another important feature: it significantly speeds up the switching action. As soon as a transition begins, the change in output voltage is fed back to the input to aid the transition. This feature, analogous to unbalancing a seesaw by walking past the center pivot, is particularly useful when the input voltage is changing slowly [Fig. 19.3(f)], and the time of threshold crossing must be determined precisely. Due to this *snap action*, the precision of transition time is limited only by the slew rate of the op amp.

In analog applications where circuit operation is sensitive to the value of the comparator output voltage (e.g., function generators), it is conventional to clamp or limit V_o using Zener diodes (Fig. 19.4) rather than letting it reach the power supply voltage $\pm V_{cc}$. This avoids a possible "latched up" state in which the op amp does not respond to the input, makes circuit behavior independent of power supply voltage fluctuation, and also speeds up the transition by avoiding saturation.

Another common comparator circuit, the *window comparator*, functions as a limit tester or voltage discriminator. The circuit gives an output high signal when the input voltage falls between preset lower and upper limits V_L and V_U. A window circuit using conventional op amps is shown in Fig. 19.5(a). The diodes not only steer the output voltages but provide a logical OR function. The output voltage V_o is low when the transistor is *on* or when V_1 is high. This can occur when the output of either A_1 or A_2 is high. A_1's output is high if $V_{in} > V_U$. V_2's output is high if $V_{in} < V_L$ [see waveform sketch in Fig. 19.5(a)]. As a result, V_o is high only when the input voltage falls within the limits of the window V_L and V_U.

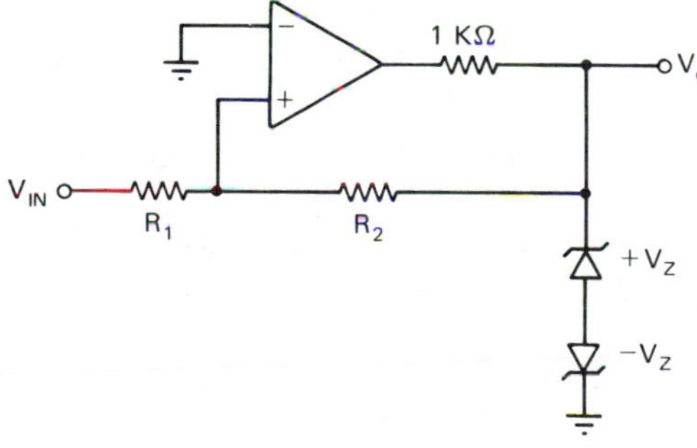

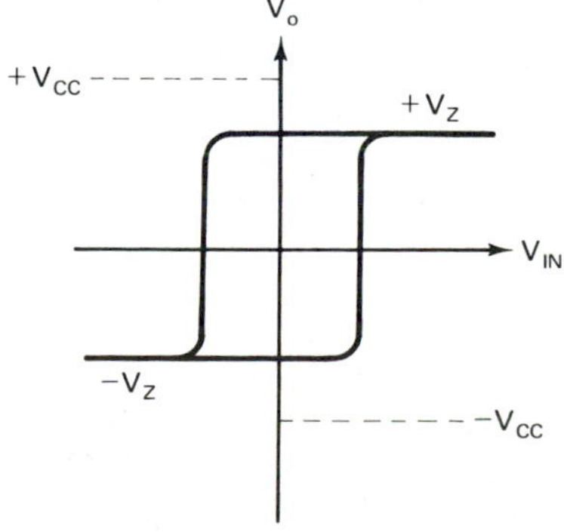

Figure 19.4 Zener diode clamping of output voltage.

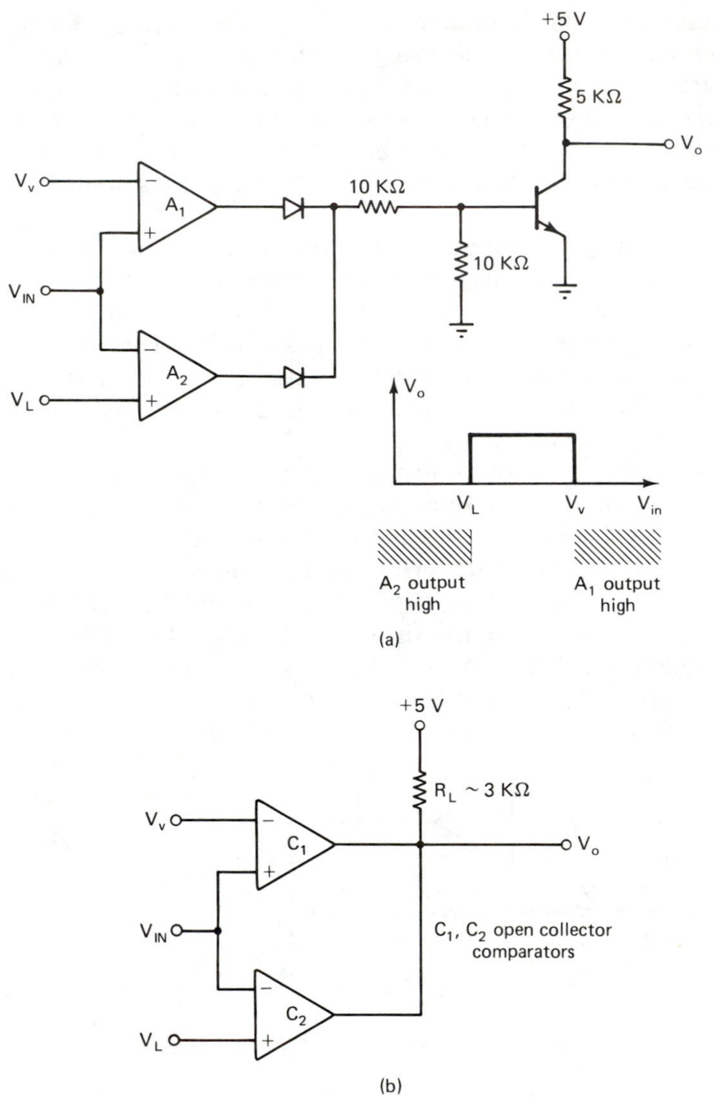

Figure 19.5 Window comparators with (a) conventional output and (b) open collector output comparators.

The circuit is simpler if comparator op amps with open-collector outputs are used [Fig. 19.5(b)], since open-collector outputs can perform the logical OR function directly [Fig. 19.2(e)].

All features of the window comparator are now available in a single comparator, the 3098, optimized for direct connection to transducers whose impedance may range from 100 Ω to many megohms, and with output driving capability up to 150 mA. A built-in flip flop provides memory of the most recent window crossing, giving hysteresis equivalent to the circuit of Fig. 19.3(a), but without the need for external resistor adjustment. A programming input makes possible low microwatt standby power dissipation when large power handling is not needed.

19.2 CLOSED-LOOP TRACKING CIRCUITS

Some op amp circuits display switching behavior even though no switch is apparent. Such circuits typically have a comparator or subtractor in series with an integrator, whose output is fed back as one of the inputs of the comparator. The sense of the feedback loop tries to keep the capacitor from charging. Switching operation begins when an op amp output reaches a bound so linear operation is no longer possible.

19.2.1 Slew-Rate Limiting Filter

Suppose a superconducting solenoid is controlled by a computer though a D/A converter. Sudden changes in the set point cause a large voltage to appear across the solenoid because of its large inductance. This could lead to a dangerous "quench" or transition to the nonsuperconducting state, avoidable by a hardware rate limit. Audio design also employs circuits which faithfully follow an input signal except when noise spikes occur. Pop and hiss eliminators are an example. Although linear low-pass filters can integrate out sudden changes, they produce an equally long lag for both small and large amplitude signals, slow down system response, and lose fidelity. A *nonlinear* rate limiter, by contrast, passes signals precisely until a certain bound is exceeded.

One version of a rate limit circuit [Fig. 19.6(a)] is based on a diode bounding circuit (Chapter 15). This is a linear amplifier except when Zener diode breakdown puts a bound on the output voltage. To limit rate dV_o/dt rather than V_o, feed the diode circuit with the derivative of the input, and follow this with an integrator to recover the original signal. Although adequate for audio signals, this circuit suffers from long-term stability problems for dc control signals due to the integrator drift.

An improved impulse filter [Fig. 19.6(b)] has integrator drift canceled by negative feedback. Although apparently a low-pass filter, the circuit actually tracks input signals faithfully when in its linear mode of operation. To understand the operation qualitatively, suppose A_1 is connected as a high-gain amplifier. The integrator output will try to keep its summing-point voltage zero. In addition to the normal feedback path through C, the integrator has a second feedback path through A_1 which opposes the capacitor charging up. The integrator feeds back V_o which follows the input V_i so the subtractor can keep voltage V_1 a minimum at the integrator input. Nonlinear limiting action begins when the output of A_1 reaches full scale ($\pm V_{cc}$ unless otherwise bounded). The integrator then generates a linear ramp

$$V_o = \pm V_{cc}t/RC \quad \text{(slew rate limiting)} \tag{19.5}$$

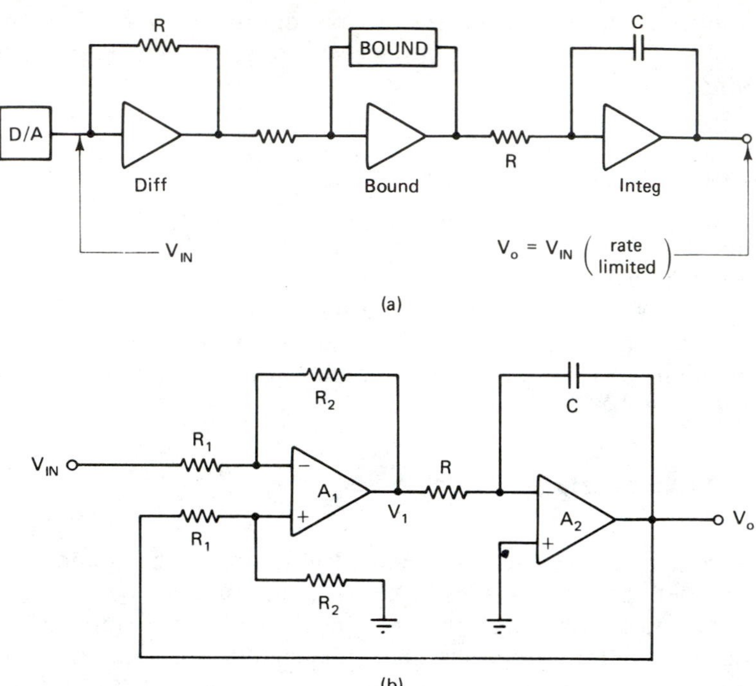

Figure 19.6 (a) Nonlinear rate limiting filter utilizing diode bound circuit. (b) Slew rate limiting filter. Use BiFET or BiMOS op amps for low bias current and high slew rate.

until V_1 comes out of limits and the feedback circuit can again function. In response to an overloading pulse input [Fig. 19.7(a)], the output slews at the above rate rather than follow the sharp steps of the pulse. As the input amplitude increases, no further increase occurs in the output triangular pulse. This feature makes it possible for the circuit to greatly attenuate large noise spikes on top of another signal [Fig. 19.7(b)]. The output accurately tracks the signal waveform, with no noticeable phase shift or attenuation. The signal-to-noise ratio has been improved by more than a factor of 10 in this example. As the signal frequency increases, a point is reached where the rate of change exceeds the slew rate limit, and waveform distortion occurs [Fig. 19.7(c)]. Component values must be selected so most noise is attenuated, yet signals of interest fall below the upper limit frequency, which is close to $RC/2\pi$.

Actual circuit operation is slightly more complicated. The transfer function in the linear mode is

$$V_o/V_i = 1/(1 + RC\omega G^{-1}) \tag{19.6}$$

where G is the subtractor gain R_2/R_1. For large G, this is frequency-independent, but for a low gain subtractor, some attenuation and phase shift of the signal will occur near $\omega = (RC)^{-1}$ because of low-pass filter action. The transition to the limiting mode occurs when V_1 reaches the power supply voltage V_{cc}. V_1 is the derivative of V_o. For sine wave signals, limiting occurs when

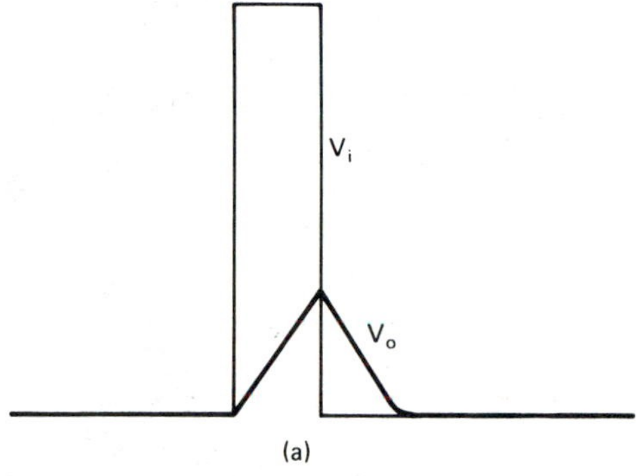

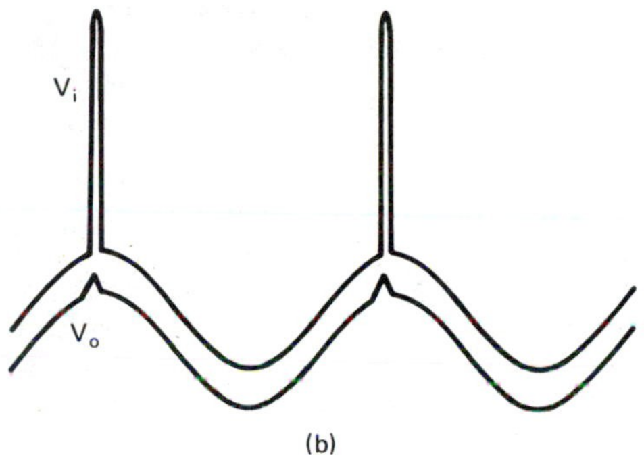

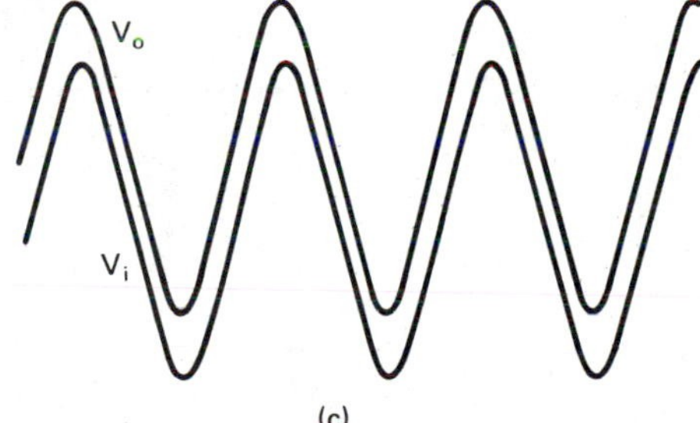

Figure 19.7 Slew-rate limiting filter waveforms.

Input (upper trace)	*Output (lower trace)*
(a) Large square pulse	Small triangular pulse
(b) Sine wave with noise pulse	Sine wave with most noise removed
(c) Sine wave above cutoff frequency	Distorted sine wave

$$V_1 = -RC\omega V_i/(1 + RC\omega G^{-1}) = V_{cc} \tag{19.7}$$

Solve for the value of ω in two cases. For large G,

$$\omega\,(\text{limit}) = (V_{cc}/V_i)(RC)^{-1} \tag{19.8}$$

and for $G = 1$,

$$\omega\,(\text{limit}) = (RC)^{-1} \tag{19.9}$$

Normally, one chooses large G to avoid signal distortion near $\omega = (RC)^{-1}$. With this choice, ω (limit) is considerably larger than $(RC)^{-1}$ since input signals are ordinarily small compared to the power supply voltage.

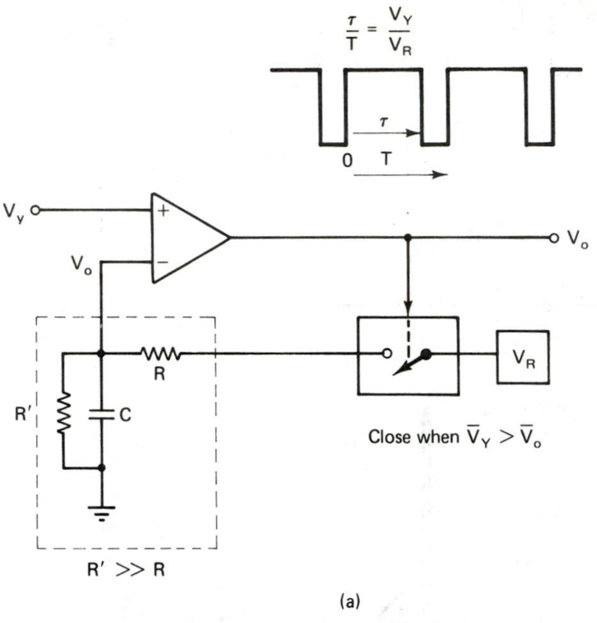

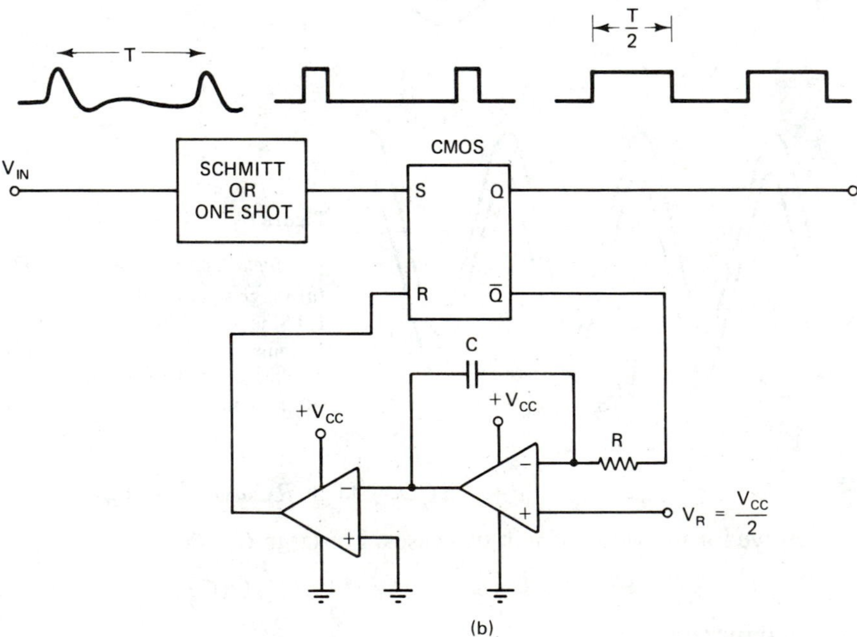

Figure 19.8 (a) The voltage-to-pulse-width converter generates an asymmetric square wave whose duty cycle is proportional to V_Y. (b) Closed loop tracking integrator can convert a periodic pulse string into a symmetric square wave.

19.2.2 Other Closed-Loop Tracking Circuits

The heart of a pulse width multiplier (Chapter 15) is a *voltage-to-pulse width converter* [Fig. 19.8(a)]. The output wave has its duty cycle proportional to the input V_Y. Initially, V_Y brings the comparator output high, closing the switch. This connects a constant voltage V_R to an RC combination. The capacitor charges up until it causes the comparator to change its state, opening the switch. The process repeats itself, creating an output waveform whose average value equals V_Y. As a result, the duty cycle $\tau/T = V_Y/V_R$ is proportional to the input voltage V_Y.

Lock-in amplifiers need to generate a reference signal which is a square wave, even if the available reference is not square. For a fixed frequency, this is easily done: adjust the period of a one-shot to equal half the period of the reference signal. But a fixed frequency circuit is a limited solution. The circuit shown in Fig. 19.8(b) generates a symmetric square wave from an input wave of *any* frequency. The input reference signal is made digital through either a Schmitt trigger or a one-shot and *sets* a flip flop. The FF output drives an integrator, whose output can *reset* the flip flop. The feedback loop tries to keep the integrator's input equal to V_R, on the average, but does so by ramping the integrator up and down an equal amount of time. The Q output is therefore a symmetric square wave. The biasing shown assumes a CMOS FF whose high and low states are V_{cc} and 0 V. Otherwise, it is necessary to interpose an FET switch which sets the integrator input at well-defined voltage values.

The *delta modulator* [Fig. 19.9(a)], related to the voltage-to-frequency converter of Chapter 9, is a superior A/D method at high frequencies and thus popular in communications. It uses the same building blocks as the previous examples, but creates a pulse string whose pattern is related to the input voltage. By analogy with previous examples, it should be clear that the integrator output on the average follows V_{in}. The sign of the comparator output lets through the FF more or less clock pulses as needed to keep the loop in balance. The *demodulator* on the other end combines the same components in a different way (left for Prob. 19.3). The output of the demodulator is a segmented approximation to the original waveform. The algorithm is quite different from a V/F converter. A *low* pulse frequency (mostly 1's or mostly 0's) commands the demodulator's integrator to charge up or down. Peaks of the input signal generate *high*-frequency data streams corresponding to the modulator integrator charging up and down for equal intervals. The delta modulation technique achieves wide dynamic range and good bandwidth because the pulse stream indicates *changes* in the signal (hence the term *delta*) rather than trying to include enough bits to reproduce each point on the waveform all over again.

The closed-loop integrator tracking circuit is also the basis for the lock-in circuit of Fig. 18.16 which subtracts off constant ambient light so the amplifier can detect the modulated component without being overwhelmed. The analysis is left for Prob. 19.4.

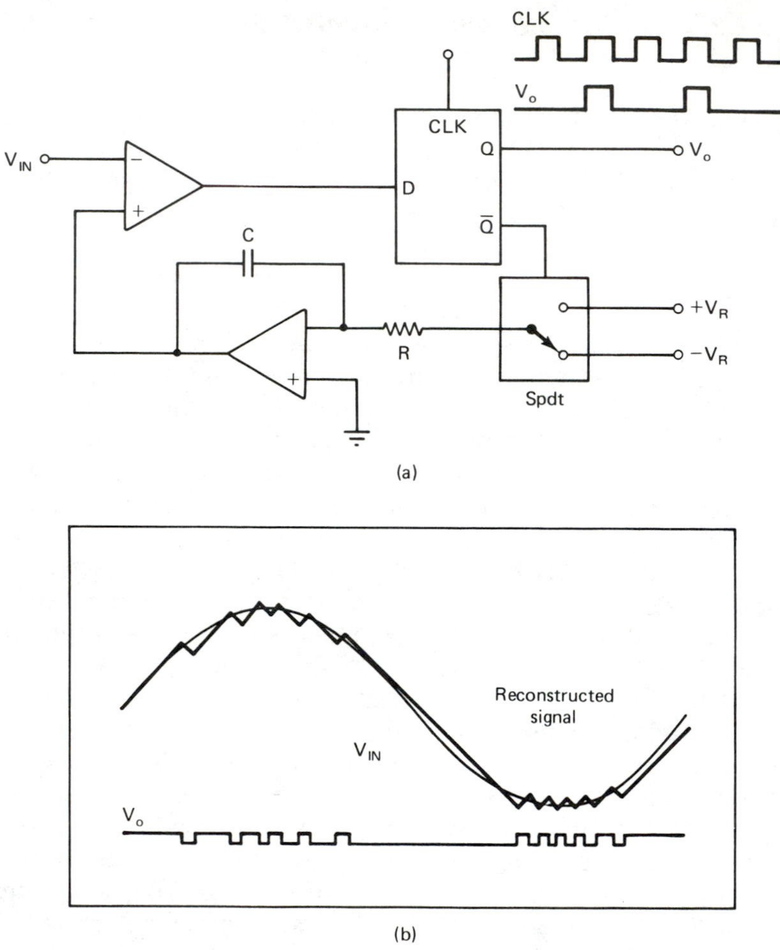

Figure 19.9 The delta modulator (a) circuit and (b) waveforms. This is an efficient technique for asynchronous serial communications.

19.3 FET ANALOG SWITCHES

The FET analog switch replaces the mechanical switch. It facilitates electronic control, switches much faster than the mechanical switch, and is about 10 times cheaper. For example, an electronic switch is generally preferred in resetting an integrator, because its speed eliminates indeterminate initial conditions. Otherwise, op amp bias current can induce drift during the slow switching interval of a mechanical switch.

Manufacturer's specification sheets shows a simple block diagram [Fig. 19.10(a)]: a switch driven by the *control* input through a buffer amplifier. The version shown is single pole–single throw (spst). For every conventional switch, the corresponding analog switch exists (single pole–double throw, double pole–single throw, etc.). Inside the switch are one or more MOSFETs [Fig. 19.10(b)]. The

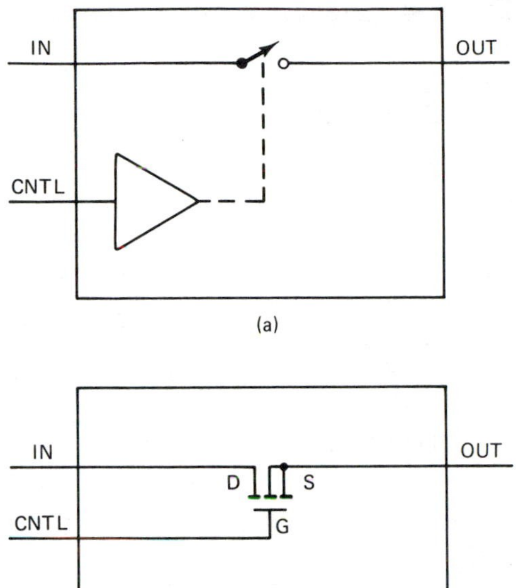

Figure 19.10 The FET analog switch. (a) Circuit symbol. (b) Internal circuit with a single MOSFET.

MOSFET switch is a voltage-variable resistor, whose resistance varies by many orders of magnitude.

	MOSFET switch	Mechanical switch
On-state resistance	$10^2\ \Omega$	$10^{-3}\ \Omega$
Off-state resistance	$>10^{10}\ \Omega$	$>10^{9}\ \Omega$
Speed	$<10^{-7}$ s	10^{-3} s

The MOSFET lacks the low on-state resistance of a mechanical switch. But in electronics, a switch is adequately "On" if its resistance is small compared to other circuit components. For example, there is no disadvantage if the switch which resets an integrator capacitor has a 100-Ω on-state resistance. More often, it is the off-state resistance which matters, and this is adequately high in a MOSFET switch. The *ratio* of off- to on-state resistance is very high ($> 10^8$) for a MOSFET switch, much higher than a JFET or bipolar transistor. This ratio could be infinite for a mechanical switch, but in practice surface leakage limits the ratio to a value comparable to the MOSFET. In any case, the mechanical switch or relay has speed far too slow for most electronic applications.

Because of its on-state resistance, a MOSFET switch dissipates power within it when passing signal current. Safe circuit design limits current flow in these fragile devices to tolerable levels. Given an on-state resistance of 100 Ω, 100 mA could flow if 10 V were put across the switch, giving a power dissipation of 1 W. Since this is in excess of the 500-mW typical upper limit, switch circuits generally include external series resistance as protection, limiting current flow to a few milliamps.

TABLE 19.2 CMOS ANALOG SWITCH EXAMPLES

Device[a, b]	Switch configuration
DG200	2 SPST
DG201, 4066, (4016)	4 SPST
DG303/307	2 SPDT, TTL/CMOS input compatible
DG302/306	2 DPST, TTL/CMOS input compatible
4053*	Triple 1 of 2 multiplexer
4052*	Dual 1 of 4 multiplexer
4051*, DG508	1 of 8 multiplexer
4097, DG507	Dual or differential 1 of 8 multiplexer
4067, DG506	1 of 16 multiplexer

[a]Devices listed in parentheses are largely superseded by newer models.

[b]Devices listed with an asterisk include internal logic level conversion (see text).

19.3.1 CMOS Analog Switch IC's

Examples of analog switch IC's are listed in Table 19.2. Figure 19.11 displays the switch combinations possible for a selection of devices.

Important analog switch specifications and typical values are listed in Table 19.3. The actual values vary considerably with power supply voltages, which can be as small as 5 V or as large as ±15 V. For example, the *propagation delay* t_P at the control input is the time between applying a control signal and having the signal output change in response. For a 4051, this time is 120 ns for $V_{DD} = 15$ V but slows to 500 ns for $V_{DD} = 5$ V. The control voltage input levels also depend upon V_{DD}. V_{IL} is the maximum value which will be interpreted as a *low*, and V_{IH} is the minimum value which will be interpreted as a *high* logic signal. The values listed give a very wide range of acceptable signals, consistent with the high noise immunity of CMOS IC's. These values are intended for control by CMOS IC outputs. If a TTL IC drives a control input, either use a pull-up resistor or select an analog switch whose control input levels are optimized for typical TTL output voltage levels (i.e., a *high* may be as low as 3.4 V). The picoamp value listed in Table 19.2 for control input current is correct only at 25°C. This current is very temperature-dependent, rising to 1 μA at 125°C. Although MOSFET gates are insulated, the steep increase is due to reverse leakage current in the protection diodes incorporated as protection for the fragile oxide.

A switch such as the 4066 has two power supplies, V_{DD} and V_{SS}. With $V_{DD} = +5$ V and $V_{SS} = -5$ V, the circuit can steer input signals of ±5 V range with 0–5 V control signals. Single power supply operation of IC's designed for ac analog signals may require offsetting the ground reference and the control input level to half the power supply voltage. An example is shown in Fig. 19.12(a) for the

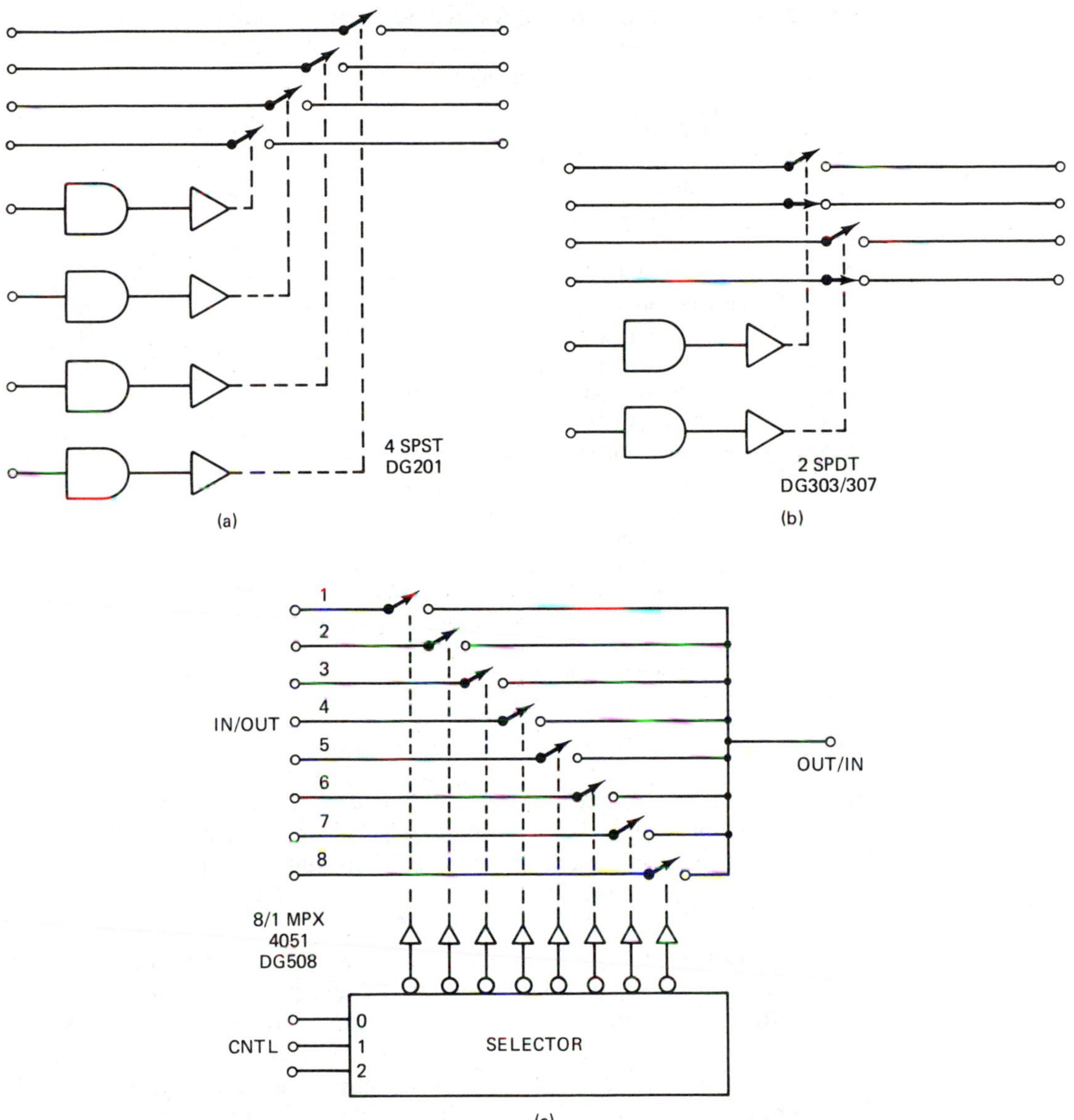

Figure 19.11 Equivalent switch circuits for several common FET analog switch packages. (a) Quad single pole-single throw. (b) Dual single pole-double throw. (c) 8-channel multiplexer.

DG200. Other IC's such as the DG 300 are optimized for single polarity operation directly [Fig. 19.12(b)]. Analog multiplexer devices such as the 4051 have additional internal level shifting called *logic level conversion*. An extra power supply V_{EE} adjusts the offset between single-ended digital signals and dual polarity analog signals. With $V_{DD} = +15$ V, $V_{SS} = 0$ V, and $V_{EE} = -15$ V, "typical" 0-V to 5-V digital signals can steer "typical" ± 15-V analog signals. The signal voltage range

TABLE 19.3 TYPICAL ANALOG SWITCH SPECIFICATIONS

Symbol	Definition	Typical value
Signal inputs and outputs		
R_{on}	On-state resistance	100 Ω
I_{is}	In-out leakage current (switch Off)	100 pA
ΔV_{in}	Signal voltage range	min $V_{EE} - 0.5$ V max $V_{DD} + 0.5$ V
t_P	Propagation delay, signal In to signal Out	10 ns
Control inputs and outputs		
V_{IL}	Input Low voltage (max)	0.4 V_{DD}
V_{IH}	Input High voltage (min)	0.5 V_{DD}
t_P	Propagation delay, control In to signal Out	150 ns
I_{in}	Input current	10 pA
Supply voltage, signal voltage, and power ranges		
V_{DD}	Positive supply voltage (typical)	+3 to +15 V
V_{SS}	Negative supply voltage (typical) for digital side	0 to −5 V
$V_{DD} - V_{SS}$	Supply voltage range (typical)	3 to 15 V
	Supply voltage range (max)	18 V
V_{EE}	Negative supply voltage (typical) (internal offset devices only)	0 to −15 V
P_D	Power dissipation, per package (max)	500 mW

ΔV_{in} must be carefully limited so as not to exceed the power supply "rails" by more than 0.5 V, or else device burnout can occur as the channel substrate pn junction becomes forward-biased. Signal levels between V_{DD} and V_{SS} are safe in either unipolar or dual-polarity operation.

Switch example. The 4066 bilateral switch typifies devices intended for transmision or multiplexing of either analog or digital signals. The circuit is shown in Fig. 19.13(a). Although a single MOS transistor could function as a switch, complementary pMOS-nMOS pairs have far less variation of on-state resistance over the analog signal range. Such a switch is highly linear, with less than 0.4% distortion. The gates of the pMOS and nMOS switches are driven by complementary signals from the control input.

Multiplexer example. The 4051 digitally controlled analog switch typifies devices optimized as analog multiplexer/demultiplexers [Fig. 19.13(b)]. The line marked *Common* and the eight lines marked *Channel* can function either as inputs or outputs, respectively. Internal decoding allows eight channel control with only three control lines. The box marked *Logic level conversion* provides the level shifting mentioned above so unipolar control signals can steer dual-polarity analog signals.

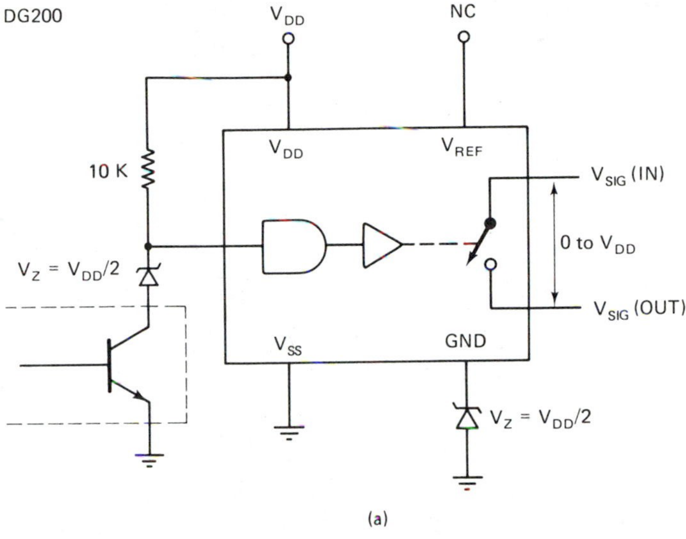

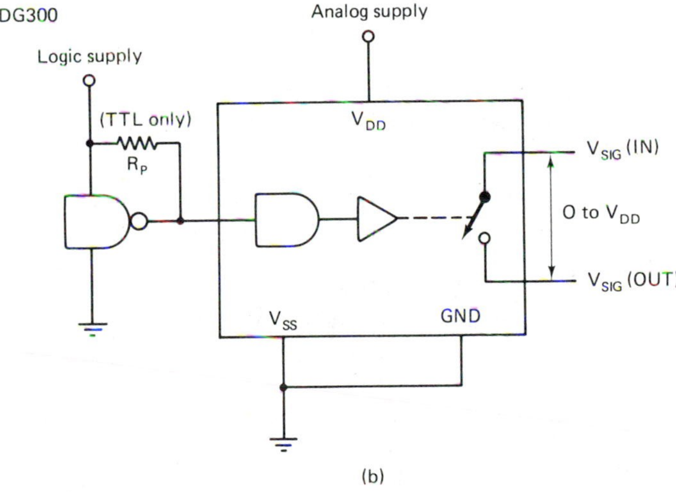

Figure 19.12 Power supply connections for single-ended operation of analog switches. (a) Bias shifting of an IC ordinarily intended for ac signals. (b) Direct connection of an IC compatible with single-ended operation.

19.4 ANALOG MULTIPLEXER APPLICATIONS

Suppose a computer is measuring signals with an A/D converter. There may be 10 or more locations or *channels* in the experiment where voltages must be measured. An analog multiplexer functions as a selector switch, connecting one voltage at a time to the A/D. This is preferable in most cases to dedicating one A/D

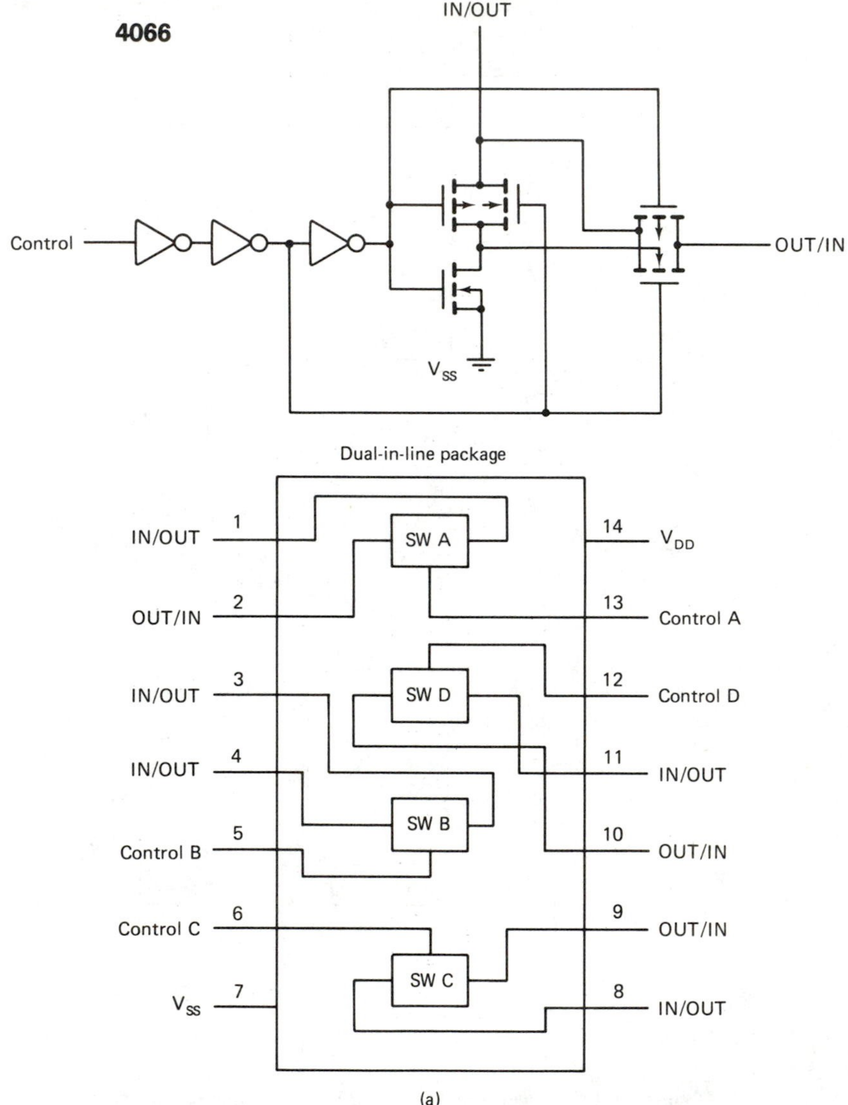

Figure 19.13 Internal circuits of (a) 4066 quad analog switch and (b) 4051, 8 to 1 analog multiplexer.

per channel, since a high resolution A/D is a relatively expensive component, and since each A/D occupies an I/O port of the computer. In the simplest example of two input voltages [Fig 19.14(a)], two analog switches have outputs connected to the A/D. Their control inputs are fed by a binary signal and its complement, so only one of the switches is closed at any time. Multiplexer IC's with more than two switches have a binary selector built in [Fig. 19.14(b)]. A binary *channel select* word brings one selector output low, connecting the selected signal to the A/D.

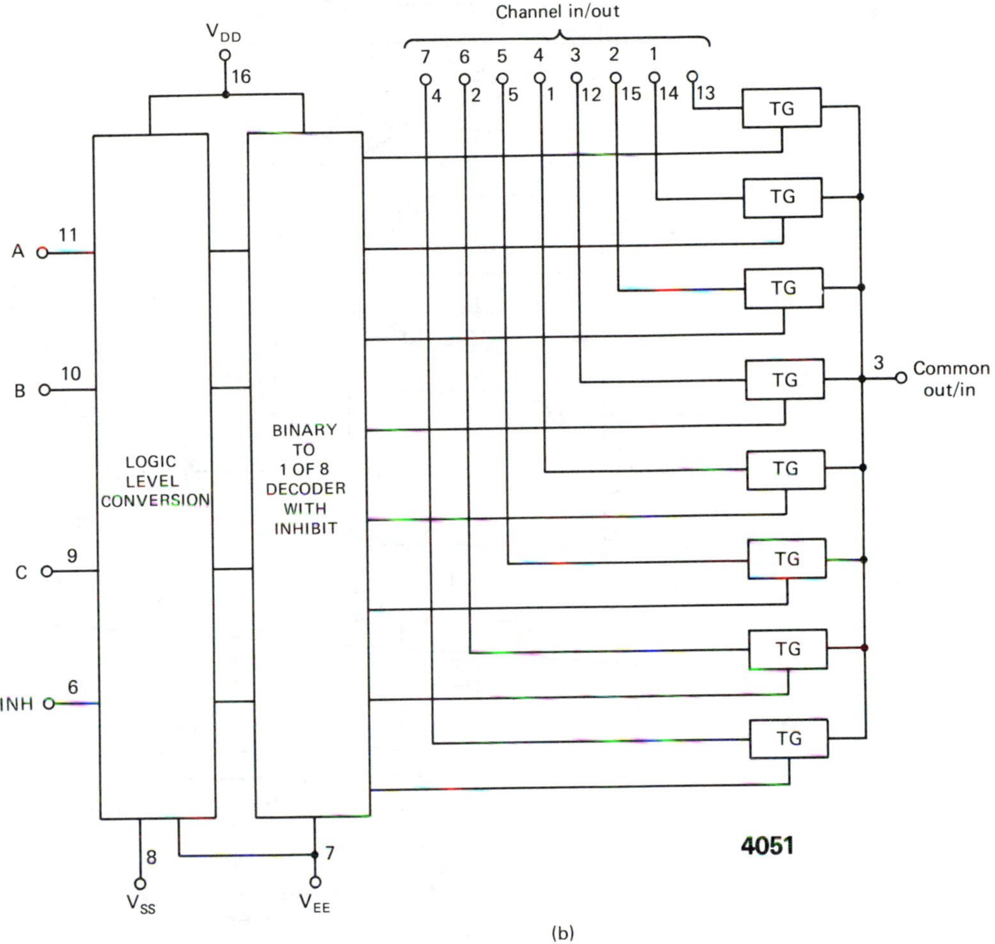

Figure 19.13 Continued

The analog switch is *bidirectional*, because the analog signal passes through a bar of semiconductor whose resistance is altered by the digital control signal. Current can flow equally well in either direction. The same IC wired as a multiplexer in Fig. 19.14(b) can be wired backwards to serve as a demultiplexer [Fig. 19.14(c)] to distribute an analog signal to numerous locations. The bidirectional property is specific to analog switches; a digital multiplexer will not work backwards.

Since the analog signals are independent, there will be transient noise when changing from one channel to another. Typically, the *select* control signals come from a computer. The computer program must wait long enough after a given channel is selected for noise on the line to settle before the A/D can be read reliably.

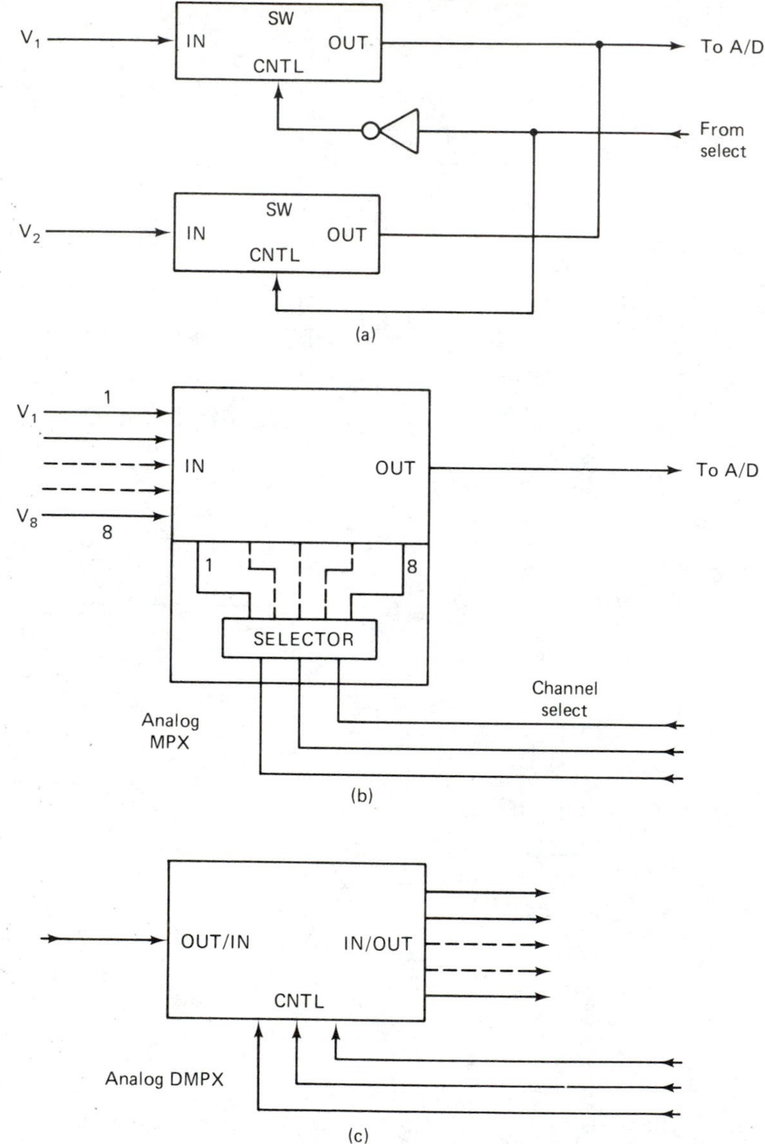

Figure 19.14 Analog multiplexing. (a) Two input voltages plus two switches wired as a spdt switch. (b) 8-channel multiplexer. (c) The same multiplexer IC wired backwards as a distributor.

19.4.1 Programmable Gain Op Amp

Digital control of op amp gain is accomplished using analog switches to select resistor values. For example, a programmable gain op amp [Fig. 19.15(a)] uses a 4051 to select one of resistors $R_a \ldots R_n$ to adjust the value of R_2 in the feedback loop. The resistor values are selected to give the desired gain range. An alterna-

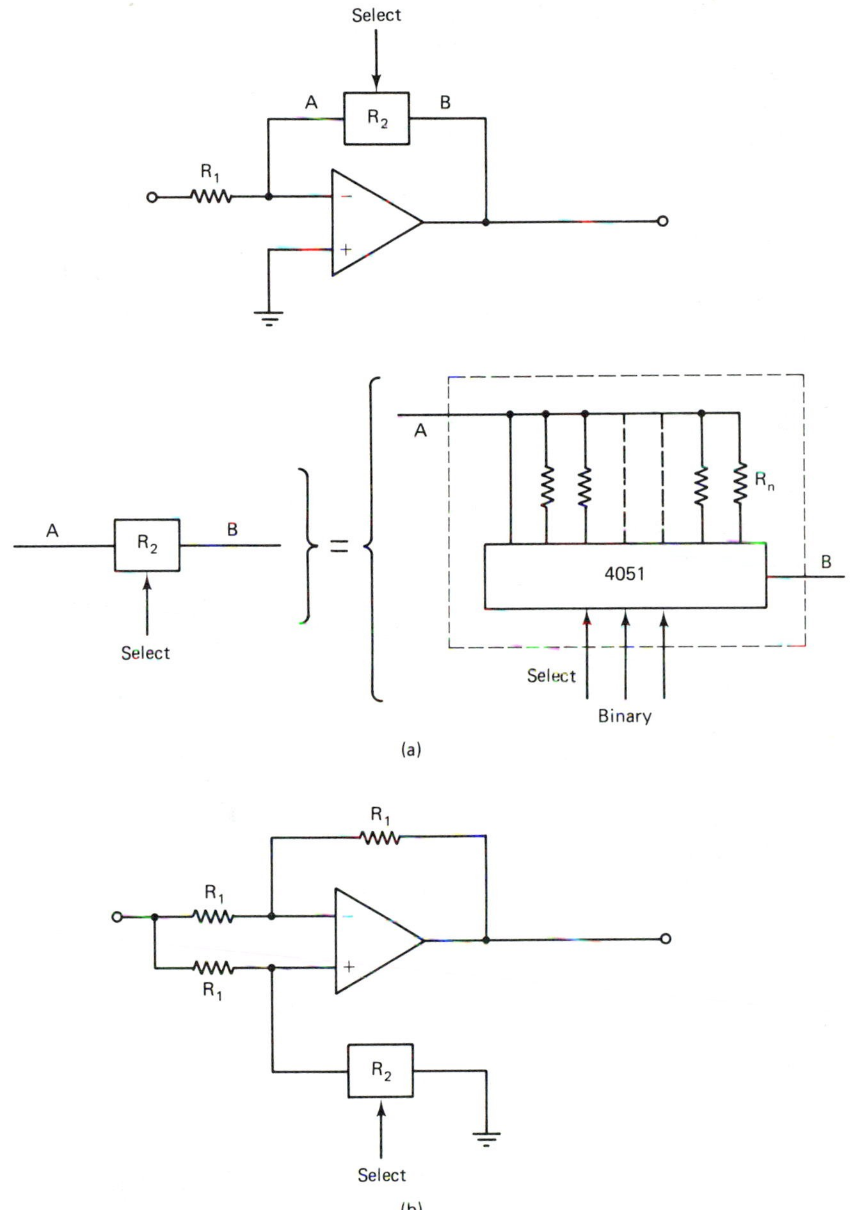

Figure 19.15 (a) Programmable gain op amp, with R_2 selected by a 4051. (b) Programmable attenuator, with R_2 of the subtractor selected by a 4051.

tive is the digital attenuator [Fig. 19.15(b)], a subtractor with both inputs connected together. Analysis shows this to be a digital multiplier or digitally controlled attenuator. The ratio V_o/V_i ranges from 1 to 0 as R_2 is varied from 0 to R_1. These circuits are equivalent to multiplying D/A converters, except that the resistor values are not limited to powers of 2. Resistor ratios related to one another logarithmically, for example, yield wide dynamic range attenuation measured directly in decibels.

19.4.2 Other Programmable Circuits

Practically any op amp circuit can be made programmable with analog switches. The bilateral current source of Fig. 19.16 has a range from ± 1 nA to ± 1 ma. The current value is $I_o = V_{in}/R_{out}$, where R_{out} is selected by the 4051. Resistor values smaller than 10 KΩ may have to be trimmed to include the on-state resistance of the switch.

Programming can be varied with time to generate the envelope of an arbitrary signal. Fig. 19.17 shows such a digital envelope generator based on the attenuator circuit of Fig. 19.15(b). As the clock cycles the counter through its range, the effective value of resistor R_2' is set by the selected resistor A ... G plus the selected resistor H ... N in series. The resistor values shown generate a bell-like decaying envelope; other choices can give arbitrary attack, sustain, and decay combinations.

19.4.3 Multiplexed Keyboard Scanning

The analog switch's ability to act as a transmission path is useful even when the signal is not linear but digital. The keyboard encoder of Fig. 19.18 is an example from electronic music. Key switches (not shown) are wired at row–column intersections of a matrix. The more significant bits 3–5 of the 4024 counter scan the row lines. While a single row line is selected, the less significant bits 0–2 scan through all column lines. The *data* lines thus sequence through the addresses of all keys. If no key is pressed, the *strobe* line remains low. If a key is pressed, a

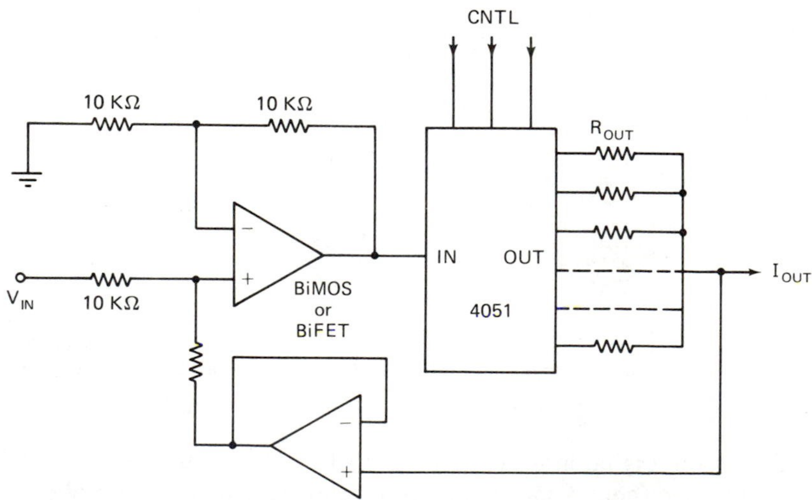

Figure 19.16 Bilateral programmable current source. (*Adapted from* A. Olesin, *Electronics.* (Sept. 18, 1975): 95.)

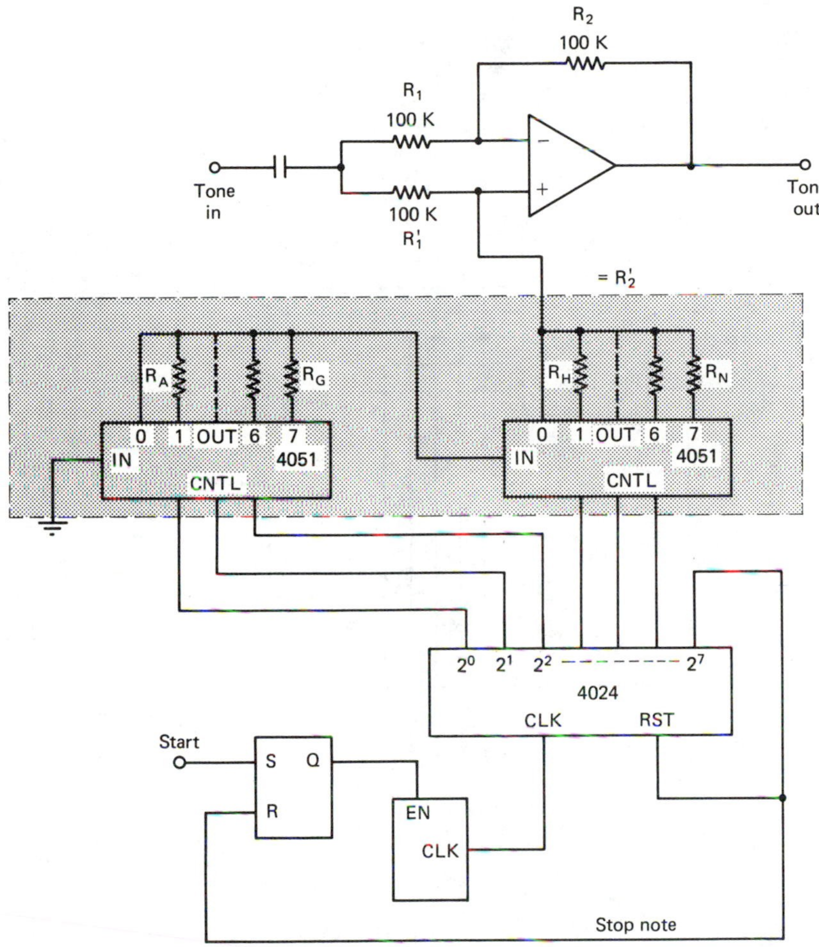

Figure 19.17 Programmable envelope generator, with resistor values selected to give bell-like decaying tone. Here, R_A–R_G = n(1.5K), n = 1–7, and R_H–R_N = m(12.5K), m = 1–7. (*Adapted from* K. Dugan, *Electronics.* (Sept. 28, 1978): 126.)

strobe signal is generated at the precise time when that key's address appears on the data lines. The *high* signal wired to the *Z* terminal of the upper 4051 finds a pathway out the *Y* terminals and through the matrix intersection closed by the switch, then back into the *Y* terminal of the lower 4051 and out its *Z* terminal, pulling the *strobe* line high. This scheme is superior to diode matrix encoding schemes whose OR logic can give address errors if more than one key is pressed at a time.

19.5 SAMPLE AND HOLD

This circuit allows a relatively slow device such as an A/D converter to encode much faster signals. The basic sample-and-hold circuit [Fig. 19.19(a)] connects a capacitor to the input signal upon command. When disconnected, the sampled

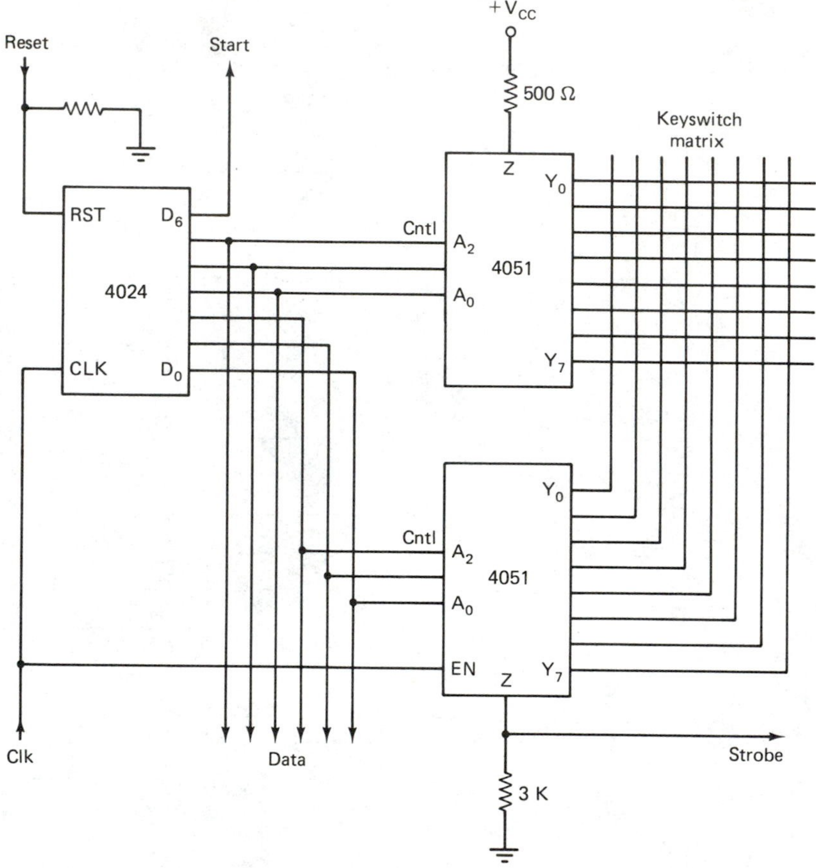

Figure 19.18 Scanning keyboard encoder for up to 64 key switches at the intersections of the matrix. (John Simonton, Jr., *Polyphony*, 1978; *courtesy* PAiA Electronics, Inc.)

voltage is held on the capacitor and read out at leisure. The high off-state resistance and low leakage of the MOSFET analog switch make this possible. The op amp follower allows readout without loss of charge on the capacitor. However, low-leakage polystyrene dielectric capacitors must be used, as well as a high-impedance BiMOS or BiFET op amp. Changing the voltage stored in this circuit requires current flow back through the signal source. This may limit circuit response speed if the source impedance is high. Improved circuits [Fig. 19.19(b)] isolate the signal source with a follower and put the capacitor in the feedback loop of an integrator. The negative feedback loop back to the input comparator requires the integrator to follow input signal changes, limited only by op amp speed.

A sample-and-hold circuit is characterized by the parameters illustrated in Fig. 19.19(c):

Hold Time. The time t_h during which the switch is open and the charge is held on the capacitor.

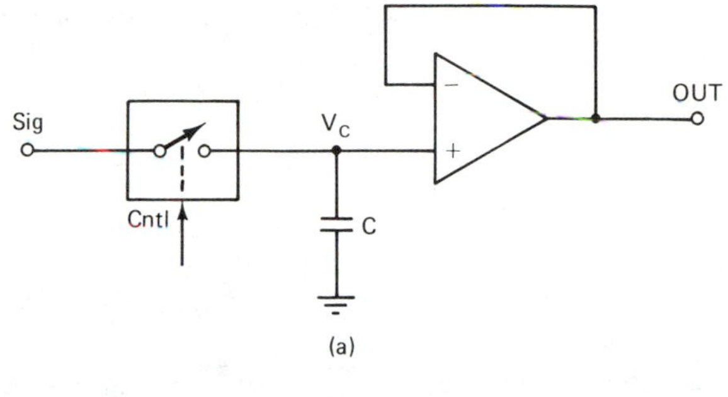

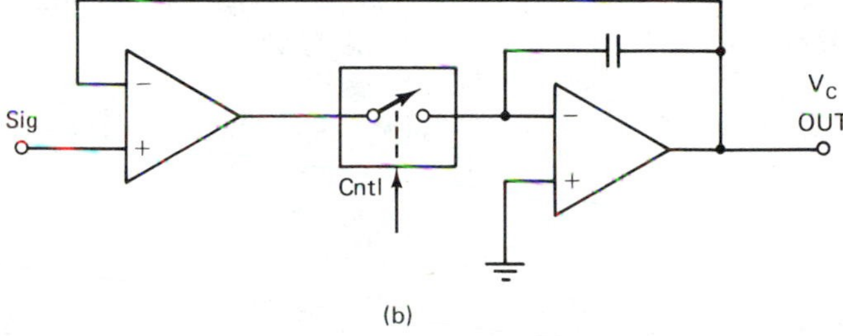

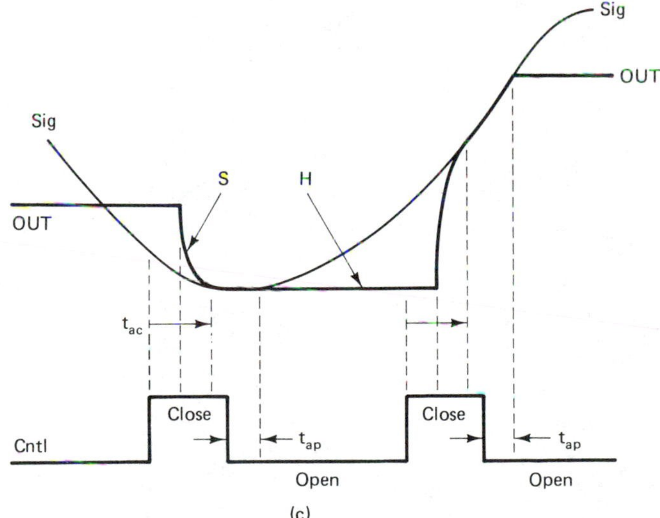

Figure 19.19 Sample and hold circuits. (a) Basic version. (b) Improved version. (c) Sample and hold waveforms, showing definition of key specifications.

Droop Rate. The rate of decay of stored voltage, dV_c/dt. Droop rate is determined by capacitor and switch leakage and op amp bias current, and can be limited to 1 mV/s with careful design.

Aperture Time. The interval t_{ap} between applying the control signal and opening of the switch to hold the sample. Available MOSFET switches allow aperture times as short as 10 ns.

Acquisition Time. The time t_{ac} required for the capacitor to charge to a specified percentage of the analog signal value, including the time it takes the switch to close. Acquisition time is limited primarily by the capacitor value C and the on-state resistance R_{sw} of the switch. With $C = 100$ pF and $R_{sw} = 100\ \Omega$, the acquisition time is about 10 ns.

19.5.1 Sample-and-Hold Applications

The sample-and-hold circuit has numerous other applications. In addition to use with A/D converters (Chapter 9), it makes possible the *sampling oscilloscope*, which can display picosecond events on a much slower oscilloscope. A picture of a repetitive waveform is built up by gradually advancing the moment of sampling. A *sampling circuit* for precise measurement of voltage values on fast pulses is shown in Fig. 19.20(a). A trigger signal from the input waveform begins the cycle [Fig. 19.20(b)]. A pulse generator with adjustable pulse delay drives the switch and also intensifies the waveform on the scope to show which point is being measured. The sampled signal is held until the digital voltmeter (DVM) can measure its amplitude.

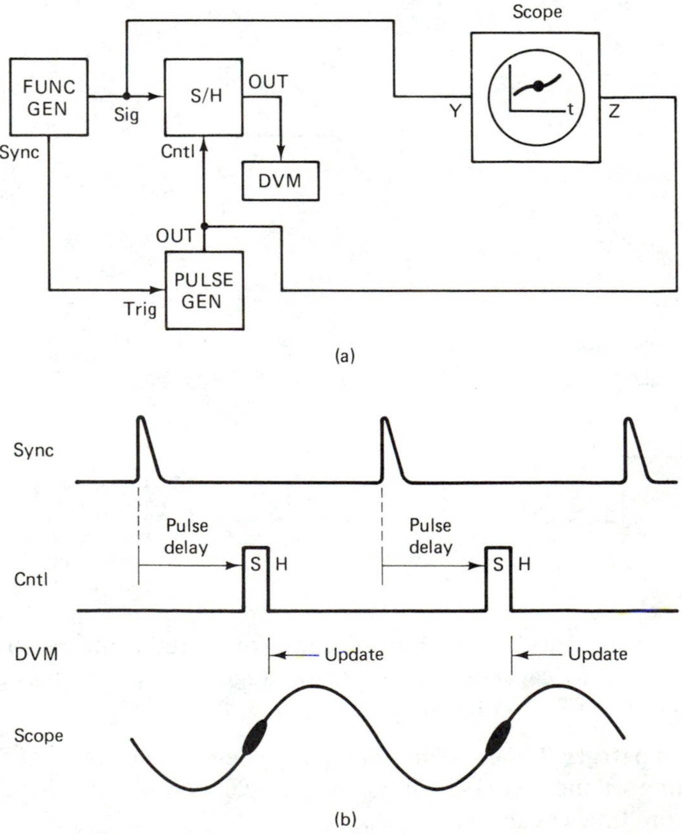

Figure 19.20 Sampling the waveform on a scope for precise digital voltage measurement. (a) Circuit. (b) Timing waveforms.

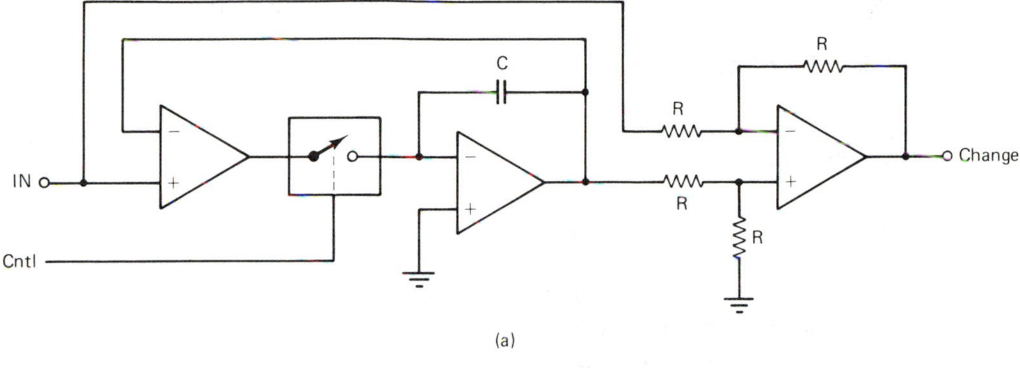

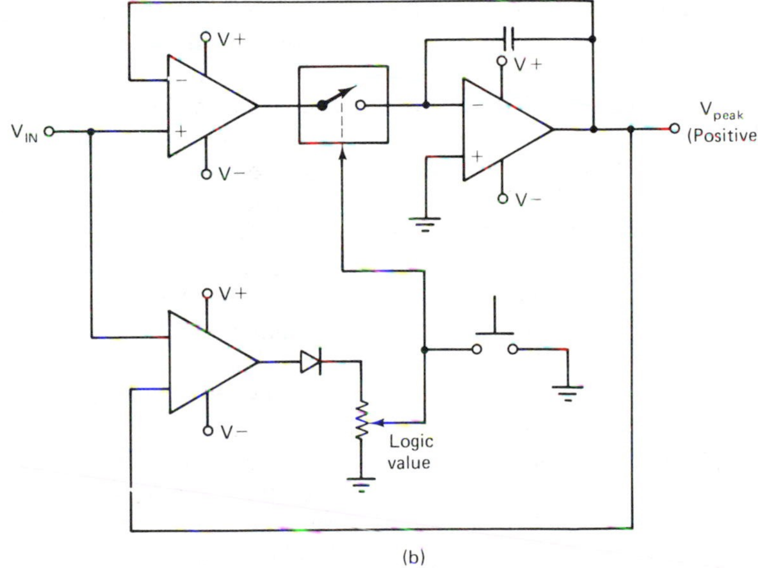

Figure 19.21 Other sample and hold applications. (a) Sampled difference circuit. (b) Peak follower circuit.

The circuit of Fig. 19.21(a) retains the most recent measurement as a reference voltage for *comparisons.* The output of the subtractor is the difference between the two most recent measurements. A straightforward switching modification (not shown) provides the *auto-zero* feature of digital voltmeters. A switch alternately connects the circuit to the signal source and to ground. A sample-and-hold circuit connects these to a subtractor so the signal fed to the DVM is automatically compensated for offset and drift of preamp components.

A *peak follower* is constructed from a sample-and-hold plus comparator [Fig. 19.21(b)]. The output is the largest value of the input voltage since the last reset signal. This circuit eliminates the error due to diode forward drop which is present in *passive* diode-plus-capacitor peak followers.

19.6 DISCRETE TIME ANALOG SIGNAL PROCESSING

19.6.1 Electronic Digital Filters

Active filter design is subtle if one needs to go beyond second-order sections (Chapter 16). The open-loop gain of an op amp is frequency dependent; phase shifts move poles around and can lead to instability. An alternative technique, called electronic *digital filtering* or *discrete time signal processing*, stores samples of the signal at a periodic rate. The characteristic frequencies of a digital filter are set by the clock frequency. This provides advantages over purely analog filtering. Filter characteristics become programmable without change of passive component values. Stability is enhanced, with no drift-susceptible components, and is limited only by the stability of the clock.

The digital filter concept is illustrated in Fig. 19.22. The input signal V_o is applied sequentially to a series of identical passive filters (low-pass RC filters in this example). With a sine wave input of frequency identical to the clock frequency f_o, each capacitor will eventually charge to a voltage equal to the average value of V_{in} at the sampled point. The output voltage V_o is also sequentially switched along the capacitors and displays a discrete sample of the input sine wave [Fig. 19.22(c)]. If the input frequency is not exactly commensurate with f_o, less average charge will accumulate on the capacitors. For large frequency differences, the stored charge will average to zero. The circuit is thus a bandpass filter symmetric about f_o. The concept is an extension of the lock-in amplifier (Chapter 18), but now the emphasis is on processing signals which need not be coherent with f_o.

A practical version of this circuit makes use of a 4051 analog multiplexer (Fig. 19.23). The switches inside connect one capacitor at a time to ground. Since only the connected capacitor makes a completed circuit, only it will change its charge and contribute an output voltage. The effect is equivalent to the circuit of Fig. 19.22, but the number of switches has been reduced by a factor of 2. The filter behavior is identical, although this simpler version is limited to single-pole filters requiring only one capacitor switching connection. Analysis shows that this is a bandpass filter with center frequency f_o and

$$Q = \pi f_o N / RC \tag{19.10}$$

The shape of the bandpass response is identical to the low-pass response but translated upwards by the *commutating* action [Fig. 19.23(b)]. The sharpness of the bandpass can be increased by increasing the number N of capacitors, since the bandwidth $\Delta\omega = \omega_o / Q = 2RC/N$. The center frequency can be shifted without changing the bandwidth, since f_o is independent of component values.

Actually, there are other peaks at multiples of f_o [Fig. 19.23(c)]; this is really a *comb* filter. Signals at frequencies $2f_o$, $3f_o$, ... also store nonzero average charge on the capacitors. This result can also be viewed as a consequence of sampling theory. Data at frequencies $f_d > f_o$ cannot be distinguished from data at a frequency $f_d - f_o$ [see Fig. 9.13(b)]. The higher-order peaks are of reduced amplitude.

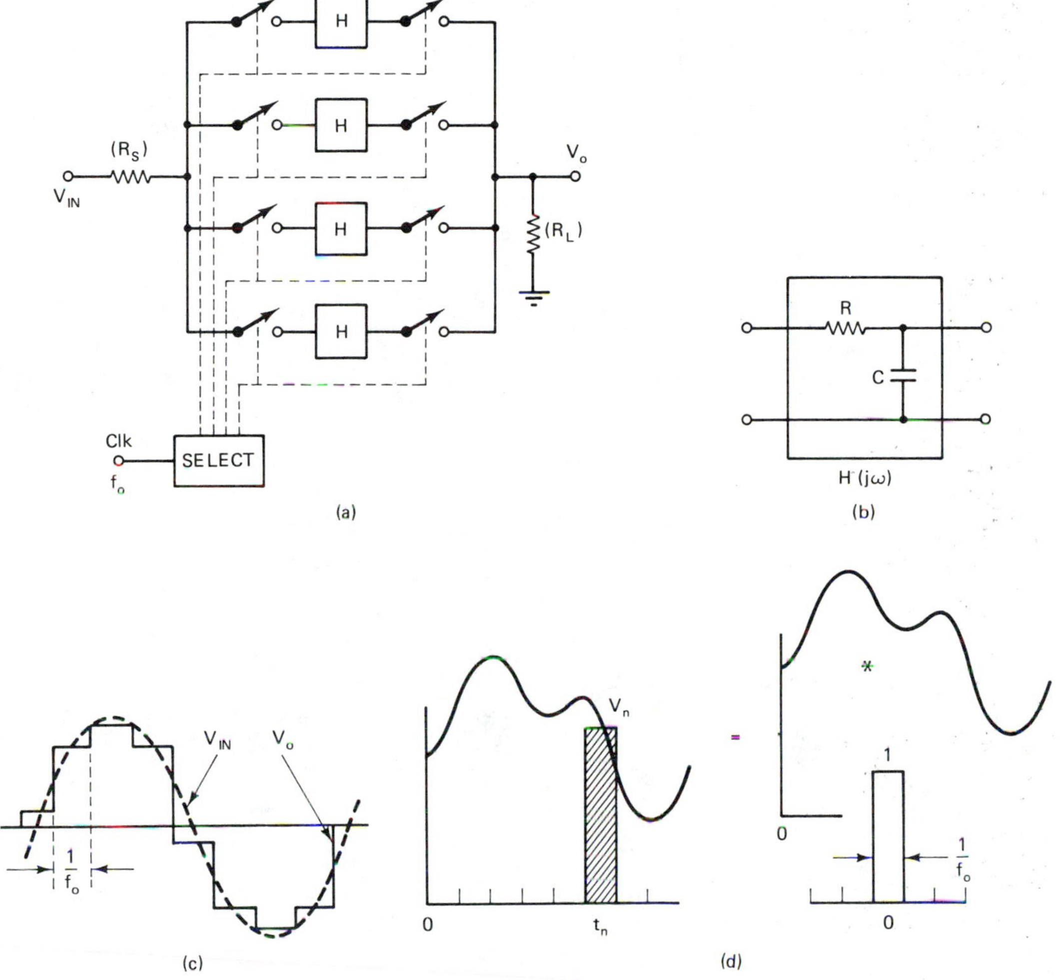

Figure 19.22 Digital filter. (a) Circuit. (b) Passive low-pass sampling element. (c) Sampled waveform. (d) Single sample (shaded bar) equals convolution of V_{in} with a pulse.

$$|A(n)|^2 = [sin^2(n\pi/N)]/[n\pi/N]^2 \tag{19.11}$$

A higher harmonic of the sine wave of Fig. 19.22(c) finds itself more averaged out by the smaller number of samples per cycle. More rigorously, the sample stored on a given capacitor is the convolution of the input signal with a pulse of width $1/f_o$ (Fig. 19.22(d)]. The frequency response is therefore the product of the signal's Fourier transform with the Fourier transform of a pulse, $(sin\ x)/x$, where $x = \pi f/f_o$).

If harmonic response is undesired, the signal is fed through a conventional low-pass filter with $f_c < f_o$ prior to reaching the switching filter. This is ordinarily advisable in data acquisition to avoid *Nyquist images* of high-frequency noise

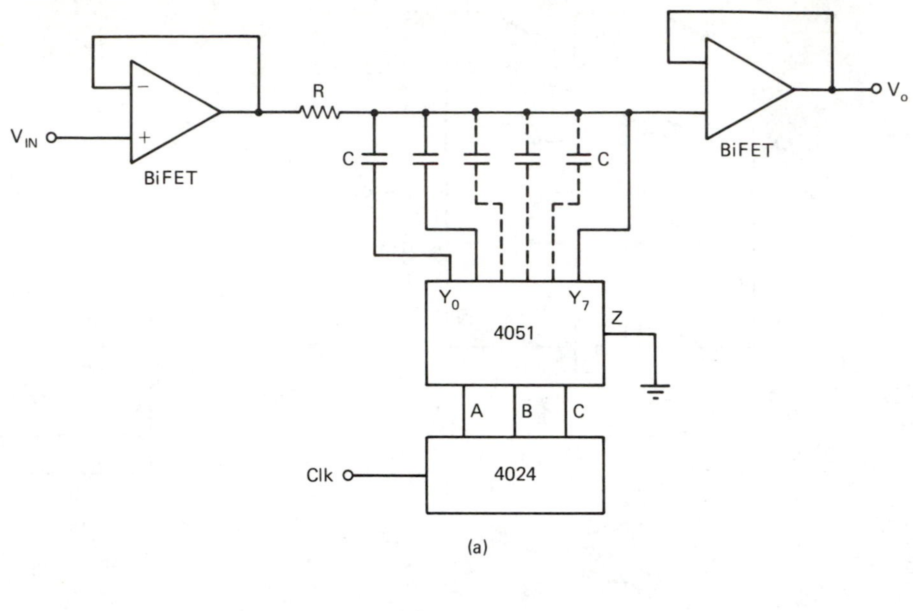

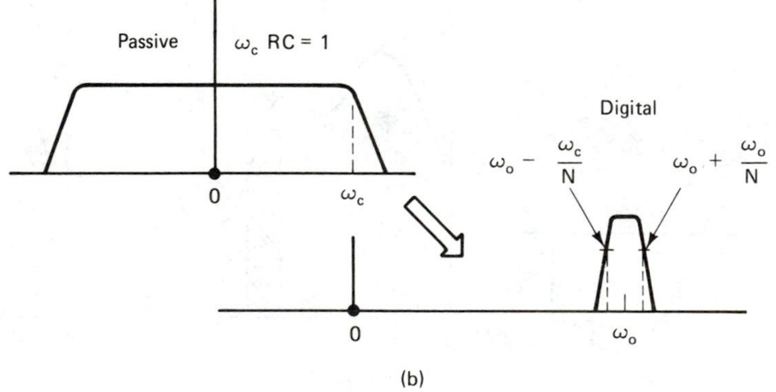

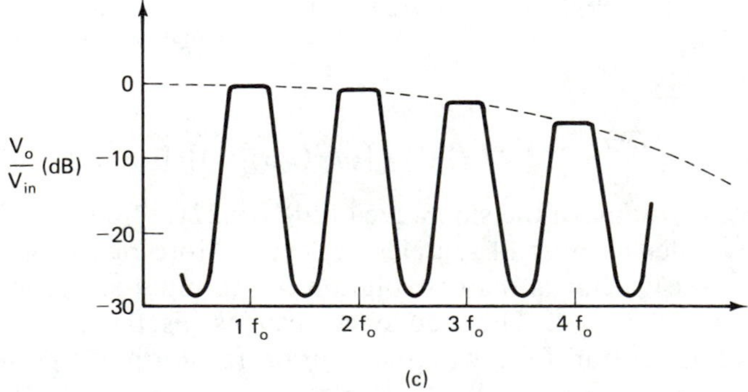

Figure 19.23 Practical digital bandpass filter. (a) Circuit. (b) Transfer function. (c) Bandpass shape as translation of low pass.

translated down into the frequency region below f_o. However, signal processing sometimes uses the comb response to advantage.

19.6.2 Signal Averaging

A signal averager (Fig. 19.24) resembles a digital filter with the number of switch-capacitor combinations increased greatly to build a finer detail picture of the input waveform. The switched storage is driven by a sequencer or ring counter

Figure 19.24 Signal averager. (a) Circuit. (b) Waveform example.

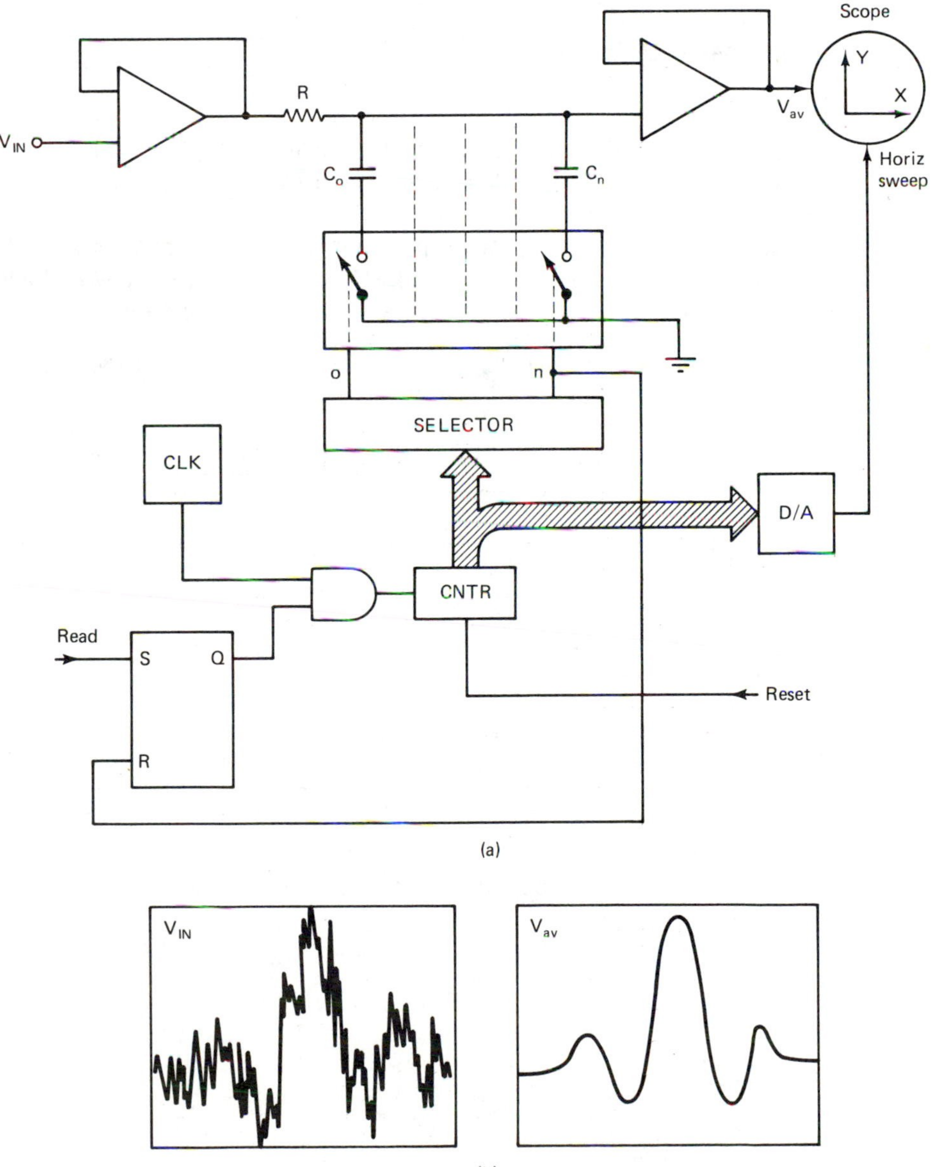

with additional logic which initiates a scan and then stops the counter's clock when all N capacitors have been updated. The *read* signal which initiates a scan is derived from the input signal to provide a coherent sample. Repeated scans build up a picture with incoherent noise averaged out [Fig. 19.24(b)]. Signal levels soon reach their average value, but noise samples decrease as $(1/M)^{1/2}$, where M is the number of scans completed.

The signal averager of Fig. 19.24 is an analog shift register. The voltage resolution can be very high because of the linear nature of the storage. Purely digital signal averagers store samples of the input signal in a series of registers or memory locations. The signal builds up as M, the noise builds up as $M^{1/2}$, so the signal-to-noise ratio also improves as $M^{1/2}$.

19.6.3 Analog Delay Lines

This IC is an analog shift register with only serial inputs and outputs. As with digital shift registers, many pieces of information can thus be stored in a package with few pins. An example is the SAD-1024 (Fig. 19.25), which stores 1024 pieces of information in a 16-pin package. Because the storage is linear, resolution is high; the dynamic range is >70 dB, corresponding to 12 binary bits per stored voltage. This is called a *charge-coupled device* (CCD) or *bucket brigade device.* Each bucket contains a switch and capacitor which stores the information as a charge. The

Figure 19.25 The analog delay line. (a) Circuit principle. (b) Bucket brigade idea. (*Adapted from* SAD-1024 application notes; *courtesy* Reticon.)

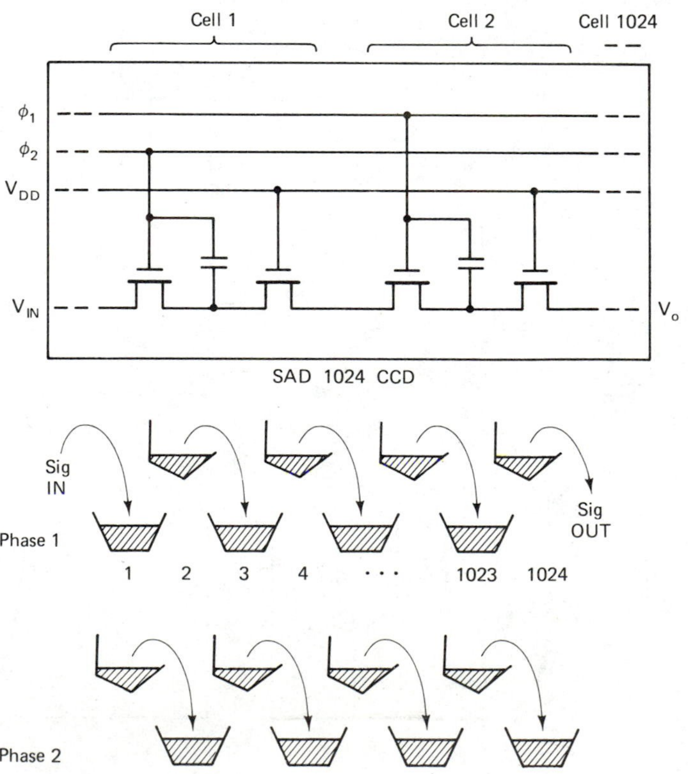

switches are MOS transistors and the capacitors are back-biased pn junctions. Charge is moved from one bucket to the next as a two-phase clock alternately opens and closes adjacent transmission gates. The physics of CCD operation involves shifting adjacent potential wells to give charge packets a preferred direction in which to move.

Compared to purely digital techniques, the strong points of the CCD as a signal processing element are the lack of A/D and D/A requirements and the high resolution. The binary equivalent of a 1024-channel CCD such as the SAD 1024 would require at least $(2^{10} \times 21^{12}) = 2^{22}$ bits or 4K bytes of dedicated storage, in addition to a 12-bit A/D and 12-bit D/A. The weak points of the CCD are speed and eventual fidelity loss. Speed is limited on the short end by maximum clock frequency which gives a minimum device delay of $\sim 200\ \mu s$. This sets an upper limit of about 10^5 Hz to the signal processing range of this device, which is more than adequate for audio signal processing. Other CCD IC's, however, are optimized for higher-frequency operation such as video image signal processing. The longest delay practical is about 1 s, set by the eventual decay of information on the capacitor (about 1% decay in 100 ms). Fidelity loss is due to charge quantization. If a given bucket stores 10^4 electrons, one electron left behind gives a transfer inefficiency of 1×10^{-4}. Unlike purely digital systems, this transfer inefficiency is cumulative, eventually blurring the information. This limitation is not serious for real-time signal processing where the throughput is continuous, but requires that other applications such as high-resolution analog memory make provision for refreshing.

To see how useful the analog delay line can be, consider the problem of slowing down or speeding up speech on a tape recording. Slowing down is useful in learning a foreign language, and speeding up for rapid reading by ear. Changing motor speed alone shifts the pitch, and leads rapidly to unintelligibility. Consider playing an instrument along with taped music. To alter the speed of your playing, you'd have to retune your instrument! A product called *variable speech control* (Fig. 19.26) adds real-time pitch shifting in the opposite direction to compensate, leaving overall pitch independent of playback speed. Two analog shift registers receive the input signal. Their clock is driven more slowly as the tape recorder motor is rotated faster. The output comes alternately from each SR content, glued together electronically. The signal is restored in speeded playback by throwing away portions of the stretched samples which overlap. Slowed playback is trickier. There will be missing sections if glued directly. But the repetitiveness of phonemes in human speech allows the missing information to be approximated by inserting a bit of a section over again, as in a tape loop. The gluing process is done at zero crossings to minimize pops which would otherwise come from sudden changes in amplitude.

19.6.4 Digital Filters with Delay Lines

The delay line lends itself to filters not possible with purely analog techniques. The delayed output may be added to or subtracted from the signal, or fed back around to the input. The techniques developed below are general ones, even

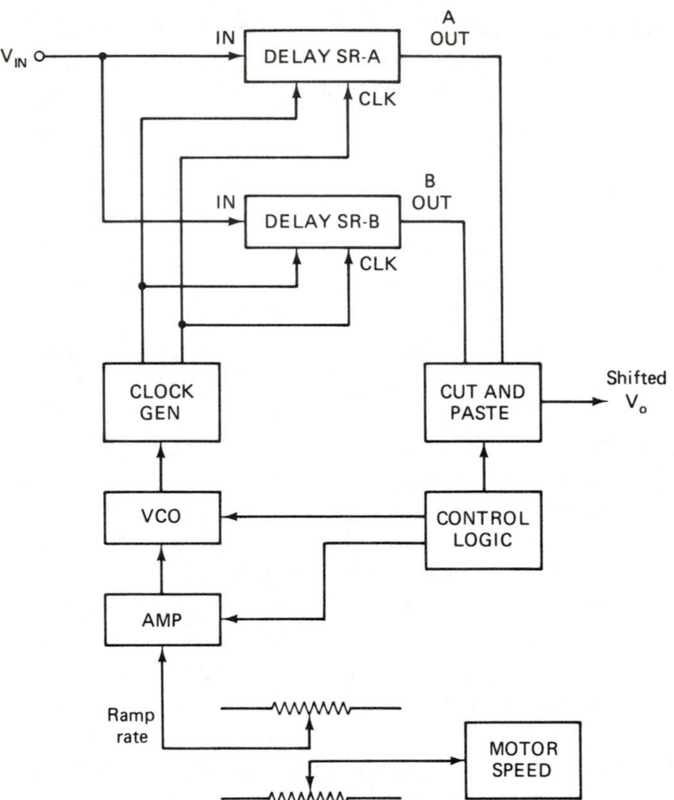

Figure 19.26 Variable speech controller simplified block diagram. (*Adapted from* M. Shiffman, *Electronics.* (Aug. 22, 1974): 87.)

though illustrated for analog delay line circuits, and are the basis for purely digital filtering.

To see the concept, first consider the amplitude and phase of a signal which has been fed through a delay of length τ [Fig. 19.27(a)]. The amplitude is of course constant as a function of frequency, but the phase exhibits wild oscillations. This happens because, when fed through fixed *time* delay, the output is 180° out of phase relative to the input each time the period $1/f$ is an odd multiple of τ. The phase lag has no physical consequence unless the output is allowed to interfere with the input. If output and input are added [Fig. 19.27(b)], the resultant amplitude will exhibit destructive interference when half-cycle phase lags occur, or at signal frequencies f given by

$$\tau = (2n + 1)T/2 = (2n + 1)/(2f) \qquad n = 0, 1, 2, \ldots$$

$$f = (n + \tfrac{1}{2})/\tau \tag{19.12}$$

Thus, for a 1-ms delay, zeros in amplitude occur at 500, 1500, 2500 ... Hz. By a similar argument, peaks occur at 1, 2, 3, ... KHz. If output and input are subtracted instead [Fig. 19.27(c)], the zeros and peaks are interchanged, with peaks at odd multiples and zeros at even multiples of $1/\tau$.

The additive and subtractive circuits resemble low-pass and high-pass filters, respectively, if signal frequencies are kept below $1/(2\tau)$. However, their transfer

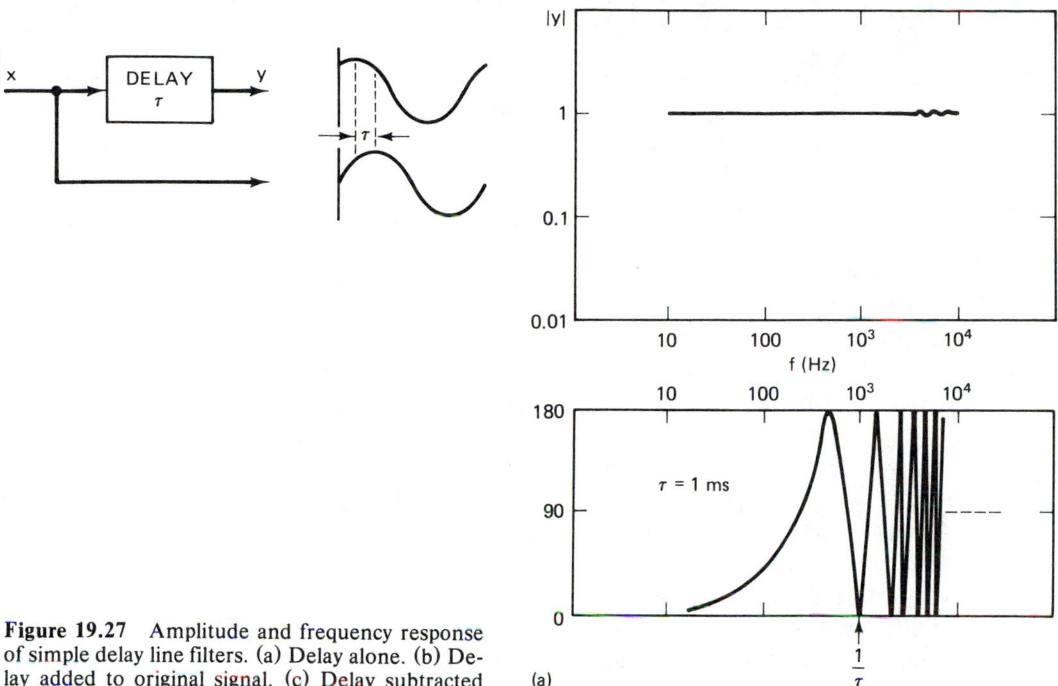

Figure 19.27 Amplitude and frequency response of simple delay line filters. (a) Delay alone. (b) Delay added to original signal. (c) Delay subtracted from original signal.

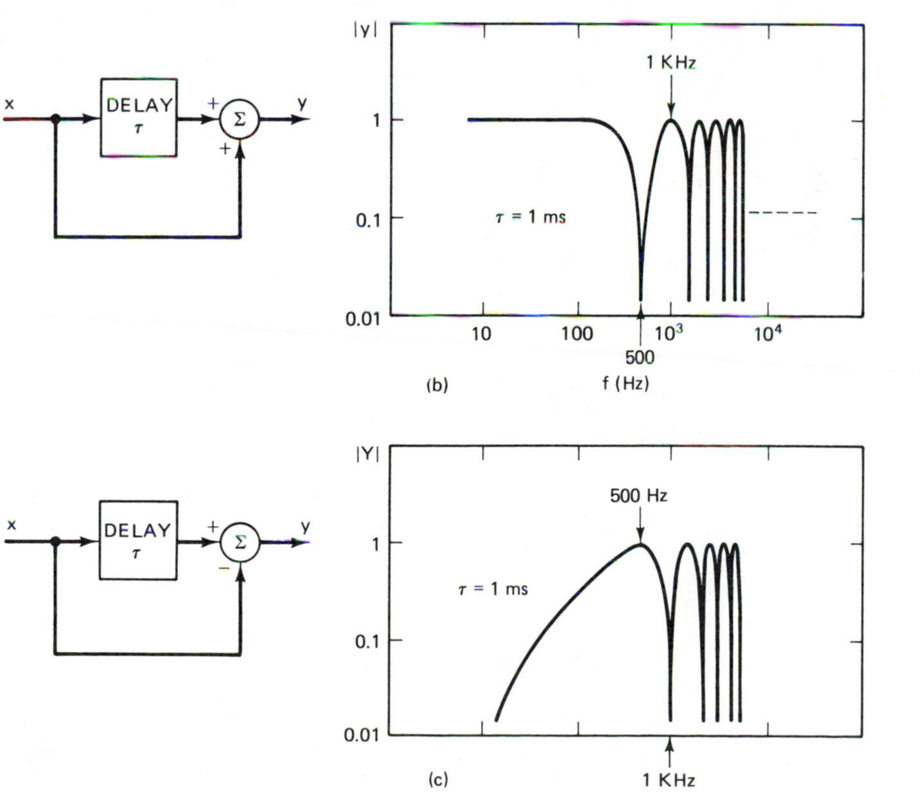

functions actually behave as cosine and sine functions. (The log distortion of Fig. 19.27 makes this less than obvious.) These are known as *comb filters.* An audio signal fed through such a filter exhibits unusual sounds known as phasing or "flanging" as the delay time is varied, due to cancellation of various frequency components in the signal. Slight modifications allow reverberation or echo effects, plus many other signal-processing filters. These examples illustrate that delay line filters have transfer functions greatly different from familiar passive or active filters, with cosine and sine form rather than simple polynomials. The explanation depends upon the mathematics of digital filters (see References).

19.6.5 Correlation and Autocorrelation

Suppose two signals are thought to be related to one another. An example is a search for gravity waves from heavy stars in space [Fig. 19.28(a)]. Gravity wave detectors are heavy masses instrumented with sensitive vibration detectors. If a gravity wave comes along, the mass is set into vibration. Of course, many other vibrations such as people walking by, nearby trucks, and mild earthquakes will create signals far greater than the sought-for gravity wave. Noise has to be averaged out by stationing two detectors far enough away from one another that coincidental disturbances of earthly origin are unlikely. The two signals $x(\tau)$ and $y(\tau)$ are *correlated* with one another in the search for gravity wave events. The correlation function $z(\tau)$ is defined as

$$z(\tau) = \int y(t)x(t - \tau)\,dt \tag{19.13}$$

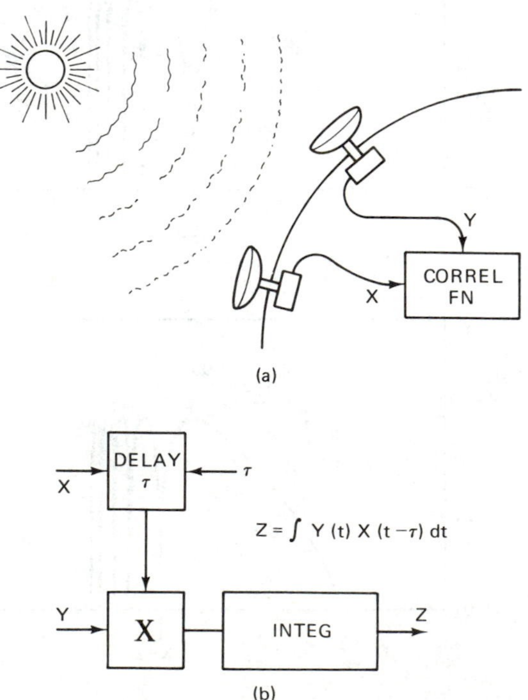

Figure 19.28 (a) A correlation experiment: the search for gravity waves. (b) Correlation function circuit (simplified).

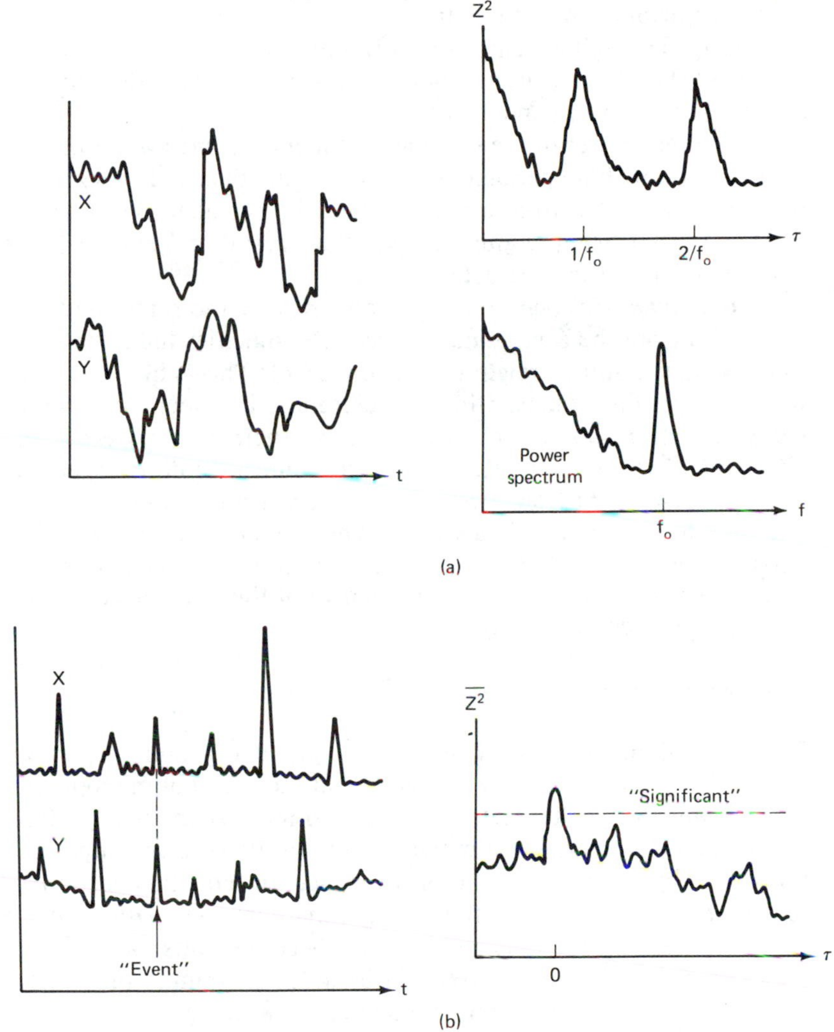

Figure 19.29 Examples of correlation function data as a function of time, delay, and frequency. (a) Data with hidden coherent sine wave. (b) Data with occasional coincident pulse.

The correlation function principle is illustrated by the circuit of Fig. 19.28(b). The circuit resembles a lock-in amplifier plus a delay line. The two signals by themselves display little visible relationship. The strength of the correlation function approach is that it lends itself to statistical evaluation of the significance of possible correlation between events. Suppose the two signals each have an identical sine wave of frequency f_o hidden in the noise. The output of the correlation function circuit will display a peak at $\tau = 1/f_o$ [Fig. 19.29(a)]. Other peaks will occur at multiples of this delay, as the two sine waves again line up. Uncorrelated noise will largely average out, except at the low-frequency end where geophysical events contribute. A Fourier transform of $z(\tau)$ will display a single peak at frequency f_0. This follows since the correlation function is equivalent to a convolution of $x(t)$ and $y(t)$, so the Fourier transform $Z(f)$ is the product of the individual transforms $X(f)$ and $Y(f)$.

The real power of the correlation function comes when the events are occasional pulses with no periodicity. Viewed individually [Fig. 19.29(b)], there is no way to tell an event from a noise spike. The correlation function, however, will display a statistically significant peak at $\tau = 0$ if genuine correlation exists between events at the two detectors.

The *autocorrelation* function is obtained by the same method but with both inputs fed from the same signal source. Periodicities hidden by noise will display as peaks in the autocorrelation function $Z(f)$. This is like coherent detection but with the signal itself generating the reference. This technique is often used in the study of noise sources, characterized by their *power spectral density* function $|Z(f)|^2$. Familiar examples include white noise (uniform spectral density), and "pink" noise ($1/f$) from electronic devices. A noise source can be used to excite and look for resonances in a system. The human voice and a violin are examples where resonances would be visible directly in the response. Resonances in noisier systems show up in the Fourier transform of the autocorrelation function, which is the power spectral density.

19.7 DIGITAL FILTERS

Real-time signal processing is increasingly being done digitally. The cost of A/D's, D/A's, and memory has fallen and the speed of microprocessors has risen to the point that digital techniques cost no more than analog ones, and have the advantage of programmability. Discrete time filtering encompasses both the analog switching methods of the previous section and this section's purely digital techniques. The common ground is the mathematics. Differential equations become difference equations and integrals become sums, since the computation is on discretely sampled data. We will be able to examine only a few applications, leaving rigorous development to the references cited.

Example: The Running Average. Consider the data shown in Fig. 19.30(a). The example plotted happens to be the price of the Japanese yen relative to the dollar for a certain month, but could just as well be someone's weight each day, the price of a particular stock, or measurements of cosmic-ray intensity during a period of solar flares. A three-point running average, also shown, tends to smooth out fluctuations so long-term trends can be spotted. Longer-term averages might bring out different features, but would tend to eliminate significant cyclic trends. This section will show how the discrete running average is related to the low-pass filter, and how a search for cyclic trends leads to a digital bandpass filter.

The three-point running average is computed by adding to sample x_n the previous two samples x_{n-1} and x_{n-2} and dividing the result by 3. This may be computed in real time with the circuit of Fig. 19.30(b). The unit delay length is set equal to the time between samples. More generally,

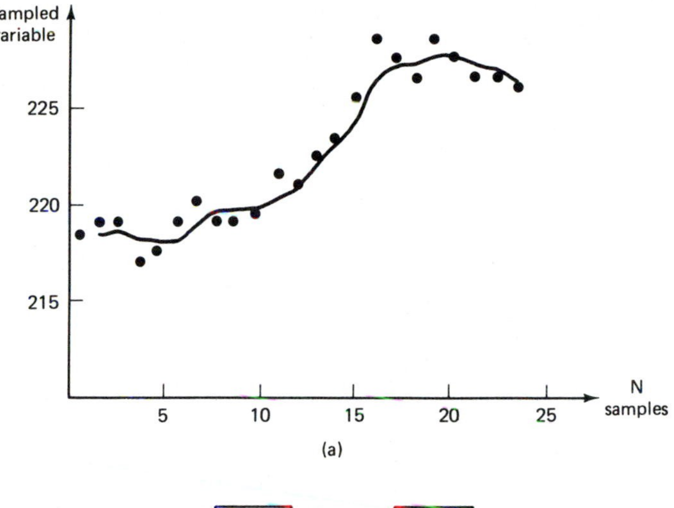

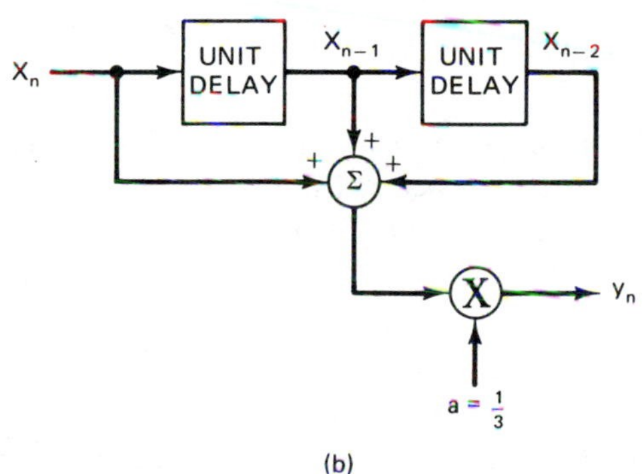

Figure 19.30 (a) Example of sampled data. The curve is a three point running average. (b) Digital filter circuit to perform the three point running average.

$$y_n = \sum_k b_k x_{n-k} \tag{19.14}$$

This defines a *nonrecursive* digital filter. Some digital filters also incorporate samples of recent outputs.

$$y_n = \sum_k [b_k x_{n-k} + a_k y_{n-k}] \tag{19.15}$$

This is called a *recursive* digital filter. Recursive filters tend to require fewer terms for a given result than nonrecursive ones. The three-point running average is nonrecursive.

It is natural to develop analytical tools built around the unit delay, z^{-1}, the basic digital filter building block. For the three-point running average, it can be shown that the transfer function takes on the form

$$H(z) = [1 + z^{-1} + z^{-2}]/3 \tag{19.16}$$

where $z = \exp(s\tau)$: $s = j\omega$ as usual, and τ is the sampling interval. We recognize this as precisely the operation performed by the three-point running average circuit of Fig. 19.30(b) if z^{-1} is identified as the *unit delay operator*.

The frequency response of the three-point running average looks like:

$$|H(\omega)|^2 = [3 + 4\cos\omega\tau + 2\cos 2\omega\tau]/9 \tag{19.17}$$

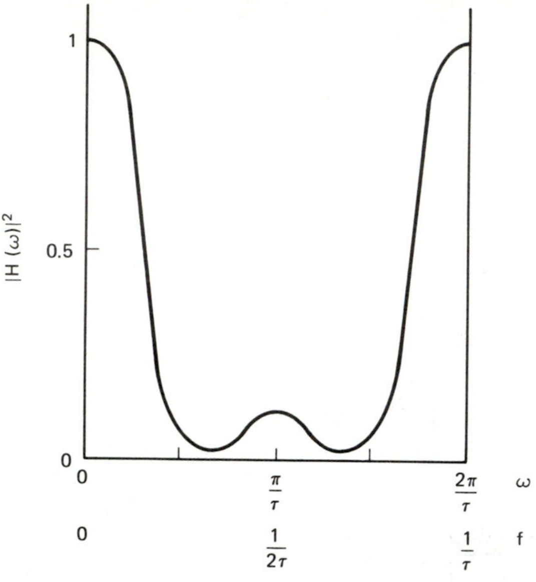

Figure 19.31 Transfer function of the three point running average in the frequency domain.

This is shown plotted in Fig. 19.31. The curve repeats itself at intervals of the sampling frequency, as expected. At low frequencies, the response looks something like a low-pass filter. Trends slower than $f_s/4$ get through with little attenuation. This makes sense; a trend which lasts for more than three samples should get through a three-point average.

19.7.1 Bandpass Digital Filter

Consider how to synthesize a digital filter which carries out the second-order bandpass function. Two delays will be required, because the corresponding differential equation is second-order (transfer function quadratic in s). A basic second-order section is shown in Fig. 19.32. The transfer function is arrived at by calculating how $V_o(z)$ is related to the voltage at intermediate points (V_1, V_2, and V_3), and how these voltages are related to the input signal and (recursively) to one another.

$$H(z) = \frac{z^2 + b_1 z + b_o}{z^2 - a_1 z - a_o} \tag{19.18}$$

The resemblance to the transfer function of the familiar analog second-order bandpass is no accident. The filter's characteristics are determined by locating the poles in the z-plane:

$$z_o = -a_1 \pm [(a_1/2)^2 - a_o]^{1/2} \tag{19.19}$$

The determination of natural frequency and Q follow familiar patterns (Chapter 16), except that the delay time τ enters as a hidden variable. An example of the filter frequency response (Fig. 19.33) shows close correspondence with the analog version, except that the periodicity of the digital filter's response (*imaging*) weakens the falloff above the cutoff frequency.

Digital filter techniques make possible both real-time synthesis and real-time analysis of complicated signals. An example is speech synthesis. Formerly accessible only with dedicated high-speed computers ($100,000 in one leading la-

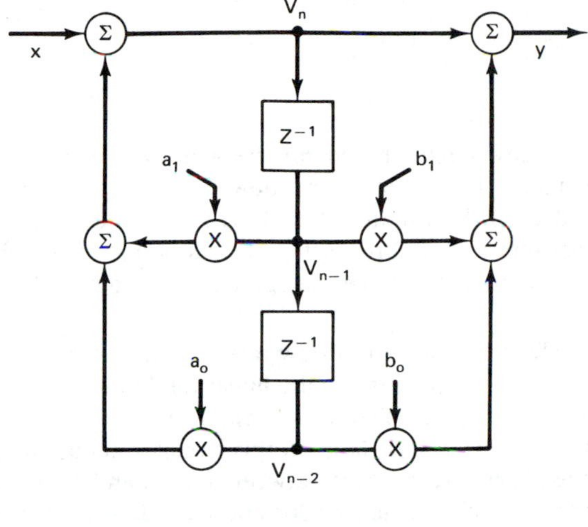

Figure 19.32 Recursive filter. Second order bandpass synthesized from two stages of delay.

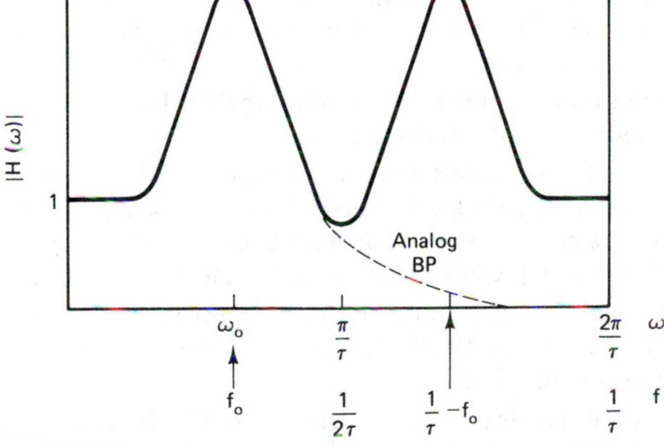

Figure 19.33 Digital bandpass filter. Transfer function as a function of frequency. The dashed line is the corresponding analog filter.

boratory), speech synthesis has now been accomplished in a \$50 consumer product, the Texas Instruments *Speak and Spell* (Fig. 19.34). Digital bandpass filters simulate the human vocal tract with its formant frequencies. The filter coefficients are changed in real time to alter the pitch and harmonic content of the sound. The words are quite recognizable, and the technology opens up wide horizons such as language translation.

Figure 19.34 Voice synthesis as an application of real-time digital filtering (the principle behind the Texas Instruments' *Speak and Spell*).

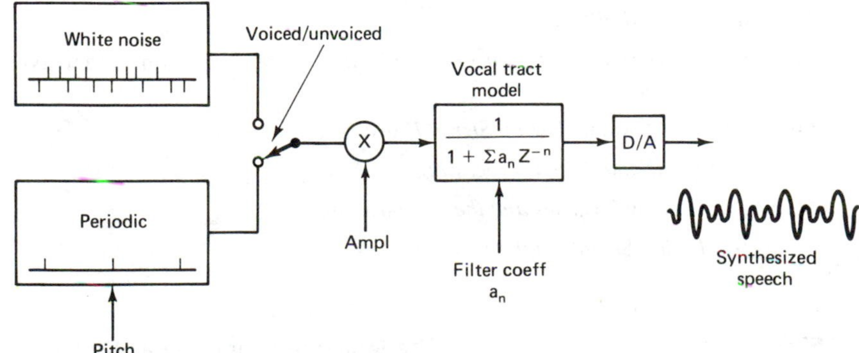

PROBLEMS

19.1. Design an indicating window circuit based on the window comparator of Fig. 19.5. A green LED lights for V_{in} within the window. Separate red LED's for V_{in} below and above the edges of the window. The design is simplest if you connect LED's with pull-up resistors to 5 V directly to selected points of the circuit. Be prepared to reverse op amp inputs to have the lighted state indicate the intended condition.

19.2. **(a)** Show that the transfer function of the slew rate limiting filter [Fig. 19.6(b)] is given by Eq. (19.6) when operating in the linear mode.

(b) Is it correct to think that negative feedback keeps V_1 near 0?

(c) Normally one chooses $G = R_2/R_1$ large in this circuit. Suppose instead that $G = 1$. Explore the phase relationships between V_i, V_o, and V_1 as a function of frequency. Under what circumstances does output follow input faithfully?

19.3. Rearrange the components of the delta modulator to make a demodulator. Given the sampled output of Fig. 19.9(b), the demodulator should generate the waveform labeled *reconstructed audio.*

19.4. Explain the operation of the ambient light-cancelling circuit of Fig. 18.16. Use the principle of the closed-loop integrator tracking circuit.

19.5. Design a revised version of the sampled difference circuit [Fig. 19.21(a)] to perform the *auto-zero* operation for a DVM. A preamp input is to be alternately connected to the signal voltage and to ground. The difference voltage is sampled and presented to the DVM, but with preamp offset and drift automatically cancelled.

19.6. Compare a two-point running average with the three-point version in the text. Sketch the circuit, analogous to Fig. 19.30(b), and calculate and plot the transfer function $H(\omega)$ analogous to Fig. 19.31.

19.7. Show how the second-order bandpass filter circuit of Fig. 19.32 has the transfer function of Eq. (19.18). **Hint:** Calculate Y_n in terms of intermediate results V_n, V_{n-1}, V_{n-2}, and then calculate V_n in terms of x_n, etc.

REFERENCES

More complete bibliographic information for the books listed below appears in the annotated bibliography at the end of the book.

BRACEWELL, *The Fourier Transform and its Applications*

HAMMING, *Digital Filters*

HIGGINS, *Experiments with Integrated Circuits*, project labs on signal averaging and on digital filtering

OPPENHEIM & SHAFER, *Digital Signal Processing*

PELED & LIU, *Digital Signal Processing*

SILICONIX, *Analog Switches and their Applications*

STEARNS, *Digital Signal Analysis*

APPENDIX 1

Powers of 2

2^N	N
1	0
2	1
4	2
8	3
16	4
32	5
64	6
128	7
256	8
512	9
1 024	10
2 048	11
4 096	12
8 192	13
16 384	14
32 768	15
65 536	16
131 072	17
262 144	18
524 288	19
1 048 576	20
2 097 152	21
4 194 304	22
8 388 608	23
16 777 216	24
33 554 432	25

2^N						N
		67	108	864		26
		134	217	728		27
		268	435	456		28
		536	870	912		29
	1	073	741	824		30
	2	147	483	648		31
	4	294	967	296		32
	8	589	934	592		33
	17	179	869	184		34
	34	359	738	368		35
	68	719	476	736		36
	137	438	953	472		37
	274	877	906	944		38
	549	755	813	888		39
1	099	511	627	776		40
2	199	023	255	552		41
4	398	046	511	104		42
8	796	093	022	208		43
17	592	186	044	416		44
35	184	372	088	832		45
70	368	744	177	664		46
140	737	488	355	328		47
281	474	976	710	656		48
562	949	953	421	312		49

APPENDIX 2

Some Guidelines for Beginners on Oscilloscope Use

Though most textbooks discuss the basic principles of the oscilloscope or "scope", there is a gap between the principles and competent oscilloscope operation. We outline a few practical guidelines. These guidelines will also be useful with many other instruments. The advantage of learning them with an oscilloscope is that:

1. With an oscilloscope, you can *see* when you are doing something wrong.
2. When something goes wrong in a circuit or a measurement, problem-tracing is speeded by checking point-by-point with an oscilloscope.

A2.1 TURNING IT ON

There are lots of knobs and switches; don't get worried. First find the power switch to turn it on! Sooner or later, try all of the controls; you cannot hurt anything. There is one exception. The INTENSITY CONTROL, which sets the brightness, can burn out a point in the screen if left sitting anywhere at too high an intensity. If it's too bright for you, it's too bright for the scope.

Both the horizontal and the vertical deflections are the result of a voltage which *you* have to give it. Often, the *mode* is

---

Portions of this section are adapted from: *A Primer of Waveforms and Their Oscilloscope Displays*, Tektronix, Inc.

y-axis: displays voltage

x-axis: displays time

The time comes from a sweep circuit which generates a *voltage ramp* at a rate set by the *sweep rate* control, moving the electron beam a fixed distance in an adjustable time interval.

A2.2 INTERPRETING THE OSCILLOSCOPE DISPLAY

A2.2.1 Voltage vs. Time Mode

The display on an oscilloscope is a graph [Fig. A2.1 (a)] of the instantaneous voltage of a wave vs. time. Elapsed time is indicated by horizontal distance across the cathode-ray-tube (CRT) screen. The instantaneous voltage of the waveform is measured vertically on the screen.

To find the *elapsed time* between two points on the graph (such as points A and B), multiply the horizontal distance between these points, measured with the scale on the scope face (*major graticule* divisions), by the setting of the TIME/DIV. control. This control sets the horizontal sweep rate of the scope. In Fig. A2.1(a), the distance between points A and B is 4.4 major divisions. If the TIME/DIV. control is set 100 μs per division, then the elapsed time between points A and B is $4.4 \times 100 = 440\ \mu$s. In general,

$$\text{Elapsed time} = (\text{horizontal distance in divisions}) \times (\text{TIME/DIV. setting}) \tag{A2.1}$$

If a MULTIPLIER control is associated with the TIME/DIV. control, multiply the above result by the setting of the MULTIPLIER. If a MAGNIFIER is in operation, divide the result by the amount of magnification.

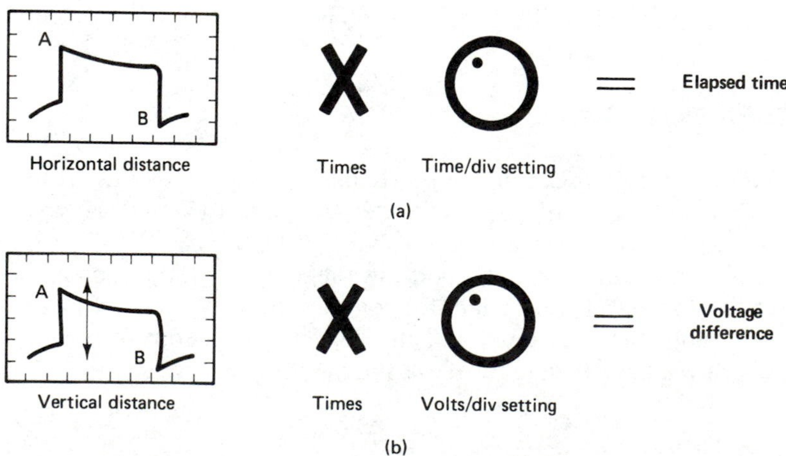

Figure A2.1 Oscilloscope measurements of (a) elapsed time and (b) voltage.

To find the *voltage difference* between any two points on the graph, such as points A and B, multiply the vertical distance between these points, measured in major divisions on the screen, by the setting of the VOLTS/DIV. control, which sets the vertical deflection factor or sensitivity of the oscilloscope. In Fig. A2.1(b), the vertical distance between points A and B is 3.6 divisions. If the VOLTS/DIV. control is set at 0.5 V per division, then the voltage difference between points A and B must be $3.6 \times 0.5 = 1.8$ V. In general,

$$\text{Voltage difference} = (\text{vertical distance in divisions}) \times (\text{VOLTS/DIV. setting}) \tag{A2.2}$$

The oscilloscope can also draw pictures of quantities other than voltages. To measure an electric *current* waveform, send the current through a small series resistor and observe the voltage across the resistor with the oscilloscope. Other quantities such as temperature, pressure, strain, speed, and acceleration can be translated into voltages by suitable transducers, and then viewed on the oscilloscope.

A2.2.2 XY Mode

Another useful scope mode utilizes two input voltages, one on the vertical and one on the horizontal axis. Uses of the XY mode include

1. *Graphing one variable as a function of another.* An example is plotting transistor characteristic curves [Fig. A2.2(a)].
2. *Determining phase relationships.* If the two signals are sinusoidal

   $$Y = A \sin(\omega t + \phi) \tag{A2.3}$$
   $$X = B \sin(\omega t)$$

   then the resulting ellipse may be used to *measure* the relative phase ϕ. This is useful in determining the phase shift between input and output in a network or a physical variable [Fig. A2.2(b)]:

   $$\phi = \sin^{-1}\left(\frac{A}{B}\right) \tag{A2.4}$$

3. *Determining frequency relationships.* When the two frequencies are different, a more complex pattern (sometimes called a *Lissajous figure*) results [Fig. A2.2(c)]. If the pattern is stationary, the two sources are *coherent.* If the pattern rotates, the rate gives information on the frequency difference.

A2.3 HOW THE OSCILLOSCOPE WORKS

Fig. A2.3 is a block diagram of a typical oscilloscope, omitting power supplies. An *electron gun*, consisting of a heated filament plus accelerating potential, shoots an electron beam towards the center of the screen. The term cathode-ray tube ori-

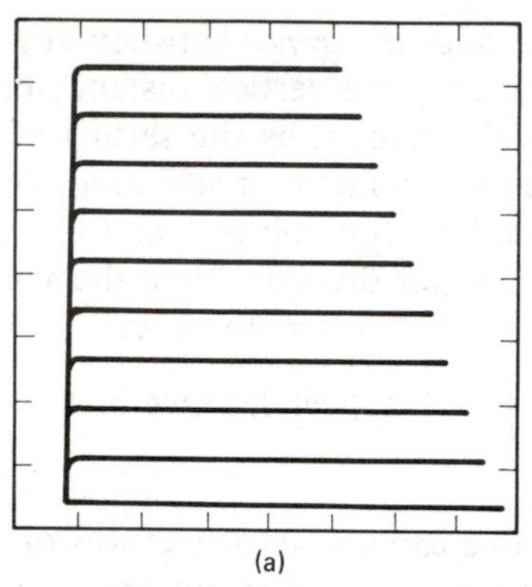

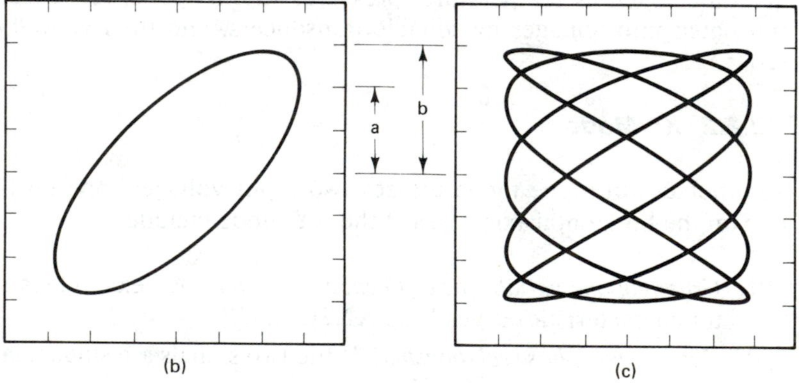

Figure A2.2 *XY* measurements on an oscilloscope of (a) transistor characteristic curves; Lissajous figures of two sine waves of (b) the same frequency and (c) differing frequency.

ginated when an electron beam was called a cathode ray. Metal plates part-way towards the screen provide a means to deflect the beam and draw the picture.

The waveform (A) is fed into the vertical-amplifier input, whose gain is set by the calibrated VOLTS/DIV. control. The output (B and C) of the vertical amplifier is fed to the vertical-deflection plates of the CRT.

The time-base generator or *sweep generator* develops a sawtooth wave (E) used as a horizontal-deflection voltage. The rising part of this sawtooth, called the run-up portion, is linear, rising through a given number of volts during each unit of time. This rate of rise is set by the calibrated TIME/DIV. control. The sawtooth voltage is fed to the time-base amplifier, which includes a phase inverter so two output sawtooth waveforms (G) and (J) are supplied simultaneously. The positive-going sawtooth is applied to the right-hand horizontal-deflection plate of the CRT, and the negative-going sawtooth is applied to the left-hand deflection plate. The electron beam is swept horizontally through a given number of graticule divisions each unit of time, with the sweep rate controlled by the TIME/DIV. control.

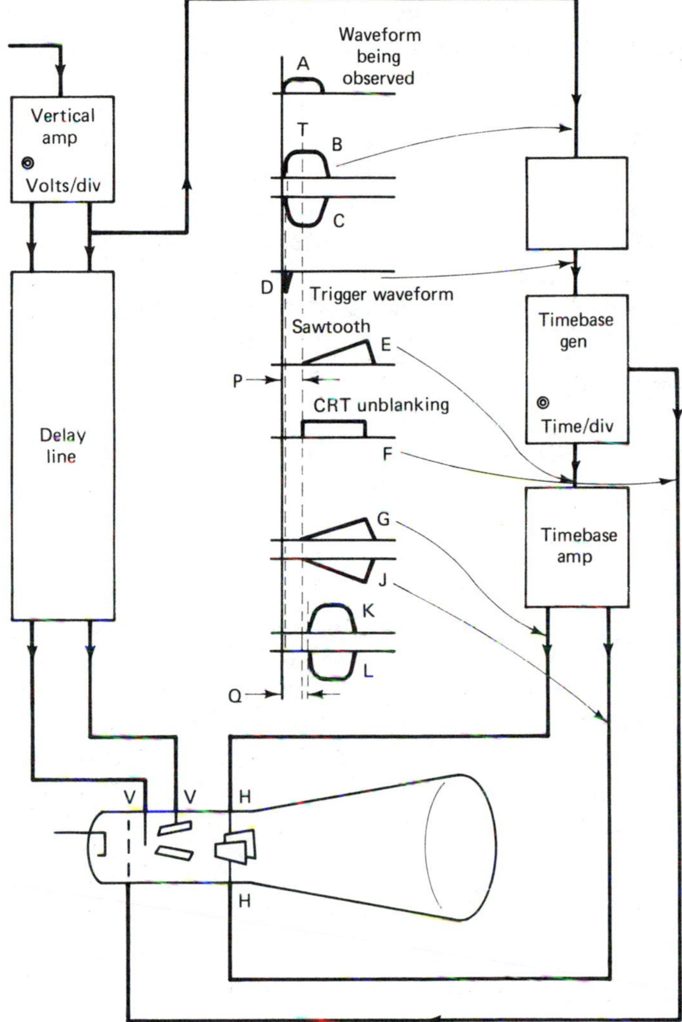

Figure A2.3 Oscilloscope block diagram.

To maintain a stable display on the CRT screen, each horizontal sweep must start at the same point on the displayed waveform. A sample of the waveform is fed to a *trigger circuit* that gives a pulse (D) at some selected point on the displayed waveform. This triggering pulse starts the run-up portion of the time-base sawtooth. As far as the display is concerned, triggering is synonymous with starting the horizontal sweep.

A rectangular *unblanking wave* (F) derived from the time-base generator is applied to the grid of the cathode-ray tube. The positive part of this wave occurs during the run-up part of the time-base output, and switches the beam *on* during its left-to-right travel, and then off again (*blanked*) during its right-to-left retrace.

In the case shown, the leading edge of the waveform displayed actuates the trigger circuit. We may want to observe this leading edge on the screen, yet the triggering and unblanking operations require a measurable time interval P of perhaps 0.1 μs. To permit us to see the leading edge, a somewhat longer delay Q is introduced by the delay line in the vertical-deflection channel, after the point where the vertical signal is fed to the trigger circuit. The delay line retards the application of the waveform to the vertical-deflection plates until the trigger and time-base circuits have the unblanking and horizontal-sweep operations under way. Thus, we can view the entire waveform, even though its leading edge triggers the horizontal sweep. If the delay line were not used, we would see only that portion of the waveform following time (T) in waveform (A).

A2.4 TRIGGERING

In the voltage vs. time mode, you must give it a *trigger* to start the *sweep* circuit. There are several *trigger modes.*

Automatic: The sweep is free running, or self-starting. It's a good mode for seeing if you have any signal at all.

Internal: The sweep is started by your signal. The circuit looks for a change, and starts when it sees it.

Line: The sweep circuit is connected to the ac line and begins every 1/60 s. Especially good for suspected line pickup, which will be synchronous.

External: You tell it when to start with an external signal related to your signal.

A2.5 SOME VERTICAL AXIS SUBTLETIES

A2.5.1 Grounding

When you connect the probe to the point in the circuit, pay attention to *grounding*; otherwise, 60 Hz *pickup* may dominate the observations. A ground makes a completed circuit for current flow. *You have to give it one.* Giving it more than one can cause an error called a *ground loop.* Implicit connections to ground through the power line may already exist in your circuit. But since the path between these two grounds may be a long one, there may be a nonnegligible resistance, and a potential difference may build up. Closing a second ground path may yield an unexpected IR drop (and error) through your probe, because of IR drops due to circulating currents.

To see the effect of no ground, touch the scope probe to your finger. A large 60 Hz signal is usually seen, since you act as an antenna. If you see a similar pattern in a measurement, the chances are you have forgotten a ground. A summary of grounding procedures is given in Fig. A2.4.

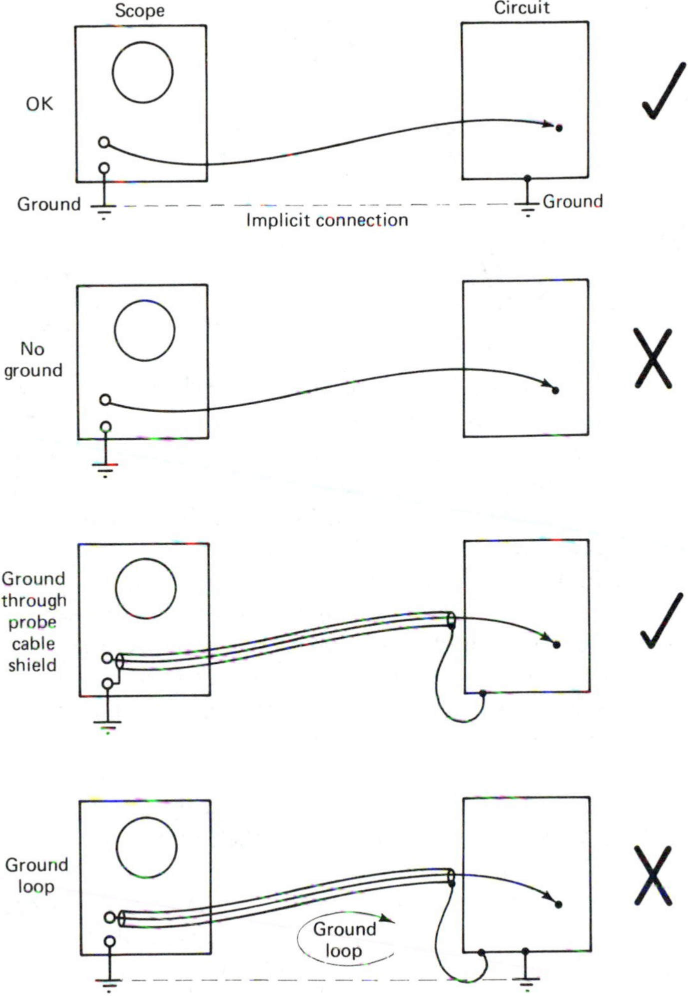

Figure A2.4 Oscilloscope grounding.

A2.5.2 Using Square Waves for Testing Frequency Response

A periodic nonsinusoidal wave is equivalent to the sum of a *fundamental* wave—a sine wave of frequency equal to that of the original wave; and a series of *harmonic components*—sine waves whose frequencies are multiples of the fundamental frequency (see Appendix 4). This fact helps us understand the problems of generating, amplifying, and displaying complicated waveforms.

If we were to reproduce a nonsinusoidal waveform by adding together its sine-wave components, each component would have to be correct in amplitude, frequency, and phase in order to reproduce the wave faithfully.

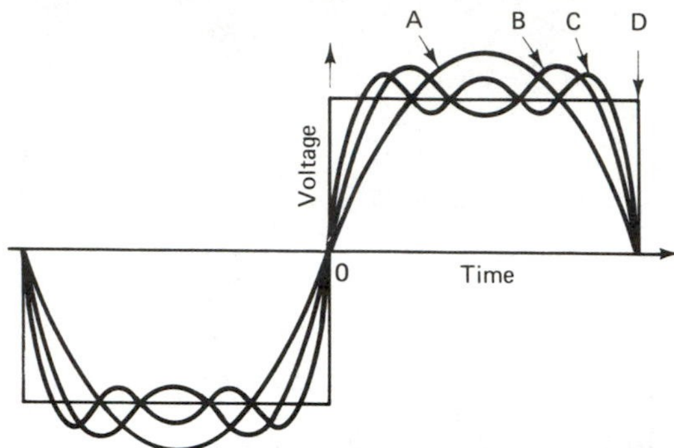

Figure A2.5 Addition of successively higher-order harmonics to a fundamental sine wave to produce a close approximation of a square wave.

An example is a rectangular or "square" waveform, which is made up of a fundamental sine wave plus an infinite series of odd harmonic sine waves (see Appendix 4). The amplitudes of the harmonics vary inversely with the frequencies. That is, the third harmonic is $\frac{1}{3}$ as strong as the fundamental; the fifth harmonic is $\frac{1}{5}$ as strong, etc. The combining of these sine-wave components to make up the original square-wave is shown in Fig. A2.5.

The first few harmonic components approach the shape of the actual square wave. Additional harmonics of higher frequencies cause the leading edge of the wave to rise more rapidly, and produce sharper corners. It requires an infinite range of harmonics to produce truly vertical edges and sharp corners. Though this is physically impossible, waves can be generated that are very close to this ideal.

Information regarding the amplitude and phase relationships of the *higher* harmonics is contained in the leading edge steepness and in the sharpness of the corners. If *low-frequency* components are not present in the proper amount and correct phase, the slope or curvature of the flat, top part of the square wave will be affected (Fig. A2.6).

Square waves are convenient for testing equipment because the *nature* of a defect, rather than simply its presence, is suggested by the distortion that occurs. By observing square wave response, we can tell if it is the transmission of high or low frequencies which is affected.

If two linear devices give identical response to square wave inputs, they can be expected to give responses similar to each other with other waveforms.

A2.5.3 AC vs. DC Coupling

Most scopes may be either ac or dc coupled to the circuit under test. In the *dc-coupled* mode, the probe is connected directly to the vertical amplifier, whose response extends to dc. In the ac-coupled mode, a capacitor is inserted between probe and amplifier. The ac-coupled mode makes it possible to look at a small ac part of a signal with a large dc component, such as the ripple in a dc power supply [Fig. A2.6(a)]. Without the capacitor, increasing the gain will push the ac waveform off the screen. The dc-coupled mode is also essential for accurately observing any signal with low-frequency components. An example is shown in Fig. A2.6(b). If a supposedly square wave of low frequency looks tilted, make sure the scope is dc-coupled.

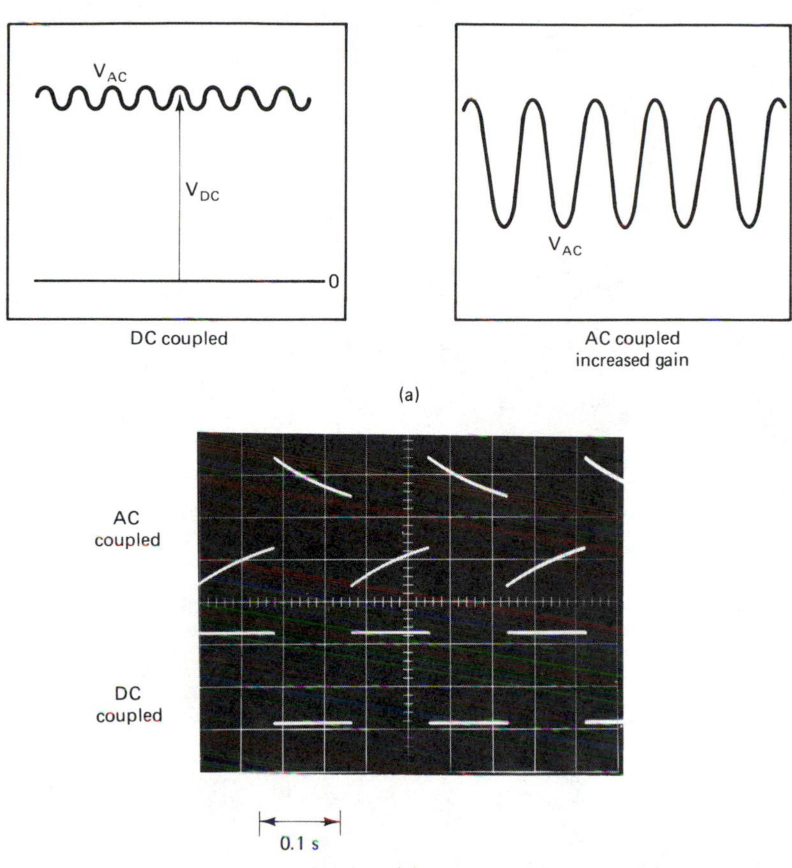

Figure A2.6 DC coupling can (a) increase the dynamic range in the measurement of an ac signal on top of a dc background, (b) remove waveform distortion in a low-frequency signal.

A2.5.4 Step Response and Waveform Distortion

Although a scope's bandwidth may be very high (up to several hundred MHz), another effect can cause waveform distortion even at audio frequencies. The bandwidth is a measure of the beam deflection electronics' ability to "follow" an input signal. Normally, a scope has a moderately high input impedance to avoid loading a source. There is also unavoidable input capacitance, due to stray capacitance in the input amplifier [Fig. A2.7(a)]. For example, $R_i = 1\ \text{M}\Omega$, $C_i =$ 100 pF. Even though C_i is small, the large value of R_i leads to a fairly low cutoff frequency. In this example, the cutoff frequency is $1/(R_i C_i) = 10^4$ Hz. A square wave in this frequency range will be distorted.

The problem is not a result of fundamental bandwidth limitations, and can be eliminated by a clever trick, which is usually built into the *10*-X *probe* [see Fig. A2.7(b)]. In addition to a 10:1 attenuator, a built-in *variable* capacitor allows the probe and scope time constants to match. The result is a frequency-independent

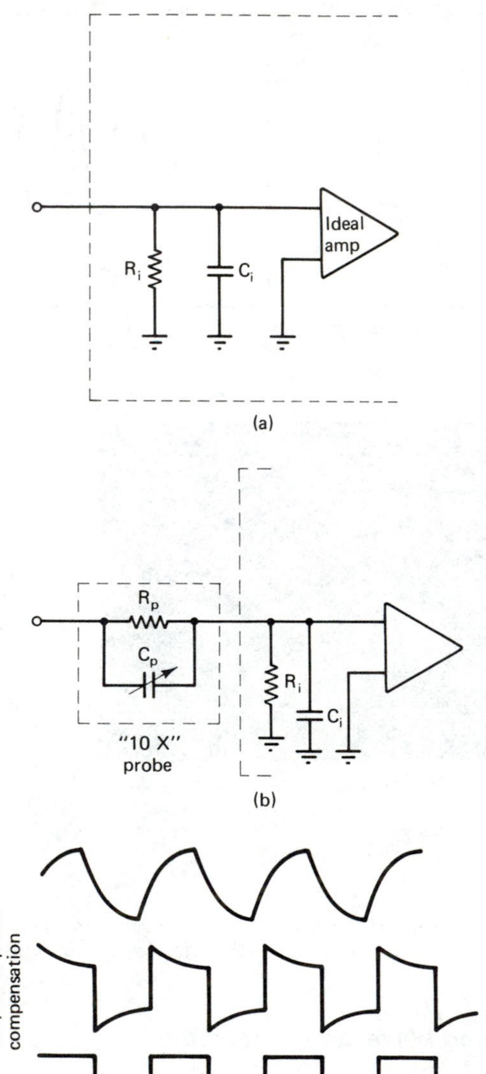

Figure A2.7 The compensated scope probe. (a) Equivalent circuit of a scope's vertical amplifier. (b) Input equivalent circuit using a compensated scope probe. (c) Examples of probe compensation: undercompensated, overcompensated, and just right.

attenuation, presenting to the scope deflection electronics a faithful representation of the input signal.

$$\frac{V_2}{V_1} = \frac{Z_s}{Z_s + Z_p} = \frac{R_s \exp (i\ \phi_s)}{R_s \exp (i\ \phi_s) + R_p \exp (i\ \phi_p)} \tag{A2.5}$$

Since

$$\tan \phi_s = R_i C_i \tag{A2.6}$$
$$\tan \phi_p = R_p C_p$$

adjusting C_p such that $\phi_s = \phi_p$ makes the phase factor drop out, leaving

$$\frac{V_2}{V_1} = \frac{R_i}{R_i + R_p} = \frac{1}{10} \quad \text{(customarily)} \tag{A2.7}$$

The attenuation is frequency-independent, as desired, and the unavoidable input capacitance of the vertical amplifier has been eliminated. This adjustment is normally performed using a built-in calibration waveform.

Normally, $R_p \sim 10\ R_i$ and $C_p \sim 0.1\ C_i$, which reduces by a factor of 10 both the resistive and capacitive loading of the circuit under test. This avoids waveform distortion due to frequency dependence of the *source* impedance. Since the signal presented to the scope is attenuated by a factor of 10, you must multiply all oscilloscope voltages by 10 to get the correct value.

C_p is made variable, and adjusted using a square wave. If C_p were absent or too small, some of the high-frequency components of the square wave would be by-passed around the oscilloscope input terminals by the input capacitance C_i, and the leading edge of the displayed square wave would be less steep [Fig. A2.7(c)].

If we adjust the probe capacitor C_p to the correct value, a compensating amount of high-frequency information will be by-passed around the probe resistor R_p to make up for the loss through C_i, and the leading edge of the displayed square wave will be restored [Fig. A2.7(e)]. But if C_p is made too large, the high-frequency response is overcompensated and too much high-frequency information is applied to the oscilloscope input. This results in an overshoot in the displayed waveform [Fig. A2.7(d)]. C_p is adjusted to its correct value by displaying the voltage calibrator signal built in to the oscilloscope, adjusting for the flattest top in the displayed waveform [Fig. A2.7(e)].

A2.5.5 Differential Inputs

Some scopes can look *differentially* at the voltage *difference* across a circuit element. This is essential when neither end of that element is grounded (use of a grounded scope will change the circuit, and perhaps even destroy it!). Two probes are required, connected to two points in the circuit under test. Differential amplifiers can tolerate a large amount of *common-mode* signal without letting it creep through to the display. A large common-mode signal may exist, for example, if the circuit is floating or has a different ground than the scope. The ability of a scope to tolerate such a signal is specified by its *common-mode rejection ratio* (CMRR), where

$$\text{CMRR} = (\text{differential signal gain})/(\text{common mode signal gain})$$

CMRR values of 90 dB are typical, and 140 dB is possible. If the common-mode signal is large enough to overload the amplifier, completely spurious waveshapes will be seen, even though the differential signal on the screen does not appear excessively large. To check for this condition, reduce the gain by one step. If the *character* of the waveform changes, suspect an overload distortion due to the common-mode signal.

APPENDIX 3

Complex Numbers

A3.1 COMPLEX NUMBERS

The manipulation of complex numbers is essential, for example, in evaluating the impedances of networks used in op amp circuits. This appendix includes the basic rules needed in the text. Complex numbers originate in arithmetic when one takes a square root of a negative number. That number may be rewritten as the square root of the positive number times j, where

$$j = \sqrt{-1}, \; j^2 = -1 \tag{A3.1}$$

The symbol j is used in electronics rather than the usual symbol i to avoid confusion because i is often used for current. Computation with numbers which have both a real and imaginary part may be carried out as long as care is taken to keep track of the real and imaginary parts separately. The complex number

$$Z = a + jb \tag{A3.2}$$

may be viewed as a vector, as shown in Fig. A3.1. The size of the real part is measured along the horizontal or real axis, and the size of the imaginary part is measured along the vertical or imaginary axis. A given point in this *complex plane* uniquely represents a complex number.

Magnitude. The length of the vector Z is evaluated from the components by the usual rule for triangles

$$|Z| = (a^2 + b^2)^{1/2} = R \tag{A3.3}$$

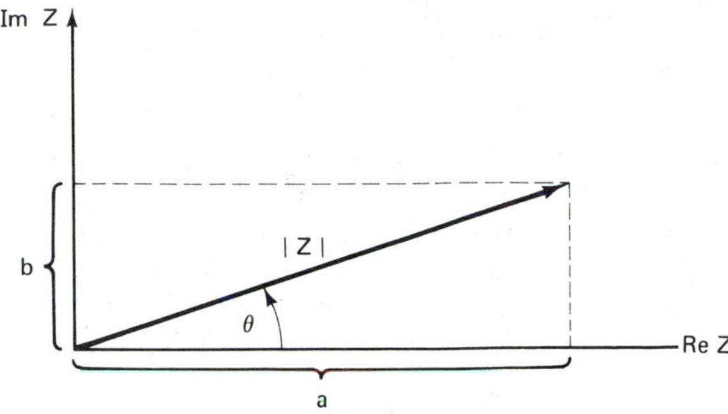

Figure A3.1 Complex quantity represented as a vector.

Phase. The phase angle of the vector Z depends upon the ratio of the components

$$\tan \theta = \frac{b}{a} \tag{A3.4}$$

To test your understanding, what is the phase of j?

Exponential notation. The manipulation of complex algebra is often simplified if use is made of magnitude and phase notation

$$Z = a + jb = \mathrm{R}e^{j\theta} \tag{A3.5}$$

where the exponential function is defined as

$$e^{j\theta} = \cos \theta + j \sin \theta \tag{A3.6}$$

The complex exponential may be viewed as a vector oriented at angle θ on a circle of unit radius (Fig. A3.2). These two representations are of course equivalent, since the cosine and sine project out the horizontal and vertical components

$$Z = R\ (\cos \theta + j \sin \theta\ ) = a + jb \tag{A3.7}$$

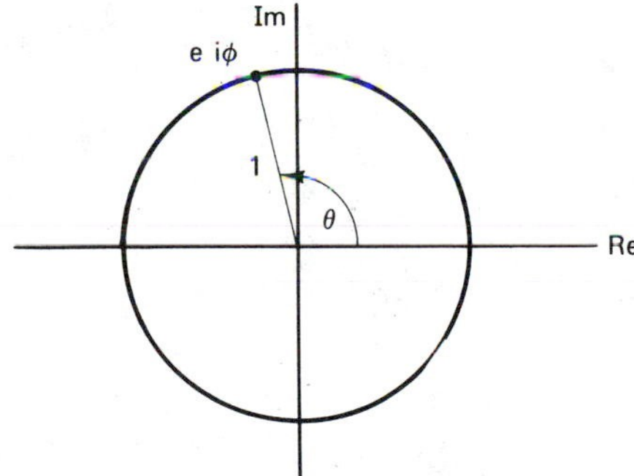

Figure A3.2 The complex exponential exp $(j\theta)$ lies on the unit circle.

A3.2 COMPLEX NUMBER MANIPULATION

Addition. In adding two complex numbers, the components are added separately. If

$$X = a + jb$$
$$Y = c + jd$$

Then

$$Z = X + Y = (a + c) + j(b + d) \qquad \text{(A3.8)}$$

This follows from the vector representation of complex numbers. The length of the sum of two vectors of length 7 and length 3 is not simply 10 (Fig. A3.3). A similar rule holds for subtraction.

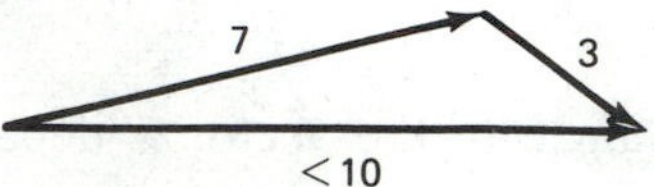

Figure A3.3 Complex numbers add like vectors.

Equalities. For two complex numbers to be equal, both the real and imaginary must separately be equal. For $X = a + jb$ and $Y = c + jd$, the equality $X = Y$ requires that

$$a = c \quad and \quad b = d \qquad \text{(A3.9)}$$

Thus, a statement of equality of complex numbers is really two equations.

Multiplication. In multiplying two complex numbers, the magnitudes multiply, and the phases add.

$$Z = X \cdot Y \qquad \text{(A3.10)}$$
$$Z = (R \cdot S)e^{j(\alpha + \theta)} = |Z|e^{j\phi}$$

where $X = Re^{j\alpha}$, $Y = Se^{j\theta}$, $|Z| = RS$, $\phi = \alpha + \theta$. This follows from the rules for multiplication of exponentials.

Division. In dividing two complex numbers, the magnitudes multiply, and the phases subtract.

$$Z = \frac{X}{Y}$$
$$Z = \frac{\text{R}e^{j\alpha}}{Se^{j\theta}} = \frac{R}{S} e^{j(\alpha - \theta)} \qquad \text{(A3.11)}$$

To test your understanding, what is the phase of $1/j$?

Complex conjugate. The complex conjugate Z^* of a number Z is a number with the same components, but with the sign of the imaginary part reversed.

$$\text{If } Z = a + jb$$
$$\text{then } Z^* = a - jb \qquad \text{(A3.12)}$$

The vector picture of this is shown in Fig. A3.4. This quantity is particularly use-

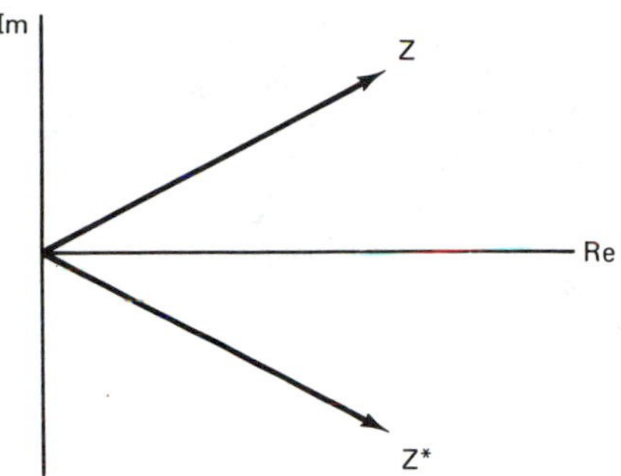

Figure A3.4 The complex conjugate Z^*.

ful in evaluating magnitudes, since the product of a number and its conjugate equals the square of its magnitude.

$$\begin{aligned} ZZ^* &= a^2 - j^2b^2 - abj + abj \\ &= a^2 + b^2 = |Z|^2 \end{aligned} \tag{A3.13}$$

To test your understanding, show that this is true using the magnitude and phase representation rather than the component representation of Z. Another application of the complex conjugate is described next.

Rationalizing. The operation of rationalizing is defined as bringing any complex expression into the standard form of Eq. (A3.2). It is useful to be able to bring the result of some calculation using complex numbers into standard form $Z = a + jb$, in order to graph its magnitude and phase. Once the expression for Z has been brought into the form of a ratio of numerator to denominator, the rule for bringing it into standard form is to multiply both numerator and denominator by the complex conjugate of the denominator.

$$Z = \frac{N}{D}\frac{D^*}{D^*} = \frac{ND^*}{|D|^2} \tag{A3.14}$$

The denominator will no longer have an imaginary part, since it is a magnitude. An example is given below.

$$\begin{aligned} Z &= \frac{c + jd}{e + jf} \\ &= \frac{c + jd}{e + jf}\,\frac{e - jf}{e - jf} = \frac{ce - j^2df + jde - jcf}{e^2 + f^2} \\ &= \frac{ce + df}{e^2 + f^2} + j\,\frac{de - cf}{e^2 + f^2} \\ &= a + jb \end{aligned} \tag{A3.15}$$

APPENDIX 4

Fourier Spectral Analysis

A4.1 FOURIER ANALYSIS

Any periodic signal with period T can be represented as a sum of sine and cosine waves, called the *Fourier series* representation.

$$f(t) = A_0/2 + \sum_{n=1}^{\infty} [A_n \cos(2n\pi/T)t + B_n \sin(2n\pi/T)t] \quad \text{(A4.1)}$$

The coefficients are obtained from the integrals

$$A_n = (2/T) \int_{-T/2}^{T/2} f(t) \cos(2n\pi t/T)\, dt \quad \text{(A4.2)}$$

$$B_n = (2/T) \int_{-T/2}^{T/2} f(t) \sin(2n\pi t/T)\, dt$$

$$n = 0,\ 1,\ 2,\ \cdots$$

Fourier coefficients for common waveforms are summarized in Table A4.1. Here are some comments about the more important ones.

1. The square wave has only odd harmonics; contributions to the even harmonic coefficients in Eq. (A4.2) cancel by symmetry.
2. The harmonic component amplitudes for the square wave fall off as $1/n$, while those for the triangular wave fall off as $1/n^2$. This follows because the

TABLE A4.1 SOME FOURIER SERIES[a]

Waveform	Fourier series
Square wave 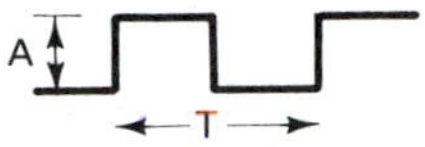	$\frac{2A}{\pi}\sum_{n=1}^{\infty}\frac{1}{n}\sin(n\omega t), \quad n = \text{odd}$
Triangular wave 	$\frac{2A}{\pi}\sum_{n=1}^{\infty}\frac{1}{n^2}\cos(n\omega t), \quad n = \text{odd}$
Sawtooth wave 	$\frac{A}{\pi}\sum_{n=1}^{\infty} -\frac{-1^n}{n}\sin(n\omega t), \quad \text{all } n$
Half-wave rectified Sine wave 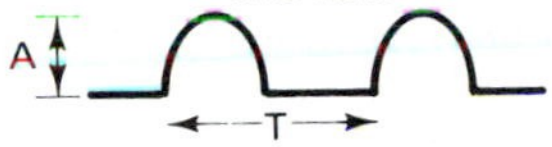	$\frac{2A}{\pi}\{\frac{\pi}{4}\sin\omega t + \sum_{n=2,4,6\cdots}^{\infty}\frac{1}{n^2-1}\cos n\omega t\}$
Full-wave rectified Sine wave 	$\frac{4A}{\pi}\sum_{n=2,4,6\ldots}^{\infty}\frac{-(-1)^n}{n^2-1}\cos(n\omega t)$
Rectangular pulse 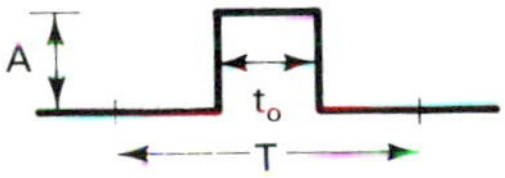	$\frac{2At_o}{T}\sum_{n=1}^{\infty}\frac{\sin(n\pi t_o/T)}{n\pi t_o/T}\cos(n\omega t)$

[a] $\omega = 2\pi/T$

triangular wave is the integral of the square wave; integrating any sine wave component brings out an extra $1/n$ factor.

3. While the triangular wave also has only odd harmonics, the sawtooth wave has all harmonics. This is why the sawtooth is preferred as a harmonic-rich signal source, as in electronic music.
4. The rectangular wave narrowed to a pulse-like limit has all harmonics of equal amplitude. The shorter the pulse, the more harmonics it takes to represent it adequately.
5. The fundamental frequency of the input sine wave is absent in the Fourier coefficients of the full-wave rectified waveform. You can see it by sketching in the lowest frequency component by hand; the lowest one in the full-wave signal has twice the input frequency. This makes a full-wave rectifier easier to filter in a power supply than a half-wave rectifier.

The Fourier series faciltates circuit analysis of complex waveforms, since the analysis can be performed on sine waves and then added up proportional to the weighting of the harmonic amplitudes V_n. For example, feed a square wave of peak-to-peak amplitude A through a low-pass filter [Fig. 1.8]. The output is

$$V_o = \sum_{n=1}^{\infty} [j\omega_n RC + 1]^{-1} V_n \simeq \sum_{n=1}^{\infty} [j\omega_n RC]^{-1} V_n \tag{A4.3}$$

where

$$V_n = (2A/n\pi)\cos(\omega_n t)$$

and

$$\omega = 2n\pi/T$$

As a result,

$$V_o \simeq \sum_{n=1}^{\infty} (2A/n\pi)^2 \cos(\omega_n t)$$

This is the Fourier series for a triangular wave; the low-pass filter approximately integrates the input signal.

A4.2 THE FOURIER TRANSFORM

Even if the signal $f(t)$ is not periodic, its spectral composition can be evaluated with the *Fourier transform* $F(\omega)$.

$$F(\omega) = \int_{-\infty}^{\infty} f(t)\exp(-j\omega t)\,dt \tag{A4.4}$$

The Fourier transform will display peaks wherever the signal has frequency components. If the transform is performed over a finite interval (as when data is recorded over a finite *window*), the peaks will be broadened and may display sidebands. The finite window of width τ is equivalent to multiplying an infinite sample by a rectangular pulse (1 within the window, 0 elsewhere), whose Fourier transform is

$$F(\omega) = \frac{\sin(\omega\tau/2)}{\omega/2} \tag{A4.5}$$

The $(\sin x)/x$ shape appears at each spectral peak because the Fourier transform $G(\omega)$ of a product of two signals is the *convolution* of the individual Fourier transforms $F(\omega)$ and $G(\omega)$.

$$G(\omega) = \int_{-\infty}^{\infty} F(\omega')G(\omega' - \omega)\,d\omega'$$

Undesirable sidebands are eliminated by multiplying the data instead by a *window weighting* function with less sharp corners. The Fourier transform of a Gaussian

$\exp[-(t/T)^2]$, for example, is also a Gaussian, and has no sidebands. This choice may broaden the transform peaks undesirably, and other window weighting functions may be a preferred compromise.

A4.3 RELATIONSHIP BETWEEN FOURIER ANALYSIS AND FOURIER SPECTRUM

The two ideas are closely related. If one evaluates the Fourier transform [Eq. (A4.4)] of a periodic waveform with an integer number of cycles fitting within the window, the spectrum displays a row of peaks which correspond exactly to the coefficients of a Fourier analysis by Eq. (A4.1). The relationship is even more apparent if the transform is evaluated numerically by computer, so the integral is approximated by a discrete sum over samples x_k.

$$G(\omega_r) = \sum_{k=0}^{N-1} x_k \exp(-2\,jrk/N)$$

$$r = 0,\ 1,\ 2,\ \ldots,\ N-1$$

This is the most common situation in digitized or *discretely sampled* data. In this case, no meaningful spectral information is contained beyond $\pm f_s/2$, where f_s is the sampling frequency. The spectrum, if extended beyond f_s, will display an exact replica or *Nyquist image.* Data subject to discrete Fourier processing must be filtered to remove components above $f_s/2$ to prevent a spillover of the image which could not be distinguished from real spectral components.

The subject of spectral analysis is a fascinating one, which we have barely touched on here in the context of electronics applications. The reader is referred to the References for further discussion.

REFERENCES

More complete bibliographic information for the books listed below appears in the annotated bibliography at the end of the book.

BRACEWELL, *The Fourier Transform and Its Applications*

HIGGINS, R. J., "The Fast Fourier Transform ...," *American Journal of Physics* (1976), vol. 44, p. 766

ITT, *Reference Data for Radio Engineers*, Chapter 44

Annotated Bibliography

In addition to the standard texts, look into Manufacturer's Data Books. In a fast-moving field, the most up-to-date application ideas will be found here. Some of the best are from: Analog Devices; Burr Brown; Motorola; National; RCA; Signetics; Texas Instruments.

ALLEY, CHARLES L., and KENNETH W. ATWOOD, *Electronic Engineering* (3rd ed.) New York: John Wiley & Sons, Inc., 1973.

______ and ______ , *Semiconductor Devices and Circuits.* New York: John Wiley & Sons, Inc., 1971. This text, set at junior college EE level, has a very clear approach, explicit diagrams, and solved examples to bring the reader close to practical implementation.

Analog Devices, Inc., Engineering Staff, *Analog-Digital Conversion Handbook*, ed. D. H. Sheingold. Norwood, Mass.: Analog Devices, 1972. Everything you ever wanted to know about A/D and D/A.

______ , *Nonlinear Circuits Handbook*, Norwood, Mass.: Analog Devices, 1974. Nonlinear function modules; log converters; multipliers and dividers.

BENEDICT, R. RALPH, *Electronics for Scientists and Engineers* (2nd ed.) Englewood Cliffs, N.J.: Prentice-Hall, Inc., 1975. A survey with adequate level of mathematical depth.

BOYLSTEAD, ROBERT L., *Introductory Circuit Analysis* (3rd ed.) Columbus, Ohio: Charles E. Merrill Publishing Co., 1977. Need some background in network analysis? Here's a place to go without feeling as if you're in the nineteenth century!

BRACEWELL, RONALD N., *The Fourier Transform and its Applications* (2nd ed.) New York: McGraw-Hill Book Company, 1978. A very graphical pictorial point of view. Emphasis on practical applications.

BROPHY, JAMES J., *Basic Electronics for Scientists* (3rd ed.) New York: McGraw-Hill Book Company, 1977. A lightly updated (the section labeled "microcomputers" actually describes a pocket calculator!) version of the original (1966) survey. Very readable, though the emphasis is more on discrete circuits than on IC's.

BROWN, PAUL B., BRUCE W. MAXFIELD, and HOWARD MORAFF, *Electronics for Neurobiologists.* Cambridge: MIT Press, 1973. Many disciplines develop specialized techniques and feel the need to break their students in without much background.

CRAWFORD, ROBERT H., *MOSFET in Circuit Design.* Texas Instruments Electronics Series. New York: McGraw-Hill Book Company, 1967. Background in how MOSFET devices are fabricated and how they operate.

DAVIS, SIDNEY A., *Feedback and Control Systems.* New York: Simon & Schuster, Inc., 1974. A "Tech Outline" of solutions to problems plus some basic theory. Nice tables of transfer functions, etc.

DEMPSEY, JOHN A., *Basic Digital Electronics with MSI Applications.* Reading, Mass.: Addison-Wesley Publishing Co., Inc., 1977. This book strikes a nice balance. Though accessible, with nice pictures, and wide-ranging, it is far from superficial, with many practical design details.

DIEFENDERFER, A. JAMES, *Principles of Electronic Instrumentation* (2nd ed.) Philadelphia: W. B. Saunders Company, 1979. Includes both text and lab manual. Although the content is fairly conventional, it is up-to-date and sensible. The writing style is very clear (perhaps a little low level) with excellent illustrations. The lab manual is a model of gentle guidance.

FLOYD, THOMAS L., *Digital Logic Fundamentals.* Columbus, Ohio: Charles E. Merrill Publishing Co., 1977. Fairly standard contents, at the junior college technical level. The coverage of analytical techniques such as Karnaugh maps is therefore quite understandable.

FRANKS, ROGER C., *Modeling and Simulation in Chemical Engineering.* New York: John Wiley & Sons, Inc., 1972. Clever diagramming of modeling. The earlier edition has better coverage of analog modeling.

GARDNER, FLOYD M., *Phaselock Techniques.* New York: John Wiley & Sons, Inc., 1979.

GIBBONS, JAMES F., *Semiconductor Electronics.* New York: McGraw-Hill Book Company, 1966. An extremely well written introductory treatment for electrical engineers. Main weakness is a complete lack of coverage of dc differential amplifiers.

GLASER, ARTHUR, and GERALD E. SHUBAK-SHARPE, *Integrated Circuit Engineering: Fabrication, Design, Application.* Reading, Mass.: Addison-Wesley Publishing Co., Inc., 1979. One of the newer books for EE's on how IC's work.

GRAEME, JERALD G., *Application of Operational Amplifiers: Third Generation Techniques.* New York: McGraw-Hill Book Company, 1973.

______ , *Designing with Operational Amplifiers: Applications, Alternatives.* New York: McGraw-Hill Book Company, 1977.

______, GENE E. TOBEY, and L. P. HUELSMAN, eds., *Operational Amplifiers: Design and Applications.* See Tobey.

GROVE, A. S., *Physics and Technology of Semiconductor Devices.* New York: John Wiley & Sons, Inc., 1967. The basic reference (in spite of its age) on how devices work.

HAMMING, R. W., *Digital Filters.* Englewood Cliffs, N.J.: Prentice-Hall, Inc., 1977. Clear yet elegant brief introductory text on discrete time signal processing.

HENRY, RICHARD W., *Electronic Systems and Instrumentation.* New York: John Wiley & Sons, Inc., 1978. Though brief, this book goes deep. Unusual approach, *beginning* with chapters on impulse response and Laplace transform in the time domain and Fourier transform in the frequency domain.

HIGGINS, RICHARD J., *Experiments with Integrated Circuits.* Englewood Cliffs, N.J.: Prentice-Hall, Inc., 1982. The lab book which accompanies this text, with experiments closely keyed to the chapters.

HOENIG, STEWARD A., and F. LELAND PAYNE, *How to Build and Use Electronic Devices Without Frustration, Pain, Mountains of Money, or an Engineering Degree.* Boston: Little, Brown & Company, 1973. The title tells all. A delightful user's guide, though limited in scope to op amp applications (i.e., little digital).

HUNTER, BILL, *CMOS Databook.* Blue Ridge Summit, Pa.: Tab Books, 1978. A wide ranging, down-to-earth survey. Particularly strong on CMOS noise immunity and interfacing rules.

Intel Corp., *Component Data Catalog.* Santa Clara, Calif.: Intel Corp., updated periodically. Here's a source for memory, interface, and microcomputer IC specifications.

———, *Memory Design Handbook.* Santa Clara, Calif.: Intel Corp., updated periodically. Design information on internal organization and applications of the latest RAM's and ROM's.

International Telephone and Telegraph Corp., *Reference Data for Radio Engineers* (6th ed.), ed. H. Westman, Indianapolis: Howard W. Sams & Co., Inc., 1975. If you wanted one classic reference encyclopedia on electronics, this would probably be the one.

JACKSON, A. S., *Analog Computation.* New York: McGraw-Hill Book Company, 1960. The best introduction I have found, in spite of its age.

JUNG, WATER G., *IC Op Amp Cookbook.* Indianapolis: Howard W. Sams & Co., Inc., 1974. Practical circuits, with all component values given. Valuable (though slightly dated) cross-comparison of internal workings of op amps.

KOHONEN, TEUVO, *Digital Circuits and Devices.* Englewood Cliffs, N.J.: Prentice-Hall, Inc., 1972.

KORN, GRANINO A., *Microprocessors and Small Digital Computer Systems for Engineers and Scientists.* New York: McGraw-Hill Book Company, 1977. Nice handbook. Compares various architectures of specific mini- and microcomputers.

———, and ———, *Electronic Analog and Hybrid Computers* (2nd ed.) New York: McGraw-Hill Book Company, 1972.

LANCASTER, DON, *CMOS Cookbook.* Indianapolis: Howard W. Sams & Co., Inc., 1977. Resembles the classic *TTL Cookbook* in format and level. Many applications are a rehash, though some are new.

———, *TTL Cookbook.* Indianapolis: Howard W. Sams & Co., Inc., 1975. A turning-point cookbook which made TTL IC's and circuits accessible to anyone. Many clever applications, plus an invaluable one-page-per-device specification summary.

LARSEN, DAVID G., and PETER R. RONY, *Logic and Memory Experiments using TTL Integrated Circuits.* Indianapolis: Howard W. Sams Co., Inc., 1978. Successor to the classic Bugbooks I and II. Very much a cookbook, so it's a painless introduction to actually doing electronics in the lab. The step-by-step instructions, helpful at first, are eventually redundant.

LION, KURT S., *Elements of Electrical & Electronic Instrumentation.* New York: McGraw-Hill Book Company, 1975.

______ , *Instrumentation in Scientific Research (Electrical Input Transducers).* New York: McGraw-Hill Book Company, 1959. Few books deal with how to make a physical or chemical quantity into an electrical signal. This book, despite its age, is still a classic.

MAGRAB, EDWARD B., and DONALD S. BLOMQUIST, *The Measurement of Time-Varying Phenomena.* New York: John Wiley & Sons, Inc., 1971. Periodic and random signals; filters; amplifiers; voltage detectors; recorders; signal generators; digital systems.

MALMSTADT, HOWARD V., CHRISTIE G. ENKE, and STANLEY R. CROUCH, *Electronic Measurements for Scientists.* Hunter, N.Y.: W. R. Benjamin, 1974. Though flashy and topical, material is scattered through four sections or "modules". For example, Fourier analysis and also impedances do not appear until quite late. The titles of sections are sometimes misleading; e.g., "Resistive Devices for Measurement and Control" is a section on dc circuits, including resistive transducers.

MARKUS, JOHN, *Electronic Circuits Manual.* New York: McGraw-Hill Book Company, 1974. A collection of recipes garnered from manufacturers' application notes and magazine articles.

MELEN, ROGER, and HARRY GARLAND, *Understanding IC Operational Amplifiers* (2nd ed.) Indianapolis: Howard Sams & Co., Inc., 1978. As with many of Sams' books, this is a good place to look to get started with practical how-to-do-it ideas and circuits.

MILLMAN, JACOB, *Microelectronics: Digital and Analog Circuits and Systems.* New York: McGraw-Hill Book Company, 1979. A first course for EE's; updated rewrite of the classic Millman & Halkias Integrated Electronics. How devices and IC's work; digital and analog circuits and systems. An encyclopedic (almost 900-page) work.

MORRIS, ROBERT L., and JOHN R. MILLER, eds., and Texas Instruments Corp. Applications Staff, *Design with TTL Integrated Circuits.* New York: McGraw-Hill Book Company, 1971. A classic, still quite useful for TTL circuits.

MORRISON, RALPH, *Grounding and Shielding Techniques in Instrumentation* (2nd ed.) New York, John Wiley & Sons, Inc., 1977. Got a tricky measuring problem? This book explains how to get around ground loops and noise pickup.

MUELLER, RICHARD S., and THEODORE J. KAMINS, *Device Electronics for Integrated Circuits.* New York: John Wiley & Sons, Inc., 1977. One of the best recent introductory treatments of the physics of device operation. Treats bipolar, MOS, and also Schottky (metal-semiconductor) devices.

National Semiconductor Corp., *Linear Databook.* Santa Clara, Calif.: National Semiconductor Corp., published periodically.

______ , *MOS/LSI Databook.* Santa Clara, Calif.: National Semiconductor Corp., published periodically.

______ , *CMOS Databook.* Santa Clara, Calif.: National Semiconductor Corp., published periodically.

______ , *Linear Applications Handbook.* Santa Clara, Calif.: National Semiconductor Corp., published periodically.

______ , *Voltage Regulators Handbook.* Santa Clara, Calif.: National Semiconductor Corp., published periodically.

——— , *Data Acquisition Handbook.* Santa Clara, Calif.: National Semiconductor Corp., published periodically.

OPPENHEIM, ALAN V., ed., *Applications of Digital Signal Processing.* Englewood Cliffs, N.J.: Prentice-Hall, Inc., 1978. Especially chapters on telecommunications, audio, speech, image processing, radar, sonar, and geophysics.

——— , and RONALD W. SHAFER, *Digital Signal Processing.* Englewood Cliffs, N.J.: Prentice-Hall, Inc., 1975. This is the "bible" for background in discrete time signals, Fourier transforms, and digital filters.

PEATMAN, JOHN B., *Microcomputer-Based Design.* New York: McGraw-Hill Book Company, 1977. Emphasis on *instrument* design using microprocessors.

PELED, ABRAHAM, and BEDE LIU, *Digital Signal Processing.* New York: John Wiley & Sons, Inc., 1976. Brief (300-page) intro at the undergraduate level. Best intro text I've seen.

Philbrick Editorial Staff, *Philbrick Applications Manual for Computing Amplifiers.* Dedham, Mass.: Teledyne-Philbrick Inc., 1966. A useful compilation of circuits for computing, measurement, and simulation, along with practical advice on how to live with the nonideal operational amplifier. Sadly, out of print.

ROBERGE, JAMES K., *Operational Amplifiers, Theory & Practice.* New York: John Wiley & Sons, Inc., 1975. By far the best op amp book at the deep analytical level for EE's. Really two books, since extensive applications plus IC op amp design are included. Warning: not an easy book.

RONY, PETER R., DAVID G. LARSEN, and JONATHAN A. TITUS, *Introductory Experiments in Digital Electronics and 8080A Microcomputer Programming and Interfacing.* Indianapolis: Howard W. Sams Co., Inc., 1978. Successor to the classic Bugbooks V and VI. If you wanted an introductory lab exposure to both digital logic and microcomputers, this is one way to get it. Suffers from the same strengths and limitations as other Bugbooks (see Larsen and Rony).

RUTKOWSKI, GEORGE B., *Handbook of Integrated Circuit Operational Amplifiers.* Englewood Cliffs, N.J.: Prentice-Hall, Inc., 1975. Not a deep book, but has extensive practical applications.

SCHWARTZ, MISCHA, and LEONARD SHAW, *Signal Processing: Discrete Spectral Analysis, Detection, and Estimation.* New York: McGraw-Hill Book Company, 1975. The math is tempered by some nice graphical data examples which help get the point across.

Scientific American, *Microelectronics.* San Francisco: W. H. Freeman & Company Publishers, 1977. A "gee-whiz" tour, from a special issue of *Scientific American.* Articles written by industry and academic leaders, with great illustrations. Covers from devices through microcomputer applications.

SHEINGOLD, DANIEL H., ed., *Transducer Interfacing Handbook.* Norwood, Mass.: Analog Devices, Inc., 1980. Modern transducers and how to interface them to microcomputers.

Signetics Corp., *Analog Data Manual.* Sunnyvale, Calif.: Signetics Corp., 1979.

——— , *Analog Applications Manual.* Sunnyvale, Calif.: Signetics Corp., 1979. The phase-locked loop discussion is the best available, and typical of the excellent quality to be found here.

Siliconix, Inc., *Analog Switches and their Applications.* Santa Clara, Calif.: Siliconix, Inc., 1976. Best guide I've found on how analog switches work, with many applications to multiplexing, sample and hold, and signal processing.

SIMPSON, ROBERT E., *Introductory Electronics for Scientists and Engineers.* Boston: Allyn & Bacon, Inc., 1974. A wide-ranging survey which begins with dc and ac circuits. Strong on discrete circuits and instrumentation, but weak on integrated circuits.

SOUCEK, BRANKO, *Microcomputers and Microprocessors.* New York: John Wiley & Sons, Inc., 1976.

STEARNS, SAMUEL D., *Digital Signal Analysis.* Rochelle Park, N.J.: Hayden Book Co., Inc., 1975. A most understandable book on digital signal processing. Sampled data, FFT, digital filters, correlation. Appendix includes digital filter design programs.

STOUT, DAVID F., and MILTON KAUFFMAN, eds., *Handbook of Op Amp Circuit Design.* New York: McGraw-Hill Book Company, 1976. A real cookbook: design handbook without telling you *why.*

SZE, S. M., *Physics of Semiconductor Devices.* New York: John Wiley & Sons, Inc., 1969. Background for how all kinds of devices work, from a solid-state physics point of view.

TOBEY, G. E., J. G. GRAEME, and L. P. HEULSMAN, eds., *Operational Amplifiers.* New York: McGraw-Hill Book Company, 1971. This is the place to go for both the insides and outsides of op amps. Compiled by Burr-Brown, it represents a distillation of earlier application manuals.

——, *Operational Amplifiers.* Vol. 2: *Second Generation Applications.* McGraw-Hill Book Company, 1971.

VASSOS, BASIL H., and GALEN W. EWING, *Analog and Digital Electronics for Scientists* (2nd ed.) New York: John Wiley & Sons, Inc., 1979. A nice collection of up-to-date ideas at the "how-to-do-it" level. On the brief side. Good problems.

WESTMAN, H., ed., *Reference Data for Radio Engineers.* See under ITT.

WHALEN, ANTHONY D., *Detection of Signals in Noise.* New York: Academic Press, Inc., 1971.

WILLIAMS, GERALD E., *Digital Technology.* Chicago: Science Research Associates, 1977. A wide-ranging introductory survey, from logic families to microprocessors.

Index

A

B

C

D

E

F

G

H

I

J

K

L

M

N

O

P

Q

R

S